Microconstituents in the Environment

Microconstituents in the Environment

Occurrence, Fate, Removal, and Management

Edited by
Rao Y. Surampalli,
Tian C. Zhang,
Chih-Ming Kao,
Makarand M. Ghangrekar,
Puspendu Bhunia,
Manaswini Behera,
and Prangya R. Rout

WILEY

This edition first published 2023

Registered Office(s)
John Wiley & Sons, Inc., 111 River Street, Hoboken, NJ 07030, USA
John Wiley & Sons Ltd, The Atrium, Southern Gate, Chichester, West Sussex, PO19 8SQ, UK

For details of our global editorial offices, customer services, and more information about Wiley products visit us at www.wiley.com.

Wiley also publishes its books in a variety of electronic formats and by print-on-demand. Some content that appears in standard print versions of this book may not be available in other formats.

Library of Congress Cataloging-in-Publication Data
Names: Surampalli, Rao Y., editor.
Title: Microconstituents in the environment : occurrence, fate, removal and management / edited by Rao Y. Surampalli [and six others].
Description: Hoboken, NJ : John Wiley & Sons Ltd, 2023. | Includes bibliographical references and index.
Identifiers: LCCN 2022055640 (print) | LCCN 2022055641 (ebook) | ISBN 9781119825258 (hardback) | ISBN 9781119825265 (pdf) | ISBN 9781119825272 (epub) | ISBN 9781119825289 (ebook)
Subjects: LCSH: Pollution. | Water–Pollution. | Pollution prevention.
Classification: LCC TD174 .M52 2023 (print) | LCC TD174 (ebook) | DDC 363.739/4–dc23/eng/20230126
LC record available at https://lccn.loc.gov/2022055640
LC ebook record available at https://lccn.loc.gov/2022055641

Cover image: © TA BLUE Capture/Shutterstock
Cover design by Wiley

Set in 9.5/12.5pt STIXTwoText by Integra Software Services Pvt. Ltd, Pondicherry, India

Contents

Preface

Microconstituents or contaminants of emerging concern (CECs) refer to any pollutants that have not previously been detected or regulated under current environmental laws, or may cause known or suspected adverse ecological and/or human health effects even at insignificant levels. They consist of pesticides, industrial chemicals, surfactants, pharmaceutical and personal care products, cyanotoxins, nanoparticles, and flame retardants, among others, that are consistently being found in groundwater, surface water, municipal wastewater, drinking water, and food sources. The presence of CECs in treated effluents and its long-term impact are to be evaluated considering their environmental partitioning and bioaccumulation potential in the aquatic species. There is an urgent need not only to develop reliable and cost-effective methods to analyze a wide range of ECs, but also to find techno-economically feasible options for their efficient removal from different ecosystems.

This book is intended to provide the readers with an understanding of the occurrence and fate of microconstituents in the environment and possible management strategies. The main topics are organized into five core parts with subdivisions of each. Part I deals with the fundamental ideas regarding microconstituents in the environment and consists of four chapters. Chapter 1 introduces the microconstituents and explores their various classifications, properties, and sources, as well as their impact on environmental ecosystems and human health. The presence of microconstituents in environmental samples and the detection methodology are discussed in Chapter 2. The sampling protocols, quantification, and analysis of microconstituents are discussed in Chapter 3. Chapter 4 deals with the toxicity assessment, including acute and chronic toxicity and dose-responses studies. Part II covers the fate and transportation of microconstituents in various environmental domains, including mathematical transport systems of microconstituents (Chapter 5), groundwater contamination by microconstituents (Chapter 6), microconstituent transport in surface water (Chapter 7), fate and transport of microconstituents in wastewater treatment plants (Chapter 8), atmospheric transport of microconstituents (Chapter 9), and modeling microconstituents based on remote sensing and GIS techniques (Chapter 10). Part III encompasses details of the various physicochemical treatment techniques of microconstituents with five chapters. Chemical precipitation (Chapter 11), adsorption (Chapter 12), ion exchange (Chapter 13), filtration and membrane separation (Chapter 14), and advanced oxidation (Chapter 15) are covered in this part. The removal of microconstituents via biological treatment techniques is discussed in Part IV. Aerobic biological treatment (Chapter 16), anaerobic biological treatment (Chapter 17), bioelectrochemical systems (Chapter 18), and hybrid treatment solutions (Chapter 19) are presented in this part. Finally, Part V focuses on the aspects of sustainability and environmental management, including regulatory framework (Chapter 20), laboratory to field application (Chapter 21), and sustainability outlook (Chapter 22).

We hope this book will be of interest to students, scientists, engineers, government officers, process managers, and practicing professionals. As a reference, this book will help the readers readily find the information they are looking for.

The editors gratefully acknowledge the hard work and patience of all authors who have contributed to this book. The views or opinions expressed in each chapter of this book are those of the authors and should not be construed as opinions of the organizations they work for.

Rao Y. Surampalli
Tian C. Zhang
Chih-Ming Kao
Makarand M. Ghangrekar
Puspendu Bhunia
Manaswini Behera
Prangya R. Rout

List of Contributors

Mansi Achhoda
Department of Biotechnology
Thapar Institute of Engineering & Technology
Patiala
Punjab
India

Lavanya Adagadda
CSIR-National Environmental Engineering Research Institute
Nagpur
Maharashtra
India

Department of Civil Engineering
M.S. Ramaiah Institute of Technology
Bangalore
Karnataka
India

Richa Babbar
Department of Civil Engineering
Thapar Institute of Engineering & Technology
Patiala
Punjab
India

Manoj Baranwal
Department of Biotechnology
Thapar Institute of Engineering & Technology
Patiala
Punjab
India

Manaswini Behera
School of Infrastructure
Indian Institute of Technology
Bhubaneswar
Odisha
India

Sailesh N. Behera
Air Quality Laboratory
Department of Civil Engineering
Shiv Nadar University
Delhi-NCR
Greater Noida
Gautam Buddha Nagar
Uttar Pradesh
India

Centre for Environmental Sciences and Engineering
Shiv Nadar University
Delhi-NCR

Greater Noida
Gautam Buddha Nagar
Uttar Pradesh
India

Puspendu Bhunia
School of Infrastructure
Indian Institute of Technology
Bhubaneswar
Odisha
India

Indrajit Chakraborty
Department of Civil Engineering
Indian Institute of Technology
Kharagpur
West Bengal
India

Sasmita Chand
Centre of Sustainable Built
Environment
Manipal School of Architecture and
Planning
Manipal Academy of Higher Education
Manipal
Karnataka
India

Hung-Hsiang Chen
Department of Civil Engineering
National Chi Nan University
Puli
Nantou County
Taiwan

Ku-Fan Chen
Department of Civil Engineering
National Chi Nan University
Puli
Nantou County
Taiwan

Younggyun Choi
Department of Environmental & IT
Engineering
Chungnam National University
Daejeon
Republic of Korea

Sanket Dey Chowdhury
School of Infrastructure
Indian Institute of Technology
Bhubaneswar
Odisha
India

Sovik Das
Department of Civil Engineering
Indian Institute of Technology
Kharagpur
West Bengal
India

Rajesh Roshan Dash
School of Infrastructure
Indian Institute of Technology
Bhubaneswar
Odisha
India

H.N.P. Dayarathne
School of Engineering and
Mathematical Sciences
La Trobe University
Bendigo
Australia

Thakshila Nadeeshani Dharmapriya
Institute of Environmental
Engineering
National Sun Yat-sen University
Kaohsiung
Taiwan

Manikanta M. Doki
Department of Civil Engineering
Indian Institute of Technology
Kharagpur
West Bengal
India

Brajesh Kumar Dubey
Department of Civil Engineering
Indian Institute of Technology
Kharagpur
West Bengal
India

Makarand M. Ghangrekar
Department of Civil Engineering
Indian Institute of Technology
Kharagpur
West Bengal
India

School of Environmental Science and Engineering
Indian Institute of Technology
Kharagpur
West Bengal
India

Sanjoy Gorai
School of Energy & Environment
Thapar Institute of Engineering & Technology
Punjab
India

Nirmalya Halder
Department of Biotechnology
Thapar Institute of Engineering & Technology
Patiala
Punjab
India

K.S. Hariprasad
Department of Civil Engineering
Indian Institute of Technology Roorkee
India

Po-Jung Huang
Department of Chemical and Materials Engineering
National Central University
Taoyuan
Taiwan

Nagireddi Jagadeesh
Department of Civil Engineering
National Institute of Technology
Andhra Pradesh
Tadepalligudem
India

Sivaraman Jayaraman
Department of Biotechnology & Medical Engineering
National Institute of Technology
Rourkela
Odisha
India

Chih-Ming Kao
Institute of Environmental Engineering
National Sun Yat-sen University
Kaohsiung
Taiwan

Civil and Environmental Engineering Department
College of Engineering
University of Nebraska
Lincoln
Omaha, NE
USA

Nageshwari Krishnamoorthy
Department of Biotechnology &
Medical Engineering
National Institute of Technology
Rourkela
Odisha
India

Department of Biotechnology
Saveetha School of Engineering
Saveetha Institute of Medical and
Technical Sciences
Chennai
India

Vishnu Kumar
Air Quality Laboratory
Department of Civil Engineering
Shiv Nadar University
Delhi-NCR
Greater Noida
Gautam Buddha Nagar
Uttar Pradesh
India

Tsai Chi Kuo
National Taiwan University of Science
and Technology
Taipei
Taiwan

Fang-Yu Liang
Institute of Environmental
Engineering
National Sun Yat-sen University
Kaohsiung
Taiwan

Jun Wei Lim
HICoE-Centre for Biofuel and
Biochemical Research
Institute of Self-Sustainable Building
Department of Fundamental and
Applied Sciences
Universiti Teknologi PETRONAS
Seri Iskandar
Perak Darul Ridzuan
Malaysia

Wei-Han Lin
School of Environment
Tsinghua University
Beijing
PR China

Pu-Fong Liu
Institute of Environmental
Engineering
National Sun Yat-sen University
Kaohsiung
Taiwan

Bandita Mainali
School of Engineering
Faculty of Science and Engineering
Macquarie University
Sydney
Australia

Challa Mallikarjuna
School of Infrastructure
Indian Institute of Technology
Bhubaneswar
Odisha
India

Thi-Mahn Nguyen
Department of Civil Engineering
National Chi Nan University
Puli
Nantou County
Taiwan

C.S.P. Ojha
Department of Civil Engineering
Indian Institute of Technology Roorkee
India

Jiun-Hau Ou
Institute of Environmental Engineering
National Sun Yat-sen University
Kaohsiung
Taiwan

Balasubramanian Paramasivan
Department of Biotechnology & Medical Engineering
National Institute of Technology
Rourkela
Odisha
India

Aditya Parihar
Department of Civil Engineering
Thapar Institute of Engineering & Technology
Patiala
Punjab
India

Monali Priyadarshini
School of Environmental Science and Engineering
Indian Institute of Technology
Kharagpur
West Bengal
India

Rajat Pundlik
School of Infrastructure
Indian Institute of Technology
Bhubaneswar
Odisha
India

Rishabh Raj
School of Environmental Science and Engineering
Indian Institute of Technology
Kharagpur
West Bengal
India

Dwarikanath Ratha
Department of Civil Engineering
Thapar Institute of Engineering & Technology
Patiala
Punjab
India

Prangya Ranjan Rout
Department of Biotechnology
Thapar Institute of Engineering and Technology
Patiala
Punjab
India

Department of Biotechnology
Dr. B. R. Ambedkar National Institute of Technology Jalandhar
Punjab
India

Chinmayee Sahoo
Department of Biotechnology & Medical Engineering
National Institute of Technology
Rourkela
Odisha
India

Naresh Kumar Sahoo
Department of Chemistry
Institute of Technical Education and Research
Bhubaneswar

Odisha
India

S.M. Sathe
Department of Civil Engineering
Indian Institute of Technology
Kharagpur
West Bengal
India

Muhammad Kashif Shahid
Research Institute of Environment &
Biosystem
Chungnam National University
Daejeon
Republic of Korea

Meena Kumari Sharma
Department of Civil Engineering
Manipal University Jaipur
Rajasthan
India

Anoop Kumar Shukla
Manipal School of Architecture and
Planning
Manipal Academy of Higher Education
Manipal
Karnataka
India

Satyavati Shukla
Key Laboratory of Geospatial
Informatics
Guilin University of Technology
Guilin
PR China

Baranidharan Sundaram
Department of Civil Engineering
National Institute of Technology
Andhra Pradesh
Tadepalligudem
India

Rao Y. Surampalli
Global Institute for Energy,
Environment,
and Sustainability
Lenexa, KS
USA

Mudit Yadav
Air Quality Laboratory
Department of Civil Engineering
Shiv Nadar University
Delhi-NCR
Greater Noida
Gautam Buddha Nagar
Uttar Pradesh
India

Zong-Han Yang
Institute of Environmental
Engineering
National Sun Yat-sen University
Kaohsiung
Taiwan

Ying-Liang Yu
Institute of Environmental
Engineering
National Sun Yat-sen University
Kaohsiung
Taiwan

Civil and Environmental Engineering
Department
College of Engineering
University of Nebraska
Lincoln
Omaha, NE
USA

Alisha Zaffar
Department of Biotechnology &
Medical Engineering
National Institute of Technology
Rourkela
Odisha
India

Tian C. Zhang
Civil and Environmental Engineering
College of Engineering
University of Nebraska
Lincoln
Omaha, NE
USA

About the Editors

Dr. Rao. Y. Surampalli, Ph.D., P.E., Dist.M.ASCE, received his M.S. and Ph.D. degrees in Environmental Engineering from Oklahoma State and Iowa State Universities, respectively. He is a Registered Professional Engineer in the branches of Civil and Environmental Engineering, and also a Board Certified Environmental and Water Resources Engineer (BCEE, D.WRE) of the American Academy of Environmental Engineers (AAEE) and American Academy of Water Resources Engineers. His career in private practice, government, university and applied research has given him the opportunity to experience and appreciate the varied interests and challenges of the environmental engineering profession, and has conducted applied research on over fifty (50) environmental science and engineering topics. He is an Adjunct Professor in five (5) universities and Distinguished/ Honorary Visiting Professor in eight (8) universities. Currently, he serves, or has served, on more than 75 national and international committees, review panels, or advisory boards including the ASCE National Committee on Energy, Environment and Water Policy. A Distinguished Engineering Alumnus of both the Oklahoma State and Iowa State Universities, Dr. Surampalli has received over 30 national awards and honors from ASCE, WEF, IWA, AAEE, NSPE, AAES; and is an elected Fellow of the American Association for the Advancement of Science (F.AAAS), an elected Member of the European Academy of Sciences and Arts (EASA), an elected Member of the Russian Academy of Engineering (RAE), an elected Distinguished Fellow of the International Water Association (Dist.F.IWA) and Fellow of the Water Environment Federation (F.WEF), an elected Member of the U.S. National Academy of Construction (NAC) and recognized as a Distinguished Member of the American Society of Civil Engineers (Dist.M.ASCE) – the highest honor of ASCE. He also is Editor-in-Chief of the ASCE *Journal of Hazardous, Toxic, and Radioactive Waste*, and past Vice-Chair of Editorial Board of *Water Environment Research* journal. He has authored more than 700 technical publications in journals and conference proceedings, including more than 370 refereed journal articles, 22 patents, 25 books, and 171 book chapters.

Dr. Tian C. Zhang, Ph.D., P.E., is a professor in the Department of Civil Engineering at the University of Nebraska–Lincoln (UNL). He received his Ph.D. in environmental engineering from the University of Cincinnati in 1994 and joined the UNL faculty in August 1994. Professor Zhang teaches courses related to water/wastewater treatment, remediation of hazardous wastes, and non-point pollution control. Professor Zhang's research involves fundamentals and applications of nanotechnology and conventional technology for water, wastewater, and stormwater treatment and management, remediation of contaminated environments, and detection/control of emerging contaminants in the environment. Professor Zhang has published more than 190 peer-reviewed journal papers, 82 books chapters, and 17 books since 1994. Professor Zhang is a Diplomate of Water Resources Engineer (D.WRE) of the American Academy of Water Resources Engineers, a Board Certified Environmental Engineer (BCEE) of the American Academy of Environmental Engineers, an elected Distinguished Member of the American Society of Civil Engineers (Dist.M.ASCE), an elected Fellow of American Association for the Advancement of Science (F.AAAS), and an elected member of European Academy of Sciences and Arts (EASA). Professor Zhang is an Associate Editor of *Journal of Environmental Engineering* (since 2007), *Journal of Hazardous, Toxic, and Radioactive Waste* (since 2006), *and Water Environment Research* (since 2008). He has been a registered professional engineer in Nebraska since 2000.

Dr. Chih-Ming Kao, Ph.D., P.E., BCEE, D.WRE, F.IWA, F.WEF, F.ASCE, F.AAAS is a Distinguished chair professor in the Institute of Environmental Engineering at National Sun Yat-sen University, Taiwan. Prof. Kao is also the Coordinator of Environmental Engineering Program at Ministry of Science and Technology, past President of The Chinese Institute of Environmental Engineering, and former President of The Taiwan Association of Soil and Groundwater Environmental Protection. Prof. Kao received his M.S. and Ph.D. degrees in Civil and Environmental Engineering from North Carolina State University in 1989 and 1993, respectively. He is a fellow member of International Water Association (IWA), American Society of Civil Engineers (ASCE), an Academician of European Academy of Sciences and Arts (EASA), a fellow member of American Association for the Advancement of Science (AAAS), a fellow member of Environment and Water Resource Institute (EWRI), a Registered Professional Engineer in the branch of Civil Engineering, a Certified Ground Water Professional, and a Professional Hydrologist in the United States. He is also a Diplomate of the American Academy of Environmental Engineers and Diplomate of American Academy of Water Resources Engineers. Prof. Kao received the "Distinguished Researcher Award" from Taiwan Ministry of Science and Technology in 2011 and 2015. He is also the receiver of the "Distinguished Engineer Professor Award" from Chinese Institute of Engineers in 2012, and receiver of the "Distinguished Honor Award" from C.T. Ho Foundation in 2013. He also received several awards from ASCE including the State of the Art of Civil Engineering Award in 2013, Hering Medal, Samuel Arnold Greely Award in 2012, and distinguished theory-oriented paper award in 2008 and 2015. He has over 350 refereed publications.

Prof. Makarand M. Ghangrekar, Fellow INAE, MASCE, is Institute Chair Professor in the Department of Civil Engineering and, and Heading two academic units, School of Environmental Science and Engineering and PK Sinha Centre for Bioenergy and renewables, and also Professor In-Charge, Aditya Choubey Centre for Re-Water Research at Indian Institute of Technology Kharagpur. He had been visiting Scientist to Ben Gurion University, Israel and University of Newcastle upon Tyne, UK under Marie Curie Fellowship by European Union and had stint as faculty of various capacities in renowned engineering colleges and research institutes. He has been working in the areas of anaerobic wastewater treatment, bioenergy recovery during wastewater treatment using microbial fuel cell and bio-electrochemical systems. He is recognized worldwide in scientific community for his research contribution in the development of bio-electrochemical processes and his research group stands among the top five research laboratories in the world in terms of scientific publications. The first of its kind MFC based onsite toilet waste treatment system 'Bioelectric toilet' developed by him received wide publicity in electronic and print media. He has successfully completed multinational collaborative projects with European Countries and few of the projects are ongoing. He has also provided design of industrial wastewater and sewage treatment plants in India and abroad. He has been working on setting up wastewater treatment plants to produce reusable quality treated water at affordable cost. He has guided 21 Ph.D. Research Scholars and 50 Master student's projects. He has contributed 204 research papers in journals of international repute, out of these 138 papers are on microbial fuel cell and also contributed 44 book chapters. His research work has been presented in more than 250 conferences. He has delivered invited lectures in the many reputed universities in the world.

Dr. Puspendu Bhunia, Ph.D., is presently holding the Professor position at the School of Infrastructure, Indian Institute of Technology Bhubaneswar, India. He received his B.E. degree in Civil Engineering from Indian Institute of Engineering Science and Technology, Shibpur, India in 2002, his M.Tech. and Ph.D. degree in Environmental Engineering from Indian Institute of Technology Kharagpur, India in 2004, and in 2008, respectively. He joined the Indian Institute of Technology Bhubaneswar faculty in July 2009. Dr. Bhunia teaches courses related to water/wastewater treatment, and remediation of hazardous wastes. His research interest includes sustainable natural treatment technologies of wastewater, nutrient removal, and green technologies for waste remediation. Dr.

Bhunia has authored 50 technical publications in refereed journals, book chapters and conference proceedings. He has presented several expert talks at different technical conferences organized nationally and internationally. Dr. Bhunia's research work has been recognized, including the Best Practice Oriented Paper award from ASCE. Dr. Bhunia is member of several professional organizations and also serves as an Associate Editor of ASCE Journal of Hazardous, Toxic, and Radioactive Waste and is reviewer for more than 30 international peer reviewed journals.

Dr. Manaswini Behera, Ph.D., is an Associate Professor of Environmental Engineering in the School of Infrastructure, Indian Institute of Technology, Bhubaneswar, which is among the top 20 Indian institutions. She has received her Ph.D. in Environmental Engineering from Indian Institute of Technology Kharagpur and master's degree in environmental engineering and management from Indian Institute of Technology Delhi. She has joined IIT Bhubaneswar in 2014. She is an Associate Editor of the ASCE Journal of Hazardous, Toxic and Radioactive Waste and and is reviewer for more than 40 international peer reviewed journals. She has published 33 refereed articles in well-known journals, 31 international conference presentations and proceedings, and 21 refereed book chapters. Her area of research is bioenergy recovery during treatment of industrial wastewater and solid waste in bioelectrochemical systems, development of separators for microbial fuel cells, wastewater treatment and reuse. She has successfully completed three sponsored research projects. She is the principal investigator for the ongoing project, SARASWATI-2.0 (INR 12 million) jointly funded by the European Union and the Department of Science and Technology, Government of India. She has guided 2 PhD students and is at present supervising six PhD research scholars. She has filed a patent on using ceramic separator as a cost-effective alternative to expensive polymeric membrane in microbial fuel cells.

Dr. Prangya Ranjan Rout, PhD., is presently serving as an Assistant Professor in the Department of Biotechnology, Thapar Institute of Engineering and Technology (TIET), Patiala, Punjab India. He holds B.Tech and M.Tech degrees in Biotechnology and a Ph.D. in Environmental Engineering. His research interest lies in the domain of bioreactor design, anaerobic digestion, bioconversion of wastes to wealth, emerging contaminants removal, membrane technology, resource recovery and reuse, and wastewater treatment. He has authored over 80 publications, including refereed journal articles (33), edited books (2), book chapters (24), national (12) and international conference (11) presentations, technical notes (1), and a granted patent to his credit. Some of the awards he has received include Odisha Young Scientist Award 2017, Best Practice Oriented Paper 2019 by American Society of Civil Engineering (ASCE), and Outstanding Reviewer 2019 by ASCE. He is an Associate Editor of the ASCE Journal of Hazardous, Toxic, and Radioactive Wastes and has served as a guest editor of a special issue of the journal. He is an Editorial Board Member of Journal of Water Process Engineering, Elsevier. He is also actively involved in Editing contributed book volumes for internationally renowned publishers like CRC Press, John Wiley & Sons, Elsevier, ASCE, etc.

Part I

Fundamental Ideas Regarding Microconstituents in the Environment

1

Introduction to Microconstituents

Manaswini Behera[1], Prangya Ranjan Rout[2,3], Puspendu Bhunia[1], Rao Y. Surampalli[4], Tian C. Zhang[5], Chih-Ming Kao[5,6], and Makarand M. Ghangrekar[7,8]

[1]School of Infrastructure, Indian Institute of Technology Bhubaneswar, Odisha, India
[2]Department of Biotechnology, Thapar Institute of Engineering and Technology, Patiala, Punjab, India
[3]Department of Biotechnology, Dr. B. R. Ambedkar National Institute of Technology Jalandhar, Punjab, India
[4]Global Institute for Energy, Environment, and Sustainability, Lenexa, KS, USA
[5]Civil and Environmental Engineering, College of Engineering, University Nabraska, Lincoln, Omaha, NE, USA
[6]Institute of Environmental Engineering, National Sun Yat-sen University, Kaohsiung, Taiwan
[7]Department of Civil Engineering, Indian Institute of Technology Kharagpur, West Bengal, India
[8]School of Environmental Science and Engineering, Indian Institute of Technology Kharagpur, West Bengal, India

1.1 Introduction

Microconstituents (MCs) are a comparatively new group of unregulated natural and man-made substances, including elements and inorganic and organic chemicals, detected within water and the environment that can cause detrimental effects to aquatic environment and human health. Humans, aquatic organisms, and other wildlife can be exposed to these compounds through environmental contact and consumption of foods and water that are contaminated with MCs. The term "microconstituents" has been developed by the Water Environment Federation (WEF) (Cleary 2008). These MCs can make their way into the environment through a variety of routes such as effluent from industries, wastewater treatment plant (WWTP) effluent, agricultural run-off, run-off from feedlot operations, and other nonpoint sources that are more difficult to quantify.

Microconstituents are also called Micropollutants, Emerging Contaminants, or Contaminants of Emerging Concern (CEC) (Bhandari et al. 2009). The absence of sensitive analytical methods was the major reason for non-detection of the existence of MCs in environmental samples before the late 1990s, as they are present at comparatively low concentrations, typically from a few ng/L to a few hundred μg/L in the aquatic environment. However, even at low concentration levels, many of these pollutants have the ability to cause substantial ecological and/or human health risks. They are considered as CECs because they still remain unregulated or are currently undergoing a regularization process. However, the directives and legal frameworks are not yet in place.

The presence of MCs in the environment can have deleterious effects on aquatic and human life via interference with the endocrine system of living organisms, antimicrobial

Microconstituents in the Environment: Occurrence, Fate, Removal, and Management, First Edition.
Edited by Rao Y. Surampalli, Tian C. Zhang, Chih-Ming Kao, Makarand M. Ghangrekar, Puspendu Bhunia, Manaswini Behera, and Prangya R. Rout.

resistance, and accumulation in soil. Uptake of MCs by plants from contaminated soils and their accumulation in the food chain can be another instrumental problem for the ecosystem and human health. Although most of the MCs have been detected at concentrations of μg/L in water supplies, often lower than their toxic concentrations, they can have longer-term effects than previously believed as some of them may be mutagenic, carcinogenic, and very persistent, with a tendency to bioaccumulate. Conventional WWTPs are not designed to remove these micropollutants because they are either transformed or remain unchanged as they enter the conventional WWTPs. Therefore, many of these MCs appear in the effluents and, in most cases, find their way into surface water and then into drinking water, exposing us to these substances and their possible effects. Since 2000, increased awareness of the risks posed by emerging contaminants to human health has raised concerns for water quality improvement. The presence of MCs in water, even at very low concentrations, has raised concerns among stakeholders, such as drinking water regulators, governments, water suppliers, and the public, regarding the potential risks to human health from exposure to traces of these pollutants via drinking water. Hence, the development of strategies and technologies for their removal and risk management should be of prime importance (Bhandari et al. 2009; Salamanca et al. 2021).

The existence of CECs in the environment is not a new phenomenon and can be dated back to 2000 years ago with the emergence of the oldest global contaminant, lead, due to over exploitation of lead deposits by the Romans and Greeks (Rout et al. 2021). Subsequently, the presence of CECs gradually swept through the traditional contaminants to the present-day nanomaterials, pharmaceuticals, personal care products, disinfection by-products, etc. The first documented awareness of emerging contaminants should probably be attributed to Rachel Carson for her 1962 book "Silent Spring" that commended the link between widespread usage of dichloro-diphenyl-trichloroethane (DDT) and environmental hazards (Carson 2002; Sauvé and Desrosiers 2014). However, extensive investigation of the environmental occurrence of CECs and their detection have been extensively carried out only during the last two decades. The slow development of sensitive analytical techniques to detect the very low concentrations of CECs present in environmental samples is the major cause of the time lag in their detection (Noguera-Oviedo and Aga 2016). Conventionally, gas chromatography with mass spectrometry (GC/MS) is used for the analysis of CECs; however, the majority of CECs are not responsive to this technique. The introduction of liquid chromatography with mass spectrometry (LC/MS) could help in realizing the ubiquitous nature of emerging contaminants (ECs) in the environment (Kolpin et al. 2002). Further improvement in mass accuracy and resolving power of quadrupole time of flight (Q-TOF) MS and orbitrap MS have accelerated ECs-related research exponentially in recent years (Rout et al. 2021). Bhandari et al. (2009) have presented an overview of CECs and their classification and have summarized the analytical methods used for separation and quantification of ECs and describes molecular biology approaches to identify organisms capable of degrading these chemicals.

Many technologies for the removal of CECs have been developed, and they can be categorized broadly into natural attenuation, conventional, and advanced treatment processes. It has been reported that even conventional WWTPs are able to remove some CECs efficiently, although they are not designed to eliminate these pollutants at low concentrations (Alvarino et al. 2018). Conventional treatment technologies, such as membrane filtration, activated carbon mediated adsorption, ozonation, etc., have also exhibited effective removal of ECs (Pesqueira et al. 2020). Advanced treatment technologies like advanced oxidation processes, constructed wetlands, bioelectrical systems, enzymatic treatment, etc., have been proposed in the past few years. Despite the availability of these treatment technologies, the removal efficiency of most of the employed technologies for CECs is not satisfactory, depending mainly on the physico-chemical properties of ECs and treatment conditions (Rout et al. 2021). The CECs will continue to be an enduring target of the scientific community due to ever increasing synthesis of a large variety of new chemicals in proportion to industrial developments and technological advancements.

The objective of this chapter is to introduce the MCs and explore their various classifications, properties, and sources, as well as their impact on the environmental ecosystem and human health. The description of all the other chapters of the book is presented in Section 1.6.

1.2 Classification of Microconstituents

The wide range of compounds that comprise MCs continues to expand as new chemicals are being identified as part of this category due to advances in analytical procedures. A broad variety of compounds can be included in the MCs, including pharmaceuticals and personal care products, pesticides/herbicides, food additives, artificial sweeteners, disinfection by-products, flame/fire retardants, and surfactants. The classification of MCs is presented in Figure 1.1

1.2.1 Pharmaceuticals and Personal Care Products

Pharmaceuticals have been receiving increasing attention as potential bioactive chemicals in the environment over the last two decades. Pharmaceuticals are continuously entering into the environment and are prevalent at small concentrations (Kolpin et al. 2002). They affect water quality and potentially impact drinking water supplies, human

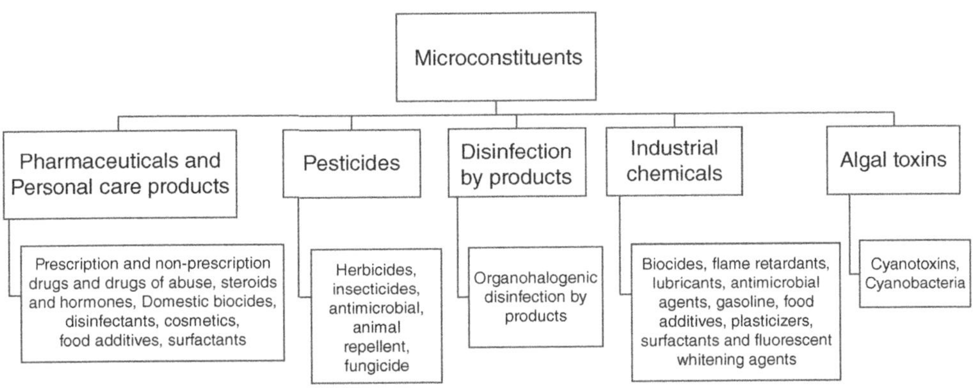

Figure 1.1 Classification of microconstituents.

health, and ecosystems (Sirés and Brillas 2012). Although pharmaceuticals have been present in water for decades, their concentration in the environment has only recently begun to be quantified and recognized as potentially unsafe to ecosystems (Rivera-Utrilla et al. 2013).

Pharmaceutical compounds most commonly detected in water bodies or WWTP effluents are active ingredients of prescription and non-prescription drugs for human and veterinary use, and illicit drugs such as the following: i) anti-inflammatories and analgesics (paracetamol, acetylsalicylic acid, ibuprofen, diclofenac); ii) antibiotics (tetracyclines, macrolides, penicillins, quinolones, sulfonamides, fluoroquinolones, chloramphenicol, iimidazole derivatives); iii) antiepileptics (carbamazepine); iv) lipid-lowering drugs (fibrates); v) β-blockers (atenolol, propanolol, metoprolol); vi) antiulcer drugs and antihistamines (ranitidine, famotidine); vii) antidepressants (benzodiazepines); and viii) other substances (cocaine, barbiturates, methadone, amphetamines, opiates, heroin, and other narcotics). These pharmaceutical compounds have been detected in the surface and groundwaters of Brazil (Ternes et al. 1999), Canada (Miao et al. 2004), China (Yuan et al. 2013), Germany (Hirsch et al. 1999; Ternes and Hirsch 2000), Holland (Belfroid et al. 1999), India (Mutiyar and Mittal 2014), Italy (Castiglioni et al. 2004), Spain (Carballa et al. 2005), Switzerland (Soulet et al. 2002), and the United States (Drewes et al. 2001), among others. As shown in Table 1.1, the concentrations of pharmaceuticals are typically $<0.1\,\mu g/L$ (or 100 ng/L) in surface waters, groundwater, and partially treated water and are generally $<0.05\,\mu g/L$ (or 50 ng/L) in treated water (WHO 2012). Several hormones, such as estrone (E1), 17-oestradiol (E2), 17-ethinylestradiol (EE2), estriol (E3), equilin, 17-oestradiol, norethindrone, equilenin, and mestranol, are listed as priority drinking water contaminants based on their health effects and environmental occurrence.

Personal care products (PCPs) comprise a diverse group of common household materials used for health, beauty, and cleaning purposes. PCPs include a large number of synthetic chemical compounds in daily use, including products such as soaps, lotions, toothpaste, fragrances, cosmetics, and sunscreens (Brausch and Rand 2011). These also include disinfectants, insect repellents, preservatives, and UV filters, among others. Some of these PCPs are considered as CECs due to their presence and negative impact on aquatic ecosystems.

Table 1.1 The presence of selected microconstituents in water and wastewater samples.

Emerging Contaminant	Sample	Concentration (ng/L)	Reference
Pharmaceuticals and Personal Care Products			
Sulfamethoxazole	Effluent of WWTP.	374.3–1309.5	Archundia et al. 2017
Cefalexin	Hospital WW.	2×10^8	Estrada et al. 2012
Ofloxacin	Effluent of WWTP.	5×10^3	Klamerth et al. 2012
Aminopyrine	Drinking water.	0.17–0.64	Cai et al. 2015
Methyl dihydrojasmonate	Effluent of WWTP.	3500	Reyes-Contreras et al. 2011
Tonalide	Urban WW.	540	Ávila et al. 2015

Table 1.1 (Continued)

Emerging Contaminant	Sample	Concentration (ng/L)	Reference
Galaxolide	Effluent of WWTP.	3500	Reyes-Contreras et al. 2011
Estrogen			
Estrone	River water.	26 ± 0.3–124 ± 15	Williams et al. 2019
Estrone	Drinking water.	0.2–0.6	Williams et al. 2003
17β-Estradiol	Surface water.	0.1–0.6	Dorabawila and Gupta 2005
Pesticides			
Diuron	Effluent of WWTP.	1×10^3	Prieto-Rodríguez et al. 2013
Atrazine	Effluent of WWTP.	1	Reungoat et al. 2012
Triclosan	Effluent of WWTP.	5×10^3	Klamerth et al. 2012
2,4-D	WW.	421	Nguyen et al. 2013
Endocrine Disrupting Substances			
Estradiol	Effluent of WWTP.	0–123	Lin et al. 2020
Estriol	Effluent of WWTP.	0–48	Lin et al. 2020
17α-Ethynylestradio	Effluent of WWTP.	0–33	Lin et al. 2020
Estriol	River water.	0–53	Lin et al. 2020
Surfactants			
Linear alkylbenzene sulphonic acid	Surface water.	4.16×10^5	Ying 2006
Alkyl ethoxysulphate	Effluent of WWTP.	3×10^3–12×10^3	Sibila et al. 2008
Quaternary ammonium compound	Sewage effluent.	6.2×10^4	Versteeg et al. 1997
Alkylphenol ethoxylate	Sewage effluent.	3.32×10^5	Ying 2006
Artificial Sweeteners			
Sucralose	Effluent of WWTP.	3.88×10^4	Subedi and Kannan 2014
Aspartame	Effluent of WWTP.	140	Subedi and Kannan 2014
Saccharin	Effluent of WWTP.	420	Subedi and Kannan 2014
Acesulfame	Effluent of WWTP.	2.14×10^3	Subedi and Kannan 2014
Plasticizers			
Bisphenol A	Effluent of WWTP.	4.8–258	Loraine and Pettigrove 2006
Tri-isobutylphosphate	Effluent of WWTP.	160	Reemtsma et al. 2008
Tributyl phosphate	Effluent of WWTP.	520	Reemtsma et al. 2008

PCPs are found in water bodies and WWTP effluents in concentrations ranging from ng/L to μg/L (Table 1.1). Fragrances, antiseptics, and sunscreens have been the most reported groups of PCPs in water bodies. The largest number of PCPs documented as CECs are found in WWTP effluents, which evidence the anthropological contribution of PCPs to water bodies. These compounds are found on all continents (Montes-Grajales et al. 2017).

1.2.2 Pesticides

Pesticides are substances that are used in agriculture to control insects, nematodes, rodents, weeds, and fungi. The term "pesticide" includes the following: herbicides, insecticides, nematicides, molluscicides, piscicides, avicides, rodenticides, bactericides, insect repellents, animal repellents, antimicrobials, and fungicides (Randall 2014). Herbicides are the most common pesticide, which account for ~80% of all pesticides use. Currently, >1100 pesticides (insecticides, acaricides, herbicides, fungicides, rodenticides) are marketed for public use in European countries (Affum et al. 2018; Burri et al. 2019). Although application of pesticides is required to achieve high yields of agricultural products, their application has resulted in food and water contamination (El-Nahhal 2020) and continuous damage to ecosystems because >90% of applied pesticides do not reach the target species (Llorent-Martínez et al. 2011).

Different pesticides and their transformed compounds are detected in surface and groundwater in many countries. Organochlorine (OC) insecticide residues (aldrin, chlordane, DDT, dieldrin, endrin, α-endosulfan, β-endosulfan, HCH, heptachlor, lindane, methoxychlor, toxaphene, γ-chlordane, γ-HCH (α-HCH, β-HCH, γ-HCH, δ-HCH)), and four metabolites (1,1-dichloro-2,2-bis(p-chlorophenyl) ethylene (DDE), 1,1-dichloro-2,2-bis(p-chloro-phenyl) ethane (DDD), heptachlor-epo, endrin aldehyde) have been found in water samples in many countries. OC residues were found in water samples in the concentration range of 0.001–2.65 μg/L in China (Zhou et al. 2008; Li et al. 2012; Wu et al. 2014; Liu et al. 2016), 0.01–0.34 μg/L in India (Mohapatra et al. 1995; Kaushik et al. 2012), 0.007–0.159 μg/L in Turkey (Bulut et al. 2010), 0.02–12.82 μg/L in Lebanon, 0.01–0.03 μg/L in South Africa (Fatoki and Awofolu 2004), 0.003–0.09 μg/L in Mexico (Díaz et al. 2009), 0.01–0.04 μg/L in Ghana (Fosu-Mensah et al. 2016; Affum et al. 2018), 0.02–0.74 μg/L in the Philippines (Navarrete et al. 2018), 0.10–6.00 μg/L in Egypt (Derbalah et al. 2014), 0.0004–0.22 μg/L in Canada (Woudneh et al. 2009), Ireland (McManus et al. 2013), and the USA (Eitzer and Chevalier 1999). Approximately 27 organophosphate (OP) insecticide residues were detected in water samples from many countries (El-Nahhal and El-Nahhal 2021). OP residues include azinphos-methyl, chlorfenvinphos, chlorpyrifos, cyanophos, diazinon, dichlofenthion, dichlorvos, dicrotophos, dimethoate, ethion, ethoprophos, fenitrothion, fenthion, isofenphos-oxon, malathion, mephosfolan, methamidophos, monocrotophos, omethoate, parathion-ethyl, parathion-methyl, phosphate, photoset, profenofos, pyridafenthion, thionazin, and triazophos.

1.2.3 Disinfection By-Products

Disinfection by-products (DBPs) are a kind of secondary pollutant produced during water disinfection treatment. They are frequently detected in the urban water cycle and reported to have (eco)toxicological impacts on aquatic systems and human health. Thus, DBPs are considered as global environmental CECs. Disinfectants are powerful oxidants that oxidize organic matter and bromide naturally present in most source waters (rivers, lakes, many groundwaters) and form DBPs that kill the harmful microorganisms. Chlorine, ozone, chlorine dioxide, and chloramines are the most common disinfectants in use, and each produces its own group of chemical DBPs in treated water.

When chlorine is used as a disinfectant, the most commonly found DBPs are the trihalomethanes (THM), halogenated acetic acids, halogenated acetonitriles, chloral hydrate, and chlorinated phenols. Alternative disinfectants other than chlorine used to treat drinking water are chlorinated furanone MX (multispectral), halopicrins, cyanogen halides, haloketones, and haloaldehydes. Bromo-trihalonitromethanes, iodo-trihalomethanes, dihaloaldehydes, MX (3-chloro-4-(dichloromethyl)-5-hydroxy-2(5H)-furanone), and brominated forms of MX are formed at higher levels. Specific DBPs of emerging toxicological interest include brominated and iodinated compounds, such as bromonitromethanes, iodo-trihalomethanes, iodo-acids, and brominated forms of MX, as well as nitrosodimethylamine (NDMA) (Richardson 2003).

1.2.4 Industrial Chemicals

Chemical compounds used in industries, such as perfluoroalkyl compounds, flame retardants (FRs), plasticizers, and preservatives, have been recently identified as CECs. Many of them are environmentally persistent compounds. The recent development of highly selective and sensitive analytical techniques has been key in identifying and bringing these organic contaminants into focus. Many different synthetic organic compounds are produced and used in large quantities worldwide for different purposes. Thousands of compounds are used as intermediates in the chemical industry (plasticizers, dyes, resins) or as food additives, antioxidants, surfactants, and detergents.

Industrial chemicals such as corrosion inhibitors (1,2,3-Benzotriazole, benzothiazol-2-sulfonic acid), perfluoroalkyl compounds (perfluorooctane sulfonic acid and its derivatives), flame retardants (polybrominated diphenyl ethers (PBDE)), 3,3′, 5,5′ tetrabromobisphenol A (TBBPA), C10–C13 polychlorinated alkanes, tris(2-chloroethyl) phosphate, plasticizers (bisphenol A, phthalates, e.g., di-2-ethylhexyl phthalate), alkylphenol ethoxylate, surfactants (nonylphenols (4-nonylphenol), octylphenol (4-(1,1′, 3,3′-tetramethyl-butyl)-phenol)), polycyclic aromatic hydrocarbons (PAHs) (naphthalene, benzo[a]pyrene), volatile organic compounds (benzene, CCl_4, 1,2-dichloroethane, 1,2-dichloroethene, dichloromethane, styrene, tetrachloroethene or perchloroethylene (PCE), toluene, $CHCl_3$, xylenes), gasoline additives (methyl tert-butyl ether (MTBE), dialkyl ethers), antifouling compounds (dibutyl tin ion, irgarol), antioxidants (2,6-Di-tert-butylphenol), and many others are identified as CECs (Calvo-Flores et al. 2018). These compounds are often found in surface waters, which receive treated or untreated industrial effluent as a discharge.

1.2.5 Algal Toxins

In the last few decades, there has been a growing interest in the occurrence of cyanotoxins and their potential toxicity in the aquatic environment. Harmful algal blooms are overgrowths of algae in water. Some produce dangerous toxins in fresh and marine water. Algal toxins are extremely potent neurotoxins or hepatotoxins that are produced from dinoflagellates, diatoms, or cyanobacteria (blue-green algae). The saxitoxins, anatoxins, microcystins, nodularins, and cylindrospermopsin produced by cyanobacteria are considered as CECs and are currently on the US EPA's contaminant candidate list (CCL) list (Richardson 2003).

1.3 Source of Microconstituents

The production and use of MCs have increased as a result of the continuous development of anthropogenic activities (industry, agriculture, health). The presence of CECs in the environment is the result of uncontrolled urbanization, development of industry, healthcare activities essential to support human well-being, agriculture, and transportation activities. CECs include a wide range of substances, produced by humans, considered indispensable for modern society. The release of MCs into the environment depends on different sources and pathways. An understanding of origins and sources of these contaminants is required so that knowledge gaps can be filled, and future research or actions can be more effectively undertaken to manage the concerns due to these contaminants.

1.3.1 Source of Pharmaceutical and Personal Care Products (PPCPs) in the Environment

In the past three decades, pharmaceutical residues have been discovered in almost all environmental matrices on every continent. This includes surface waters (lakes, rivers, streams), sea water, groundwaters, wastewater treatment plants (influents, effluents), soils, and sludges. Once PPCP residues enter water and soil, they are also taken up the plants growing in these soils or waters (Figure 1.2).

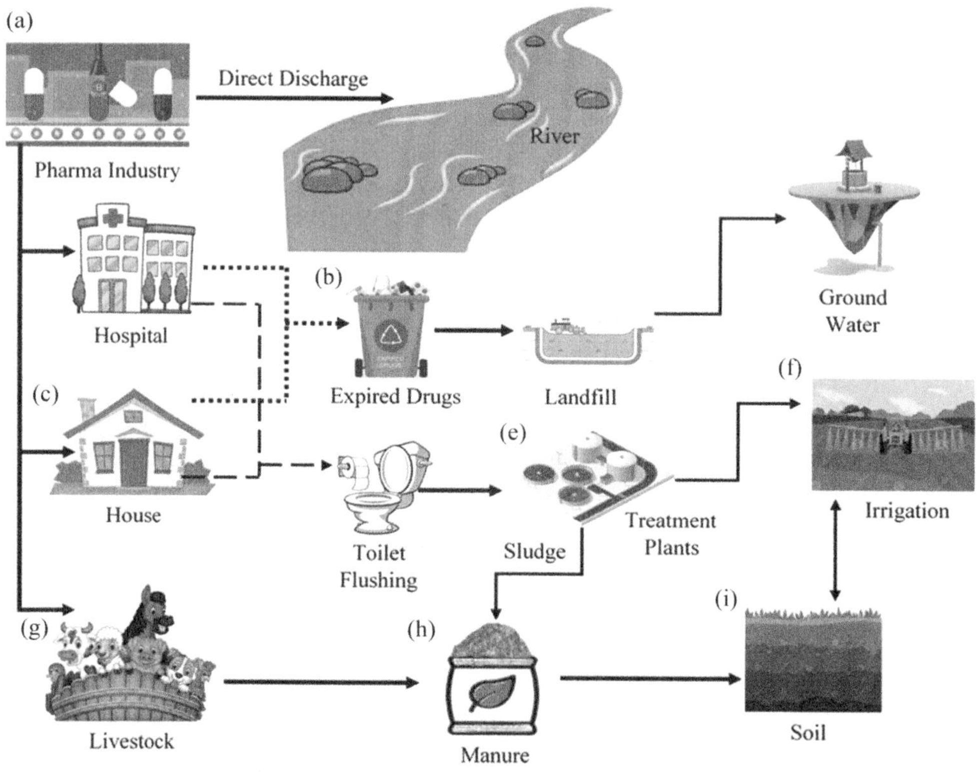

Figure 1.2 Source and pathway of pharmaceuticals and personal care products in the environment. *Sources:* (a) lucadp/Adobe Stock, (b) Derariad/Adobe Stock, (c) Oleksandr/ Adobe Stock, (e) Weenee/ Adobe Stock, (f) Seahorsevector/Adobe Stock, (g) tigatelu/Adobe Stock and (i) Taras/Adobe Stock.

Urban wastewater is the dominant emission pathway for PPCPs. The PPCPs used in households are discharged into the wastewater and reach the WWTPs. WWTPs have never been designed for complete removal of PPCPs; instead, they are generally designed to handle easily and moderately degradable organics in the mg/L range. Remediation efficiencies can be <10% in the case of such PPCPs. The residues are left behind in the effluent and find their way into the surface water.

The second most commonly recorded discharge source for the PPCPs is hospitals and healthcare facilities. A multitude of different pharmaceuticals are found in the sewage effluent of healthcare facilities, often at high concentrations. Most hospitals do not have on-site sewage treatment plants and are usually connected directly to urban sewage systems. Another important source for pharmaceuticals is commercial animal husbandry. Veterinary pharmaceuticals are often administered to livestock to treat or prevent diseases. Veterinary drugs and their metabolites excreted in the animals' urine and feces reach surface waters and groundwater in runoff after precipitation events. Other important sources that can cause confined increases of pharmaceutical concentrations in environmental matrices are pharmaceutical manufactures, aquaculture facilities, and wastewater used to irrigate field crops. Pharmaceutical production industries in particular can cause high concentrations in the range of milligrams per liter in rivers adjacent to production facilities, especially in developing and emerging countries. The source and fate of PPCPs in the environment can be summarized as follows:

i) Metabolic excretion (unmetabolized drug, drug conjugates, bio-active metabolites) by humans and pets after individual use of drugs;
ii) Disposal of unused and expired drugs into the sewage system;
iii) Release of treated/untreated hospital wastewater into the sewage system;
iv) Treated effluent from domestic WWTPs discharged to surface waters, re-injected to recharge aquifers, recycled/reused for irrigation or domestic use;
v) Discharge of untreated sewage directly to surface waters (straight-piping from homes);
vi) Use of sewage solids for soil amendments and fertilization and release from agricultural run-off to surface water or percolation through soil to reach groundwater;
vii) Release to surface water through aquaculture, medicated feed, and subsequent excreta; and
viii) Disposal to dump yards and landfills as domestic refuse and leaching from the dump yards and poorly engineered landfills to reach surface water and groundwater.

1.3.2 Source of Pesticides in the Environment

Pesticide use has grown in order to supply food to the exploding population. Different pesticides (herbicides, insecticides, fungicides, etc.) are being used worldwide to control insects, weeds, and other organisms that attack food crops. Water is one of the main ways that pesticides are transported (Figure 1.3) from the areas where they are applied to other locations where they may cause health problems. Pesticides, like most other water contaminants, enter the hydrological system from widely dispersed nonpoint sources and from point sources that are associated with specific points of release. As the dominant sources of

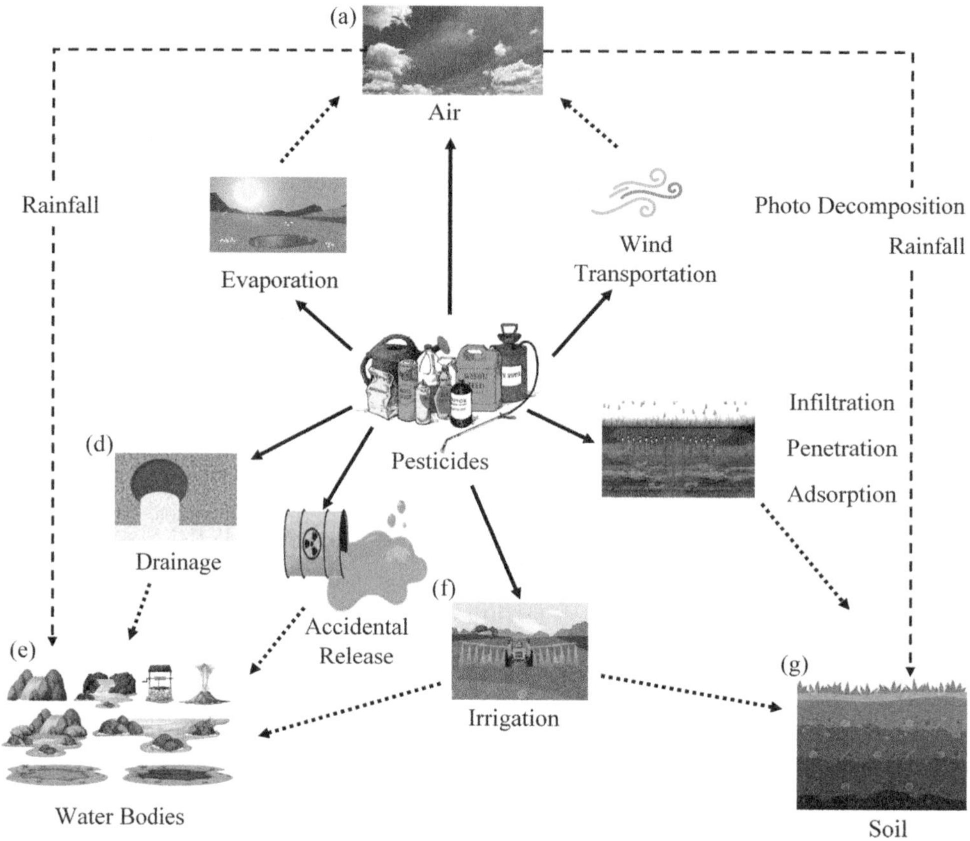

Figure 1.3 Source and pathways of pesticides in the environment. *Sources:* (a) Magann/Adobe Stock, (b) anuwat/Adobe Stock, (e) Blueringmedia/ Adobe Stock, (f) Seahorsevector/Adobe Stock (g) Taras / Adobe Stock.

pesticides found in streams and groundwater, nonpoint sources include runoff to streams from agricultural and urban land, rights-of-way, turf areas, seepage to groundwater in areas where pesticides are used, and deposition of pesticides from the atmosphere. Probable point sources of pesticides include pesticide manufacturing plants, mixing-and-loading facilities, spills, wastewater recharge facilities (wells or basins), waste disposal sites, and sewage treatment plants. Point-source pollution originates from specific, identifiable locations, such as the following: i) pesticide manufacturing plants; ii) wastewater recharge facilities (wells or basins); iii) waste disposal sites; iv) sewage treatment plants; v) a pesticide spill entering a storm sewer; vi) back-siphoning of pesticides into water supplies; vii) contaminated surface water entering sinkholes, viii) repeated pesticide spills at mixing and loading sites; ix) careless spilling of wash water at equipment cleanup sites; x) improper handling of spills and leaks at storage sites; and xi) improper disposal of containers and excess pesticides.

Once pesticides and the new compounds, formed by the transformation of a pesticide by chemical or biological reactions, reach the atmosphere, streams, or groundwater, they move through the hydrological system with air, water, or particles, depending on the chemical and physical properties of the compounds. The pathways of pesticides in different environmental matrices can be summarized as follows:

i) Pesticides are transported to the atmosphere during and after application by the process of evaporation or may travel as spray droplets, vapors, dusts, or solid particles, and even on blown soil particles by wind drift. The pesticides return to earth via precipitation and dry deposition and can reach streams and groundwater.
ii) Surface waters are vulnerable to pesticide contamination because runoff from agricultural and urban areas flows directly into surface water bodies along with both dissolved and particle associated pesticides.
iii) Pesticides can reach water-bearing aquifers below ground from their applications onto crop fields, seepage of contaminated surface water, accidental spills and leaks, improper disposal, and even through injection of waste material into wells.
iv) Groundwater can be an important source of pesticides to some streams during low-flow periods, when groundwater contributes a major portion of the streamflow.

1.3.3 Source of Disinfection By-Products in the Environment

Disinfection is a crucial step in drinking water treatment and is usually performed to protect public health from pathogenic microorganisms and waterborne diseases. Along with this, a sufficient disinfectant residual needs to be maintained within a water distribution system for prevention of the regrowth of unwanted pathogens. Harmful pathogens present in water are destroyed by the use of disinfectants such as chlorine, chlorine dioxide, chloramine, ozone, and ultraviolet light. However, some naturally occurring organic matter, anthropogenic contaminants, bromide, and iodide are also present in water, and when a chemical disinfectant such as chlorine is added to that water, it tends to react with organic matter to form undesired DBPs, which are known to be cytotoxic, genotoxic, and/or carcinogenic. The formation of DBPs at WWTPs and in distribution networks (Figure 1.4) is a function of several factors such as the pH, temperature, source water characteristics, type of disinfectant, and residence time.

The reason for formation of DBPs in water and wastewater treatment plants can be summarized as follows:

i) Natural organic matter present in water bodies and anthropogenic contaminants are the primary precursor of DBPs in water treatment plants. Among these precursors, phenolic structures identified in humic acids (HAs) and fulvic acids have been recognized as the principal precursors of trihalomehanes (THMs) and haloacetic acids (HAAs), while amino acids, oligopeptides, and nitrogen-containing contaminants serve as the major nitrogenous DBP (N-DBP) precursors (Gallard and von Gunten 2002; Lee and Westerhoff 2006; Chu et al. 2010; Yang et al. 2010; Bond et al. 2012).
ii) Organic polyelectrolytes have been widely used as coagulants or coagulant aids for enhancing the coagulation efficiency in water treatment, such as polyacrylamide (PAM), anion PAM (APAM), cation PAM (CPAM), poly (diallyldimethylammonium chloride) (PDADMAC), and epichlorohydrin/dimethylamine (ECH/DMA) (Bolto and Gregory 2007). However, the amide, quaternary amine, and primary amine groups of these compounds and their monomers can react with several disinfectants to produce toxic N-DBPs.

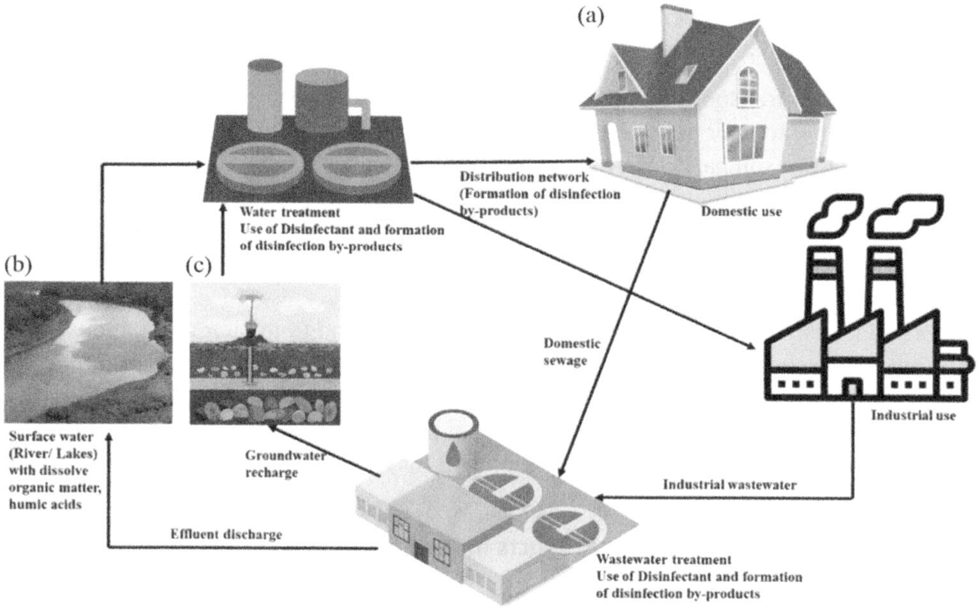

Figure 1.4 Source and pathways of disinfection by-products in the environment. *Sources:* (a) Viktoriia Protsak/Adobe Stock, (b) Keller Jo / U.S. Fish and Wildlife Service and (c) Kajani/Adobe Stock.

iii) Certain treatment steps in WWTPs have the potential to transform existing DBPs into more toxic compounds. For example, activated carbon (AC), ion-exchange resins, or membrane treatments can increase the ratios of brominated DBPs (Br-DBPs) to chlorinated DBPs (Cl-DBPs), resulting in greater mammalian cell cytotoxicity and genotoxicity, even though the overall DBP concentrations decrease.

iv) The formation of biofilm in the water distribution systems acts as a major precursor for DBP if residual disinfectants are present. Often the concentration of THMs at water distribution systems and the points of use are higher than those in the final water, while chlorine is rapidly consumed within the pipe system (Toroz and Uyak 2005; Pieri et al. 2014).

v) Effluents from WWTPs contain DBPs of health concerns when the effluent water is discharged to surface water or used for aquifer recharge. The pattern of DBP formation is strongly affected by whether or not the WWTP achieved good nitrification. Chlorine addition to poorly nitrified effluents formed low levels of halogenated DBPs, except for (in some cases) dihalogenated acetic acids, but often with substantial amounts of *N*-nitrosodimethyamine (NDMA). Chlorination of well-nitrified effluent typically resulted in substantial formation of halogenated DBPs but much less NDMA (Krasner et al. 2009).

1.3.4 Source of Industrial Chemicals in the Environment

Many different synthetic organic compounds are produced and used in large quantities worldwide for different purposes. Thousands of compounds are used as intermediates in the chemical industry (plasticizers, dyes, resins) or as food additives, antioxidants,

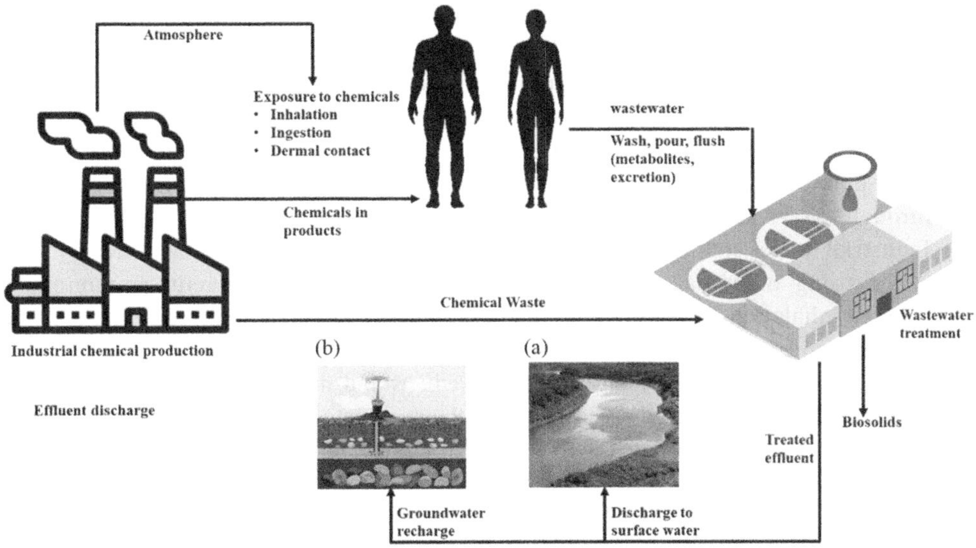

Figure 1.5 Source and pathways of industrial chemicals in the environment. *Sources:* (a) Keller Jo/U.S. Fish and Wildlife Service and (b) Kajani/Adobe Stock.

surfactants, and detergents. Consequently, these substances are continuously released into the environment, mainly due to industrial and domestic wastewater effluent discharges (Figure 1.5).

The volatile organic compound (VOC) have been reported to enter into the environment from industrial activities over the past four or five decades, whereas the discovery of other classes of synthetic organic chemicals (e.g., illicit drugs, perfluoroalkyl compounds, flame retardants, plasticizers, preservatives, food additives, gasoline additives, lubricants, andantimicrobial agents) has been more recent. Many of these industrial chemicals are environmentally persistent compounds. Organohalogen compounds (e.g., polychlorinated naphthalenes, chlorinated paraffins, polychlorinated biphenyls) are well known environmental contaminants. Chloroform is a common groundwater pollutant, in both urban and rural areas, that has both a natural and an anthropogenic origin. Perchlorate has been designated as an important environmental contaminant after its discovery in a number of water supplies. The source and pathways of industrial chemicals can be summarized as follows:

i) Many MCs are emitted from industrial activities during the production of chemicals into the atmosphere in the form of gaseous emissions or particulates. Among emitted pollutants are products of incomplete combustion that include a large number of organic compounds. Many emitted contaminants are chemically and biologically stable in the environment and resist typical environmental degradation processes, including atmospheric photooxidation, direct photolysis, and hydrolysis. Hence, these chemicals are extremely persistent in the environment, and some of them bioaccumulate in the environment. These contaminants settle to the ground through wet or dry deposition.

ii) When the contaminants disperse into the air from industrial activities, workers and people living in close vicinity to that facility might be exposed directly through inhalation or indirectly through consumption of food or water contaminated by deposition of the pollutants to soil and vegetation. Other people can be exposed through a different mix of environmental pathways after the contaminants travel some distance in the atmosphere, go through various chemical and physical transformations, or move through soil, water, or food.
iii) Industrial wastes containing various MCs reach the wastewater treatment facility.
iv) While using industrial chemicals, people are exposed to different contaminants through inhalation, ingestion, or dermal contact. The MCs are removed from the body through urine and excreta, and reach the wastewater treatment plants.
v) Many MCs are transformed into other by-products and many cannot be treated in the WWTPs. The MCs in treated wastewater find their way into surface water or groundwater.
vi) MCs are found in soil irrigated with reclaimed wastewater and subsequently enter into the food chain.
vii) The biosolids from the WWTPs may contain different MCs and enter into the environment upon disposal.

1.3.5 Source of Algal Toxins in the Environment

A harmful algal bloom occurs when toxin-producing algae grow excessively in a body of water. Algae are always present in natural bodies of water like oceans, rivers, and lakes, but only a few types of algae produce toxins. Toxin production in algae is stimulated by environmental factors such as light, warmer temperatures in summer, and excessive nutrients from fertilizers or sewage waste deposited by runoff or effluent discharge (Figure 1.6).

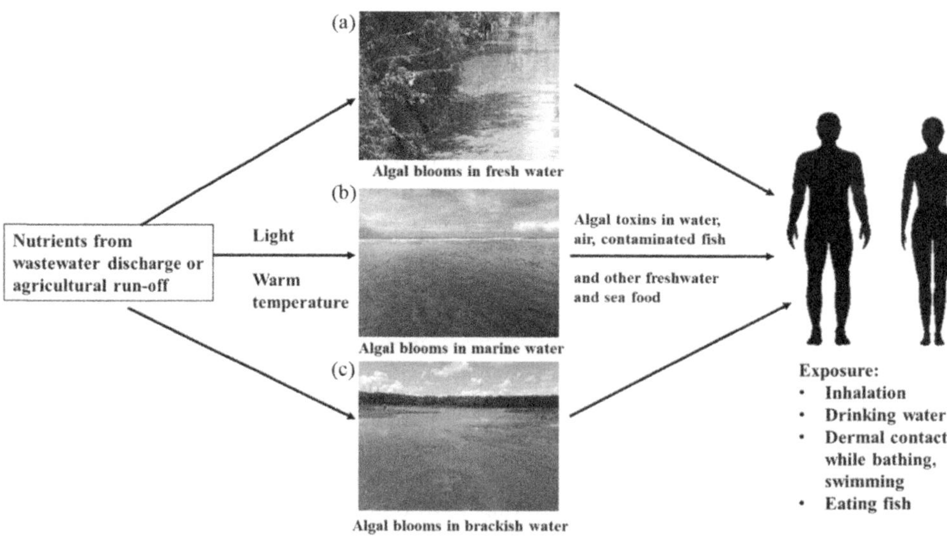

Figure 1.6 Source and pathway of algal toxins in the environment. *Sources:* (a) United States Environmental Protection Agency/Public Domain, (b) Buteo/Adobe Stock and (c) Justin / Adobe Stock.

Algal blooms can harm people, animals, and the environment when they produce toxins, become too dense, use up the oxygen in the water, and release harmful gases. These harmful blooms can be caused by many types of phytoplankton. However, there are three main types of phytoplankton tbat cause most blooms that make people and animals sick: cyanobacteria (blue-green algae), dinoflagellates (microalgae or red tide), and diatoms (microalgae or red tide). Cyanobacterial blooms are mostly found in fresh waters, whereas dinoflagellates and diatoms mostly grow in marine water and brackish water.

Harmful algae and cyanobacteria can affect ecosystems in many ways. Toxins can remain in the cells of phytoplankton or can be released into water. Common toxins produced by cyanobacteria are microcystin, cylindrospermopsin, anatoxin, guanitoxin, saxitoxin, nodularin, and lyngbyatoxin. The toxins produced by dinoflagellates are brevetoxin, azaspiracid, ciguatoxins, okadic acid, saxitoxin, and dinophysistoxin, whereas diaotoms produce domoic acid. People can be exposed to toxins from the fish they catch and eat, from swimming in or drinking the water, and even from the air they breathe.

1.4 Physical and Chemical Properties of Microconstituents

The detection and quantification of MCs in water or wastewater have been developed as an important subject of study, necessitating extremely sophisticated analytical methods to detect concentrations ranging from micrograms to nanograms per liter. Even though various studies have addressed the harmful consequences of MCs and advised effective interventions for their control, several crucial research gaps remain. Suitable analytical methodologies and advanced instruments for detecting MCs and their metabolites are being investigated. The appropriate reference materials for numerous CECs are presently undefined. The environmental fate, toxicity, and reactivity of MCs and their long-term (chronic) health impacts are yet unclear (Naidu et al. 2016). There is a major necessity for an in-depth understanding of MCs, including information on applying diverse analytical techniques and developing rapid and successful screening methods to detect these chemicals. Because CEC concentrations are often in the μg/L range, sample pretreatment (extraction, pre-concentration, washing) is important for accurate measurement. Solid-phase extraction, which was proposed to replace liquid–liquid extraction, is the most widely described method for extracting CECs from aqueous environmental samples. The liquid–liquid extraction technique is less favorable since it is resource-intensive, difficult to automate, and consumes a substantial amount of solvents (Moliner-Martínez et al. 2011). Due to the intrinsic complexity of CECs, chromatography (gas/liquid) in conjunction with mass spectrometry is used as the predominant analytical method. Environmental samples, particularly surface water containing WWTP effluents, are relatively complex in nature. Even with sophisticated mass spectrometers, accurate identification of compounds can be extremely difficult. Co-eluting compounds, molecules with equivalent mass-to-charge ratios, shifts in retention time, and matrix features such as ion suppression can contribute to compound misinterpretation. Furthermore, access to pure standard compounds is limited. It is also critical to identify the primary molecule in the environmental sample by elucidating the degradation mechanism of multiple CECs (Alvarez and Jones-Lepp 2010; Rodriguez-Narvaez et al. 2017).

The physico-chemical properties of these pollutants, such as solubility, polarity, structural complexity, functional groups, sorption coefficients, and lipophilic characteristics, might impact their environmental degradation. The octanol-water partitioning coefficient (K_{OW}), which represents the hydrophobic partitioning of CECs, can influence these properties. Changes in this coefficient directly impact the degradability, bioaccumulation, and solubility of CECs. Because these qualities alter depending on the CEC, the degree of the potential threat varies. Some toxins degrade rapidly, while others remain in the environment for a longer period. CECs having log $K_{OW} > 3$ are considered hydrophobic and thus have a higher potential for bioaccumulation, either in soil matrix/sediment or in the tissues of organisms. Contaminants having a log K_{OW} of 1 are hydrophilic (water-soluble) and biodegradable; therefore, the bioaccumulation rate is relatively low. The K_{OW} values of the CECs must be considered while selecting the appropriate treatment technology for the contaminants. CECs with lower K_{OW} values could be treated efficiently with methods involving the degradation of pollutants. In contrast, treatment procedures that include pollutant adsorption on various matrices can achieve greater outcomes with contaminants with higher K_{OW} values (Rempel et al. 2021). A brief list of CECs, their concentration detected in environmental samples, and their characteristics are described in Table 1.2.

In spite of the availability of various treatment methods for CECs, the pollutant removal efficacy of most of the technologies is insufficient and largely dependent on the physico-chemical characteristics of the CECs and operational parameters. For example, hydrophobic CECs such as miconazole and bisphenol A are generally present in sludge due to their propensity to partition onto the particulate phase, while CECs such as ciprofloxacin are adsorbed onto sludge via electrostatic interaction. The phase sorption encompasses both the absorption of CECs onto the lipid component of primary sludge via hydrophobic interactions and the adsorption of CECs onto the surface of sludge particles, which is mostly accomplished through electrostatic attractions (Alvarino et al. 2018; Tran et al. 2018). The physico-chemical parameters of CECs (size, presence of the functional group, polarity, acid dissociation constant (pKa), K_{OW}), the properties of the adsorbent (mineral content, size, surface area, pore-volume), and the operational conditions (pH, temperature, contact time) influence CEC adsorption. The biodegradation of CECs also depends on physico-chemical properties like structural complexity and the availability of specific functional groups. CECs with electron-donating functional groups are more readily biodegraded than CECs with electron-drawing functional groups (Luo et al. 2014). The removal of CECs in membrane bioreactors also depends on physico-chemical characteristics such as particle size, biodegradability, and hydrophobic interaction with the membrane. Nonpolar CECs were eliminated primarily through size exclusion and adsorption onto the membrane/biofilm layer, while the polar CECs were removed mainly via biodegradation and limited adsorption (Goswami et al. 2018).

1.5 Impact on Human Society and Ecosystem

The MCs present in water, wastewater, soil, and air can harm humans as well as the other living entities in the ecosystem by bioaccumulation, toxicity, or by altering natural metabolic activities.

Table 1.2 Physico-chemical properties of microconstituents.

Emerging Contaminants	Compounds	Sample Detected	Conc. (μg/L)	Log K_{OW}	pK_a	Reference
Pharmaceuticals	Azythromycin	Surface water.	0.06–0.17	4.02	8.50	Rodriguez-Mozaz et al. 2015
		Wastewater.	0.66–1.68			Fatta-Kassinos et al. 2011
	Amoxicillin	Surface water.	0.2–0.3	0.87	3.20	Watkinson et al. 2009
		Wastewater.	0.33–0.69			Mirzaei et al. 2019
	Ciprofloxacin	WWTP effluent.	0.09	0.28	6.09	Rout et al. 2021
	Erythromycin	WWTP effluent.	0.042	3.06	8.88	
	Diclofenac	Groundwater.	0.04	4.15	4.51	Cabeza et al. 2012
		Surface water.	0.14–0.31			Tran et al. 2014
		Wastewater.	0.6			Rosal et al. 2010
	Ibuprofen	WWTP effluent.	0.4	3.50	4.85	Rout et al. 2021
Personal Care Products	Triclosan	Wastewater.	0.18–4.4	4.76	7.90	Yu et al. 2013
	Bisphenol A	River water.	0.09–0.3	3.32	9.60	Williams et al. 2019
	Benzotriazole	River water.	0.04–0.52	1.44	8.37	
	Galaxolide	Wastewater.	25	5.9	5.7	Rosal et al. 2010
Pesticides	Carbofuran	Surface water.	0.95–1.67	2.32	11.9	Chowdhury et al. 2012
	Carbendazim	Surface water.	0.1–1.8	1.52	4.48	Peng et al. 2018
	Atrazine	Groundwater.	0.2–0.7	2.61	1.68	Espíndola and Vilar 2020
		Surface water.	0.1–0.8			

(Continued)

Table 1.2 (Continued)

Emerging Contaminants	Compounds	Sample Detected	Conc. (μg/L)	Log K_{OW}	pK_a	Reference
Hormones	Estrone	Wastewater.	0.13	3.13	10.33	He et al. 2013
	Progesterone	Surface water.	0.2	3.87	18.92	Kolpin et al. 2002
	17β-Estradiol	Wastewater.	1.1–1.2	4.01	10.46	He et al. 2013
Disinfection by-products	Trichloromethane	River water.	7.8–315.3	1.97	15.7	Heng et al. 2021
	Trichloro acetaldehyde	River water.	12.5–19.8	0.99	9.66	
	Trichloro acetonitrile	River water.	0.12–0.52	2.09	–	
Industrial Chemicals	Caffeine	Surface water.	6	−0.07	10.4	Kolpin et al. 2002
		Wastewater.	66			Rosal et al. 2010
	Nicotine	Wastewater.	1.1–14.6	1.17	8.5	Kumar et al. 2019
Algal Toxins	Microcystin	Surface water.	12.5	−0.7	12.4	Richardson 2003
	Anatoxin	Drinking water.	8.5	1.12	9.4	
	Cylindro-spermopsin	Drinking water.	8.1–97.1	-	10.26	

1.5.1 Impact on Human Health

The CECs have the potential to induce a large range of acute and long-term effects (e.g., endocrine disruption, immunotoxicity, neurological disorders, cancers) on human health. The adverse effects on human health are shown in Table 1.3 with respect to different types of MCs based on conclusions of the comprehensive epidemic literature and representative case reports relevant to emerging contaminants.

1.5.2 Impact on the Ecosystem

Several studies have described alterations in animal behavior due to exposure to some MCs. For example, tadpoles (*Bufo arabicus*) exposed to fluoxetine (at 3 μg/L) were more susceptible to predation from dragonfly larvae (*Anax imperator*) (Barry 2014). Brodin et al. (2013) described changes in the behavior of European perch (increased activity – number of swimming bouts for 10 min; increased boldness – the inverse of latency to enter a novel area during the total trial time; reduced sociality – cumulative time (in seconds) spent close to a group of co-specifics and reduced feeding rate) after exposure to oxazepan. Many other literatures have reported the effects of CECs at trace concentrations in animal behavior. CECs are also known to alter microbial communities and function, and may be responsible for spreading antibiotic resistance (Wang et al. 2019). Studies have described the effects of different CECs exposure on the microbiota of water-related environments (rivers, marine environment, wetlands, etc.), soil, and in engineered systems (e.g., activated sludge from WWTPs). Proia et al. (2013) evaluated the effects of pharmaceuticals and pesticides on fluvial biofilms in a river in the Mediterranean region. They observed an increase in autotrophic biomass and in peptidase, and a decrease in phosphatase and photosynthetic efficiency, when biofilms were moved to areas with higher concentrations of CECs. The effect of CECs on ecosystems are shown in Table 1.4.

Table 1.3 Impact of microconstituents on human health.

Emerging Contaminant	Intake Route	Effects	Reference
Pharmaceuticals			
Cytotoxic anticancer agents.	Oral	Toxicity in humans.	Yadav et al. 2021
17α-thinylestradiol	Oral	Toxicity in humans.	Fent et al. 2006
A-fluoro-b-alanine	Oral	Toxicity after bio-accumulation.	Česen et al. 2016
Personal Care Products			
Bisphenols	Oral	These chemicals can bind with sex hormone-binding globulin (SHBG) at high affinity, thereby disrupting natural hormones' steroid-binding function.	Sheikh et al. 2017
Alkylphenol polyethoxylates.	Oral	Act as endocrine disruptors.	Sheikh et al. 2017

(Continued)

Table 1.3 (Continued)

Emerging Contaminant	Intake Route	Effects	Reference
Triclosan	Oral	Cause hepatotoxicity and hepatocarcinogenicity.	Zhang et al. 2019
Fragrances	Dermal	Increase asthmatic symptoms and dermal irritation and contact dermatitis.	Bickers et al. 2003
Parabens	Oral	Affect male reproductive health. Increase risk of breast cancer.	Parada et al. 2019
Pesticides			
Organophosphorus chemicals.	Oral	Inhibitors of the human metabolism of both testosterone and estradiol.	Hernández et al. 2013
Diethyl alkylphosphate.	Oral	Chronic degenerative diseases Neurodevelopmental deficits and cancer.	London et al. 2012
Disinfection by-products (DBPs)			
Chlorination DBPs	Oral	Cause cancer of the bladder, rectum and colon, birth defects.	Malcolm et al. 1999
N-nitrosodimethylamine (by-product of wastewater chlorination)	Oral	Carcinogenic compound.	Malcolm et al. 1999
Ozonation DBP.	Oral	Renal cell cancer and infections with *Cryptosporidium Parvum.*	Havelaar et al. 2000
Trichloromethane	Oral	Cancer.	Jo et al. 1990
Industrial Chemicals			
Microplastics	Oral	obstruction of the intestinal tract, reticence of gastric enzyme discharge, condensed feeding stimulus, and diminish steroid hormone levels, interruption in ovulation and malfunction to reproduce.	Fendall and Sewell 2009; Wright et al. 2013
1,4-Dioxane	Oral	Carcinogenic compound.	Kumar et al. 2020
Naphthenic acids	Oral	Toxic compound.	Kumar et al. 2020
Benzotriazoles (anticorrosive)	Oral	Toxic compound.	Kumar et al. 2020
Alkylphenol ethoxylate	Oral	Toxic compound.	Kumar et al. 2020
Algal Toxins			Kumar et al. 2020
Cyanotoxins	Oral	Liver and nervous system damage.	Kumar et al. 2020
Cyanobacteria	Oral	Inhibition of protein synthesis causing a reduction in body weight.	Humpage and Falconer 2003

Table 1.4 Effect of microconstituents on flora and fauna.

Emerging Contaminant	Contact Source	Effects	Reference
Pharmaceuticals and Personal Care Products			
Triclosan	Wastewater and solid waste.	Microbial resistance, endocrine disruption in small animals.	Kumar et al. 2009; Dhillon et al. 2015
17β-Estradiol	Wastewater.	Disruption of fish population dynamics and endocrine system at low concentration.	Casey et al. 2003; Kunz et al. 2014
Caffeine	Wastewater.	Not shown any carcinogenicity or change in fertility or lactation when evaluated of rats and monkeys.	IARC 1991
Pesticides			
Pyethroids and pyrethrins (Deltamethrin)	Agriculture wastewater.	Accumulation in the aquatic ecosystem and adverse effects on invertebrates and fish.	Bjorling-Poulsen et al. 2008; Pérez-Fernández et al. 2010
Dithiocarbamates (Maneb)	Agriculture wastewater.	Toxicity to aquatic organisms.	Van Wezel and Van Vlaardingen 2004; Belpoggi et al. 2006
Phenylureas (Tebuthiuron)	Agriculture wastewater.	Increase phase-I biotransformation enzymes in small and large fish and genotoxicity in small fish.	Caux et al. 1997
Disinfection by-products (DBPs)			
Trichloroacetic acid (TCAA)	Wastewater.	The DNA damage was analyzed in the roots of *Allium cepa*.	Ranjan et al. 2019
Tribromomethane (TBM)	Wastewater.	Depletion of root growth appears similar to the effects of arsenic.	Ranjan et al. 2019
Trihalomethanes (THM)	Water.	Cancer observed in animals (primarily liver cancer).	Ranjan et al. 2019
Trichloromethane (TCM)	Wastewater.	Decrease in root growth.	Ranjan et al. 2019
Industrial Chemicals			
Tetrabromodiphenyl ether – BDE 47	Industrial wastewater.	Damage cell mitochondria; hormone disruption and impaired energy metabolism during growth phase of fish.	Pazin et al. 2014

(Continued)

Table 1.4 (Continued)

Emerging Contaminant	Contact Source	Effects	Reference
Pentabromodiphenyl ether – BDE 100	Industrial wastewater.	Damage cell mitochondria; affect neurodevelopment in fauna.	Huang et al. 2010; Pereira et al. 2013
Linear alkylbenzene sulfonic acid (LAS)	Industrial wastewater.	Link to phospholipids on the cell membrane and proteins, to increase permeability causing cell death.	Pereira et al. 2015
Sodium lauryl sulfate (SLS)	Industrial wastewater.	Bind to bioactive macromolecules such as peptides, enzymes, and DNA, to modify their biological function through changes in the polypeptide chain folding and the surface charge of a molecule.	Ivanković and Hrenović 2010
Algal Toxins			
Cyanotoxins	Wastewater.	Poisoning of shellfish; death of large fish;	Richardson 2003
Cyanobacteria	Wastewater.	paralytic shellfish poisoning.	Richardson 2003

1.6 The Structure of the Book

This book is intended to provide readers with an understanding of the occurrence and fate of microconstituents (MCs) in the environment and possible management strategies. The main topics are organized into five core parts with subdivisions of each. Part I deals with the fundamental ideas regarding MCs in the environment and consists of four chapters. Chapter 1 introduces the MCs and explores their various classifications, properties, and sources, as well as their impact on the environmental ecosystem and human health. The presence of MCs in diverse environmental samples is discussed in Chapter 2. This chapter also describes the detection methodology of MCs, the roadblocks in identification, and suggests appropriate methods for conducting surveys. The sampling protocols, quantification, and analysis of different MCs are discussed in Chapter 3, which includes a wide range of monitoring and screening methods to analyze these pollutants. The source tracking technique is described with the help of case studies. This chapter also examines the application of remote sensing and geographical information systems (GIS) for monitoring MCs in groundwater and surface water. Chapter 4, the last chapter of the first part, deals with the toxicity assessment, including acute and chronic toxicity and dose-responses studies for the determination of safe levels of MCs.

The fate and transportation of MCs in various environmental domains are covered in Part II, which consists of six chapters. Chapter 5 discusses the fundamentals of pollutant transport modeling and different software available for mathematical modeling. The mathematical and computational limitations of transport modeling are also included in this chapter. Chapter 6 covers the major MCs found in groundwater as well as the mechanisms of groundwater contamination. The modeling of pollutant transport in groundwater and its limitations are also examined. The major MCs present in surface water as well as sources and pathways of surface water contamination are covered in Chapter 7. The fate of pollutants and transport modeling of MCs for the coastal marine environment and the water cycle studies and shortcomings of mathematical tools are also addressed. Chapter 8 deals with the fate and transport of MCs in WWTPs. This chapter elucidates on the biotransformation of MCs in WWTPs and the characteristics of generated biosolids. Different models for estimating sediment and water quality, as well as the fate of the MCs, are also discussed. Chapter 9 explores the transport of MCs in the atmosphere. Contaminant control approaches, as well as the significance of atmospheric conditions on pollutant transport, are mentioned. This chapter also discusses the fate of atmospheric MCs and modeling methodologies. Several aspects of remote sensing and GIS technique-based modeling of MCs are described in Chapter 10. The emerging application, as well as uncertainty in environmental modeling, are also briefly discussed.

Part III of the book encompasses details of the various physico-chemical treatment of MCs with five chapters. Chapter 11 summarizes the chemical precipitation technique in view of the removal of MCs. The limitations regarding sludge disposal are addressed in this chapter. Chapter 12 deals with the adsorption mechanisms and their advantages and limitations. It includes several conventional and emerging adsorbents and desirable adsorbent properties. The techniques for the modification of adsorbent characteristics are explained. Finally, the disposal techniques of adsorbents are discussed. Chapter 13 discusses the treatment techniques related to ion exchange. Different engineering aspects of designing ion exchange resins and limitations of this method are also discussed. Membrane technology is an alternative physico-chemical approach to the removal of MCs. Membranes are made of various materials that create unique filtering characteristics and retain different kinds of pollutants. Chapter 14 briefly discusses several features and properties of membranes for the retention of MCs. Membrane fouling and other operational challenges are also included in the discussion. Chapter 15 includes the management of MCs via advanced oxidation processes. It discusses different oxidizing agents capable of producing extremely reactive oxygenated species with low selectivity (e.g., hydroxyl radicals (OH·)), which oxidize almost all organic compounds, generating either less toxic compounds, or mineralizing them entirely to CO_2, H_2O, and inorganic compounds. The energy utilization, economic aspect, and practicability of this process are also explained in this chapter.

The removal of MCs via biological treatment techniques is discussed in Part IV. Several aerobic and anaerobic biological processes have been investigated for MCs removal. Chapter 16 explains the aerobic biological treatment processes which include the suspended and attached growth systems. The effectiveness of these techniques and the challenges regarding handling of MCs are mentioned in this chapter. Chapter 17 deals with the anaerobic treatment process and explains the key contributing factors related to the efficacy of the process. Chapter 18 introduces the concept of bio-electrochemical systems (BES). The various types of BES, their configurations, and electrode and membrane materials are included in the chapter. It also reviews a few scale-up studies. The challenges and future prospects of this technology are further discussed. Several innovative hybrid treatment units have been examined in recent years, since the traditional treatment techniques are inadequate for removing a wide variety of MCs effectively. Different types of hybrid treatment options are explored in Chapter 19. The future direction of these methods is also briefly discussed.

Finally, Part V focuses on the aspects of sustainability and environmental management. During designing a treatment process to address a certain environmental problem, it must be understood that these processes will introduce additional environmental pressures through resource consumption and energy utilization. These need to be reduced to a minimum to prevent the emergence of other environmental problems stemming from the original. The requirements of legislative measures and existing regulatory frameworks are incorporated in Chapter 20. The guidelines for pollution control at the source are also briefly discussed. Chapter 21 deals with case studies of lab- and field-scale applications for the treatment of MCs. The complexity of field applications and the sustainability of large-scale treatment systems are also included. The last chapter of the book (Chapter 22) emphasises sustainable approaches regarding the removal of MCs. The importance of systematic information sharing between all stakeholders regarding health and environmental risks is discussed.

1.7 Conclusions

While the present generation is struggling to keep the watercourses clean by remediating and minimizing the conventional pollutants present in the huge amount of wastewater generated due to population growth, the microconstituents (MCs) or the contaminants of emerging concern (CECs) are now warranting attention. Wastewater effluents are a major source for many of the CECs due to use of CEC-containing products in households, and from pharmaceuticals, personal care products, detergents, foods and beverages, food packaging, and fire retardants. After use, these chemicals are released into wastewater. Many of the MCs are incompletely removed by wastewater treatment and enter the rivers, lakes, and drinking water supplies. Surface and agricultural run-offs can also be important sources of MCs, such as herbicides, insecticides, and fungicides, into the environment. Most of the MCs can become transformed in the environment to different forms by microbial degradation, photolysis, and hydrolysis, and can also react with disinfectants in water or WWTPs to form disinfection by-products. Major concerns of the MCs include widespread occurrence, persistence, bioaccumulation, toxicity, and adverse impacts on human health and ecosystems. Many of the MCs are identified as responsible for probable endocrine disruption, antibiotic resistance, cytotoxicity, carcinogenicity, and genotoxicity.

The environmental fate, toxicity, and reactivity of MCs and their long-term (chronic) health impacts are yet unclear. There is a major necessity for an in-depth understanding of MCs, including information on applying diverse analytical techniques and developing rapid and successful screening methods to detect these chemicals. Due to the difficulty in detecting and handling CECs, the concentration reduction at the source must be emphasised through policies and regulations, public awareness, and imposing discharge limit of CECs in the environment. This poses a key challenge because it involves the improvement of the existing framework and the implementation of new cost-effective technologies that can permanently remove a wide range of pollutants. This book gives an overview of CECs, their fate and transportation in the environmental media, and the different treatment technologies such as physico-chemical and biological used for their removal. Different types of hybrid technologies adopted for the treatment of CECs are discussed. The sustainability aspects together with environmental management of different treatment technologies along with techno economic assessments and life-cycle assessment are covered. Lab- and field-scale case studies for the removal of CECs are also presented. There is an overall progression through the book from the introduction to MCs, and their detection strategies, followed by treatment methods to the examples and case studies of practical applications considering sustainable management.

References

Affum, A.O., Acquaah, S.O., Osae, S.D., and Kwaansa-Ansah, E.E. (2018). Distribution and risk assessment of banned and other current-use pesticides in surface and groundwaters consumed in an agricultural catchment dominated by cocoa crops in the Ankobra Basin, Ghana. *Science of the Total Environment* 633: 630–640.

Alvarez, D.A. and Jones-Lepp, T.L. (2010). Sampling and analysis of emerging pollutants. In: *Water Quality Concepts, Sampling, and Chemical Analysis* (eds Li, Y. and Migliaccio, K.). 199–226. CRC Press.

Alvarino, T., Suarez, S., Lema, J., and Omil, F. (2018). Understanding the sorption and biotransformation of organic micropollutants in innovative biological wastewater treatment technologies. *Science of the Total Environment* 615: 297–306.

Archundia, D., Duwig, C., Lehembre, F. et al. (2017). Antibiotic pollution in the Katari subcatchment of the Titicaca Lake: major transformation products and occurrence of resistance genes. *Science of the Total Environment* 576: 671–682.

Ávila, C., Bayona, J.M., Martín, I. et al. (2015). Emerging organic contaminant removal in a full-scale hybrid constructed wetland system for wastewater treatment and reuse. *Ecological Engineering* 80: 108–116.

Barry, M.J. (2014). Fluoxetine inhibits predator avoidance behavior in tadpoles. *Toxicological & Environmental Chemistry* 96(4): 641–649.

Belfroid, A.C., Van der Horst, A., Vethaak, A.D. et al. (1999). Analysis and occurrence of estrogenic hormones and their glucuronides in surface water and waste water in the Netherlands. *Science of the Total Environment* 225 (1–2): 101–108.

Belpoggi, F., Soffritti, M., Guarino, M. et al. (2006). Results of long-term experimental studies on the carcinogenicity of ethylene-bis-dithiocarbamate (mancozeb) in rats. *Annals of the New York Academy of Sciences* 982: 123–136.

Bhandari, A., Surampalli, R., Adams, C.D. et al. (2009). *Contaminants of Emerging Environmental Concern*. Reston, VA: ASCE.

Bickers, D.R., Calow, P., Greim, H.A. et al. (2003). The safety assessment of fragrance materials. *Regulatory Toxicology and Pharmacology* 37 (2): 218–273.

Bjorling-Poulsen, M., Andersen, H.R., and Grandjean, P. (2008). Potential developmental neurotoxicity of pesticides used in Europe. *Environmental Health* 7: 1–22.

Bolto, B. and Gregory, J. (2007). Organic polyelectrolytes in water treatment. *Water Research* 41 (11): 2301–2324.

Bond, T., Templeton, M.R., and Graham, N. (2012). Precursors of nitrogenous disinfection by-products in drinking water: a critical review and analysis. *Journal of Hazardous Materials* 235: 1–16.

Brausch, J.M. and Rand, G.M. (2011). A review of personal care products in the aquatic environment: environmental concentrations and toxicity. *Chemosphere* 82 (11): 1518–1532.

Brodin, T., Fick, J., Jonsson, M., and Klaminder, J. (2013). Dilute concentrations of a psychiatric drug alter behavior of fish from natural populations. *Science* 339 (6121): 814–815.

Bulut, S., Erdogmus, S.F., Konuk, M., and Cemek, M. (2010). The organochlorine pesticide residues in the drinking waters of Afyonkarahisar, Turkey. *Ekoloji* 19 (74): 24–31.

Burri, N.M., Weatherl, R., Moeck, C., and Schirmer, M. (2019). A review of threats to groundwater quality in the anthropocene. *Science of the Total Environment* 684: 136–154.

Cabeza, Y., Candela, L., Ronen, D., and Teijon, G. (2012). Monitoring the occurrence of emerging contaminants in treated wastewater and groundwater between 2008 and 2010. The Baix Llobregat (Barcelona, Spain). *Journal of Hazardous Materials* 239: 32–39.

Cai, M.Q., Wang, R., Feng, L., and Zhang, L.Q. (2015). Determination of selected pharmaceuticals in tap water and drinking water treatment plant by high-performance liquid chromatography-triple quadrupole mass spectrometer in Beijing, China. *Environmental Science and Pollution Research* 22 (3): 1854–1867.

Calvo-Flores, F.G., Isac-García, J., and Dobado, J.A. (2018). *Emerging Pollutants: Origin, Structure, and Properties*. John Wiley & Sons.

Carballa, M., Omil, F., and Lema, J.M. (2005). Removal of cosmetic ingredients and pharmaceuticals in sewage primary treatment. *Water Research* 39 (19): 4790–4796.

Carson, R. (2002). *Silent Spring*. Houghton Mifflin Harcourt.

Casey, F.X., Larsen, G.L., Hakk, H., and Šimůnek, J. (2003). Fate and transport of 17β-estradiol in soil—water systems. *Environmental Science & Technology* 37 (11): 2400–2409.

Castiglioni, S., Fanelli, R., Calamari, D. et al. (2004). Methodological approaches for studying pharmaceuticals in the environment by comparing predicted and measured concentrations in River Po, Italy. *Regulatory Toxicology and Pharmacology* 39 (1): 25–32.

Caux, P.Y., Kent, R.A., Bergeron, V. et al. (1997). Canadian water quality guidelines for tebuthiuron. *Environmental Toxicology Water Quality* 12: 61–95.

Česen, M., Eleršek, T., Novak, M. et al. (2016). Ecotoxicity and genotoxicity of cyclophosphamide, ifosfamide, their metabolites/transformation products and their mixtures. *Environmental Pollution* 210: 192–201.

Chowdhury, A.Z., Jahan, S.A., Islam, M.N. et al. (2012). Occurrence of organophosphorus and carbamate pesticide residues in surface water samples from the Rangpur district of Bangladesh. *Bulletin of Environmental Contamination and Toxicology* 89 (1): 202–207.

Chu, W.H., Gao, N.Y., Deng, Y., and Krasner, S.W. (2010). Precursors of dichloroacetamide, an emerging nitrogenous DBP formed during chlorination or chloramination. *Environmental Science & Technology* 44 (10): 3908–3912.

Cleary, J. (2008). Microconstituents – what are they and what is their fate through wastewater treatment. *Proceedings of the Water Environment Federation* 2008: 3444–3460.

Derbalah, A., Ismail, A., Hamza, A., and Shaheen, S. (2014). Monitoring and remediation of organochlorine residues in water. *Water Environment Research: A Research Publication of the Water Environment Federation* 86 (7): 584–593.

Dhillon, G.S., Kaur, S., Pulicharla, R. et al. (2015). Triclosan: current status, occurrence, environmental risks and bioaccumulation potential. *International Journal of Environmental Research and Public Health* 12: 5657–5684.

Díaz, G., Ortiz, R., Schettino, B. et al. (2009). Organochlorine pesticides residues in bottled drinking water from Mexico City. *Bulletin of Environmental Contamination and Toxicology* 82 (6): 701–704.

Dorabawila, N. and Gupta, G. (2005). Endocrine disrupter, estradiol, in Chesapeake Bay tributaries. *Journal of Hazardous Materials* 120 (1–3): 67–71.

Drewes, J.E., Fox, P., and Jekel, M. (2001). Occurrence of iodinated X-ray contrast media in domestic effluents and their fate during indirect potable reuse. *Journal of Environmental Science and Health, Part A* 36 (9): 1633–1645.

Eitzer, B.D. and Chevalier, A. (1999). Landscape care pesticide residues in residential drinking water wells. *Bulletin of Environmental Contamination and Toxicology* 62 (4): 420–427.

El-Nahhal, Y. (2020). Pesticide residues in honey and their potential reproductive toxicity. *Science of the Total Environment* 741: 139953.

El-Nahhal, I. and El-Nahhal, Y. (2021). Pesticide residues in drinking water, their potential risk to human health and removal options. *Journal of Environmental Management* 299: 113611.

Espíndola, J.C. and Vilar, V.J. (2020). Innovative light-driven chemical/catalytic reactors towards contaminants of emerging concern mitigation: a review. *Chemical Engineering Journal* 394: 124865.

Estrada, A.L., Li, Y.Y., and Wang, A. (2012). Biodegradability enhancement of wastewater containing cefalexin by means of the electro-Fenton oxidation process. *Journal of Hazardous Materials* 227: 41–48.

Fatoki, O.S. and Awofolu, O.R. (2004). Levels of organochlorine pesticide residues in marine-, surface-, ground- and drinking waters from the Eastern Cape Province of South Africa. *Journal of Environmental Science and Health, Part B* 39 (1): 101–114.

Fatta-Kassinos, D., Hapeshi, E., Achilleos, A. et al. (2011). Existence of pharmaceutical compounds in tertiary treated urban wastewater that is utilized for reuse applications. *Water Resources Management* 25 (4): 1183–1193.

Fendall, L.S. and Sewell, M.A. (2009). Contributing to marine pollution by washing your face: microplastics in facial cleansers. *Marine Pollution Bulletin* 58 (8): 1225–1228.

Fent, K., Weston, A.A., and Caminada, D. (2006). Ecotoxicology of human pharmaceuticals. *Aquatic Toxicology* 76 (2): 122–159.

Fosu-Mensah, B.Y., Okoffo, E.D., Darko, G., and Gordon, C. (2016). Assessment of organochlorine pesticide residues in soils and drinking water sources from cocoa farms in Ghana. *SpringerPlus* 5 (1): 1–13.

Gallard, H. and von Gunten, U. (2002). Chlorination of natural organic matter: kinetics of chlorination and of THM formation. *Water Research* 36 (1): 65–74.

Goswami, L., Kumar, R.V., Borah, S.N. et al. (2018). Membrane bioreactor and integrated membrane bioreactor systems for micropollutant removal from wastewater: a review. *Journal of Water Process Engineering* 26: 314–328.

Havelaar, A.H., De Hollander, A.E., Teunis, P.F. et al. (2000). Balancing the risks and benefits of drinking water disinfection: disability adjusted life-years on the scale. *Environmental Health Perspectives* 108 (4): 315–321.

He, Y.J., Chen, W., Zheng, X.Y. et al. (2013). Fate and removal of typical pharmaceuticals and personal care products by three different treatment processes. *Science of the Total Environment* 447: 248–254.

Heng, B., Zhang, R., Wang, Z. et al. (2021). Occurrence and risk assessment of volatile halogenated disinfection by-products in an urban river supplied by reclaimed wastewater. *Ecotoxicology and Environmental Safety* 211: 111912.

Hernández, A.F., Parrón, T., Tsatsakis, A.M. et al. (2013). Toxic effects of pesticide mixtures at a molecular level: their relevance to human health. *Toxicology* 307: 136–145.

Hirsch, R., Ternes, T., Haberer, K., and Kratz, K.L. (1999). Occurrence of antibiotics in the aquatic environment. *Science of the Total Environment* 225 (1–2): 109–118.

Huang, S.C., Giordano, G., and Costa, L.G. (2010). Comparative cytotoxicity and intracellular accumulation of five polybrominated diphenyl ether congeners in mouse cerebellar granule neurons. *Toxicological Sciences* 114 (1): 124–132.

Humpage, A.R. and Falconer, I.R. (2003). Oral toxicity of the cyanobacterial toxin cylindrospermopsin in male Swiss albino mice: determination of no observed adverse effect level for deriving a drinking water guideline value. *Environmental Toxicology: An International Journal* 18 (2): 94–103.

International Agency for Research on Cancer World Health Organization – IARC (1991). IARC monographs on the evaluation of carcinogenic risks to humans: coffee, tea, methylzanthines and methylglyoxal, v. 51.

Ivanković, T. and Hrenović, J. (2010). Surfactants in the environment. *Arhiv za higijenu rada i toksikologiju* 61 (1): 95–109.

Jo, W.K., Weisel, C.P., and Lioy, P.J. (1990). Chloroform exposure and the health risk associated with multiple uses of chlorinated tap water. *Risk Analysis* 10 (4): 581–585.

Kaushik, C.P., Sharma, H.R., and Kaushik, A. (2012). Organochlorine pesticide residues in drinking water in the rural areas of Haryana, India. *Environmental Monitoring and Assessment* 184 (1): 103–112.

Klamerth, N., Malato, S., Agüera, A. et al. (2012). Treatment of municipal wastewater treatment plant effluents with modified photo-Fenton as a tertiary treatment for the degradation of micro pollutants and disinfection. *Environmental Science & Technology* 46 (5): 2885–2892.

Kolpin, D.W., Furlong, E.T., Meyer, M.T. et al. (2002). Pharmaceuticals, hormones, and other organic wastewater contaminants in US streams, 1999–2000: a national reconnaissance. *Environmental Science and Technology* 36: 1202–1211.

Krasner, S.W., Westerhoff, P., Chen, B. et al. (2009). Occurrence of disinfection byproducts in United States wastewater treatment plant effluents. *Environmental Science & Technology* 43 (21): 8320–8325.

Kumar, M., Borah, P., and Devi, P. (2020). Priority and emerging pollutants in water. In: *Inorganic Pollutants in Water* (eds Devi, P., Singh, P., and Kansal, S.). 33–49. Elsevier.

Kumar, R., Tscharke, B., O'Brien, J. et al. (2019). Assessment of drugs of abuse in a wastewater treatment plant with parallel secondary wastewater treatment train. *Science of the Total Environment* 658: 947–957.

Kumar, V., Chakraborty, A., Kural, M.R., and Roy, P. (2009). Alteration of testicular steroidogenesis and histopathology of reproductive system in male rats treated with triclosan. *Reproductive Toxicology* 27: 177–185.

Kunz, P.Y., Kienle, C., Carere, M. et al. (2014). *In vitro* bioassays to screen for endocrine active pharmaceuticals in surface and waste waters. *Journal of Pharmaceutical and Biomedical* 15: 107–115.

Lee, W. and Westerhoff, P. (2006). Dissolved organic nitrogen removal during water treatment by aluminum sulfate and cationic polymer coagulation. *Water Research* 40 (20): 3767–3774.

Li, B., Qu, C., and Bi, J. (2012). Identification of trace organic pollutants in drinking water and the associated human health risks in Jiangsu Province, China. *Bulletin of Environmental Contamination and Toxicology* 88 (6): 880–884.

Lin, X., Xu, J., Keller, A.A. et al. (2020). Occurrence and risk assessment of emerging contaminants in a water reclamation and ecological reuse project. *Science of the Total Environment* 744: 140977.

Liu, J., Qi, S., Yao, J. et al. (2016). Contamination characteristics of organochlorine pesticides in multimatrix sampling of the Hanjiang River Basin, southeast China. *Chemosphere* 163: 35–43.

Llorent-Martínez, E.J., Ortega-Barrales, P., Fernández-de Córdova, M.L., and Ruiz-Medina, A. (2011). Trends in flow-based analytical methods applied to pesticide detection: a review. *Analytica Chimica Acta* 684 (1–2): 30–39.

London, L., Beseler, C., Bouchard, M.F. et al. (2012). Neurobehavioral and neurodevelopmental effects of pesticide exposures. *Neurotoxicology* 33 (4): 887–896.

Loraine, G.A. and Pettigrove, M.E. (2006). Seasonal variations in concentrations of pharmaceuticals and personal care products in drinking water and reclaimed wastewater in southern California. *Environmental Science & Technology* 40 (3): 687–695.

Luo, Y., Guo, W., Ngo, H.H. et al. (2014). A review on the occurrence of micropollutants in the aquatic environment and their fate and removal during wastewater treatment. *Science of the Total Environment* 473: 619–641.

Malcolm, M.S., Weinstein, P., and Woodward, A.J. (1999). Something in the water? A health impact assessment of disinfection by-products in New Zealand. *The New Zealand Medical Journal* 112 (1098): 404–407.

McManus, S.L., Coxon, C.E., Richards, K.G., and Danaher, M. (2013). Quantitative solid phase microextraction–gas chromatography mass spectrometry analysis of the pesticides lindane, heptachlor and two heptachlor transformation products in groundwater. *Journal of Chromatography A* 1284: 1–7.

Miao, X.S., Bishay, F., Chen, M., and Metcalfe, C.D. (2004). Occurrence of antimicrobials in the final effluents of wastewater treatment plants in Canada. *Environmental Science & Technology* 38 (13): 3533–3541.

Mirzaei, R., Mesdaghinia, A., Hoseini, S.S., and Yunesian, M. (2019). Antibiotics in urban wastewater and rivers of Tehran, Iran: consumption, mass load, occurrence, and ecological risk. *Chemosphere* 221: 55–66.

Mohapatra, S.P., Kumar, M., Gajbhiye, V.T., and Agnihotri, N.P. (1995). Ground water contamination by organochlorine insecticide residues in a rural area in the Indo-Gangetic plain. *Environmental Monitoring and Assessment* 35 (2): 155–164.

Moliner-Martínez, Y., Ribera, A., Coronado, E., and Campíns-Falcó, P. (2011). Preconcentration of emerging contaminants in environmental water samples by using silica supported Fe_3O_4 magnetic nanoparticles for improving mass detection in capillary liquid chromatography. *Journal of Chromatography A* 1218 (16): 2276–2283.

Montes-Grajales, D., Fennix-Agudelo, M., and Miranda-Castro, W. (2017). Occurrence of personal care products as emerging chemicals of concern in water resources: a review. *Science of the Total Environment* 595: 601–614.

Mutiyar, P.K. and Mittal, A.K. (2014). Occurrences and fate of selected human antibiotics in influents and effluents of sewage treatment plant and effluent-receiving River Yamuna in Delhi (India). *Environmental Monitoring and Assessment* 186 (1): 541–557.

Naidu, R., Espana, V.A.A., Liu, Y., and Jit, J. (2016). Emerging contaminants in the environment: risk-based analysis for better management. *Chemosphere* 154: 350–357.

Navarrete, I.A., Tee, K.A.M., Unson, J.R.S., and Hallare, A.V. (2018). Organochlorine pesticide residues in surface water and groundwater along Pampanga River, Philippines. *Environmental Monitoring and Assessment* 190 (5): 1–8.

Nguyen, L.N., Hai, F.I., Kang, J. et al. (2013). Removal of emerging trace organic contaminants by MBR-based hybrid treatment processes. *International Biodeterioration & Biodegradation* 85: 474–482.

Noguera-Oviedo, K. and Aga, D.S. (2016). Lessons learned from more than two decades of research on emerging contaminants in the environment. *Journal of Hazardous Materials* 316: 242–251.

Parada, H., Jr, Gammon, M.D., Ettore, H.L. et al. (2019). Urinary concentrations of environmental phenols and their associations with breast cancer incidence and mortality following breast cancer. *Environment International* 130: 104890.

Pazin, M., Pereira, L.C., and Dorta, D.J. (2014). Toxicity of brominated flame retardants, BDE-47 and BDE-99 stems from impaired mitochondrial bioenergetics. *Toxicol Mech Methods* 25: 34–41.

Peng, Y., Fang, W., Krauss, M. et al. (2018). Screening hundreds of emerging organic pollutants (EOPs) in surface water from the Yangtze River Delta (YRD): occurrence, distribution, ecological risk. *Environmental Pollution* 241: 484–493.

Pereira, L.C., de Souza, A.O., and Dorta, D.J. (2013). Polybrominated diphenyl ether congener (BDE-100) induces mitochondrial impairment. *Basic & Clinical Pharmacology & Toxicology* 112: 418–424.

Pereira, L.C., de Souza, A.O., Bernardes, M.F.F. et al. (2015). A perspective on the potential risks of emerging contaminants to human and environmental health. *Environmental Science and Pollution Research* 22 (18): 13800–13823.

Pérez-Fernández, V., Garcia, M.A., and Marina, M.L. (2010). Characteristics and enantiomeric analysis of chiral pyrethroids. *Journal of Chromatography A* 1217: 968–989.

Pesqueira, J.F., Pereira, M.F.R., and Silva, A. (2020). Environmental impact assessment of advanced urban wastewater treatment technologies for the removal of priority substances and contaminants of emerging concern: a review. *Journal of Cleaner Production* 261: 121078.

Pieri, P., Andra, S.S., Charisiadis, P. et al. (2014). Variability of tap water residual chlorine and microbial counts at spatially resolved points of use. *Environmental Engineering Science* 31 (4): 193–201.

Prieto-Rodríguez, L., Oller, I., Klamerth, N. et al. (2013). Application of solar AOPs and ozonation for elimination of micropollutants in municipal wastewater treatment plant effluents. *Water Research* 47 (4): 1521–1528.

Proia, L., Osorio, V., Soley, S. et al. (2013). Effects of pesticides and pharmaceuticals on biofilms in a highly impacted river. *Environmental Pollution* 178: 220–228.

Randall, C. (2014). Pest management. In: *National Pesticide Applicator Certification Core Manual*, 2nd edn. Washington, DC: National Association of State Departments of Agriculture Research Foundation.

Ranjan, J., Mandal, T., and Mandal, D.D. (2019). Environmental risk appraisement of disinfection by-products (DBPs) in plant model system: *Allium cepa*. *Environmental Science and Pollution Research* 26 (9): 8609–8622.

Reemtsma, T., Quintana, J.B., Rodil, R. et al. (2008). Organophosphorus flame retardants and plasticizers in water and air. I: Occurrence and fate. *TrAC Trends in Analytical Chemistry* 27 (9): 727–737.

Rempel, A., Gutkoski, J.P., Nazari, M.T. et al. (2021). Current advances in microalgae-based bioremediation and other technologies for emerging contaminants treatment. *Science of the Total Environment* 772: 144918.

Reungoat, J., Escher, B.I., Macova, M. et al. (2012). Ozonation and biological activated carbon filtration of wastewater treatment plant effluents. *Water Research* 46 (3): 863–872.

Reyes-Contreras, C., Matamoros, V., Ruiz, I. et al. (2011). Evaluation of PPCPs removal in a combined anaerobic digester-constructed wetland pilot plant treating urban wastewater. *Chemosphere* 84 (9): 1200–1207.

Richardson, S.D. (2003). Disinfection by-products and other emerging contaminants in drinking water. *TrAC Trends in Analytical Chemistry* 22 (10): 666–684.

Rivera-Utrilla, J., Sánchez-Polo, M., Ferro-García, M.Á. et al. (2013). Pharmaceuticals as emerging contaminants and their removal from water: a review. *Chemosphere* 93 (7): 1268–1287.

Rodriguez-Mozaz, S., Chamorro, S., Marti, E. et al. (2015). Occurrence of antibiotics and antibiotic resistance genes in hospital and urban wastewaters and their impact on the receiving river. *Water Research* 69: 234–242.

Rodriguez-Narvaez, O.M., Peralta-Hernandez, J.M., Goonetilleke, A., and Bandala, E.R. (2017). Treatment technologies for emerging contaminants in water: a review. *Chemical Engineering Journal* 323: 361–380.

Rosal, R., Rodríguez, A., Perdigón-Melón, J.A. et al. (2010). Occurrence of emerging pollutants in urban wastewater and their removal through biological treatment followed by ozonation. *Water Research* 44 (2): 578–588.

Rout, P.R., Zhang, T.C., Bhunia, P., and Surampalli, R.Y. (2021). Treatment technologies for emerging contaminants in wastewater treatment plants: a review. *Science of the Total Environment* 753: 141990.

Salamanca, M., López-Serna, R., Palacio, L. et al. (2021). Study of the rejection of contaminants of emerging concern by a biomimetic aquaporin hollow fiber forward osmosis membrane. *Journal of Water Process Engineering* 40: 101914.

Sauvé, S. and Desrosiers, M. (2014). A review of what is an emerging contaminant. *Chemistry Central Journal* 8 (1): 1–7.

Sheikh, I.A., Tayubi, I.A., Ahmad, E. et al. (2017). Computational insights into the molecular interactions of environmental xenoestrogens 4-tert-octylphenol, 4-nonylphenol, bisphenol A (BPA), and BPA metabolite, 4-methyl-2, 4-bis (4-hydroxyphenyl) pent-1-ene (MBP) with human sex hormone-binding globulin. *Ecotoxicology and Environmental Safety* 135: 284–291.

Sibila, M.A., Garrido, M.C., Perales, J.A., and Quiroga, J.M. (2008). Ecotoxicity and biodegradability of an alkyl ethoxysulphate surfactant in coastal waters. *Science of the Total Environment* 394 (2–3): 265–274.

Sirés, I. and Brillas, E. (2012). Remediation of water pollution caused by pharmaceutical residues based on electrochemical separation and degradation technologies: a review. *Environment International* 40: 212–229.

Soulet, B., Tauxe, A., and Tarradellas, J. (2002). Analysis of acidic drugs in Swiss wastewaters. *International Journal of Environmental & Analytical Chemistry* 82 (10): 659–667.

Subedi, B. and Kannan, K. (2014). Fate of artificial sweeteners in wastewater treatment plants in New York State, USA. *Environmental Science & Technology* 48 (23): 13668–13674.

Ternes, T.A. and Hirsch, R. (2000). Occurrence and behavior of X-ray contrast media in sewage facilities and the aquatic environment. *Environmental Science & Technology* 34 (13): 2741–2748.

Ternes, T.A., Stumpf, M., Mueller, J. et al. (1999). Behavior and occurrence of estrogens in municipal sewage treatment plants. I: Investigations in Germany, Canada and Brazil. *Science of the Total Environment* 225 (1–2): 81–90.

Toroz, I. and Uyak, V. (2005). Seasonal variations of trihalomethanes (THMs) in water distribution networks of Istanbul City. *Desalination* 176 (1–3): 127–141.

Tran, N.H., Urase, T., and Ta, T.T. (2014). A preliminary study on the occurrence of pharmaceutically active compounds in hospital wastewater and surface water in Hanoi, Vietnam. *CLEAN–Soil, Air, Water* 42 (3): 267–275.

Tran, N.H., Reinhard, M., and Gin, K.Y.H. (2018). Occurrence and fate of emerging contaminants in municipal wastewater treatment plants from different geographical regions; a review. *Water Research* 133: 182–207.

Van Wezel, A.P. and Van Vlaardingen, P. (2004). Environmental risk limits for antifouling substances. *Aquatic Toxicology* 66: 427–444.

Versteeg, D.J., Stanton, D.T., Pence, M.A., and Cowan, C. (1997). Effects of surfactants on the rotifer, *Brachionus calyciflorus*, in a chronic toxicity test and in the development of QSARs. *Environmental Toxicology and Chemistry: An International Journal* 16 (5): 1051–1058.

Wang, Y., Lu, J., Mao, L. et al. (2019). Antiepileptic drug carbamazepine promotes horizontal transfer of plasmid-borne multi-antibiotic resistance genes within and across bacterial genera. *The ISME Journal* 13 (2): 509–522.

Watkinson, A.J., Murby, E.J., Kolpin, D.W., and Costanzo, S.D. (2009). The occurrence of antibiotics in an urban watershed: from wastewater to drinking water. *Science of the Total Environment* 407 (8): 2711–2723.

WHO (World health Organisation) (2012). Pharmaceuticals in drinking water. ISBN 978 92 4 150208 5.

Williams, M., Kookana, R.S., Mehta, A. et al. (2019). Emerging contaminants in a river receiving untreated wastewater from an Indian urban centre. *Science of the Total Environment* 647: 1256–1265.

Williams, R.J., Johnson, A.C., Smith, J.J., and Kanda, R. (2003). Steroid estrogens profiles along river stretches arising from sewage treatment works discharges. *Environmental Science & Technology* 37 (9): 1744–1750.

Woudneh, M.B., Ou, Z., Sekela, M. et al. (2009). Pesticide multiresidues in waters of the Lower Fraser Valley, British Columbia, Canada. Part II: Groundwater. *Journal of Environmental Quality* 38 (3): 948–954.

Wright, S.L., Thompson, R.C., and Galloway, T.S. (2013). The physical impacts of microplastics on marine organisms: a review. *Environmental Pollution* 178: 483–492.

Wu, C., Luo, Y., Gui, T., and Huang, Y. (2014). Concentrations and potential health hazards of organochlorine pesticides in shallow groundwater of Taihu Lake region, China. *Science of the Total Environment* 470: 1047–1055.

Yadav, A., Rene, E.R., Mandal, M.K., and Dubey, K.K. (2021). Threat and sustainable technological solution for antineoplastic drugs pollution: review on a persisting global issue. *Chemosphere* 263: 128285.

Yang, X., Fan, C., Shang, C., and Zhao, Q. (2010). Nitrogenous disinfection byproducts formation and nitrogen origin exploration during chloramination of nitrogenous organic compounds. *Water Research* 44 (9): 2691–2702.

Ying, G.G. (2006). Fate, behavior and effects of surfactants and their degradation products in the environment. *Environment International* 32 (3): 417–431.

Yu, Y., Wu, L., and Chang, A.C. (2013). Seasonal variation of endocrine disrupting compounds, pharmaceuticals and personal care products in wastewater treatment plants. *Science of the Total Environment* 442: 310–316.

Yuan, S., Jiang, X., Xia, X. et al. (2013). Detection, occurrence and fate of 22 psychiatric pharmaceuticals in psychiatric hospital and municipal wastewater treatment plants in Beijing, China. *Chemosphere* 90 (10): 2520–2525.

Zhang, H., Shao, X., Zhao, H. et al. (2019). Integration of metabolomics and lipidomics reveals metabolic mechanisms of triclosan-induced toxicity in human hepatocytes. *Environmental Science & Technology* 53 (9): 5406–5415.

Zhou, R., Zhu, L., and Chen, Y. (2008). Levels and source of organochlorine pesticides in surface waters of Qiantang River, China. *Environmental Monitoring and Assessment* 136 (1): 277–287.

2

Occurrence

Prangya Ranjan Rout[1,2], Manaswini Behera[3], Puspendu Bhunia[3], Tian C. Zhang[4], and Rao Y. Surampalli[5]

[1]*Department of Biotechnology, Thapar Institute of Engineering & Technology, Patiala, Punjab, India*
[2]*Department of Biotechnology, Dr. B. R. Ambedkar National Institute of Technology Jalandhar, Punjab, India*
[3]*School of Infrastructure, Indian Institute of Technology Bhubaneswar, Odisha, India*
[4]*Civil and Environmental Engineering, College of Engineering, University of Nebraska, Lincoln, Omaha, NE, USA*
[5]*Global Institute for Energy, Environment and Sustainability, Lenexa, KS, USA*

2.1 Introduction

The environmental occurrence of diversified newly identified natural as well as man-made compounds in the past couple of decades emerges as a matter of increasing environmental concern and has drawn global attention. The environmental occurrence of this vast and expanding array of compounds is in the range of μg/L to ng/L, hence termed as microconstituents (MCs) or micropollutants (MPs) (Luo et al. 2014). The MCs, which are predominantly organic in nature, are otherwise known as trace organic compounds (TrOCs). As per the United States Geological Survey (USGS) definition, MCs are any natural/synthetic chemicals or microbial contaminants like antibiotic resistance genes (ARGs)/bacteria, which are not commonly monitored in the environment but have potentially harmful effects on the environmental and/or human health (USGS 2017). The environmental occurrence of these unregulated trace compounds dates back 2000 years but they are only now being detected in recent times due to the development of sensitive and advanced analytical techniques such as gas chromatography-mass spectrometry (GC-MS) and liquid chromatography-mass spectrometry (LC-MS), etc., hence also called emerging pollutants (EPs) or contaminants of merging concern (CECs) (Sauvé and Desrosiers 2014; Rout et al. 2021). Therefore, as per the definition of the United States Environmental Protection Agency (USEPA), MCs are noticeably new compounds that have damaging effects on the ecology and health of various life forms (de Oliveira et al. 2019). The MCs include industrial chemicals, personal care products, pharmaceuticals, steroid hormones, nanomaterials, drugs of abuse, pesticides, and many other newly detected compounds.

The environmental occurrence of MCs has been linked to a range of harmful consequences, including endocrine disrupting effects, microbial antibiotic resistance, and short- and long-term toxicity. Due to their low concentration, the short-term toxic effects of MCs

Microconstituents in the Environment: Occurrence, Fate, Removal, and Management, First Edition.
Edited by Rao Y. Surampalli, Tian C. Zhang, Chih-Ming Kao, Makarand M. Ghangrekar, Puspendu Bhunia, Manaswini Behera, and Prangya R. Rout.

are not significant but their continual release into the environment, recalcitrant nature, complex forming tendency, and unwanted synergistic effects, etc., are the key contributors to chronic health effects of living beings to their long-term exposure. Depending on the dose and length of time of exposure, MCs can induce both female and male sexual disorders, such as polycystic ovaries, lower male fertility, longer conception time, greater rates of miscarriage, breast malignancies, reduction in aquatic biodiversity, and harmful effects on human and wild life endocrine systems, etc. (Stackelberg et al. 2004; de Oliveira et al. 2019). The potential health risks associated with MCs are assessed using the risk quotient (RQ), which is the ratio of the predicted environmental concentrations (PEC) and the predicted no-effect concentrations (PNECs) of MCs (Eq. 2.1). PNEC is the threshold concentration of MCs exposure which will not have any harmful effect on living beings. To predict the environmental concentration of specific MCs, PEC can be done based on the reported literature, whereas PNECs can be calculated from the LC_{50} values of MCs, that is the concentration level of MCs lethal to 50% of population of a specific living being. According to the RQ values, MCs are categorized as negligible risk, $RQ < 0.01$, low risk, $0.01 > RQ < 0.1$, medium risk, $0.1 > RQ < 1$, and high risk, $RQ > 1$ (Díaz-Garduño et al. 2017).

$$\mathrm{RQ} = \frac{PEC}{PENC} \tag{2.1}$$

MCs are known to be released into different environmental matrices via multiple pathways, including domestic discharges, industrial discharges, sewage treatment plants, agricultural runoff, hospital discharges, aquaculture effluent, improper manufacturer disposal, landfill leachates, water treatment plants, effluents from livestock farming, etc. The released MCs may either retain their original structures and concentrations or become transformed into other inactive/active forms during their lifespan in environmental matrices (Yang et al. 2017). The existing technologies for MCs removal can be broadly categorized into natural, conventional, and advanced processes. All the natural MCs reduction processes like sorption, dilution, photolysis, volatilization, biodegradation, etc., are simple and cost-effective processes but are less efficient (Rizzo et al. 2019). Conventional processes like adsorption by activated carbon, membrane-based filtration, and ozonation are proved to be effective in MCs removal but formation of oxidation by-products (ozonation), concentrate discharge (membrane filtration), disposal of MCs saturated adsorbent (adsorption), etc., are major problems to be tackled (Rout et al. 2015; Pesqueira et al. 2020; Shahid et al. 2020). Advanced processes, such as application of bioelectrical systems, constructed wetlands, advanced oxidation processes, enzymatic treatment, etc., are demonstrating promising MCs removal potential but the scalability of these advanced technologies in a cost-effective way still remains an unanswered question. Therefore, the removal of MCs continues to be a major challenge to the global scientific community in spite of the availability of many treatment technologies. Additionally, the existing wastewater treatment plants (WWTPs) are not yet adequate to remove MCs from wastewater of diverse origin. Consequently, many of these MCs escape the treatment steps of WWTPs and end up in different environmental matrices (Bolong et al. 2009).

To date, most MCs have no regulated environmental discharge limits. MCs like pharmaceuticals and personal care products (PPCPs) have been tightly regulated in order to reduce their use (Daughton 2002). However, due to the beneficial usage of PPCPs for humans and animals, these items are unlikely to be restricted (Jones et al. 2005). A small number of MCs have been regulated by some countries. For example, as per Directive 2008/105/EC, environmental quality criteria was set for a small number of MCs like bisphenol A, nonylphenol, diuron, etc., by the European Union (European Parliament and The Council 2008). The Canadian government has likewise declared nonylphenol ethoxylates and nonylphenol to be hazardous chemicals (Canadian Environmental Protection Act 1999). Other MCs are yet to be added to the list of regulated chemicals. Moreover, in most WWTPs, safeguards and monitoring activities for MCs are not adequately established (Bolong et al. 2009). Therefore, the scientific community and regulatory bodies need to obtain an understanding of not just the impact of individual and additive negative environmental impacts of MCs but also their occurrence and transport to different environmental matrices. Further research on acute and chronic impacts of MCs, their major sources, and migration pathways in the environmental matrices, is critical for setting regulatory limits. Table 2.1 presents an outline of MC classifications and their sources of origin. This chapter emphasises the recent occurrence of MCs in terms of their concentration in different environmental matrices.

Table 2.1 General classification and sources of MCs.

Major Classes	Representative MCs	Probable Sources	Adverse Impact
Pharmaceuticals	Diclofenac, Ciprofloxacin, Carbamazepine, Diazepam, Metoprolol Testosterone.	Hospital discharges; Pharmaceutical industry discharge; Discharges from aquaculture and livestock farming and domestic discharges.	Antibiotic resistance; Microbial growth inhibition; Feminization of males; Reduced fertility.
Industrial chemicals	Dimethyl adipate (DMAD), Tris (1-chloro-2-propyl) phosphate (TCPP).	Induatrial discharges; Improper manufacturer disposal.	Affects nervous, endocrine, and reproductive systems.
Biocides	Butachlor, Epoxiconazole, Metaldehyde	Surface water; Aquaculture discharge; Agricultural runoff.	Affects respiratory systems and wild life.
Personal care products	*N,N'*-diethyltoluamide (DEET), Galaxolide, 4-benzophenone.	Wastewater treatment plant discharge; Landfill leachate; Surface water.	Toxic to aquatic organisms; Affects human nervous system; Carcinogenic to rodents.
Surfactants	Polysorbates, Sodium lauryl sulfate (SLS), Sodium dodecyl sulfate (SDS).	Industrial discharge; Domestic discharge.	Toxic to environment; Carcinogenic; Absorption of pollutants in water.
Endocrine disrupting compounds (EDCs)	Bisphenol A, Xenoestrogen, Dioctyl phthalate (DOP).	Soil/Sediments; Surface water; Secondary sludge; Drinking water.	Affects endocrine system, causes developmental delays and birth defects.
Perfluorinated alkylated substances (PFASs)	Perfluorooctanoic acid (PFOA), Perfluorooctanesulfonate (PFOS).	Wastewater; Surface water; Sediments; Ground water.	Causes cardiovascular diseases; Carcinogenic; Poor fetal development.
Artificial Sweeteners	Saccharin, Aspartame, Sucralose.	Landfill leachate; Sewage treatment plant discharge.	Damage to algae and water fleas; Forms recalcitrant by-products.

2.2 Goals of Occurrence Survey

The occurrence survey of MCs is vital in order to understand how these pollutants enter and are transported in different environmental matrices worldwide. However, achieving this goal is challenging due to lack of information on how MCs occur in different natural environmental components like surface water, deposition, groundwater, etc., and man-made environmental components such as industrial wastewater, municipal wastewater, etc. This lack of information on MCs can be attributed to the unavailability of sensitive detection techniques capable of quantifying low concentrations of MCs in environmental samples (Noguera-Oviedo and Aga 2016). Though the traditional approach for MCs analysis is gas chromatography-mass spectrometry (GC-MS), a significant number of MCs are non-detectable using this technique, whereas liquid chromatography-mass spectrometry (LC-MS) techniques can detect a dominant group of MCs (Kolpin et al. 2002). There was a gap of 27 years between the first quantification of MCs in surface water using GC/MS (1975) and the widespread detection of MCs in US streams using LC-MS (2002). The current MCs detection practices include solid-phase extraction (SPE) followed by liquid chromatography-tandem mass spectrometry (LC-MS/MS). In addition to quantification challenges, research on MCs has typically focused on one or few selected environmental components, making it difficult to perform rigorous, interdisciplinary research that examines the significance of many emission pathways. Therefore, in the last few decades, the monitoring of MCs has become a matter of growing concern to make a complete dataset of their occurrence in diversified environmental compartments. Data from monitoring studies have regularly established the existence of hundreds of MCs in environmental matrices worldwide. Recently, long-term monitoring of MCs occurrence at the small scale, such as the watershed level, has been explored to recognize key insights into their sources (Carpenter and Helbling 2018). For instance, multivariate and mass balance analyses revealed three types of MC sources, sewage treatment plant (STP) outfalls, diffuse runoff, and mixed pathways, in a Minnesota river (Fairbairn et al. 2016). Likewise, a geospatial analysis of PFAS revealed the occurrence of MCs at higher concentrations in urban areas, particularly when connected to metal smelting industries, textile mills, airports, etc. (Zhang et al. 2016). These examples validate the importance of occurrence surveys and powerful ways in which MCs occurrence data can be used to progress our understanding of MC sources.

2.3 Environmental Occurrence of Microconstituents

Environmental occurrence of MCs is attributed to their diversified sources including effluents from industries, hospitals, domestic discharges, and runoff from animal farming and agriculture (Figure 2.1). Leachates from landfills, leakages from sewage treatment plants (STPs), discharges from aquaculture, and reclaimed water mediated irrigations are some of the additional contributors of MCs to the environment (Yang et al. 2017). Effluents from industries like chemical, biocides, pharmaceuticals, personal care products (PCPs), etc. are the dominant contributors of MCs to the environments (Barbosa et al. 2016). Hospital effluents are the sources of MCs like drugs, antibiotic resistance microbes, radioactive elements, etc., whereas domestic discharge is the prime source of PCPs such as sunscreen creams, shampoos, bodywash, toothpastes, etc. (Tiwari et al. 2017). Steroid hormones and

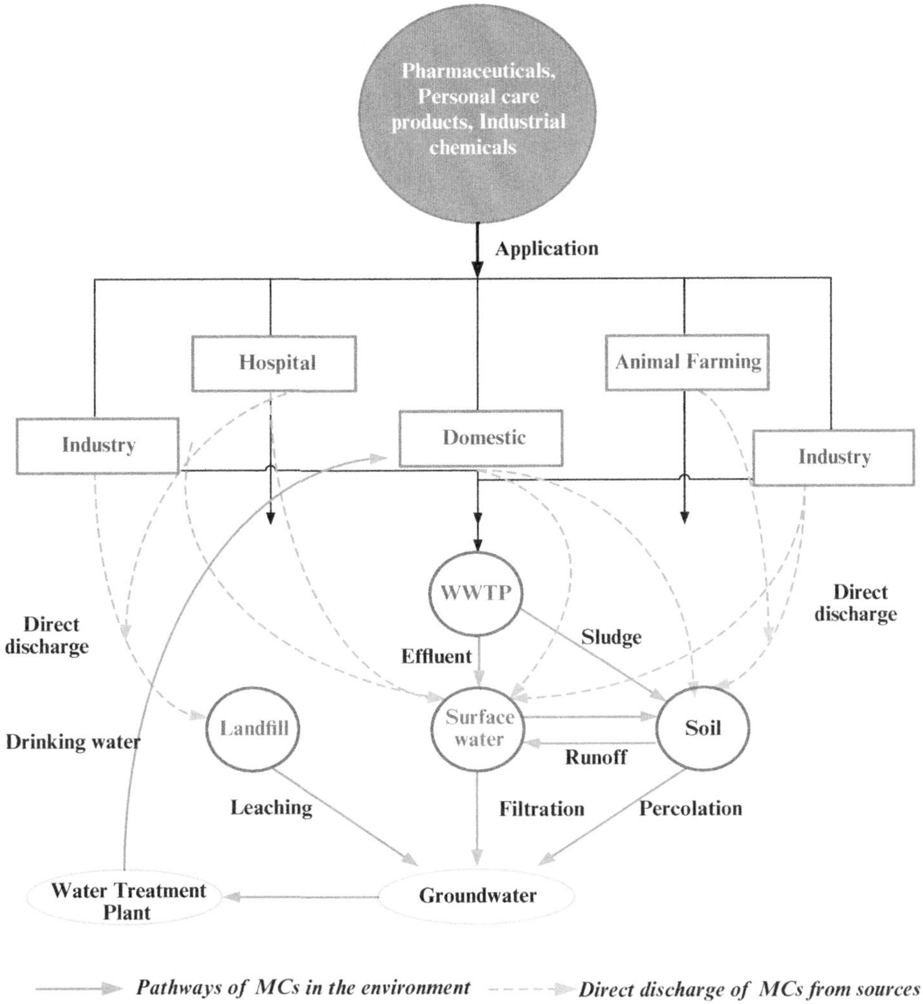

Figure 2.1 Major pathways of MCs in the environmental matrices.

pesticides are the major MCs from animal farming and agricultural activities, respectively. After environmental appearance, MCs further drift to different environmental matrices like WWTPs, sludge, soil, surface water (e.g., rivers and lakes), groundwater, estuaries and marine environments, drinking water, landfills, etc., through various paths (Figure 2.1). Environmental factors, such as dispersion, dilution, precipitation, etc., further cause concentration variations of MCs in different environmental components (Luo et al. 2014).

2.3.1 Occurrence of Microconstituents in Groundwater

The occurrence of MCs in groundwater has been properly characterized, mostly in some parts of the US and Europe. It has been observed that groundwater is less contaminated with MCs, whereas reservoirs are pointedly more contaminated, displaying higher concentrations of MCs, particularly in China (Yan et al. 2015). Based on studies conducted in 14 countries across Asia, America, the Middle East, and Europe, the most commonly detected MCs in groundwater include carbamazepine, ibuprofen, caffeine, diclofenac, salicylic acid, paracetamol, and sulfamethoxazole because of their wide and frequent consumption (Peng et al. 2014). But the detected concentration of most of the MCs are at levels comparable to the PENCs, <100 ng/L, as presented in Table 2.2; therefore, without

Table 2.2 Concentration of selected MCs in different environmental matrices.

MCs Types	WWTP Effluent Concentration (ng/L)	WTP Effluent Concentration (ng/L)	Groundwater Concentration (ng/L)	Surface Water Concentration (ng/L)	Soil Concentration (μg/Kg)	Sludge Concentration (ng/g Dry Weight)
Antibiotics						
Ciprofloxacin	96.3	–	0.28	77	–	1400–4800
Erythromycin	42	1.3–2	–	–	–	6–79
Trimethoprim	100	–	18	145	3.1–60.1	BDL-13
NSAIDs						
Diclofenac	647	BDL	24	794	0.35–1.16	BDL-133
Ibuprofen	394	1.3	3.97	203	1.51–5.03	BDL-3988
Ketoprofen	86	20–37	2886	293	–	BDL-58.4
Naproxen	462	BDL	204	322–328	–	BDL-1022
Lipid regulators						
Bezafibrate	139	–	60.8	–	–	17–64
Gemfibrozil	137.7	BDL	41	163	–	BDL-1192
Hormones						
Estrone	15	<0.15	14	69.1	–	BDL-17.5
Ethinylestradiol	2	<0.2	1.9	230	–	BDL-17.5
Beta-blockers						
Atenolol	843	–	10	34–100	–	BDL-86
Metoprolol	166	BDL	12	–	–	BDL-226
Anticonvulsants						
Carbamazepine	482	1–4	420	735	2.6–7.5	BDL-50
Insect repellents						
DEET	678.1	49	–	–	–	2–6
Biocides						
Triclosan	74.8	<0.1	39.8	124–220	BDL-16.7	354–15600
Psychoactive drugs						
Caffeine	191.1	–	4500	568	–	BDL-805

BDL = Below detection limit; NSAIDs = Nonsteroidal anti-inflammatory drugs; DEET = *N,N'*-diethyltoluamide.
Note: The data in this table are summarized primarily from Rout et al. 2021; Das et al. 2017; Huerta-Fontela et al. 2011; Tran et al. 2018; Tiwari et al. 2017; Petrie et al. 2015; Luo et al. 2014; Li 2014; Yang et al. 2011; Wang and Wang 2016; Yang et al. 2017. http://www.drugbank.ca, https://pubchem.ncbi.nlm.nih.gov, and https://pubchem.ncbi.nlm.nih.gov/source/ChemIDplus.

substantial environmental effects (Luo et al. 2014). Nevertheless, noticeably high concentrations of ketoprofen (2886 ng/L), naproxen (204 ng/L), carbamazepine (420 ng/L), and caffeine (4500 ng/L) were reported in groundwater of some European countries, California, Canada, etc. (Li 2014). Being the most commonly used analgesic, carbamazepine was found in 42% of groundwater samples taken from 164 sites across 23 different European countries. Likewise, in Guangzhou (China), salicylic acid was monitored with 98% detection frequency in the concentration range of 43.7 to 2014.7 ng/L (Peng et al. 2014). Similarly, steroid hormones were discovered in the groundwater at a level 2 to 3 times higher than PENCs at a land application site in the US, probably due to the application of wastewater effluent for irrigation (Luo et al. 2014). In comparison to nonsteroidal anti-inflammatory drugs (NSAIDs) and antibiotics, lipid regulators (bezafibrate, gemfibrozil), hormones (estrone, ethinylestradiol), and beta-blockers (atenolol, metoprolol) were detected in groundwater less frequently.

MCs in groundwater are the result of industrial, agricultural, livestock farming, and anthropogenic activities. Today, potential causes and ongoing sources of groundwater contamination include leachate from landfills, bank filtration, infiltration from wastewater sources, groundwater recharge utilizing reclaimed water, percolation from agricultural land, and seepage from sewage systems (Sui et al. 2015). Additionally, these channels have a major impact on groundwater quality, especially in areas with shallow groundwater tables and high transmissivity aquifers. Soil is another main conduit for some MCs to pollute groundwater, in particular through the application of sludge as a fertilizer or soil conditioner. Similarly, burying dead animals on land greatly contributes to the MCs in soil that eventually reach groundwater. Another significant issue and noticeable source of groundwater MC contamination is medication to prevent disease. The majority of veterinary medications used are often expelled (50–100%) through urine and feces, which penetrate the surrounding systems and seriously endanger groundwater (Kim et al. 2011). Furthermore, MCs of surface water are also subject to natural attenuation, eventually finding their way into the groundwater through the hyporheic zone, either by lateral hydraulic exchange or vertical hydraulic exchange. The concentrations of MCs from other sources (such as leachate from landfills and seepage from sewage systems) to groundwater can typically be reduced by the mechanisms driving subsurface flow and transport, such as adsorption to aquifer material, transit duration, degradation, and dilution. However, as the groundwater is used for direct or indirect potable water reuse, MC prevalence within it might be of serious concern (Luo et al. 2014).

2.3.2 Occurrence of Microconstituents in Surface Water

Effluents from WWTPs or STPs are the principal pathway through which MCs enter surface water like rivers and lakes and accumulate in the aquatic environment. Diverse natural processes, such as sorption into sediments and suspended solids, dilution, photolysis, and degradation, are used to attenuate MCs in surface water bodies. MCs may be present at concentrations at least one order of magnitude lower than the WWTPs effluent levels in rivers and lakes because of their dilution. MCs are far more common in rivers and lakes during dry weather than during wet weather. Similarly, the concentrations of MCs in surface water are found to be lower during summer than in winter, which could be due to enhanced biodegradation of MCs at warmer temperatures during dry summers and increased dilution during wetter summers (Gómez et al. 2012). Furthermore, rainfall helps to dilute MCs in surface water, whereas it contributes additional MCs to the surface water via chemical leaching from pavement materials, building and roof coatings, sewage overflow, and flash floods mediating increased runoff (K'oreje et al. 2020). In general, MC concentration in natural water bodies in densely populated areas is typically more severe due to the widespread use of these chemicals by a large population.

NSAIDs, antibiotics, and hormones are the most prevalent MCs found in surface water, as presented in Table 2.2, because of their high consumption rate. Especially, diclofenac, carbamazepine, and caffeine were detected at alarmingly high levels with respective concentrations of 794, 735, and 538 ng/L, respectively. Beta-blockers that are used to treat vasodilation, bronchodilation, and the relaxation of visceral smooth muscles were detected regularly in surface waters at concentrations of up to 34–100 ng/L. Pesticides like MCs occurrence in rivers and lakes is governed by various factors like soil characteristics, crop types, climatic conditions (rainfall, temperature, wind, moisture), aquatic body characteristics (flow rate, depth), and land features near aquatic bodies (slope, soil usage, water bodies distance). K'oreje et al. (2020) reported the occurrence of pesticides including herbicides, insecticides, and fungicides in the ranges of 0.2–14, 0.06–69, and 0.1–9 ng/L, respectively, in African surface waters (K'oreje et al. 2020). The release of hospital effluents and other highly contaminated waters was the main cause of the high concentrations of MCs that were discovered in the surface water of Costa Rica (Luo et al. 2014). WWTPs being the predominant source of MCs to the surface water, these findings reporting the widespread occurrence of MCs in surface water confirm that these recalcitrant MCs are difficult to remove by conventional WWTPs.

2.3.3 Occurrence of Microconstituents in Marine Water

Estuaries and marine environments receive MCs through atmospheric deposition, direct discharges of domestic and industrial wastewater, and riverine sources like harbors, shipping, port activities, aquaculture, and oil exploration (Ghekiere et al. 2013). Amusement beaches in particular are vulnerable to contamination, through sailing, windsurfing, and swimming, like marine activities mediated absorption of MCs into water. Additionally, marine outfalls from coastal cities can seriously pollute recreational beaches with MCs. Also, MCs that commonly occur in fresh water are released on a large scale into marine ecosystems. Consequently, a diverse group of MCs like biocides, pharmaceuticals, corrosion inhibitors, and stimulants accumulate in the marine environment (Nödler et al. 2014). The accumulation of MCs in marine and, in particular, coastal environments, primarily depends on the persistent nature of the MCs and the rate of water exchange and evaporation in the receiving marine water. A comparatively lower concentration of an antifouling agent in the North Sea than that of the Baltic Sea of the German marine environments was attributed to the higher water exchange rate of the North Sea (Biselli et al. 2000). The hierarchical structures of the MCs and hydrometeorological variables like rain fall, tide levels, etc., are other important factors that impact the change in MCs concentrations in a marine environment (Baek et al. 2022). Marine water samples were collected from the shorelines of the San Francisco Bay, Baltic Sea, Pacific

Ocean, Mediterranean Sea, Northern Adriatic Sea, Balearic Sea, Aegean Sea, etc., and analyzed for diverse MCs. During long-term monitoring of MCs concentrations at the marine outfall, occurrences of high concentration of MCs were reported when the tide level was low after rainfall (Baek et al. 2022). More than half of all the samples contained caffeine, atrazine, theobromine, paraxanthine, tolyltriazole, etc. Given that untreated wastewater was the chief source of caffeine, the prevalence of raw sewage in marine ecosystems is the reason behind the elevated concentration of caffeine in marine environments. Carbamazepine, sulfamethoxazole, iopamidol, paracetamol, and atenolol had detection frequencies between 20 and 32%, whereas detection of pesticide compounds exhibited an irregular trend (Nödler et al. 2014). Mevinphos, telmisartan, and methiocarb are reported to be some of the most ecotoxicologically hazardous MCs, posing a high risk to marine ecosystems (Baek et al. 2022). Taking into account the apparent negative impacts on the marine environment created by the complex mixture of MCs, extensive global monitoring studies encompassing a wide range of compounds representing numerous MCs sources to marine environments are implied. The Marine Strategy Framework Directive (MSFD) was established by the European Union to create a framework in which the member states will be responsible to take the necessary steps to achieve and maintain good marine environmental status by developing strategies to monitor and reduce MC inputs in each marine region. Consequently, the first-phase MSFD report highlights that the pollution in the marine environment has decreased in some places, but levels of some MCs are still beyond permissible limits (Nödler et al. 2014). Therefore, it is recommended that additional efforts should be made to achieve the targeted goal of good marine environmental status.

2.3.4 Occurrence of Microconstituents in Drinking Water

Since most MCs concentrations are below the limit of detection in drinking water, very few studies have reported the presence of MCs in this medium. A number of variables, such as the kind of water source and seasons, affect the degree of occurrence of MCs. Additionally, the technologies used for water treatment play a critical part in their elimination. Erythromycine, ibuprofen, ketoprofen, estrone, ethinylestradiol, carbamazepine, *N,N'*-diethyltoluamide (DEET), and triclosan are the most often reported MCs in drinking water. However, all of these MC concentrations were below 100 ng/L, as presented in Table 2.2. According to Kleywegt et al. (2011), carbamazepine concentration in some of the drinking water sources was around 600 ng/L, which is roughly 6–10 times more than the concentration of the majority of MCs. In Valencia, Spain, tap and mineral water were examined by Carmona et al. (2014) for the presence of MCs. Salicylic acid was the most prevalent of the MCs in tap water samples; however, ibuprofen and diclofenac were also present in higher concentrations. Salicylic acid is the most frequently discovered MC in drinking water samples, according to Vulliet et al. (2011), who also demonstrated the presence of 25 MCs in drinking water. Although they do exist, the majority of MC concentrations in drinking water are present at safe levels. However, due to direct consumption of drinking water, much care must be taken for routine monitoring of the MCs owing to their negative effects on human and animal health.

2.3.5 Occurrence of Microconstituents in WWTPs Effluent and Sludge

Concentrations of MCs in WWTPs vary depending on how they are used. For instance, taking into account inter-day variation, a peak in antibiotic load was seen in the influent and effluent of WWTPs between 7:00 and 9:00 am. The trend is because of the ingested medications building up in the body through metabolization during sleep and subsequent excreted through urine and feces to the wastewater stream (Coutu et al. 2013). Therefore, rate of excretion plays a vital role in controlling the introduction of MCs, pharmaceuticals in particular, to the wastewater. Similarly, considering inter-day variability, a higher concentration of cocaine was detected in the wastewater samples of European countries. The observed trend is attributed to the recreational use of cocaine over the weekend (Rout et al. 2021). Likewise, seasonal variations of MCs in WWTPs are also reported in the literature. For example, high concentrations of sunscreen compounds are detected in wastewater during summer seasons, whereas compounds of cough syrup like pholcodine are observed in elevated concentrations in wastewater during winter seasons (Petrie et al. 2015). Some of the reported MC concentrations in WWTPs are <100 ng/L (e.g., ciprofloxacin erythromycin, ketoprofen, estrone, ethinylestradiol, triclosan). However, most of the MCs are detected in concentrations of >100 ng/L including trimethoprim, diclofenac, ibuprofen, naproxen, atenolol, metoprolol, DEET, caffeine, bezafibrate, and gemfibrozil (Figure 2.2). Compounds like diclofenac, ibuprofen, and carbamazepine have low excretion rates, but are still present in wastewater in very high concentrations, possibly because of the enormous usage of these compounds, which counterbalance the low rates of excretion of the compounds. The plentiful occurrence of caffeine is

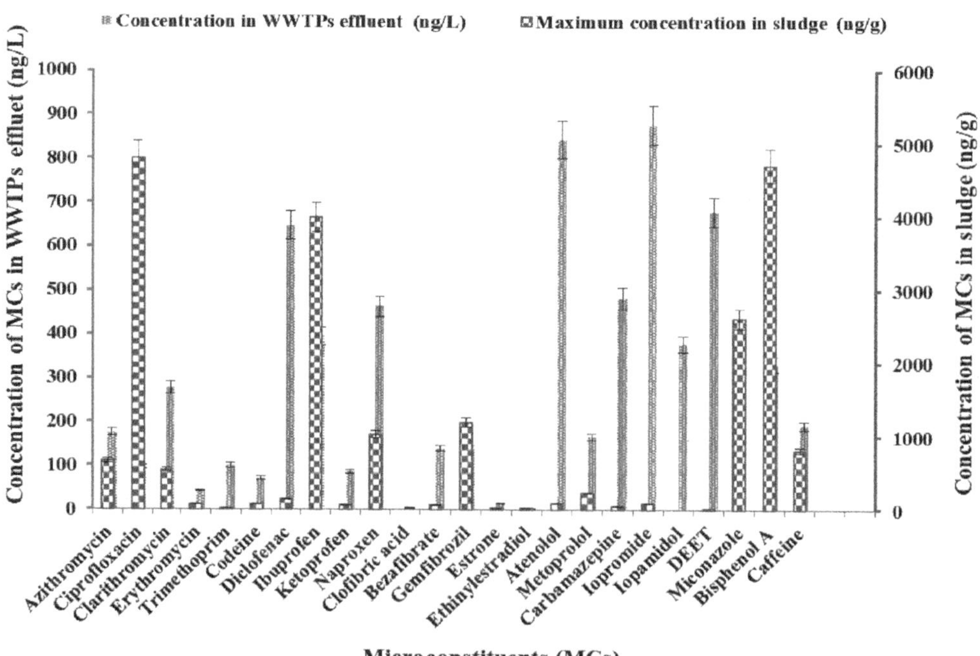

Figure 2.2 Concentrations of selected MCs in WWTPs effluent and sludge.

probable linked with the high rate of consumption of coffee, soft drinks, and tea worldwide. Hormones, in particular steroids, commonly show less detected concentrations; however, on account of the high estrogenic effect of the steroid hormones, their incidence, even at small concentrations, is of concern (Luo et al. 2014).

Unlike the extensive studies undertaken pertinent to the occurrence of MCs in their dissolved phase, studies related to MC occurrence in the particulate phase has rarely been reported in literature. The limited availability of reports on MC occurrence in sludge is perhaps due to the complex sludge matrices and nonexistence of precise analytical techniques to detect MCs in the sludge matrix (Tran et al. 2018). Comparatively higher concentrations of MCs existence in sludge than in wastewater have been testified to in past studies. So, the data of the incidence and distribution of MCs in sludge is imperative, not only from the sludge management approach point of view, but also from the prospective of controlling its usage in agricultural applications as a fertilizer, since the recalcitrant MCs in sludge can contaminate soil and groundwater (Rout et al. 2021). The variable concentrations of selected MCs in the range below a detection limit of 15,600 ng/g sludge dry weight, as per their occurrence in sewage sludge, are presented in Figure 2.2. The wide range of variations in MC concentrations in sludge are reliant on their usage patterns hydrophobicity like physico-chemical properties (Tran et al. 2018). Particularly, the octanol-water partitioning coefficient (K_{OW}), that implies the separation of MCs between dissolved and particulate phases based on their hydrophobic nature, governs the extent of retention of MCs in sludge. In principle, $logK_{OW}$ <2.5 indicates high mobility of MCs; $logK_{OW}$ >4 specifies retention of MCs in sludge matrix (low mobility), and $logK_{OW}$ values in between 2.5 and 4 designate medium mobility of the MCs (Rogers 1996). For instance, $logK_{OW}$ of DEET and trimethoprim of <1.5 means they are less likely to be retained in the sludge matrix, hence reported in the least concentrations, 2–6 and BDL 13 ng/g sludge dry weight (Table 2.2), respectively in a sludge sample. However, ciprofloxacin and triclosan have $logK_{OW}$; therefore, they are mostly retained in the sludge matrix in respective concentrations of 1400–4800 and 354–15,600 ng/g sludge dry weight (Table 2.2) (Petrie et al. 2015).

2.3.6 Occurrence of Microconstituents in Soil

Monitoring of MCs in soil is not a commonly practiced approach. On the other hand, behaving as a pollutant reservoir, soils play a vital role in MC transfer to different environmental components. Various physico-chemical processes that take place in soil, including adsorption, precipitation, dissipation, etc., facilitate accumulation of MCs and their subsequent transfer to aquatic environments like groundwater, rivers, lakes, etc. (Gasperi et al. 2018). The application of sewage sludge as a soil conditioner or fertilizer in agricultural fields is the chief source of MC contamination in soil. Additionally, other factors caused by humans primarily associated to land usage and industrialization as well as river basin network activities are also responsible for soil contamination with MCs. Soil, manured with sewage sludge, becomes severely polluted with antibiotics, steroids, and hormones. The concentrations of some of the MCs, like PFASs, polycyclic aromatic hydrocarbons (PAHs), phthalates, polychlorinated biphenyls (PCBs), etc., were detected in the range of 0.2–55,000 μg/kg dry weight soil (Gasperi et al. 2018). However, it was reported that MCs like PCBs and PAHs accumulated in the soil just after spreading of the sewage sludge, but a couple of months later the concentration of MCs reduced to the original level as before the spreading of sludge. As per an investigative report by the Netherlands National Baseline Survey of Soil Quality, the majority of the MC concentrations were below the limit of detection in soil samples (Brus et al. 2009). In fact, physicochemical properties of MCs like distribution coefficient (K_d), acid dissociation constant (pKa), and octanol-water partitioning coefficient (K_{OW}), etc., govern the extent of retention of MCs in soil, and K_d in particular plays the most significant role. K_d is the ratio of MCs concentration in the solid phase (soil) to the concentration in liquid phase (water), as presented in Eq. (2.2).

$$K_d = \frac{C_s}{C_l} \tag{2.2}$$

where C_s is the contaminant concentration on solid (μg/L); and C_l is the contaminant concentration in liquid (μg/L). For MCs with a K_d value <300–500 L/kg/soil, their adsorption mediated retention in soil is insignificant (Das et al. 2017). Out of all the MCs presented in Table 2.2, except for ciprofloxacin (K_d = 19,952.6–38,900), estrone (K_d = <100–645), ythinylestradiol (K_d = 330–750), and triclosan (K_d = 1905–9550), the MCs have a K_d value <300–500 L/kg/soil and so minimum retention in the soil matrix can be expected (Goswami et al. 2018).

2.3.7 Occurrence of Microconstituents in Foods and Vegetables

One of the primary means by which living beings are exposed to emerging contaminants is through food. The consumption of food containing diverse organic micropollutants accounts for 90 to 98% of average human exposure. Because of their use in agricultural and food processing, PCBs and organochlorine pesticides (OCPs), like MCs are frequently being detected in foods, particularly in vegetables (Khan et al. 2022). A study by Sabourin et al. (2012) reported the uptake of MCs like hormones, parabens, and pharmaceuticals by the vegetables, such as tomatoes, carrots, potatoes, and sweetcorn grown in soils fertilized with municipal biosolids at a controlled application rate. The detected concentration was in the range of 0.33 to 6.25 ng/g dry weight for some of the MCs. However, the authors found low uptake potential of MCs by the crops under normal farming conditions. A recent study also discussed a similar finding where uptake and metabolization of pharmaceutical and personal care products were observed in the roots and leaves of lettuce grown in treated wastewater (Bigott et al. 2022). Out of the 14 MCs investigated, maximum concentrations of sulfamethoxazole, sucralose, and citalopram were observed in the roots, whereas acesulfame, sucralose, and carbamazepine had the highest concentrations in the leaves. Another significant obstacle to providing food security for these highly populated places is the lack of agricultural land in urban and peri-urban settlements. The anaerobic digestate from food waste, a dominant contributor of MCs to agricultural activities, is increasingly used as a nutrient medium in hydroponics for food production for highly populated places lacking agricultural land (Lee et al. 2020, 2021). Golovko et al. (2022) investigated MCs concentrations in anaerobic digestate mediated biofertilizers and observed 48 out of 113 investigated MCs were detected at least once, and caffeine, pyroxidine, theobromine, and nicotine occurrence were in the highest concentrations. Therefore, it would be highly essential to have RQ mediated concentration limits of MCs in foods and vegetables for their safe consumption.

2.4 Challenges and Future Prospective in Occurrence Survey

Due to the lack of knowledge on MC distribution and migration in various environmental matrices, MC occurrence surveys are relatively difficult. The lack of sensitive detection techniques capable of measuring low concentrations of MCs in environmental samples until the early 1990s can be blamed for this lack of information. Furthermore, it is challenging to conduct thorough, interdisciplinary research that investigates the impact of numerous emission pathways, since research on MCs has traditionally concentrated on one or a small number of selected environmental components. As a result of the scarcity of information regarding the persistence and toxicity of MC transform products, they are not typically monitored. The extraction of MCs or their transformation products from complicated environmental matrices like sludge and their quantification/identification should be the main goals of the development of new procedures or protocols. Due to the challenges associated with the detection, quantifications, and treatment of MCs, it is important to focus on reduction of their concentration at the source itself through regulation, and raising public awareness, thereby limiting MC release into the environment. An additional prospective would be to study the mechanisms causing the attenuation or persistence of pollutants in solid or particulate phases like soils and sludge. Since these processes are driven by the pedological and structural properties of the soils and sludges, an interdisciplinary study integrating the structural and physico-chemical properties of the soil as well as sludge is of great importance.

2.5 Conclusions

This chapter discusses the occurrence of microconstituents (MCs) in different environmental components including groundwater, surface water, drinking water, marine water, wastewater treatment plants, sewage sludge, and soil. The chapter concludes that the occurrence, distribution, and migration of MCs can be nearly predicted based on their physico-chemical features like distribution coefficient (K_d), octanol-water partitioning coefficient (K_{OW}), acid dissociation constant (pKa), etc. The MCs with K_d of <300–500 and $\log K_{OW}$ of >4 have better hydrophobicity mobility in the liquid phase, so can be observed in higher concentrations in aqueous environments; therefore, the said MCs will be less retained in the solid and particulate phases. The environmental occurrence of MCs has been linked to harmful consequences such as endocrine disrupting effects and microbial antibiotic resistance. The potential health risks associated with MCs are assessed using risk quotient (RQ): $RQ < 0.01$ negligible risk, $0.01 > RQ < 0.1$ low risk, $0.1 > RQ < 1$ medium risk, and $RQ > 1$ high risk. This chapter advocates the importance of an MCs occurrence survey and prevailing ways in which the occurrence data can be explored to progress our understanding of MCs sources.

References

Baek, S.S., Yun, D., Pyo, J. et al. (2022). Analysis of micropollutants in a marine outfall using network analysis and decision tree. *Science of the Total Environment* 806: 150938.

Barbosa, M.O., Moreira, N.F., Ribeiro, A.R. et al. (2016). Occurrence and removal of organic micropollutants: an overview of the watch list of EU Decision 2015/495. *Water Research* 94: 257–279.

Bigott, Y., Gallego, S., Montemurro, N. et al. (2022). Fate and impact of wastewater-borne micropollutants in lettuce and the root-associated bacteria. *Science of the Total Environment* 831: 154674.

Biselli, S., Bester, K., Hühnerfuss, H., and Fent, K. (2000). Concentrations of the antifouling compound Irgarol 1051 and of organotins in water and sediments of German North and Baltic Sea marinas. *Marine Pollution Bulletin* 40 (3): 233–243.

Bolong, N., Ismail, A.F., Salim, M.R., and Matsuura, T. (2009). A review of the effects of emerging contaminants in wastewater and options for their removal. *Desalination* 239 (1–3): 229–246.

Brus, D.J., Lame, F.P.J., and Nieuwenhuis, R.H. (2009). National baseline survey of soil quality in the Netherlands. *Environmental Pollution* 157 (7): 2043–2052.

Canadian Environmental Protection Act. (1999). Priority substances list assessment report, "nonylphenol and its ethoxylates."

Carmona, E., Andreu, V., and Picó, Y. (2014). Occurrence of acidic pharmaceuticals and personal care products in Turia River Basin: from waste to drinking water. *Science of the Total Environment* 484: 53–63.

Carpenter, C.M. and Helbling, D.E. (2018). Widespread micropollutant monitoring in the Hudson River estuary reveals spatiotemporal micropollutant clusters and their sources. *Environmental Science & Technology* 52 (11): 6187–6196.

Coutu, S., Wyrsch, V., Wynn, H.K. et al. (2013). Temporal dynamics of antibiotics in wastewater treatment plant influent. *Science of the Total Environment* 458: 20–26.

Das, S., Ray, N.M., Wan, J. et al. (2017). Micropollutants in wastewater: fate and removal processes. In: *Physico-Chemical Wastewater Treatment and Resource Recovery* (eds Farooq, R. and Ahmad, Z.), 75–117.

Daughton, C.G. (2002). Environmental stewardship and drugs as pollutants. *Lancet (London, England)* 360 (9339): 1035–1036.

de Oliveira, M., Frihling, B.E.F., Velasques, J. et al. (2019). Pharmaceuticals residues and xenobiotics contaminants: occurrence, analytical techniques and sustainable alternatives for wastewater treatment. *Science of the Total Environment* 705: 135568.

Díaz-Garduño, B., Pintado-Herrera, M.G., Biel-Maeso, M. et al. (2017). Environmental risk assessment of effluents as a whole emerging contaminant: efficiency of alternative tertiary treatments for wastewater depuration. *Water Research* 119: 136–149.

European Parliament and The Council (2008). Directive 2008/105/EC of the European Parliament and of the Council of 16 December 2008 on environmental quality standards in the field of water policy, amending and subsequently repealing Council Directives 82/176/EEC, 83/513/EC, 84/156/EEC, 84/491/EEC, 86/280/EEC and amending Directive 2000/60/EC of the European Parliament and of the Council. *Official Journal of Europian Union* L348: 84–97.

Fairbairn, D.J., Arnold, W.A., Barber, B.L. et al. (2016). Contaminants of emerging concern: mass balance and comparison of wastewater effluent and upstream sources in a mixed-use watershed. *Environmental Science & Technology* 50 (1): 36–45.

Gasperi, J., Ayrault, S., Moreau-Guigon, E. et al. (2018). Contamination of soils by metals and organic micropollutants: case study of the Parisian conurbation. *Environmental Science and Pollution Research* 25 (24): 23559–23573.

Ghekiere, A., Verdonck, F., Claessens, M. et al. (2013). Monitoring micropollutants in marine waters, can quality standards be met? *Marine Pollution Bulletin* 69 (1–2): 243–250.

Golovko, O., Ahrens, L., Schelin, J. et al. (2022). Organic micropollutants, heavy metals and pathogens in anaerobic digestate based on food waste. *Journal of Environmental Management* 313: 114997.

Gómez, M.J., Herrera, S., Solé, D. et al. (2012). Spatio-temporal evaluation of organic contaminants and their transformation products along a river basin affected by urban, agricultural and industrial pollution. *Science of the Total Environment* 420: 134–145.

Goswami, L., Kumar, R.V., Borah, S.N. et al. (2018). Membrane bioreactor and integrated membrane bioreactor systems for micropollutant removal from wastewater: a review. *Journal of Water Process Engineering* 26: 314–328.

Huerta-Fontela, M., Galceran, M.T., and Ventura, F. (2011). Occurrence and removal of pharmaceuticals and hormones through drinking water treatment. *Water Research* 45 (3): 1432–1442.

Jones, O.A., Voulvoulis, N., and Lester, J.N. (2005). Human pharmaceuticals in wastewater treatment processes. *Critical Reviews in Environmental Science and Technology* 35 (4): 401–427.

Khan, N.H., ur Rahman, A., Zuljalal, F. et al. (2022). Food additives as environmental micropollutants. In: *Environmental Micropollutants* (eds Hashmi, M.Z., Wang, S., and Ahmed, Z.), 63–79. Elsevier.

Kim, K.R., Owens, G., Kwon, S.I. et al. (2011). Occurrence and environmental fate of veterinary antibiotics in the terrestrial environment. *Water, Air, & Soil Pollution* 214 (1): 163–174.

Kleywegt, S., Pileggi, V., Yang, P. et al. (2011). Pharmaceuticals, hormones and bisphenol A in untreated source and finished drinking water in Ontario, Canada: occurrence and treatment efficiency. *Science of the Total Environment* 409 (8): 1481–1488.

Kolpin, D.W., Furlong, E.T., Meyer, M.T. et al. (2002). Pharmaceuticals, hormones, and other organic wastewater contaminants in US streams, 1999–2000: a national reconnaissance. *Environmental Science & Technology* 36 (6): 1202–1211.

K'oreje, K.O., Okoth, M., Van Langenhove, H., and Demeestere, K. (2020). Occurrence and treatment of contaminants of emerging concern in the African aquatic environment: literature review and a look ahead. *Journal of Environmental Management* 254: 109752.

Lee, E., Rout, P.R., and Bae, J. (2021). The applicability of anaerobically treated domestic wastewater as a nutrient medium in hydroponic lettuce cultivation: nitrogen toxicity and health risk assessment. *Science of the Total Environment* 780: 146482.

Lee, E., Rout, P.R., Kyun, Y., and Bae, J. (2020). Process optimization and energy analysis of vacuum degasifier systems for the simultaneous removal of dissolved methane and hydrogen sulfide from anaerobically treated wastewater. *Water Research* 182: 115965.

Li, W.C. (2014). Occurrence, sources, and fate of pharmaceuticals in aquatic environment and soil. *Environmental Pollution* 187: 193–201.

Luo, Y., Guo, W., Ngo, H.H. et al. (2014). A review on the occurrence of micropollutants in the aquatic environment and their fate and removal during wastewater treatment. *Science of the Total Environment* 473: 619–641.

Nödler, K., Voutsa, D., and Licha, T. (2014). Polar organic micropollutants in the coastal environment of different marine systems. *Marine Pollution Bulletin* 85 (1): 50–59.

Noguera-Oviedo, K. and Aga, D.S. (2016). Lessons learned from more than two decades of research on emerging contaminants in the environment. *Journal of Hazardous Materials* 316: 242–251.

Peng, X., Ou, W., Wang, C., Wang, Z. et al. (2014). Occurrence and ecological potential of pharmaceuticals and personal care products in groundwater and reservoirs in the vicinity of municipal landfills in China. *Science of the Total Environment* 490: 889–898.

Pesqueira, J.F., Pereira, M.F.R., and Silva, A.M. (2020). Environmental impact assessment of advanced urban wastewater treatment technologies for the removal of priority substances and contaminants of emerging concern: a review. *Journal of Cleaner Production* 261: 121078.

Petrie, B., Barden, R., and Kasprzyk-Hordern, B. (2015). A review on emerging contaminants in wastewaters and the environment: current knowledge, understudied areas and recommendations for future monitoring. *Water Research* 72: 3–27.

Rizzo, L., Malato, S., Antakyali, D. et al. (2019). Consolidated vs. new advanced treatment methods for the removal of contaminants of emerging concern from urban wastewater. *Science of the Total Environment* 655: 986–1008.

Rogers, H.R. (1996). Sources, behaviour and fate of organic contaminants during sewage treatment and in sewage sludges. *Science of the Total Environment* 185 (1–3): 3–26.

Rout, P.R., Bhunia, P., and Dash, R.R. (2014). Modeling isotherms, kinetics and understanding the mechanism of phosphate adsorption onto a solid waste: ground burnt patties. *Journal of Environmental Chemical Engineering* 2 (3): 1331–1342.

Rout, P.R., Bhunia, P., and Dash, R.R. (2015). A mechanistic approach to evaluate the effectiveness of red soil as a natural adsorbent for phosphate removal from wastewater. *Desalination and Water Treatment* 54 (2): 358–373.

Rout, P.R., Zhang, T.C., Bhunia, P., and Surampalli, R.Y. (2021). Treatment technologies for emerging contaminants in wastewater treatment plants: a review. *Science of the Total Environment* 753: 141990.

Sabourin, L., Duenk, P., Bonte-Gelok, S. et al. (2012). Uptake of pharmaceuticals, hormones and parabens into vegetables grown in soil fertilized with municipal biosolids. *Science of the Total Environment* 431: 233–236.

Sauvé, S. and Desrosiers, M. (2014). A review of what is an emerging contaminant. *Chemistry Central Journal* 8 (1): 1–7.

Shahid, M.K., Kashif, A., Rout, P.R. et al. (2020). A brief review of anaerobic membrane bioreactors emphasizing recent advancements, fouling issues and future perspectives. *Journal of Environmental Management* 270: 110909.

Stackelberg, P.E., Furlong, E.T., Meyer, M.T. et al. (2004). Persistence of pharmaceutical compounds and other organic wastewater contaminants in a conventional drinking-water-treatment plant. *Science of the Total Environment* 329 (1–3): 99–113.

Sui, Q., Cao, X., Lu, S. et al. (2015). Occurrence, sources and fate of pharmaceuticals and personal care products in the groundwater: a review. *Emerging Contaminants* 1 (1): 14–24.

Tiwari, B., Sellamuthu, B., Ouarda, Y. et al. (2017). Review on fate and mechanism of removal of pharmaceutical pollutants from wastewater using biological approach. *Bioresource Technology* 224: 1–12.

Tran, N.H., Reinhard, M., and Gin, K.Y.H. (2018). Occurrence and fate of emerging contaminants in municipal wastewater treatment plants from different geographical regions: a review. *Water Research* 133: 182–207.

USGS (2017). Contaminants of emerging concern in the environment. Environmental Health: Toxic Substances Hydrology Program. US Geological Survey.

Vulliet, E., Cren-Olivé, C., and Grenier-Loustalot, M.F. (2011). Occurrence of pharmaceuticals and hormones in drinking water treated from surface waters. *Environmental Chemistry Letters* 9 (1): 103–114.

Wang, J. and Wang, S. (2016). Removal of pharmaceuticals and personal care products (PPCPs) from wastewater: a review. *Journal of Environmental Management* 182: 620–640.

Yan, Q., Zhang, Y.X., Kang, J. et al. (2015). A preliminary study on the occurrence of pharmaceutically active compounds in the river basins and their removal in two conventional drinking water treatment plants in Chongqing, China. *Clean–Soil, Air, Water* 43 (6): 794–803.

Yang, X., Flowers, R.C., Weinberg, H.S., and Singer, P.C. (2011). Occurrence and removal of pharmaceuticals and personal care products (PPCPs) in an advanced wastewater reclamation plant. *Water Research* 45 (16): 5218–5228.

Yang, Y., Ok, Y.S., Kim, K.H. et al. (2017). Occurrences and removal of pharmaceuticals and personal care products (PPCPs) in drinking water and water/sewage treatment plants: a review. *Science of the Total Environment* 596: 303–320.

Zhang, X., Lohmann, R., Dassuncao, C. et al. (2016). Source attribution of poly-and perfluoroalkyl substances (PFASs) in surface waters from Rhode Island and the New York Metropolitan Area. *Environmental Science & Technology Letters* 3 (9): 316–321.

3

Sampling, Characterization, and Monitoring

Mansi Achhoda[1,‡], *Nirmalya Halder*[1,‡], *Lavanya Adagadda*[2,3], *Sanjoy Gorai*[4], *Meena Kumari Sharma*[5], *Naresh Kumar Sahoo*[6], *Sasmita Chand*[7], *and Prangya RanjanRout*[1,8]

[1]Department of Biotechnology, Thapar Institute of Engineering & Technology, Patiala, Punjab, India
[2]CSIR-National Environmental Engineering Research Institute, Nagpur, Maharashtra, India
[3]Department of Civil Engineering, M.S. Ramaiah Institute of Technology, Bangalore, Karnataka, India
[4]School of Energy & Environment, Thapar Institute of Engineering & Technology, Punjab, India
[5]Department of Civil Engineering, Manipal University Jaipur, Rajasthan, India
[6]Department of Chemistry, Institute of Technical Education and Research, Bhubaneswar, Odisha, India
[7]Centre of Sustainable Built Environment, Manipal School of Architecture and Planning, Manipal Academy of Higher Education, Manipal, Karnataka, India
[8]Department of Biotechnology, Dr. B. R. Ambedkar National Institute of Technology Jalandhar, Punjab, India
‡ These authors contributed equally

3.1 Introduction

The detrimental and destructive effects of microconstituents (MCs), otherwise known as micropollutants (MPs), emerging pollutants, trace organic chemicals, etc., have been scientifically proven so far by the result of many years of study and research conducted by scientists from various nations. It is critical to identify and monitor the extent of environmental pollution in order to reduce the threat that MCs provide to both public health and the ecological hierarchy (Rout et al. 2021c). In order to guarantee the security of crops and food worldwide, it is crucial to establish new, accurate, and dependable measuring and control technologies for MCs. Sampling is the process of choosing a portion of material that is appropriately representative of the area of the environment being sampled and is small enough in volume to be transported and handled in the lab. Representativeness and integrity are the two fundamental challenges in sampling. Sampling procedures serve as the first step in the characterization and monitoring of MCs. It is frequently preferable to enhance the time window and the resolution by taking more samples in environments where the pollutant concentrations may change over time. With such fictitious time-integrated water sampling, whether automatic or manual (Stuer-Lauridsen 2005), particularly when measuring trace contamination, sample planners and analytical scientists need to optimize the entire measurement process including the sampling procedures. A sample plan should be created after the goal of the analysis has been determined in order to fulfil it. Sampling is the first step in the most typical process for determining pollution levels and monitoring them. Next comes extraction, then instrumental analysis. This methodology has been approved for use in regulation and legislation because it is well-established and validated. Long-term monitoring is necessary in addition to input regulation methods to assess the possible risk posed by MCs.

Microconstituents in the Environment: Occurrence, Fate, Removal, and Management, First Edition.
Edited by Rao Y. Surampalli, Tian C. Zhang, Chih-Ming Kao, Makarand M. Ghangrekar, Puspendu Bhunia, Manaswini Behera, and Prangya R. Rout.

Since monitoring has been ongoing for several decades, it has become clear that MCs are present everywhere in the environment (Roose and Udo 2005). Predictive and real-time monitoring can be utilized to detect anomalies, such as increases or decreases in MC concentration (Stacklin 2009). Identification of the number of MCs that are present in the aquatic environment is steadily growing as a result of mass spectrometry's constantly improved sensitivity. An increasing number of MCs are being monitored and quantified at the same time using multi-residue technologies based on liquid chromatography-mass spectrometry (Hermes et al. 2018). This chapter discusses the sampling protocols of MCs, the methods of MCs detection and analysis in different environmental samples, techniques employed to track their major sources, and application of remote sensing as well as the Geographical Information System (GIS) for the monitoring of MCs in different environmental matrices.

3.2 Sampling Protocols of Different Microconstituents

Analysis of the occurrence and removal of MCs from the environment demands highly specialized equipment and analytical professionals with extensive training. Determining what types of MCs there are and their quantity present in waterbodies is the first step in understanding their possible biological consequences on the environment. To do this, sampling, extraction, and analysis of the samples through advanced technologies must be used to confirm the identity of targeted and unknown compounds in minute quantities present in environmental samples (Alvarez and Jones-Lepp 2011).

3.2.1 Sample Preparation

The sample ought to be a true representation of the original environmental matrix (air, water, sediment, biota, etc.) and ought to be free of contamination during sample collection and transfer to the testing laboratory. The foundation of any effective sampling strategy is a well-thought-out plan and defined procedures. While organizing the sampling trip, it is important to consider these vital areas of inquiry. First, choose the sampling technique for collection of a sample, and second, calculate the sample size needed to satisfy the analytical method's minimal quantitation requirements.

One of the most crucial considerations is sample size since, if it is too small, it may be necessary to adjust the study's overall design or the manner in which the data is processed, which could completely affect the sampling strategy. In order to overcome any bias created by the sample collection, quality control (QC) measures must be identified, safety precautions must be taken, and sampling goals must be determined for each sample. Temperature, flow, pH, turbidity, and other measurements of water quality may be necessary, depending on the study's design and the characteristics of the chemicals being studied. It is important to indicate the methods utilized to collect the samples, the sites where they were taken, the time and date, and the QC techniques that are employed. The approaches present in Figure 3.1 can be used to collect the samples.

Transport, preservation, storage conditions, and processing times for samples are all important parameters because sample deterioration may lead to data inaccuracy. Despite the fact that deterioration can occur with any sample matrix approach, it is more specific to active matrices like wastewater and wastewater sludge. Users of the data believe that

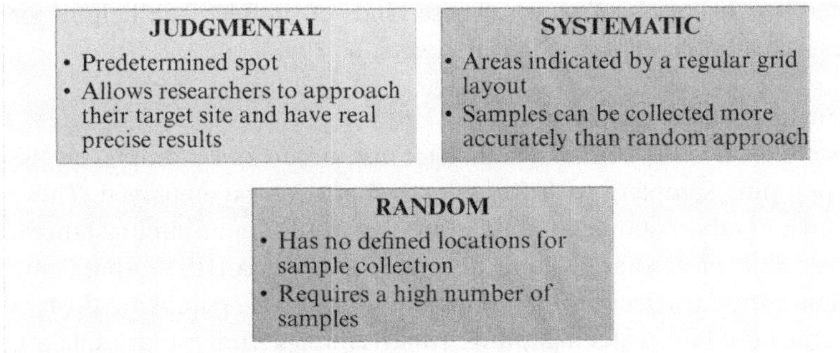

Figure 3.1 Different approach for collection of samples.

going over the holding times will lead to reduced analyte concentration levels. It is possible that this is not always true. The analyte usually starts to disintegrate into various analytical parts. Due to interferences brought on by sample degradation, this not only leads to a highly varied breakdown of analyte concentration but also could result in extremely variable recovery rates (Ohman et al. 2009).

3.2.1.1 Traditional Sampling Techniques

In order to take samples of surface waters, the sample vial is frequently filled directly. Hand-picking or the use of a net held upstream are the two methods used to harvest samples in shallow waters. These techniques are no longer effective for water layers below 0.5 m. Hence specialized water samplers are employed. Sample vials are dropped on a rope or steel cable which can be remotely controlled. Utilizing pumps is another choice. For instance, peristaltic pumps can be used in conjunction with in-line filtration to collect higher volumes of water, preventing air dust contamination in the field (Madrid and Zayas 2007). The Kemmerer, Van Dorn, and double check-valve bailer designs, as shown in Figure 3.2, are common samplers, and they all feature a tube or bottle that gathers the water sample.

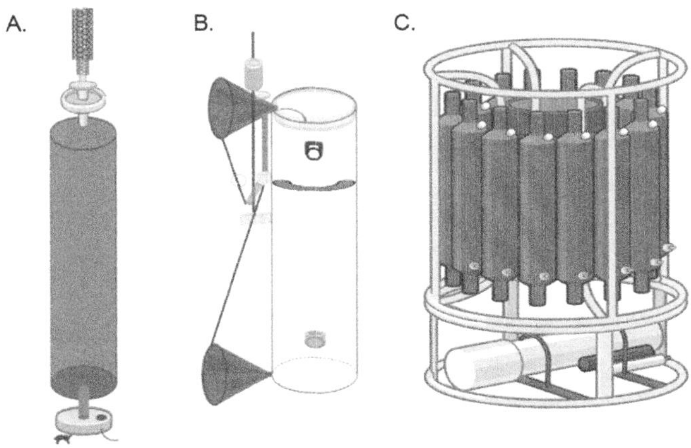

Figure 3.2 Different samplers for wastewater sampling: **A.** Kemmerer sampler; **B.** Van Dorn sampler; **C.** Niskin sampler.

The sample is restricted by caps or check valves. These kinds of samplers are helpful for taking precise samples from certain depths (Gibs and Iwatsubo 1998).

3.2.1.2 Automatic Samplers and Pumps

For the collection of samples from ephemeral, small streams, urban storm drains, or distant sites, automatic pumping samplers with fixed-depth input can be employed. These samplers can be set up to gather samples under a number of different circumstances, including precipitation, stage, or discharge. Point-integrated samples are those that come from automatic samplers or pumps. Suction-lift pumps and submersible pumps are the two main categories for pumps used for water sampling. Water samples from estuaries, lakes, and reservoirs can be obtained using pumps (Ferahtia et al. 2021).

3.2.1.3 Pore-Water Sampling

Pore-water has a significant impact on the bioavailability of MCs and nutrients, making it a crucial matrix for the monitoring of sediment quality (Rout et al. 2021b). Research conducted on the leaching of chemicals into the water column typically collects and analyzes pore-water. For sampling sediment pore-water, there are two main categories of procedures: *in situ* methods and *ex situ* methods (Figure 3.3). *Ex situ* methods are based on the study of sediment cores taken at the sediment–water interface of any water body (Anschutz and Charbonnier 2021), typically by pneumatic pressure or centrifugation, though extraction is also an option. *In situ* methods, on the other hand, involve removing the sediment of interest from its natural setting and isolating the pore-water elsewhere. The benefit of using *in situ* procedures is that they minimize the disturbances that come with removing materials from their environment (Viollier et al. 2003).

3.2.2 Extraction of Microconstituents

Due to the relatively low concentrations and complicated matrices that are often found in environmental fluids, sample treatment and enrichment methods are essential during analytical processing of the sample. The target analytes must first be extracted and pre-concentrated to make them identifiable in ambient samples before separating and identifying them. The technique of liquid–liquid extraction (LLE), which is based on the relative

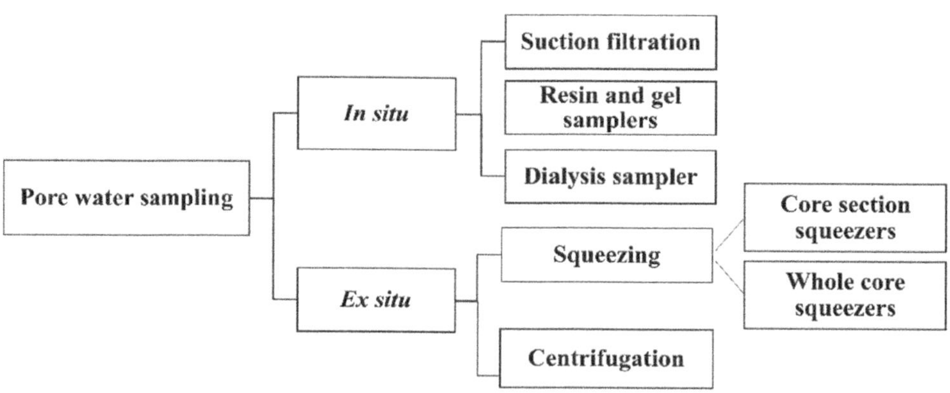

Figure 3.3 Pore-water sampling method.

solubilities of the analyte and the solvent, transfers the analyte from an aqueous sample to a significant amount of the latter (Rout et al. 2016). The most well-known and often used separation method is LLE, which is likely the oldest method still in use (Figure 3.4A). Despite the fact that LLE has a lot of benefits, solid-phase extraction has become very popular in recent years. Compounds that are suspended or dissolved in a liquid matrix can be removed according to their chemical and physical attributes using the sample technique known as solid-phase extraction (SPE), as demonstrated in Figure 3.4B. Polymeric or

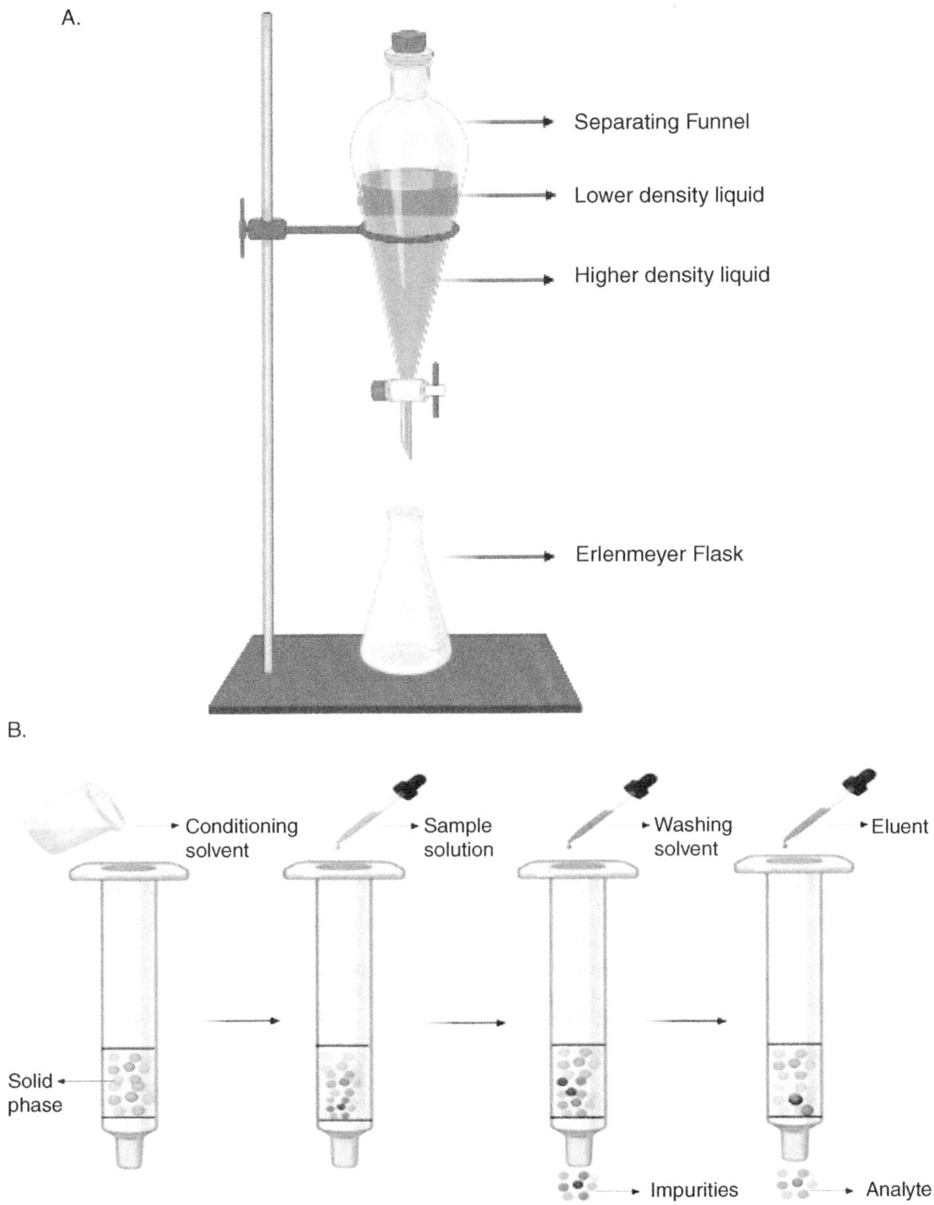

Figure 3.4 **A.** Liquid–liquid extraction of MCs, **B.** Solid phase extraction of MCs.

silica-based sorbents are employed for reversed phase SPE. Compounds are mostly held on the sorbent in both situations through hydrophobic interactions (Rout et al. 2017; Lee et al. 2019). The interaction between the analyte and the sorbent can be broken up by eluting the analyte with an organic solvent (Danaceau et al. 2017). A modern extraction method was created for parabens, alkyl-phenolic compounds, and phenyl-phenol employing hydrophilic-lipophilic-balanced sorbent in SPE and liquid chromatography–mass spectrometry. In the process of liquid–liquid extraction, particles are separated according to their relative solubilities using immiscible solvents. Constituents are extracted using immiscible solvents according to their solubility. When the solvents are physically separated from one another, the partitioned MCs are separated and extracted from field samples. MCs like bisphenol A, alkylphenol ethoxylates, nonylphenol ethoxylates, and the majority of steroids and hormones that contain hydroxyl groups are extracted from water using LLE with subsequent detection by GC-MS. The main benefits of SPE are the following: i) the effective use of time and labor; ii) the significant reduction in the amount of organic solvents used, which lowers the cost of analysis; iii) the decrease in waste generated, which contributes to their environmentally safe nature; and (iv) the potential for automation (Saha et al. 2018).

3.2.3 Passive Sampling

The assessment of unbound dissolved concentrations of pollutants in water via passive sampling is a sensitive and time integrative process, making it a promising method for sampling MCs. Due to the strong affinity of the analyzed compounds or the sorbent material in the sampler, organic contaminants are diffused from the sampling media to the sampler. Since passive samplers are a well-defined sampling medium with a known absorption capacity, they have a significant advantage over alternative matrices used for trend monitoring, such as sediments or biota for a particular sample. Table 3.1 presents a list of

Table 3.1 Types of different passive samplers and their use.

Sampler	Typical Deployment	Construction	Analytes	References
Ceramic dosimeter.	Up to 1 year	Ceramic tube (5.1 cm) filled with a solid-phase sorbent material, closed with PTFE lids.	PAHs, BTEX, chlorinated hydrocarbons.	Bopp et al. 2005
Chemcatcher.	14 days–1 month	A housing made of inert plastic (e.g., PTFE), containing a disk of solid receiving phase bound in a porous polymer, and a disk of diffusion-modulating membrane.	Polar and nonpolar organics.	Booij and Chen 2018
Dosimeter.	Up to 2 months	Activated carbon receiving phase in a perforated acrylic housing.	BTEX and atrazine.	DiGiano et al. 1988
LDPE and silicone strips.	1 month	Low-density polyethylene or silicone strips.	Hydrophobic organic compounds.	Booij et al. 2002

Table 3.1 (Continued)

Sampler	Typical Deployment	Construction	Analytes	References
PDB (Passive diffusion bag sampler).	2 weeks	Dialysis membrane or a low-density polyethylene bag filled with distilled water.	Polar organic compounds, VOCs, metals, trace elements.	Vroblesky and Hyde 1997
POCIS (Polar organic chemical integrative sampler)	Up to 2 months	Solid sorbent receiving phase material enclosed in a polyethersulphone membrane.	Herbicides and pharmaceuticals with logK_{OW} <3.	Rosen et al. 2018
SPMD (Semi-permeable membrane devices).	1 month	Flat tube of LDPE filled with triolein.	Hydrophobic semi-volatile organic compounds.	Creusot et al. 2013
DGT (Diffusion gradients in thin films).	1 week.	Two layers of acrylamide gel mounted in a holder device, one containing an acceptor phase, the other acting as a thin diffusion layer.	Metallic elements including the common heavy metals, phosphorous, sulphide.	Schintu et al. 2008
PIMS (Passive integrative mercury sampler).	Weeks to months.	LDPE lay-flat tubing.	Neutral Hg species.	Brumbaugh et al. 2002
PLM (Permeation liquid membrane).	Hours.	Microporous hydrophobic support separating test solution from receiving solution.	Cu, Pb.	Slaveykova et al. 2004
SLMD (Stabilized liquid membrane device).	Days to weeks.	LDPE lay-flat tubing containing an acidic solution with high affinity for the target elements.	Divalent metal ions.	Brumbaugh et al. 2002

different passive samplers and their use in different MC analyses. The time-weighted average (TWA) water concentration can be calculated using the concentration found in a passive sampler device over a long period of time (Vrana et al. 2014). Chemical TWA values are frequently used to quantify exposure and form the basis of an ecological risk assessment procedure for chemical stressors. Materials utilized in the fabrication of passive samplers have a consistent composition as well as diffusion and partition properties, which enables reproducible sampling, quantification, and comparison of the quantities of hydrophobic organic molecules throughout time, geography, and various environmental conditions (Taylor et al. 2021). In aquatic systems, contaminant concentrations are frequently fluctuating or take the form of irregular pulses. TWAs throughout the exposure period are represented by the concentrations reported by passive samplers. To support reliable concentration estimates needed for the field use of passive samplers, it is necessary to quantify the absorption in passive samplers under discontinuous exposure situations (Vrana et al. 2005). The capacity of passive sampling devices to gather analytes and subsequently quantify low and extremely low concentrations, as well as their simplicity of usage in the field without the need for electrical power or specialized equipment, are additional benefits. This is a significant benefit compared to spot sampling techniques, where analyte concentrations are frequently below the limit of quantification and limit of detection of the selected analytical method, preventing the identification of ultra-trace MPs in the environment (Thomatou et al. 2011).

3.2.4 Quality Assurance and Quality Control

Quality assurance is a set program for laboratory operations that outlines the steps necessary to provide reliable data with known precision and accuracy. Quality control (QC) samples are typically used to determine the precision and bias that are specific to the laboratory or an analyst, or to evaluate the performance of the entire measurement system or just a part of it. The analytical QC program aims to evaluate the capability and state of analytical facilities in concerned laboratories, to pinpoint the significant obstacles, both random and systematic that laboratories face while operating, and also to offer the necessary support that is required to verify the data from the water quality monitoring (Bhushan and Basu 2017). With each batch of samples that is evaluated, QC samples should be run. Generally, QC samples should make up 10 to 20% of any analytical run.

3.2.5 Internal vs. External Quality Control

The internal QC entails internal practices for ongoing operation monitoring and rigorous day-to-day evaluation of the produced data to determine whether it is trustworthy enough to be disseminated. The processes essentially check the accuracy of the data using duplicate analyses of test samples and/or control samples, as well as the bias of the data using control samples. These processes happen at the batch level. The first step in creating QC samples is to create field replicates. Throughout the remaining sample handling and analysis procedures, they are handled as independent samples. These samples are used to evaluate error related to sampling methodology, analytical procedures, and sample heterogeneity. When calculating total error (precision) for crucial samples with contamination concentrations at or close to the action level, field replicates may be employed. Background samples, which are regarded as "clean" samples, determine the aqueous matrix's inherent chemical makeup. They give a starting point for comparing pollutant concentration levels with samples taken on the spot. In order to prepare spike samples, the indicator being analyzed is added to the sample at a specified concentration. The concentration in the sample should rise in a predictable way. It is employed to evaluate the method's precision. The external QC entails involvement in national and/or international inter-laboratory sample and data exchange programs as well as reference assistance from other laboratories. External QC evaluations are helpful as an unbiased test of the internal QC methods because they can spot mistakes, biased standards, or corrupted data that might escape detection in an internal QC system. The purpose of QC is to guarantee that there is a reasonable amount of checking at different stages of the data gathering process to ensure that the quality of the data is maintained and, if necessary, that it is discovered with a minimum loss of data.

3.3 Quantification and Analysis of Microconstituents

An essential component for micropollutant (MPs) removal is the monitoring of water quality and the analytical techniques used to find MPs. In the fight to eliminate and manage MPs, the identification of pollutants and their measurement at trace amounts remain the major obstacles. The majority of analytical methods combine mass spectrometry (MS) with chromatography. The principles of isolation, ionization, signal detection, signal translation, identification, and reporting underlie these analytical procedures. To elicit the analyte's distinctive signal response, various ionization procedures, including electrospray ionization and atmospheric pressure ionization, are used. A heterogeneous composite of environmental samples comprising additional interfering molecules usually surrounds the sample analytes of interest. These contaminants lower the signal-to-noise ratio and increase background noise, which lowers the analyte's quality of detection. It is necessary for all researchers to have easy access to a safe, quick, dependable, and sensitive approach for the robust and accurate detection of MPs. MCs concentrations can occasionally range from ng/L to mg/L. Sometimes the traditional detection methods are unable to distinguish them in these low quantities. New enhanced methods of MC detection with superior abilities to identify MPs at extremely low levels have been made accessible owing to the ongoing efforts of research teams in environmental analytical chemistry. Identification of MPs is difficult due to the fact that they are constantly evolving and are not thoroughly comprehended. The presence of a known list of MPs can be determined by all regulatory tests, but mostly because environmental and physicochemical conditions are dynamic, MPs can exist in a variety of forms (derivatives, metabolites, transformation products). Due to the lack of understanding regarding various types of MPs, it is challenging to evaluate water samples effectively and consider them safe for use. Researchers have used anticipatory approaches and simulated mechanisms of the possible changes that an MP may undergo to address the problem of the unpredictable nature of MPs.

3.3.1 Detection Techniques

Gathering collection of adequate data by detection with the existing analytical tools is a barrier in the characterization of MCs and their intermediates. Traditionally high-performance liquid chromatograph, liquid chromatography tandem electrochemical detection, liquid chromatography tandem mass spectrometry, capillary electrophoresis, gas chromatography, and gas chromatography tandem mass spectrometry are the most frequently used techniques for determining various MCs (Figure 3.5). These analytical techniques include experimental validation using reference molecules. A key identification tool to identify the transformation by-products of the desired contaminants would also include information currently present in the literature and databases (NIST, Wiley, etc.). However, because of a number of issues with addressing unidentified structures, detecting transformation products is a difficult challenge. They often occur in small quantities in the samples, and it is rarely possible to obtain standard material for structure elucidation.

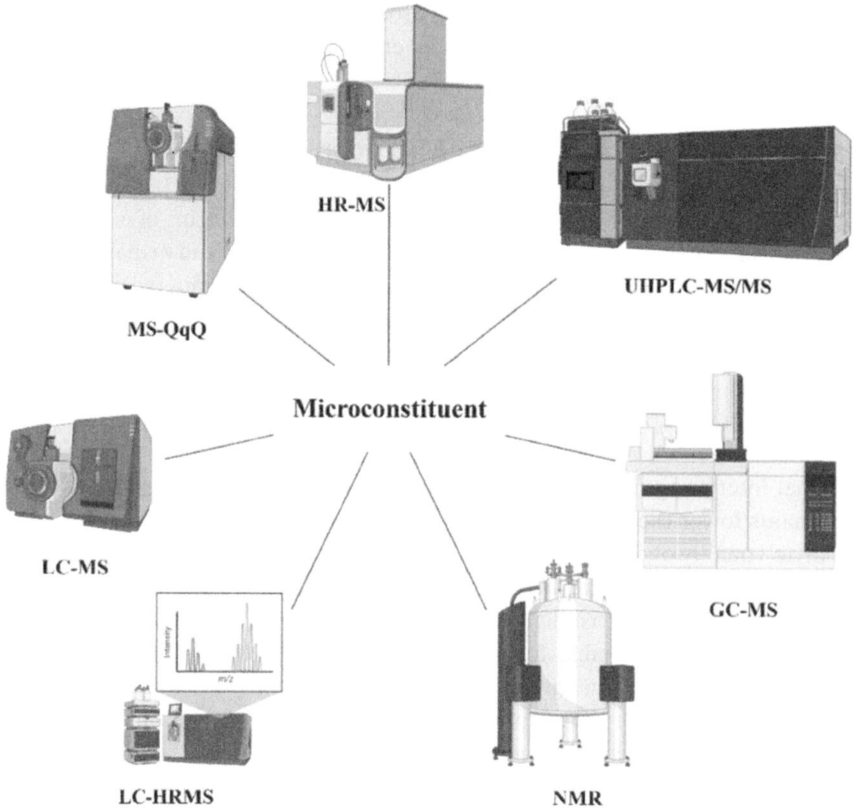

Figure 3.5 Traditional methods for detection of MCs.

3.3.2 UV-Visible Optical Methods

UV-visible spectroscopy belongs among the first techniques to be integrated with other techniques of traditional detections and analysis methods of MCs. The chromatographic separation was further integrated with UV-visible spectrometry for countering the limit of detection concentration. Along with single wavelength detection, absorption spectrum scanning of samples can provide substantial information about the effluents. Electronic transitions between molecular orbitals occur as a result of the electromagnetic spectrum's visible and ultraviolet energy absorption. After chromatographic separation, the spectral scanning of the specimen or of each individual allied co-product are typically measured, as listed in Table 3.2. UV-visible spectrum of reference material is used to identify degradation products. The availability of the specific functional groups has a direct bearing on the presence and absence of the absorbance maxima. The absorption maxima are primarily caused by unsaturated groups, or chromophores, which are most relevant for identification and quantitative study. Typically, UV-visible necessitates relatively low analyte concentrations. Specific UV absorption decreases with the removal of a large series of MCs. However,

Table 3.2 Microconstituents detected by UV-Visible spectrometry.

Sl. No.	Compound	Used Wavelength	Reference
1.	4-Formyl-aminoantipyrine, Methyl-benzotriazole, Metoprolol, Sulfa-methoxazole, Valsartan.	254 nm	Altmann et al. 2016
2.	Diclofenac, Carbamazepine, Estrone, Naproxen, Phenazone, Bezafibrate, Ibuprofen, Ketoprofen, Clofibric acid, Iopromide, Iopamidol, Iohexol.	254 nm	Bahr et al. 2007
3.	Benzotriazole, Iomeprol, Diclofenac, Carbamazepine, Sulfamethoxazole, Primidone, Bezafibrate.	254 nm	Altmann et al. 2014
4.	Atenolol, Trimethoprim, Primidone, Sulfamethoxazole, Meprobamate, Diphenhydramine, Diltiazem, Carbamazepine, Gemfibrozil, Triclocarban.	254 nm	Anumol et al. 2015
5.	Acetaminophen, Atenolol, Bezafibrate, Carbamazepine, Ciprofloxacin, Diclofenac, Erythromycin, 17-α-estradiol, 17-β-estradiol, Estrone, Furosemide, Gabapentin, Lorazepam, Naproxen, Norfloxacin, Ofloxacin, Oxazepam, Propranolol, Roxithromycin, Sulfadiazine, Sulfadimethoxine, Sulfamethoxazole, Sulfathiazole, Tetracycline, Trimethoprim, AMPA, Diuron, Glyphosate, Mecoprop, Terbutryn, OP1EO, OP2EO, Phthalates, DEHP, DEP, DnBP, Triclocarban, Triclosan, Acesulfame, Saccharin, Sucralose, Benzotriazole, Bisphenol A, PFOA, PFOS.	254 nm	Mailler et al. 2016
6.	Clothianidin, KHP, Metaldehyde, Thiamethoxam, Tryptophan, Uric acid.	200–360 nm	Spangenberg et al. 2021
7.	Clarithomycin, Venlafaxin.	254 nm	Wittmer et al. 2015

the UV-visible spectra for several compounds that contain the same chromophore will be the same. In consideration of this, even though UV-visible is useful for estimating degradation products, more sophisticated analytical techniques are required to identify potential reactive intermediates.

3.3.3 NMR Spectroscopy

Nuclear magnetic resonance (NMR) spectroscopy has become increasingly important as a tool for identifying the chemical and physical characteristics of a variety of organic substances. Because atomic nuclei absorb energy at specific frequencies when exposed to a magnetic field, such an analysis is made possible. The most beneficial and prevalent nuclei for NMR measurements are those with finite magnetic moments and spin quantum numbers of $I = ½$ such as 1H, 11B, 13C, 15N, 17O, 19F, and 31P. Since the most frequent nuclei in organic molecules are hydrogen and carbon, the ability to investigate these nuclei by NMR is crucial for determining the organic structure. Unfortunately, compared to other techniques like UV/Visible spectroscopy, NMR is rather insensitive since the nuclear transition energy is significantly lower (usually of the order of 10^4) than the electronic transition. Additionally, a variety of factors, including the magnetic field strength of the instrument (B_0), sample concentration, acquisition time (NS), and measurement temperature, affect the signal-to-noise ratio (S/N) in NMR. However, because it may offer clear information on a target molecule, NMR is possibly the most conservative analytical method currently in use. NMR can reveal chemical structures and offer extremely intricate proof that a molecule is what it purports to be. NMR is also a quantitative method because the strength of the signal is inversely proportional to the number of nuclear spins. A list of MCs detected by NMR is summarized in Table 3.3.

Chromatography and NMR can be coupled in a number of ways. The on-line linking of LC to SPE and then to NMR became possible more recently, which addressed some of the analytical limitations of prior approaches. The chromatographic spots are trapped in SPE cartridges of this arrangement, and by multi-trapping within the same cartridge, they can be concentrated up to several times. Recently, LC-UV-SPE-NMR-MS was used to separate 2-phenyl-1,4-benzopyrone and phenolic acids from Greek oregano extract. The chemicals were separated by LC, trapped in SPE cartridges, and then eluted for acquisition by NMR and MS. The study showed that two similar compounds may be easily differentiated by MS and NMR when co-eluting in the LC (and consequently trapped within the same cartridge). Since this method enables the separation, concentration, and NMR acquisition of the analyte together within a single system, it is appropriate for the analysis of unusual or/and unknown chemicals in heterogeneous mixtures. Additionally, simultaneous interpretation of the UV, MS, and NMR data collected for each trapped peak resulted in a clear attribution of their structure. Different techniques, such as MS or NMR, can help with the detection of MC by-products, but frequently, the whole chemical description of a molecule can only be attained by combining the compound information obtained from numerous sources.

Advanced analytical tools use MS to identify the MS fragmentation pathways (cleaved moieties), whereas 1H and 13C NMR are used to establish structural integrity. As a result, one of the most effective methods for identifying an unidentified molecule is the integration of MS and NMR. Compound identification has been done using the extended hyphenation of LC-NMR-MS, which combines the benefits of NMR spectroscopy with MS spectrometry. Utilizing the benefits of both technologies simultaneously or, if possible, online, is the most effective method to do it. However, it is still more usual to conduct investigations by LC-MS and LC-(SPE)-NMR separately due to their complicated analytical set-up.

Table 3.3 Microconstituents detected by NMR.

Sl. No.	Compound	Reference
1.	1-methoxy-1-methyl-3-phenyl urea	Haroon et al. 2022
2.	Telmisartan	Shah and Singh 2010
3.	Irbesartan	Shah et al. 2010
4.	Linuron	Haroon et al. 2021

3.3.4 Chromatographic Methods Tandem Mass Spectrometry

A variety of detection techniques, such as HPLC/NMR, UV/MS, etc., can be used in the approach of comprehensive chemical characterization of MCs and its by-products. Traditionally, mass-based analysis is used as part of the most common methodology for analyzing these MCs. In general, compared to other approaches, mass-based procedures using mass spectrometry (MS) exhibit relatively low detection limits. When using MS, a spectrometric technique, it is possible to identify molecules with a high mass-to-charge ratio that can be used to determine the molecular mass (MM) of a chemical. In mass spectrometry, energy at specific frequencies is absorbed. A molecule is intentionally split during the MS analyses, the pieces are identified by their masses, and the original structure is then deduced from the fragments. Gas chromatography-mass spectrometry (GC-MS and GC-MS/MS), a technique that requires multistep sample preparation and has low selectivity, has historically been the leading technique used to monitor contamination levels. With the ability to detect a variety of polar and non-volatile substances, liquid chromatography with UV/fluorescence tandem mass spectrometry (LC-MS/MS) is now a common analytical tool and a crucial method for environmental analysis. Adoption of this approach has minimized sample preparation while increasing the specificity and reliability of identification. In order to perform direct analysis by LC-MS/MS detection, or analysis without the requirement to concentrate ultra-trace-level pollutants in the water sample, sample preparation must be reduced. Several mass analyzers that are capable of identifying, quantifying, and resolving ambiguities by choosing the proper ionization and acquisition characteristics can be added to liquid chromatography mass spectrometers. Even though the chemical characteristics of water pollutants vary greatly, atmospheric pressure ionization (API) techniques can ionize the majority of them. Electrospray ionization (ESI), the most popular ionization technique for residue analysis by LC-MS/MS, is appropriate for the analysis of the majority of the constituents that pollute water. Higher ionization energy is needed to separate less polar molecules using reverse phase chromatography. Time-of-flight (TOF) mass analyzers are highly suited for the study of chemical contaminants in water samples, primarily for analytical reasons, owing to their inherent qualities of high resolution, mass accuracy, and good sensitivity in scanning broad mass ranges. The hybrid quadrupole-TOF (QqTOF) is one of the most effective contemporary lab-based techniques for identifying unknown compounds in aqueous systems because it enables the pre-selection of a precursor ion in the quadrupole filter and the measurement of the full-scan product ion spectra with high mass accuracy. For instance, the QqTOF system was implemented to identify 22 triazole compounds in water with a detection limit of 0.5 ppb, along with primary fungicides and by-products. An extensive list of MCs detected by chromatographic tandem mass spectrometry are presented in Table 3.4.

3.3.5 Biological Assay for Detection

Through the data generated on the toxicity, estrogenicity, and antibacterial activity of the released wastewater, biological evaluation of the MC by-products supplements chemical analysis (Rout et al. 2021c). In order to evaluate and assess the potential effects of these pollutants, biological entities need to be employed. An environmental contaminant is a

Table 3.4 Microconstituents detected by chromatographic tandem mass spectrometry.

Sl. No.	Compound	Technique Used	Reference
1.	Diclofenac, Ibuprofen 4-AAA, 4-FAA, 10,11-DHC, Lamotrigine, Gabapentin, Bezafibrate, Iopamidol, Iomeprol, Iopromide, Iohexol, Carbamazepine, Metoprolol, Sotalol, Acetyl-Sulfamethoxazole, Sulfamethoxazole, Hydrochlorothiazide, 1H-benzotriazole, 4-MBT+5-MBT, Isoproturon, Chloridazon, Atrazine, Desethylatrazine, Simazine, Terbuthylatrazine, Metazachlor, BH479-4, BH479-8, Metolachlor, Metolachlor-C, Metolachlor-S, Amidotrizoic acid, Metazachlor metabolite, Chlorotoluron, Diuron.	Liquid Chromatography-tandem Mass Spectrometry (LC-MS/MS).	Posselt et al. 2018
2.	Tetraethyleneglycol, Laureth-5 and di-(2-ethylhexyl) phosphoric acid)), Hydrochlorothiazide, Theophyline.	Ultra-performance liquid chromatography tandem Mass Spectrometry (UPLC-MS/MS).	
3.	1,4-dichlorobenzene, Atrazine, Prometon, Tributyl phosphate, Nicotine, Tetrachloroethene.	High-Performance Liquid Chromatography/tandem Mass Spectrometry (HPLC-MS/MS).	Elliott et al. 2018
4.	Androstanedione, Benadryl, Benzedrex, Bupivacaine, Cathine, Chloroxylenol, Clindamycin Clorophene, Escitalopram, Etilefrine, Flucanazole, Hexestrol, Isosorbide, Ketamine, Levomenthol, Lidocaine, Mequinol, Methadone, Midazolam, Polyalkylene glycol monoalkyl ethers, Prasterone, Propofol, Ropivacaine, Testosterone, Cypionate, Thiopental, Triclosan, Valproic acid, Venlafaxine.	Two-Dimensional Gas Chromatography coupled with Time Of Flight Mass Spectrometry (GC×GC-TOFMS).	Meza et al. 2020
5.	1,3-Dimethylurate, Anileridin, Atenolol acid, Azithromycin, Caffeine, Celecoxib, Ciprofloxacin, Coumarin, Ephedrine, Fexofenadine, Fluconazole, Gabapentin, Glucosamine, Irbesartan, Lamotrigine, Levocetirizine, Metformin, Methotrexate, Morphine-3-glucuronide, Mycophenolate, Nevirapine, Rifaximin, Safranin, Testosterone Propionate, Voriconazole.	Ultra High Performance Liquid Chromatography coupled with Time Of Flight Mass Spectrometry (UHPLC-TOFMS).	Meza et al. 2020
6.	Alfuzosin, Alverine, Amisulprideb, Amitriptyline, Atenolol, Atorvastatin, Betamethasone, 17 valerate, Bezafibrate, Bisoprolol, Carbamazepine, Cefalexin, Celiprololb, Cetirizine, Chlorpheniramine, Citalopram, Clarithromycinb, Clopidogrel, Cyclizine, Diclofenac, Diltiazem (cis), Dipyridamole, Doxazosin, Erythromycinb, Fexofenadineb, Flecainideb, Fluconazoleb, Furosemide, Gliclazide, Ibuprofena, Indapamidea, Irbesartanb, Ketoconazole, Ketoprofen, Labetalol, Lamotrigine, Lansoprazole, Lidocaine, Loperamide, Loratadine, Losartan, Mefenamic acida, Metoclopramide, Metoprolol, Mirtazapine, Naproxena, Nifedipine, Omeprazole, Oxprenololb, Pantoprazole, Phenytoin, Piroxicam, Procyclidine, Propranolol, Quinine, Ranitidine, Salbutamol, Salicylic acida, Sertraline, Sotalolb, Sulfasalazine, Sumatriptanb, Tamsulosin, Telmisartan, Terbinafine, Timolol, Tramadolb, Trazodone, Trimethoprim, Valsartanb, Venlafaxine, Verapamil, Warfarin.	UHPLC/ Q-ToF-MS.	Gravell et al. 2020

7.	Amphetamine, Amphetamine-D8, Methamphetamine, Methamphetamine-D8, MDEA, MDEA-D5, Cocaethylene, Cocaethylene-D8, Norcocaine, Norcocaine-D3, EME, Benzoylecgonine and norcocain, Benzoylecgonine Benzoylecgonine-D8, Morphine, 6-Monoacetylmorphine, Methadone, Methadone-D9, EDDP, EDDP-D3, Codeine, Hydrocodone, Oxymorphone, Buprenorphine, Heroin, Oxycodone, Oxycodone-D6, Alprazolam, Alprazolam-D5, Diazepam-D5, Lorazepam-D4, LSD, Hydroxy-LSD, Chlorpromazine, Clozapine, Norclozapine, Olanzapine, Risperidone, Aripiprazole-D8, Lidocaine, Fentanyl, Norfentanyl, Ketamine, Norketamine, Carbamazepine, Carbamazepine-D10, Lamotrigine, Topiramate, Levetiracetam, Primidone, Nortryptiline, Clomipramine, Imipramine, Doxepine, Mirtazapine, Desmethylmirtazapine, Citalopram, Citalopram-D6, Fluoxetine, Paroxetine, Sertraline, Norsetraline, Desmethylcitalopram, Venlafaxine, Venlafaxine-D6, Pseudoephedrine, Pseudoephedrine & ephedrine, Bupropion, 11-Hydroxy-THC, Cannabinol, THC, THC and cannabidiol, THC-D3, Clopidogrel carboxylic acid, Clopidogrel, Diphenhydramine, Diphenhydramine-D3, Atenolol, Atenolol-D7, Caffeine, Caffeine-13C3, Paraxanthine, Nicotine, Cotinine, Cotinine-D3, Zopiclone, Mephedrone, PB22, JWH073, JWH018, Memantine.	LC-MS/MS.	Asimakopoulos et al. 2017
8.	Fluoxetine, Chlorpyrifos.	GC-MS.	Caglak et al. 2019
9.	Caffeine, Benzophenone-3, Carbamazepine.	GC-MS.	Lempart et al. 2018
10.	Naphthalene, Acenaphthylene, Acenaphthene, Fluorene, Phenanthrene, Anthracene, Fluoranthene, Pyrene, Benz[α]anthracene, Chrysene, Benzo[α]fluoranthene, Benzo[*k*]fluoranthene, Benzo[α]pyrene, Indenol[1,2,3-cd] pyrene, Dibenz[*a,h*]anthacene, Benzo[*ghi*]perylene, α -BHC, Lindane, β-BHC, Heptachlor, Aldrin, Dieldrin, Endrin, Endosulfan, p,p′ – DDD, p,p′ – DDT, Endrin aldehyde, Methoxychlor.	GC-MS.	Basheer et al. 2003
11.	4-nonylphenol (4-NP), 4-octylphenol (4-OP), anthracene (Ant), alachlor (Ac), heptachlor (Hc), heptachlor epoxide (Hce), and bis(2-ethylhexyl)phthalate (DEHP).	GC-MS.	Król and Dudziak 2019
12.	Methylparaben, Ethylparaben, Propylparaben, butylparaben.	GC-FID.	Amin et al. 2019

substance that, in a specific quantity and chemical form, poses a threat to the ecosystem's inhabitants (bioindicators) and has harmful or hazardous consequences (Rout et al. 2021a; Tadda et al. 2021). The physico-chemical features and qualities of an MC molecule, in addition to its following toxicity, are determined by its structure. Similar structures correlate to similar characteristics, which can result in identical toxicity in molecules. Scientists established a model based on eight molecular structural features as characteristic parameters using a K-closest neighbor pattern identification method to discuss how chemical structures are connected to one another. Over 90% of the variance explained in the 8 variables was explained by K-nearest neighbors. Ecotoxicological tests can be performed on samples collected using any degree of biological organization, including molecular, whole, populations, colonies, or combinations of species (Rout et al. 2018). However, the examination of toxicity with standardized test protocols will only be accurate if numerous essential features are taken into account. Due to the necessity to biologically analyze numerous environmental samples in a brief amount of time, miniature tests for toxicity or "microbiotests" are devised for fast and reliable testing. These have gained popularity in recent times, which are also known as alternative tests and second-generation tests. However, the measurement of cytotoxicity with standardized test protocols will only be accurate if many individual elements are taken into account. In order for environmental samples to be tested using bioassays, they ought to be reflective, and the collection, storage, and preparation processes must not alter the sample's toxicity. Additionally, a test's variance should be measured, and the influence of factors (i.e., pH, dissolved solids, EC) must also be determined. To assess the cytotoxicity of certain compounds, a variety of ecotoxicity assays have been devised. Contaminated medium assays and/or bioassays were developed as a result of the implementation of these techniques for the assessment of the toxicity of environmental samples. In order to test a particular chemical impact on a biological system, a biological assay, termed a bioassay, is used. Bioassays can give a qualitative or quantitative assessment of a chosen system. In this type of scenario, the analysis frequently includes estimating a particular chemical concentration or potency by the measurement of particular biological reactions.

A toxicity test can be thought of as a bioassay that enables damage measurement. The advantages of adopting quick, sensitive, repeatable, and inexpensive bacterial assays, including bioluminescent bacteria (BLB), have recently come to light. *Vibrio fisheri* (NRRL B-11177) is the most well bioluminescent bacteria, emits light as a by-product of their regular metabolism (~10% of their metabolic energy), and the intensity of the light released is a measure of their metabolic activity. The light emission is decreased when such bacteria are subjected to a toxin, giving an indication of the sample's acute toxicity. Its benefits over alternatives include speed, affordability, and ease of use. Additionally, this standardized microorganism has undergone extensive research, and a variety of systems are market accessible. These techniques provide great specificity and good sensitivity, but they are unsuitable for simultaneous real-time detection and processing of multiple samples rapidly. This process requires highly skilled operators, along with lengthy procedures and complicated pre-treatment procedures. Equipment used for this process is specialized and costly. Additionally, the techniques are unsuited for field research and *in-situ* sample management. Although immunochemical methods, such as enzyme-linked immunosorbent assays (ELISA), can also be employed to identify endocrine disrupting compounds (EDCs), they are less effective than chromatographic methods due to the lower stability of the biomaterials used in the assays (Rout et al. 2021c). These assays also involve complex multistage processes which may require expensive equipment. Although these techniques provide good selectivity and detection limits, the required specific antibodies or proteins must be produced via recombinant technologies. ELISA-based techniques are challenging to apply in non-specialized labs and in the field environment. They require time-consuming processes like repeated washing and incubation as well as an enzymatic process for the generation of the final output. Additionally, ELISA-based techniques are particular to a single type of constituent or, at best, compounds that are structurally similar. Thus, a single type of ELISA is unsuitable for the detection and analysis of multiple analytes. Furthermore, MCs are structurally versatile and new types of MCs are continuously introduced into the environment as they are linked with the market-driven evolution of novel chemicals. A reliable and easy analytical technique for the detection and quantification of MCs is urgently required. The summary of MCs detected by bioassays are presented in Table 3.5.

Table 3.5 Microconstituents detected by battery of bioassays.

Sl. No.	Compound	Activity Checked	Outcome	Reference
1.	Trichloroethane.	Methanogenic activity.	Reduction in bioluminescence inhibition *V. fischeri*.	Erasin et al. 1994
2.	Heterogenous sample.	***Battery of microbiotests*** *Vibrio fischeri* (Microtox test). *Daphnia magna* (Daphtoxkit F). *Tetrahymena thermophila* (Protoxkit F).	Mobility reduction in *D. magna*. Growth inhibition of *T. thermophila*.	Kungolos 2004
3.	NA	***Phytotoxicity assay*** Green algae *Pseudokirchneriella subcapitata*. ***Microbiotests*** *Vibrio fischeri*.	PSII inhibition I-PAM assay. Bioluminescence inhibition.	Macova et al. 2011
4.	Heterogenous sample.	*Vibrio fischeri*. *Photobacterium phosphoreum*.	Bioluminescence inhibition.	Escher et al. 2014
5.	17β-estradiol Bisphenol A.	Chinese hamster ovary cells (Hsp 47 assay).	Stress response.	Etteieb et al. 2016
6.	• Triclosan • Acridine • 17α-ethinylestradiol (EE2) • 3-nitrobenzanthrone (3-NBA).	***Bioassay battery*** Unicellular green algae growth inhibition (Algae) Daphnid immobilization (Daphnia). Zebrafish embryo lethality and morphological effects (FET). Mechanism-specific assays evaluated 140 estrogenicity (ER-Luc and YES) Mutagenicity (Ames).		Di Paolo et al. 2016

3.3.6 Sensors and Biosensors for Detection

A biosensor is made of receptor-transducer apparatus which employs identification components of biological origin, for delivering both selectively quantifiable or partially-quantifiable analytical information. These consist of a biological component (i.e., an enzyme or an antibody) that is immobilized on the surface of the transducer (electrode). When these come into contact with the analyte of interest like an MC, this in turn causes a modification which is quantifiable. Bioassays and biosensors for detection of MCs are presented in Figure 3.6. This process happens within the vicinity of the transducer surface, enabling further to convert a (bio)chemical process into a quantifiable digital signal. The mass-transport kinetics of solutes and metabolites, as well as the loading of the sensing molecule, are the primary parameters that affect the biosensor response. The transduction element (i.e., electrochemical, optical, piezoelectrical, thermal) or the biological recognition basis are typically used to classify biosensors (e.g., enzymatic, immunoaffinity recognition, whole-cell sensor, DNA). For a variety of biological and commercial processes, biosensors provide concise, robust, quick, and reagentless real-time monitoring. Significant advancements in

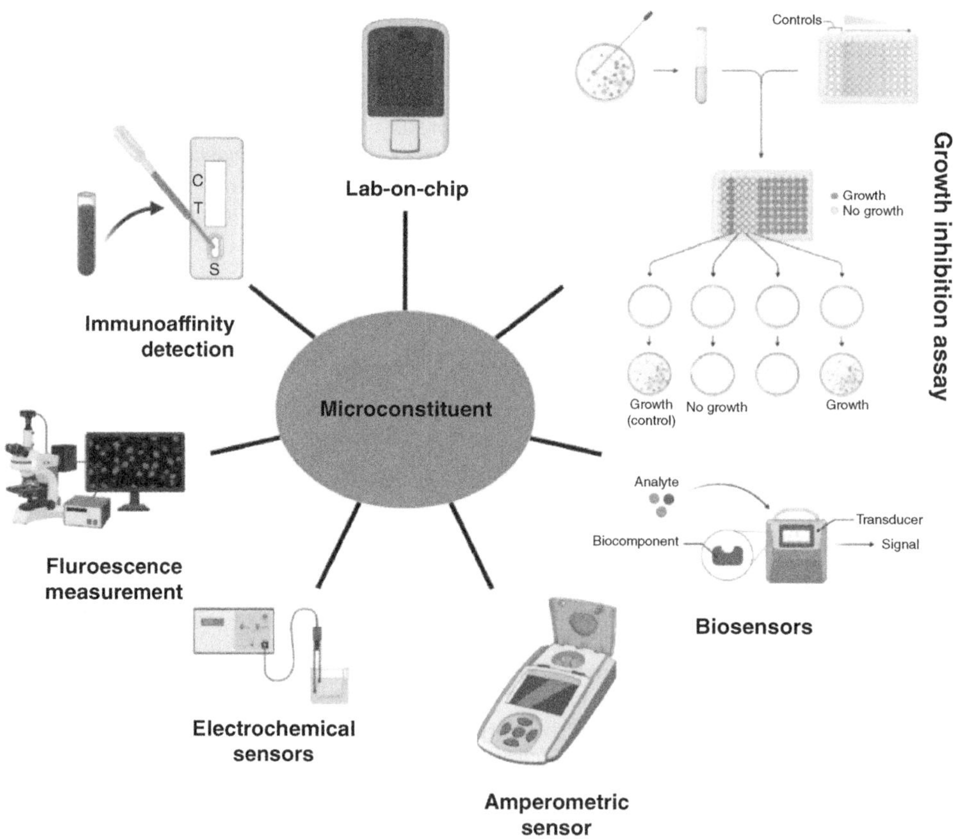

Figure 3.6 Bioassays and biosensors for detection of MCs.

the field of biosensor technology have been seen recently. This includes breakthroughs in material technologies, signal transduction techniques, and sophisticated computer software to operate devices.

3.4 Source Tracking Techniques

3.4.1 Performance Criteria

Anthropogenic activities harness the existence of various MCs in freshwater courses, which impact on water quality. Similarly, MCs enter into the water bodies through various routes like direct or indirect emancipation, or discharge of treated effluents from wastewater treatment plants (WETPs). Correspondingly, stormwater runoff, illicit connections, poor pipe materials, and inadequate treatment infrastructure also confer the presence of MCs in water courses. Therefore, it is imperative to identify the relevant source so that adequate intervention measures can be adopted to control the entry of MCs into water courses. Tracers are supportive in source apportionment. Moreover, tracers are valuable compounds, which are imperative to identify a contaminant within a specific source, particularly when ambiguous transportation pathways exist (Lim et al. 2017). Furthermore, the criteria for an ideal tracer should be in such a way that it is source specific, and released into the environment in ubiquitous quantities and readily available during the analysis. Various kinds of tracers such as microbial, chemical, and a combination of both can be extensively used. Microbial markers like *Escherichia coli* (*E. coli*) and *E. enterococci* are usually employed to identify the potential pathogen contamination and to determine source and extent of contamination. Additionally, microbial source tracking techniques include denaturing gradient and temperature gradient gel electrophoresis, which contribute to detect and distinguish the source of contamination. Similarly, an extensive variety of chemicals are also utilized as tracers to identify the source of contamination to surface and as well as for ground water. Recently, MCs like artificial sweeteners (acesulfame and sucralose), pharmaceutical and personal care products (caffeine and carbamazepine), and fecal sterols/stanols have been utilized as chemical markers. A summary of different MCs used as tracers is given in Table 3.6.

3.4.2 Tracer Selection

The selection of a tracer depends on several parameters including geology of location, land use pattern, level and type of MC, and environmental circumstances. In order to choose an appropriate and realistic tracer, it is essential to comprehend its characteristics. Furthermore, usage of multiple tracers is worthwhile rather than using a single marker for collecting adequate and sufficient information. The ratio of coprostanol/cholesterol is used to assess the level of sewage pollution. Likewise, a high ratio of sterols/stanols indicate the contamination by fecal matter, especially from herbivorous animals like cow and sheep. Additionally, measurement of physico-chemical parameters like conductivity will further guide to decide the source of contamination. The steps that need to be followed for selecting appropriate tracer are the following: i) site survey; ii) monitoring the incidence of tracers; iii) study of fate and transportation of tracer; and iv) validation (Lim et al. 2017).

Table 3.6 Summary of different MCs used as tracers.

Category of Microconstituent	Compound	Reference
Artificial sweeteners.	Sucralose	Ferrer and Thurman 2010
	Cyclamate	Zirlewagen et al. 2016
	Acesulfame.	Wu et al. 2014
Pharmaceuticals and personal care products.	Caffeine	Young et al. 2008
	Diclofenac	Ekklesia et al. 2015
	Carbamazepine	Sauvé et al. 2012
	Ibuprofen	Pongmala et al. 2015
	Acetaminophen	Sun et al. 2016
	Triclocarbon	Sun et al. 2016
	Atenolol	de Sousa et al. 2014
	Propanolol.	de Sousa et al. 2014
Sterols/stanols.	Coprostanol	Derrien et al. 2012
	Epicoprostanol	Derrien et al. 2012
	Campestanol	Furtula et al. 2012
	Cholesterol	Furtula et al. 2012
	Cholestanol.	Furtula et al. 2012

i) *Site survey*: Generally, contaminants are source- and site-specific. As mentioned earlier, anthropogenic activities, land use patterns, and environmental conditions influence the type, occurrence, fate, and transportation of contaminants. Hence the identification of a potential tracer starts with the site survey and characterization of the site includes hydrology, type of soil, population, land use, and environmental conditions of that particular site. Additionally, the degree and type of MC can be evaluated based on consumption rate, sorption rate, stability, and biodegradation, which depend on fate and transportation through soil.
ii) *Monitoring the incidence of tracers*: Monitoring in wastewater and receiving water courses is a tedious and challenging task due to dilution during transport and intrusion of matrix constituents in water. The quantification limit of tracers needs to be at sub-ppt levels in order to detect the targeted compounds in receiving water courses. During the monitoring program, the sample needs to be collected at source (wastewater), transportation (canal), and at receiving water courses (ground water, surface water). Also, the sampling should be collected under different weather conditions (wet and dry), at different time intervals (seasonally), and also temporal and spatial distribution. Subsequently, additional samples, which are not impacted by the source, need to be collected for comparison. Later, with the aid of indices (detection ratio and detection frequency) and analysis, the potential marker is to be identified. Fenech et al. (2012) investigated the suitability of caffeine, acetaminophen, and carbamazepine as a tracer due to their high detection frequencies.

iii) *Study of fate and transport*: Fate and transport are the vital factors contributing to the choice of an appropriate tracer. Transportation is majorly influenced by different parameters including dilution and natural attenuation. However, adsorption and biodegradation are the superior mechanisms in the subsurface system (Rout et al. 2014; Altaf et al. 2021). Badruzzaman et al. (2013) illustrated the identification of a potential marker based upon the transportation process.
iv) *Validation*: After the analysis of samples collected during monitoring, a low quantity of tracers should be present in samples that are not impacted by source, and a greater quantity of it is usually observed in samples that are collected at source. Therefore, samples need to be collected at various locations for better evaluation and validating the tracer.

3.4.3 Different Source Tracking Methods

Tracing the source of contamination will significantly help to curtail the pollution levels in freshwater courses. Tracer ratio analysis is the crucial parameter for source identification and apportionment. Detection of sterols/stanols as a tracer will be an added advantage to distinguish human and non-human contamination. Evershed and Bethell (1996) used the ratio of COP/5β-stigmasterol to classify the source of contamination. If the ratio of COP/5β-stigmasterol is >1.5, it indicates that the contamination is by fecal matter. Kuroda et al. (2014) utilized different analysis include tritium, principal component analysis, multiple regression, and a combination of these analysis to identify the source of perfluoroalkyl acids in Japanese groundwater samples. Later they evaluated that sewage (44.6%) is the primary source of contamination, followed by point sources and street runoff. An overview of different sterols/stanols ratios used to determine wastewater contamination is provided in Table 3.7.

Table 3.7 Sterols/Stanols ratio used to determine wastewater contamination.

	Degree of Contamination			
Sterols/Stanols Ratio	**High**	**Normal**	**No.**	**Reference**
$\frac{\text{Coprostanol} + \text{Epicoprostanol}}{\text{Coprostanol} + \text{Epicoprostanol} + \text{Cholestanol}}$	>0.7	0.3 – 0.7	<0.3	Reeves and Patton 2005
$\frac{\text{Coprostanol}}{\text{Cholestanol} + \text{cholesterol}}$	>0.06	–	–	Bujagić et al. 2016
$\frac{\text{Epicoprostanol}}{\text{Coprostanol}}$	<0.2	0.2 – 0.8	>0.8	Froehner et al. 2009
$\frac{\text{Coprostanol}}{\text{Cholesterol}}$	>1.0	–	<1.0	Reeves and Patton 2005
$\frac{\text{Coprostanol}}{\text{Epicoprostanol}}$	>1.5	–	<1.5	Reeves and Patton 2005

3.4.4 Statistical Approaches in Source Tracking Modeling

Measurement of various physico-chemical parameters during tracer monitoring significantly contributes to support the findings of source contamination with the aid of statistical approaches including principal component analysis (PCA) and multiple linear regression (MLR).

3.4.4.1 Principal Component Analysis (PCA)

PCA is mainly used to determine the spatial and temporal variations of physico-chemical parameters and MCs as tracers for indicating a potential and promising tracer (de Sousa et al. 2014). PCA is a multivariate technique used to evaluate the correlation among diverse variables that are monitored. Principal components (PCs) are created among the variables with an aim to maximize the variation by each PC (Vieira et al. 2012). After identifying the PCs, remaining variables are omitted without the substantial loss of data. A loading plot of the PCA depicts the correlation or no correlation among the variables by measuring the distance between them. A lower distance between variables in the loading plot indicate that they are correlated. Additionally, the score for each variable has been provided on the loading plot. If the scores are the same, it shows that the samples are similar.

3.4.4.2 Multiple Linear Regression (MLR)

In MLR modeling, four types of equations are developed, as in Eqs (3.1) to (4.4) (Cho and Lee 2018).

$$Y = \alpha_0 + \sum_{i=1}^{n} \alpha_i X_i \tag{3.1}$$

$$Ln(Y) = \alpha_0 + \sum_{i=1}^{n} \alpha_i X_i \tag{3.2}$$

$$Ln(Y) = \alpha_0 + \sum_{i=1}^{n} (\alpha_i Ln(X_i)) \tag{3.3}$$

$$Ln(Y) = \alpha_0 + \sum_{i=1}^{m} (\alpha_i Ln(X_i)) + \sum_{i=m+1}^{n} (\alpha_i X_i) \tag{3.4}$$

where α_0 = regression constant; and α_i = regression coefficient corresponding to variable X_i

Original variables are used to develop Eq. (3.1). Similarly dependent variables are used to develop Eq. (3.2). In Eq. (3.3), all the dependent variables are $\log^{-e}$ transformed. In Eq. (3.4), dependent and explanatory variables are $\log^{-e}$ transformed. The fitness of the above-mentioned equations is evaluated based on coefficient values. Collinearity of the predicted variables that are generated using the MLR model is verified by computing the variance inflation factor (VIF) (Cho et al. 2009). Collinearity exists if the largest VIF value is >10 or the average VIF value is >1. The performance of the MLR model is validated using the jackknife method. With this method, each site is deleted and MLR is performed for the remaining sites with the same dependent variables as that of the deleted site. Pollutant discharge of the deleted site is computed with the help of the model that is developed using the remaining sites. The same process is repeated and one site is deleted each time.

3.5 Remote Sensing and GIS Applications for Monitoring

3.5.1 Basic Concepts and Principles

Remote sensing (RS) is the technology of collecting data from the earth's surface without physical contact, whereas the Geological Information System (GIS) analyses the remotely sensed collected data in a computer system. Remote sensing can collect data beyond the range of human vision. The working principle of remote sensing is the emission or reflection of different wavelength electromagnetic spectra from objects or materials. Satellites are the devices that collect remote sensing data. Satellite-produced datasets are freely or commercially available to download from different organization's websites. Some satellite data providers are Google Earth, Sentinel Hub, USGS Satellite imagery, NOAA, Earth on AWS, NASA Worldview, etc. The satellite data collected from these sources have a wide application in agriculture (Piekarczyk 2014), water quality monitoring (Seyhan and Dekker 1986; Ramadas and Samantaray 2018; Wang and Yang 2019), flood management, forest fire mapping (Mohajane et al. 2021), mineral resource mapping (Rajan Girija and Mayappan 2019), water resource management (Giardino et al. 2010; Wang and Xie 2018; Duan et al. 2021), water quality modeling (Wang et al. 2011; El Ouali et al. 2021), and many more.

GIS is a spatial information system that creates, manages, analyzes, and maps all data types. A spatial map provides a vision of the geographical feature or other spatial phenomena by graphically conveying information about location and attributes. GIS can help in the integration of multiple resources data, recording of how events change over space and time, effective communication of results, and modeling of various geographical features. The combined application of GIS and RS is an important tool in environmental monitoring, especially in the field of water quality monitoring (Dekker et al. 2002). In groundwater analysis, GIS provides a clear understanding by estimating groundwater pollution vulnerability, groundwater movement modeling, and solute transport modeling (Nas and Berktay 2010). The basic groundwater and surface water quality of various parameters such as pH, electrical conductivity, total dissolved solids, total suspended solids, dissolved oxygen, chemical oxygen demand, biological oxygen demand, sodium ion, potassium ion, calcium, total hardness, iron, potassium, magnesium, and many more parameters of different locations is monitored using GIS techniques (Mantzafleri et al. 2009; Tiwari et al. 2015; Kaviarasan et al. 2016; Ramadas and Samantaray 2018; El Behairy et al. 2021; Oseke et al. 2021; Vasistha and Ganguly 2022).

3.5.2 Measurement and Estimation Techniques

The spatial maps of any features are either remotely sensed data or site-based interpolated data. Remote sensed images or data are directly collected from various features such as elevation, water bodies, land use, land cover, built-up, etc. Beyond the remotely sensed dataset, spatial maps are also created for those sensors that have yet to be developed. The interpolation technique has created spatial maps of on-site collected data. The basic interpolation techniques are Inverse Distance Weighting (IDW), Spline, Kriging, Thiessen Polygons, etc. (Naoum and Tsanis 2004; Albrecht 2007). IDW is the most direct implemented method of Tobbler's First law. The IDW provides the most probable value of an unknown place concerning some known values at locations where we have a measurement (Albrecht 2007). The unknown value is calculated with the weighted average value of the known points. IDW is the most popular interpolation technique widely used in water quality monitoring. Spline interpolation is a type of interpolation where the interpolant is a special type of piece-wise polynomial. Spline interpolation fits low-degree polynomials to a small subset of the values, instead of fitting single and high-degree polynomial simultaneously (Hall and Meyer 1976; Albrecht 2007). Kriging interpolation techniques provide the best linear unbiased estimation based on the semivariogram (Lin and Chen 2004). The structure of spatial variability is used for describing semivariograms. There are two types of Kriging methodology: ordinary and universal. Ordinary Kriging methodology has five types of semivariogram methodology based on statistical equations. These are the Spherical semivariogram model, Circular semivariogram model, Exponential semivariogram model, Gaussian or normal distribution semivariogram model, and the Linear semivariogram model. In contrast, the universal Kriging methodology has two types of semivariogram models. These are the Universal Kriging with linear drift and Universal Kriging with quadratic drift. All these Kriging methods can be performed in ArcGIS software. The Thiessen polygon method is widely used in hydrological analysis. The Thiessen polygon defines the hydrometeorological zone of any location, differentiating from the nearest locations. The interpolation uses the Thiessen polygon method by directly inputting the value of a known place which is meant to be the same as that for the hydrometeorological zone. However, of all the interpolation techniques, the IDW is most widely used in water quality monitoring (Mantzafleri et al. 2009; El Behairy et al. 2021; Oseke et al. 2021; Singh et al. 2022; Vasistha and Ganguly 2022).

3.5.3 Applications for Microconstituents Monitoring

The MCs in water bodies are pharmaceuticals and personal care products, detergents, steroid hormones, industrial chemicals, pesticides, microplastics, etc. (Rousseau et al. 2021; Das et al. 2022; Rout et al. 2022). The sources of these pollutants in surface water bodies are domestic households, industry, hospitals, animal farms, and agriculture. In contrast, groundwater sources are the transportation from these sources via soil infiltration. The literature on the physico-chemical parameter of water quality were widely studied (Mantzafleri et al. 2009; Tiwari et al. 2015; Kaviarasan et al. 2016; Ramadas and Samantaray 2018; El Behairy et al. 2021; Oseke et al. 2021; Vasistha and Ganguly 2022). There was no literature on the application of GIS on MCs monitoring in ground and surface water during our review. Therefore, a case study is incorporated considering microplastic data to simulate the monitoring of MCs over a certain area (Samandra et al. 2022). The study was conducted at Bacchus Marsh Irrigation District (Victoria, Australia) and water samples were collected from six locations. The microplastic from these samples was analyzed and it was found that the microplastic constituents of the region lay in the range of 16 to 97 microplastics/L, with an average number of 38 ± 8 microplastics/L (Samandra et al. 2022). The most common interpolation methodology of IDW was used for spatial plotting of microplastics over the study area (Figure 3.7). Figure 3.7 predicted the microplastic level at an unknown place within the area concerning the known places. The spatial map generated in Figure 3.7 is for microplastic pollutants. Similarly, other MCs can also be monitored using the interpolation techniques of GIS, and further extensive research in this field is required.

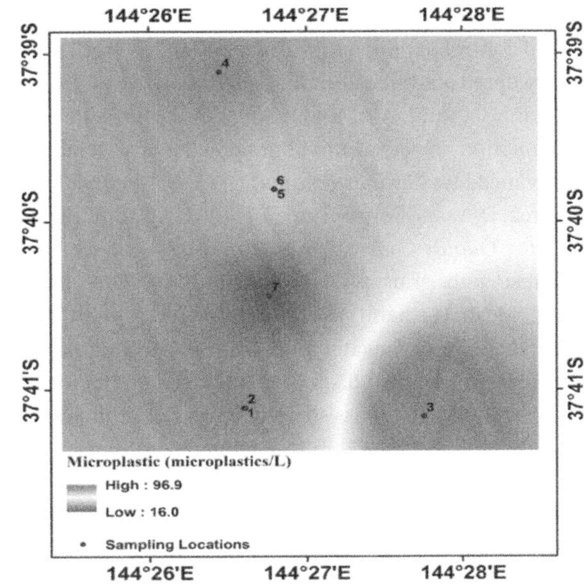

Figure 3.7 IDW interpolation of microplastics in Bacchus Marsh Irrigation District. *Source:* Adapted from (Victoria, Australia).

3.6 Conclusions

In light of the potential for the accumulation of residual metabolites that can have a negative impact on the environment, it is crucial to understand how MCs behave in the environment. A strategic screening approach to identify transformation products of MCs is necessary. In order to assess the possible impacts posed by the degradation and transitional products of MCs, an integration of both chemical and biological analyses is required. The extraction and identification of target chemicals from immensely complex composites at very low quantities is now partly attributable to advancements in analytical instrumental methods, such as the advancement in chromatographic resolution, detection sensitivity, and specificity. The establishment of proper methods to detect and measure unidentified degradation products is also crucial. Increasingly advanced reactivity–structure relationships are currently being explored that employ molecular orbital energy modeling to forecast patterns for moderate MCs particle oxidation by the oxidant (e.g., hydroxyl radical). Although it is generally advised to use standard compounds and databases, the lack of research for many of the products created during the alteration of MCs has necessitated the development of identification approaches.

References

Albrecht, J. (2007). *Key Concepts and Techniques in GIS*. Sage.

Altaf, R., Lin, X., Zhuang, W.Q. et al. (2021). Nitrilotrismethylenephosphonate sorption from wastewater on zirconium-lanthanum modified magnetite: reusability and mechanism study. *Journal of Cleaner Production* 314: 128045.

Altmann, J., Massa, L., Sperlich, A. et al. (2016). UV254 absorbance as real-time monitoring and control parameter for micropollutant removal in advanced wastewater treatment with powdered activated carbon. *Water Research* 94: 240–245.

Altmann, J., Ruhl, A.S., Zietzschmann, F., and Jekel, M. (2014). Direct comparison of ozonation and adsorption onto powdered activated carbon for micropollutant removal in advanced wastewater treatment. *Water Research* 55: 185–193.

Alvarez, D.A. and Jones-Lepp, T.L. (2011). Sampling and analysis of emerging pollutants. In: *Water Quality Concepts, Sampling, and Analyses* (eds Migliaccio, K. and Li, Y.), 199–226. https://pubs.er.usgs.gov/publication/70157305

Amin, M.M., Hashemi, M., Ebrahimpour, K., and Chavoshani, A. (2019). Determination of parabens in wastewater and sludge in a municipal wastewater treatment plant using microwave assisted dispersive liquid-liquid microextraction coupled with gas chromatography-mass spectrometry. Environmental Health Engineering and Management Journal 6 (3): 215–224. https://ehemj.com/browse.php?a_id=530&sid=1&slc_lang=en.

Anschutz, P. and Charbonnier, C. (2021). Sampling pore-water at a centimeter resolution in sandy permeable sediments of lakes, streams, and coastal zones. *Limnology and Oceanography: Methods* 19 (2): 96–114.

Anumol, T., Sgroi, M., Park, M. et al. (2015). Predicting trace organic compound breakthrough in granular activated carbon using fluorescence and UV absorbance as surrogates. *Water Research* 76: 76–87.

Asimakopoulos, A.G., Kannan, P., Higgins, S., and Kannan, K. (2017). Determination of 89 drugs and other micropollutants in unfiltered wastewater and freshwater by LC-MS/MS: an alternative sample preparation approach. *Analytical and Bioanalytical Chemistry* 409 (26): 6205–6225.

Badruzzaman, M., Oppenheimer, J.A., and Jacangelo, J.G. (2013). Impact of environmental conditions on the suitability of microconstituents as markers for determining nutrient loading from reclaimed water. *Water Research* 47 (16): 6198–6210.

Bahr, C., Schumacher, J., Ernst, M. et al. (2007). SUVA as control parameter for the effective ozonation of organic pollutants in secondary effluent. *Water Science and Technology* 55 (12): 267–274.

Basheer, C., Balasubramanian, R., and Lee, H.K. (2003). Determination of organic micropollutants in rainwater using hollow fiber membrane/liquid-phase microextraction combined with gas chromatography–mass spectrometry. *Journal of Chromatography A* 1016 (1): 11–20.

Bhushan, C. and Basu, D.D. (2017). *Gauging the Ganga: Guidelines for Sampling and Monitoring Water Quality*. New Delhi: Centre for Science and Environment.

Booij, K. and Chen, S. (2018). Review of atrazine sampling by polar organic chemical integrative samplers and Chemcatcher. *Environmental Toxicology and Chemistry* 37 (7): 1786–1798.

Booij, K., Smedes, F., and Van Weerlee, E.M. (2002). Spiking of performance reference compounds in low density polyethylene and silicone passive water samplers. *Chemosphere* 46 (8): 1157–1161.

Bopp, S., Weiß, H., and Schirmer, K. (2005). Time-integrated monitoring of polycyclic aromatic hydrocarbons (PAHs) in groundwater using the Ceramic Dosimeter passive sampling device. *Journal of Chromatography A* 1072 (1): 137–147.

Brumbaugh, W.G., Petty, J.D., Huckins, J.N., and Manahan, S.E. (2002). Stabilized liquid membrane device (SLMD) for the passive, integrative sampling of labile metals in water. *Water, Air, and Soil Pollution* 133 (1): 109–119.

Bujagić, I.M., Grujić, S., Jauković, Z., and Laušević, M. (2016). Sterol ratios as a tool for sewage pollution assessment of river sediments in Serbia. *Environmental Pollution* 213: 76–83.

Caglak, A., Chormey, D.S., Onkal Engin, G., and Bakirdere, S. (2019). Determination of micropollutants in wastewater matrix using gas chromatography–mass spectrometry after optimization of dispersive liquid–liquid microextraction. *International Journal of Environmental Science and Technology* 16 (11): 7285–7292.

Cho, J.H. and Lee, J.H. (2018). Multiple linear regression models for predicting nonpoint-source pollutant discharge from a highland agricultural region. *Water* 10 (9): 1156.

Cho, K.H., Kang, J.H., Ki, S.J. et al. (2009). Determination of the optimal parameters in regression models for the prediction of chlorophyll. Part a: A case study of the Yeongsan Reservoir, Korea. *Science of the Total Environment* 407 (8): 2536–2545.

Creusot, N., Tapie, N., Piccini, B. et al. (2013). Distribution of steroid-and dioxin-like activities between sediments, POCIS and SPMD in a French river subject to mixed pressures. *Environmental Science and Pollution Research* 20 (5): 2784–2794.

Danaceau, J., Haynes, K., and Chambers, E. (2017). A comprehensive comparison of solid phase extraction (SPE) vs. solid liquid extraction (SLE) vs. liquid liquid extraction (LLE) sample prep techniques in bioanalysis and forensic toxicology analyses. Waters Corporation.

Das, A.A., Mishra, S., Pradhan, P. et al. (2022). Recent advancements in microplastics treatments: characteristics, occurrence, and removal technologies. *Materials Today: Proceedings* 67: 1211–1217.

de Sousa, D.N.R., Mozeto, A.A., Carneiro, R.L., and Fadini, P.S. (2014). Electrical conductivity and emerging contaminant as markers of surface freshwater contamination by wastewater. *Science of the Total Environment* 484: 19–26.

Dekker, A.G., Vos, R.J., and Peters, S.W.M. (2002). Analytical algorithms for lake water TSM estimation for retrospective analyses of TM and SPOT sensor data. *International Journal of Remote Sensing* 23 (1): 15–35.

Derrien, M., Jardé, E., Gruau, G. et al. (2012). Origin of fecal contamination in waters from contrasted areas: stanols as Microbial Source Tracking markers. *Water Research* 46 (13): 4009–4016.

DiGiano, F.A., Elliot, D., and Leith, D. (1988). Application of passive dosimetry to the detection of trace organic contaminants in water. *Environmental Science & Technology* 22 (11): 1365–1367.

Di Paolo, C., Ottermanns, R., Keiter, S. et al. (2016). Bioassay battery interlaboratory investigation of emerging contaminants in spiked water extracts: towards the implementation of bioanalytical monitoring tools in water quality assessment and monitoring. *Water Research* 104: 473–484.

Duan, W., Maskey, S., Chaffe, P.L. et al. (2021). Recent advancement in remote sensing technology for hydrology analysis and water resources management. *Remote Sensing* 13 (6): 1097.

Ekklesia, E., Shanahan, P., Chua, L.H.C., and Eikaas, H.S. (2015). Associations of chemical tracers and faecal indicator bacteria in a tropical urban catchment. *Water Research* 75: 270–281.

El Behairy, R.A., El Baroudy, A.A., Ibrahim, M.M. et al. (2021). Modelling and assessment of irrigation water quality index using GIS in semi-arid region for sustainable agriculture. *Water, Air, & Soil Pollution* 232 (9): 1–19.

El Ouali, A., El Hafyani, M., Roubil, A. et al. (2021). Modeling and spatiotemporal mapping of water quality through remote sensing techniques: a case study of the Hassan Addakhil Dam. *Applied Sciences* 11 (19): 9297.

Elliott, S.M., Erickson, M.L., Krall, A.L., and Adams, B.A. (2018). Concentrations of pharmaceuticals and other micropollutants in groundwater downgradient from large on-site wastewater discharges. *PLoS One* 13 (11): e0206004.

Erasin, B.R., Turner, A.P., and Wheatley, A.D. (1994). A fixed film bioassay for the detection of micropollutants toxic to anaerobic sludges. *Analytica Chimica Acta* 298 (1): 1–10.

Escher, B.I., Allinson, M., Altenburger, R. et al. (2014). Benchmarking organic micropollutants in wastewater, recycled water and drinking water with *in vitro* bioassays. *Environmental Science & Technology* 48 (3): 1940–1956.

Etteieb, S., Kawachi, A., Han, J. et al. (2016). Assessment of organic micropollutants occurrence in treated wastewater using heat shock protein 47 stress responses in Chinese hamster ovary cells and GC/MS-based non-target screening. *Water Science and Technology* 74 (10): 2407–2416.

Evershed, R.P. and Bethell, P.H. (1996). Application of multimolecular biomarker techniques to the identification of fecal material in archaeological soils and sediments. In: *Archaeological Chemistry Organic, Inorganic, and Biochemical Analysis* (ed. Orna, M.V.), 157–172. New Rochelle, NY: ACS Publications. https://pubs.acs.org/doi/book/10.1021/bk-1996-0625.

Fenech, C., Rock, L., Nolan, K. et al. (2012). The potential for a suite of isotope and chemical markers to differentiate sources of nitrate contamination: a review. *Water Research* 46 (7): 2023–2041.

Ferahtia, A., Halilat, M.T., Mimeche, F., and Bensaci, E. (2021). Surface water quality assessment in semi-arid region (El Hodna watershed, Algeria) based on water quality index (Wqi). *Studia Universitatis Babes-Bolyai, Chemia* 66 (1).

Ferrer, I. and Thurman, E.M. (2010). Analysis of sucralose and other sweeteners in water and beverage samples by liquid chromatography/time-of-flight mass spectrometry. *Journal of Chromatography A* 1217 (25): 4127–4134.

Froehner, S., Martins, R.F., and Errera, M.R. (2009). Assessment of fecal sterols in Barigui River sediments in Curitiba, Brazil. *Environmental Monitoring and Assessment* 157 (1): 591–600.

Furtula, V., Osachoff, H., Derksen, G. et al. (2012). Inorganic nitrogen, sterols and bacterial source tracking as tools to characterize water quality and possible contamination sources in surface water. *Water Research* 46 (4): 1079–1092.

Giardino, C., Bresciani, M., Villa, P., and Martinelli, A. (2010). Application of remote sensing in water resource management: the case study of Lake Trasimeno, Italy. *Water Resources Management* 24 (14): 3885–3899.

Gibs, J. and Iwatsubo, R.T. (1998). *Selection of Equipment for Water Sampling* (ed. Wilde, F.D.). US Government Printing Office.

Gravell, A., Fones, G.R., Greenwood, R., and Mills, G.A. (2020). Detection of pharmaceuticals in wastewater effluents: a comparison of the performance of Chemcatcher® and polar

organic compound integrative sampler. *Environmental Science and Pollution Research* 27 (22): 27995–28005.
Hall, C.A. and Meyer, W.W. (1976). Optimal error bounds for cubic spline interpolation. *Journal of Approximation Theory* 16 (2): 105–122.
Haroon, M., Iali, W., and Al-Saadi, A.A. (2021). Conformational analysis and concentration detection of linuron: spectroscopic NMR and SERS study. *Spectrochimica Acta Part A: Molecular and Biomolecular Spectroscopy* 263: 120223.
Haroon, M., Iali, W., Saleh, T.A., and Al-Saadi, A.A. (2022). NMR evidence for hydrogen bonding stabilized anti-conformation of 1-methoxy-1-methyl-3-phenylurea and the concentration detection by SERS. *Journal of Molecular Liquids* 357: 119096.
Hermes, N., Jewell, K.S., Wick, A., and Ternes, T.A. (2018). Quantification of more than 150 micropollutants including transformation products in aqueous samples by liquid chromatography-tandem mass spectrometry using scheduled multiple reaction monitoring. *Journal of Chromatography A* 1531: 64–73.
Kaviarasan, M., Geetha, P., and Soman, K.P. (2016). GIS-based ground water quality monitoring in Thiruvannamalai District, Tamil Nadu, India. In: *Proceedings of the International Conference on Soft Computing Systems* (eds Suresh, L.P. and Panigrahi, B.K.), 685–700. New Delhi: Springer. https://link.springer.com/book/10.1007/978-81-322-2671-0
Król, M. and Dudziak, M. (2019). Determination of organic micropollutants in water using gas chromatography-mass spectrometry. *Multidisciplinary Digital Publishing Institute Proceedings* 16 (1): 57.
Kungolos, A. (2004). Testing industrial effluents with bioassays in Greece. *Waste Management and the Environment*. II: *WIT Transactions on Ecology and the Environment* 78: 521–528. https://www.witpress.com/elibrary/wit-transactions-on-ecology-and-the-environment/78/14404.https://www.witpress.com/books/978-1-85312-738-0.
Kuroda, K., Murakami, M., Oguma, K. et al. (2014). Investigating sources and pathways of perfluoroalkyl acids (PFAAs) in aquifers in Tokyo using multiple tracers. *Science of the Total Environment* 488: 51–60.
Lee, E., Rout, P.R., Shin, C., and Bae, J. (2019). Effects of sodium hypochlorite concentration on the methanogenic activity in an anaerobic fluidized membrane bioreactor. *Science of the Total Environment* 678: 85–93.
Lempart, A., Kudlek, E., and Dudziak, M. (2018). Determination of micropollutants in water samples from swimming pool systems. *Water* 10 (8): 1083.
Lim, F.Y., Ong, S.L., and Hu, J. (2017). Recent advances in the use of chemical markers for tracing wastewater contamination in aquatic environment: a review. *Water* 9 (2): 143.
Lin, G.F. and Chen, L.H. (2004). A spatial interpolation method based on radial basis function networks incorporating a semivariogram model. *Journal of Hydrology* 288 (3–4): 288–298.
Macova, M., Toze, S., Hodgers, L. et al. (2011). Bioanalytical tools for the evaluation of organic micropollutants during sewage treatment, water recycling and drinking water generation. *Water Research* 45 (14): 4238–4247.
Madrid, Y. and Zayas, Z.P. (2007). Water sampling: traditional methods and new approaches in water sampling strategy. *TrAC Trends in Analytical Chemistry* 26 (4): 293–299.
Mailler, R., Gasperi, J., Coquet, Y. et al. (2016). Removal of a wide range of emerging pollutants from wastewater treatment plant discharges by micro-grain activated carbon in fluidized bed as tertiary treatment at large pilot scale. *Science of the Total Environment* 542: 983–996.

Mantzafleri, N., Psilovikos, A., and Blanta, A. (2009). Water quality monitoring and modeling in Lake Kastoria, using GIS. Assessment and management of pollution sources. *Water Resources Management* 23 (15): 3221–3254.

Meza, L.C., Piotrowski, P., Farnan, J. et al. (2020). Detection and removal of biologically active organic micropollutants from hospital wastewater. *Science of the Total Environment* 700: 134469.

Mohajane, M., Costache, R., Karimi, F. et al. (2021). Application of remote sensing and machine learning algorithms for forest fire mapping in a Mediterranean area. *Ecological Indicators* 129: 107869.

Naoum, S. and Tsanis, I.K. (2004). Ranking spatial interpolation techniques using a GIS-based DSS. *Global Nest* 6 (1): 1–20.

Nas, B. and Berktay, A. (2010). Groundwater quality mapping in urban groundwater using GIS. *Environmental Monitoring and Assessment* 160 (1): 215–227.

Ohman, K., Chen, M., Knezivic, Z. et al. (2009). The sampling protocol and measurement of micro-constituents recent learnings. In: *WEFTEC 2009*, 750–763. Water Environment Federation.

Oseke, F.I., Anornu, G.K., Adjei, K.A., and Eduvie, M.O. (2021). Assessment of water quality using GIS techniques and water quality index in reservoirs affected by water diversion. *Water-Energy Nexus* 4: 25–34.

Piekarczyk, J. (2014). Application of remote sensing in agriculture. *Geoinformatica Polonica – The Journal of the Polish Academy of Arts and Sciences* 13 (1): 69–75. https://access.portico.org/Portico/auView?auId=ark:%2F27927%2Fpgjm39xdr.

Pongmala, K., Autixier, L., Madoux-Humery, A.S. et al. (2015). Modelling total suspended solids, *E. coli* and carbamazepine, a tracer of wastewater contamination from combined sewer overflows. *Journal of Hydrology* 531: 830–839.

Posselt, M., Jaeger, A., Schaper, J.L. et al. (2018). Determination of polar organic micropollutants in surface and pore-water by high-resolution sampling-direct injection-ultra high-performance liquid chromatography-tandem mass spectrometry. *Environmental Science: Processes & Impacts* 20 (12): 1716–1727.

Rajan Girija, R. and Mayappan, S. (2019). Mapping of mineral resources and lithological units: a review of remote sensing techniques. *International Journal of Image and Data Fusion* 10 (2): 79–106.

Ramadas, M. and Samantaray, A.K. (2018). Applications of remote sensing and GIS in water quality monitoring and remediation: a state-of-the-art review. In: *Water Remediation, Remediation* (eds Bhattacharya, S., Gupta, A.B., Gupta, A. and Pandey, A.), 225–246. Springer. https://link.springer.com/chapter/10.1007/978-981-10-7551-3_13. https://link.springer.com/book/10.1007/978-981-10-7551-3

Reeves, A.D. and Patton, D. (2005). Faecal sterols as indicators of sewage contamination in estuarine sediments of the Tay Estuary, Scotland: an extended baseline survey. *Hydrology and Earth System Sciences* 9 (1/2): 81–94.

Roose, P. and Udo, A.T. (2005). Monitoring organic microcontaminants in the marine environment: principles, programmes and progress. *TrAC Trends in Analytical Chemistry* 24 (11): 897–926.

Rosen, G., Lotufo, G.R., George, R.D. et al. (2018). Field validation of POCIS for monitoring at underwater munitions sites. *Environmental Toxicology and Chemistry* 37 (8): 2257–2267.

Rousseau, D., Louage, F., Wang, Q., and Zhang, R. (2021). Constructed wetlands for urban wastewater treatment: an overview. *Encyclopedia of Inland Waters*. In: *Reference Module in Earth Systems and Environmental Sciences*. (eds Chapman, D., Irvine, K., Warner, S. et al.), 3rd edn, vol. 4, 272–284. Germany, Elsevier. https://www.sciencedirect.com/science/article/pii/B9780128191668001080?via%3Dihub. https://www.sciencedirect.com/referencework/9780128220412/encyclopedia-of-inland-waters

Rout, P.R., Bhunia, P., and Dash, R.R. (2014). Modeling isotherms, kinetics and understanding the mechanism of phosphate adsorption onto a solid waste: ground burnt patties. *Journal of Environmental Chemical Engineering* 2 (3): 1331–1342.

Rout, P.R., Bhunia, P., and Dash, R.R. (2017). Assessing possible applications of waste organic solid substances as carbon sources and biofilm substrates for elimination of nitrate toxicity from wastewater. *Journal of Hazardous, Toxic, and Radioactive Waste* 21 (3): 04016027.

Rout, P.R., Dash, R.R., and Bhunia, P. (2016). Nutrient removal from binary aqueous phase by dolochar: highlighting optimization, single and binary adsorption isotherms and nutrient release. *Process Safety and Environmental Protection* 100: 91–107.

Rout, P.R., Dash, R.R., Bhunia, P. et al. (2021a). Comparison between a single unit bioreactor and an integrated bioreactor for nutrient removal from domestic wastewater. *Sustainable Energy Technologies and Assessments* 48: 101620.

Rout, P.R., Dash, R.R., Bhunia, P., and Rao, S. (2018). Role of *Bacillus cereus* GS-5 strain on simultaneous nitrogen and phosphorous removal from domestic wastewater in an inventive single unit multi-layer packed bed bioreactor. *Bioresource Technology* 262: 251–260.

Rout, P.R., Mohanty, A., Sharma, A. et al. (2022). Micro-and nanoplastics removal mechanisms in wastewater treatment plants: a review. *Journal of Hazardous Materials Advances* 6: 100070. https://www.sciencedirect.com/science/article/pii/S2772416622000262.

Rout, P.R., Shahid, M.K., Dash, R.R. et al. (2021b). Nutrient removal from domestic wastewater: a comprehensive review on conventional and advanced technologies. *Journal of Environmental Management* 296: 113246.

Rout, P.R., Zhang, T.C., Bhunia, P., and Surampalli, R.Y. (2021c). Treatment technologies for emerging contaminants in wastewater treatment plants: a review. *Science of the Total Environment* 753: 141990.

Saha, S., Singh, A.K., Keshari, A.K. et al. (2018). Modern extraction techniques for drugs and medicinal agents. In: *Ingredients Extraction by Physicochemical Methods in Food* (eds Grumezescu, A.M. and Holban, A.M.), 65–106. Academic Press. https://www.sciencedirect.com/science/article/pii/B9780128115213000028. https://www.sciencedirect.com/book/9780128115213/ingredients-extraction-by-physicochemical-methods-in-food

Samandra, S., Johnston, J.M., Jaeger, J.E. et al. (2022). Microplastic contamination of an unconfined groundwater aquifer in Victoria, Australia. *Science of the Total Environment* 802: 149727.

Sauvé, S., Aboulfadl, K., Dorner, S. et al. (2012). Fecal coliforms, caffeine and carbamazepine in stormwater collection systems in a large urban area. *Chemosphere* 86 (2): 118–123.

Schintu, M., Durante, L., Maccioni, A. et al. (2008). Measurement of environmental trace-metal levels in Mediterranean coastal areas with transplanted mussels and DGT techniques. *Marine Pollution Bulletin* 57 (6–12): 832–837.

Seyhan, E. and Dekker, A. (1986). Application of remote sensing techniques for water quality monitoring. *Hydrobiological Bulletin* 20 (1): 41–50.

Shah, R.P. and Singh, S. (2010). Identification and characterization of a photolytic degradation product of telmisartan using LC–MS/TOF, LC–MSn, LC–NMR and on-line H/D exchange mass studies. *Journal of Pharmaceutical and Biomedical Analysis* 51 (55): 1037–1046. doi: 10.1086/j.jpba.2009.11.008.

Singh, G., Gorai, S., Ratha, D. et al. (2022). Monitoring groundwater quality status using geo-spatial technique: a case study on Malwa region Punjab, India. *International Journal of Energy and Water Resources* 6 (2): 253–266.

Slaveykova, V.I., Parthasarathy, N., Buffle, J., and Wilkinson, K.J. (2004). Permeation liquid membrane as a tool for monitoring bioavailable Pb in natural waters. *Science of the Total Environment* 328 (1–3): 55–68.

Spangenberg, M., Bryant, J.I., Gibson, S.J. et al. (2021). Ultraviolet absorption of contaminants in water. *Scientific Reports* 11 (1): 1–8.

Stacklin, C. (2009, January). Predictive source control: application of advanced numerical methods and surrogate monitoring for microconstituents in water reclamation. In: *WEFTEC 2009*, 723–742. Water Environment Federation.

Stuer-Lauridsen, F. (2005). Review of passive accumulation devices for monitoring organic micropollutants in the aquatic environment. *Environmental Pollution* 136 (3): 503–524.

Sun, Q., Li, Y., Li, M. et al. (2016). PPCPs in Jiulong River estuary (China): spatiotemporal distributions, fate, and their use as chemical markers of wastewater. *Chemosphere* 150: 596–604.

Tadda, M.A., Altaf, R., Gouda, M. et al. (2021). Impact of Saddle-Chips biocarrier on treating mariculture wastewater by moving bed biofilm reactor (MBBR): mechanism and kinetic study. *Journal of Environmental Chemical Engineering* 9 (6): 106710.

Taylor, A.C., Fones, G.R., Vrana, B., and Mills, G.A. (2021). Applications for passive sampling of hydrophobic organic contaminants in water: a review. *Critical Reviews in Analytical Chemistry* 51 (1): 20–54.

Thomatou, A.A., Zacharias, I., Hela, D., and Konstantinou, I. (2011). Passive sampling of selected pesticides in aquatic environment using polar organic chemical integrative samplers. *Environmental Science and Pollution Research* 18 (7): 1222–1233.

Tiwari, A.K., De Maio, M., Singh, P.K., and Mahato, M.K. (2015). Evaluation of surface water quality by using GIS and a heavy metal pollution index (HPI) model in a coal mining area, India. *Bulletin of Environmental Contamination and Toxicology* 95 (3): 304–310.

Vasistha, P. and Ganguly, R. (2022). Water quality assessment in two lakes of Panchkula, Haryana, using GIS: case study on seasonal and depth wise variations. *Environmental Science and Pollution Research* 29 (28): 43212–43236.

Vieira, J.S., Pires, J., Martins, F.G. et al. (2012). Surface water quality assessment of Lis river using multivariate statistical methods. *Water, Air, & Soil Pollution* 223 (9): 5549–5561.

Viollier, E., Rabouille, C., Apitz, S.E. et al. (2003). Benthic biogeochemistry: state of the art technologies and guidelines for the future of *in situ* survey. *Journal of Experimental Marine Biology and Ecology* 285: 5–31.

Vrana, B., Klučárová, V., Benická, E. et al. (2005). Passive sampling techniques for monitoring pollutants in water. *TrAC Trends in Analytical Chemistry* 24 (10): 845–868.

Vrana, B., Klučárová, V., Benická, E. et al. (2014). Passive sampling: an effective method for monitoring seasonal and spatial variability of dissolved hydrophobic organic contaminants and metals in the Danube river. *Environmental Pollution* 184: 101–112.

Vroblesky, D.A. and Hyde, W.T. (1997). Diffusion samplers as an inexpensive approach to monitoring VOCs in ground water. *Groundwater Monitoring & Remediation* 17 (3): 177–184.

Wang, X., Fu, L., and He, C. (2011). Applying support vector regression to water quality modelling by remote sensing data. *International Journal of Remote Sensing* 32 (23): 8615–8627.

Wang, X. and Xie, H. (2018). A review on applications of remote sensing and geographic information systems (GIS) in water resources and flood risk management. *Water* 10 (5): 608.

Wang, X. and Yang, W. (2019). Water quality monitoring and evaluation using remote sensing techniques in China: a systematic review. *Ecosystem Health and Sustainability* 5 (1): 47–56.

Wittmer, A., Heisele, A., McArdell, C.S. et al. (2015). Decreased UV absorbance as an indicator of micropollutant removal efficiency in wastewater treated with ozone. *Water Science and Technology* 71 (7): 980–985.

Wu, M., Qian, Y., Boyd, J.M. et al. (2014). Direct large volume injection ultra-high performance liquid chromatography-tandem mass spectrometry determination of artificial sweeteners sucralose and acesulfame in well water. *Journal of Chromatography A* 1359: 156–161.

Young, T.A., Heidler, J., Matos-Pérez, C.R. et al. (2008). *Ab initio* and *in situ* comparison of caffeine, triclosan, and triclocarban as indicators of sewage-derived microbes in surface waters. *Environmental Science & Technology* 42 (9): 3335–3340.

Zirlewagen, J., Licha, T., Schiperski, F. et al. (2016). Use of two artificial sweeteners, cyclamate and acesulfame, to identify and quantify wastewater contributions in a karst spring. *Science of the Total Environment* 547: 356–365.

4

Toxicity Assessment of Microconstituents in the Environment

Nagireddi Jagadeesh[1], *Baranidharan Sundaram*[1], *and Brajesh Kumar Dubey*[2]

[1]*Department of Civil Engineering, National Institute of Technology Andhra Pradesh, Tadepalligudem, India*
[2]*Department of Civil Engineering, Indian Institute of Technology Kharagpur, West Bengal, India*

4.1 Introduction

The emerging pollutants/contaminants are synthetic or naturally occurring chemicals detected in water that adversely affect human and ecological health. During the last decade, the research for identifying the source, occurrence, and adsorption of emerging pollutants, such as pharmaceuticals and personal care products (PPCPs), persistent organic pollutants (POPs), steroid hormones, nanomaterials, cosmetic products, endocrine-disrupting chemicals (EDCs), and perfluorinated compounds, has increased significantly (Visanji et al. 2018). Nanoparticles (NPs), active substances or medicinal goods, flame retardants, pigments, stabilizers, and other chemical additives may be included with these items. In the past several decades, a broad variety of novel chemical pollutants have been discovered in aquatic habitats due to the increased consumption, lack of regulation, and inadequate waste management of commercial items. These pollutants are now regarded of health and ecological significance (López et al. 2022).

Throughout the world there is an interest in nanoscience and nanotechnology that is quickly becoming a top priority for the research community (Bystrzejewska-Piotrowska et al. 2009). Natural, semi-synthetic, and synthetic chemical compounds, including nanomaterials (MPs/PNPs) and metal nanoparticles (Me-NPs), as well as sub-products formed from the transformation and/or degradation of these pollutants in nature, constitute emerging contaminants (López et al. 2022). It is vital to go through the most frequent nanomaterials (NMs) utilized in environmental analysis before moving on to their applications, including carbon NMs, AuNPs, magnetic nanoparticles, and quantum dots (QDs) (Liu et al. 2014). Due to their high surface-to-volume ratio and tiny particle sizes, nanomaterials may fill microscopic holes and cracks with remarkable characteristics. As a result, nanoparticles have become more common in the production of nanocomposites. Various inorganic

Microconstituents in the Environment: Occurrence, Fate, Removal, and Management, First Edition.
Edited by Rao Y. Surampalli, Tian C. Zhang, Chih-Ming Kao, Makarand M. Ghangrekar, Puspendu Bhunia, Manaswini Behera, and Prangya R. Rout.

nanoparticles are used to reinforce and fill ceramic composites. Ceramic matrices with these nanoparticles have improved mechanical, chemical, and thermal stability, and antibacterial and fire retardant qualities, as well as elasticity, self-healing, and biocompatibility (Mantilaka et al. 2019).

The wide range of applications of nanomaterials are mainly in the treatment of wastewater, removal of hazardous materials, corrosion control and protection from the automobile industry, and emission control and technology (Brar et al. 2010). When nanomaterials are extensively produced and utilized, nano-sized particles are constantly released into the environment, posing a danger to both the occupying population and the general public. In several studies, particle exposure has been linked to cardiovascular disease. There have also been several instances where nanoparticles may have infiltrated cardiac tissue, built up, and subsequently caused harm (Cheng et al. 2021). There are considerable worries regarding the environmental impact of emerging contaminants (ECs) and the development of antibiotic resistance as a result of the discharge of medicines, personal care products, and pesticides (Okoye et al. 2022). When taken as a whole, pharmaceuticals and personal care products (PPCP) are seen as a potential source of environmental contamination with pharmaceuticals. More than ever before, PPCPs are being manufactured and used in ever larger quantities. Even though there has been a significant increase in the manufacture and usage of PPCPs, there has been a lack of therapy for the PCs released by PPCPs (Samal et al. 2022). The presence of medicines and personal care items in wastewater treatment plant effluents, and surface, ground, and drinking water has increased significantly in recent years.

Water and wastewater treatment facilities are the primary source of PPCPs entering the aquatic environment. Since the vast majority of medicine taken by patients is excreted in their urine and feces, it is sent to the wastewater treatment plant unaltered (Khraisheh et al. 2014). PPCPs elimination is a hot topic right now because of the growing concerns about new toxins like these. Due to their widespread use as painkillers or to reduce inflammation and fever, non-steroidal anti-inflammatory medications (NSAIDs) are the most often found PPCPs in environmental samples (Czech et al. 2020). PCPs such as anti-inflammatory medicines, antimicrobials, and heavy metals are included in the list of environmental pollutants of concern. Wastewater may include active or inactive PCP remains, depending on the source of that waste (Buta et al. 2021). Pollution from human activities such as mining, smelting and other metal-based industries; waste disposal; excretion; animal manure; runoffs; cars and roadworks; and other sources have all contributed to the emergence of heavy metal pollution (López et al. 2022). Pesticides, insecticides, fertilizers, and other agricultural chemicals, many of which include heavy metals, have become secondary sources of heavy metal contamination (Briffa et al. 2020).

Surface water is contaminated when the emerging contaminants such as antibiotics, pesticides, hormones, microplastics (MPs), and PPCPs become mixed within it (Galindo-Miranda et al. 2019). The addition of emerging pollutants to the wastewater has been growing every year and makes these wastewater characteristics more toxic than the parent water source (Agüera et al. 2013). The USEPA (United States Environmental Protection Agency), along with many research articles, has stated that there are unfamiliar compounds in the wastewater and the aquatic environments (Deblonde et al. 2011). There was no research until 2005 in developing countries due to the lack of knowledge on emerging pollutants and their effect on the ecosystem and human health. Since 2005, research on emerging pollutants began and rose gradually in the aquatic environment in these countries, but there has been a lack of proper technology and research publication in emerging pollutants removal from the wastewater (Visanji et al. 2018). Even though the knowledge and research options have increased in removing emerging pollutants from wastewater, their availability is still a risk to terrestrial animals, aquatic life, and humans.

This potential risk can lead to decreased male fertility, including congenital disabilities, changes in the reproduction system, and occasionally causing cancer. The monitoring of emerging pollutants in wastewater is difficult because their trace concentrations vary from nanograms per liter (ng/L) to milligrams per liter (mg/L) and because of the unavailability of expensive, sophisticated, and high analytical instruments (Visanji et al. 2018). Many physical, chemical, and biological methods were investigated to remove emerging pollutants from wastewater (Dhangar and Kumar 2020). Among these, the reduction of emerging contaminants using the adsorption, ultra/nano filtration, advanced oxidation process, and reverse osmosis techniques is the most welcoming methods. This study summarizes the review of fate, behavior, and transportation of MPs in the environment. MPs concentration in the environment, especially in the aquatic environment, focuses more on its toxic effects on various aquatic species as well as humans.

4.2 Microplastics in the Environment

Microplastics may enter the marine water column, sediment biogeochemical pathways, and food webs due to their tiny particle size and extensive spread (Bharath K et al. 2021). Although our knowledge of MPs is still immature, it is critical in determining the plastic's environmental destiny and its remediation in these channels. Recent years have seen an enormous amount of study across several sectors, which has improved our knowledge, but it is clear that the results need to be streamlined and consolidated. The biogeochemical significance of microbe–plastic interactions is examined in this section, based on information from laboratory and field research. In particular, new methodologies and analytical approaches are being focused on in order to better understand how microorganisms and MP litter interact in a two-way process and the resulting environmental consequences (Rogers et al. 2020). Plastics are raising concern as a growing threat to the environment, which contaminate air, soil, water, and food. On average, a person consumes 5 g of plastic through various ingestion pathways, such as drinking water, consumption of various aquatic species, and inhalation of air (Dalberg Advisors et al. 2019).

With the use of online databases such as Science Direct, Google Scholar, and Research Gate, a comprehensive literature evaluation was undertaken. Suitable keywords have been used for exploring the literature studies such as "microplastics," "behavior," "fate," "toxic effects," "environment," "India," "microplastics," and "world" in the online data bases in various components in the environment. Figure 4.1 shows the research articles published from 2016 to date on MPs worldwide, including India, which provides ideas on the current research trends. It was observed that the research on MPs in India still lacks the necessity to accelerate as much as possible by conducting more studies on MPs in various components of the environment, such as air, water, biota, sediments, salt, and aquatic species.

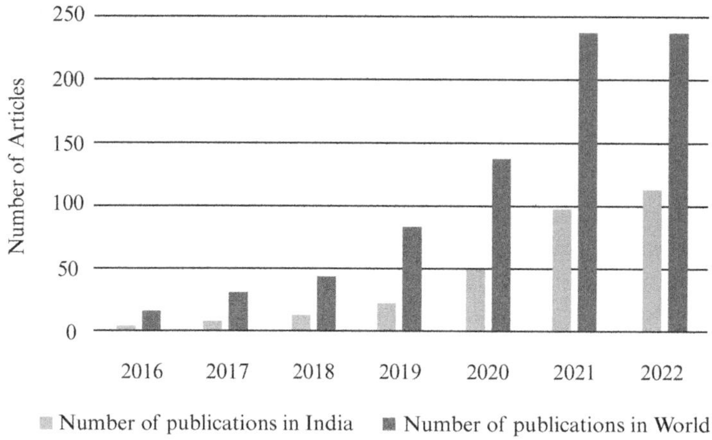

Figure 4.1 Research articles on MPs in the environment (India vs. World).

4.3 Microplastics Pathways, Fate, and Behavior in the Environment

Plastics were the great innovation of the twentieth century, and have been used in a variety of industries because of their high strength-to-weight ratio and low cost. The global data for plastic consumption has increased exponentially from 1.5 million tons (1950) to 360 million tons (2018), as per Statista's statistical data (Bui et al. 2020). The basic classification of plastics is based on their size range; MPs whose size is >2.5 cm, followed by mesoplastics 2.5 to 5 mm, and MPs in the range of 0.001 to 5 mm (Jeyasanta et al. 2020). MPs have become a global issue as they are found in almost all pharmaceuticals, paints, cleaning products, and personnel care products, including facial creams, soaps, body lotions, and toothpaste (Westphalen and Abdelrasoul 2018).

MPs are distributed worldwide and frequently identified in soil, marine waters, freshwater sediments, and wastewater (Ziajahromi et al. 2016; Lares et al. 2018; Shen et al. 2020; Veerasingam et al. 2020; Daniel et al. 2020a). MPs have been identified in aquatic and terrestrial ecosystems and are being ingested by various means to end up in the human body (Karuppasamy et al. 2020). Ingestion and entanglement of MPs in aquatic and terrestrial ecosystems through the food chain cause potential health risks such as starvation and immune functional issues (Xu et al. 2020). Also, the interaction of MPs with organic and inorganic pollutants in the environment results in leaching of chemicals and heavy metals, which ultimately affect the organisms residing in the respective ecosystems (Talvitie et al. 2017; Naidu 2019; Narmadha et al. 2020; Padervand et al. 2020; Veerasingam et al. 2020; Wu et al. 2021). These are new challenges to society, and the majority of reports have come from MPs concerning pollution, which has become of public concern in the last decade. Figure 4.2 shows the pathway of MPs into the environment through various phases that ultimately affect humans.

According to a survey by the Un-Plastic Collective (UPC), the amount of waste plastic produced in India is 9.5 million tonnes per year. Of this amount, 43% of the plastic is utilized as packaging material of different thicknesses, and 40% of the garbage is deemed to be

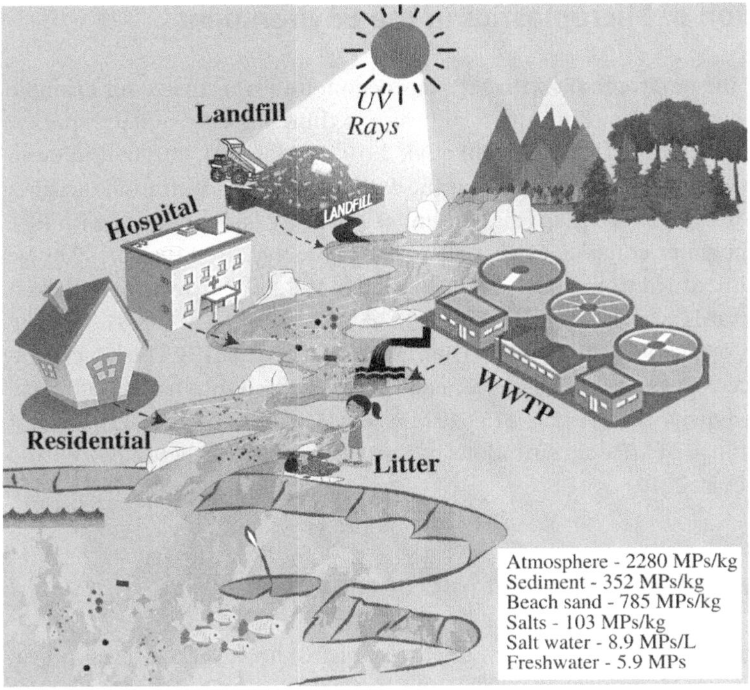

Figure 4.2 Pathways of MPs into the environment.

littering waste. Recent studies on the pollution caused by MPs, as well as their presence, destiny, and behavior in India, has become of increasing concern for many academics. This is mostly due to the exponential growth of MPs and the improper handling of plastic waste, which results in contamination in the majority of water bodies, including rivers, lakes, seas, and soil. This causes considerable environmental disturbance, which has to be addressed as soon as possible (Fu and Wang 2019). Recently, there has been a focus on MPs in the Indian environment, and most of the studies that have been conducted have reported on marine, fresh water, sediments, salt, atmospheric dust, and aquatic species.

Despite this, research is still being conducted, and MPs in the environment, especially in wastewater, are not being taken into consideration (Naidu et al. 2017; Kumar et al. 2018; Sarkar et al. 2019; Karuppasamy et al. 2020; Narmadha et al. 2020; Veerasingam et al. 2020). The primary source of MPs released into the environment is WWTPs, where the wastewater with MPs is being treated and the effluents are disposed of back into the environment (Sol et al. 2020). Though the MPs from wastewater have been efficiently removed by WWTPs (up to 99.5%), it is still observed that there is a lack of mass balance in influent and effluent concentration of MPs after disposal to land or water (Koutnik et al. 2021). It has also been observed that the huge influent load of wastewater with MPs into the WWTPs results in escape of MPs into the environment through effluents and sludge (Wei et al. 2020). However, the existence of MPs and their behavior in Indian wastewater has not yet been discussed and the objective of this study is to summarize the various studies performed on MPs in the Indian environment and identification of research gaps in wastewater management.

4.4 Concentration of Microplastics in the Environment

The contamination of the environment with MPs has been found worldwide, for example in mountains, atmospheric dust, marine and fresh water, sediments, salts, aquatic species, and wastewater (Rezania et al. 2018; Ambrosini et al. 2019; Daniel et al. 2020b; Sridharan et al. 2021; Yozukmaz 2021). It was reported that the MPs average concentration in atmospheric dust in China was 1.5 to 120 g/kg (indoor) and 0.2 to 9.02 g/kg (outdoor), with PET and PC as the major polymer complex (Liu et al. 2019). It was reported that the average concentration of MPs discarded from homes and observed in the soil, earthworms, chicken feces, and chicken gizzards was 0.87 ± 1.9 MPs/g, 14.8 ± 28.8 MPs/g, 129.8 ± 82.3 MPs/g, and 10.2 ± 13.8 MPs/g, respectively, in Mexico (Huerta Lwanga et al. 2017). The concentrations of MPs in surface waters of the US and Canada were estimated to be 466,000 MPs/km^2, with an average density of $43{,}157 \pm 115{,}519$ MPs/km^2 (Eriksen et al. 2013). Also, it was reported that the range of MP concentrations in European lakes was between 0.82 and 4.42 MPs/m^3 (Fischer et al. 2016).

4.5 Influence of Microplastics on Microorganisms

In WWTPs, MPs interact with a wide range of microorganisms; however, the most prevalent contact is with fungus (Mammo et al. 2020; Wang et al. 2020); algae (Zhang et al. 2017; Liu et al. 2020; Reichelt and Gorokhova 2020; Nava and Leoni 2021); and bacteria (Oberbeckmann et al. 2015; Viršek et al. 2017; Rummel et al. 2017; Michels et al. 2018; Parrish and Fahrenfeld 2019; Mammo et al. 2020; Tu et al. 2020; Nava and Leoni 2021). The interaction of MPs with these microorganisms causes the attachment of germs to the surfaces of the MPs, which in turn changes the function of the MP cells, which then affects the gene of the bacteria (Tuson and Weibel 2013). Because the findings of a meta-analysis study on algal species showed that MPs hindered algae development, the researchers came to the conclusion that it is risky to expose microorganisms to MPs (Reichelt and Gorokhova 2020).

Despite the fact that microalgae are the principal producers of food for aquatic creatures, their development is stunted when they are exposed to MPs, which thus has a deleterious effect on the ecology of aquatic environments (Liu et al. 2020). The surfaces of MPs act as a source of biofilm formation, which shields bacteria (physical-mechanical). As a source of biofilm development, the surfaces of MPs protect bacteria against oxidative stress and infection (physical-mechanical). The size, polymer type, and hydrophobic nature of MPs all impact the biofilm's ability to hold bacteria (Oberbeckmann et al. 2015). The size of MPs had no effect on the adhesion of microbes to their surfaces, which was surprising. A few northern ocean bacterium species have been discovered to adhere to the surfaces of polyethylene and polyvinyl chloride (PVC) MPs, according to this study (Mammo et al. 2020). Because of the aggregation and attachment of bacterial colonies on MPs, these particles have a propensity to sink in bodies of water. This gives the impression that they are suitable food for aquatic creatures, but they can cause major health problems for these animals (Michels et al. 2018).

4.6 Toxicity Mechanisms

Physical harm, such as internal abrasions and obstructions, will occur once the MP is deposited in the organism. Additionally, leaching ingredient pollutants, such as monomers and plastic additives, may cause carcinogenesis and endocrine disruption in humans (Veerasingam et al. 2020). In addition to genotoxicity, MPs have been shown to affect the endocrine and neurotransmission systems of marine organisms. Both vertebrates and insects have been reported to have MPs in their digestive systems owing to the presence of bisphenols in fish liver and muscle (Sharma et al. 2021).

In order to better understand the fate of MPs in WWTPs, researchers apply a variety of treatment extensions. MPs in WWTPs have an influence on the land, aquatic life, soil, and human health via their behaviors during treatment and discharge (Wang et al. 2020). When MPs are released into fresh water after being held in sludge or being filtered via the WWTP, they contaminate the water to cause severe water pollution (Wang et al. 2020). In freshwater lakes, MPs persist for extended periods of time, ultimately being deposited with sediments as a result of their lengthy retention duration (Miranda et al. 2020). MPs found in wastewater are often deposited in sludge and ultimately reach the soil, according to several accounts in the literature (Bilal and Iqbal 2020).

4.6.1 Effect on Aquatic Ecosystem

In order for the aquatic ecosystem to work effectively, there must be interactions between biotic and abiotic factors. Since 2020, 77.5% of research publications on marine ecosystems have been published, 16.2% in freshwater systems, and 7.6% in estuaries, according to Meng et al. (2020). An increasing number of studies are looking at the ecological effects of MPs on aquatic species, and a lot of effort is being put into gaining a thorough picture of the effects (Schmidt et al. 2020). Plastic pollution, which has accumulated over the last 15 years, is expected to decimate 15 to 35% of aquatic flora and fauna by 2050 (Chandra et al. 2020). Storms and cyclones may transport pollutants at lower concentrations into the aquatic environment than by rivers and streams, resulting in MP contamination (Westphalen and Abdelrasoul 2018; Wang et al. 2020). Increasing usage of plastics, water pollution, and poor management of aquatic systems may affect the concentrations of MPs in aquatic systems (Wang et al. 2020).

In Scotland, effluent discharge from WWTPs discharged 6.5×10^7 MPs into the aquatic environment, with tertiary treatment at the WWTP having a successful removal rate of 98% (Bretas Alvim et al. 2020). WWTPs discharge an average of 7000 micro beads per day into the aquatic environment, according to US research, and found an annual release of 8×10^9 microbeads (Wu et al. 2019). Toxic pollution in aquatic ecosystems is mostly caused by the accumulation of different plastic bits and synthetic MPs like PVC and PET after they are thrown away (Ngo et al. 2019). This is because plastics in aquatic systems take far longer to decompose under the influence of UV light radiation than they do on a terrestrial surface (Parrish and Fahrenfeld 2019). MPs absorb harmful metals from wastewater treatment plants and release them into waterways, producing ecotoxicity. This is because MPs have a very hydrophobic surface (Westphalen and Abdelrasoul 2018).

According to reports, MPs have also been consumed by zooplankton, marine fish, mammalian species, sea birds, and sea turtles. Plastic debris entanglement and subsequent asphyxiation may occur when substantial volumes of MPs are inhaled by aquatic food species because of their long stay in the aquatic environment and their form and size (Environment Protection Authority 2016).

4.6.2 Effect on Human Health

Plastics are being used at a far higher rate than ever before, and as a result they are having an effect not only on aquatic ecosystems but also on human health. This is a major cause for concern for the majority of researchers, as well as the regulatory authorities (Miranda et al. 2020). The data has shed light on the worldwide spatial distribution of MPs, which have been discovered in great amounts in remote locations, such as the ocean bottom, the ice of the Arctic, and the Antarctic, and are known to have negative impacts on the surrounding ecosystem (Wang et al. 2020). For a long time, researchers have been investigating the physical, chemical, and biological interactions of MPs. Despite this, additional research is urgently required to investigate the effect that MPs have on both the aquatic environment and human health (Bilal and Iqbal 2020). Due to the existence of MPs in the environment and human exposure, there is an increased danger to human health due to pollution of the food chain and water bodies (Bui et al. 2020). MPs and their toxicity are being studied as a consequence of the eating and inhalation of tainted food, which presents major health hazards to the general population (Miranda et al. 2020).

WWTPs are a critical point of entry for MPs into the environment (Bakaraki Turan et al. 2021). Sustainable soil health may be jeopardized if MPs in WWTP sludge leak into groundwater and contaminate it (Li et al. 2018). With WWTP sludge dumping into the environment, a considerable volume of MPs affects the natural aquatic ecosystem to a larger degree than other pollutants (Li et al. 2018). Sludge from WWTPs may provide an environmental concern to people and animals because it can penetrate freshwater sources and hence pose a threat to human health (Li et al. 2020). Research on MPs has been continuing; however, there is an even greater need to analyze their dynamic behavior in the context of WWTPs, their controversial implications on human health, and their detrimental effects on stream networks (Schmidt et al. 2020).

4.6.3 Toxicity Testing

Traces of metals and possibly hazardous organic compounds may be found in MPs. Upon entering the body, these harmful compounds can be absorbed by the plastic surface. MPs have been linked to a variety of health problems in humans. These concerns pertain to the chemical toxicity of plastics (e.g., bisphenol A, phthalates, styrene) or the particle toxicity of MPs owing to their tiny size (Seth and Shriwastav 2018). Even tap water contains MPs. Plastic pieces may also transport disease-causing organisms and serve as a vector for environmental illnesses. Soil fauna may also be affected by MPs, which can influence their health and the functions of the soil (He et al. 2018). MPs have the potential to be toxic to aquatic organisms such as zooplankton, shrimps, crabs, and fish, which may then be transported to higher trophic levels of the food chain (López et al. 2022).

The end point of the MPs is ultimately mostly the aquatic environment, and as a result, the toxicity is also significant in this environment as a result of the massive build-up of MPs. We have come up with two methods that may assist in identifying the toxicity of MPs in aquatic environments. It was determined that bifenthrin, a common wastewater-derived MP, had acute harmful effects on freshwater species, whether in the presence or absence of other MPs. The sorption of bifenthrin on MPs may be affected by the presence of dissolved organic carbon (DOC) in surface water. DOC has been shown to interact with bifenthrin and the potential toxicity of both bifenthrin and MPs was evaluated in laboratory and more realistically environmental situations using studies in both synthetic water and river water. Bifenthrin's adsorption to MPs in water was also studied in a table top laboratory experiment in order to better understand how the two interact. Toxicity bioassays were carried out in accordance with the Organisation for Economic Cooperation and Development (OECD) guidelines using species from aquatic environment and cultivating of their eggs (Wang et al. 2020). Following the normal procedure indicated in the OECD procedures, egg masses were taken from culture tanks and introduced into 600 mL glass beakers containing 500 mL moderately hard water (MHW). In accordance with the USEPA's approved procedures, MHW was made using analytical grade reagents, the cultures were incubated at 21°C with an L:D 16:8 h, and gently aerated until the larvae emerged. In order to carry out the bioassays, the second instar larvae (b5 d post-hatch) were taken from the culture beakers.

4.6.3.1 Test Without PE MPs

Using copper (II) sulphate pentahydrate ($CuSO_4{\cdot}5H_2O$) at concentrations ranging from 31.25 to 500 g/L as a reference toxicant, a 48 h acute test was carried out to guarantee the health and adequate sensitivity of freshwater species for bioassay purposes. In two 48 h acute bioassays, one with moderately hard water (MHW) and the other with river water, OECD methods were used to perform the two experiments.

HPLC-quality methanol was used to create bifenthrin stock solutions (analytical grade, Sigma-Aldrich; N99.9%, Sigma-Aldrich). Based on preliminary range-finding studies, the six nominal bifenthrin concentrations used in each bioassay ranged from 0.1 to 3.2 micrograms per litre. All of the water samples were tested for HPLC-grade methanol concentrations of 0.01 and 0.01%, respectively. Beakers filled with bifenthrin stock were placed in the fume hood and allowed to completely evaporate. After complete evaporation, each beaker was filled with 25 ml of MHW (or river water) and swirled with a Pasteur pipette to provide the solvent with a maximum concentration of 0.01% (HPLC-grade methanol). At 21°C, L:D 16:8 h, 4 replicates were used for all treatments. Images were taken after 48 h of exposure and the immobility (i.e., death) of the exposed species was recorded. During the acute exposure period of 48 h, each duplicate was checked for the remaining aquatic species using a stereo microscope. This process continues until the exposure is complete.

4.6.3.2 With Microbeads

Using a stock solution of 100 g/mL of a combination of synthetic pyrethroids in water, the sorption experiments were carried out in order to investigate the potential interactions between MPs and the synthetic pyrethroids. Ultrapure water (18.2 MΩ/cm) was utilized for the sorption experiments, which were carried out in the presence and absence of microscopic plastic particles according to the same protocol as the toxicity bioassays. There were three duplicates of each treatment, which were sealed in beakers and incubated at 21°C in a shaker (60 rpm). A Whatman 0.4 m glass microfiber filter (Whatman) was used to remove MPs from the solution, and dichloromethane was used to extract synthetic pyrethroids from the solution. Note that the trials were used to examine the probable loss of pyrethroids to the beaker and/or filter. A gas chromatograph-mass spectrometer (GC–MS) was used to compare chemical concentrations in water with and without MPs.

4.6.3.3 Analysis Limitations

Adsorbed organic pollutants, as well as microplastics themselves, may be an issue. In recent years, scientists studying microplastic toxicity have started looking at how harmful it might be in combination with other chemicals (Park and Park 2021). Organs might be poisoned if they are exposed to both microplastics and PPCPs in a combined form. Study design flaws include the size, shape, and concentration of the tested microplastics not relevant to the environment; the lack of detailed test particle characterization, such as the size distribution, density, and assessment of chemicals that may have already been absorbed before exposure; the use of nonstandard endpoints or biomarkers and the lack of appropriate covariates; and the use of nonstandard endpoints or biomarkers and the lack of appropriate covariates.

The lack of standard testing for microplastic ecotoxicity has resulted in large data gaps, and EC50 data would be beneficial for regulatory risk assessment as a consequence. When possible, OECD test recommendations should be followed in the design of a study. Monitoring and impact studies must disclose results in concentrations that are similar, which is of critical importance. Units of particle number per volume, surface area per volume, and mass per volume have all been employed in particle toxicity research. The use of a variety of standard units may be suitable while conducting these investigations. The most important aspect of ecotoxicity studies is that researchers adequately define the test particles utilized and provide these data in order to make comparisons to exposure data possible. Employing nonlinear regression and log-log dose-response curves utilized to generate the LC_{50} and LC_{10} values at 95% confidence interval will not provide accurate results. In order to determine whether or not the data warrant further investigation in light of the control samples, the Dunnett's multi comparison test (also known as ANOVA) was carried out. On the other hand, the findings did not change, even when considering samples with a greater number of participants ($n > 30$).

4.7 Risk Assessment

Environmental hazards presented by members of congress have been assessed using two previously developed indices: the hazard index (HI) and the pollutant load index (PLI). Supplementary data shows whether or not the two procedures are scientifically and ethically acceptable (Picó et al. 2021). The transfer of risk was initiated from fresh water, air, wind, and wastewater to the aquatic animals, where the aquatic animals consume MPs as food and these MPs are ingested in their stomach. Aquatic animals/species, such as fish and crab, are consumed by humans and this can also cause death in humans. The amount of risk related to the consumption of MPs was calculated using the following Eq. (4.1) (Yozukmaz 2021).

$$R = C * I \tag{4.1}$$

where R is the estimated risk due to MPs ingestion; C is the concentration of MPs identified in the sample; and I is the rate of ingestion of MPs.

Although human and ecological risk evaluations for microplastics are ongoing, more study is required for a standard of concentration and accuracy.

4.8 Future Challenges in Quantification of the Environment

As the use of MPs in everyday life has risen, no adequate corrective action or evaluation has been taken, which has had a worldwide impact (Boucher and Friot 2017). It is vital to look into the presence of MPs in wastewater and their route to WWTPs. Many studies have been done on MPs, their fate and pathways, and the effects they have on WWTPs, ecosystems, and humans. There are still several years until large-scale, thorough examination, and removal of MPs from WWTPs can be achieved (Freeman et al. 2020). In order to sample, pre-treat, or characterize wastewater MPs, there is no one specified method. There are a variety of ways to sample, pre-treat, purify, and identify MPs in wastewater (Iyare et al. 2020). More research is needed on the effects of sludge disposal with MPs on land animals, since the majority of MPs in WWTPs are trapped in biological sludge (Bretas Alvim et al. 2020). There has been a growing awareness of the long-term dangers of plastic use and the global need to minimize plastic use and promote eco-friendly materials that may disintegrate in the environment after their disposal by researchers, the general public, and politicians worldwide. Different investigations on the incidence, fate, and transportation of MPs in WWTP effluent samples have been conducted; however, no unique kinetics for eliminating MPs from WWTP effluents have been identified.

4.9 Conclusions

Environmental pollutants/contaminants have harmed human and ecological health. Pharmaceuticals and personal care products (PPCPs), persistent organic pollutants (POPs), steroids, nanomaterials, cosmetic items, and endocrine-disrupting compounds (EDCs) are common environmental contaminants. These contaminants threaten human health and the environment. Nanoscience and nanotechnology are gaining prominence worldwide. Emerging pollutants include natural or semi-synthetic toxic chemicals and their transition or breakdown in nature, such as metal nanoparticles. MPs are increasingly being used in goods, which raises the likelihood that they may end up in water and wastewater worldwide. Discharging MPs into wastewater has a number of negative consequences. Toxicological studies have shown that water containing minimal amounts of MPs may be recycled and reused after being removed from wastewater. However, if the concentration of MPs is high, MPs reduce the effectiveness of WWTPs by inhibiting microbial development. One of the greatest threats to aquatic ecosystems, microbes, and human health is the widespread presence of MPs in the environment. Even though there are many other sources of MPs, WWTPs are the primary source of MPs released into the aquatic environment. In addition to their removal and treatment, it is important to study the effects of these pollutants on aquatic creatures, such as fish, bacteria, and other zooplankton in freshwater bodies. Research on MPs worldwide is reviewed in this chapter.

References

Agüera, A., Martínez Bueno, M.J., and Fernández-Alba, A.R. (2013). New trends in the analytical determination of emerging contaminants and their transformation products in environmental waters. *Environmental Science and Pollution Research* 20: 3496–3515. doi: 10.1007/s11356-013-1586-0.

Ambrosini, R., Azzoni, R.S., Pittino, F. et al. (2019). First evidence of microplastic contamination in the supraglacial debris of an alpine glacier. *Environmental Pollution* 253: 297–301. doi: 10.1016/j.envpol.2019.07.005.

Bakaraki Turan, N., Sari Erkan, H., and Onkal Engin, G. (2021). Microplastics in wastewater treatment plants: occurrence, fate and identification. *Process Safety and Environmental Protection* 146: 77–84. doi: 10.1016/j.psep.2020.08.039.

Bharath, K., M., Srinivasalu, S., Natesan, U. et al. (2021). Microplastics as an emerging threat to the freshwater ecosystems of Veeranam Lake in south India: a multidimensional approach. *Chemosphere* 264: 128502. doi: 10.1016/j.chemosphere.2020.128502.

Bilal, M. and Iqbal, H.M.N. (2020). Transportation fate and removal of microplastic pollution: a perspective on environmental pollution. *Case Studies in Chemical and Environmental Engineering* 2: 100015. doi: 10.1016/j.cscee.2020.100015.

Boucher, J. and Friot, D. (2017). Primary microplastics in the oceans, marine environmental research. *IUCN Library System* 111.

Brar, S.K., Verma, M., Tyagi, R.D., and Surampalli, R.Y. (2010). Engineered nanoparticles in wastewater and wastewater sludge: evidence and impacts. *Waste Management* 30: 504–520. doi: 10.1016/j.wasman.2009.10.012.

Bretas Alvim, C., Mendoza-Roca, J.A., and Bes-Piá, A. (2020). Wastewater treatment plant as microplastics release source: quantification and identification techniques. *The Journal of Environmental Management* 255: 109739; https://doi.org/10.1016/j.heliyon.2020.e04691.

Briffa, J., Sinagra, E., and Blundell, R. (2020). Heavy metal pollution in the environment and their toxicological effects on humans. *Heliyon* 6: pe04691. http://doi.org/10.1016/j.heliyon.2020.e04691.

Bui, X.T., Vo, T.D.H., Nguyen, P.T. et al. (2020). Microplastics pollution in wastewater: characteristics, occurrence and removal technologies. *Environmental Technology & Innovation* 19: 101013. doi: 10.1016/j.eti.2020.101013.

Buta, M., Hubeny, J., Zieliński, W. et al. (2021). Sewage sludge in agriculture: the effects of selected chemical pollutants and emerging genetic resistance determinants on the quality of soil and crops: a review. *Ecotoxicology and Environmental Safety* 214: 112070. 10.1016/j.ecoenv.2021.112070.

Bystrzejewska-Piotrowska, G., Golimowski, J., and Urban, P.L. (2009). Nanoparticles: their potential toxicity, waste and environmental management. *Waste Management* 29: 2587–2595. https://doi.org/10.1016/j.ecoenv.2021.112070.

Chandra, P., Enespa, and Singh, D.P. (2020). Microplastic degradation by bacteria in aquatic ecosystem, microorganisms for sustainable environment and health. In: *Microorganisms for Sustainable Environment and Health,* 431–467. Elsivier. https://doi.org/10.1016/B978-0-12-819001-2.00022-X.

Cheng, Y., Chen, Z., Yang, S. et al. (2021). Nanomaterials-induced toxicity on cardiac myocytes and tissues, and emerging toxicity assessment techniques. *The Science of the Total Environment* 800: 149584. doi: 10.1016/j.scitotenv.2021.149584.

Czech, B., Shirvanimoghaddam, K., Trojanowska, E., and Naebe, M. (2020). Sorption of pharmaceuticals and personal care products (PPCPs) onto a sustainable cotton based adsorbent. *Sustainable Chemistry and Pharmacy* 18: 100324. https://doi.org/10.1016/j.scp.2020.100324.

Dalberg Advisors, de Wit, W., and Bigaud, N. (2019). No plastic in nature: assessing plastic ingestion from nature to people (an Analysis for WWF). *World Wide Fund Nature.* 7.

Daniel, D.B., Ashraf, P.M., and Thomas, S.N. (2020a). Abundance, characteristics and seasonal variation of microplastics in Indian white shrimps (*Fenneropenaeus indicus*) from coastal waters off Cochin, Kerala, India. *The Science of the Total Environment* 737: 139839. doi: 10.1016/j.scitotenv.2020.139839.

Daniel, D.B., Thomas, S.N., and Thomson, K.T. (2020b). Assessment of fishing-related plastic debris along the beaches in Kerala Coast, India. *Marine Pollution Bulletin* 150: 110696. doi: 10.1016/j.marpolbul.2019.110696.

Deblonde, T., Cossu-Leguille, C., and Hartemann, P. (2011). Emerging pollutants in wastewater: a review of the literature. *International Journal of Hygiene and Environmental Health* 214: 442–448. doi: 10.1016/j.ijheh.2011.08.002.

Dhangar, K. and Kumar, M. (2020). Tricks and tracks in removal of emerging contaminants from the wastewater through hybrid treatment systems: a review. *The Science of the Total Environment* 738: 140320. doi: 10.1016/j.scitotenv.2020.140320.

Environment Protection Authority (2016). Plastic shopping bags – options paper.

Eriksen, M., Mason, S., Wilson, S. et al. (2013). Microplastic pollution in the surface waters of the Laurentian Great Lakes. *Marine Pollution Bulletin* 77: 177–182. doi: 10.1016/j.marpolbul.2013.10.007.

Fischer, E.K., Paglialonga, L., Czech, E., and Tamminga, M. (2016). Microplastic pollution in lakes and lake shoreline sediments: a case study on Lake Bolsena and Lake Chiusi (central Italy). *Environmental Pollution* 213: 648–657. doi: 10.1016/j.envpol.2016.03.012.

Freeman, S., Booth, A.M., Sabbah, I. et al. (2020). Between source and sea: the role of wastewater treatment in reducing marine microplastics. *Journal of Environmental Management* 266: 110642. doi: 10.1016/j.jenvman.2020.110642.

Fu, Z. and Wang, J. (2019). Current practices and future perspectives of microplastic pollution in freshwater ecosystems in China. *The Science of the Total Environment* 691: 697–712. doi: 10.1016/j.scitotenv.2019.07.167.

Galindo-Miranda, J.M., Guízar-González, C., Becerril-Bravo, E.J. et al. (2019). Occurrence of emerging contaminants in environmental surface waters and their analytical methodology: a review. *Water Science and Technology: Water Supply* 19: 1871–1884. doi: 10.2166/ws.2019.087.

He, D., Luo, Y., Lu, S. et al. (2018). Microplastics in soils: analytical methods, pollution characteristics and ecological risks. *TrAC – Trends in Analytical Chemistry* 109: 163–172. doi: 10.1016/j.trac.2018.10.006.

Huerta Lwanga, E., Mendoza Vega, J., Ku Quej, V. et al. (2017). Field evidence for transfer of plastic debris along a terrestrial food chain. *Scientific Reports* 7: 1–7. doi: 10.1038/s41598-017-14588-2.

Iyare, P.U., Ouki, S.K., and Bond, T. (2020). Microplastics removal in wastewater treatment plants: a critical review. *Environmental Science: Water Research & Technology* 6: 2664–2675. doi: 10.1039/d0ew00397b.

Jeyasanta, K.I., Sathish, N., Patterson, J., and Edward, J.K.P. (2020). Macro-, meso- and microplastic debris in the beaches of Tuticorin district, southeast coast of India. *Marine Pollution Bulletin* 154: 111055. doi: 10.1016/j.marpolbul.2020.111055.

Karuppasamy, P.K., Ravi, A., Vasudevan, L. et al. (2020). Baseline survey of micro and mesoplastics in the gastro-intestinal tract of commercial fish from southeast coast of the Bay of Bengal. *Marine Pollution Bulletin* 153: 110974. doi: 10.1016/j.marpolbul.2020.110974.

Khraisheh, M., Kim, J., Campos, L. et al. (2014). Removal of pharmaceutical and personal care products (PPCPs) pollutants from water by novel TiO_2-Coconut Shell Powder (TCNSP) composite. *Journal of Industrial and Engineering Chemistry* 20: 979–987. doi: 10.1016/j.jiec.2013.06.032.

Koutnik, V.S., Alkidim, S., Leonard, J. et al. (2021). Unaccounted microplastics in wastewater sludge: where do they go? *ACS ES&T Water* 1: 1086–1097. doi: 10.1021/acsestwater.0c00267.

Kumar, V.E., Ravikumar, G., and Jeyasanta, K.I. (2018). Occurrence of microplastics in fishes from two landing sites in Tuticorin, southeast coast of India. *Marine Pollution Bulletin* 135: 889–894. doi: 10.1016/j.marpolbul.2018.08.023.

Lares, M., Ncibi, M.C., Sillanpää, M., and Sillanpää, M. (2018). Occurrence, identification and removal of microplastic particles and fibers in conventional activated sludge process and advanced MBR technology. *Water Research* 133: 236–246. doi: 10.1016/j.watres.2018.01.049.

Li, X., Chen, L., Mei, Q. et al. (2018). Microplastics in sewage sludge from the wastewater treatment plants in China. *Water Research* 142: 75–85. doi: 10.1016/j.watres.2018.05.034.

Li, Y., Li, W., Jarvis, P. et al. (2020). Occurrence, removal and potential threats associated with microplastics in drinking water sources. *The Journal of Environmental Chemical Engineering* 8: 104527. doi: 10.1016/j.jece.2020.104527.

Liu, C., Li, J., Zhang, Y. et al. (2019). Widespread distribution of PET and PC microplastics in dust in urban China and their estimated human exposure. *Environment International* 128: 116–124. doi: 10.1016/j.envint.2019.04.024.

Liu, G., Jiang, R., You, J. et al. (2020). Microplastic impacts on microalgae growth: effects of size and humic acid. *Environmental Science & Technology* 54: 1782–1789. doi: 10.1021/acs.est.9b06187.

Liu, Q., Zhou, Q., and Jiang, G. (2014). Nanomaterials for analysis and monitoring of emerging chemical pollutants. *TrAC – Trends in Analytical Chemistry* 58: 10–22. doi: 10.1016/j.trac.2014.02.014.

López, A.D.F., Fabiani, M., Lassalle, V.L. et al. (2022). Critical review of the characteristics, interactions, and toxicity of micro/nanomaterials pollutants in aquatic environments. *Marine Pollution Bulletin* 174: 113276. https://doi.org/10.1016/j.marpolbul.2021.113276.

Mammo, F.K., Amoah, I.D., Gani, K.M. et al. (2020). Microplastics in the environment: interactions with microbes and chemical contaminants. *The Science of the Total Environment* 743: 140518. doi: 10.1016/j.scitotenv.2020.140518.

Mantilaka, M.M.M.G.P.G., Wijesinghe, W.P.S.L., Dissanayake, D.M.S.N. et al. (2019). Current review on the utilization of nanoparticles for ceramic matrix reinforcement. In: *Interfaces in Particle and Fibre Reinforced Composites: Current Perspectives on Polymer, Ceramic, Metal and Extracellular Matrices*, 345–367. Woodhead Publishing. https://doi.org/10.1016/B978-0-08-102665-6.00013-3.

Meng, Y., Kelly, F.J., and Wright, S.L. (2020). Advances and challenges of microplastic pollution in freshwater ecosystems: a UK perspective. *Environmental Pollution* 256: 113445. doi: 10.1016/j.envpol.2019.113445.

Michels, J., Stippkugel, A., Lenz, M., Wirtz, K., and Engel, A. (2018). Rapid aggregation of biofilm-covered microplastics with marine biogenic particles. *Proceedings of the Royal Society B Biological Sciences* 285: 1885. https://doi.org/10.1098/rspb.2018.1203.

Miranda, M.N., Silva, A.M.T., and Pereira, M.F.R. (2020). Microplastics in the environment: a DPSIR analysis with focus on the responses. *The Science of the Total Environment* 718: 134968. doi: 10.1016/j.scitotenv.2019.134968.

Naidu, S.A. (2019). Preliminary study and first evidence of presence of microplastics and colorants in green mussel, *Perna viridis* (Linnaeus, 1758), from southeast coast of India. *Marine Pollution Bulletin* 140: 416–422. doi: 10.1016/j.marpolbul.2019.01.024.

Naidu, S.A., Ranga Rao, V., and Ramu, K. (2017). Microplastics in the benthic invertebrates from the coastal waters of Kochi, southeastern Arabian Sea. *Environmental Geochemistry and Health* 40: 1377–1383. doi: 10.1007/s10653-017-0062-z.

Narmadha, V.V., Jose, J., Patil, S. et al. (2020). Assessment of microplastics in roadside suspended dust from urban and rural environment of Nagpur, India. *International Journal of Environmental Research* 14: 629–640. doi: 10.1007/s41742-020-00283-0.

Nava, V. and Leoni, B. (2021). A critical review of interactions between microplastics, microalgae and aquatic ecosystem function. *Water Research* 188: 116476. doi: 10.1016/j.watres.2020.116476.

Ngo, P.L., Pramanik, B.K., Shah, K., and Roychand, R. (2019). Pathway, classification and removal efficiency of microplastics in wastewater treatment plants. *Environmental Pollution* 255: 113326. doi: 10.1016/j.envpol.2019.113326.

Oberbeckmann, S., Löder, M.G.J., and Labrenz, M. (2015). Marine microplastic-associated biofilms: a review. *Environmental Chemistry* 12(5): 551–562. https://doi.org/10.1071/EN15069.

Okoye, C.O., Okeke, E.S., Okoye, K.C. et al. (2022). Occurrence and fate of pharmaceuticals, personal care products (PPCPs) and pesticides in African water systems: a need for timely intervention. *Heliyon* 8: e.09143. https//doi,org/10.1016/j.heliyon.2022.e09143.

Padervand, M., Lichtfouse, E., Robert, D., and Wang, C. (2020). Removal of microplastics from the environment: a review. *Environmental Chemistry Letters* 18: 807–828. doi: 10.1007/s10311-020-00983-1.

Park, H. and Park, B.S. (2021). Review of microplastic distribution, toxicity, analysis methods, and removal technologies. *Water (Switzerland)* 2021: 2736. https://doi.org/10.3390/w13192736

Parrish, K. and Fahrenfeld, N.L. (2019). Microplastic biofilm in fresh- and wastewater as a function of microparticle type and size class. *Environmental Science: Water Research & Technology* 5: 495–505. doi: 10.1039/c8ew00712h.

Picó, Y., Soursou, V., Alfarhan, A.H. et al. (2021). First evidence of microplastics occurrence in mixed surface and treated wastewater from two major Saudi Arabian cities and assessment of their ecological risk. *The Journal of Hazardous Materials* 416: 125747. https//doi.org/10.1016/j.jhazmat.2021.125747.

Reichelt, S. and Gorokhova, E. (2020). Micro- and nanoplastic exposure effects in microalgae: a meta-analysis of standard growth inhibition tests. *Frontiers in Environmental Science* 8: 1–17. doi: 10.3389/fenvs.2020.00131.

Rezania, S., Park, J., Md Din, M.F. et al. (2018). Microplastics pollution in different aquatic environments and biota: a review of recent studies. *Marine Pollution Bulletin* 133: 191–208. doi: 10.1016/j.marpolbul.2018.05.022.

Rogers, K.L., Carreres-Calabuig, J.A., Gorokhova, E., and Posth, N.R. (2020). Micro-by-micro interactions: how microorganisms influence the fate of marine microplastics. *Limnology and Oceanography Letters* 5: 18–36. doi: 10.1002/lol2.10136.

Rummel, C.D., Jahnke, A., Gorokhova, E. et al. (2017). Impacts of biofilm formation on the fate and potential effects of microplastic in the aquatic environment. *Environmental Science & Technology Letters* 4: 258–267. doi: 10.1021/acs.estlett.7b00164.

Samal, K., Mahapatra, S., and Hibzur Ali, M. (2022). Pharmaceutical wastewater as Emerging Contaminants (EC): treatment technologies, impact on environment and human health. 6: 100076 https://doi.org/10.1016/j.nexus.2022.100076.

Sarkar, D.J., Das Sarkar, S., Das, B.K. et al. (2019). Spatial distribution of meso and microplastics in the sediments of River Ganga at eastern India. *The Science of the Total Environment* 694: 133712. doi: 10.1016/j.scitotenv.2019.133712.

Schmidt, C., Kumar, R., Yang, S., and Büttner, O. (2020). Microplastic particle emission from wastewater treatment plant effluents into river networks in Germany: loads, spatial patterns of concentrations and potential toxicity. *The Science of the Total Environment* 737: 139544. doi: 10.1016/j.scitotenv.2020.139544.

Seth, C.K. and Shriwastav, A. (2018). Contamination of Indian sea salts with microplastics and a potential prevention strategy. *Environmental Science and Pollution Research* 25: 30122–30131. doi: 10.1007/s11356-018-3028-5.

Sharma, S., Basu, S., Shetti, N.P. et al. (2021). Microplastics in the environment: occurrence, perils, and eradication. *The Chemical Engineering Journal* 408: 127317. doi: 10.1016/j.cej.2020.127317.

Shen, M., Song, B., Zhu, Y. et al. (2020). Removal of microplastics via drinking water treatment: current knowledge and future directions. *Chemosphere* 251: 126612. doi: 10.1016/j.chemosphere.2020.126612.

Sol, D., Laca, A., Laca, A., and Díaz, M. (2020). Approaching the environmental problem of microplastics: importance of WWTP treatments. *The Science of the Total Environment* 740: 140016. doi: 10.1016/j.scitotenv.2020.140016.

Sridharan, S., Kumar, M., Singh, L. et al. (2021). Microplastics as an emerging source of particulate air pollution: a critical review. *Journal of Hazardous Materials* 418: 126245. doi: 10.1016/j.jhazmat.2021.126245.

Talvitie, J., Mikola, A., Koistinen, A., and Setälä, O. (2017). Solutions to microplastic pollution: removal of microplastics from wastewater effluent with advanced wastewater treatment technologies. *Water Research* 123: 401–407. doi: 10.1016/j.watres.2017.07.005.

Tu, C., Zhou, Q., Zhang, C. et al. (2020). Biofilms of microplastics biofilms of microplastics. In: *Handbook of Environmental Chemistry*, vol. 95: 299–317. Springer Science and Business Media Deutschland GmbH. https://doi.org/10.1007/698_2020_461.

Tuson, H.H. and Weibel, D.B. (2013). Bacterior–surface interactions. In: *Soft Matter* May 2013: 4368–4380. https://doi.org/10.1039/c3sm27705d.

Veerasingam, S., Ranjani, M., Venkatachalapathy, R. et al. (2020). Microplastics in different environmental compartments in India: analytical methods, distribution, associated contaminants and research needs. *TrAC – Trends in Analytical Chemistry* 133: 116071. doi: 10.1016/j.trac.2020.116071.

Viršek, M.K., Lovšin, M.N., Koren, Š. et al., 2017. Microplastics as a vector for the transport of the bacterial fish pathogen species *Aeromonas salmonicida*. Marine Pollution Bulletin 125(1–2): 301–309. https://doi.org/10.1016/j.marpolbul.2017.08.024.

Visanji, Z., Sadr, S.M.K., Johns, M. et al. (2018). Emerging pollutants in developing countries. *Optimising Sustainable Treatment Solutions* 3: 2204–2191. doi: 10.29007/txmw.

Wang, C., Zhao, J., and Xing, B. (2020). Environmental source, fate, and toxicity of microplastics. *The Journal of Hazardous Materials* 124357. doi: 10.1016/j.jhazmat.2020.124357.

Wang, F., Zhang, X., Zhang, S. et al. (2020). Interactions of microplastics and cadmium on plant growth and arbuscular mycorrhizal fungal communities in an agricultural soil. *Chemosphere* 254: 126791. doi: 10.1016/j.chemosphere.2020.126791.

Wang, W., Ge, J., and Yu, X. (2020). Bioavailability and toxicity of microplastics to fish species: a review. *Ecotoxicology and Environmental Safety* 189: 109913. doi: 10.1016/j.ecoenv.2019.109913.

Wei, S., Luo, H., Zou, J. et al. (2020). Characteristics and removal of microplastics in rural domestic wastewater treatment facilities of China. *The Science of the Total Environment* 739: 139935. doi: 10.1016/j.scitotenv.2020.139935.

Westphalen, H. and Abdelrasoul, A. (2018). Challenges and treatment of microplastics in water. In: *Water Challenges of an Urban World. IntechOpen.* https//doi.org/10.5772/intechopen.71494.

Wu, M., Tang, W., Wu, S. et al. (2021). Fate and effects of microplastics in wastewater treatment processes. *The Science of the Total Environment* 757: 143902. doi: 10.1016/j.scitotenv.2020.143902.

Wu, P., Huang, J., Zheng, Y. et al. (2019). Environmental occurrences, fate, and impacts of microplastics. *Ecotoxicology and Environmental Safety* 184: 109612. doi: 10.1016/j.ecoenv.2019.109612.

Xu, S., Ma, J., Ji, R. et al. (2020). Microplastics in aquatic environments: occurrence, accumulation, and biological effects. *The Science of the Total Environment* 703: 134699. doi: 10.1016/j.scitotenv.2019.134699.

Yozukmaz, A. (2021). Investigation of microplastics in edible wild mussels from İzmir Bay (Aegean Sea, Western Turkey): a risk assessment for the consumers. *Marine Pollution Bulletin* 171: 112733. doi: 10.1016/j.marpolbul.2021.112733.

Zhang, K., Xiong, X., Hu, H. et al. (2017). Occurrence and characteristics of microplastic pollution in Xiangxi Bay of Three Gorges Reservoir, China. *Environmental Science & Technology* 51: 3794–3801. doi: 10.1021/acs.est.7b00369.

Ziajahromi, S., Neale, P.A., and Leusch, F.D.L. (2016). Wastewater treatment plant effluent as a source of microplastics: review of the fate, chemical interactions and potential risks to aquatic organisms. *Water Science and Technology* 74: 2253–2269. doi: 10.2166/wst.2016.414.

Part II

The Fate and Transportation of Microconstituents

5

Mathematical Transport System of Microconstituents

Dwarikanath Ratha[1], Richa Babbar[1], K.S. Hariprasad[2], C.S.P. Ojha[2], Manoj Baranwal[3], Prangya Ranjan Rout[3,4], and Aditya Parihar[1]

[1]*Department of Civil Engineering, Thapar Institute of Engineering & Technology, Patiala, Punjab, India*
[2]*Department of Civil Engineering, Indian Institute of Technology Roorkee, India*
[3]*Department of Biotechnology, Thapar Institute of Engineering & Technology, Patiala, Punjab, India*
[4]*Department of Biotechnology, Dr. B. R. Ambedkar National Institute of Technology Jalandhar, Punjab, India*

5.1 Introduction

The world cannot be sustained without water and its sustainability is posing a serious challenge for researchers. Since surface water is filtered while passing through porous media during the process of infiltration, the quality of groundwater is usually considered as better than that of surface water. But the sustainability of both groundwater and surface water are seen as under threat due to continuous degradation of their natural quality. Industrialization and agricultural activities have increased in the last three decades due to urbanization and rapid rise in the human population. Pesticides and fertilizers applied onto agricultural fields are filtered through irrigation water and degrade the soil and groundwater quality. Similarly, the wastes discharged from various industries transport through runoff/infiltered rainwater and degrade the natural quality of both surface and groundwater. According to the United States Geological Survey (USGS), any chemical of natural or synthetic origin, or microorganisms that have the potential to have a harmful effect on humans and ecology, but cannot be monitored easily in the environment, can be termed as emerging contaminants (USGS (US Geological Survey) 2017). Therefore, microbial contaminants, such as antibacterial drugs mediated antibiotic resistance genes/bacteria, can be termed as emerging microbial constituents (Rout et al. 2021).

Microconstituents (MCs) are natural or synthetic compounds which when present in drinking water can have an adverse effect on public health through toxicity or disease. The constituents that include pharmaceutical products, personal care products, endocrine disrupting compounds, herbicides, etc. (Ying and Droste 2015) are termed as microconstituents (MCs). The wastes from these industries, when dumped on land, become the source of MCs in surface water or groundwater. The presence of MCs in both surface water and groundwater has been a concern for the last three decades. The toxic substances and

Microconstituents in the Environment: Occurrence, Fate, Removal, and Management, First Edition.
Edited by Rao Y. Surampalli, Tian C. Zhang, Chih-Ming Kao, Makarand M. Ghangrekar, Puspendu Bhunia, Manaswini Behera, and Prangya R. Rout.

emerging biological contaminants generated from these MCs enter the water body through either surface runoff or the infiltration processes of industrial waste discharges, effluent of wastewater treatment plants, agricultural operations, etc. The biological contaminants have an adverse impact on the quality of water, soil, and sediments. The term "microconstituents" was initiated by the Water Environment Federation (WEF). Table 5.1 lists the various chemical and biological MCs. The types of health effects caused in humans due to the presence of biological MCs in potable water are also listed in Table 5.1. The presence of biological MCs in the potable water sources make water unsuitable for its use and poses challenges for water resources managers. The health risks caused due to the presence of MCs in water depends on their transport and survival. These contaminants are released either onto the land surface or at a certain depth below the ground surface, either naturally or by man-made activities. In certain cases, these contaminants can also be released onto the ground surface by design or accidentally. Later, these contaminants transport through unsaturated soil media and join the groundwater through various routes such as infiltration, direct migration, artificial recharge, inter aquifer exchange, etc. The major anthropogenic sources along with the reason for contamination of ground or surface water are as follows: i) Municipal (sewer leakage, liquid waste, solid waste, landfill, etc.); ii) Industrial (liquid waste, leakage, mining activity, oil field brine, etc.); iii) Agricultural (return flows, fertilizer and pesticides, animal waste, etc.); and iv) Miscellaneous (saltwater intrusion, acid rain, etc.) (Chahar 2015). Based on the geometry of the source or cause of pollution, all the sources can be categorized into point source, line source, and diffused or area source. If the contaminants originate from a single location, it is known as the point source, whereas if the contaminants originate from a linear alignment, it is known as the line source. The diffused or area source considers an undefined extensive area. The geometry of the contamination source helps in a better understanding of plume behavior and its dynamics during its transportation. Light Non-Aqueous Phase Liquids (LNAPL), Dense Non-Aqueous Phase Liquids (DNAPL), and inorganic and other dissolved constituents are the three main categories of point source groundwater contamination. The contamination of groundwater or surface water can be classified based on the mechanism of contaminant transport within a system, which further helps in selecting a specific numerical model for determination of variation of concentration with respect to space and time. They are conservative/non-conservative, single phase/multi-phase, and unstratified/stratified.

The presence of toxic substances or biological contaminants such as bacteria, viruses, and protozoa in drinking water causes adverse effects on human health. However, it is difficult to completely eliminate the source activities, and the subsurface will continue to receive an increasing quantity of contaminants. Since the source activities cannot be eliminated, it is necessary to protect the groundwater and surface water from the source of contamination. It is observed that the concentration of contaminant with respect to space and time during its movement depends on its size, adsorption characteristics, degree of reaction, and inactivation. Groundwater is believed to be of purer form than surface water but once the aquifer is contaminated due to natural or anthropogenic activities, the restoration of its quality will take decades, mainly due to long residence times. Since biological contaminants come under MCs, this chapter considers the governing equations for transportation of biological contaminants.

Table 5.1 Chemical and biological microconstituents.

Chemical	Biological		
	Bacteria (Human Health Effects)	**Virus (Human Health Effects)**	**Protozoa (Human Health Effects)**
• Alkylphenols • Antibiotics • Disinfecton by-products • Endocrine Disruptors • Estrogenic compounds • Heavy metals • NDMA • Personal care products • Pharmaceuticals • Polybrominated Diphenyl Ethers	• Aeromonas (Gastroenteritis, Septicemia) • Cyanobacteria (Gastroenteritis, Respiratory effects, Skin irritations, Allergic responses, and Liver damage). • *Helicobacter* (Peptic Ulcers, Gastritis). • *Legionella* (Severe Pneumonia, Gastro intestinal symptoms). • *Mycobacterium avium* complex (Pulmonary infection, Cervical lymphadenitis, Chrohn's disease). • Pathogenic *E. Coli.*	• Adenovirus (Gastroenteritis). • Calcivirus (Gastroenteritis). • Coxsackievirus (Meningitis, Myocarditis, Gastroenteritis and Hepatitis). • Echovirus (Diarrhea, Common cold, Otitis).	• *Cryptosporidium* (Chronic diarrhea). • *Giardia* (Chronic diarrhea). • *Microsporidia* spp. (Gastroenteritis, Eye infections, Respiratory infection, Genitourinary track infection). • Toxoplasma.

Source: Adapted summarized from Robbles et al. (2006).

The use of mathematical models is preferred to experimental simulation and also captures the attention of researchers for solving the problems in all naturally occurring phenomena. A mathematical model usually describes a system by a set of variables and equations that establish relationships between those variables. Numerical simulations of mathematical models, representing the transport equation, can be conducted to predict the concentration with respect to space and time by developing a computer program and this program can be used to carry out numerical experiments. These numerical experiments are usually carried out in parallel to the physical experiment. This is to ensure model validation and to interpret physical experiments along with some real-life aspect of experimental conditions and physics behind the transport. Moreover, the numerical simulation can be used as a design tool for carrying out the research for a complex real-life problem at minimum cost.

5.2 Need for Mathematical Models

A mathematical model is a representation of a real-life problem based on assumptions and hypotheses. Mathematical models representing subsurface hydrology are usually in the form of differential equations along with initial and boundary conditions that describe the hydraulic relations within the system. The physical behavior and mechanism of such problems can be understood through analyzing the results obtained from experimental studies. With the use of mathematical models, understanding of the mechanism of sub-surface systems is convenient and economical in comparison to conducting experiments. Also, the complex geometry, initial, and boundary conditions make the experiments a very tedious job. Therefore, the development of mathematical models for sub-surface flow problems and their solution through analytical equations or numerical simulations has gained more popularity than experimental simulations.

5.3 Fundamentals of Pollutant Transport Modeling

The movement and accumulation of pollutants in the groundwater (solvent), either in the saturated or unsaturated porous medium, is represented in the form of a partial differential equation. Given the specific initial and boundary conditions, the variation of concentration of pollutants at any time can be predicted. A conservation of the mass statement for an elementary volume is a first step toward describing the mass transport through porous media using the following partial formula.

$$\begin{aligned}&\left[\textit{Net rate of change of mass of pollutant within the volume}\right]\\&\quad=\left[\textit{Flux of pollutant out of the volume}\right]-\left[\textit{Flux of pollutant into the volume}\right]\\&\quad\pm\left[\textit{Loss or gain of pollutant due to the reactions}\right]\end{aligned}$$

The contaminant released onto the land surface, either due to natural or anthropogenic activities, starts to migrate through the upper crust of soil during rainwater infiltration and ultimately joins the groundwater table. The migration of contaminant starts initially through the unsaturated zone and its transportation through the saturated zone starts only when the contaminant reach the aquifer. The transportation mechanism in the unsaturated zone is a complex process due to non-linearity of flow in this zone (Persson and Berndtsson 1998). Since the velocity of infiltered water is a function of degree of saturation, the flow and transport equation are required to be solved one after another to determine the variation of concentration with respect to space and time. Various transport mechanisms are as follows: i) advection; ii) diffusion; iii) dispersion; iv) sorption; v) decay; vi) physical, natural, or anthropogenic activities; vii) hydrolysis, volatilization, and bio-transformation; viii) transport in aquifers with a pronounced bimodal permeability distribution; and ix) chemical reaction (Chahar 2015). The subsurface fate and transport of microbes is broadly governed by the processes of convection, hydrodynamic dispersion, sedimentation, motility and chemotaxis, growth and decay, and particle–surface interactions (Scheibe and Wood 2003).

Advection or convection is the process of contaminant movement attributed to movement of bulk of water, in which the hydraulic gradient is the driving force. This process of transport is predominant in highly permeable materials such as sand and gravel. The steady state one-dimensional (1D) mass flux of the solute per unit area per unit time (F_x) due to advection can be expressed as per Eq. (5.1).

$$F_a = v_x n C = -K \frac{dh}{dx} C \tag{5.1}$$

where v_x is the seepage velocity in the x-direction; n is the medium; K is the hydraulic conductivity of the medium; C is the contaminant concentration; and dh/dx is the hydraulic gradient in the x-direction. The advection process for unsteady flow can be represented by Eq. (5.2).

$$\frac{\partial C}{\partial t} = -v_x \frac{\partial C}{\partial x} \tag{5.2}$$

During the contaminant transportation process, the contaminants have a tendency to spread out from the expected path they would have followed due to advection. This phenomenon is known as hydrodynamic dispersion. The spreading process causes the mixing of the contaminant with water that leads to its dilution. This hydrodynamic dispersion process during contaminant transport occurs due to mechanical mixing during advection and molecular diffusion due to the thermal-kinetic energy of the solute particles (Freeze and Cherry 1997). The mechanical dispersion occurs due to difference in pore size, non-uniform flow velocity, and tortuosity of pore channels in the porous medium.

The process in which molecular constituents in a solution start moving in the direction of concentration gradient, i.e., the molecular motion in a solution occurs from a region of higher concentration to lower concentration, is called diffusion. According to Fick's first law, the mass of diffusing substance passing through a given cross-section per unit time (F) is proportional to the concentration gradient (dc/dx) and can be expressed by Eq. (5.3).

$$F_d = -D \frac{dC}{dx} \tag{5.3}$$

where D is the diffusion coefficient.

Fick's second law provides a differential equation which relates the concentration of a diffusing substance to space and time and the expression is shown in Eq. (5.4).

$$\frac{\partial C}{\partial t} = D \frac{\partial^2 C}{\partial x^2} \tag{5.4}$$

The transport of conservative (non-reactive) dissolved constituents in saturated, homogeneous, isotropic, steady state, and uniform flow can be expressed by the advection-dispersion equation. The 1D form of the advection-dispersion equation is represented by Eq. (5.5).

$$\frac{\partial C}{\partial t} = D_{hx} \frac{\partial^2 C}{\partial x^2} - v_x \frac{dC}{dx} \tag{5.5}$$

where the coefficient of hydrodynamic dispersion (D_{hx}) is the summation of mechanical dispersion and molecular diffusion that can be represented by Eq. (5.6).

$$D_{hx} = D_m + D^* = \alpha_l v_x + D^* \tag{5.6}$$

where α_l is the longitudinal dynamic dispersivity; and D^* is the diffusion coefficient.

Inactivation or decay is also a mechanism which affects the transport of non-conservative (reactive) dissolved constituents through saturated and unsaturated porous media. The first-order kinetic equation that represents the inactivation of dissolved constituents can be represented by Eq. (5.7).

$$\frac{dC}{dt} = -\lambda C \tag{5.7}$$

where λ is inactivation constant. Table 5.2 shows the inactivation rate coefficients of different types of viruses in groundwater and wastewater.

The 1D form of contaminant transport equation considering advection, dispersion, and inactivation can be represented by Eq. (5.8) (Runkel and Bencala 1995).

$$\frac{\partial C}{\partial t} = D_{hx}\frac{\partial^2 C}{\partial x^2} - v_x\frac{\partial C}{\partial x} - \lambda C \tag{5.8}$$

If there is no inactivation, Eq. (5.8) will become the advection-dispersion equation, as in Eq. (5.1).

Sorption is another contaminant transport mechanism which tends to retard the movement. Sorption is a chemical and physical process that implies an exchange of

Table 5.2 Inactivation rate coefficient of viruses in groundwater and wastewater.

^{0}C	Water	MS2	FRNAPH	PRD1	POLIO1	ECHO1	HAV
4	GRW	0.063					
7	GRW	0.0058		0.01			
7	GRW	0.10		0.10			
10	GRW	0.11		0.025	0.11		0.19
10	GRW[a]				0.010		
10	GRW-deionized				0.032		
10	GRW				0.013		
12	GRW	0.16			0.18	0.24	
13	GRW	0.22			0.20	0.25	
16	GRW(5.4 mg/l O_2)				0.21	0.28	
16	GRW(0.2 mg/ O_2)				0.069		
17	0.18				0.31		
22	GRW (0.06 mg/l O_2)				0.16		
22	GRW				0.10		
23	GRW	0.36		0.035	0.17		0.18
23	GRW	1.3		0.12			
23	GRW	0.58		0.30			
23	GRW	0.73			1.2	0.92	

^{0}C = temperature; GRW = groundwater
[a]Sterilized.
(Summarized from Schijven 2001; Ratha 2008).

contaminant mass between mobile fluid and the immobile regions existing in the porous medium (Brusseau et al. 1989; Sardin et al. 1991).

The 1D form of contaminant transport equation in the liquid phase considering advection, dispersion, sorption, and inactivation can be represented by Eq. (5.9) (Jin et al. 1997).

$$R\frac{\partial C}{\partial t} = \frac{\partial}{\partial x}\left(D_{hx}\frac{\partial C}{\partial x}\right) - v_x\frac{\partial C}{\partial x} - \lambda RC \tag{5.9}$$

where R is retardation factor

The 1D virus transport in hydraulically homogeneous, saturated porous media accounting for virus adsorption and inactivation can be expressed as Eq. (5.10) (Sim and Chrysikopoulos 1996).

$$\frac{\partial C}{\partial t} + \frac{\rho}{\theta}\frac{\partial C^*}{\partial t} = \frac{\partial}{\partial x}\left(D_{hx}\frac{\partial C}{\partial x}\right) - v_x\frac{\partial C}{\partial x} - \lambda C - \lambda^*\frac{\rho}{\theta}C^* \tag{5.10}$$

where C is the aqueous phase virus concentration; C^* is the mass of virus adsorbed on the solid matrix; D represents the hydrodynamic dispersion coefficient; v is the pore water velocity in the flow direction; ρ is the bulk density of the solid matrix; λ is the first-order inactivation rate coefficient in the aqueous viruses; λ^* is the first-order inactivation rate coefficient in the sorbed viruses; x is the Cartesian coordinate; and t is the time coordinate.

Analytical solutions of these governing equations for different initial and boundary conditions provide the exact solution and confirm the accuracy of numerical solutions developed for any complex case. Following boundary conditions at inlet may be used for numerical simulations:

a) Continuous injection of contaminant for infinite duration at the upstream boundary,
b) Continuous injection of contaminant for finite duration at the upstream boundary.

Equation (5.11) shows the initial condition of the domain and Eq. (5.12) shows the boundary condition for the case of continuous source of contaminant for an infinite duration at the upstream boundary.

The contaminant concentration may be assumed to be zero at $t = 0$.

$$C(x,0) = 0 \ for\ x \geq 0 \tag{5.11}$$

$$C(0,t) = C_0 \ for\ t \geq 0; C(\infty,t) = 0 \ for\ t \geq 0 \tag{5.12}$$

Equation (5.13) shows the analytical solution for the advection-dispersion equation considering initial and boundary conditions represented in Eqs (5.11) and (5.12) (Ogata and Banks 1961).

$$C(x,t) = \frac{C_0}{2}\left[erfc\left(\frac{x - vt}{2\sqrt{Dt}}\right) + exp\left(\frac{vx}{D}\right)erfc\left(\frac{x + vt}{2\sqrt{Dt}}\right)\right] \tag{5.13}$$

Neglecting the second term of Eq. (5.13) will introduce an error of 3% for the values of $D/vx < 0.002$ (Ogata and Banks 1961; Runkel 1996).

Similarly, Eqs (5.14) to (5.16) show the analytical solution for the non-conservative contaminant transport equation (Eq. 5.8), considering initial and boundary conditions represented in Eqs (5.11) and (5.12) (Runkel 1996).

$$C(x,t) = \frac{C_0}{2}\left\{exp\left[\frac{vx}{2D}(1-\Gamma)\right]erfc\left(\frac{x-vt\Gamma}{2\sqrt{Dt}}\right)+exp\left[\frac{vx}{2D}(1+\Gamma)\right]erfc\left(\frac{x+vt\Gamma}{2\sqrt{Dt}}\right)\right\} \quad (5.14)$$

$$\Gamma = \sqrt{1+2H} \quad (5.15)$$

$$H = \frac{2\lambda D}{v^2} \quad (5.16)$$

Since the second term of Eq. (5.14) is small relative to the first term and will introduce a very small error, it can be eliminated. Also, it can be further simplified by approximating Γ by the first two terms in a binomial expression for a very small value of H (of the order of 0.0025). Equation (5.17) shows the approximate analytical solution (Runkel 1996) for the non-conservative contaminant transport equation (Eq. 5.8) considering initial and boundary conditions represented in Eqs (5.11) and (5.12).

$$C(x,t) = \frac{C_0}{2}\left\{exp\left(\frac{-\lambda x}{v}\right)erfc\left(\frac{x-vt[1+H]}{2\sqrt{Dt}}\right)\right\}. \quad (5.17)$$

Equation (5.18) shows the boundary condition for the case of continuous source of contaminant for finite duration (t_0) at the upstream boundary.

$$C(0,t) = C_0 \; for \; t_0 \geq t \geq 0; C(0,t) = 0 \; for \; t \geq t_0; C(\infty,t) = 0 \; for \; t \geq 0. \quad (5.18)$$

The exact and approximate analytical solution of the Eq. (5.8) subjected to initial and boundary conditions given in Eqs (5.11) and (5.18) for $t > t_0$ is represented in Eqs (5.19) and (5.20) (Runkel 1996).

$$C(x,t) = \frac{C_0}{2}\left\{\begin{array}{l} exp\left[\frac{vx}{2D}(1-\Gamma)\right]\left[erfc\left(\frac{x-vt\Gamma}{2\sqrt{Dt}}\right)-erfc\left(\frac{x-v(t-t_0)\Gamma}{2\sqrt{D(t-t_0)}}\right)\right] \\ +exp\left[\frac{vx}{2D}(1+\Gamma)\right]\left[erfc\left(\frac{x+vt\Gamma}{2\sqrt{Dt}}\right)-erfc\left(\frac{x+v(t-t_0)\Gamma}{2\sqrt{D(t-t_0)}}\right)\right] \end{array}\right\}. \quad (5.19)$$

$$C(x,t) = \frac{C_0}{2}exp\left(\frac{-\lambda x}{v}\right)\left\{erfc\left(\frac{x-vt[1+H]}{2\sqrt{Dt}}\right)-erfc\left(\frac{x-v(t-t_0)[1+H]}{2\sqrt{D(t-t_0)}}\right)\right\}. \quad (5.20)$$

The analytical solution for the governing equation shown in Eq. (5.9) subjected to initial and boundary conditions given in Eqs (5.11) and (5.12) is represented in Eq. (5.21).

$$C(x,t)=\frac{C_0}{2}exp\left(\frac{xv}{2D}\right)\left[exp\left(-\frac{x}{2D}\left(\sqrt{v^2+4\lambda D}\right)\right)erfc\left(\frac{Rx-\left(\sqrt{v^2+4\lambda D}\right)t}{2\sqrt{DRt}}\right)+exp\left(\frac{x}{2D}\left(\sqrt{v^2+4\lambda D}\right)\right)erfc\left(\frac{Rx-\left(\sqrt{v^2+4\lambda D}\right)t}{2\sqrt{DRt}}\right)\right]. \quad (5.21)$$

The analytical solution for the governing equation shown in Eq. (5.9) subjected to initial and boundary conditions given in Eqs (5.11) and (5.18) is represented in Eq. (5.22).

$$C(x,t)=\frac{C_0}{2}exp\left(\frac{xv}{2D}\right)\left[\begin{array}{l}exp\left(-\frac{x}{2D}\left(\sqrt{v^2+4\lambda D}\right)\right)\left[erfc\left(\frac{Rx-\left(\sqrt{v^2+4\lambda D}\right)t}{2\sqrt{DRt}}\right)-erfc\left(\frac{Rx-\left(\sqrt{v^2+4\lambda D}\right)(t-t_0)}{2\sqrt{DR(t-t_0)}}\right)\right]\\+exp\left(\frac{x}{2D}\left(\sqrt{v^2+4\lambda D}\right)\right)\left[erfc\left(\frac{Rx-\left(\sqrt{v^2+4\lambda D}\right)t}{2\sqrt{DRt}}\right)-erfc\left(\frac{Rx-\left(\sqrt{v^2+4\lambda D}\right)(t-t_0)}{2\sqrt{DR(t-t_0)}}\right)\right]\end{array}\right]$$
(5.22)

5.4 Development of Numerical Model

5.4.1 Advective Transport

The advective transport was solved numerically by using the finite volume method based on monotone upwind schemes for conservation laws (MUSCL) proposed by Van Leer (1977). This method is globally high-order accurate and non-oscillatory (Putti et al. 1990). Equation (5.23) shows the advective transport.

$$\frac{\partial C}{\partial t}+\frac{\partial F}{\partial x}=0. \quad (5.23)$$

Figure 5.1 shows the space-time diagram. In the finite volume method, Eq. (5.23) is integrated over the solution domain Ω in space and time as

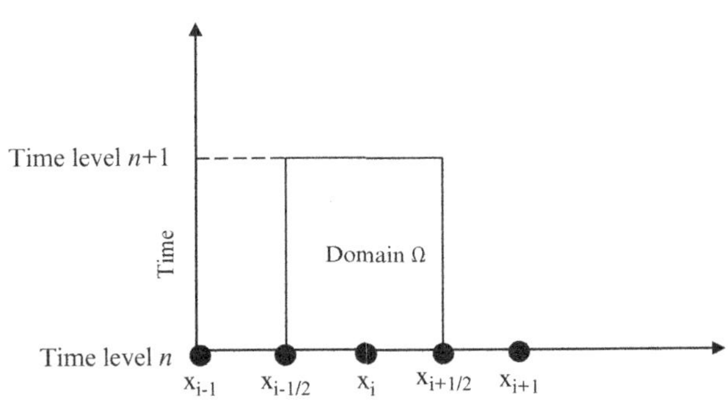

Figure 5.1 Space-time diagram.

$$\iint_{\Omega}\left(\frac{\partial C}{\partial t}+\frac{\partial F}{\partial x}\right)dxdt=0 \tag{5.24}$$

Taking the limits of the domain Ω, Eq. (5.24) can be written as

$$\int_{x_{i-1/2}}^{x_{i+1/2}} Cdx+\int_{t^n}^{t^{n+1}} Fdt=0 \tag{5.25}$$

Integration of Eq. (5.25) leads to

$$\frac{\left(\bar{C}_i^{n+1}-\bar{C}_i^n\right)}{\Delta t}+\frac{\left(\bar{F}_{i+1/2}^n-\bar{F}_{i-1/2}^n\right)}{\Delta x}=0 \tag{5.26}$$

where

$$\begin{aligned}\bar{C}_i^n&=\frac{1}{\Delta x}\int_{x_{i-1/2}}^{x_{i+1/2}} C_i^n dx\\ \bar{F}_{i+1/2}^n&=\frac{1}{\Delta t}\int_{t^n}^{t^{n+1}} F_{i+1/2}^n dt\end{aligned} \tag{5.27}$$

Equation (5.27) can be written in a discrete form by assuming the cell centered concentration C_i as cell averaged concentration as

$$\frac{\left(C_i^{n+1}-C_i^n\right)}{\Delta t}+\frac{\left(\bar{F}_{i+1/2}^n-\bar{F}_{i-1/2}^n\right)}{\Delta x}=0 \tag{5.28}$$

where C_i is the cell centered concentration and the flux $\bar{F}_{i+1/2}^n$ is the time averaged flux.

Putti et al. (1990) adopted a second-order upwind method for calculation of the advective flux. Assuming a linear distribution in the cell, the mass concentration value at the cell interface is reconstructed using an MUSCL approach as (Van Leer 1977)

$$C_{i+1/2}^L=C_i+\frac{1}{2}\delta C_i \tag{5.29}$$

$$C_{i+1/2}^R=C_{i+1}-\frac{1}{2}\delta C_{i+1} \tag{5.30}$$

where δC_i is the gradient of the concentration distribution in cell i, while L and R represent the left and right faces of the cell interface.

The value of gradient δC_i is calculated using limiters to avoid numerical oscillations. The limiter adds a certain amount of dissipation to the scheme. Superbee, Van Albada, and Minmod limiters can be employed, which can influence the accuracy of the advective transport.

The gradient of the concentration distribution in cell i for monotonicity, prescribed by the Superbee and Minmod limiters, is represented in Eq. (5.31).

$$\left(\delta C_i\right)_{mon} = ave\left(\Delta\_C, i\Delta_+ C_i\right) \tag{5.31}$$

with

$$\begin{aligned} \Delta\_C_i &= C_i - C_{i-1} \\ \Delta_+ C_i &= C_{i+1} - C_i \end{aligned} \tag{5.32}$$

Equations (5.33) and (5.34) shows the *ave(a,b)* of Eq. (5.31) for Minmod and Superbee limiters.

Minmod limiter:

$$ave(a,b) = \begin{cases} sign(a)\min\left(|a|,|b|\right) & \text{if ab} > 0 \\ 0 & \text{otherwise} \end{cases} \tag{5.33}$$

Superbee limiter:

$$ave(a,b) = \begin{cases} sign(a)\min\left(2|a|,\max\left(|a|,|b|\right),2|b|\right) & \text{if ab} > 0 \\ 0 & \text{otherwise} \end{cases} \tag{5.34}$$

Equation (5.35) shows the gradient of the concentration distribution in cell *i*, for applying the *Van Albada limiter:*

$$\delta C_i = R_i \left[\frac{1}{2}(1-k)\left(\Delta\_C_i\right) + \frac{1}{2}(1+k)\left(\Delta_+ C_i\right)\right] \tag{5.35}$$

where R_i is the Van Albada limiter and expressed in Eq. (5.36).

$$R_i = \frac{\left[2\left(\Delta_+ C_i\right)\left(\Delta\_C_i\right) + \varepsilon\right]}{\left[\left(\Delta_+ C_i\right)^2 + \left(\Delta\_C_i\right)^2 + \varepsilon\right]} \tag{5.36}$$

where ε is a very small number used to avoid occurrence of division by zero in Eq. (5.36). Equation (5.35), when $k = 1/3$, gives a third-order accurate scheme.

The fluxes at the interface are evaluated based on Eqs (5.29) and (5.30) as

$$\bar{F}^n_{i+1/2} = v_i C^L_{i+1/2} \quad \text{when } v_i > 0 \tag{5.37}$$

$$\bar{F}^n_{i+1/2} = v_{i+1} C^R_{i+1/2} \quad \text{otherwise} \tag{5.38}$$

Integration over time:

Explicit and implicit methods are commonly used for the time integration of the discretized Eq. (5.28). A choice of $\overline{F}^n$, as given in Eqs (5.37) and (5.38), results in an Euler scheme which is first-order accurate in time and is unconditionally unstable (Tai et al. 1997). To alleviate this difficulty, Hancock's Scheme (Van Albada et al. 1997) is usually employed, which is a two-step second-order accurate explicit scheme (Putti et al. 1990). The two half time steps in this method can be represented as predictor and corrector steps as follows:

Predictor step:

$$C_i^{n+1/2} = C_i^n - \frac{\Delta t}{2\Delta x} v_i \delta C_i \tag{5.39}$$

Corrector Step:

$$C_i^{n+1} = C_i^n - \frac{\Delta t}{\Delta x}\left(\overline{F}_{i+1/2}^{n+1/2} - \overline{F}_{i-1/2}^{n+1/2}\right) \tag{5.40}$$

The C_i obtained from the predictor step is used for calculation of the fluxes. The time step is limited by the courant number. For stability of this scheme, Courant number, $C_u = \frac{v\Delta t}{\Delta x}$, should be less than or equal to 1.

5.4.2 Dispersive Transport

The dispersive transport is performed on the concentrations, resulting from the advective transport in each time step.

$$\frac{\partial C}{\partial t} = D_{hx} \frac{\partial^2 C}{\partial x^2}. \tag{5.41}$$

The dispersive part is solved by a conventional, fully implicit, finite-difference scheme, which is unconditionally stable for the final concentrations. Figure 5.2 illustrates the discretization of Eq. (5.41) using the finite difference method.

5.4.3 Discretization in Space and Time

For a typical interior node i, surrounded by the adjoining nodes $i - 1$ and $i + 1$, the finite difference approximation of Eq. (5.41) can be written as

$$\frac{C_i^n}{\Delta t} = -\left(\frac{D}{\Delta x^2}\right) C_{i-1}^{n+1} + \left(\frac{2D}{\Delta x^2} + \frac{1}{\Delta t}\right) C_i^{n+1} - \left(\frac{D}{\Delta x^2}\right) C_{i+1}^{n+1} \tag{5.42}$$

where Δx and Δt are the spatial and temporal increments; n denotes the time level at which solution is known; and $n + 1$ denotes the time level where the solution is sought.

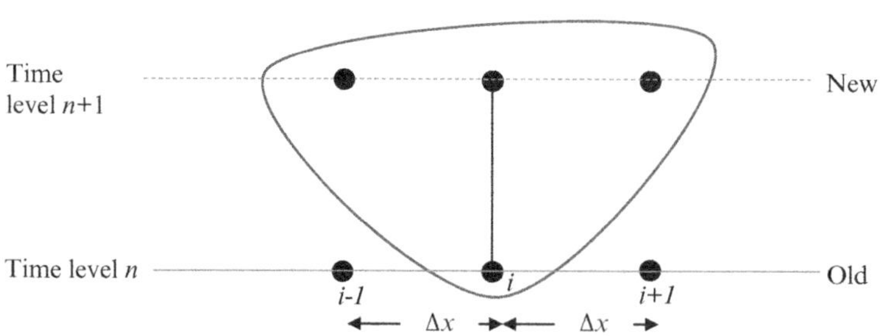

Figure 5.2 Definition sketch of finite difference discretization.

Equation (5.42) can be written in matrix form as

$$[\mathbf{A}][\mathbf{C}]=[\mathbf{B}] \tag{5.43}$$

where **A** is a tridiagonal coefficient matrix; **C** is the vector of unknown concentration C_i at time level $n+1$; and **B** is the vector of known quantities at the time level n. The tridiagonal systems shown in Eq. (5.43) are solved using the Thomas algorithm (Remson et al. 1971).

Contaminants will transport initially through an unsaturated zone before joining the groundwater table if the source of contaminant is located in unsaturated porous media. The nature of virus movement in an unsaturated zone is significantly different from that in a saturated zone (Sim and Chrysikopoulos 2000). The sorption mechanism is highly predominant during transportation through the unsaturated zone. Electrostatic double layer interaction and van der Waals forces are the major reason for sorption on a liquid–solid interface (Sim and Chrysikopoulos 2000). Physical absorption, chemical absorption, and ion exchange also cause the sorption of solute on the liquid–solid interface. It is reported in the literature that sorption increases by decreasing moisture content (Lance and Gerba 1984; Sim and Chrysikopoulos 2000). Since the moisture content in the unsaturated zone depends on the pressure head and changes with respect to space and time, the modeling of contaminant transport in the unsaturated zone is much more difficult than modeling in the saturated zone. The nonlinearity of the governing flow equation (Richards equation) needs mass conservative schemes for accurate prediction of velocities and moisture contents.

There are three standard forms of Richards equation available.

Pressure head based (ψ-based)

$$C(\psi)\frac{\partial\psi}{\partial t}=\frac{\partial}{\partial z}\left[K(\theta)\left\{\frac{\partial\psi}{\partial z}+1\right\}\right] \tag{5.44}$$

Moisture content based (Θ-based)

$$\frac{\partial\theta}{\partial t}=\frac{\partial}{\partial z}\left[D(\theta)\frac{\partial\theta}{\partial z}\right]+\frac{\partial K(\theta)}{\partial z} \tag{5.45}$$

Mixed form

$$\frac{\partial\theta}{\partial t}=\frac{\partial}{\partial z}\left\{K(\theta)\left(\frac{\partial\psi}{\partial z}+1\right)\right\} \tag{5.46}$$

where ψ is the pressure head; θ is the moisture content; z is the vertical coordinate taken positive upward; t is the time coordinate; $C=d\theta/d\Psi$ is the specific moisture capacity of the soil; K is the unsaturated hydraulic conductivity of the soil; and $D=K/C$ is the soil moisture diffusivity.

Equations (5.47) to (5.50) shows the relationship given by Van Genuchten (1980), which are used for θ–ψ and K–θ relationships.

θ–ψ Relationship:

$$\Theta = \left\{ \frac{1}{1+\left(\alpha_v |\psi|\right)^{n_v}} \right\}^{m_v} \tag{5.47}$$

where α_v and n_v are unsaturated soil parameters with

$$m_v = 1 - \frac{1}{n_v} \tag{5.48}$$

Θ is the effective saturation defined as

$$\Theta = \frac{\theta - \theta_r}{\theta_s - \theta_r} \tag{5.49}$$

where θ_s is saturated water content and θ_r is residual water content of the soil.

K–θ Relationship:

$$K(\Theta) = K_s \left\{ 1 - \left\{ 1 - \Theta^{\left(1/m_v\right)} \right\}^{m_v} \right\}^2 \Theta^{\left(1/2\right)} \tag{5.50}$$

where K_s is saturated hydraulic conductivity.

Initial condition:
Initial condition in the form of the pressure head or moisture content distribution at the beginning of the simulation used and shown in Eqs (5.51) and (5.52).

$$t = 0, \psi = \psi_0 \; 0 \le z \le L \tag{5.51}$$

Or

$$t = 0, \theta = \theta_0 \; 0 \le z \le L \tag{5.52}$$

where ψ_0 and θ_0 are the specified pressure head and moisture contents at the beginning of the simulation.

Lower boundary condition:
Depending on the presence of the water table, Eqs (5.53) to (5.54) show the two different types of lower boundary conditions.

In case the water table is very near to the ground surface, atmospheric pressure head ($\psi = 0$) is applied at the water table, i.e.,

$$t > 0, \psi = 0 \; z = 0 \tag{5.53}$$

where L is the depth of the water table.

For the case of the deeper groundwater table, a gravity drainage $\left(\frac{\partial \psi}{\partial z} = 0\right)$ boundary condition is applied at a certain depth below the ground surface, i.e.,

$$t > 0, \left(\frac{\partial \psi}{\partial z} = 0 \right), z = 0 \tag{5.54}$$

Upper boundary condition:
For infiltration under ponding conditions, a Dirichlet type boundary condition is assigned at the ground surface, i.e.,

$$t > 0 \ \psi = \psi_{top} \ z = L \tag{5.55}$$

For infiltration/evaporation with constant flux, a Neuman type condition is assigned at the ground surface, i.e.,

$$t > 0, -K\left(\frac{\partial \psi}{\partial z} + 1 \right) = q_{top}, z = L \tag{5.56}$$

where q_{top} is the infiltration/evaporation rate.

For solving the differential equation (Eq. 5.46), a finite difference grid is superimposed over the solution domain. For a typical interior node j, a fully implicit finite difference approximation of the term on the right side of Eq. (5.46), based on Picard's scheme for the nonlinear terms, can be written as (Clement et al. 1994)

$$\begin{aligned} \frac{\partial}{\partial z}\left\{ K(\theta)\left(\frac{\partial \psi}{\partial z} + 1 \right) \right\} = \frac{1}{\Delta z}\Bigg[& \left(\frac{K_j^{n+1,m} + K_{j+1}^{n+1,m}}{2} \right)\left(\frac{\psi_{j+1}^{n+1,m+1} - \psi_j^{n+1,m+1}}{\Delta z} \right) \\ & - \left(\frac{K_j^{n+1,m} + K_{j-1}^{n+1,m}}{2} \right)\left(\frac{\psi_j^{n+1,m+1} - \psi_{j-1}^{n+1,m+1}}{\Delta z} \right) \Bigg] \\ & + \frac{1}{\Delta z}\left[\left(\frac{K_j^{n+1,m} + K_{j+1}^{n+1,m}}{2} \right) - \left(\frac{K_j^{n+1,m} + K_{j-1}^{n+1,m}}{2} \right) \right] \end{aligned} \tag{5.57}$$

where n denotes the discrete time level at which the solution is known; $\Delta t = t^{n+1} - t^n$ is the time step; and $K(\theta)$ is a nonlinear function of θ, which is linearized using a Picard iteration scheme. The current and previous Picard iteration levels are denoted as $m + 1$ and m respectively. The hydraulic conductivity is arithmetically averaged between nodes. Use of the arithmetical mean is justified by the findings of Kirkland et al. (1992) that solution of the Richards equation is relatively insensitive to the interblock-averaging scheme used for hydraulic conductivity. Kirkland et al. (1992) also found that use of a Crank–Nicholson scheme on the mixed form of the Richards equation fails to reduce truncation error and is subject to potential inabilities.

5.5 Application of Models

A numerical model was applied to determine both non-conservative and conservative microbial transport in a saturated medium and the concentration was determined which varies as a function of space and time. The results obtained from numerical simulation were validated with an analytical solution. A conservative microbial source was considered that is continuously injecting at the source for a duration of 150 days. The flow velocity is considered as 0.75 m/s. The numerical model was solved for both advection dominated and dispersion dominated transport considering the Peclet number of 150 and 5, respectively. Similarly, the numerical model was simulated for Courant numbers of 0.5, 0.75, and 0.9 in order to check the effect of the Courant number on the accuracy of numerical simulation. It is observed from Figures 5.3 to 5.6 that the results obtained from the numerical model are confirmed by the results of analytical solutions for both advection and dispersion dominated transport. Also, it is observed from Figures 5.4 and 5.6 that there is no effect of Courant number on the accuracy of the results obtained from numerical simulation. Figure 5.7 shows the breakthrough curve for a case of non-conservative microbial transport

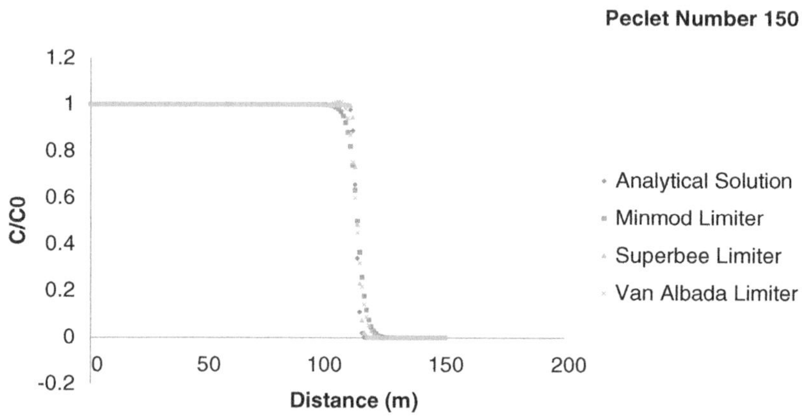

Figure 5.3 Comparison of relative concentration with distance for advection dominated conservative microbial transport using various limiters.

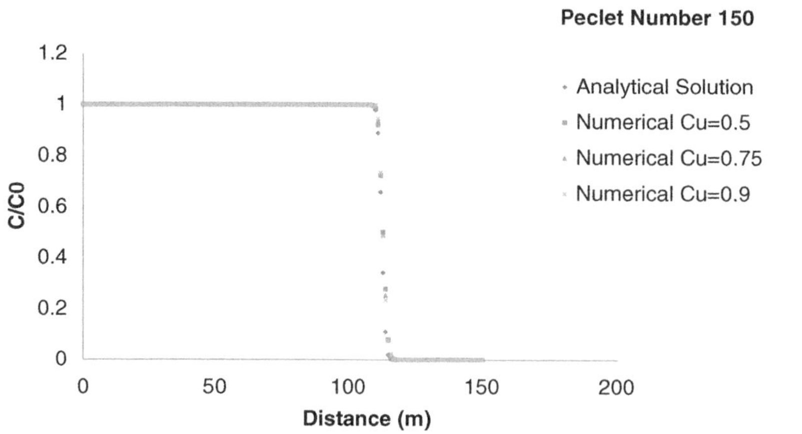

Figure 5.4 Comparison of relative concentration with distance for advection dominated conservative microbial transport using various Courant numbers.

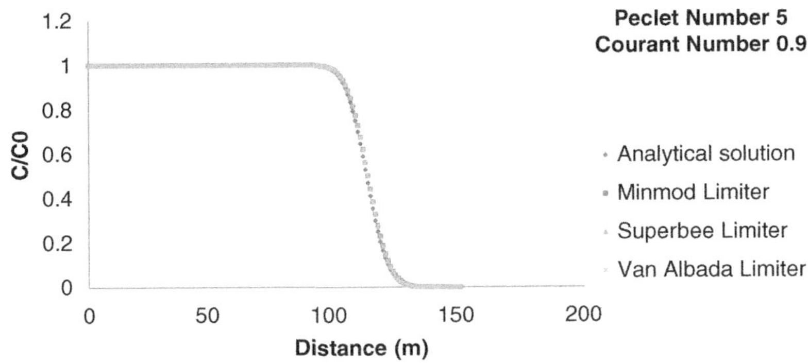

Figure 5.5 Comparison of relative concentration with distance for dispersion dominated conservative microbial transport using various limiters.

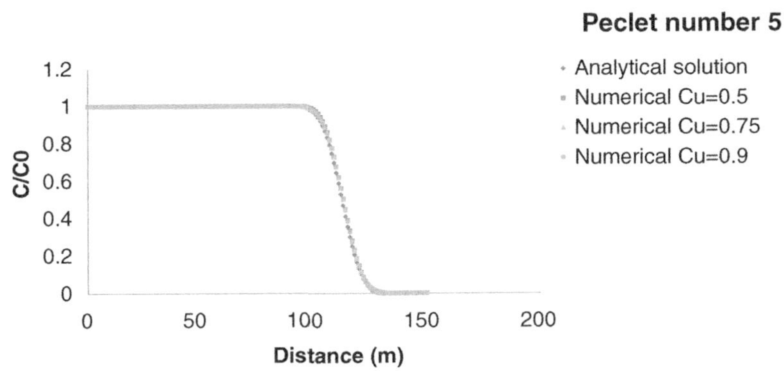

Figure 5.6 Comparison of relative concentration with distance for dispersion dominated conservative microbial transport using various Courant numbers.

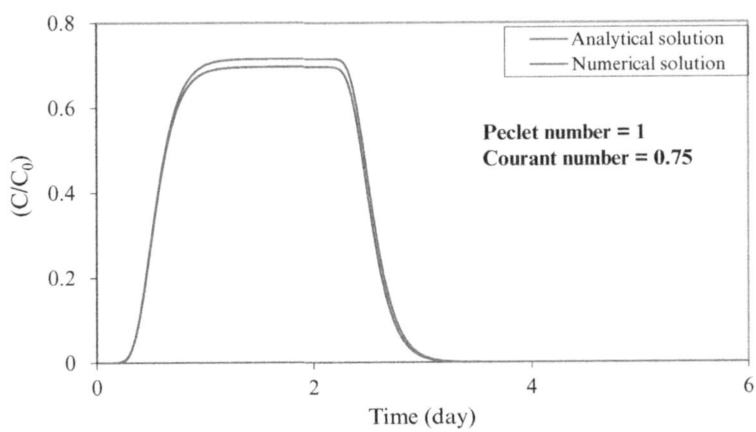

Figure 5.7 Comparison of breakthrough curves for non-conservative microbial transport obtained from analytical and numerical simulations.

through saturated porous media in which microbes were injected at the source for 2 days but the concentration at 20 cm downstream from the source was observed for 6 days through numerical simulation. The inactivation coefficient of microbes was considered as 0.58/day. The breakthrough curve obtained from numerical simulation was compared with the analytical solution and it is observed that the curve obtained through numerical simulation matches with the results obtained through analytical solution.

5.6 Softwares for Pollutant Transport

Various softwares are available to solve the governing equation of pollutant transport. This section highlights some of the aspects of Hydrus 1D software that can be used to analyze the pollutant/microbial transport in various moisture conditions of soil.

5.6.1 Hydrus Model for Pollution Transport

Hydrus 1D is a software that can simulate water movement in a heterogenous or homogenous soil medium. Hydrus 1D can also be used for analysis in soils with soil strata varying with depth. The model numerically solves the Richards equation for saturated-unsaturated water flow and Fickian-based advection dispersion equations for solute transport phenomena. Various infiltration models can be used for the analysis using Hydrus1D.

The contaminant transport in an unsaturated medium was solved using Hydrus 1D software and the concentration was determined, which varied as a function of space and time. A homogeneous soil column of 100 cm was considered in which water was flowing vertically. For infiltration under ponding conditions, Dirichlet type boundary conditions were assigned at the ground surface, i.e.,

$$\text{For } t > 0 \ \psi = \psi_{top} = -10\,\text{cm}$$

and the initial value of $\psi = -1000$ cm for all points of the soil column. A modified Van Genuchten model was chosen in which the following parametric values were considered for simulation.

Unsaturated soil parameters α_v and n_v are taken as 0.036 cm^{-1} and 1.56, respectively. The value of saturated moisture content and residual moisture content are 0.43 and 0.078, respectively. The value of saturated hydraulic conductivity of soil is taken as 1.04 cm/h. Loam type of soil is considered as a soil media through which the contaminant transportation is occurring. Figures 5.8 to 5.10 show the variation of pressure head, moisture content, and concentration with time at various observation points.

5.7 Mathematical and Computational Limitation

The mathematical analytical models for understanding the fate and transport of emerging contaminants are largely employed for well defined, saturated, homogeneous, and isotropic aquifers. These analytical models can also be used for verification of accuracy of

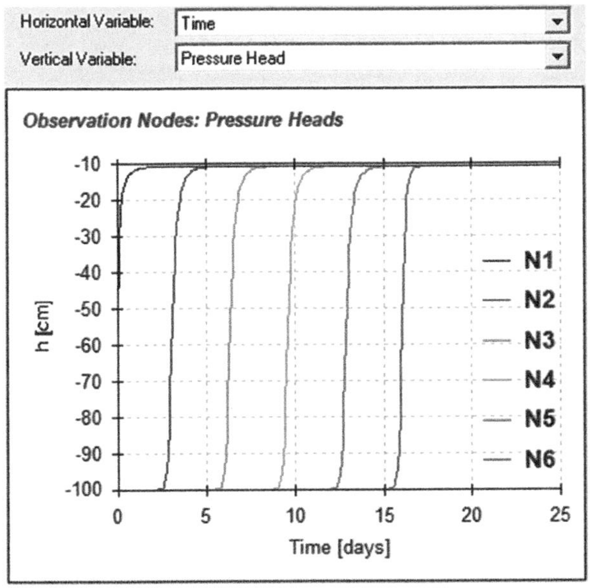

Figure 5.8 Variation of Pressure heads with time at selected observation points.

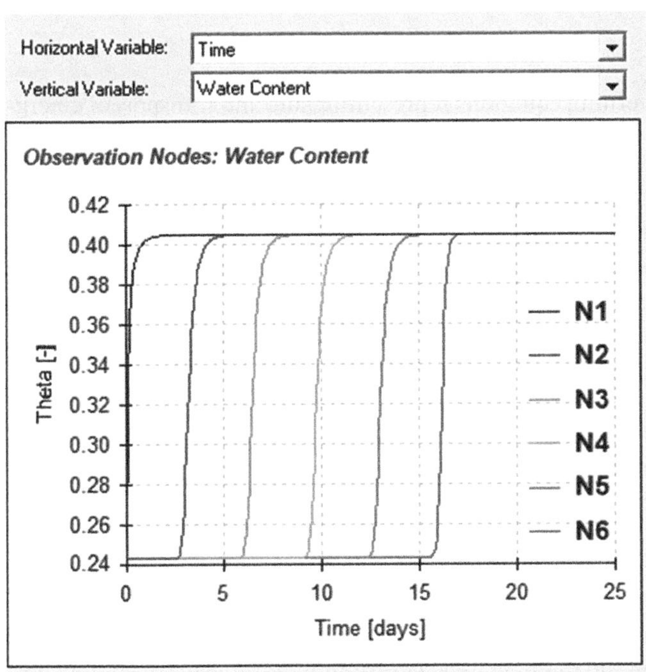

Figure 5.9 Variation of moisture content with time at selected observation points.

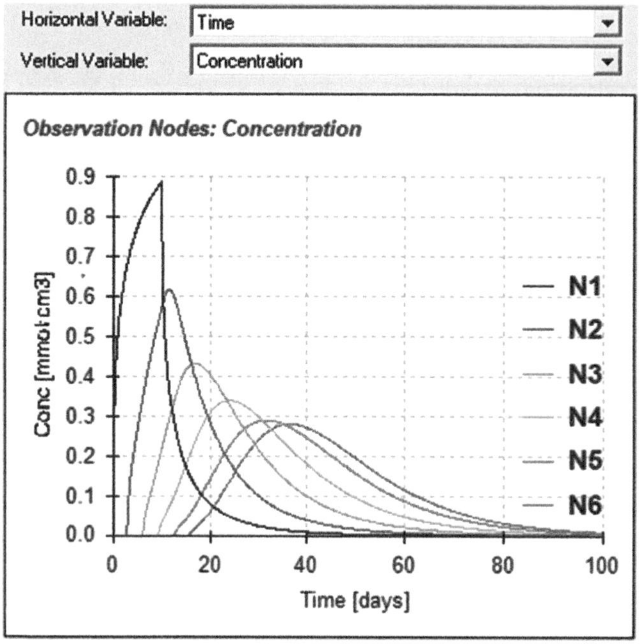

Figure 5.10 Variation of concentration with time at selected observation points.

numerical simulations. The governing equations representing fate and transport of emerging chemical contaminants through both saturated and unsaturated porous media are difficult to solve because there is a non-linear relationship existing between the contaminant concentration of liquid phase and sorbed phase (Huang et al. 1998). The non-linearity is usually converted to the linear form during numerical simulation that often introduces numerical error. Literature are available that have used the modified Picard iteration algorithm for solving non-linear transport equation in unsaturated zone (Huang et al. 1998; Kim et al. 2008; Ratha et al. 2009). The physical and hydraulic properties of the porous media may change because of accumulation of microbial biomass due to which the flow path of contaminant, microbial growth, and distribution, etc., will change (Rockhold et al. 2004). The air-water-interfacial area model developed by Cary (1994) was used by Sim and Chrysikopoulos (2000) for virus transport modeling, but their model uses some empirical parameters which are the limitation of the model. Later Kim et al. (2008) used the air-water-interfacial area model for bacterial transport. The mathematical model developed for simulating flow and transport considers a simple first-order decay reaction rather than considering the actual cell growth or substrate limitations on biodegradation rates and this is a limitation of the transport model (Rockhold et al. 2004). The non-linear attachment and detachment process in the mathematical description of transport of soluble pollutants, heavy metal, bacteria, and suspension-colloidal particles, etc., still remain as a limitation of numerical simulation and therefore Bai et al. (2019) proposed a nonlinear attachment-detachment model with hysteresis (Bai et al. 2019).

It is observed from the literature that the existing limitations of the representation of the transport phenomena in a form of a mathematical model has been continuously improved during the last two decades. Although numerical simulation of a naturally occurring phenomenon is a cost-effective method when compared to experimental simulation, there still exists certain computational limitations of numerical simulations. Several studies have confirmed that the numerical oscillation arises during the simulation of the mathematical equation when a higher-order finite difference scheme is used (Ratha et al. 2007). Although the distance between two grids and the time step can be reduced in order to eliminate these types of oscillation, it can increase the simulation cost. Since the number of iterations during the simulation increases due to decrease in the time step, higher memory capacity along with high-speed computing is required for simulation which increases the computational cost. In certain cases, the non-uniformity and unsteady boundary conditions of the real field cannot be accurately represented during the simulation. The validation of a numerical simulation requires the experimental data but in most cases the experimental data of a study area may not be available.

5.8 Conclusions

The present chapter discusses the governing equation of pollutant transport and its solution using both analytical and numerical methods. The advection-dispersion equation was solved numerically by a hybrid model using the operator split approach. The advection part was solved using the explicit finite volume model, whereas the dispersion part was solved using the implicit finite difference method. These approaches were globally second-order accurate. It is observed that this numerical model can be used for a wide range of both Peclet numbers and Courant numbers. The results obtained from numerical simulation were similar to analytical solutions. The present chapter also describes the numerical simulation of pollutant transport in unsaturated media using Hydrus 1D software. The variation of moisture content, pressure head, and concentration with time obtained from Hydrus 1D software are also represented.

Acknowledgments

Authors would like to acknowledge Sarthak Goyal, Shivam Lamba, Piyush Garg, and Udit Sharma, final year undergraduate students for simulating Hydrus 1D software.

References

Bai, B., Rao, D., Chang, T., and Guo, Z. (2019). A nonlinear attachment-detachment model with adsorption hysteresis for suspension-colloidal transport in porous media. *Journal of Hydrology* 578: 124080.

Brusseau, M.L., Rao, P.S.C., and Gillham, R.W. (1989). Sorption nonideality during organic contaminant transport in porous media. *Critical Reviews in Environmental Science and Technology* 19 (1): 33–99.

Cary, J.W. (1994). Estimating the surface area of fluid phase interfaces in porous media. *Journal of Contaminant Hydrology* 15 (4): 243–248.

Chahar, B.R. (2015). *Groundwater Hydrology*. McGraw Hill Education (India) Private Limited.

Clement, T.P., Wise, W.R., and Molz, F.J. (1994). A physically based, two-dimensional, finite-difference algorithm for modeling variably saturated flow. *Journal of Hydrology* 161 (1–4): 71–90.

Freeze, R.A. and Cherry, J.A. (1997). *Groundwater*, 604. Englewood Cliffs, NJ: Prentice Hall Inc.

Huang, K., Mohanty, B.P., Leij, F.J., and van Genuchten, M.T. (1998). Solution of the nonlinear transport equation using modified Picard iteration. *Advances in Water Resources* 21 (3): 237–249.

Jin, Y., Yates, M.V., Thompson, S.S., and Jury, W.A. (1997). Sorption of viruses during flow through saturated sand columns. *Environmental Science & Technology* 31 (2): 548–555.

Kim, M.K., Kim, S.B., and Park, S.J. (2008). Bacteria transport in an unsaturated porous media: incorporation of air–water interface area model into transport modelling. *Hydrological Processes: An International Journal* 22 (13): 2370–2376.

Kirkland, M.R., Hills, R.G., and Wierenga, P.J. (1992). Algorithms for solving Richards' equation for variably saturated soils. *Water Resources Research* 28 (8): 2049–2058.

Lance, J.C. and Gerba, C.P. (1984). Virus movement in soil during saturated and unsaturated flow. *Applied and Environmental Microbiology* 47 (2): 335–337.

Ogata, A. and Banks, R.B. (1961). A solution of the differential equation of longitudinal dispersion in porous media: fluid movement in earth materials. US Government Printing Office.

Persson, M. and Berndtsson, R. (1998). Estimating transport parameters in an undisturbed soil column using time domain reflectometry and transfer function theory. *Journal of Hydrology* 205 (3–4): 232–247.

Putti, M., Yeh, W.W.G., and Mulder, W.A. (1990). A triangular finite volume approach with high-resolution upwind terms for the solution of groundwater transport equations. *Water Resources Research* 26 (12): 2865–2880.

Ratha, D. (2008). Analysis and Parameter estimation of virus transport through sub-surface media. PhD thesis.

Ratha, D., Hari Prasad, K.S., and Ojha, C.S.P. (2007). A finite volume model for the solution of the advection-dispersion equation. *ISH Journal of Hydraulic Engineering* 13 (2): 122–135.

Ratha, D.N., Ojha, C.S.P., and Prasad, K.S. (2009). *151-A Modified Picard's Method for Virus Transport in Ground Water*. New Delhi: Allied Publishers Pvt. Limited.

Remson, I., Hornberger, G.M., and Molz, F.J. (1971). Numerical methods in subsurface hydrology.

Robles, L., Slifko, T., and Kunihiro, K. (2006). Analytical innovations for the detection of biological microconstituents. *Florida Water Resources Journal* 12: 60–65. https://fwrj.com/?page_id=163

Rockhold, M.L., Yarwood, R.R., and Selker, J.S. (2004). Coupled microbial and transport processes in soils. *Vadose Zone Journal* 3 (2): 368–383.

Rout, P.R., Zhang, T.C., Bhunia, P., and Surampalli, R.Y. (2021). Treatment technologies for emerging contaminants in wastewater treatment plants: a review. *Science of the Total Environment* 753: 141990.

Runkel, R.L. (1996). Solution of the advection-dispersion equation: continuous load of finite duration. *Journal of Environmental Engineering* 122 (9): 830–832.

Runkel, R.L. and Bencala, K.E. (1995). Transport of reacting solutes in rivers and streams. In: *Environmental Hydrology (ed. Singh, V.P.)*, 137–164. Dordrecht: Springer.

Sardin, M., Schweich, D., Leij, F.J., and Van Genuchten, M.T. (1991). Modeling the nonequilibrium transport of linearly interacting solutes in porous media: a review. *Water Resources Research* 27 (9): 2287–2307.

Scheibe, T.D. and Wood, B.D. (2003). A particle-based model of size or anion exclusion with application to microbial transport in porous media. *Water Resources Research* 39 (4).

Schijven, J.F. (2001). Virus removal from groundwater by soil passage. modeling, field, and laboratory experiments. PhD thesis.

Sim, Y. and Chrysikopoulos, C.V. (1996). One-dimensional virus transport in porous media with time-dependent inactivation rate coefficients. *Water Resources Research* 32 (8): 2607–2611.

Sim, Y. and Chrysikopoulos, C.V. (2000). Virus transport in unsaturated porous media. *Water Resources Research* 36 (1): 173–179.

Tai, C.H., Chiang, D.C., and Su, Y.P. (1997). Explicit time marching methods for the time-dependent Euler computations. *Journal of Computational Physics* 130 (2): 191–202.

USGS (US Geological Survey). (2017). Contaminants of emerging concern in the environment.

Van Albada, G.D., Leer, B.V., and Roberts, W. (1997). A comparative study of computational methods in cosmic gas dynamics. In: *Upwind and High-Resolution Schemes* (eds. Yousuff Hussaini, M., Leer, B., and Rosendale, J.), 95–103. Berlin, Heidelberg: Springer.

Van Genuchten, M.T. (1980). A closed-form equation for predicting the hydraulic conductivity of unsaturated soils. *Soil Science Society of America Journal* 44 (5): 892–898.

Van Leer, B. (1977). Towards the ultimate conservative difference scheme. III: Upstream-centered finite-difference schemes for ideal compressible flow. *Journal of Computational Physics* 23 (3): 263–275.

Ying, Z. and Droste, R.L. (2015). Sorption of microconstituents onto primary sludge. *Water Science and Technology* 72 (5): 779–784.

6

Groundwater Contamination by Microconstituents

Jiun-Hau Ou[1], *Ku-Fan Chen*[2], *Rao Y. Surampalli*[3], *Tian C. Zhang*[4], *and Chih-Ming Kao*[1,4]

[1] *Institute of Environmental Engineering, National Sun Yat-sen University, Kaohsiung, Taiwan*
[2] *Department of Civil Engineering, National Chi Nan University, Puli, Nantou County, Taiwan*
[3] *Global Institute for Energy, Environment and Sustainability, Lenexa, KS, USA*
[4] *Civil and Environmental Engineering, College of Engineering, University of Nebraska, Lincoln, Omaha, NE, USA*

6.1 Introduction

The Water Environment Federation (WEF) estimates that there are more than 100,000 chemicals currently in use by different societies and industries, with up to 1000 new chemicals being introduced each year (Rahman et al. 2018). Some of these chemicals transport and migrate to the ecosystems via different point sources (e.g., sewage sludges, wastewater effluents, landfill leachates, combined sewer overflows) and different non-point sources (NPS) (e.g., farmlands, roads). However, some of these sources have not been thoroughly investigated and characterized, and there are other routes of chemical transport that have not yet been identified.

Microconstituents (MCs) are used to describe the man-made and naturally occurring chemicals which are detected in the environment with potential harmful effects on the ecosystem and human health. MCs include elements and inorganic/organic chemicals. Increased awareness of the risks posed by MCs to human health has raised concerns for water quality improvement. Many researchers have focused on developing treatment methods for MCs removal (Salamanca et al. 2021). The main sources of these MCs contain veterinary drugs and food additives used in livestock farming, pesticides used in agriculture, and pharmaceuticals and personal care products (PPCPs). Because these MCs cannot be fully metabolized, they are usually discharged to wastewater treatment plants (WWTPs) or discharged to water bodies directly without proper treatment. Moreover, the WWTPs are not designed for MCs treatment, their capabilities on MCs removal are not effective, and thus, a significant amount of MCs may appear in the effluents, which can be eventually transported to water bodies (including the subsurface aquifer). All the MC sources can result in both point sources or NPS pollution, and part of the MCs may leach into groundwater as a result

Microconstituents in the Environment: Occurrence, Fate, Removal, and Management, First Edition.
Edited by Rao Y. Surampalli, Tian C. Zhang, Chih-Ming Kao, Makarand M. Ghangrekar, Puspendu Bhunia, Manaswini Behera, and Prangya R. Rout.

of rainfall and soil infiltration, resulting in groundwater contamination by MCs (Fatta-Kassinos et al. 2015). Due to the constant discharges of MCs to water bodies via point sources or NPS, they appear persistent for chronic toxicity to ecosystems. The potential exists for subtle effects, even at ppb levels (μg/L). Because the groundwater contamination by MCs has not been fully investigated, the risk of their impacts on subsurface environments (including groundwater systems) are unknown challenges. As groundwater is very difficult to clean up once polluted and it is also an important water resource in many countries, it is a necessity to expand our knowledge and understanding of MCs in groundwater systems.

6.2 Major Microconstituents in Groundwater

Groupings of MCs commonly found in the subsurface environment include microplastics, PPCPs, industrial chemicals (e.g., polybrominated diphenyl ethers (PBDEs), bisphenol A(BPA), perfluorooctanesulfonic acid (PFOS)) pesticides, persistent bio-accumulative toxics (PBTs), endocrine disruptors, nanomaterials, natural compounds (e.g., microbial toxins), and disinfection by-products (DBPs). Many MCs are used worldwide for various purposes. Due to their large consumption, MCs may enter the aqueous environments including subsurface aquifers at ng/L to μg/L concentration levels. Continuous exposure to low and sub-toxic concentrations of some specific MCs may result in unintended effects on non-target species and undesirable effects on ecosystems and humans (Fatta-Kassinos et al. 2015).

PPCPs are the main sources of MCs in the environment (Table 6.1) (Liu and Wong 2013), which are ubiquitous in our lives. Many pharmaceuticals become wastes when they pass

Table 6.1 Classification of PPCPs.

PPCPs/Subgroups	Representative Compounds
Pharmaceuticals	
Antibiotics.	Clarithromycin; erythromycin; sulfamethoxazole; sulfadimethoxine; ciprofloxacin; norfloxacin; chloramphenicol.
Hormones.	Estrone (E1); estradiol (E2); ethinylestradiol (EE2).
Analgesics and anti-inflammatory drugs.	Diclofenac; ibuprofen; acetaminophen; acetylsalicylic acid.
Antiepileptic drugs.	Carbamazepine; primidone.
Blood lipid regulators.	Clofibrate; gemfibrozil.
β-blockers.	Metoprolol; propanolol.
Contrast media.	Diatrizoate; iopromide.
Cytostatic drugs.	Ifosfamide; cyclophosphamide.
Personal Care Products	
Antimicrobial agents/Disinfectants.	Triclosan; Triclocarban.
Synthetic musks/Fragrances	Galaxolide (HHCB); Toxalide (AHTN).
Insect repellants.	*N,N*-diethyl-m-toluamide (DEET).
Preservatives.	Parabens (alkyl-p-hydroxybenzoates).
Sunscreen UV filters.	2-ethyl-hexyl-4-trimethoxycinnamate (EHMC); 4-methyl-benzilidine-camphor (4MBC).

their expiration dates. One pathway for PPCPs into the ecosystem is absorption of PPCPs by humans following therapeutic uses, then followed by excretion and discharge to sewage systems. After treatment in WWTPs, some of the treated wastewater and biosolids may be used for land application as fertilizers and irrigation. The manufacture factories of PPCPs also discharge the PPCPs-contained wastewater to the WWTPs, which may become another source of PPCPs to the ecosystems, including groundwater bodies (Ebele et al. 2017).

Endocrine-disrupting and pharmaceutical compounds are classified as emerging organic contaminants (EOCs). They have been widely used in food preservation, healthcare, agricultural practices, veterinary care, and industrial applications (Verlicchi et al. 2012). Microplastics are also commonly found MCs in groundwater bodies as ubiquitous pollutants (Verlicchi et al. 2012; Atugoda et al. 2021). Microplastics have been defined as any type of fibers, plastic fragments, or beads with a diameter ranging from 100 nm to <5 mm and have been considered as a source of terrestrial and aquatic MC contamination (Atugoda et al. 2021). PPCPs, EOCs, and microplastics are the most commonly found MCs in groundwater systems, and thus they have become of global concern and have resulted in intensive research in the field of MCs (Prata et al. 2018; Atugoda et al. 2021).

6.3 Mechanisms for Groundwater Contamination By Microconstituents

The fate of MCs in the ecosystems including the groundwater environment is driven by their physical/chemical/biological characteristics. MCs are complex molecules with different physical and chemical properties. They can be characterized by their molecular weights, ionic nature, and relatively low solubility. Most MCs have specific effects within humans (e.g., central nervous system, cardiovascular, anti-inflammatory, antibacterial, anticancer, immune). Many of these special properties are relevant to the assessment of their environmental fate and transport properties (Padhye and Tezel 2014). The treated wastewater discharges from the WWTP enter the water bodies, and the residual MCs can transport and transform in the environment media, causing pollution of the surface and subsurface environments (soil and groundwater) (Boxall et al. 2012).

MCs in groundwater can be closely related to human activities because most of them are man-made or synthetic products. Once the MC-contained wastewater discharges to the water bodies, the receptors (e.g., groundwater) are polluted. MCs can remain in the receptors and can be transported or transformed via natural attenuation processes. MCs can be transported to the groundwater through the hyporheic zone. The concentrations of MCs have been analyzed in subsurface environments including soils, sediments, and groundwater. The primary processes and mechanisms affecting the fate and transport of MCs in the subsurface media include adsorption, advection, diffusion, dispersion, and degradation. MCs can leach into the subsurface aquifers after their vertical transport processes through the vadose zone from the land surface (Laws et al. 2011).

Adsorption may influence the fate of MCs in subsurface environments by affecting their bioavailability, migration, and plant uptake. MCs with higher adsorption characteristics are usually less mobile in subsurface environments with limited transport and vertical transport potential. However, MCs with lower adsorption characteristics can easily migrate vertically to the water saturated zone (groundwater zone) (Lin and Gan 2011). Moreover, environmental factors (e.g., soil properties, hydrogeological conditions) may affect the fate and transport behaviors of MCs in the subsurface environment (Katz et al. 2009). Compared to soil environments, the microbial population and diversity are lower with relatively lower oxidation-reduction potential. Thus, the biodegradation potential of MCs in the groundwater system is not significant, and they may undergo incomplete degradation patterns. This may result in the production of hazardous metabolites, or may remain unchanged in groundwater for a long period of time (Lapworth et al. 2012; Daugherty et al. 2016).

Some of the MCs are polar and ionizable, so pH conditions may affect their presence or speciation in groundwater. The organic and inorganic ions, including dissolved organic matter present in the soil and groundwater system, interfere with the adsorption mechanisms occurring between MCs and soil particles. Inorganic and organic ion species in groundwater may compete with MCs to interact with the active sites of soil particles, resulting in the suppression of the adsorption process (Guo and Wang 2019). The initial concentrations of MCs also play important roles in the maximum adsorption capacity. At low MCs concentrations, the adsorption mechanism occurs rapidly because there is a high probability of MCs attaching to the binding sites of soil particles. However, with increased MCs concentrations, the active sites of the soil particles become occupied, and so adsorption reaches a steady-state equilibrium rapidly (Zhang et al. 2018).

There are several pathways for MCs to enter groundwater environments including surface runoff, agricultural activity, and flooding. The downward movement of MCs may occur through the soil pores and cracks when the size of the MC is smaller than the size of soil pore (Bläsing and Amelung 2018; Wong et al. 2020). Groundwater can be contaminated by MCs from surface water recharge (e.g., wastewater from WWTPs) (Koelmans et al. 2019). MC particles may also adsorb a variety of contaminants, such as organic pollutants and heavy metals based on their physiochemical properties. The co-transport of MCs with adsorbed pollutants may result in significant combined environmental pollution problems, which has drawn considerable attention to the soil and groundwater environment (Browne et al. 2013; Godoy et al. 2019).

6.4 Modeling Transport of Microconstituents

The Pharmaceutical Assessment and Transport Evaluation (PhATE™) model is a geographical information system-based water quality model, which can be applied to evaluate the combined risk from different MCs (e.g., 17β-estradiol, estrone, 17α-ethinyl estradiol). Researchers used the PhATE™ model for MCs transport in different environmental media (initially developed to assess the transport of pharmaceuticals in water bodies) (Cunningham et al. 2012). This kind of availability has the potential to simulate the MCs transport in the subsurface groundwater environment. Table 6.2 presents the modeling transport of MCs.

The Risk Identification and Ranking (RAIDAR) model offers an assessment of the fate of chemicals (e.g., MCs) in the environment. Results from other studies show that the model is influenced by estimates of loss processes, such as WWTP removal efficiencies and biotransformation (Gouin et al. 2012). Moreover, the quantitative molecular similarity assessment (QMSA) is a regression-based structure property model (Li and Colosi 2012). QMSA

Table 6.2 Modeling transport of MCs.

Model Name	Introduction	References
Pharmaceutical Assessment and Transport Evaluation (PhATE™).	Pharmaceutical Assessment and Transport Evaluation (PhATE™) model was developed to simulate concentrations of active pharmaceutical ingredients in 11 watersheds across the United States. The PhATE™ model is based upon mass balance equations that describe contaminant fate in each segment of the watershed. For each segment, mass enters either via inflow from upstream segments or from WWTP as point sources along the segment. Contaminants leave the segment via first-order in-stream decay, flow diversions, or outflow to a downstream segment.	(Anderson et al. 2004)
Quantitative molecular similarity assessment (QMSA).	Quantitative molecular similarity analysis (QMSA) is one alternative to traditional, regression-based structure–property models. This is an evolving technique, which tends to be particularly appealing for applications in which predictive accuracy may be hampered by limited data availability. QMSA models are based on the assumption that "similar" molecules behave "similarly," and they have been used to accurately predict various parameters related to environmental risk assessment, including mutagenicity, hepatoxicity, and skin sensitization plus various other REACH-relevant toxicity endpoints.	(Li and Colosi 2012)
Risk Identification And Ranking (RAIDAR).	Risk Identification And Ranking (RAIDAR) is a screening level risk assessment model that combines information on chemical partitioning, reactivity, bioaccumulation, exposure, effect levels, and emission rates. The RAIDAR model is structured to bring together information on chemical partitioning, reactivity, environmental fate and transport, bioaccumulation, exposure, endpoints, and emission rates in a coherent mass balance framework to yield an overall estimate of risk at a screening level.	(Arnot et al. 2006)
Princeton Ocean Model.	Princeton Ocean Model is a three-dimensional, coastal ocean circulation model. The model, originally designed to carry out coastal ocean predictions, has been extended to include estuaries as well as whole basins, like the Golf of Mexico, and to operate in both prognostic and diagnostic modes. Schimmelpfennig et al. (2012) used the model to understand the retention time of nutrients and trace pollutants in lake water and to efficiently apply regulatory measures.	(Blumberg and Mellor 1987; Schimmelpfennig et al. 2012)
Artificial Neural Networks.	Artificial Neural Networks are inspired by the early models of sensory processing by the brain. An artificial neural network can be created by simulating a network of model neurons in a computer. By applying algorithms that mimic the processes of real neurons, we can make the network "learn" to solve many types of problems. Silva et al. (2019) evaluated the use of artificial neural networks to predict soil sorption and desorption, as well as the environmental contamination potential of diuron, hexazinone, and sulfometuron-methyl herbicides in Brazilian soils.	(Krogh 2008; Silva et al. 2019)

(Continued)

Table 6.2 (Continued)

Model Name	Introduction	References
Life-Cycle Assessment.	Life Cycle Assessment is a tool to assess the environmental impacts and resources used throughout a product's life cycle, i.e., from raw material acquisition, via production and use phases, to waste management. In other researches, life-cycle assessment methodology is used as a means to quantify the potential environmental impacts on ecotoxicity and human toxicity of wastewater containing priority and emerging pollutants.	(Finnveden et al. 2009; Muñoz et al. 2008)
Targeted National Sewage Sludge Survey (TNSSS).	The EPA initiated the Targeted National Sewage Sludge Survey (TNSSS) in the mid-2000s to characterize pollutants that may be present in suspended solids (SS) generated by the nation's publicly owned treatment works (POTWs) to develop regulations on the use or disposal of SS.	(USEPA 2009)
Monte Carlo simulation models.	Monte Carlo simulation is a type of simulation that relies on repeated random sampling and statistical analysis to compute the results. This method of simulation is closely related to random experiments, experiments for which the specific result is not known in advance. In this context, Monte Carlo simulation can be considered as a methodical way of doing so-called what-if analysis.	(Raychaudhuri 2008)
Quantitative structures-activity relationship (QSAR).	Quantitative structures–activity relationship (QSAR) models were built to estimate the soil sorption coefficients (log K_{oc}) of substituted organic chemicals. QSAR is a useful technique to correlate their physical, chemical, biological, or environmental activities to their physico-chemical property descriptors. Because the experimental determination is time-consuming and expensive, estimated values based on QSAR models are now widely used.	(Nantasenamat et al. 2009)

modeling has the potential to be applied to assess the fate and transport of MCs in groundwater, the environmental fate behaviors of as-yet uncharacterized emerging contaminants of regulatory interest (Li and Colosi 2012). The hydrodynamic Princeton Ocean Model has been applied to assess the transport of pollutant and to optimize pipeline operation strategies to maintain water quality in Lake Tege in Germany, which was polluted by MCs (Schimmelpfennig et al. 2012). Artificial neural networks were applied to assess the breakthrough profiles of pharmaceuticals through floodplain filtration systems (Behera et al. 2012).

Life cycle assessment was also applied to evaluate whether a WWTP is more effective for pharmaceutical removal (Igos et al. 2012). Monte Carlo simulation models were used to assess the fate of ciprofloxacin from WWTP effluent (Chari and Halden 2012; Harris et al. 2013). Researchers also assessed the most sensitive parameters in the estimation of ecotoxicological characterization factors by Monte Carlo analysis using different modeling approaches (Morais et al. 2013). The fate and transport of 9 commonly found MCs were simulated using HYDRUS1D software in reclaimed water in 90 cm deep soil (Chen et al. 2013). de García et al. (2013) applied a quantitative structures–activity relationship (QSAR) model to evaluate the adverse impacts of 96 pharmaceutical compounds and metabolites in combination with environmental toxicological indexes. Their results show the highest levels of risk based on the PBT index included antidepressants, hormones, blood lipid regulators, personal care products, and antidepressant metabolites (de García et al. 2013).

6.5 Limitations

Routes of MCs transportation into groundwater media include transformation pathways, which are important to understand the fate of MCs in groundwater environments. Besides of the difficult selection of relevant transformation products for monitoring purposes, one of the key challenges in evaluation of transformation products of MCs in groundwater is the low detected concentrations with potential toxicity. The low concentration also limits the bacterial isolation, enrichment, and sensitive detection (Fatta-Kassinos et al. 2011). Lack of testing guidelines could be the main limitation, and recent activities have been conducted to define strategies of MCs tests and validation of existing tests for groundwater analyses (Petrovic et al. 2004).

A new approach to overcome the limitations of MCs analyses is to employ high-resolution mass spectrometric detection methods. Hybrid tandem mass spectrometers (combining two mass spectrometric techniques including one high-resolution technique) have been shown to enable reliable, sensitive, and fast detection and identification of low-molecular weight substances due to their high mass resolution and mass accuracy (Martínez Bueno et al. 2007). Full-scan chromatograms enable selective searching for the molecular ions of transformation products according to their exact mass, whereas mass spectrometry/mass spectrometry provides structural information from compound fragmentation (Hollender et al. 2010).

6.6 Concluding Remarks

This chapter introduces the basic knowledge of and common types of microconstituents (MCs) in groundwater. The chapter also deals with the assessment of their environmental fate and transport properties in the groundwater system. The modeling tools, which can be used for the fate and transport of MCs in groundwater, as well as their limitations, are also discussed. Results from many investigations imply that a variety of MCs have been discovered in groundwater systems. Thus, groundwater contamination has become of global concern, which has received increased attention to this issue. One of the most important management strategies is to formulate regulations and standards of MCs in groundwater, especially where groundwater is a source for potable water. It is also a necessity to understand the occurrence, distribution, behavior, potential toxicity, and risks of MCs before a thorough management guideline and management strategy can be established. Through detailed investigations, it is suggested and expected that favorable results can be obtained, which will be applied for the prevention and control of groundwater contamination by MCs.

References

Anderson, P.D., D'Aco, V.J., Shanahan, P. et al. (2004). Screening analysis of human pharmaceutical compounds in US surface waters. *Environmental Science & Technology* 38 (3): 838–849.

Arnot, J.A., Mackay, D., Webster, E., and Southwood, J.M. (2006). Screening level risk assessment model for chemical fate and effects in the environment. *Environmental Science & Technology* 40 (7): 2316–2323.

Atugoda, T., Vithanage, M., Wijesekara, H. et al. (2021). Interactions between microplastics, pharmaceuticals and personal care products: implications for vector transport. *Environment International* 149: 106367.

Behera, S.K., Rene, E.R., and Park, H.-S. (2012). Neural network modeling of sorption of pharmaceuticals in engineered floodplain filtration system. *Expert Systems with Applications* 39 (5): 6052–6060.

Bläsing, M. and Amelung, W. (2018). Plastics in soil: analytical methods and possible sources. *Science of the Total Environment* 612: 422–435.

Blumberg, A.F. and Mellor, G.L. (1987). A description of a three-dimensional coastal ocean circulation model. *Three-Dimensional Coastal Ocean Models* 4: 1–16.

Boxall, A.B., Rudd, M.A., Brooks, B.W. et al. (2012). Pharmaceuticals and personal care products in the environment: what are the big questions? *Environmental Health Perspectives* 120 (9): 1221–1229.

Browne, M.A., Niven, S.J., Galloway, T.S. et al. (2013). Microplastic moves pollutants and additives to worms, reducing functions linked to health and biodiversity. *Current Biology* 23 (23): 2388–2392.

Chari, B.P. and Halden, R.U. (2012). Predicting the concentration range of unmonitored chemicals in wastewater-dominated streams and in run-off from biosolids-amended soils. *Science of the Total Environment* 440: 314–320.

Chen, W., Xu, J., Lu, S. et al. (2013). Fates and transport of PPCPs in soil receiving reclaimed water irrigation. *Chemosphere* 93 (10): 2621–2630.

Cunningham, V.L., D'Aco, V.J., Pfeiffer, D. et al. (2012). Predicting concentrations of trace organic compounds in municipal wastewater treatment plant sludge and biosolids using the PhATE™ model. *Integrated Environmental Assessment and Management* 8 (3): 530–542.

Daugherty, M. (2016). Adsorption of organic pollutants to microplastics: the effects of dissolved organic matter. Master thesis, Department of Environmental Science, Northwestern University: USA.

de García, S.O., Pinto, G.P., García-Encina, P.A., and Mata, R.I. (2013). Ranking of concern, based on environmental indexes, for pharmaceutical and personal care products: an application to the Spanish case. *Journal of Environmental Management* 129: 384–397.

Ebele, A.J., Abdallah, M.A.-E., and Harrad, S. (2017). Pharmaceuticals and personal care products (PPCPs) in the freshwater aquatic environment. *Emerging Contaminants* 3 (1): 1–16.

Fatta-Kassinos, D., Manaia, C., Berendonk, T.U. et al. (2015). COST Action ES1403: new and emerging challenges and opportunities in wastewater REUSe (NEREUS). *Environmental Science and Pollution Research* 22 (9): 7183–7186.

Fatta-Kassinos, D., Meric, S., and Nikolaou, A. (2011). Pharmaceutical residues in environmental waters and wastewater: current state of knowledge and future research. *Analytical and Bioanalytical Chemistry* 399 (1): 251–275.

Finnveden, G., Hauschild, M.Z., Ekvall, T. et al. (2009). Recent developments in Life Cycle Assessment. *Journal of Environmental Management* 91 (1): 1–21. doi: 10.1016/j.jenvman.2009.06.018.

Godoy, V., Blázquez, G., Calero, M. et al. (2019). The potential of microplastics as carriers of metals. *Environmental Pollution* 255: 113363.

Gouin, T., van Egmond, R., Price, O.R., and Hodges, J.E. (2012). Prioritising chemicals used in personal care products in China for environmental risk assessment: application of the RAIDAR model. *Environmental Pollution* 165: 208–214.

Guo, X. and Wang, J. (2019). Sorption of antibiotics onto aged microplastics in freshwater and seawater. *Marine Pollution Bulletin* 149: 110511.

Harris, S., Morris, C., Morris, D. et al. (2013). Simulation model to predict the fate of ciprofloxacin in the environment after wastewater treatment. *Journal of Environmental Science and Health, Part A* 48 (7): 675–685.

Hollender, J., Singer, H., Hernando, D. et al. (2010). The challenge of the identification and quantification of transformation products in the aquatic environment using high resolution mass spectrometry. In: *Xenobiotics in the Urban Water Cycle*, 195–211. Springer

Igos, E., Benetto, E., Venditti, S. et al. (2012). Is it better to remove pharmaceuticals in decentralized or conventional wastewater treatment plants? A life cycle assessment comparison. *Science of the Total Environment* 438: 533–540.

Katz, B.G., Griffin, D.W., and Davis, J.H. (2009). Groundwater quality impacts from the land application of treated municipal wastewater in a large karstic spring basin: chemical and microbiological indicators. *Science of the Total Environment* 407 (8): 2872–2886. doi: 10.1016/j.scitotenv.2009.01.022.

Koelmans, A.A., Nor, N.H.M., Hermsen, E. et al. (2019). Microplastics in freshwaters and drinking water: critical review and assessment of data quality. *Water Research* 155: 410–422.

Krogh, A. (2008). What are artificial neural networks? *Nature Biotechnology* 26 (2): 195–197. doi: 10.1038/nbt1386.

Lapworth, D., Baran, N., Stuart, M., and Ward, R. (2012). Emerging organic contaminants in groundwater: a review of sources, fate and occurrence. *Environmental Pollution* 163: 287–303.

Laws, B.V., Dickenson, E.R., Johnson, T.A. et al. (2011). Attenuation of contaminants of emerging concern during surface-spreading aquifer recharge. *Science of the Total Environment* 409 (6): 1087–1094.

Li, C. and Colosi, L.M. (2012). Molecular similarity analysis as tool to prioritize research among emerging contaminants in the environment. *Separation and Purification Technology* 84: 22–28.

Lin, K. and Gan, J. (2011). Sorption and degradation of wastewater-associated non-steroidal anti-inflammatory drugs and antibiotics in soils. *Chemosphere* 83 (3): 240–246.

Liu, J.-L. and Wong, M.-H. (2013). Pharmaceuticals and personal care products (PPCPs): a review on environmental contamination in China. *Environment International* 59: 208–224. doi: 10.1016/j.envint.2013.06.012.

Martínez Bueno, M.J., Agüera, A., Gómez, M.J. et al. (2007). Application of liquid chromatography/quadrupole-linear ion trap mass spectrometry and time-of-flight mass spectrometry to the determination of pharmaceuticals and related contaminants in wastewater. *Analytical Chemistry* 79 (24): 9372–9384.

Morais, S.A., Delerue-Matos, C., Gabarrell, X., and Blánquez, P. (2013). Multimedia fate modeling and comparative impact on freshwater ecosystems of pharmaceuticals from biosolids-amended soils. *Chemosphere* 93 (2): 252–262.

Muñoz, I., José Gómez, M., Molina-Díaz, A. et al. (2008). Ranking potential impacts of priority and emerging pollutants in urban wastewater through life cycle impact assessment. *Chemosphere* 74 (1): 37–44. doi: 10.1016/j.chemosphere.2008.09.029.

Nantasenamat, C., Isarankura-Na-Ayudhya, C., Naenna, T., and Prachayasittikul, V. (2009). A practical overview of quantitative structure-activity relationship. *EXCLI Journal* 8: 74–88.

Padhye, L.P. and Tezel, U. (2014). Fate of environmental pollutants. *Water Environment Research* 86 (10): 1714–1773.

Petrovic, M., Eljarrat, E., De Alda, M.L., and Barceló, D. (2004). Endocrine disrupting compounds and other emerging contaminants in the environment: a survey on new monitoring strategies and occurrence data. *Analytical and Bioanalytical Chemistry* 378 (3): 549–562.

Prata, J.C., Lavorante, B.R.B.O., Montenegro Mdc, B.S.M., and Guilhermino, L. (2018). Influence of microplastics on the toxicity of the pharmaceuticals procainamide and doxycycline on the marine microalgae *Tetraselmis chuii*. *Aquatic Toxicology* 197: 143–152. doi: 10.1016/j.aquatox.2018.02.015.

Rahman, S.M., Eckelman, M.J., Onnis-Hayden, A., and Gu, A.Z. (2018). Comparative life cycle assessment of advanced wastewater treatment processes for removal of chemicals of emerging concern. *Environmental Science & Technology* 52 (19): 11346–11358.

Raychaudhuri, S. (2008). Introduction to Monte Carlo simulation. In: *Winter Simulation Conference* (eds Mason, S.J., Hill, R.R., Mönch, L. et al.), 91–100. IEEE.

Salamanca, M., López-Serna, R., Palacio, L. et al. (2021). Study of the rejection of contaminants of emerging concern by a biomimetic aquaporin hollow fiber forward osmosis membrane. *Journal of Water Process Engineering* 40: 101914. doi: 10.1016/j.jwpe.2021.101914.

Schimmelpfennig, S., Kirillin, G., Engelhardt, C., and Nützmann, G. (2012a). Effects of wind-driven circulation on river intrusion in Lake Tegel: modeling study with projection on transport of pollutants. *Environmental Fluid Mechanics* 12 (4): 321–339.

Schimmelpfennig, S., Kirillin, G., Engelhardt, C. et al. (2012b). Seeking a compromise between pharmaceutical pollution and phosphorus load: management strategies for Lake Tegel, Berlin. *Water Research* 46 (13): 4153–4163.

Silva, T.S., de Freitas Souza, M., Maria da Silva Teófilo, T. et al. (2019). Use of neural networks to estimate the sorption and desorption coefficients of herbicides: a case study of diuron, hexazinone, and sulfometuron-methyl in Brazil. *Chemosphere* 236: 124333. doi: 10.1016/j.chemosphere.2019.07.064.

USEPA (2009). Targeted national sewage sludge survey sampling and analysis technical report. Washington, DC: Office of Water.

Verlicchi, P., Al Aukidy, M., and Zambello, E. (2012). Occurrence of pharmaceutical compounds in urban wastewater: removal, mass load and environmental risk after a secondary treatment: a review. *Science of the Total Environment* 429: 123–155. doi: 10.1016/j.scitotenv.2012.04.028.

Wong, J.K.H., Lee, K.K., Tang, K.H.D., and Yap, P.-S. (2020). Microplastics in the freshwater and terrestrial environments: prevalence, fates, impacts and sustainable solutions. *Science of the Total Environment* 719: 137512.

Zhang, X., Zheng, M., Wang, L. et al. (2018). Sorption of three synthetic musks by microplastics. *Marine Pollution Bulletin* 126: 606–609.

7

Microconstituents in Surface Water

Po-Jung Huang[1], *Fang-Yu Liang*[2], *Thakshila Nadeeshani Dharmapriya*[2], *and Chih-Ming Kao*[2,3]

[1]*Department of Chemical and Materials Engineering, National Central University, Taoyuan, Taiwan*
[2]*Institute of Environmental Engineering, National Sun Yat-sen University, Kaohsiung, Taiwan*
[3]*Civil and Environmental Engineering Department, College of Engineering, University of Nebraska, Lincoln, Omaha, NE, USA*

7.1 Introduction

Microconstituents (MCs), also called emerging contaminants (ECs), in the aqueous environment have recently become a worldwide issue. The concentration of MCs is commonly in the range of a few ng/L to several μg/L. Current wastewater treatment plants (WWTPs) are not specifically designed to eliminate MCs. Therefore, they can be potentially released into surface water (e.g., rivers, lakes, creeks, wetlands, oceans). The MCs cannot only be transported from wastewater effluent but also from runoff through climatic conditions like rainfall, snowmelt, and storms. Additionally, evaporation is also another source of transport for these MCs. In this chapter, we summarize the recent occurrence of various MCs in surface water and discuss their possible fate during transportation.

7.2 Major Microconstituents in Surface Water

Due to different applications, microconstituents (MCs) can be grouped as pharmaceuticals and personal care products (PPCPs), steroid hormones, industrial additives, pesticides, and heavy metal ions. The collected data consist of reports and research from five continents from 2000 to the present.

7.2.1 Pharmaceuticals and Personal Care Products (PPCPs)

Pharmaceuticals are detected in the water environment, mainly originating from wastewater from hospitals or industrial effluents and wastewater treatment plants (WWTPs). Additionally, these pharmaceuticals and their metabolic products can also be detected in domestic effluents because of the disposal of drugs and excretions. Runoff water is another way to leach pharmaceuticals from the soil. Surprisingly, pharmaceuticals are also detected in the Arctic regions. These pharmaceutical compounds are metabolized in the human

Microconstituents in the Environment: Occurrence, Fate, Removal, and Management, First Edition.
Edited by Rao Y. Surampalli, Tian C. Zhang, Chih-Ming Kao, Makarand M. Ghangrekar, Puspendu Bhunia, Manaswini Behera, and Prangya R. Rout.

body, subsequently excreted through urine and feces to become potential contaminants. The excretion rate plays an important role in determining the frequency of pharmaceutical chemicals in surface water (Luo et al. 2014). Table 7.1 presents the excretion rate of some pharmaceutical compounds. Even though the excretion rate of these pharmaceuticals is relatively low compared with others, like norfloxacin or amoxicillin, it reflects the massive usage in hospital or household wastewater.

The occurrence of pharmaceuticals in surface water has been recently reported. Carbamazepine, caffeine, diclofenac, ibuprofen, and sulfamethoxazole are reported most frequently in surface water of many countries (Reyes et al. 2021). Carbamazepine is an anthropogenic marker to access water quality because of its persistence. In most cases, carbamazepine is poorly removed by conventional WWTPs, resulting in its release through the effluent. Additionally, carbamazepine is an anticonvulsant medication for curing neuropathic pain. Hai et al. (2018) reported that around 10% of carbamazepine can be excreted from the human body to domestic wastewater. Moreover, the metabolized carbamazepine, carbamazepine-10,11-epoxide, was also detected in surface water (Doummar 2014).

The main source of caffeine is mostly related to human activity, and only a few caffeine products are naturally contributed from caffeine-producing plants, like cocoa and coffee plants. The United States Environmental Protection Agency (USEPA) announced that caffeinated drinks like coffee or tea are the major sources, and they are transported into drainage from direct disposal of caffeinated drinks or the residues from drinking vessels. Moreover, several researchers also observed that the caffeine in surface water is more related to population density. Tamura et al. (2017) quantified that the caffeine concentration in the water stream is 5.33 μg/L in Japan, and Pico et al. (2020) monitored caffeine concentrations in Saudi Arabia reaching 21 μg/L. Compared to wastewater, caffeine concentration is relatively low in surface water.

Caffeine and acetaminophen were also found to have the highest concentrations in Africa, reaching 1214 μg/L and 343 μg/L, respectively. Additionally, nicotine of 9.34 μg/L was also observed in Nigeria's surface water, which is attributed to the presence of tobacco farms near to the sampling sites (Ebele et al. 2020). Caffeine, acetaminophen, and methylparaben were detected in South America, and their concentrations were 13 μg/L, 1.72 μg/L, and 0.66 μg /L, respectively (Chaves et al. 2020). These results indicate that pharmaceutical contamination can be significantly influenced by location and human activities (Ebele et al. 2020).

Table 7.1 Human excretion rate of pharmaceutical compounds (Ternes 1998; Huschek et al. 2004; Jjemba 2006; Luo et al. 2014).

Excretion Rate	Pharmaceutical Compound
Low (≤5%)	Aspirin (acetylsalicyclic acid), carbamazepine, gemfibrozil, ibuprofen.
Moderately low (6–39%)	Diclofenac, metroprolol, primidone, sulfamethoxazole.
Relatively high (40–69%)	Bezafibrate, norfloxacin, trimethoprim.
High (≥70%)	Amoxicillin, ciprofloxacin, tetracycline.

Diclofenac is a non-steroidal anti-inflammatory drug (NSAID) and is mostly used as a sodium salt for relieving pain and inflammation. The annual global consumption of diclofenac achieves 940 tons (Lonappan et al. 2016). Diclofenac is most frequently found in surface water and occasionally found in groundwater and drinking water. The main source of NSAIDs in surface water has been considered as the release of WWTP effluent (Kasprzyk-Hordern et al. 2009). Moreover, diclofenac is seldom detected in soil due to its low K_{OW}. Even though some studies detected the existence of diclofenac in soil, especially in sandy sediment, it can be potentially leached into groundwater or drinking water after rain events (Scheurell et al. 2009). Diclofenac is frequently detected in rivers, estuaries, and lakes in Europe, with the concentrations ranging from 2 to 260 ng/L. In Asia, the most frequent finding of diclofenac is located in river water. Scheurell et al. (2009) reported that the highest concentration was detected in Pakistan of 4.9 μg/L because of the absence of advanced WWTPs. Diclofenac also was detected in the coastal zones of Brazil and Pakistan where the concentration was 19.4 and 100 ng/L, respectively (Scheurell et al. 2009). In Costa Rica, the most frequently detected NSAIDs in surface water were carbamazepine, sulfamethoxazole, and triclosan, and their median concentrations were 82, 56, and 263 ng/L, respectively (Spongberg et al. 2011). In South Africa, three kinds of NSAIDs (naproxen, ibuprofen, and diclofenac) are detected in an aqueous environment, especially in river water, and their maximum concentrations were 6.84, 19.2, and 9.70 μg/L, respectively (Madikizela and Chimuka 2017). Their low K_{OW} (1.60 for naproxen, 2.48 for ibuprofen, and 1.90 for diclofenac) was the main reason for them to escape from WWTPs. Compared with influent/effluent of wastewater in other countries, these results demonstrated that the release of NSAIDs was caused by the absence of advanced WWTPs (Madikizela and Chimuka 2017).

The fifth most abundant pharmaceutical observed in surface water is antibiotics in both developed and developing countries (Kairigo et al. 2020). Many antibiotics groups, like enrofloxacin, ciprofloxacin, oxolinic, ofloxacin, norfloxacin, sulfapyridine, trimethoprim, metronidazole, and their metabolites are reported as having higher concentrations in hospital effluent, especially in India. The economic growth in Indian relies on the development of agriculture. The manufacturing and usage of antibiotics have rapidly developed, resulting in antibiotic contamination in surface water.

In Vietnam, surface water contains sulfamethoxazole (21 ng/L), sulfadiazine (4 ng/L), trimethoprim (17 ng/L), and enrofloxacin (12 ng/L) (Giang et al. 2015). Surface water contains ng/L grade antibiotics in China. For example, macrolides, sulfonamides, and trimethoprim were the most frequently detected antibiotics in the Dongjiang River, with the concentrations ranging from 0.9 to 67.4 ng/L (Zhang et al. 2012). The source of macrolides and sulfonamides/trimethoprim were probably from domestic wastewater and livestock agriculture wastewater, respectively. The concentration of antibiotics in river water was generally below the concentration in coastal water because of dilution, which implies that the contamination of antibiotics becomes serious once antibiotics are detected in offshore waters. Furthermore, this antibiotic contamination could be attributed to agricultural activities near the sea (Zhang et al. 2012). Zuo et al. (2011) observed that the contaminated level of antibiotics increased following an increase in population density. This confirms that the main source of antibiotics in the surface water is connected with human activities.

Personal care products frequently found in surface water were disinfectants, synthetic musks, and UV filters. Triclosan (TCS) and triclocarban (TCC) are the most frequent components in personal care products, including soaps, detergents, toothpaste, medical disinfectants, and kitchen cleaners (Sandborgh-Englund et al. 2006; Calafat et al. 2008). Therefore, TCS and TCC are commonly detected in household wastewater. In the United States, these two substances were measured in surface water ranging from 4.5 to 47.3 ng/L and 3.9 to 28.3 ng/L, respectively (Gautam et al. 2014). Yang et al. (2015) observed that TCS was detected in hospital wastewater effluent ranging from 3 to 68 ng/L in the Jiaosu River and from 2.7 to 51 ng/L in the Diao-Bao River in Taiwan. Moreover, the amount of the TCS in surface water in the wet season is higher than that in the dry season due to photo-decomposition.

Polycyclic musks occurred in the Songhua River with 1,3,4,6,7,8-hexahydro-4,6,6,7,8,8-hexamethyl-cyclopenta-[γ]-2-benzopyran (HHCB) and 7-acetyl-1,1,3,4,4,6-hexamethyl-1,2,3,4-tetrahydronapthalene (AHTN) as the two main species, and their concentrations detected in surface water were 2 to 37 ng/L and 1 to 87 ng/L, respectively (Zhang et al. 2020). HHCB detected in the Taihu Lake and the Jiaozhou Bay wetland achieved 212.04 ng/L and 208 ng/L, respectively (Guo et al. 2013; Jiang et al. 2018). In Singapore and Korea, HHCB is still the most abundant substance in the range of 5.16 to 42.9 ng/L and 0.27 to 0.342 ng/L, respectively (Lee et al. 2015; Wang and Kelly 2017). Additionally, polycyclic musks (PCMs) were also detected in the seawater samples from the Matosinhos beach. In the urban catchment of China, HHCB and AHTN are dominant substances in the range of 13.2 to 95 ng/L and 2.98 to 232 ng/L, respectively. Urban wastewaters and volatilization were identified as the main sources for PCMs (Homem et al. 2016).

UV filters usually enter the aquatic environment through industrial effluent, storm water runoff, or anthropogenic activities. BP-3, 2-ethylhexyl-4-methoxycinnamate (EHMC: also known as octinoxate), homosalate (HMS), octocrylene (OC), and octyl dimethyl-p-amino-benzoic acid (ODPABA) were most frequently found in surface water, especially in seawater (Mao et al. 2019). These UV filters in the ocean exhibit a high threat to marine species, especially coral. Mitchelmore et al. (2019) collected seawater samples from Hawaii and detected at least 8 UV-filters in the samples, with homosalate (HMS) (53.0–625.7 ng/L) being the most abundant substance followed by OS, BP-3, and OC. Tsui et al. (2017) monitored seawater near coral from the eastern Pearl River Estuary of the South China Sea and found the maximum concentrations at 4 and 8 ng/L of BP-3 and ODPABA, respectively. Thus, the concentration of these compounds was highly influenced by anthropogenic activities and seasonal variations (Astel et al. 2020). Figure 7.1 summarizes the concentration and distribution of PPCPs in surface water on different continents.

7.2.2 Endocrine-Disrupting Chemicals

Hormones existing in humans work by producing chemicals to stimulate biological responses through specific hormones acceptors. Hormones based on their structures can be classified into four groups: aslipid-derived hormones; amino acid-derived hormones; peptide hormones; and glycoprotein. In recent decades, some chemicals, like phenolic compounds used in industry or agriculture, were found to potentially interfere with health effects, especially for the endocrine system. These compounds are called endocrine-disrupting

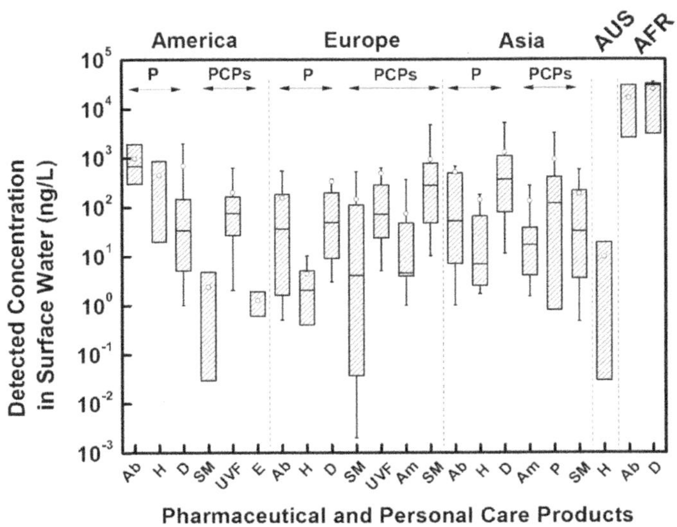

Figure 7.1 The occurrence and distribution of pharmaceutical and personal care products (PPCPs) in the surface water worldwide. (A = antibiotics; H = hormone; D = drug; SM = synthetic musks; UVF = UV filter; E = estrogen; Am = antimicrobial reagents; P = preservative).

chemicals (EDCs). The USEPA also defined an exogenous substance that caused adverse health effects in an intact organism or its progeny, consequent to endocrine function. The presence of EDCs in surface water may come from nature, pharmaceuticals, agriculture, and industry. Estrogen estrone (E1), 17β-estradiol (E2), and estriol (E3) are natural steroids excreted by humans and animals so that they are widely discovered in the aqueous environment. Yao et al. (2018) collected 217 water samples from surface water in China. E1 and E3 were the most abundant substances in river water, with a mean concentration of 1.8 and 3.7 ng/L, respectively. The concentration of E2 is 0.56 mg/L, which is lower than the two natural steroids. Some reports mentioned that E1 and E3 are the main substance excreted from human and animals. Moreover, E2 in an aqueous environment is easily degraded or oxidized through microorganisms, resulting in low concentrations in surface water. Jurgens et al. (2002) observed that the transformation of E2 to E1 occurred in an English river with half-lives of 0.2 to 9 days through microorganisms. The third most abundant estrogen in surface water is synthetic estrogen 17α-ethynylestradiol-3-sulfate (EE2), with a mean concentration of 1.7 ng/L (Zuo et al. 2013). The geographical distribution of estrogens indicated that the occurrence of estrogen was considered as the correlation of human activities in China (Yao et al. 2018). The estrogen concentration at eastern coastal areas was higher than that in other areas because of the higher population density in that area. The estrogen concentration in marine water is relatively low compared with phenol compounds such as Bisphenol A (BPA) and nonylphenol. In Italy, E1, E2, and EE2 were detected in the ranges of 3.2 to 6.7 ng/L, 3.0. to 175 ng/L, and 4.6 to 28 ng/L, respectively.

Xenoestrogen is defined as a synthetic or semisynthetic industrial chemical that can be activated biologically in an accident and then interfere with the functions of humans and animals. BPA and phthalates are the well-known xenoestrogens detected in surface water. BPA is widely used in industrial and manufacturing, as plasticizers, food containers, and antioxidants in cosmetics. Santhi et al. (2012) monitored seven points in the Langat River in Malaysia from upstream to downstream. BPA was detected in 93% of all the sample points, and its concentration increased from upstream to downstream. The highest concentration of BPA in river water was 215 ng/L. There are no industries located in the upstream area; thus, the occurrence of BPA should come from household wastewater only. Meanwhile, some medium industries and WWTPs located in the downstream area indicated a higher amount of industrial effluent containing a higher amount of BPA (Santhi et al. 2012). Shehab et al. (2020) discussed the BPA occurrence in forest, urban, and industrial regions in Malaysia. The highest concentration of BPA in river water was 5.52 ng/L and was found in urban regions, especially in the industrial region. This high concentration was attributed to rubber factories, paper mills, steel towers manufacturing, chemical manufacturing, car industries, and textile mills. It demonstrated that BPA existing in the surface water is highly correlated with industrial activities. In 1991, BPA has been demonstrated as an xenoestrogen because of its significant negative health impact on humans (Vandenberg et al. 2009). Therefore, BPA is already banned from production and use in many countries. According to the high usage of BPA in industries, biophenol S (BPS) is used as an alternative chemical for replacing BPA. BPS exhibits higher stability, indicating difficult degradation through heat, sunlight, and even microorganisms (Wu et al. 2018). However, some reports mention that the toxicity of BPS might be equal or higher than that of BPA. Therefore, it is also required to consider the occurrence of BPS in surface water for human exposure. In China, BPS, BPA, and BPF (BPA derivatives) were detected in river water, and the concentrations were 0.51–8.9, 7.9–42, and 0.5–29 ng/L, respectively (Wu et al. 2018). In Asia, the higher concentration of BPS was found in India (26.5 ng/L). Even though the BPA and its analogs are detected in surface water or other aqueous environments, the most abundant BPA occurs in sediments because of its hydrophobicity. The respective values of the normalized partition coefficient of BPS, BPA, and BPF are around 2.5, 2.8, and 3.7, respectively (Choi and Lee 2017). Hence, a higher level of these chemicals are detected in sediment.

Phthalate Esters (PAEs) are widespread compounds used as plastic additives to improve the flexibility and elasticity of plastic-based products. The toxicity of PAEs is similar to BPA, destroying the endocrine system (Luis et al. 2021). Therefore, the USEPA announced di-(2-ethylhexyl) phthalate (DEHP), di-n-octyl phthalate (DnOP), dimethyl phthalate (DMP), diethyl phthalate (DEP), dibutyl phthalate (DBP), and butyl benzyl phthalate (BBP) as priority pollutants. Also, the European Parliament and the Council of the European Union declared restriction of PAEs useage in children's appliances in 2005. He et al. (2019) collected water from Lake Chaohu in China and six phthalate esters (PAEs) (i.e., dimethyl phthalate (DMP), diethyl phthalate (DEP), diisobutyl phthalate (DIBP), dibutyl phthalate (DBP), butyl benzyl phthalate (BBP), and di(2-ethylhexyl) phthalate (DEHP)) were detected. The concentration of total PAEs ranged from 0.37 to 13.2 μg/L in lake water, and DIBP was the dominant substance. Kingsley and Witthayawirasak (2020) analyzed surface water samples from the southern part of Thailand, and the total concentration of PAEs ranged from 1.44 to 12.08 μg/L. DEHP is the dominant compound in surface water, and its concentration ranged from 1.28 to 5.28 μg/L. After DEHP, DiNP and DBP were followed in the range from ND to 3.36 μg/L and ND to 3.44 μg/L, respectively. To consider the position, the highest concentrations of PAEs were located near industrial cities in southern Thailand. The industrial effluents are dominantly attributed to PAEs contamination in Thailand.

7.2.3 Industrial Chemicals

Fire retardants (or flame retardants) are abundant additives in plastics, and the purpose of addition is to increase their heat and fire resistance. Organohalogen (tetrabromobisphenol A, TBBPA) and organophosphorus compounds (triphenyl phosphate, TPP) are the main groups of fire retardants. Since the production and use of brominated flame retardants (BFRs) are restricted, the organophosphorus compound is widely used (Segev et al. 2009). Organophosphorus compounds are composed of organophosphate triesters in the central group with their substituents, like halogenated alkyl, non-halogenated alkyl, and aryl phosphates. According to these different substituents, organophosphorus flame retardants can be classified as four groups, including chlorinated organophosphorus flame retardants (Cl-OPFRs), Br-OPFRs, alkyl-OPFRs, and aryl-OPFRs. Cl-OPFR, like tris(2-chloroisopropyl) phosphate (TCIPP) is the additive flame retardant in polyurethane (PU) for the building and automobile industries, while aryl-OPFR, like tris(phenyl) phosphate (TPHP) is added to polyvinyl chloride (PVC) and thermoplastics (van der Veen and de Boer 2012; Cristale et al. 2017). Usually, the addition of OPFRs to plastic is processed by physical blending and then manufactured under the melting point of plastics. Therefore, OPFRs readily evaporate and transport to the environment.

Several researchers reported that OPFRs have been discovered in aquatic environments, like surface water (rivers, lakes), wastewater, drinking water, and even coastal water. Moreover, OPFRs can also be observed in Antarctic fresh water and Arctic Ocean water, indicating the possibility of long-distance transportation. Zhang et al. (2021) reported that OPFRs such as tris(2-chloro-propyl) phosphate (TCIPP), tris(2-chloroethyl) phosphate (TCEP), and tri-n-butyl phosphate (TNBP) are the dominant species in surface seawater of Qinzhou Bay, with concentrations of 217.5, 144.8, and 18.3 ng/L, respectively. Xiao et al. (2021) also observed the presence of OPFSs in surface seawater of the West Pacific. Tris(2-chloroethyl) phosphate (TCEP) is the most abundant contributor, whose concentration is 10.9 ng/L, and TNBP (6.26 ng/L) and TDCIPP (4.96 ng/L) came after TCEP. Additionally, the presence of TPTP in surface seawater has a widespread distribution. The composition of OPFRs in the Pacific Ocean is similar to water from China, Korea, and Japan, indicating that the introduction of OPFRs originates from the coastal area of the east and southeast areas. In Germany, the dominant substance, TCPP was detected in surface water (River Elbe) and surface seawater (German Bight) in the range of 40 to 250 ng/L and of 3 to 28 ng/L, respectively. Others, such as triethyl phosphate (TEP), tri-iso-butyl phosphate (TiBP), tris(2-butoxyethyl), phosphate (TBEP), and triphenylphosphine oxide (TPPO) were found in surface water in concentrations of ~180 ng/L (Regnery and Puttmann 2010; Bollmann et al. 2012). Bollmann et al. (2012) also observed that the concentration of these OPFRs changed seasonally. Non-halogenated OPFSs were degraded through microorganisms or photo-decomposition. In the US, tris(2-chloro-propyl) phosphate (TCIPP) was an abundant retardant in rivers (74.6 ng/L), lakes (59.4 ng/L), tap water (11.6 ng/L), rainwater (61.8 ng/L), and surface seawater (31.9 ng/L). Kim and Kannan (2018) also monitored the seasonal change of total OPFRs and found that the concentration in surface water reached a minimum in the winter season. This implied higher temperatures might launch the evaporation of OPFRs from building materials, resulting in detecting OPFRs in rainwater and even in surface seawater through atmospheric diffusion. Figure 7.2 summarizes the occurrence and distribution of steroidal

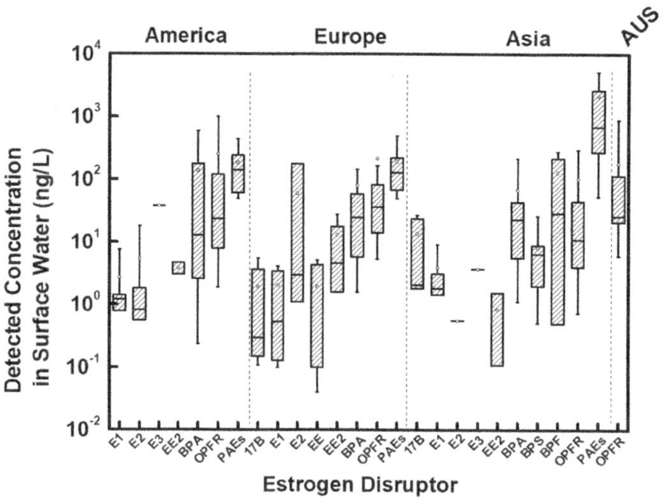

Figure 7.2 Occurrence and distribution of steroid hormone, xenoestrogens, organophosphorus flame retardants (OPFR), and phthalate ester (PhE) in surface water worldwide. (E1 = Estrogen; E2 = Estradiol; E3 = estriol; EE2 = 17α-ethynylestradiol; 17B = 17B Estradiol; BPA = bisphenol A; BPS = bisphenol sulfone; BPF = 4,4′-methylenediphenol; OPFR = organophosphorus flame retardants; PAEs = phthalate ester).

hormone, xenoestrogens, organophosphorus flame retardants (OPFR), and phthalate ester (PAE) in surface water on different continents.

7.2.4 Pesticides

Agriculture is the most important industry worldwide. It is necessary to maintain and support ever-increasing food production for human survival (Sadowski and Baer-Nawrocka 2018). However, the growth of agricultural crops is continously threatened by insects and microorganisms. Therefore, the use of chemicals, called pesticides, can effectively protect these crops. According to the target species, pesticides can be classified into three groups as herbicides, insecticides, and fungicides. Atrazine is the most common herbicide detected in surface water because of its widespread use in soybean and corn crops (de Souza et al. 2020). Ouyang et al. (2019) monitored atrazine and acetochlor in three geographical areas (i.e., estuary, coastal, bay) in the Yellow Sea region. The bay area contained the highest concentration of atrazine (76 ng/L), while the lowest concentration of atrazine was detected in the estuary area (26 ng/L). It was predicted that atrazine would deposit during transportation from agriculture sampling sites to the marine environment. However, acetochlor in the bay area was detected in lower concentrations than in the estuary, implying the marine environment has decreased its solubility. In Canada, atrazine and glyphosate were the most frequently found herbicides in surface water, and the concentrations varied from 4 to 666 ng/L and from 2 to 3000 ng/L, respectively (Lalonde and Garron 2020). Montiel-Leon et al. (2019) observed that both atrazine and glyphosate existed in south shore tributaries of the St Lawrence River, while only glyphosate was found in north shore tributaries of that river. This observation corresponds to the agricultural impact. Higher agriculture activities increase the amount of herbicide usage resulting in a higher concentration of herbicide in discharged wastewater (Montiel-León et al., 2019). Additionally, the different distributions of atrazine and glyphosate also can be attributed to the hydrological properties of the system and different biological activities (Montiel-León et al. 2019).

Metalochlor is the second most abundant herbicide in the US for its widespread use in corn and potato crops (Sun et al. 2019). The main reason for easy observation of metalochlor in surface water is due to its low degradation efficiency; its half-life in surface water is achieved in a couple of years (de Souza et al. 2020). Additionally, the undegraded metalochlor potentially starts volatilization under normal temperatures and moisture, then the evaporated metalochlor is transported for a long distance in the atmosphere (Glinski et al. 2018). Rose et al. (2018) reported that only 10% of metalochlor volatilized into the atmosphere; 0.3% of it returns as rainfall; 0.4% is moved into surface water; and 0.3% sinks into sediment or soil. This also indicated that the wide distribution was caused by the volatilization.

In Costa Rica, diuron was the most frequent herbicide observed in river water, and the measured concentration achieved 22,770 ng/L. The main reason might be due to the higher persistence in the aquatic environment (Carazo-Rojas et al. 2018). Balakrishnan et al. (2012) reported that diuron was found in surface seawater in Japan, and the concentration ranged from 10 to 62 ng/L. The sources of diuron were considered as agricultural and urban run off.

The pathway of insecticides releasing is similar to that of herbicides releasing through surface flushing, runoff, and pulverization. In Costa Rica, dimethoate was the abundant insecticide that existed in surface water. Stackpoole et al. (2021) summarized the three types of pesticides found from 2013 to 2017 in the US. Clorpyrifos and dimethoate were insecticides most frequently found in surface water, which were also highly detected in European countries. Figure 7.3 summarizes the arrangement for occurrence and distribution of pesticides (insecticides, herbicides, and fungicides) in surface water on different continents.

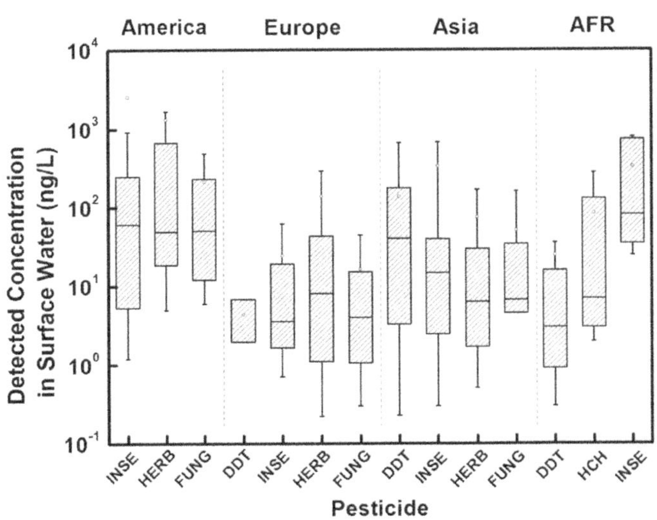

Figure 7.3 Occurrence and distribution of pesticides worldwide. (INSE = insecticide; HERB = herbicide; FUNG fungicide).

7.3 Water Cycles, Sources, and Pathways of Microconstituents, and the Applicability of Mathematical Models

In recent years, many studies have shown that our aquatic environments contain chemicals and compounds that can cause ecological damage. As detection technology procedures continue to measure compounds in microscale concentrations, additional compounds are being identified in our aquatic environments (e.g., drinking water, rivers, etc.). These products are known as microconstituents (MCs) (e.g., drugs, antibiotics, fragrances, soaps, pesticides, plasticizers). Pharmaceuticals and personal care products (PPCPs) are the most common sources of MCs in the environment. PPCPs are ubiquitous in nature and our lives, and millions of pharmaceuticals become waste products each year as products pass their expiry date. The following are examples of PPCPs and pesticides.

7.3.1 Pharmaceutical and Personal Care Products (PPCPs)

The sources of PPCPs in surface water include WWTP effluent, treated industrial effluent and tanks, municipal sewage transportation infrastructure, septic tank systems, pixel-treated sewage tanks, immersion aftercare treatment, inhalation or runoff overflow from contaminated land, runoff from agricultural fields or regions, aquaculture, land animals and particles produced, swimming in surface waters, waste from mis-designed landfills, or leachate in cemeteries, which may be released to the environment (Figure 7.4). For a river

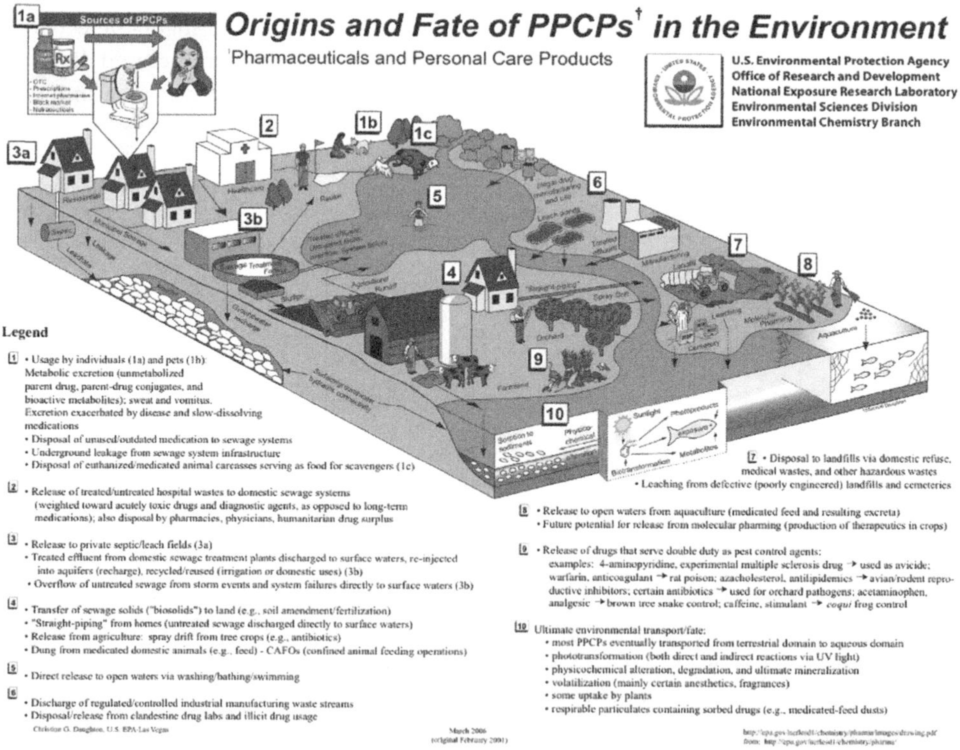

Figure 7.4 Origins and fate of PPCPs in the environment. *Source:* Daughton, 2006/U.S. Environmental Protection Agency/Public Domain.

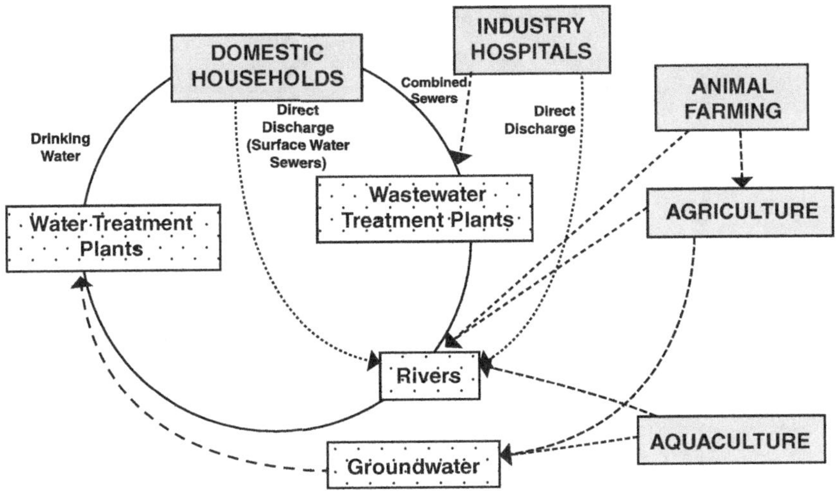

Figure 7.5 Sources and pathways of PPCPs in the urban water cycle (Ellis 2006).

basin, the occurrence and fate of PPCPs are affected by various factors, such as geography, regional characteristics, climate, population size and density, economy, and consumption habits of PPCPs.

PPCPs enter the water body in their original form or as metabolites. The main route is via the human body from where they are metabolized and excreted into the water body. The waste exudate easily enters the environment, causing difficulties in the treatment of the back-end WWTP. Therefore, the concentration of the discharged water into the natural water body is also high, leading to pollution of rivers, lakes, groundwater, and drinking water sources.

Figure 7.5 shows the main anthropogenic sources of PPCPs. Discharges directly entering receiving water bodies during manufacturing, processing, and distribution are considered negligible. However, losses from fish farms and agriculture go directly to surface water and groundwater without sewage treatment. During plant treatment, the water discharged through the surface to the underwater way is also untreated.

7.3.2 Pesticides in Surface Water

Figure 7.6 shows pesticides in the Hydrological Cycle System. The pesticides may enter surface water directly through runoff, spills, or various effluents. Pollution may also be indirect. Pesticides first enter the atmosphere or groundwater and then transfer to surface water. Pesticides are usually transferred from fields to surface water. Drainage caused by runoff, rain, or irrigation water is one of the main ways that pesticides are transported from their application areas to other areas of the environment. Once the pesticides reach streams, they can be widely dispersed in other streams, rivers, lakes, reservoirs, and oceans.

In order to simulate the fate of pesticides at the watershed scale, it is necessary to consider the spatially variable land management and landscape characteristics, the temporally variable meteorological and hydrology, and the dissipation process in the river. After pesticides

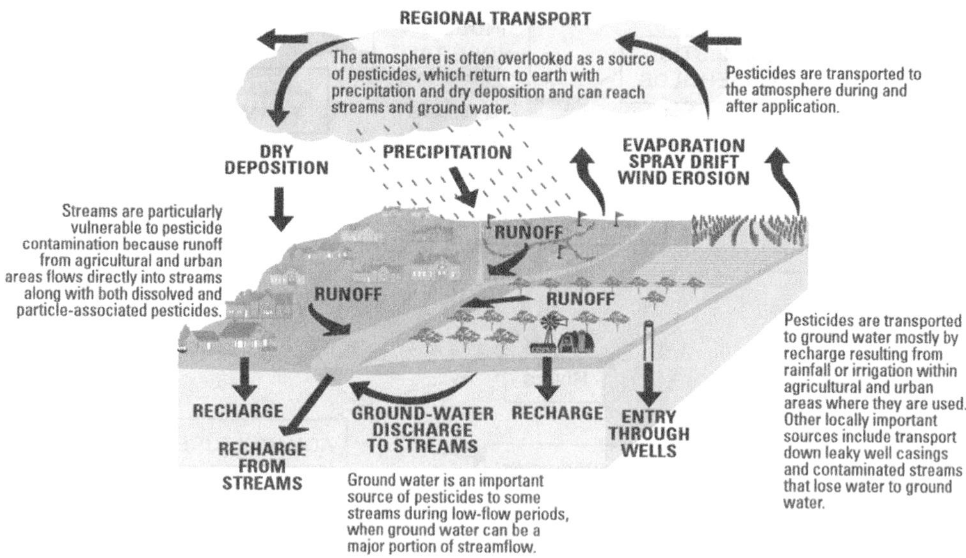

Figure 7.6 Pesticides in the Hydrological Cycle System. *Source:* Majewski and Capel, 1995/U.S. Department of the Interior/Public Domain.

are used on farmland, they enter surface water, groundwater, soil, plants, and air, and accumulate in the water through biological circulation, breaking the ecological balance, affecting people's daily drinking water, and thus having an impact on human health. Therefore, the impact of pesticides on the environment is worthy of our attention.

The existance of MCs in surface water mainly comes from the release of WWTP effluent compared with other sources. Even though the wastewater in WWTPs is treated by several processes like flocculation, sedimentation, and activated sludge treatment, there are still some limitations for removing traces of pollutants at the level of ng/L. Krzmarzick and Novak (2014) rearranged the chlorinated organic compounds, like diclofenac and triclosan, in different WWTPs, compared their concentrations in influent and effluent streams, and found that all of them are difficult to eliminate completely. Higher removal efficiency (e.g., >97%) in WWTPs usually occurs when the influent concentration is higher; the removal ability of WWTPs dramatically reduces as the influent concentration of pollutants reduces. Additionally, some pollutants, like pesticides or fire retardants, show high resistance. Therefore, these pollutants cannot be easily degraded by traditional drinking water and wastewater treatment plants. Eventually, the remaining pollutants are detected in effluent and eventually in surface water.

Rainfall and stormwater runoff potentially lead to leached pollutants from the soil into surface water. Müller et al. (2021) collected water samples during a storm event at the Ammer River and reported a different chemical compositions profile. MC-loaded soil and sediment may be released and suspended during storm and rain events, resulting in contamination of surface water. Phillips et al. (2012) also discovered that storm and flood events caused overflow and then the untreated wastewater was discharged into surface water.

Atmospheric transportation also plays an important role in impacting surface water. Spray drift is the common method to treat pesticides toward agricultural crops. The

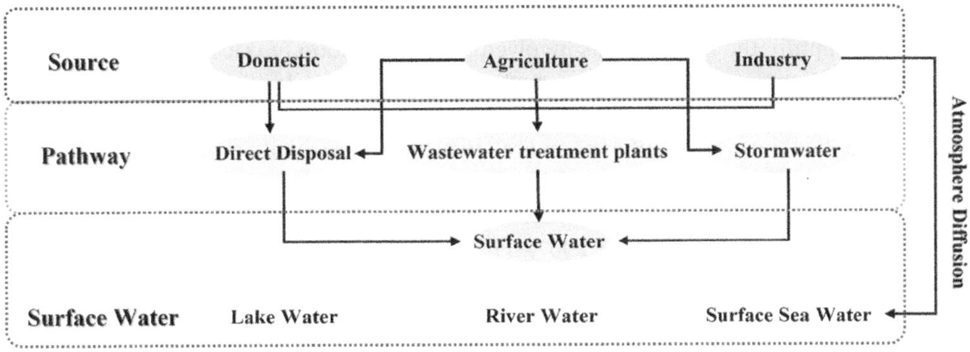

Figure 7.7 Possible sources and pathways for MCs.

generation of tiny droplets containing pesticides are easily suspended and then transported in the air. Therefore, the contamination can be caused by long-distance influences (Lalonde and Garron 2020). Volatilization is another way to transport, especially long-distance transportation from a territory to surface water and even to surface seawater. Figure 7.7 presents the possible sources and pathways for MCs.

7.3.3 The Applicability of Mathematical Models

Chemicals used daily in homes, work places, and urban environments will eventually be transported in water, whether in urban wastewater, parks, or garden runoff. Domestic wastewater is contaminated by chemicals, drugs and their metabolites, detergents, personal care products, plastics, and flame retardants. These chemicals are also known as micropollutants (MPs) and have been scientifically tested in several water bodies worldwide. The concentration of MPs is usually much lower. The way pollutants accumulate may have a serious impact on a hazard assessment. According to reports, pesticides, drugs, cosmetics, and cleaning products from agricultural irrigation and domestic sewage are usually found in lakes and rivers in the form of MPs. Despite their low concentration, these pollutants may have adverse effects on aquatic ecosystems. To optimally assess possible changes associated with these MCs, it is necessary to couple pollution models with the hydrological model, which is capable of simulating low river flows under future climate change (Michalak 2016). Scientific hypotheses are not approximations of the truth; model predictions are made by numerically solving the mathematical equations that represent parameterizations (e.g., infiltration and runoff), parameters (e.g., soil hydraulic conductivity), and observations (e.g., precipitation, solar radiation). The solving process is a form of rigorous deduction.

7.3.4 Advantages and Disadvantages of Mathematical Tools

The SWAT model has demonstrated its ability to simulate water quality and will be able to be used in the field of agricultural planning and management in the future. The issue of pesticide runoff at the watershed scale has received increasing attention in recent years, which provides opportunities for model development and application. However, it is difficult to fully capture the mechanisms involved in the migration of trace pollutants and the resulting environmental consequences, such as the leaf area index calculation of potential transpiration and potential soil evaporation. Most agricultural pesticides, especially herbicides, are applied during a significant and relatively short seasonal period. The input pathways for further diffusion of pesticides into surface water are atmospheric deposition after volatilization and aeolian deposition of soil particles containing pesticides that were previously eroded by wind (Holvoet et al. 2007). The determination of daily potential plant transpiration is considered to be the limit of the annual vegetation growth cycle in the growth module of the standard SWAT model, which can also affect the simulation of transpiration (Alemayehu et al. 2017).

Future studies using the SWAT model will require good spatial and temporal data in order to make more detailed assessments and reduce uncertainties. The main weakness of this model is the non-spatial representation of HRUs in sub-watersheds. This keeps the model simple and supports the depth of the model almost every time. The model's land use, soil, and slope heterogeneity are measured through the sub-watershed area. This subtle method suddenly changed HRU traffic and phone calls. The model needs to obtain a wide range of different data to run, and the calculation requires a large number of parameters to modify during the growth period, which can hinder the modeler's use. However, the environment is a complex system to model; decomposition or parameterization may lead to inaccuracies or even wrong results (Glavan and Pintar 2012). Table 7.2 summarizes the advantages and disadvantages of SWAT and HSPF models.

Figure 7.8 shows the global applications of SWAT in pesticide modeling (Wang et al. 2019). The majority of published SWAT pesticide research is done in the United States, primarily in the Midwest and California. The database in the SWAT model applies to North America (https://www.card.iastate.edu/swat_articles); thus, other regions may need to establish their own databases. In terms of appearance data, the watershed in North America has expanded in a large area, while the watershed in some countries like Taiwan has been built due to its small size and denseness. Future research on the SWAT model will require good spatial and temporal data in order to make more detailed assessments with reduced uncertainty.

Table 7.2 The advantages and disadvantages of SWAT and HSPF models.

Model	Advantage	Disadvantage
SWAT	It can assess the seasonal and annual changes of the non-point source pollutant load, determine the long-term water-quality change trend, and describe the land use type and topography in the actual watershed.	It is not suitable for flood simulation of a single event, and some parameters of the model's database must be modified during application.
HSPF	GIS can be used as model input and conversion, reducing the time of processing data and increasing model credibility.	When a typhoon occurs, the flow has a large error, and the pre-operation time is long.

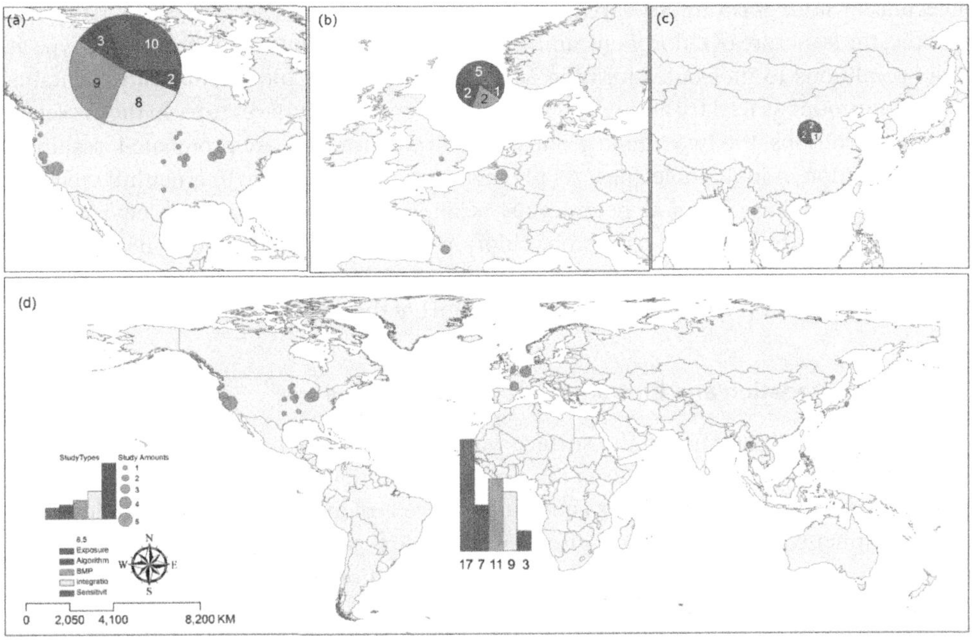

Figure 7.8 Global application of SWAT in pesticide modeling (Wang et al. 2019).

7.4 Fate and Transport of Microconstituents in Aquatic Environments

7.4.1 Adsorption of Microconstituents

The migration of microconstituents (MCs) is determined by their physical and chemical properties, which can be described by the acid dissociation constant (pK_a) and n-octanol-water partition coefficient (K_{OW}) (Rogers 1996; Schafer et al. 2011; Kim and Zoh 2016). K_{OW} is a partition coefficient to describe the distribution of MCs for octanol-water systems and is defined as the ratio of the concentration of MCs in n-octanol at water under equilibrium at a specific temperature, as shown in Eq. (7.1).

$$K_{OW} = \frac{concentration\ in\ n - octanol}{concentration\ in\ water} \tag{7.1}$$

It is indicated that hydrophilic MCs prefer to exist in water resulting in low K_{OW}, while the MCs with high K_{OW} are considered hydrophobic chemicals. Adsorption of MCs to solids or suspended particles in the surface water generally occurs through hydrophobic interaction. Therefore, Rogers (1996) provided a general rule for the prediction and estimation of sorption potential for MCs, i.e., the MC exhibited lower sorption ability if its $\log K_{OW}$ was <2.5, moderate sorption for $\log K_{OW}$ between 2.5 and 4, and high sorption for $\log K_{OW}$ >4.0.

Additionally, some MCs contain functional groups, such as amines, carboxylic groups, and phosphate, resulting in their different characterizations. The MCs exhibit either molecular or ionic form under varying pH of the environment (Schafer et al. 2011). For example, the structure of chlorpheniramine can be influenced by the pH from ionic type at acidic conditions to molecules forming at basic conditions. Chlorpheniramine contains two amine groups as tertiary amine and heterocyclic amine in its structure. In the presence of acidic conditions, the two amine groups in chlorpheniramine are protonated, resulting in the formation of ionic molecules. As pH increases from an acidic to a neutral environment, deprotonation occurs at heterocyclic amines. When the pH of the environment further increases to the basic condition, chlorpheniramine is converted to its molecular form (Lv et al. 2014). Therefore, the occurrence of MCs is predictable through the combination of their physical properties and the pH of the environment.

7.4.2 Biodegradation and Biotransformation of Caffeine

The fate of hydrophilic MCs in the surface water is potentially processed by biodegradation. Caffeine in the surface water is mainly attributed to domestic wastewater, and the concentration of caffeine achieves a couple of hundreds of milligrams in household drainage. The estimated consumption of caffeine worldwide is generally between 80 and 400 mg per person per day. The concentration and distribution of caffeine increase as the population density increases. Therefore, the amount of caffeine in surface water is highly dependent on anthropogenic activities. Caffeine has high solubility because of its extremely low octanol-water coefficient (log $K_{OW} = -0.07$). Some researchers reported that caffeine in surface water might be biodegradable during transportation. Bacterial strains (*Pseudomonas* and *Serratia*), fungal strains (*Aspergillus* and *Penicillium*), and algae (*Aiptasia* and *Pseudoterogorgia*) are discovered to biodegrade caffeine (Dash and Gummadi 2007; Edwards et al. 2015). The best microorganism, found by Dash and Gummadi (2007), *Pseudomonas sp.* NCIM 5235, can completely degrade 6.4 g/L of caffeine in 24 hours. Caffeine is transported from household drainage into surface water, like rivers or ponds, and deposited in the marine environment. The occurrence of caffeine in coastal water will potentially affect the environmental parameters, like pH and temperature, resulting in inhibition of coral growth (Vieira et al. 2022).

7.4.3 Biodegradation and Biotransformation of Steroidal Estrogen

The biotransformation of steroid estrogens can be commonly observed for natural steroidal estrogens, E1, in surface water. The major reason is caused by the higher excretion rate from microorganisms and biotransformation of E2 to E1. Jürgens et al. (2002) found that E2 was easily oxidized to E1 through microorganisms in English rivers, and its half-lives were approximately 0.2 to 9 days. However, synthetic estrogens, like EE2, have more resistance toward biodegradation or biotransformation compared with natural steroids. The half-life of EE2 was estimated to be approximately 108 days under aerobic conditions. Under anaerobic conditions, the degradation of EE2 was even longer than that in the aerobic environment. EE2 exhibits higher persistence either under aerobic or anaerobic conditions. Although EE2 is not easily decomposed through the biological route, the photodegradation process may occur under light irradiation because of the aromatic structure in EE2. Zuo et al. (2013) observed that the half-life of EE2 through sunlight irradiation was 23 hours, which was faster than microorganisms. However, the photodegradation efficiency is easily affected by environmental effects, like color or turbidity. Therefore, the photo-degradation can be classified as direct photolysis and indirect photolysis. Direct photolysis occurs when the estrogenic molecules directly adsorb light within 280 to 390 nm to launch the photochemical reaction of themselves, while the indirect photolysis occurs using hydroxyl radical generated from dissolved organic matter (DOM) in surface water. DOM, like humic substances, nitrate/nitrite, and transition metal ions, to trigger the serial reaction for generating hydroxyl and superoxide radical, leading to decomposition of EE2. The two main reactions, radical initiation and photo-decomposition, illustrate indirect photolysis of EE2 in surface water (Zuo et al. 2006).

Radical Initiation:
Dissolved Organic Matter (DOM):

$$DOM + h\nu \rightarrow DOM^{-} \tag{7.2}$$

$$DOM^{-} + O_2 \rightarrow oxidized\ DOM + O_2^{-} \tag{7.3}$$

Nitrate/Nitrite:

$$NO_2^{-} / NO_3^{-} + H_2O + h\nu \rightarrow NO / NO_2 + OH + OH^{-} \tag{7.4}$$

Transition Metal Ion:

$$Fe(III) - organic\ complex + h\nu \rightarrow Fe(II) + Org \tag{7.5}$$

$$Org + O_2 \rightarrow oxidized\ org + O_2 \tag{7.6}$$

$$2O_2^{-} + 2H^{+} \rightarrow H_2O_2 + O_2 \tag{7.7}$$

$$Fe(II) + H_2O_2 \rightarrow Fe(III) + OH + OH^{-} \tag{7.8}$$

$$H_2O_2 + h\nu \rightarrow 2OH \tag{7.9}$$

Photo-decomposition:

$$EE2 + OH^{\cdot} \rightarrow decomposed\ products \tag{7.10}$$

The photo-decomposition of MCs in the period of transportation should be predictable because of their structure. Cristale et al. (2017) observed that OPFRs potentially decomposed under sunlight irradiation through direct and indirect decomposition. Nine OPFRs classified into three groups are observed for their ability to photo-induced decompose in river water: alkyl phosphates (tri(butyl) phosphate (TNBP), tris(2-butoxyethy) phosphate (TBOEP), and tris(2-ethylhexyl) phosphate (TEHP); chloroalkyl phosphates (tris(2-chloroethyl) phosphate (TCEP), tris(2-chloroisopropyl) phosphate (TCPP), and tris(1,3-dichloropropyl) phosphate (TDCPP); as well as aryl phosphates (2-ethylhexyl diphenyl phosphate (EHDP), tris(phenyl) phosphate (TPHP); and tris(methylphenyl) phosphate (TMPP)). The experimental results showed that the photo-decomposition rate of aryl phosphate was faster than that of talkyl and chloroalkyl phosphate due to its aromatic structure. Even though acryl OPFRs are facile to adsorb sunlight, their degradation still relies on indirect photo-decomposition. Cristale et al. (2017) observed that the photo-decomposition rates of acryl OPFRs were faster in a sufficient oxygen environment, which demonstrates that the photo-induced decomposition of MCs in surface water occurs by both direct and indirect processes (Cristale et al. 2017).

Photo-degradation efficiency is also influenced by water depth. For example, lake water is stratified into three layers, epilimnion, thermocline, and hypolimnion, because of heating by sunlight. The epilimnion is the surface layer of lake water, which is typically warmer and has a higher dissolved oxygen concentration than the hypolimnion. In the thermocline layer, temperature is dramatically changed, and it will become almost constant at a specific depth, which is determined by the penetration of sunlight. The hypolimnion layer is the bottom layer, which usually receives insufficient sunlight irradiation. As shown in (Eq. 7.11), the Lambert–Beer approach can be used to describe the relationship between adsorbed photo flux (P_a) and depth (d_{th}) (Calderaro and Vione 2020).

$$Pa = \int_{\lambda} p^0(\lambda)\left[1 - 10^{-100 A_1(\lambda) DOC d_{th}}\right] d\lambda \tag{7.11}$$

where $p^0(\lambda)$ is the incident spectral photo flux density of sunlight at λ wavelength; A_1 is the specific absorbance of the lake water; DOC is the dissolved organic carbon including both sunlight-absorbing and non-absorbing; d_{th} is the depth of the thermocline; and 100 is the conversion factor between meters and centimeters. This formula also considers the effect of DOC concentrations. When DOC increases, the depth of the penetrated light decreases. Moreover, the higher wavelength of incident light leads to the exponential decay of light density in surface water. The proper wavelength is around $\lambda = 290–450\,\text{nm}$, and the specific absorbance is equal to $A_1(\lambda) = 0.45e^{-0.015\lambda}$ (Vione et al. 2010). The epilimnion lake water is able to adsorb 96 to 99.9% of incident light. As a result, most MCs are photodegraded at the epilimnion layer (Calderaro and Vione 2020).

Surfactants in WWTPs are degraded during secondary treatment processes resulting in 90 to 95% elimination efficiency of initial surfactants concentrations. Then these WWTP effluents containing different surfactants and the associated degraded products are discharged into surface water (Olkowska et al. 2014). Surfactants present in surface water can undergo sorption and aerobic/anaerobic degradation processes. Generally, the sorption process is relative to the hydrophobicity of the surfactant. The concentration in sediment or suspended solid samples increases with decreasing polarity of the group in the surfactant, like C_{13}-linear alkylbenzene sulfonate (C_{13}-LAS) and nonylphenol ethoxylate (NPEO). The higher polarity of surfactants, like C_{10}-linear alkylbenzene sulfonate (C_{10}-LAS) and nonylphenoxy carboxylic acids (NPEC), tend to exist as their dissolved forms in water (Lara-Martin et al. 2008a, 2008b). When considering the types of surfactants, higher concentrations of anionic surfactants are observed in water compared with cationic and nonionic compounds. The major component in suspended solids and sediment is silica dioxide, which exhibits a negative charge. Therefore, the repulsive force is generated between solid and anionic surfactants, while the attractive force is formed between solid and cationic and non-ionic surfactants. As a result, cationic and non-ionic surfactants much more easily attach onto solids compared with anionic ones.

Biodegradation of surfactants occurs not only in WWTPs but also in the environment. Microorganisms in the environment are able to take up these surfactants as the source of their carbon source and energies through metabolic reactions. Cationic surfactants, like Tetradecyl trimethyl ammonium chloride (TMAC) and Dodecyl trimethyl ammonium chloride (DMAC), are degraded under aerobic conditions resulting in *N*-dealkylation and *N*-demethylation. Additionally, the aerobic biodegradability of cationic surfactants from the group of quaternary ammonium is decreased with an increase in the number of non-methyl group. For example, the degradation rate of Me_4N^+ is higher than R_4N^+, where Me and R are a methyl group and a alkyl group, respectively (Garcia et al. 2001). Some researchers observed that linear alkylbenzene sulfonates (LAS) in surface water can be degraded significantly after 4 days, and the mineralization of these surfactants can be complete within 7 to 30 days. This rapid degradation efficiency of LAS relies on the group of fatty alcohol sulphates (AS), which can be facile to decompose through enzymatic cleavage (Ying 2006).

Hydrolysis and oxidation reactions play important roles in the degradation of a non-ionic surfactant. Using nonylphenol ethoxylates (NPEO) as examples, microorganisms primarily react with the ethoxylated chain through ω-oxidation and later α-,β-oxidation leading to a shorter chain. A shorter chain with a low molecular weight is more rapidly degraded than a longer chain. Typically, the mineralization of NPEO achieves from 50 to 80% (Potter et al. 1999; Staples et al. 2001).

Per- and polyfluoroalkyl substances (PFASs) are kinds of fluorinated fatty acids, and the abundant carbon–fluoride bonds in the structure of PFASs lead to unique hydrophobic and hydrophilic properties. PFASs are broadly used in various household and industrial applications, like water-resistant coating, firefighting foams, and semiconductors. Perfluorooctanoic acid (PFOA) and Perfluorooctanesulfonic acid (PFOS) are used most commonly in industry, resulting in their most common occurrence at the global level. The USEPA has set the advisory levels for PFOA and PFOS in drinking water at 70 ppb of an individual or combined concentrations because of their serious and chronic toxicity. Therefore, short-chain PFASs, such as C_6 polyfluoroalkyl substances, have been developed as alternatives in recent years. However, these fluoride compounds still exhibit potential health issues in the US.

PFASs enter the environment either by their direct production or by indirect sources, like degradation or transformation of their precursors. Some researchers report that PFOS concentration increases after secondary biological treatments, which is caused by the transformation of PFOS precursors. In over 25 countries (either developed or developing) PFASs can be detected in surface water. The concentrations of PFASs in developed and more industrialized countries are higher than those in developing countries.

7.5 Modeling of Microconstituents in Aquatic Environments

Surface Water Model Applications refers to estimation of impacts of changes in land use, climate, best management practices (BMPs) on streamflow, management practices, sediment, nutrients, etc., with the help of some surface water models, like Soil and Water Assessment Tool (SWAT), Hydrological Simulation Program Fortran (HSPF), TOPMODEL, and Variable Infiltration Capacity (VIC) (Hu et al. 2007). Countries usually have their own hydrological models. However, most national hydrological modeling standards are limited, such as model maintenance capabilities, calculation costs, and model simulation technical capabilities.

The Watershed Area Model is used to evaluate non-point source pollution and the nutrients and pollution load output to the water. Therefore, this model can effectively evaluate non-point source pollution by inputting meteorological data (rainfall, temperature, humidity, etc.), combining land use, soil characteristics, humanities, and other factors to do a single rainfall or continuous simulation and/or to calculate the non-point source pollution caused by the erosion of two transport currents, which can effectively evaluate the load of non-point source pollution in the watershed area. The data required for different watershed models are different. Some models may only apply to a single land use form, such as urban, rural, or agricultural land, while other models are suitable for watershed areas with mixed land use forms. The input data of the watershed model, such as topography, land use, rainfall station, etc., need to be processed by the Geographic Information System (GIS). Using the GIS function, the data parameters required by the model can be quickly converted into the data required by the model, and then connected to the provided numerical model for simulation calculation, and the simulation results can be returned to the GIS system function interface to query and display the results.

7.5.1 BASINS System Overview

The USEPA developed BASINS (Better Assessment Science Integrating Point and Nonpoint Sources), which is a watershed multi-objective environmental analysis system (USEPA). As shown in Figure 7.9, the BASINS 4.5 Core system combines several components to provide a comprehensive set of tools for conducting watershed and water quality analyses. This system combines GIS, watershed database, and multiple water quality simulation assessments. Tools and plug-in modes include Hydrological Simulation Program Fortran (HSPF), Soil and Water Assessment Tool (SWAT), the Water Quality Analysis Simulation Program (WASP), Storm Water Management Model (SWMM), and Pollutant Load (PLOAD). Their functions are to assist in the division of watersheds, integrate environmental data, analysis tools, etc., and support users to provide simpler modeling and analysis. PLOAD (Pollutant Load) is a simple watershed load estimation formula, which uses the Export Coefficient to estimate the annual average pollution load of non-point sources in a sub-catchment. HSPF (Hydrological Simulation Program Fortran) and SWAT (Soil and Water Assessment Tool) are watershed models that can be continuously simulated. WASP is purely a river water quality model and cannot simulate the production of non-point source pollution in the catchment area, so it needs to be coordinated with other models, while SWMM is mainly used in drainage system design, flood retention ponds, sewer overflow control strategies, and assessment of sewage and sewer leakage. Impact studies the non-point source pollution load in the wastewater load distribution, and evaluates the effectiveness of the Best Management Practices (BMP) facility to reduce the pollution load on rainy days.

BASINS is based on the concept of the watershed area. It integrates the watershed area data and the point source and non-point source pollution analysis required for total assessment control into the framework of the built-in GIS platform. BASINS is a useful tool

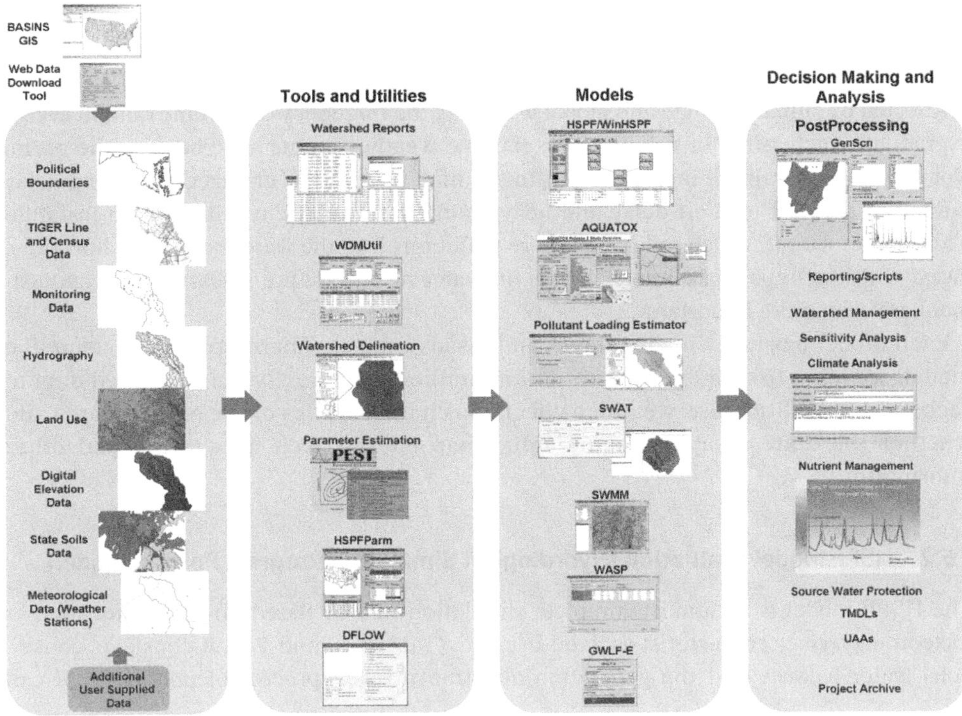

Figure 7.9 BASINS system overview. *Source:* United States Environmental Protection Agency.

for those interested in watershed management; for example, development of total maximum daily loads (TMDLs) and National Pollutant Discharge Elimination System (NDPES). This comprehensive model is suitable for the analysis and distribution calculation of TMDL. Because this non-point source pollution model adopts a continuous simulation method, it is also suitable for the development of control strategies with seasonal changes.

The BASINS non-point source pollution model is based on HSPF. Integrating hydrology, water management, and water quality into one, it can continuously simulate connection flow and pollution load, analyze point source discharge, and calculate river flow and water quality history. The HSPF model contains many modules, which can be selected to simulate and calculate various water management, water quality, or erosion parameters. The model is divided into three parts, PERLND, IMPLND, and RCHRES, which represent the simulation of permeable areas, impervious areas, and river courses in the watershed area, respectively. Descriptions of these three modules are as follows.

PERLND module: PERLND is mainly used to simulate the related reaction procedures of water quality and water quantity in the infiltration area. It is the most commonly used module in HSPF, which can simulate the movement process of water under three paths of flooding, intermediate flow, and groundwater flow.

IMPLND module: The IMPLND module is applied to the simulation of impermeable areas in the watershed area, mainly to simulate areas with little or no infiltration, such as urban areas.

RCHRES module: The RCHRES module can simulate the water quality of the infiltration area and the non-infiltration area into the river or reservoir. This module mainly simulates open channels and culverts or fully mixed lakes.

Affected by climate change and global warming, the frequency of extreme rainfall events is gradually increasing. In recent years, extreme weather events have become the norm. Not only has the rainfall increased, but the rainfall pattern has changed from long-delay rainfall in the past to short-delay and heavy rainfall nowadays, resulting in an instantaneous larger runoff, which may carry more pollutants into the water body in a short time. Faced with the impact of climate change on water resources, pre-assessment and adjustment are even more important.

Each model has its own advantages and disadvantages. Therefore, before using the model, we must first know the simulation method, complexity, and required data of each model. Additionally, we much check the characteristics of the research area and whether there are complete topographic map data as well as weather and other information.

7.5.2 HSPF Model Evaluation (Hydrological Simulation Program Fortran Model)

The HSPF model is an almost complete simulation model. Based on basic data such as meteorology, river characteristics, land use, etc. (Tables 7.3 and 7.4), it considers continuous water balance and the generation and transmission process of pollutants. It can

Table 7.3 Meteorological data required by HSPF mode.

Data Category	Profile Name	Illustrate	Unit
Meteorological	PERC	Rainfall.	In/hr
	EVAP	Evaporation capacity.	In/hr
	ATEM	Temperature.	°F
	WIND	Wind speed.	Mph
	SOLR	Sunlight.	Ly/hr
	PEVT	Evaporative potential.	In/hr
	DEWP	Dew point temperature.	°F
	CLOU	Cloud cove.r	tenths

Table 7.4 GIS data required by HSPF mode.

Data Category	Profile Name	Format
GIS	River system diagram.	Polyline
	Boundary map.	Polygon
	Land use.	Polygon
	Numerical Terrain Model (DEM).	Raster

simulate water quality, water volume, and organic and inorganic pollutants. However, its data input is complicated and belongs to a highly complex model. It can also simulate continuous rainfall and changes in the on-site rain storm (Park et al. 2020).

Results of the model simulation need to go through a parameter calibration and verification process to ensure the reliability of the simulation so that the model parameters can accurately express the local hydrology and waterquality conditions. In the process of model calibration and verification, different judgment indicators are used to evaluate the feasibility of the model (Table 7.5). Reliability and the advantages and disadvantages of different judgment indexes need to be specified. Only by fully understanding the characteristics of the judgment index can the most suitable one be selected and the credibility of the model can be effectively evaluated.

Based on the distribution of existing data, the initial performance evaluation criteria for recommended statistical performance measures, Table 7.6 shows the performance evaluation criteria for the HSPF model parameter calibration.

Table 7.5 Parameters related to flow of the HSPF model (Park et al. 2020).

Parameter	Description	Units
LZSN	Lower zone nominal storage.	inches
INFILT	Soil infiltration capacity index.	inches/h
AGWRC	Ground water recession coefficient.	none
UZSN	Upper zone nominal storage.	inches
INTFW	Interflow inflow parameter.	none
LZETP	Lower zone ET parameter.	none
DEEPFR	Fraction of groundwater inflow to deep recharge.	none
IRC	Interflow recession parameter.	none

Source: Park et al., 2020/MDPI/Public Domain CC BY 4.0.

Table 7.6 Performance evaluation criteria for the HSPF model parameter calibration.

			Performance Evaluation Criteria			
Measure	Output Response	Temporal Scale[a]	Very Good	Good	Satisfactory	Not Satisfactory
R^2	Flow[b]	D-M-A	$0.85 < R^2 \leq 1$	$0.75 < R^2 \leq 0.85$	$0.60 < R^2 \leq 0.75$	$R^2 \leq 0.60$
NSE			$0.80 < NSE \leq 1$	$0.70 < NSE \leq 0.80$	$0.50 < NSE \leq 0.70$	$NSE \leq 0.50$
PBIAS (%)			$PBIAS < \pm 5$	$\pm 5 \leq PBIAS \leq \pm 10$	$\pm 10 \leq PBIAS \leq \pm 15$	$PBIAS \geq \pm 15$

[a] D, M, and A denote daily, monthly, and annual temporal scales, respectively.
[b] includes stream flow, surface runoff, base flow, and tile flow, as appropriate, for watershed- and field-scale models.
Source: Park et al., 2020/MDPI/Public Domain CC BY 4.0.

7.5.3 Fundamental Mechanisms of SWAT Pesticide Modeling

The SWAT (Soil and Water Assessment Tool) model was developed in the early 1990s by a team led by Dr Jeff Amold at the Agricultural Research Center of the United States Department of Agriculture (USDA-ARC), with the SWRRB (Simulator for Water Resources in Rural Basics) model being its predecessor. The SWAT model is a tool that allows us to perform a comprehensive analysis of the water balance within a watershed area. The SWAT model is the only pollution simulation model that can predict the impact of land changes on water quality, water volume, sediment yield, pollutant transmission, etc., and can match the actual soil, land use, and management methods, combined with many models developed by the ARC over the years. The model takes into account several meteorological parameters (precipitation, wind, humidity, temperature, solar radiation) and information about agricultural practices (plowing, sowing, irrigation, fertilization, etc.). The model also allows for agricultural simulations including forecasting yields, the amount of nutrients taken up by plants, or the amount of nutrients leached from the field. Therefore, the SWAT model is also used to optimize agricultural practices.

7.5.3.1 SWAT Model Description

The SWAT model is a continuous, long-term, physically-based semi-distributed model developed to assess impacts of climate and land management on hydrological processes, sediment loading, and pollution transport in watersheds. SWAT has continuity in the simulation time and uses the time of day as the time simulation scale. It is a long-term simulation model and is not suitable for event flood simulation. It uses days as the basic time units to simulate hydrology, silt, surface crop growth, land use, nutrient cycle, and pesticide transmission, and evaluates the impact of water in watershed areas on hydrology, silt, nutrients, and pesticides under management operations of different land uses on soil properties. Spatially, the model subdivides the watershed area into smaller sub-watersheds, which then can be further subdivided into several Hydrological Response Units (HRUs) to reflect different land use, soil properties, and slopes, and can specify the flow direction of HRU water flow.

The model uses HRU as the main calculation unit to estimate the output of flow, silt, nutrients, and pesticides in the HRU. The total output of each HRU in the sub-watershed area can be obtained (Abbaspour et al. 2015). The output is calculated along the river course until it reaches the watershed area exit. Watershed hydrology is simulated in two phases: the land phase, which controls the amount of water, sediment, nutrient, and pesticide loadings to the main channel in each sub-basin; and the routing phase, which controls the movement of water, sediments, and so on through the streams of the sub-basins to the outlets (Abbaspour et al. 2015). The SWAT model simulates the movement of water. As shown in Figure 7.10, its hydrological calculation includes surface runoff, evapotranspiration, soil water, and groundwater. When precipitation reaches the ground in the watershed area, the water on the ground infiltrates into the soil layer. If the water volume exceeds the infiltration water volume, the water on the ground will form surface runoff into the water body, and the surface runoff transmission and water body evaporation will cause water loss.

In the model, the soil is divided into 10 layers. If the water content in the soil layer exceeds its field capacity, and the next soil layer is unsaturated, the water in the soil layer will penetrate into the next soil layer, and finally, pass through the bottom layer. The soil

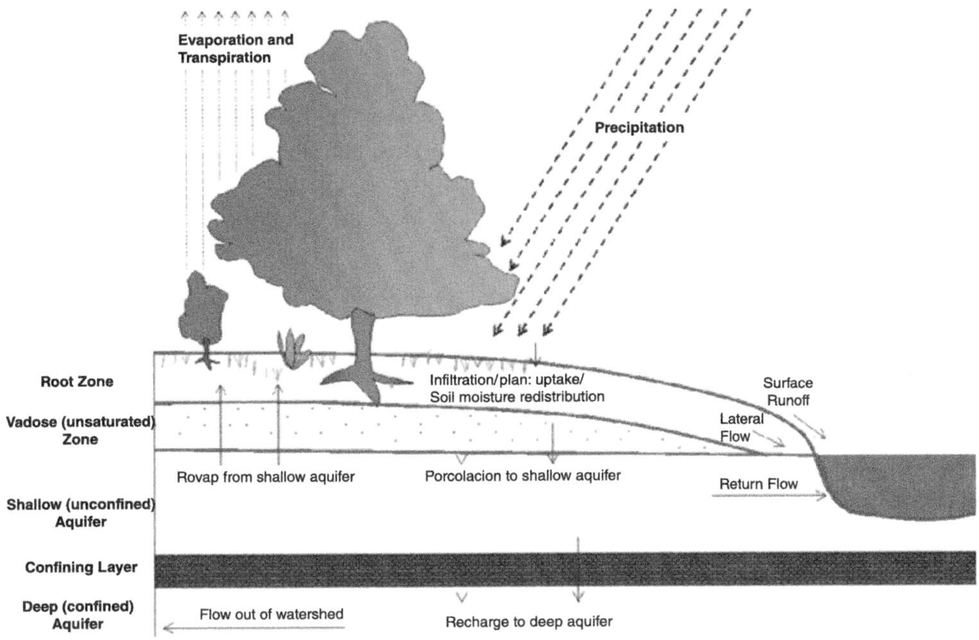

Figure 7.10 Schematic representation of the hydrological cycle.

layer is refilled into the aquifer, so if the water content in the soil layer is saturated, osmosis will be inhibited, and water will gather in the soil profile, causing lateral flow and converging into the main river. Water in the soil layer is lost due to plant uptake and soil evaporation.

In the aquifer part, the model is mainly divided into the shallow aquifer and deep aquifer. The water in the shallow aquifer will return to the main river, forming the base flow of the main river, or permeate into the deep aquifer. In the shallow aquifer, the water volume will be reduced due to re-evaporation. The water after the river course calculation flows into the river course of the next sub-watershed area. This cyclic process is the movement of water in the SWAT model.

The input data required for SWAT simulation are as follows: numerical elevation model DEM, soil data, land use data, and meteorological data. In order to make the model more accurately represent the characteristics of the study area, users can also input concentrated management information, pesticides, herbicides, fertilizers, or compost data.

7.5.3.2 SWAT Model Set-Up

The hydrological component of the model at each HRU simulates the hydrological balance using Eq. (7.12) (Abbasi et al. 2019).

$$\mathbf{SWt} = \mathbf{SWo} + \sum_{t=1}^{t} (\mathbf{Rday} - \mathbf{Q}\boldsymbol{surf} - \mathbf{Ea} - \mathbf{W}\boldsymbol{seep} - \mathbf{Q}\boldsymbol{gw}) \tag{7.12}$$

where S*Wt* is final soil moisture content (mm); S*Wo* is initial water content (mm); *t* is time in days; R*day* is precipitation (mm); Q*surf* is surface runoff (mm); E*a* is evapotranspiration (mm); *Wseep* is the amount of water seeping into the soil profile (mm); and *Qgw* is the amount of return flow (mm). The SWAT model simulates the process of silt production in the watershed area, which is mainly divided into two parts: surface erosion and silt transport in the river channel. The silt on the surface is scoured by surface runoff and enters the river channel with the surface runoff; the silt in the river channel will also show sedimentation and erosion transport behavior.

$$\text{Sed} = 11.8 * \left(Q_{surf} * q_{peak} * \text{area}_{hru}\right)^{0.56} * K_{USLE} * C_{USLE} * P_{USLE} * LS_{USLE} * C_{FRG} \tag{7.13}$$

where Sed is the sediment yield (metric tons) of an HRU; Q_{surf} is the volume of surface runoff (mm/10^4 m^2); q_{peak} is peak runoff rate (m^3/s); area_{hru} is the area of the HRU (10^4 m^2); K_{USLE} is Universal Soil Loss Equation (USLE) soil erodibility factor; C_{USLE} is USLE cover and management factor; P_{USLE} is USLE support practice factor; LS_{USLE} is USLE topographical factor; and C_{FRG} is coarse fragment factor.

Surface runoff transports the adsorbed pesticides in sediment and into the main channels. Equation 7.14 can be used to calculate pesticide loads that have been adsorbed.

$$pst_{sed} = 0.001 \frac{C_{solidphase} * sed}{\text{Area}_{hru} * \varepsilon pst_{sed}} \tag{7.14}$$

where pst_{sed} is sediment with adsorbed pesticide in the main channel (kg/pesticide/10^4/m^2); $C_{solidphase}$ is concentration of sediment on the topsoil layer; *sed* is daily sediment yield (metric tonnes); $Area_{hru}$ is the area of the HRU (10^4/m^2); and εpst_{sed} is enrichment ratio of the sediment. The total amount of pesticides is calculated by adding the amounts that have been adsorbed and those that have been dissolved.

7.5.4 Model Sensitivity Analysis, Calibration, and Validation

The following introduces the more commonly used judgment indicators in the re-evaluation of hydrological and water quality models, the method of statistical standard regression (including a coefficient of determination), the method of dimensionless analysis (including Nash–Sutcliffe efficiency), and the method classified as error-index (including percent bias). The advantages and disadvantages of each determination index are described as follows.

The coefficient of determination (R^2), Nash–Sutcliffe efficiency (NSE), and percent bias (PBIAS) were used Eqs (7.15) to (7.17) to evaluate the model.

7.5.4.1 Coefficient of Determination, R^2

R^2 evaluated the correlation between the observed and simulated values. When the relationship between the independent variable and the variable factor can be represented by regression, the goodness of fit of the regression formula can be described by R^2. Therefore, R^2 is often used to express the variable explanatory ability in regression analysis. It was expressed as a value between 0 and 1, which indicated that there was a completely linear relationship. When the R^2 of this linear regression is closer to 1, it means that the simulated value is closer to the measured value, and the simulation result is better. The R^2 equation is as in Eq. (7.15).

$$R^2 = \left(\frac{\sum_{i=1}^{n} (o_i - \bar{o})(s_i - \bar{s})}{\sqrt{\sum_{i=1}^{n} (o_i - \bar{o})^2} \sqrt{\sum_{i=1}^{n} (s_i - \bar{s})^2}} \right)^2 \tag{7.15}$$

where o_i is the ith observation; s_i is the sth analog value; is the mean observation; average simulated value; and n is the total number of samples.

7.5.4.2 Nash-Sutcliffe Efficiency Coefficient, NSE

The essential factor of hydrological models is used to perform the Nash–Sutcliffe Efficiency coefficient. For the non-linear models, NSE values can be negative. When the model performance is unsuitable, the NSE indicates negative. When efficiency is equal to 1 it represents a perfect match of estimated discharge with the observed data, whereas an efficiency of 0 suggests that the predictions of the model are as accurate as the observed data. If the NSE value is <0.5 it can be described that the model line is inadequate, and when the NSE value is 0.5–0.65 it can be described that the model line is satisfactory. When the NSE value is 0.65–0.75 it can be described that the model line is great, and when the NSE value is 0.75–1 it can be depicted that the model line perfectly fits the data, while when the NSE value is 0 it is indicated that the line does not fit the data at all.

$$NSE = 1 - \frac{\sum_{i=1}^{n} (o_i - s_i)^2}{\sum_{i=1}^{n} (o_i - \bar{o})^2} \tag{7.16}$$

where o_i is the ith observation; s_i is the sth analog value; is the mean observation; and n is the total number of samples.

PBIAS calculates the average tendency of the simulated data to be greater or less than the observed data, and the best value is 0. The lower the value, the lower the deviation of the simulation. If PBIAS is positive, the simulation value is underestimated, and the negative value means the model is overestimated.

$$PBIAS = 100 \times \frac{\sum_{i=1}^{n} (s_i - o_i)}{\sum_{i=1}^{n} (o_i)} \tag{7.17}$$

where o_i is the ith observation; s_i is the sth analog value; and n is the total number of samples.

Sensitivity analysis mainly discusses the degree of influence on the model simulation results when the input parameters change. When the simulation results are extremely sensitive to the changes of a certain input parameter, it means that a slight change in the parameter value will cause a large model output value. Therefore, sensitivity analysis can be used to understand the degree of influence of various input parameters on output values such as flow, silt, and nutrients. Parameters that are not sensitive or whose sensitivity approaches zero can be excluded from the verification procedure.

To identify the most sensitive parameters in the modeling, the calibration and uncertainty procedures (SWAT-CUP) tools were used. It is expressed by the uncertainty factor P-factor (usually represented by 95 PPU). This value is calculated at the 2.5 and 97.5% levels of an output variable, excluding 5% extremely bad simulations.

The range of P-factor is 0 to 1, and the range of R-factor is 0 to infinity. When the P-factor and R-factor are 1 and 0 respectively, the simulation results are consistent with the measured data. There are many parameters in the SWAT mode. It is difficult to make each input parameter reach an accurate value, and only the important parameter values can be made as correct as possible. Through the analysis results, the parameters with lower sensitivity are eliminated to improve the convenience and accuracy of inputting parameter values.

7.5.5 Basin Level Modeling (Pesticide Transport)

As shown in Table 7.7, many watershed models have been developed in recent years to simulate the flux of pesticides from land to surface water on a watershed scale. Here, we take the SWAT model as an example (Wang et al. 2019) to illustrate the basic mechanism of pesticide exposure simulation, the fate of pesticides at the basin level, and the transportation model. With the SWAT model, the algorithm for calculating pesticide penetration solves the loss of pesticides in surface runoff and lateral flow.

The loss of part of the pesticides applied to farmland is mainly due to wind. Part of the remaining pesticides is trapped by crop leaves or reaches the soil surface. The amount of pesticides reaching the soil surface is represented by Eq. (7.18).

$$pst_{surf} = \left(pest * ap_{ef}\right) \bullet \left(1 - \frac{1.99532 - erfc\left[1.333 * LAI - 2\right]}{2.1}\right) \tag{7.18}$$

where $pest_{surf}$ is the amount of pesticide applied to the soil surface (kg/ha); *pest* is the actual amount of pesticide applied (kg/ha); ap_{ef} is the pesticide application efficiency; *LAI* is the leaf area index, which affects ground cover; and *erfc* is the complementary error function. With an increase in LAI, the interception of growing leaves reduces the amount of pesticides on the soil surface. Pesticides in the soil environment will be further distributed between the interstitial water and soil particles.

SWAT uses a chromatography module (King and McCarty 1968) to express the phase distribution of pesticides, and the distribution ratio is determined by the soil adsorption

Table 7.7 Watershed models for pesticide fate.

Country	Watershed Models	Reference
Belgium	SEPTWA	(Beernaerts et al. 2005)
Germany	DRIPS	(Röpke et al. 2004)
USA	SWAT	(Singh et al. 2005)
USA	HSPF	(Singh et al. 2005)

coefficient (K_p). The higher the K_p value, the more easily the pesticide binds to the soil particles than to the soil water.

$$K_p = \frac{C_{solidphase}}{C_{solution}} = K_{OC} \bullet \frac{o\gamma gC}{100} \tag{7.19}$$

where K_p is soil adsorption coefficient (m^3/ton); $C_{solidphase}$ (mg/kg) and $C_{solution}$ (mg/L) are the concentration of pesticides in solid and liquid phases of soil; K_{oc} is the soil organic carbon content normalized soil adsorption coefficient (m^3/ton); and *oygC* is the percentage of organic carbon in the soil.

Pesticides in the soil environment will also be degraded via different processes (e.g., volatilization, photodegradation, hydrolysis, biological decomposition). SWAT uses lumped half-life (hlife_s) to represent the overall degradation of pesticides in the soil environment (soil water and particles).

$$pst_{s,ly,t} = \left(pst_{s,ly,t}\right) \cdot exp\left(\frac{0.693}{hlife\_s} \cdot t\right) \tag{7.20}$$

where $pst_{s,ly,t}$ and $pst_{s,ly,o}$ are the amounts of pesticides in the soil layer at time t and 0 (kg/ha); and *hlife_s* is the half-life of the pesticide in the soil (days). Pesticides distributed in the liquid phase can be transported through soil water fluxes (surface runoff, lateral flow, infiltration), as represented by Eq. (7.21).

$$pst_{flow} = \left(pst_{s,ly,t}\right) \cdot \left(1 - exp\left[\frac{-\omega_{mobile}}{SAT_{ly} + K_p \cdot \rho_b \cdot depth_{ly}}\right]\right) \tag{7.21}$$

where Pst_{flow} the pesticide content in soil water flux(kg pst/ha); ω_{mobile} the mount of mobile water (mm); SAT_{ly} is the amount of water in the soil layer at saturation (mm); ρ_b is the bulk density of the soil layer (Mg/m^3); and $depth_{ly}$ is the depth of the soil layer (mm). The amount of pesticides attached to soil particles and transported by surface runoff during soil erosion is calculated by Eq. (7.22).

$$pst_{sed} = 0.001 \cdot \frac{100 \cdot K_p \cdot pst_{s,ly,t}}{\left(SAT_{ly} + K_p \cdot \rho_b \cdot depth_{ly}\right)} \cdot \frac{sed}{area_{hur}} \cdot \epsilon_{pst,sed} \tag{7.22}$$

where pst'_{sed} is the amount of sorbed pesticide ready to be transported by surface runoff via soil erosion (kg pst/ha); $area_{hru}$ is the hydrological response unit (HRU) area (ha); *sed* is the sediment production on a certain day (tons); and $\varepsilon_{pst,sed}$ is the pesticide enrichment ratio.

After simulating the fate of pesticides on a field scale, the mobile pesticides (pst_{flow} and pst_{sed}) produced by the field can be transported to the main channel. Soluble pesticides (pst_{flow}) can be transported with the amount calculated by surface runoff, lateral flow, or seepage flow. The SWAT algorithm used to calculate pesticide infiltration can simultaneously solve the loss of pesticides in surface runoff and lateral flow. The GLEAM mode can simulate the hydrological cycle of pesticides at the terrestrial stage.

$$pst'_{surf} = \beta_{pst} \cdot \frac{pst_{flow}}{\omega_{mobile}} \cdot Q_{surf} \tag{7.23}$$

$$pst'_{lat} = \begin{cases} \beta_{pst} \cdot \dfrac{pst_{flow}}{\omega_{mobile}} \cdot Q_{lat,surf} & \textit{for top } 10mm \\ \dfrac{pst_{flow}}{\omega_{mobile}} \cdot Q_{lat,ly} & \textit{for lower layers} \end{cases} \tag{7.24}$$

$$pst'_{per,ly} = \frac{pst_{flow}}{\omega_{mobile}} \cdot \omega_{prec,ly} \tag{7.25}$$

where pst'_{surf}, pst'_{lat}, and pst'_{perly} are the amounts of pesticides transported to the lower layer through surface runoff, lateral flow, and permeation per day (kg/ha); Q_{surf} is the surface runoff produced on a certain day (mm); $Q_{lat.surf}$ and $Q_{lat,ly}$ are the amounts of water discharged from the lateral flow of the top 10 mm layer and the lower soil layer (mm); β_{pst} is the pesticide percolation coefficient; and $\omega_{prec,ly}$ is the amount of water percolating to the underlying layer (mm).

The lag process exists between the pesticides produced in the field and the pesticides delivered by the channel. For large watersheds where the concentration time exceeds 1 day, only part of the water/sediment produced can be transported to the main river channel.

$$pst'_{surf} = \left(pst_{surf} + pst_{surstor\,\iota-1}\right) \cdot \left(1 - exp\left[\frac{-surlag}{t_{conc}}\right]\right) \tag{7.26}$$

$$pst'_{lat} = \left(pst_{lat} + pst_{lasstor\,\iota-1}\right) \cdot \left(1 - exp\left[\frac{-1}{TT_{lat}}\right]\right) \tag{7.27}$$

$$pst'_{sed} = \left(pst_{sed} + pst_{sedstor\,\iota-1}\right) \cdot \left(1 - exp\left[\frac{-surlag}{t_{conc}}\right]\right) \tag{7.28}$$

where pst_{surf} and pst_{lat} are the amounts of dissolved pesticides discharged into the main channel through surface runoff and lateral flow in a given number of days (tons); pst_{sed} is the amount of pesticide adsorbed by the sediment discharged into the main channel through runoff; $pst_{surstor,l-1}$, $pst_{lasstor,l-1}$, and $pst_{sedstor,l-1}$ are the amounts of pesticides stored or lagging in surface runoff, lateral flow, and sediments the previous day; *surlag* is the surface runoff lag coefficient; TT_{lat} is the lateral flow travel time (days); and t_{conc} is the time of concentration for the HRU (h).

7.6 Conclusions

Microconsitutuents (MCs) in surface water are complex and can easily affect ecoenvironments under various environmental conditions, like climate and physical properties of the surface water. Hydrophobic MCs, like bisphenol A or organophosphate esters (OPFRs), prefer to attach to suspended particles or sediments, while hydrophilic MCs are decomposed or metabolized through microorganisms and photolysis, resulting in various metabolites in surface water. These metabolites might potentially bring more serious environmental impact and threat to human health. The physical properties of MCs, like $\log K_{ow}$, pK_a, and vapor pressure, can effectively help us to predict the distribution of MCs, trace possible pollution points, and plan remediation activities. Soil and Water Assessment Tool (SWAT) and Hydrological Simulation Program Fortran (HSPF) are useful models to gain a better understanding of the relationship between land use activities and surface water quality occurring in the watershed. The SWAT model can process and simulate the MCs at different seasons and annual changes over a long period. Additionally, this model can integrate the land type and topography to predict surface water quality. HSPF introduces the lumped-parameter model to simulate the transportation of MCs in the aquatic environments. However, large-scale water flow, like flood or typhoon, will challenge predictions through SWAT and HSPF models. Additionally, some pollutants, like pesticides, can be transported not only in an aqueous environment but also in the atmosphere (via diffusion). Thus, evaluation of MCs with atmospheric models may compensate for the possible bias of the SWAT and HSPF models in some special cases.

References

Abbasi, Y., Mannaerts, C.M., and Makau, W. (2019). Modeling pesticide and sediment transport in the Malewa River Basin (Kenya) using SWAT. *Water* 11: 87.

Abbaspour, K.C., Rouholahnejad, E., Vaghefi, S. et al. (2015). A continental-scale hydrology and water quality model for Europe: calibration and uncertainty of a high-resolution large-scale SWAT model. *Journal of Hydrology* 524: 733–752.

Alemayehu, T., Griensven, A.V., Woldegiorgis, B.T., and Bauwens, W. (2017). An improved SWAT vegetation growth module and its evaluation for four tropical ecosystems. *Hydrology and Earth System Sciences* 21: 4449–4467.

Astel, A., Stec, M., and Rykowska, I. (2020). Occurrence and distribution of UV filters in beach sediments of the Southern Baltic Sea Coast. *Water* 12: 3024.

Balakrishnan, S., Takeda, K., and Sakugawa, H. (2012). Occurrence of Diuron and Irgarol in seawater, sediments and planktons of Seto Inland Sea, Japan. *Geochemical Journal* 46: 169–177.

Beernaerts, S., Debongnie, P., G¢rard, M. et al. (2005). Evaluation of crop-protection-product losses into surface waters with the SEPTWA system. *International Journal of Environmental Analytical Chemistry* 85: 41–50.

Bollmann, U.E., Moler, A., Xie, Z.Y. et al. (2012). Occurrence and fate of organophosphorus flame retardants and plasticizers in coastal and marine surface waters. *Water Research* 46: 531–538.

Calafat, A.M., Ye, X., Wong, L.Y. et al. (2008). Urinary concentrations of triclosan in the US population: 2003–2004. *Environmental Health Perspectives* 116: 303–307.

Calderaro, F. and Vione, D. (2020). Possible effect of climate change on surface-water photochemistry: a model assessment of the impact of browning on the photodegradation of pollutants in lakes during Summer Stratification Epilimnion vs. Whole-Lake Phototransformation. *Molecules* 25: 2795.

Carazo-Rojas, E., P¢rez-rojas, G., P¢rez-villanueva, M. et al. (2018). Pesticide monitoring and ecotoxicological risk assessment in surface water bodies and sediments of a tropical agro-ecosystem. *Environmental Pollution* 241: 800–809.

Chaves, M.D., Barbosa, S.C., Malinowski, M.D. et al. (2020). Pharmaceuticals and personal care products in a Brazilian wetland of international importance: occurrence and environmental risk assessment. *Science of the Total Environment* 734: 139374.

Choi, Y.J. and Lee, L.S. (2017). Partitioning behavior of Bisphenol alternatives BPS and BPAF compared to BPA. *Environmental Science & Technology* 51: 3725–3732.

Cristale, J., Dantas, R.F., De Luca, A. et al. (2017). Role of oxygen and DOM in sunlight induced photodegradation of organophosphorous flame retardants in river water. *Journal of Hazardous Materials* 323: 242–249.

Dash, S.S. and Gummadi, S.N. (2007). Enhanced biodegradation of caffeine by *Pseudomonas* sp. using response surface methodology. *Biochemical Engineering Journal* 36: 288–293.

Daughton, C. (2006). Origins and fate of PPCPs in the environment. US Environmental Protection Agency.

de Souza, R.M., Seibert, D., Quesada, H.B. et al. (2020). Occurrence, impacts and general aspects of pesticides in surface water: a review. *Process Safety and Environmental Protection* 135: 22–37.

Doummar, J. (2014). Carbamazepine breakthrough as indicator for specific vulnerability of karst springs: application on the Jeita spring, Lebanon. *Applied Geochemistry* 47: 150–156.

Ebele, A.J., Oluseyi, T., Drage, D.S. et al. (2020). Occurrence, seasonal variation and human exposure to pharmaceuticals and personal care products in surface water, groundwater and drinking water in Lagos State, Nigeria. *Emerging Contaminants* 6: 124–132.

Edwards, Q.A., Kulikov, S.M., and Garner-O'Neale, L.D. (2015). Caffeine in surface and wastewaters in Barbados, West Indies. *Springerplus* 4.

Ellis, J.B. (2006). Pharmaceutical and personal care products (PPCPs) in urban receiving waters. *Environmental Pollution* 144: 184–189.

Garcia, M.T., Ribosa, I., Guindulain, T. et al. (2001). Fate and effect of monoalkyl Quaternary ammonium surfactants in the aquatic environment. *Environmental Pollution* 111: 169–175.

Gautam, P., Carsella, J.S., and Kinney, C.A. (2014). Presence and transport of the antimicrobials triclocarban and triclosan in a wastewater-dominated stream and freshwater environment. *Water Research* 48: 247–256.

Giang, C.N.D., Sebesvari, Z., Renaud, F. et al. (2015). Occurrence and dissipation of the antibiotics Sulfamethoxazole, Sulfadiazine, Trimethoprim, and Enrofloxacin in the Mekong Delta, Vietnam. *Plos One* 10: e0131855.

Glavan, M. and Pintar, M. (2012). Strengths, weaknesses, opportunities and threats of catchment modelling with soil and water assessment tool (SWAT) model. In: *Water Resources Management and Modeling*, 39–64. IntechOpen.

Glinski, D.A., Purucker, S.T., Van Meter, R.J. et al. (2018). Analysis of pesticides in surface water, stemflow, and throughfall in an agricultural area in South Georgia, USA. *Chemosphere* 209: 496–507.

Guo, G.H., Wu, F.C., He, H.P. et al. (2013). Screening level ecological risk assessment for synthetic musks in surface water of Lake Taihu, China. *Stochastic Environmental Research and Risk Assessment* 27: 111–119.

Hai, F.I., Yang, S.F., Asif, M.B. et al. (2018). Carbamazepine as a possible anthropogenic marker in water: occurrences, toxicological effects, regulations and removal by wastewater treatment technologies. *Water* 10: 107.

He, Y., Wang, Q.M., He, W., and Xu, F.L. (2019). The occurrence, composition and partitioning of phthalate esters (PAEs) in the water-suspended particulate matter (SPM) system of Lake Chaohu, China. *Science of the Total Environment* 661: 285–293.

Holvoet, K.M., Seuntjens, P., and Vanrolleghem, P.A. (2007). Monitoring and modeling pesticide fate in surface waters at the catchment scale. *Ecological Modelling* 209: 53–64.

Homem, V., Alves, A., Alves, A., and Santos, L. (2016). Ultrasound-assisted dispersive liquid-liquid microextraction for the determination of synthetic musk fragrances in aqueous matrices by gas chromatography-mass spectrometry. *Talanta* 148: 84–93.

Hu, H., Wang, G., Bi, X., et al. eds. (2007). Application of two hydrological models to Weihe River basin: a comparison of VIC-3L and SWAT. In: *Geoinformatics 2007: Geospatial Information Technology and Applications*, 643–652. SPIE.

Huschek, G., Hansen, P.D., Maurer, H.H. et al. (2004). Environmental risk assessment of medicinal products for human use according to European Commission recommendations. *Environmental Toxicology* 19: 226–240.

Jiang, S.J., Wang, L., Zheng, M.G. et al. (2018). Determination and environmental risk assessment of synthetic musks in the water and sediments of the Jiaozhou Bay wetland, China. *Environmental Science and Pollution Research* 25: 4915–4923.

Jjemba, P.K. (2006). Excretion and ecotoxicity of pharmaceutical and personal care products in the environment. *Ecotoxicology and Environmental Safety* 63: 113–130.

Jurgens, M.D., Holthaus, K.I.E., Johnson, A.C. et al. (2002). The potential for estradiol and ethinylestradiol degradation in English rivers. *Environmental Toxicology and Chemistry* 21: 480–488.

Kairigo, P., Ngumba, E., Sundberg, L.-R. et al. (2020). Contamination of surface water and river sediments by antibiotic and antiretroviral drug cocktails in low and middle-income countries: occurrence, risk and mitigation strategies. *Water* 12: 1376.

Kasprzyk-Hordern, B., Dinsdale, R.M., and Guwy, A.J. (2009). The removal of pharmaceuticals, personal care products, endocrine disruptors and illicit drugs during wastewater treatment and its impact on the quality of receiving waters. *Water Research* 43: 363–380.

Kim, M.K. and Zoh, K.D. (2016). Occurrence and removals of micropollutants in water environment. *Environmental Engineering Research* 21: 319–332.

Kim, U.J. and Kannan, K. (2018). Occurrence and distribution of organophosphate flame retardants/plasticizers in surface waters, tap water, and rainwater: implications for human exposure. *Environmental Science & Technology* 52: 5625–5633.

King, P.H. and McCarty, P.L. (1968). A chromatographic model for predicting pesticide migration in soils. *Soil Science* 106: 248–261.

Kingsley, O. and Witthayawirasak, B. (2020). Occurrence, ecological and health risk assessment of phthalate esters in surface water of U-Tapao Canal, Southern, Thailand. *Toxics* 8: 58.

Krzmarzick, M.J. and Novak, P.J. (2014). Removal of chlorinated organic compounds during wastewater treatment: achievements and limits. *Applied Microbiology and Biotechnology* 98: 6233–6242.

Lalonde, B. and Garron, C. (2020). Temporal and spatial analysis of surface water pesticide occurrences in the maritime region of Canada. *Archives of Environmental Contamination and Toxicology* 79: 12–22.

Lara-Martin, P.A., Gomez-Parra, A., and Gonzalez-Mazo, E. (2008a). Reactivity and fate of synthetic surfactants in aquatic environments. *TrAC-Trends in Analytical Chemistry* 27: 684–695.

Lara-Martin, P.A., Gomez-Parra, A., and Gonzalez-Mazo, E. (2008b). Sources, transport and reactivity of anionic and non-ionic surfactants in several aquatic ecosystems in SW Spain: a comparative study. *Environmental Pollution* 156: 36–45.

Lee, S., Kim, S., Park, J. et al. (2015). Synthetic musk compounds and benzotriazole ultraviolet stabilizers in breast milk: occurrence, time-course variation and infant health risk. *Environmental Research* 140: 466–473.

Lonappan, L., Pulicharla, R., Rouissi, T. et al. (2016). Diclofenac in municipal wastewater treatment plant: quantification using laser diode thermal desorption-atmospheric pressure chemical ionization-tandem mass spectrometry approach in comparison with an established liquid chromatography-electrospray ionization-tandem mass spectrometry method. *Journal of Chromatography A* 1433: 106–113.

Luis, C., Algarra, M., Camara, J.S., and Perestrelo, R. (2021). Comprehensive insight from phthalates occurrence: from health outcomes to emerging analytical approaches. *Toxics* 9: 157.

Luo, Y.L., Guo, W.S., Ngo, H.H. et al. (2014). A review on the occurrence of micropollutants in the aquatic environment and their fate and removal during wastewater treatment. *Science of the Total Environment* 473: 619–641.

Lv, G.C., Wu, L.M., Li, Z.H. et al. (2014). Binding sites of chlorpheniramine on 1:1 layered kaolinite from aqueous solution. *Journal of Colloid and Interface Science* 424: 16–21.

Madikizela, L.M. and Chimuka, L. (2017a). Occurrence of naproxen, ibuprofen, and diclofenac residues in wastewater and river water of KwaZulu-Natal Province in South Africa. *Environmental Monitoring and Assessment* 189: 1–12, 348.

Majewski, M.S. and Capel, P.D. (1995). Pesticides in the atmosphere: distribution, trends, and governing factors. *Pesticides in the Hydrologic System* 1: 214.

Mao, F.J., He, Y.L., and Gin, K.Y.H. (2019). Occurrence and fate of benzophenone-type UV filters in aquatic environments: a review. *Environmental Science: Water Research & Technology* 5: 209–223.

Michalak, A.M. (2016). Study role of climate change in extreme threats to water quality. *Nature* 535: 349–350.

Mitchelmore, C.L., He, K., Gonsior, M. et al. (2019). Occurrence and distribution of UV-filters and other anthropogenic contaminants in coastal surface water, sediment, and coral tissue from Hawaii. *Science of the Total Environment* 670: 398–410.

Montiel-León, J.M., Munoz, G., Duy, S.V. et al. (2019). Widespread occurrence and spatial distribution of glyphosate, atrazine, and neonicotinoids pesticides in the St Lawrence and tributary rivers. *Environmental Pollution* 250: 29–39.

Muller, M.E., Zwiener, C., and Escher, B.I. (2021). Storm event-driven occurrence and transport of dissolved and sorbed organic micropollutants and associated effects in the Ammer River, Southwestern Germany. *Environmental Toxicology and Chemistry* 40: 88–99.

Olkowska, E., Ruman, M., and Polkowska, Z. (2014). Occurrence of surface active agents in the environment. *Journal of Analytical Methods in Chemistry* 2014: 769708.

Ouyang, W., Zhang, Y., Gu, X. et al. (2019). Occurrence, transportation, and distribution difference of typical herbicides from estuary to bay. *Environment International* 130: 104858.

Park, J., Jung, E., Jung, I., and Cho, J. (2020). Does future climate bring greater streamflow simulated by the HSPF model to South Korea? *Water* 12: 1884.

Phillips, P.J., Chalmers, A.T., Gray, J.L. et al. (2012). Combined sewer overflows: an environmental source of hormones and wastewater micropollutants. *Environmental Science & Technology* 46: 5336–5343.

Pico, Y., Alvarez-Ruiz, R., Alfarhan, A.H. et al. (2020). Pharmaceuticals, pesticides, personal care products and microplastics contamination assessment of Al-Hassa irrigation network (Saudi Arabia) and its shallow lakes. *Science of the Total Environment* 701: 135021.

Potter, T.L., Simmons, K., Wu, J. et al. (1999). Static die-away of a nonylphenol ethoxylate surfactant in estuarine water samples. *Environmental Science & Technology* 33: 113–118.

Regnery, J. and Puttmann, W. (2010). Occurrence and fate of organophosphorus flame retardants and plasticizers in urban and remote surface waters in Germany. *Water Research* 44: 4097–4104.

Reyes, N.J.D.G., Geronimo, F.K.F., Yano, K.A.V. et al. (2021). Pharmaceutical and personal care products in different matrices: occurrence, pathways, and treatment processes. *Water* 13: 1159.

Rogers, H.R. (1996). Sources, behaviour and fate of organic contaminants during sewage treatment and in sewage sludges. *Science of the Total Environment* 185: 3–26.

Röpke, B., Bach, M., and Frede, H.-G. (2004) DRIPS: a DSS for estimating the input quantity of pesticides for German river basins. *Environmental Modelling & Software* 19: 1021–1028.

Rose, C.E., Coupe, R.H., Capel, P.D., and Webb, R.M.T. (2018). Holistic assessment of occurrence and fate of metolachlor within environmental compartments of agricultural watersheds. *Science of the Total Environment* 612: 708–719.

Sadowski, A. and Baer-Nawrocka, A. (2018). Food and environmental function in world agriculture: interdependence or competition? *Land Use Policy* 71: 578–583.

Sandborgh-Englund, G., Adolfsson-Erici, M., Odham, G., and Ekstrand, J. (2006). Pharmacokinetics of triclosan following oral ingestion in humans. *Journal of Toxicology and Environmental Health Part A: Current Issues* 69: 1861–1873.

Santhi, V.A., Sakai, N., Ahmad, E.D., and Mustafa, A.M. (2012). Occurrence of bisphenol A in surface water, drinking water and plasma from Malaysia with exposure assessment from consumption of drinking water. *Science of the Total Environment* 427: 332–338.

Schafer, A.I., Akanyeti, I., and Semiao, A.J. (2011). Micropollutant sorption to membrane polymers: a review of mechanisms for estrogens. *Advances in Colloid and Interface Science* 164: 100–117.

Scheurell, M., Franke, S., Shah, R.M., and Huhnerfuss, H. (2009). Occurrence of diclofenac and its metabolites in surface water and effluent samples from Karachi, Pakistan. *Chemosphere* 77: 870–876.

Segev, O., Kushmaro, A., and Brenner, A. (2009). Environmental impact of flame retardants (persistence and biodegradability). *International Journal of Environmental Research and Public Health* 6: 478–491.

Shehab, Z.N., Jamil, N.R., and Aris, A.Z. (2020). Occurrence, environmental implications and risk assessment of Bisphenol A in association with colloidal particles in an urban tropical river in Malaysia. *Scientific Reports* 10: 20360.

Singh, J., Knapp, H.V., Arnold, J., and Demissie, M. (2005). Hydrological modeling of the Iroquois river watershed using HSPF and SWAT 1. *JAWRA Journal of the American Water Resources Association* 41: 343–360.

Spongberg, A.L., Witter, J.D., Acuna, J. et al. (2011). Reconnaissance of selected PPCP compounds in Costa Rican surface waters. *Water Research* 45: 6709–6717.

Stackpoole, S.M., Shoda, M.E., Medalie, L., and Stone, W.W. (2021). Pesticides in US rivers: regional differences in use, occurrence, and environmental toxicity, 2013 to 2017. *Science of the Total Environment* 787: 147147.

Staples, C.A., Naylor, C.G., Williams, J.B., and Gledhill, W.E. (2001). Ultimate biodegradation of alkylphenol ethoxylate surfactants and their biodegradation intermediates. *Environmental Toxicology and Chemistry: An International Journal* 20: 2450–2455.

Sun, Y., Zhao, L., Li, X. et al. (2019). Stimulation of earthworms (*Eisenia fetida*) on soil microbial communities to promote metolachlor degradation. *Environmental Pollution* 248: 219–228.

Tamura, I., Yasuda, Y., Kagota, K. et al. (2017). Contribution of pharmaceuticals and personal care products (PPCPs) to whole toxicity of water samples collected in effluent-dominated urban streams. *Ecotoxicology and Environmental Safety* 144: 338–350.

Ternes, T.A. (1998). Occurrence of drugs in German sewage treatment plants and rivers. *Water Research* 32: 3245–3260.

Tsui, M.M.P., Lam, J.C.W., Ng, T.Y. et al. (2017). Occurrence, distribution, and fate of organic UV filters in coral communities. *Environmental Science & Technology* 51: 4182–4190.

van der Veen, I. and de Boer, J. (2012). Phosphorus flame retardants: properties, production, environmental occurrence, toxicity and analysis. *Chemosphere* 88: 1119–1153.

Vandenberg, L.N., Maffini, M.V., Sonnenschein, C. et al. (2009). Bisphenol-A and the great divide: a review of controversies in the field of endocrine disruption. *Endocrine Reviews* 30: 75–95.

Vieira, L.R., Soares, A.M.V.M., and Freitas, R. (2022). Caffeine as a contaminant of concern: a review on concentrations and impacts in marine coastal systems. *Chemosphere* 286: 131675.

Vione, D., Das, R., Rubertelli, F. et al. (2010). Modelling the occurrence and reactivity of hydroxyl radicals in surface waters: implications for the fate of selected pesticides. *International Journal of Environmental Analytical Chemistry* 90: 260–275.

Wang, Q. and Kelly, B.C. (2017). Occurrence and distribution of synthetic musks, triclosan and methyl triclosan in a tropical urban catchment: influence of land-use proximity, rainfall and physicochemical properties. *Science of the Total Environment* 574: 1439–1447.

Wang, R., Yuan, Y., Yen, H. et al. (2019). A review of pesticide fate and transport simulation at watershed level using SWAT: current status and research concerns. *Science of the Total Environment* 669: 512–526.

Wu, L.H., Zhang, X.M., Wang, F. et al. (2018). Occurrence of bisphenol S in the environment and implications for human exposure: a short review. *Science of the Total Environment* 615: 87–98.

Xiao, K.Y., Lu, Z.B., Yang, C. et al. (2021). Occurrence, distribution and risk assessment of organophosphate ester flame retardants and plasticizers in surface seawater of the West Pacific. *Marine Pollution Bulletin* 170: 112691.

Yang, G.C., Tsai, H.J., and Chang, F.K. (2015). Occurrence of triclosan in the tropical rivers receiving the effluents from the hospital wastewater treatment plant. *Environmental Monitoring and Assessment* 187: 151.

Yao, B., Li, R., Yan, S.W. et al. (2018). Occurrence and estrogenic activity of steroid hormones in Chinese streams: a nationwide study based on a combination of chemical and biological tools. *Environment International* 118: 1–8.

Ying, G.G. (2006). Fate, behavior and effects of surfactants and their degradation products in the environment. *Environment International* 32: 417–431.

Zhang, H.D., Bu, Q.W., Wu, D.K., and YU, G. (2020). Polycyclic musks in surface water and sediments from an urban catchment in the megacity Beijing, China. *Environmental Pollution* 263: 114548.

Zhang, L., Lu, L., Zhu, W.J. et al. (2021). Organophosphorus flame retardants (OPFRs) in the seawater and sediments of the Qinzhou Bay, Northern Beibu Gulf: occurrence, distribution, and ecological risks. *Marine Pollution Bulletin* 168: 112368.

Zhang, X., Zhang, D.D., Zhang, H. et al. (2012). Occurrence, distribution, and seasonal variation of estrogenic compounds and antibiotic residues in Jiulongjiang River, South China. *Environmental Science and Pollution Research* 19: 1392–1404.

Zou, S.C., Xu, W.H., Zhang, R.J. et al. (2011). Occurrence and distribution of antibiotics in coastal water of the Bohai Bay, China: impacts of river discharge and aquaculture activities. *Environmental Pollution* 159: 2913–2920.

Zuo, Y., Zhang, K., and Deng, Y. (2006). Occurrence and photochemical degradation of 17α-ethinylestradiol in Acushnet River Estuary. *Chemosphere* 63: 1583–1590.

Zuo, Y., Zhang, K., and Zhou, S. (2013). Determination of estrogenic steroids and microbial and photochemical degradation of 17α-ethinylestradiol (EE2) in lake surface water, a case study. *Environmental Science: Processes & Impacts* 15: 1529–1535.

8

Fate and Transport of Microconstituents in Wastewater Treatment Plants

Zong-Han Yang[1], *Po-Jung Huang*[2], *Ku-Fan Chen*[3], *and Chih-Ming Kao*[1,4]

[1]*Institute of Environmental Engineering, National Sun Yat-sen University, Kaohsiung, Taiwan*
[2]*Department of Chemical and Materials Engineering, National Central University, Taoyuan, Taiwan*
[3]*Department of Civil Engineering, National Chi Nan University, Puli, Nantou County, Taiwan*
[4]*Civil and Environmental Engineering Department, College of Engineering, University of Nebraska, Lincoln, Omaha, NE, USA*

8.1 Introduction

8.1.1 The Sources of Microconstituents in Wastewater Treatment Plants

Anthropogenic and natural compounds existing in the environment under trace concentrations are named microconstituents (MCs), and have a negative effect on the environment and human health. These are also known as micropollutants (MPs) (Oppenheimer et al. 2012; Ahmad et al. 2019). Amounts of MCs are sourced from medicines and industrial chemicals, containing pharmaceutically active compounds (PhACs), personal care products (PCPs), polyfluorinated chemicals (PFCs), endocrine-disrupting chemicals (EDCs), pesticides, and micro/nanoparticles (i.e., microplastics) (Wang and Wang 2016; Lutterbeck et al. 2020; Viveknand et al. 2021). Based on their trace concentrations ranging from ngL^{-1} to μgL^{-1} and their influence on the biota (Sörengård et al. 2019), the detection, quantification, and treatment of the MCs become challenging for conventional wastewater treatment plants (WWTPs) (Peña-Guzmán et al. 2019; García et al. 2020; Mpatani et al. 2021a).

Figure 8.1 shows the sources of MCs in the environment. The PhACs and PCPs commonly originate from hospital and domestic liquid wastes. Increased PhAC and persistent chemical residue concentrations are found in WWTPs through municipal sewer networks due to high drug usage in human healthcare and incomplete conversion of drugs in the human body (Wang and Wang 2016). PCPs usually include analgesics, stimulant drugs, antibiotics, antiseptics, antimicrobials, sunscreen agents, fragrances, cosmetics, and their metabolites (Daughton 2016; Antonopoulou et al. 2021). These PCPs are associated with increased antimicrobial resistance and decreased reproductive health, and are likely to result in increased vulnerability of human health and ecosystems (Daughton 2016; Madikizela et al. 2018; Dong et al. 2019; Quesada et al. 2019; Lutterbeck et al. 2020).

Another major MC source from hospitals are endocrine-disrupting chemicals (EDCs), which are ubiquitous and widely used in many products of our daily lives. Bisphenol A (BPA) is one of the most common EDCs and has been used in the manufacture of polycarbonate

Microconstituents in the Environment: Occurrence, Fate, Removal, and Management, First Edition.
Edited by Rao Y. Surampalli, Tian C. Zhang, Chih-Ming Kao, Makarand M. Ghangrekar, Puspendu Bhunia, Manaswini Behera, and Prangya R. Rout.

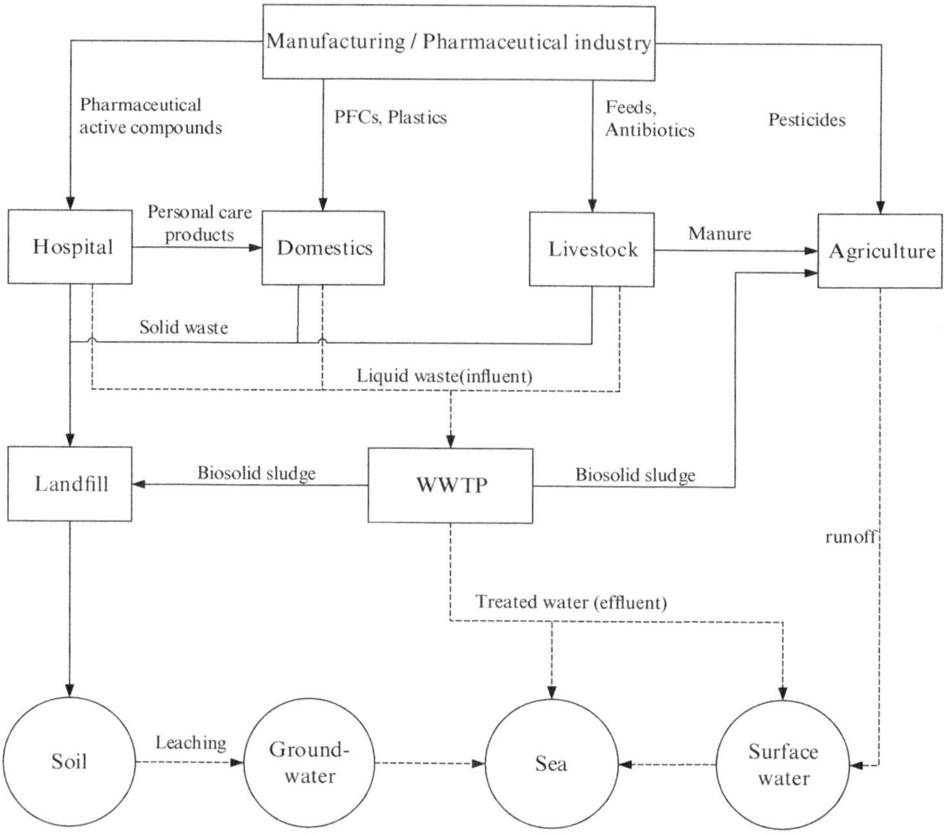

Figure 8.1 Schematic diagram showing the source of MPs to WWTP.

plastics and epoxy resins for many years. EDCs have been implicated in causing prostate cancer, endometriosis, breast cancers, and testicular diseases, which endanger the reproductive health of animals and humans (Kawa et al. 2021; Martín-Pozo et al. 2021).

In many countries, livestock is one of the main economic activities, and farm management strategies and environmental regulations for the generated livestock wastes and wastewater vary. Because of its nutrient contents, livestock wastewater can be used to supplement soil and applied to meet agricultural needs (Hamid and Eskicioglu 2012; Golwala et al. 2021).

MCs in the aquatic environment come from point sources (e.g., WWTP effluents, spills, leachates from landfills, etc.) (Majumder et al. 2021) or nonpoint sources (diffuse sources), such as runoff from fields (Couto et al. 2019). However, existing conventional WWTPs are not designed to treat all types of MCs, nor can they remove MCs and their microbial metabolites effectively. This will cause an undesirable release of MCs to water bodies including rivers, streams, groundwater, and seawater. Additionally, the sludge from WWTPs is characterized by high concentrations of organic matter and solids, containing a variety of MCs, which pose risks to human health and ecosystems (Arvaniti and Stasinakis 2015; Wu et al. 2015).

MCs may enter subsurface environments (soils and groundwater) via surface waters (Quesada et al. 2019), hyporheic interaction, and infiltration (Gogoi et al. 2018). Other routes for MCs to enter groundwater contain artificial recharge of treated wastewater (Khan et al. 2020a), direct exfiltration from sewer systems or leaking pipes (Buta et al. 2021), and single-event contaminations from spills, accidents, etc. (Citulski and Farahbakhsh 2012; Arvaniti and Stasinakis 2015).

8.1.2 The Behavior of Microconstituents

Some MCs can be metabolized in human/animals or transformed in the environment or during wastewater treatment processes to specific transformation products (TPs) (Harb et al. 2019). MCs and their metabolites/TPs possess different physico-chemical properties, which allow them to interact in numerous ways in the environment. These interactions usually occur in specific paths, coexisting redox conditions, and temperature gradients, and so valuable information of transport and fate and environmental indicators can be obtained through these interactions (Taboada-Santos et al. 2020).

The biological and chemical stability of MCs are the important parameters, because it represents the fate of an MC in a WWTP and in the environment (Grandclément et al. 2017). For example, the attenuation mechanism can be caused by biological or nonbiological degradation processes (e.g., hydrolysis, photolysis) (Faria et al. 2020; Shakerian et al. 2020). In the water system, the abiotic degradation via photolysis strongly depends on the system characteristics (Kulandaivelu et al. 2020; Khan et al. 2021; Mpatani et al. 2021a). Biological transformation within WWTPs can be the dominant attenuation process for many MCs (Lim et al. 2019; Faria et al. 2020; Choi et al. 2021). Recent studies indicate that degradation capabilities of ecosystems may change over time or space in/along surface water bodies (Olasupo and Suah 2021). Acesulfame, an artificial sweetener, was assumed to be persistent during wastewater treatment for the last two decades, but now shows a remarkable degradation to sulfamic acid (Taboada-Santos et al. 2020; Saidulu et al. 2021). Table 8.1 presents the physical and chemical characteristics of different category of MCs in WWTP influents.

The degradation patterns of MCs can be changed due to the adaption of microbial consortia to new compounds or due to altered redox conditions (Ma et al. 2020). The degradation of MCs depends on the presence of some specific microorganisms (Shakerian et al. 2020). Anthropogenic or natural boundary conditions can change ecosystems (Gutiérrez et al. 2021; Sossalla et al. 2021). The variations in geochemical or hydrological conditions in WWTPs or other treatment facilities limit the occurrence of the respective compound in the environment.

8.2 The Fate of Microconstituents in WWTPs

8.2.1 Traditional Wastewater Treatment Process

The traditional wastewater treatment process is presented in Figure 8.2. The WWTP process is designed to remove solids, organic compound loadings, and pathogens. Although several treatment steps in a wastewater treatment plant can partially remove MCs, complete removal has proved to be a challenge due to the wide variety of MCs, their extremely low concentrations,

Table 8.1 Physico-chemical characteristics of common MPs.

Categories	ECs	Molecular Formula[1]	Molecular Weight (g mol^{-1})[1]	Solubility (mg L^{-1}@25°C)[1]	pKa[1]	Log K_{ow}[1]	K_{oc}[1]	Henry's Constant atm (L mole^{-1})[1]	BCFs[1]
PhACs[2]	Ibuprofen	$C_{13}H_{18}O_2$	206.28	21	4.91	3.97	3400	3.4×10^{-4}	3(estimated)
	Diclofenac	$C_{14}H_{11}Cl_2NO_2$	296.1	2.37	4.15	4.51	245	4.7×10^{-9}	3(estimated)
	Naproxen	$C_{14}H_{14}O_3$	230.26	15.9	4.15	3.18	330	3.4×10^{-7}	3(estimated)
	Ketoprofen	$C_{16}H_{14}O_3$	254.28	51	4.45	4.24	–	–	–
	ciprofloxacin	$C_{17}H_{18}FN_3O_3$	331.34	<1	6.09	0.28	61,000	–	3(estimated)
PCPs[2]	Salicylic acid	$C_7H_6O_3$	138.12	2240	2.98	2.26	404	7.3×10^{-6}	3(estimated)
	Propylparaben	$C_{10}H_{12}O_3$	180.20	500	8.5	3.04	290	4.3×10^{-6}	50
PFCs[2]	Perflurooctanoic acid, PFOA	$C_8H_{15}O_2$	414.07	3300	−0.5 ~ 4.2	4.81	83 ~ 389	90.8	100 ~ 230
	Perflurooctane sulfonix acid, PFOS	$C_8H_{17}O_3S$	500.13	570	<1.0	4.49	126 ~ 39,800	4.3×10^{-4}	200 ~ 1500
EDCs[2]	4-Nonylphenol (4-NP)	$C_{15}H_{24}O$	220.35	7	−5.4, 10.31	5.76	6900 ~ 53,300	3.4×10^{-2}	380.19
	Bisphenol-A	$C_{15}H_{16}O_2$	228.29	300	9.6	3.32	251 ~ 3886	4.0×10^{-8}	5.1 ~ 73.4
Pesticides	Mecoprop	$C_{10}H_{11}ClO_3$	214.64	880	3.1 ~ 3.78	3.13	5 ~ 43	3.8×10^{-6}	3(estimated)
	Triclosan	$C_{12}H_7Cl_3O_2$	289.5	10	7.9	4.76	3500 ~ 20,000	2.1×10^{-5}	2.7 ~ 90

[1] These data are collected from PubChem (https://pubchem.ncbi.nlm.nih.gov/).
[2] PhACs = pharmaceutical activated compounds; PCPs = personal care products; PFCs = polyfluorinated chemicals; EDCs = endocrine-disrupting chemicals (EDCs).

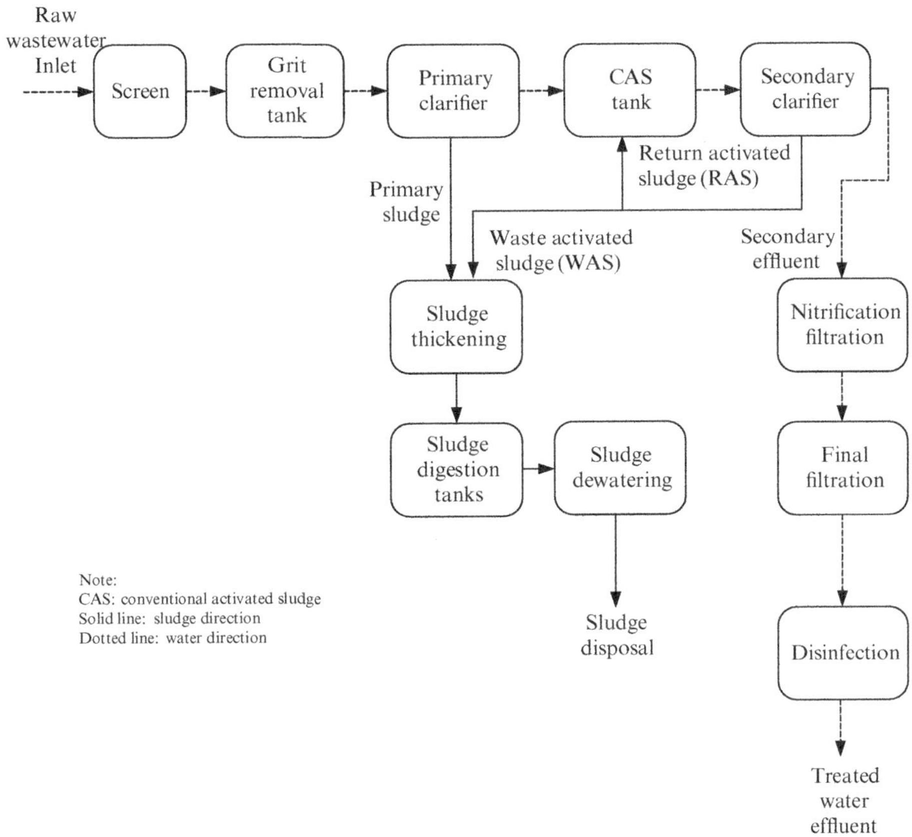

Figure 8.2 Schematic representation of conventional WWTP processes.

and their unique characteristics (Rout et al. 2021). In conventional WWTPs, the main treatment units (e.g., aerobic and anaerobic processes) are not specifically designed for removal of MCs (Abtahi et al. 2018). Most MCs are not easily degraded by conventional biological processes. Thus, domestic wastewater becomes the main source of original or partially transformed MCs in surface water bodies (Grandclément et al. 2017; Faria et al. 2020; Fundneider et al. 2021).

MCs can reduce and inhibit the abundance of nitrifying and denitrifying bacteria, with nitrogen and phosphorus removal functions and phenolic compounds degrading bacteria in WWTPs (Song et al. 2018; Kora et al. 2020; Vo et al. 2020; Sossalla et al. 2021). Moreover, the digestion process is a commonly applied technique for sludge treatment. However, sludge digestion of MCs may affect microbial communities, and thus inhibit sludge hydrolysis and anaerobic biodegradation processes including acetic acid accumulation, hydrogen generation, and methane production (Arvaniti and Stasinakis 2015; Buta et al. 2021).

8.2.2 The Fate of MCs in WWTPs

Removal of MCs in conventional WWTPs can be attributed to biotransformation and sorption mechanisms (Tran et al. 2018). However, sorption is not a viable removal mechanism in anaerobic systems due to the relatively longer solid retention time (SRT) applied

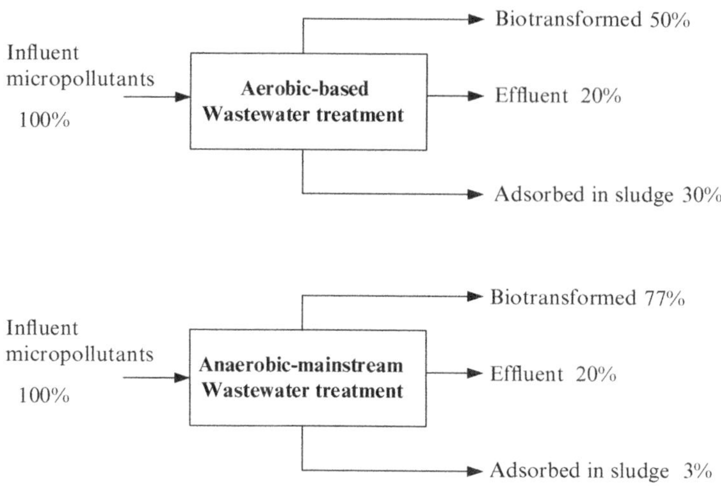

Figure 8.3 Comparative MCs mass balance across conventional aerobic activated sludge and anaerobic wastewater treatment systems.

(>100 days). The calculated liquid–solid partitioning coefficients (K_d) for MCs are in a wide range (10 to >1000 L kg^{-1}) in anaerobic sludges (Wijekoon et al. 2015; Berthod et al. 2016). Longer SRTs are required for removal of MCs in most anaerobic studies (>70 days) (Xiao et al. 2017; Shakerian et al. 2020; Fundneider et al. 2021; Zhang et al. 2021), and this indicates that sorption cannot be used solely for MCs removal, and so biotransformation should be the main mechanism responsible for MCs removal.

The results are in contrast to aerobic treatment systems, which discharge a significant amount of MCs in wasted activated sludge (Petrie et al. 2015). However, shorter SRTs (3–10 days) are used for aerobic systems. Thus, the environmental impact of MCs associated with wasted aerobic sludge is significant, especially when land application is adopted for sludge disposal or reuse (Monsalvo et al. 2014; Thomaidi et al. 2016). The potential for MC discharges to the environment from aerobic-based treatment is greater because of the higher rates of sludge wasting (Figure 8.3) (Venegas et al. 2021). This also reveals that differences in environmental loadings of MCs via sludge wasting could be more significant between aerobic and anaerobic systems (Monsalvo et al. 2014; Song et al. 2018).

8.2.3 Biodegradation of Microconstituents

The biodegradation of MCs is affected by their toxicity factors (Saidulu et al. 2021). Table 8.2 shows the ranges of biomass normalized rate constant (K_{bio}) for MCs. Based on K_{bio} values, the biodegradability of a compound can be predicted. When K_{bio} of a compound is higher than 10, it implies a higher biodegrading tendency of this compound. When K_{bio} of a compound is lower than 0.01, it implies a lower biodegrading tendency of this compound (Tran et al. 2018; Saidulu et al. 2021). When MCs are non-toxic to microorganisms, the biotransformation pathway can occur (Quintana et al. 2005; Nsenga Kumwimba and Meng 2019). Researchers report that in the metabolic biodegradation of some non-toxic MCs,

Table 8.2 The solid–water distribution coefficient and the biomass normalized rate constant of common microconstituents.

		K_d	K_{bio}
Categories	**Compounds**	**L/g^{-1} SS**	**L/g^{-1}/MLSS/d^{-1}**
PhACs	Ibuprofen	0.006 ~ 0.102	1.956 ~ 70
	Diclofenac	0.016 ~ 0.723	0.619 ~ 1.38
	Naproxen	0.013 ~ 0.03	0.005 ~ 1.9
	Ketoprofen	0.016 ~ 0.226	0.24 ~ 3.36
	Ciprofloxacin	19.9 ~ 38.9	0.081 ~ 0.51
	Azithromycin	0.034 ~ 2.156	0.24 ~ 0.75
	Atenolol	0.0059 ~ 0.095	0.75 ~ 18.4
	Propranolol	0.46 ~ 1.12	0.098 ~ 18.24
PCPs	Salicylic acid	–	0.6 ~ 9
Pesticides	Atrazine	0.003 ~ 0.146	–
	Triclosan	3.61	0.024 ~ 14.73
EDCs	Nonylphenol	–	0.7 ~ 13.3
	Bisphenol-A	0.314 ~ 0.505	0.48 ~ 12.2
	Estrone-1	0.645	< 200

Note: "-" = data not found; K_d = solid-water distribution coefficient; K_{bio} = biomass normalized rate constant.

only heterotrophic bacteria are responsible (Tran et al. 2013; Xiao et al. 2017), whereas a growth substrate (e.g., ammonia, acetate) is required to transform the non-growth substrate during the cometabolic biodegradation processes (Figure 8.4). (Sarmah and Northcott, 2008; Sathyamoorthy et al. 2013; Men et al. 2017).

The anaerobic process is another way to treat MCs due to its less sludge production (Reyes-Contreras et al., 2020; Song et al. 2018). However, researchers found operational conditions, such as hydraulic retention time (HRT), temperature, acclimatization period of sludge, and additional growth substrate could influence the biodegradation and sorption of MCs in an anaerobic treatment system (Stasinakis 2012). For example, <15% removal of ibuprofen was observed in the up-flow anaerobic sludge blanket (UASB), which was influenced by the methanogenic rate, upward velocity, and HRT (Alvarino et al. (2014). In another similar anaerobic system, complete microbial transformation of naproxen occurred, but was not effective for diclofenac (Lahti and Oikari (2011). However, HRT could increase the reaction time for microbe transformation, but it seemed to also contribute to a rise in the concentration of nonylphenol in the effluent wastewater if there were negative removals of MCs at high HRT (Queiroz et al. (2012). Besides, SRT and temperature do not seem have a significant influence on the removal of PhACs and PCPs, because higher than 85% removal efficiencies were both observed for naproxen under middle (37 °C) and high temperature (55 °C) conditions Carballa et al. (2007).

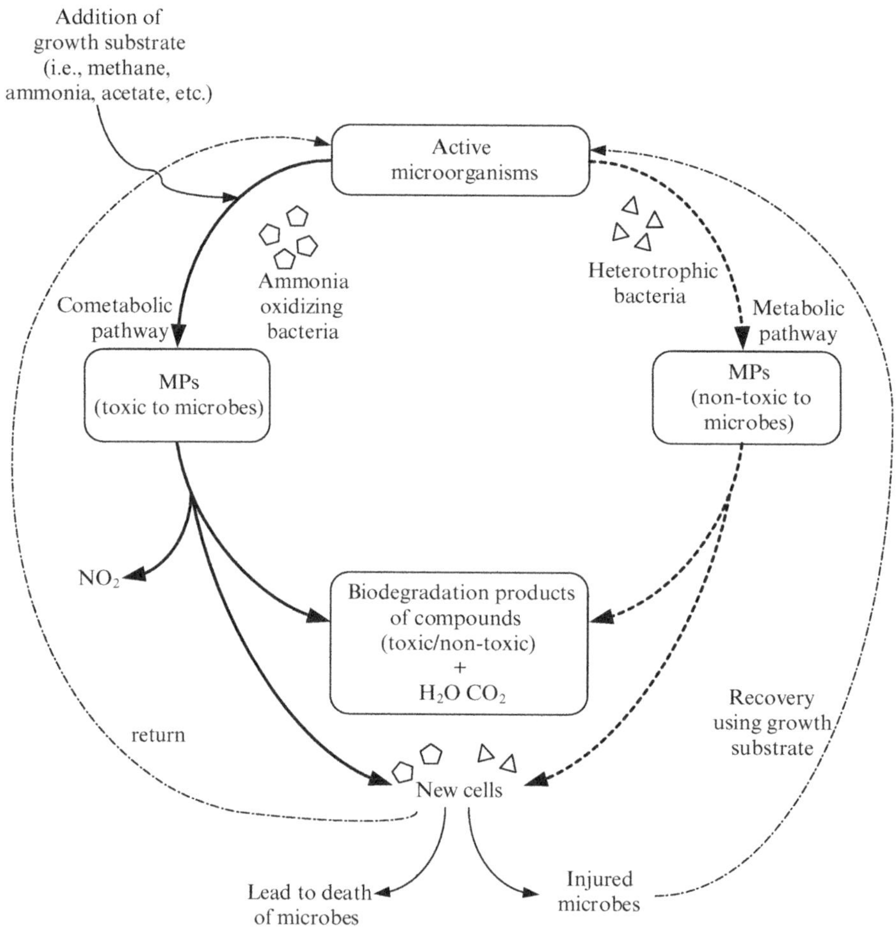

Figure 8.4 The major removal mechanisms of MCs biodegradation.

8.2.4 Sorption Onto Sludge Solids in WWTPs

Another significant mechanism controlling the fate of MCs in WWTP is sorption. There are several sorption mechanisms presented in Figure 8.5 which include physical and chemical mechanisms such as hydrophobic interaction, electrostatic attraction, van der Waal forces, hydrogen bonding, inner-sphere complexation, ion-exchange, pore diffusion, and π–π interaction (Czajka and Londry 2006). Researchers observed that physical sorption is the main pathway to organic pollutant removal, whereas with metallic ions, chemical sorption plays the criticle role for MC removal (Faria et al. 2020; Rios Miguel et al. 2020; Gonzalez-Gil et al. 2018). In general, sorption of an organic contaminant depends on its physico-chemical properties (log K_{OW} and pKa values), solid–water distribution coefficient (K_d), and organic carbon fraction present on the sludge solids (Buta et al. 2021).

Several studies have reported that high log K_{OW} (>3.5) implies MCs has high sorption affinity to sediment particles in water (Tiwari et al. 2017). For example, 4-nonylphenolin EDCs and most of PhAc compounds, tend to have high sorption affinity onto sludge solids due to their relatively high log K_{OW} values (Table 8.1), while the charge of the compounds in the aqueous

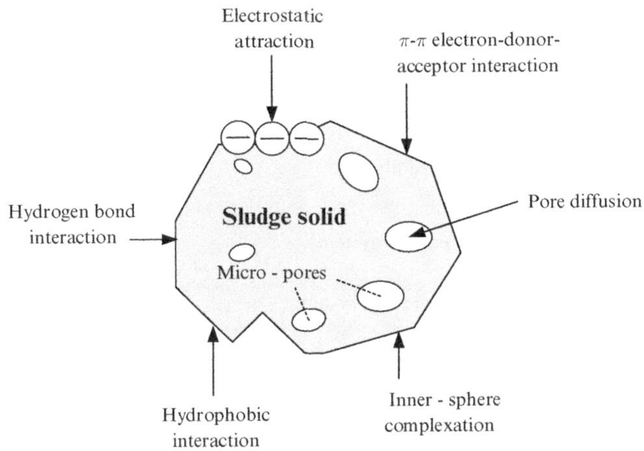

Figure 8.5 Major removal mechanisms sorption of ECs onto sludge solids.

medium can be assessed based on their pKa value, and charged compounds may hinder the mechanisms of biological removal techniques. Another key parameter is K_d, where low K_d (<0.1 L/g-1/SS) showed lower sorption potential of MCs under pH 7 conditions (Stevens-Garmon et al. (2011. However, the K_d value is influenced by several parameters, such as characteristics of solid phase and process conditions (Tran et al. 2009). Moreover, Xue et al. (2010) have observed that the absence of nitrate favors the sorption of MCs onto activated sludge.

8.3 Treatment Methods for Microconstituents Removal

The first step in pretreatment is to cut up the solid fraction in the wastewater, while dissolved organic carbon (DOC), nutrients, and a few PhACs are removed by natural means. Only a few MCs are removed in the pretreatment stage, with some MCs also originating from the conventional WWTP effluent, as WWTPs are ineffective in treating recalcitrant MCs. Therefore, advanced technologies should be implemented to prevent the discharge of MCs into aquatic environments. Several technologies were used in the existing WWTP to enhance the removal of MCs (Figure 8.6). Microbial metabolism is the most used technology in WWTPs for MCs removal, such as activated sludge process (ASP), membrane bioreactor (MBR), moving bed biofilm reactor (MBBR), and trickling filter (TF), owing to their lower operation costs than chemical treatments. This scheme implementation will also be dependent upon adequate technologies and operational conditions. The minor influential parameters include environmental constraint, local legislation, and economics of that particular area.

8.3.1 Activated Sludge Process (ASP)

ASP is a type of biological treatment process for wastewater using aeration and a biological floc composed of bacteria (Xue et al. 2010; Tran et al. 2013). The ASP for removing MCs begins with an aeration tank where air (or oxygen) is injected into the wastewater and some part of the settled sludge is recycled from the secondary sedimentation tank to the

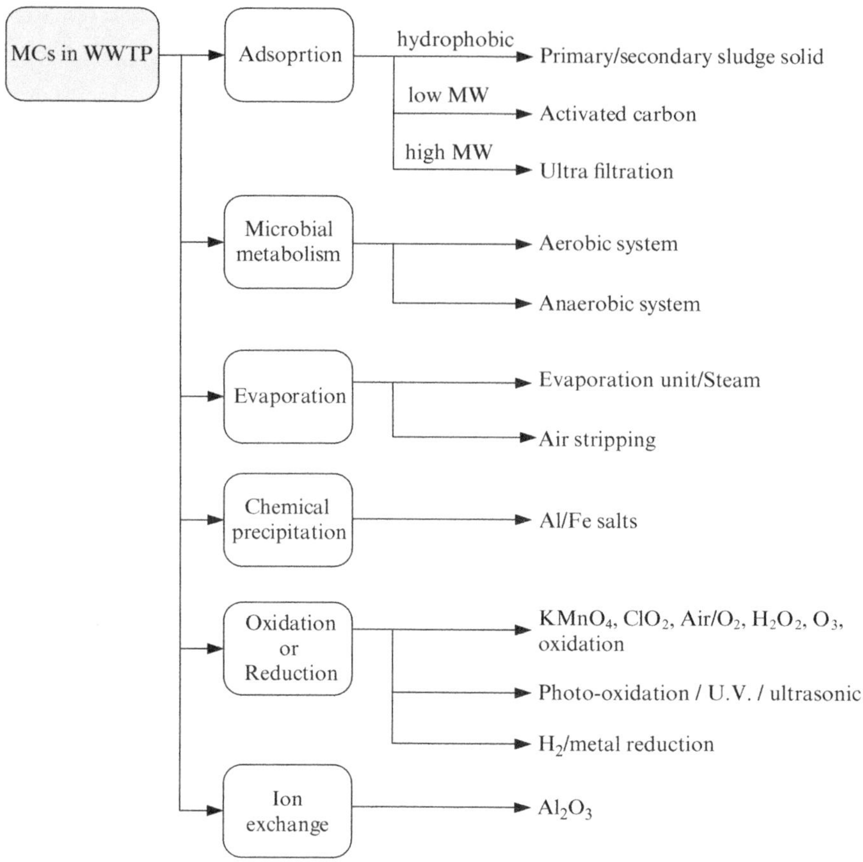

Figure 8.6 Fate during sewage and industrial effluent treatment.

aeration tank to balance the food to microorganisms (F/M) ratio (Mpatani et al. 2021b). Research revealed that the degradation of PhACs (i.e., ibuprofen, naproxen, ketoprofen) in ASP was >80% (Martínez-Alcalá et al. (2017); Melo-Guimarães et al. (2013) and degradation of EDCs (estron-1) were >98% using ASP (Petrie et al. (2015). However, ASP was not satisfactory for nonylphenol and triclosan (<50%) degradation due to its lower biodegradability (Behera et al. 2011). Using a cationic flocculant (Zetag 8140-Polyacrylamide) could improve the biodegradability of PhAC compounds (Melo-Guimarães et al. 2013), but were not efficient at biodegrading beta-blockers (i.e., atenolol, propranolol)

8.3.2 Membrane Bioreactor (MBR)

The membrane bioreactor is a combination of membrane processes such as microfiltration or ultrafiltration with ASP. MBR is an efficient treatment method for organic compound removal in WWTPs and has a high effluent quality due to its complete biodegradation retention in the reactor (Arola et al. 2017; Goswami et al. 2018). The removal efficiency of MCs in the MBR process is mainly affected by several parameters, such as influent wastewater characteristics (e.g., pH, organic matter concentration), membrane characteristics (pore size, chemical properties of the membrane), and operational conditions (e.g., redox conditions, HRT, SRT, temperature) (Khan et al. 2020b). Aerobic reactions are most favorable for the biodegradation of PhACs (ibuprofen, ketoprofen, naproxen, atenolol), but biogradation of other PhACs (diclofenac) was not satisfactory using thr MBR process compared with ASP (Kosek et al. 2020). MBR also has shown higher removal efficiencies for EDCs (bisphenol-A and estrone-1) and PCPs (salicylic acid), as most of these compound can be effectively removed (>90%) by the MBR process (Tadkaew et al. 2011).

8.3.3 Moving Bed Biofilm Reactor (MBBR)

The MBBR system consists of an aeration tank with special carriers that provide enough surface for biofilm growing, causing the removal of MCs along with nutrients from different wastewater sources (Casas et al. 2015; Kora et al. 2020). MBBRs exhibit some inherent benefits, such as uniform distribution of biomass, less volume requirement, capacity to handle high OLR, lower HRT, eliminating dead space that causes clogging and channeling, less sludge production, and generally no strict recirculation requirement (Abu Bakar et al. 2018). In an MBBR system, the electron-donating functional group of different compounds plays a key role in the variation of observed removal efficiencies (Zupanc et al. 2013). Therefore, MBBR has shown >85% removal efficiency for naproxen, salicylic acid, ibuprofen, ketoprofen and bisphenol-A. However, a lower 45% of removal efficiency has been bserved for diclofenac (Casas et al. 2015; Krzeminski et al. 2019).

8.3.4 Trickling Filter

Trickling Filter (TF) is an aerobic treatment system that utilizes microorganisms attached to a medium to remove organic matter from wastewater. TFs are mainly composed of four major components: a filtermedium such as stones, plastic shapes, or wooden slats; an enclosure to hold the liquid; a distribution system; and an under-drain system. TF is easy to operate, with low energy consumption, self-cleansing mechanism, and better sludge thickening. The removal efficiency of PhACs were higher than 70% using the TF system (Lin et al. 2009). Researchers also found that the average removal efficiency of TF was much lower than that of ASP (Kasprzyk-Hordern et al. 2009). However, it was found to give a better performance in terms of removing the microbial loads and showed lower energy requirements compared to ASP when TF was modified to include aerobic as the upper part and anaerobic as the lower part (Lamba and Ahammad (2017).

8.4 Critical Parameters in WWTP Operation for MCs

8.4.1 ASP Operation

Table 8.3 shows the different critical parameters for WWTP operation. The HRT is controlling the duration of liquid phase in the reactor that has a significant influence on the removal efficiencies for MCs. The removal efficiencies of some MCs (i.e., ibuprofen, naproxen, ketoprofen, 2,4-D, Estrone-1) were increased with an increase in HRT (Castillo Meza et al. 2020; Mojiri et al. 2020). This is because of more contact opportunities between

Table 8.3 Critical parameters, advantages, and limitations of biological treatment methods to remove MCs.

Methods	Critical Parameter	Advantages	Limitations
ASP	HRT: 5–15 h SRT: 5–14 days F/M: 0.25–0.5 $g\,BOD \cdot gML/SS^{-1}d^{-1}$ pH: 6.5–7.8 Temp.: 22–25 °C	• Basic process and to operate. • Lower capital and operational costs compared to chemical treatments.	• Unsuitable for high influent COD. • Huge sludge production containing large amount of ECs.
MBR	HRT: 6–24 h SRT: 15–30 days pH: 7–7.5 Nominal pore size: 0.4–1 μm	• Effective for removal of recalcitrant pollutants. • High removal efficiencies achieved for ECs.	• Large amount of every requirement. • Membrane fouling.
MBBR	HRT: 6–12 h pH: 7–8 Filling ratio: 30–50% Temp.: 20–25 °C	• Lower HRT as compared other biological methods. • Simple and robust design.	• Need skilled personnel. • Periodic monitoring required. • Long start-up period.
TF	HRT: 36–48 h pH: 7–7.9	• Large energy requirement and self-cleaning mechanism. • Withstands shock loads.	• Need modification for removing large varieties of ECs. • Bed clogging. • Long start-up period.

MCs and microbes when an increased HRT is implemented. However, there was not enough removal efficiency for some recalcitrant MCs (diclofenac and mecoprop) to be biodegraded under high HRT conditions (He et al. 2021). Moreover, higher HRT results in large volume requirements for the reactor, thereby increasing capital and operational expenditures. Therefore, it is essential to choose the optimum HRT to maintain an economically feasible system (Petrie et al. 2015).

SRT has another significant impact on biodegradation and sorption of contaminants during ASP due to an increase in metabolic activities with increased SRT (Falås et al. 2016). SRT impacts the bacterial population adaptability and thereby enhances their potential to metabolize the targeted MCs. Some studies observed that an average 10 days of SRT is sufficient to obtain lower effluent concentrations of MCs through ASP (Banihashemi and Droste 2014; Falås et al. 2016; Gallé et al. 2019). Moreover, higher SRT, along with suitable operational conditions, will be sufficient to remove most of the MCs (up to 80%) (Behera et al, 2011). However, increasing SRT can also change the sorption affinity of contaminants Gani et al. (2019).

The F/M ratio also influences the microbial growth, sludge settling, and removal of MCs. The optimal operation of F/M is 0.25 to 0.5 $g/BOD/g^{-1}$ $MLSS/d^{-1}$ in ASP. It was observed that the reduction of F/M could increase the removal efficiency of Di(2-ethylhexyl) phthalate (DEHP) (Gani et al. 2019). This may be due to the reduction of COD concentration in the aeration tank with the decreasing F/M, thus increasing the substrate consumption by microorganisms (Li et al. 2011; Hamza et al. 2018).

8.4.2 MBR Operation

HRT and SRT are two important parameters for MBR operation. The optimal HRT for MBR operation is in the range of 6 to 24 h (Wijekoon et al. 2013). Up to 100% of some MCs can be removed when the HRT in MBR reaches 24 h (Wijekoon et al. 2013). SRT influences membrane fouling during the wastewater filtration process. Some studies recommend that a longer SRT (65 days) is required to obtain a better PhAC removal efficiency and most of the MCs can be removed effectively with an SRT of 15 to 30 days (Kimura et al. 2007). Membrane material and its nominal pore size play important roles in MCs removal in MBR. Operational results show that polysulfone and polypropylene membranes can result in better MCs removal efficiencies (Babu et al. 2018). The surface charge of the membrane also plays an essential role in MCs removal efficiencies. The nominal range for the membrane pore size applied for MCs treatment is from 0.004 to 1 μm in most studies (Zamri et al. 2021; Mpatani et al. 2021a).

8.4.3 MBBR Operation

HRT can also affect the interaction time between biofilm of the carrier and targeted MCs in MBBR (Luo et al. 2014a). Moreover, HRT can influence the MCs mass transfer from its surrounding fluid to biofilm (Tang et al. 2017). According to Derakhshan et al. (2018), the atrazine removal efficiency was reduced from 73 to 64% when the HRT was reduced from 24 to 12 h. Moreover, the MBBR showed better removal efficiency of some MCs under relatively low HRTs compared to other similar processes (Jiang et al. 2018). The optimal range of HRT in the MBBR system to remove some MCs is in the range of 6 to 12 h (Jiang et al. 2018).

Both biofilm structure and performance of MBBR are affected by the aeration rate (Barwal and Chaudhary 2015). Ahmadi et al. (2015) reported that the removal of phthalates was increased with an increase in aeration rate; however, the removal efficiency was not improved when the rate was higher than 60 L/h^{-1} (Ahmadi et al. 2015). This could be due to the fact that biofilm was washed out at higher rates. The aeration rate will also affect the sludge dewatering characteristics. Different plastic carriers with different specific surface areas and shapes have been applied in the MBBR for the growth of attached microorganisms. The carrier type and filling ratio play important roles in the biofilm stability and reactor homogeneity. To maintain the free floatation of bio-carriers, the filling ratio is usually up to 70% (Barwal and Chaudhary 2015). The oxygen transfer rate, mixing intensity, HRT, and energy consumption are also affected by the filling ratio of the bio-carrier. Casas et al. (2015) reported that up to 98% of ibuprofen could be removed with a filling ratio of 50% using 500 Anox Kaldnes as the bio-carrier. Results from another study showed 93% of ibuprofen removal with a filling ratio of 20% using polyurethane sponge as the bio-carrier under the same HRT (Luo et al. 2014a). Up to 90% of PCPs removal could be obtained with the filling ratios ranging from 20 to 50% (Ahmadi et al. 2015; Jiang et al. 2018).

8.4.4 TF Operation

In TF systems, nitrification and denitrification are affected by HRT (Lim et al. 2019). Researchers concluded that the removal efficiency of PhACs correlated with HRT (Lim et al. 2019). The optimal HRT to remove MCs in TF is in the range from 36 to 60 h. Results from Brandt et al. (2013) showed that the TF operated with a low HRT tended to lower the removal efficiency for EDCs and PhACs, especially when packed material had low solid retention (Brandt et al. 2013). In TF, packed media can be a crucial component, which can enhance the microbial attachment and subsequent microbial growth. The selection of packed media depends on target MCs, influent concentrations, temperature ranges, and level of treatment required (Brandt et al. 2013).

8.5 Conclusions

This chapter emphasises the occurrences and toxicological effects of some commonly found Microcomstituents (MCs), and main wastewater treatment technologies applied for their removal. Researchers conclude that the risks caused by MCs are significantly high, although they are usually found at low levels. Among them, pesticides, PhACs, PCPs, and EDCs are ubiquitous MCs in water bodies causing significant toxic effects on ecosystems. Thus, it is necessary to develop a more reliable and effective treatment system for toxic MCs removal from domestic wastewater, which is a main source of MCs. The biological treatment process is usually selected for MCs removal because of the advantages of cost effectiveness and technical simplicity in wastewater management. During the operation of the applied wastewater treatment system, HRT and SRT play important roles in achieving higher removal efficiencies. Moreover, ASP and MBR have been approved as suitable biological processes for MCs treatment. However, the overall removal efficiencies for MCs are around 60 to 70%, and this indicates that biological processes are not sufficient to obtain complete MCs removal. Thus, the application of an advanced chemical oxidation process or a membrane filtration system as a post polishing unit can be an option to obtain a higher MCs removal efficiency.

References

Abtahi, S.M., Ilyas, S., Joannis Cassan, C. et al. (2018). Micropollutants removal from secondary-treated municipal wastewater using weak polyelectrolyte multilayer-based nanofiltration membranes. *Journal of Membrane Science* 548: 654–666.

Abu Bakar, S.N.H., Abu Hasan, H., Mohammad, A.W. et al. (2018). A review of moving-bed biofilm reactor technology for palm oil mill effluent treatment. *Journal of Cleaner Production* 171: 1532–1545.

Ahmad, J., Naeem, S., Ahmad, M. et al. (2019). A critical review on organic micropollutants contamination in wastewater and removal through carbon nanotubes. *Journal of Environmental Management* 246: 214–228.

Ahmadi, E., Gholami, M., Farzadkia, M. et al. (2015). Study of moving bed biofilm reactor in diethyl phthalate and diallyl phthalate removal from synthetic wastewater. *Bioresource Technology* 183: 129–135.

Alvarino, T., Suarez, S., Lema, J.M., and Omil, F. (2014). Understanding the removal mechanisms of PPCPs and the influence of main technological parameters in anaerobic UASB and aerobic CAS reactors. *Journal of Hazardous Materials* 278: 506–513.

Antonopoulou, M., Kosma, C., Albanis, T., and Konstantinou, I. (2021). An overview of homogeneous and heterogeneous photocatalysis applications for the removal of pharmaceutical compounds from real or synthetic hospital wastewaters under lab or pilot scale. *Science of the Total Environment* 765: 144163.

Arola, K., Hatakka, H., Mänttäri, M., and Kallioinen, M. (2017). Novel process concept alternatives for improved removal of micropollutants in wastewater treatment. *Separation and Purification Technology* 186: 333–341.

Arvaniti, O.S. and Stasinakis, A.S. (2015). Review on the occurrence, fate and removal of perfluorinated compounds during wastewater treatment. *Science of the Total Environment* 524–525: 81–92.

Babu, V.S., Padaki, M., D'Souza, L.P. et al. (2018). Effect of hydraulic coefficient on membrane performance for rejection of emerging contaminants. *Chemical Engineering Journal* 334: 2392–2400.

Banihashemi, B. and Droste, R.L. (2014). Sorption–desorption and biosorption of bisphenol A, triclosan, and 17α-ethinylestradiol to sewage sludge. *Science of the Total Environment* 487: 813–821.

Barwal, A. and Chaudhary, R. (2015). Impact of carrier filling ratio on oxygen uptake and transfer rate, volumetric oxygen transfer coefficient and energy saving potential in a lab-scale MBBR. *Journal of Water Process Engineering* 8: 202–208.

Behera, S.K., Kim, H.W., Oh, J.-E., and Park, H.-S. (2011). Occurrence and removal of antibiotics, hormones and several other pharmaceuticals in wastewater treatment plants of the largest industrial city of Korea. *Science of the Total Environment* 409 (20): 4351–4360.

Berthod, L., Roberts, G., Sharpe, A. et al. (2016). Effect of sewage sludge type on the partitioning behaviour of pharmaceuticals: a meta-analysis. *Environmental Science: Water Research & Technology* 2 (1): 154–163.

Brandt, E.M.F., de Queiroz, F.B., Afonso, R.J.C.F. et al. (2013). Behaviour of pharmaceuticals and endocrine disrupting chemicals in simplified sewage treatment systems. *Journal of Environmental Management* 128: 718–726.

Buta, M., Hubeny, J., Zieliński, W. et al. (2021). Sewage sludge in agriculture: the effects of selected chemical pollutants and emerging genetic resistance determinants on the quality of soil and crops: a review. *Ecotoxicology and Environmental Safety* 214: 112070.

Carballa, M., Omil, F., Ternes, T., and Lema, J.M. (2007). Fate of pharmaceutical and personal care products (PPCPs) during anaerobic digestion of sewage sludge. *Water* 41: 2139–21500. https://doi.org/10.1016/j.watres.2007.02.012.

Casas, M.E., Chhetri, R.K., Ooi, G. et al. (2015). Biodegradation of pharmaceuticals in hospital wastewater by staged moving bed biofilm reactors (MBBR). *Water Research* 83: 293–302.

Castillo Meza, L., Piotrowski, P., Farnan, J. et al. (2020). Detection and removal of biologically active organic micropollutants from hospital wastewater. *Science of the Total Environment* 700: 134469.

Choi, S., Yoom, H., Son, H. et al. (2021). Removal efficiency of organic micropollutants in successive wastewater treatment steps in a full-scale wastewater treatment plant: bench-scale application of tertiary treatment processes to improve removal of organic micropollutants persisting after secondary treatment. *Chemosphere* 132629. https://doi.org/10.1016/j.chemosphere.2021.132629.

Citulski, J. and Farahbakhsh, K. (2012). Overcoming the toxicity effects of municipal wastewater sludge and biosolid extracts in the yeast estrogen screen (YES) assay. *Chemosphere* 87 (5): 498–503.

Couto, C.F., Lange, L.C., and Amaral, M.C.S. (2019). Occurrence, fate and removal of pharmaceutically active compounds (PhACs) in water and wastewater treatment plants: a review. *Journal of Water Process Engineering* 32: 100927.

Czajka, C.P. and Londry, K.L. (2006). Anaerobic biotransformation of estrogens. *Science of the Total Environment* 367 (2): 932–941.

Daughton, C.G. (2016). Pharmaceuticals and the environment (PiE): evolution and impact of the published literature revealed by bibliometric analysis. *Science of the Total Environment* 562: 391–426.

Derakhshan, Z., Ehrampoush, M.H., and A.H. Mahvi, A.H. (2018). Evaluation of a moving bed biofilm reactor for simultaneous atrazine, carbon and nutrients removal from aquatic environments: modeling and optimization. *Journal of Indian Engineering Chemistry* 67: 219–230. https://doi.org/10.1016/j.jiec.2018.06.032.

Dong, W., Jin, Y., Zhou, K. et al. (2019). Efficient degradation of pharmaceutical micropollutants in water and wastewater by FeIII-NTA-catalyzed neutral photo-Fenton process. *Science of the Total Environment* 688: 513–520.

Falås, P., Wick, A., Castronovo, S. et al. (2016). Tracing the limits of organic micropollutant removal in biological wastewater treatment. *Water Research* 95: 240–249.

Faria, C.V., Ricci, B.C., Silva, A.F.R. et al. (2020). Removal of micropollutants in domestic wastewater by expanded granular sludge bed membrane bioreactor. *Process Safety and Environmental Protection* 136: 223–233.

Fundneider, T., Acevedo Alonso, V., Wick, A. et al. (2021). Implications of biological activated carbon filters for micropollutant removal in wastewater treatment. *Water Research* 189: 116588

Gallé, T., Koehler, C., Plattes, M. et al. (2019). Large-scale determination of micropollutant elimination from municipal wastewater by passive sampling gives new insights in governing parameters and degradation patterns. *Water Research* 160: 380–393.

Gani, K.M., Bux, F., and Kazmi, A.A. (2019). Diethylhexyl phthalate removal in full scale activated sludge plants: effect of operational parameters. *Chemosphere* 234: 885–892.

García, J., García-Galán, M.J., Day, J.W. et al. (2020). A review of emerging organic contaminants (EOCs), antibiotic resistant bacteria (ARB), and antibiotic resistance genes (ARGs) in the environment: increasing removal with wetlands and reducing environmental impacts. *Bioresource Technology* 307: 123228.

Gogoi, A., Mazumder, P., Tyagi, V.K. et al. (2018). Occurrence and fate of emerging contaminants in water environment: a review. *Groundwater for Sustainable Development* 6: 169–180.

Golwala, H., Zhang, X., Iskander, S.M., and Smith, A.L. (2021). Solid waste: an overlooked source of microplastics to the environment. *Science of the Total Environment* 769: 144581.

Gonzalez-Gil, L., Mauricio-Iglesias, M., Serrano, D. et al. (2018). Role of methanogenesis on the biotransformation of organic micropollutants during anaerobic digestion. *Science of the Total Environment* 622–623: 459–466.

Gonzalez-Gil, L., Papa, M., Feretti, D. et al. (2016). Is anaerobic digestion effective for the removal of organic micropollutants and biological activities from sewage sludge? *Water Research* 102: 211–220.

Goswami, L., Vinoth Kumar, R., Borah, S.N. et al. (2018). Membrane bioreactor and integrated membrane bioreactor systems for micropollutant removal from wastewater: a review. *Journal of Water Process Engineering* 26: 314–328.

Grandclément, C., Seyssiecq, I., Piram, A. et al. (2017). From the conventional biological wastewater treatment to hybrid processes, the evaluation of organic micropollutant removal: a review. *Water Research* 111: 297–317.

Gutiérrez, M., Grillini, V., Mutavdžić Pavlović, D., and Verlicchi, P. (2021). Activated carbon coupled with advanced biological wastewater treatment: a review of the enhancement in micropollutant removal. *Science of the Total Environment* 790: 148050.

Hamid, H. and Eskicioglu, C. (2012). Fate of estrogenic hormones in wastewater and sludge treatment: a review of properties and analytical detection techniques in sludge matrix. *Water Research* 46 (18): 5813–5833.

Hamza, R.A., Sheng, Z., Iorhemen, O.T. et al. (2018). Impact of food-to-microorganisms ratio on the stability of aerobic granular sludge treating high-strength organic wastewater. *Water Research* 147: 287–298.

Harb, M., Lou, E., Smith, A.L., and Stadler, L.B. (2019). Perspectives on the fate of micropollutants in mainstream anaerobic wastewater treatment. *Current Opinion in Biotechnology* 57: 94–100.

He, W., Wang, Q., Zhu, Y. et al. (2021). Innovative technology of municipal wastewater treatment for rapid sludge sedimentation and enhancing pollutants removal with nano-material. *Bioresource Technology* 324: 124675.

Jiang, Q., Ngo, H.H., Nghiem, L.D. et al. (2018). Effect of hydraulic retention time on the performance of a hybrid moving bed biofilm reactor-membrane bioreactor system for micropollutants removal from municipal wastewater. *Bioresource Technology* 247: 1228–1232.

Kasprzyk-Hordern, B., Dinsdale, R.M., and . Guwy, J. (2009). The removal of pharmaceuticals, personal care products, endocrine disruptors and illicit drugs during wastewater treatment and its impact on the quality of receiving waters. *Water Resources* 43: 363–380. https://doi.org/10.1016/j.watres.2008.10.047.

Kawa, I.A., Akbar, M., Fatima, Q. et al. (2021). Endocrine disrupting chemical Bisphenol A and its potential effects on female health. *Diabetes & Metabolic Syndrome: Clinical Research & Reviews* 15 (3): 803–811.

Khan, H.K., Rehman, M.Y.A., and Malik, R.N. (2020a). Fate and toxicity of pharmaceuticals in water environment: an insight on their occurrence in South Asia. *Journal of Environmental Management* 271: 111030.

Khan, M.T., Shah, I.A., Ihsanullah, I. et al. (2021). Hospital wastewater as a source of environmental contamination: an overview of management practices, environmental risks, and treatment processes. *Journal of Water Process Engineering* 41: 101990.

Khan, N.A., Khan, S.U., Ahmed, S. et al. (2020b). Recent trends in disposal and treatment technologies of emerging pollutants: a critical review. *TrAC Trends in Analytical Chemistry* 122: 115744.

Kimura, K., Hara, H., and Watanabe, Y. (2007). Elimination of selected acidic pharmaceuticals from municipal wastewater by an activated sludge system and membrane bioreactors. *Environmental Science & Technology* 41 (10): 3708–3714.

Kora, E., Theodorelou, D., Gatidou, G. et al. (2020). Removal of polar micropollutants from domestic wastewater using a methanogenic: aerobic moving bed biofilm reactor system. *Chemical Engineering Journal* 382: 122983.

Kosek, K., Luczkiewicz, A., Fudala-Książek, S. et al. (2020). Implementation of advanced micropollutants removal technologies in wastewater treatment plants (WWTPs): examples and challenges based on selected EU countries. *Environmental Science & Policy* 112: 213–226.

Krzeminski, P., Tomei, M.C., Karaolia, P. et al. (2019). Performance of secondary wastewater treatment methods for the removal of contaminants of emerging concern implicated in crop uptake and antibiotic resistance spread: a review. *Science of the Total Environment* 648: 1052–1081.

Kulandaivelu, J., Choi, P.M., Shrestha, S. et al. (2020). Assessing the removal of organic micropollutants from wastewater by discharging drinking water sludge to sewers. *Water Research* 181: 115945.

Lahti, M. and Oikari, A. (2011). Microbial transformation of pharmaceuticals naproxen, bisoprolol, and diclofenac in aerobic and anaerobic environments. *Archives of Environmental Contamination and Toxicology* 61 (2): 202–210.

Lamba, M. and Ahammad, S.Z. (2017). Performance comparison of secondary and tertiary treatment systems for treating antibiotic resistance. *Water Resources* 127: 172–182, https://doi.org/10.1016/j.watres.2017.10.025.

Li, A.-J., Li, X.-Y., and Yu, H.-Q. (2011). Effect of the food-to-microorganism (F/M) ratio on the formation and size of aerobic sludge granules. *Process Biochemistry* 46 (12): 2269–2276.

Lim, M., Ahmad, R., Guo, J. et al. (2019). Removals of micropollutants in staged anaerobic fluidized bed membrane bioreactor for low-strength wastewater treatment. *Process Safety and Environmental Protection* 127: 162–170.

Lin, A.Y.-C., Yu, T.-H., and Lateef, S.K. (2009). Removal of pharmaceuticals in secondary wastewater treatment processes in Taiwan. *Journal of Hazardous Materials* 167 (1): 1163–1169.

Luo, Y., Guo, W., Ngo, H.H. et al. (2014a). Removal and fate of micropollutants in a sponge-based moving bed bioreactor. *Bioresource Technology* 159: 311–319.

Luo, Y., Guo, W., Ngo, H.H. et al. (2014b). A review on the occurrence of micropollutants in the aquatic environment and their fate and removal during wastewater treatment. *Science of the Total Environment* 473–474: 619–641.

Lutterbeck, C.A., Colares, G.S., Dell'Osbel, N. et al. (2020). Hospital laundry wastewaters: a review on treatment alternatives, life cycle assessment and prognosis scenarios. *Journal of Cleaner Production* 273: 122851.

Ma, X.Y., Dong, K., Tang, L. et al. (2020). Investigation and assessment of micropollutants and associated biological effects in wastewater treatment processes. *Journal of Environmental Sciences* 94: 119–127.

Madikizela, L.M., Ncube, S., and Chimuka, L. (2018). Uptake of pharmaceuticals by plants grown under hydroponic conditions and natural occurring plant species: a review. *Science of the Total Environment* 636: 477–486.

Majumder, A., Gupta, A.K., Ghosal, P.S., and Varma, M. (2021). A review on hospital wastewater treatment: a special emphasis on occurrence and removal of pharmaceutically active compounds, resistant microorganisms, and SARS-CoV-2. *Journal of Environmental Chemical Engineering* 9 (2): 104812.

Martín-Pozo, L., Gómez-Regalado, M.D.C., Moscoso-Ruiz, I., and Zafra-Gómez, A. (2021). Analytical methods for the determination of endocrine disrupting chemicals in cosmetics and personal care products: a review. *Talanta* 234: 122642.

Martínez-Alcalá, I., Guillén-Navarro, J.M., and Fernández-López, C. (2017). Pharmaceutical biological degradation, sorption and mass balance determination in a conventional activated-sludge wastewater treatment plant from Murcia, Spain. *Chemical Engineering Journal* 316: 332–340. https://doi.org/10.1016/j.cej.2017.01.048.

Melo-Guimarães, A., Torner-Morales, F.J., Durán-Álvarez, J.C., and Jiménez-Cisneros, B.E. (2013). Removal and fate of emerging contaminants combining biological, flocculation and membrane treatments. *Water Science Technology* 67 (4): 877–885.

Men, Y., Achermann, S., Helbling, D.E. et al. (2017). Relative contribution of ammonia oxidizing bacteria and other members of nitrifying activated sludge communities to micropollutant biotransformation. *Water Research* 109: 217–226.

Mojiri, A., Zhou, J., Vakili, M., and Van Le, H. (2020). Removal performance and optimisation of pharmaceutical micropollutants from synthetic domestic wastewater by hybrid treatment. *Journal of Contaminant Hydrology* 235: 103736.

Monsalvo, V.M., McDonald, J.A., Khan, S.J., and Le-Clech, P. (2014). Removal of trace organics by anaerobic membrane bioreactors. *Water Research* 49: 103–112.

Mpatani, F.M., Aryee, A.A., Kani, A.N. et al. (2021a). A review of treatment techniques applied for selective removal of emerging pollutant-trimethoprim from aqueous systems. *Journal of Cleaner Production* 308: 127359.

Mpatani, F.M., Han, R., Aryee, A.A. et al. (2021b). Adsorption performance of modified agricultural waste materials for removal of emerging micro-contaminant bisphenol A: a comprehensive review. *Science of the Total Environment* 780: 146629.

Nsenga Kumwimba, M. and Meng, F. (2019). Roles of ammonia-oxidizing bacteria in improving metabolism and cometabolism of trace organic chemicals in biological wastewater treatment processes: a review. *Science of the Total Environment* 659: 419–441.

Olasupo, A. and Suah, F.B.M. (2021). Recent advances in the removal of pharmaceuticals and endocrine-disrupting compounds in the aquatic system: a case of polymer inclusion membranes. *Journal of Hazardous Materials* 406: 124317.

Oppenheimer, J.A., Badruzzaman, M., and Jacangelo, J.G. (2012). Differentiating sources of anthropogenic loading to impaired water bodies utilizing ratios of sucralose and other microconstituents. *Water Research* 46 (18): 5904–5916.

Peña-Guzmán, C., Ulloa-Sánchez, S., Mora, K. et al. (2019). Emerging pollutants in the urban water cycle in Latin America: a review of the current literature. *Journal of Environmental Management* 237: 408–423.

Petrie, B., Barden, R., and Kasprzyk-Hordern, B. (2015). A review on emerging contaminants in wastewaters and the environment: current knowledge, understudied areas and recommendations for future monitoring. *Water Research* 72: 3–27.

Queiroz, F.B., Brandt, E.M.F., Aquino, S.F.et al. (2012). Occurrence of pharmaceuticals and endocrine disruptors in raw sewage and their behavior in UASB reactors operated at different hydraulic retention times. *Water Science. Technology* 66: 2562–2569. https://doi.org /10.2166/wst.2012.482.

Quesada, H.B., Baptista, A.T.A., Cusioli, L.F. et al. (2019). Surface water pollution by pharmaceuticals and an alternative of removal by low-cost adsorbents: a review. *Chemosphere* 222: 766–780.

Quintana, J.B., Weiss, S., and Reemtsma, T. (2005). Pathways and metabolites of microbial degradation of selected acidic pharmaceutical and their occurrence in municipal wastewater treated by a membrane bioreactor. *Water Research* 39 (12): 2654–2664.

Reyes-Contreras, C., Neumann, P., Barriga, F. et al. (2020). Organic micropollutants in sewage sludge: influence of thermal and ultrasound hydrolysis processes prior to anaerobic stabilization. *Environmental Technology* 41 (11): 1358–1365.

Rios Miguel, A.B., Jetten, M.S.M., and Welte, C.U. (2020). The role of mobile genetic elements in organic micropollutant degradation during biological wastewater treatment. *Water Research X* 9: 100065.

Rout, P.R., Zhang, T.C., Bhunia, P., and Surampalli, R.Y. (2021). Treatment technologies for emerging contaminants in wastewater treatment plants: a review. *Science of the Total Environment* 753: 141990.

Saidulu, D., Gupta, B., Gupta, A.K., and Ghosal, P.S. (2021). A review on occurrences, eco-toxic effects, and remediation of emerging contaminants from wastewater: special emphasis on biological treatment based hybrid systems. *Journal of Environmental Chemical Engineering* 9 (4): 105282.

Sarmah, A.K. and Northcott, G.L. (2008). Laboratory degradation studies of four endocrine disruptors in two environmental media. *Environmental Toxicology and Chemistry* 27 (4): 819–827.

Sathyamoorthy, S., Chandran, K., and Ramsburg, C.A. (2013). Biodegradation and cometabolic modeling of selected beta blockers during ammonia oxidation. *Environmental Science & Technology* 47 (22): 12835–12843.

Shakerian, F., Zhao, J., and Li, S.-P. (2020). Recent development in the application of immobilized oxidative enzymes for bioremediation of hazardous micropollutants: a review. *Chemosphere* 239: 124716.

Song, X., Luo, W., McDonald, J. et al. (2018). An anaerobic membrane bioreactor–membrane distillation hybrid system for energy recovery and water reuse: removal performance of organic carbon, nutrients, and trace organic contaminants. *Science of the Total Environment* 628–629: 358–365.

Sörengård, M., Campos-Pereira, H., Ullberg, M. et al. (2019). Mass loads, source apportionment, and risk estimation of organic micropollutants from hospital and municipal wastewater in recipient catchments. *Chemosphere* 234: 931–941.

Sossalla, N.A., Nivala, J., Reemtsma, T. et al. (2021). Removal of micropollutants and biological effects by conventional and intensified constructed wetlands treating municipal wastewater. *Water Research* 201: 117349. https://doi.org/10.1016/j.watres.2021.117349).

Stasinakis, A.S. (2012). Review on the fate of emerging contaminants during sludge anaerobic digestion. *Bioresource Technology* 121: 432–440.

Stevens-Garmon, J., Drewes, J.E., Khan, S.J. et al. (2011). Sorption of emerging trace organic compounds onto wastewater sludge solids. *Water Resources* 45: 3417–3426. https://doi.org/10.1016/j.watres.2011.03.056.

Taboada-Santos, A., Behera, C.R., Sin, G. et al. (2020). Assessment of the fate of organic micropollutants in novel wastewater treatment plant configurations through an empirical mechanistic model. *Science of the Total Environment* 716: 137079.

Tadkaew, N., Hai, F.I., McDonald, J.A. et al. (2011). Removal of trace organics by MBR treatment: the role of molecular properties. *Water Research* 45 (8): 2439–2451.

Tang, K., Ooi, G.T.H., Litty, K. et al. (2017). Removal of pharmaceuticals in conventionally treated wastewater by a polishing moving bed biofilm reactor (MBBR) with intermittent feeding. *Bioresource Technology* 236: 77–86.

Thomaidi, V.S., Stasinakis, A.S., Borova, V.L., and Thomaidis, N.S. (2016). Assessing the risk associated with the presence of emerging organic contaminants in sludge-amended soil: a country-level analysis. *Science of the Total Environment* 548–549: 280–288.

Tiwari, B., Sellamuthu, B., Ouarda, Y. et al. (2017). Review on fate and mechanism of removal of pharmaceutical pollutants from wastewater using biological approach. *Bioresource Technology* 224: 1–12.

Tran, N.H., Reinhard, M., and Gin, K.Y.-H. (2018). Occurrence and fate of emerging contaminants in municipal wastewater treatment plants from different geographical regions: a review. *Water Research* 133: 182–207.

Tran, N.H., Urase, T., and Kusakabe, O. (2009). The characteristics of enriched nitrifier culture in the degradation of selected pharmaceutically active compounds. *Journal of Hazardous Materials* 171 (1): 1051–1057.

Tran, N.H., Urase, T., Ngo, H.H. et al. (2013). Insight into metabolic and cometabolic activities of autotrophic and heterotrophic microorganisms in the biodegradation of emerging trace organic contaminants. *Bioresource Technology* 146: 721–731.

Venegas, M., Leiva, A.M., Reyes-Contreras, C. et al. (2021). Presence and fate of micropollutants during anaerobic digestion of sewage and their implications for the circular economy: a short review. *Journal of Environmental Chemical Engineering* 9 (1): 104931.

Viveknand, A.C., Mohapatra, S., and Tyagi, V.K. (2021). Microplastics in aquatic environment: challenges and perspectives. *Chemosphere* 282: 131151. https://doi.org/10.1016/j.chemosphere.2021.131151.

Vo, H.N.P., Ngo, H.H., Guo, W. et al. (2020). Micropollutants cometabolism of microalgae for wastewater remediation: effect of carbon sources to cometabolism and degradation products. *Water Research* 183: 115974.

Wang, J. and Wang, S. (2016). Removal of pharmaceuticals and personal care products (PPCPs) from wastewater: a review. *Journal of Environmental Management* 182: 620–640.

Wijekoon, K.C., Hai, F.I., Kang, J. et al. (2013). The fate of pharmaceuticals, steroid hormones, phytoestrogens, UV-filters and pesticides during MBR treatment. *Bioresource Technology* 144: 247–254.

Wijekoon, K.C., McDonald, J.A., Khan, S.J. et al. (2015). Development of a predictive framework to assess the removal of trace organic chemicals by anaerobic membrane bioreactor. *Bioresource Technology* 189: 391–398.

Wu, X., Dodgen, L.K., Conkle, J.L., and Gan, J. (2015). Plant uptake of pharmaceutical and personal care products from recycled water and biosolids: a review. *Science of the Total Environment* 536: 655–666.

Xiao, Y., Yaohari, H., De Araujo, C. et al. (2017). Removal of selected pharmaceuticals in an anaerobic membrane bioreactor (AnMBR) with/without powdered activated carbon (PAC). *Chemical Engineering Journal* 321: 335–345.

Xue, W., Wu, C., Xiao, K. et al. (2010). Elimination and fate of selected micro-organic pollutants in a full-scale anaerobic/anoxic/aerobic process combined with membrane bioreactor for municipal wastewater reclamation. *Water Research* 44 (20): 5999–6010.

Yang, Y., Ok, Y.S., Kim, K.-H. et al. (2017). Occurrences and removal of pharmaceuticals and personal care products (PPCPs) in drinking water and water/sewage treatment plants: a review. *Science of the Total Environment* 596–597: 303–320.

Zamri, M.F.M.A., Bahru, R., Suja, F. et al. (2021). Treatment strategies for enhancing the removal of endocrine-disrupting chemicals in water and wastewater systems. *Journal of Water Process Engineering* 41: 102017.

Zhang, S., Lei, Z., Dzakpasu, M. et al. (2021). Removal of trace organic contaminants in municipal wastewater by anaerobic membrane bioreactor: efficiencies, fates and impact factors. *Journal of Water Process Engineering* 40: 101953.

Zupanc, M., Kosjek, T., M. Petkovˇsek, M. et al. (2013). Removal of pharmaceuticals from wastewater by biological processes, hydrodynamic cavitation and UV treatment. Ultrasonic Sonochemistry 20: 1104–1112. https://doi.org/10.1016/j.ultsonch.2012.12.003.

9

Various Perspectives on Occurrence, Sources, Measurement Techniques, Transport, and Insights Into Future Scope for Research of Atmospheric Microplastics

Sailesh N. Behera[1,2], *Mudit Yadav*[1], *Vishnu Kumar*[1], *and Prangya Ranjan Rout*[3,4]

[1]*Air Quality Laboratory, Department of Civil Engineering, Shiv Nadar University, Delhi-NCR, Greater Noida, Gautam Buddha Nagar, Uttar Pradesh, India*
[2]*Centre for Environmental Sciences and Engineering, Shiv Nadar University, Delhi-NCR, Greater Noida, Gautam Buddha Nagar, Uttar Pradesh, India*
[3]*Department of Biotechnology, Thapar Institute of Engineering and Technology, Patiala, Punjab, India*
[4]*Department of Biotechnology, Dr. B. R. Ambedkar National Institute of Technology Jalandhar, Punjab, India*

9.1 Introduction

Studying various aspects of plastic production and its associated pollution issues started from the time when the first synthetic polymer was formulated, and manufactured commercially for its further uses (Dris et al. 2015a; Lehner et al. 2019; Millican and Agarwal 2021). Successively numerous developments occurred worldwide in manufacturing and application of several plastic products. Such uses and applications have created plastic debris, known as plastic wastes in the environment (Barnes et al. 2009; Evode et al. 2021). Due to more dependency on plastic products meant for our use, with various purposes during the last three decades, exceptional rise in plastic pollution has occurred worldwide, with an annual growth in plastic production of 3% (Plastic Europe 2018; Enyoh et al. 2019). For example, the global production of plastics was 348 million tons during 2017, and it was predicted that there would be a 4-fold increase in production from 311 million tons during 2014 to 1.2 billion tons by 2050 (Ryan 2015; Mbachu et al. 2020). Hence, it is expected that an increase in production of plastics will cause a rise in the generation of plastic wastes. Previous estimations showed that 10% of plastic produced worldwide ended up as waste and it was assessed that only 3% of waste was recycled during 2016, leaving 7% of waste unattended without proper management for disposal (Thompson 2006; Verla et al. 2017).

Because of its light weight, ability to be made into numerous desirable forms, and the viability in increasing the production volume of plastics, the materials produced from plastics have been gaining versatility in everyday life applications including household items, personal care products, infrastructure items, machinery accessories, various tools, and many others (Awasthi et al. 2017; Idumah and Nwuzor 2019). On the other hand, improper waste management with illegal waste disposal of plastic waste and slower processes of

Microconstituents in the Environment: Occurrence, Fate, Removal, and Management, First Edition.
Edited by Rao Y. Surampalli, Tian C. Zhang, Chih-Ming Kao, Makarand M. Ghangrekar, Puspendu Bhunia, Manaswini Behera, and Prangya R. Rout.

disintegration of waste under usual environmental conditions have caused adverse effects on the environment, aquatic life, animals, and human health (Alam and Ahmade 2013; Wichai-Utcha and Chavalparit 2019; Yousefi et al. 2021). Therefore, the emphasis on by-products created from plastic waste has become famously known as microplastics. Microplastics, being highly mobile, persistent, and easy to be taken up by various organisms, can have serious effects on the environment and health of humans, animals, and aquatic life (Gasperi et al. 2018; Huang et al. 2021a; Vivekanand et al. 2021). For these reasons, microplastics are believed to be emerging environmental pollutants that can cause potential hazards to many ecosystems covering multiple spectrums in the global ecosphere that include water, air and soil, and human health.

From the definition point of view, a microplastic can be defined as a plastic particle whose size may vary in range from 100 nm to 5 mm (Thompson et al. 2004; Masura et al. 2015; Zhang et al. 2020a). As an update from the later concept in the standard international unit nomenclature (SI units), the size of microplastics varies from 1 μm to 5 mm (Hartmann et al. 2019; Zhang et al. 2020a). Research studies on plastic particles are increasing day by day and new concepts are evolving. In the perspectives of size, nanoplastics are also gaining attention because they are believed to be as ubiquitous as their bulk counterpart of microplastics (Alimi et al. 2018; Junhao et al. 2021; Gigault et al., 2021). A plastic particle with size of <1 μm is normally classified as a nanoplastic particle and this particle has significant importance with respect to seafood safety and enhancement of contaminant transport in the environment and possible implications for human health (Smith et al. 2018; Bank and Hansson 2019; Zhou et al. 2022). Nevertheless, nanoplastics present in the environment are yet to be identified in a more defined way, classified systematically, and quantified widely with scientific standardized methodologies. Therefore, microplastics are the targeted plastic particles for scientific and technical communities to further explore their occurrence and fate in different components of the environment.

From the perspectives of source of origin, microplastics in the environment are classified into primary and secondary microplastics (Wang et al. 2019; Kabir et al. 2021). Primary microplastics are generated from primary sources of plastics, originally manufactured products at small sizes that include exfoliating plastic particles in toiletries (washing milk, sanitizer, microbeads, etc.) (Browne et al. 2007; Junhao et al. 2021); whereas, secondary microplastics are generated from secondary sources such as breakdown of large plastic debris and degradation of macroplastics that include fibers from synthetic textiles (Can-Güven 2021; Zhang et al. 2022). Recently, research on microplastics in water and soil has become well understood due to their larger presence in the aquatic and terrestrial ecosystems and effecting various plant and anmial life (da Costa et al. 2018; Prata et al. 2019; Yuan et al. 2022). It is noteworthy that marine plants, and other flora and fauna in the marine ecosystems, are severely affected by illegal discharge of plastic debris and wastes, having microplastics as constituents in the aquatic bodies of oceans and rivers, and other aquatic bodies (Hammer et al. 2012; Peng et al. 2020). Similarly, when plastic debris containing microplastics is dumped in the soil, it affects natural water flow in the pores of soil and growth of organisms causing reduction in yield of plants (Boots et al. 2019; Zhang et al. 2020b). Compared to these two components of the environment (water and soil), research studies of microplastics related to the air or atmospheric matrices are less reported and various conceptual outlines are less understood in the literature.

Atmospheric deposition was inferred to as an important responsible source of microplastics to fresh water for the first time through their occurrence in atmospheric fallout and dust (Dris et al. 2015a). Since that particular study, there have been successive reports on characteristics and occurrence of atmospheric microplastics in road dust and ambient air at various locations worldwide, including Tehran (Iran), Dongguan (China), Da Nang (Vietnam), Kathmandu (Nepal), Kusatsu (Japan), and Paris (France) (Cai et al. 2017; Dehghani et al. 2017; Yukioka et al. 2020). Subsequently, occurrence of airborne microplastics are treated as the prevalent pollutants in the atmosphere or air medium. Allen et al. (2019) reported the occurrence of a large number of microplastics with a quantity of 365 items/day/m^2 in ambient air of the pristine mountainous area of the French Pyrenees. The occurrence of atmospheric microplastics and their potential risks through inhalation into the respiratory tract of humans were estimated using various ecological remedial investigation (RI) models, and respiratory threat models suitable for the human body (Liu et al. 2019; Vianello et al. 2019; Sun et al. 2021).

From the perspectives of transport and fate of microplastics after their emission from water and soil mediums, their behavior in the atmosphere has recently gained attention, as these airborne particles undergo transport and are finally deposited into water or soil mediums (Huang et al. 2020; Zhang et al. 2020c). Similarly, loose, small size, and lower density of lighter microplastics from water and soil mediums have the potential to be transferred to the air medium followed by their transportation along the wind direction (Allen et al. 2019; Zhang et al. 2020b). The sizes of microplastic particles in the atmosphere are smaller compared to water and soil mediums, and so atmospheric microplastics are suspended and re-suspended in air (Dris et al. 2015b; Liu et al. 2019). Once emitted from the source of origin, microplastics can undergo long-range transport to reach remote regions and cause pollution issues while suspended in the air medium and finally affect aquatic and soil mediums after deposition on their surfaces (Wright et al. 2020; Huang et al. 2021b).

Atmospheric microplastics have the tendency to be directly and continuously inhaled by humans and enter into their respiratory systems. Such exposure to microplastics through the inhalation process poses harmful health effects to humans (Chen et al. 2020; Rahman et al. 2021). Through their ability to be inhaled and combined with other toxic and hazardous pollutants (arsenic, chromium, nickel, polycyclic aromatic hydrocarbons (PAHs), etc.), it is believed that atmospheric microplastics have become emergent components of air pollution (Wright and Kelly 2017; Rochman et al. 2019; Tourinho et al. 2019; Sridharan et al. 2021). Therefore, atmospheric microplastics can cause respiratory ailments and related diseases, and carcinogenic and non-carcinogenic health effects in humans through exposure on a short-term and/or a long-term basis.

From the aforementioned literature review, it is clear that atmospheric microplastics are found in all components of the atmosphere. There are several aspects of these microplastics that are not yet well understood. Assessing the needs of today and providing a compact documented material with systematic compilation of pertinent findings and information from previous studies, this chapter has synthesized knowledge-based information under one umbrella. The specific objectives of this chapter on atmospheric microplastics are summarized as follows: i) providing pertinent information on classification and properties of atmospheric microplastics; ii) compilation of information on responsible sources of emissions;

iii) synthesizing various methods in sampling, sample preparation, and characterization in detection of these pollutants for measurement and quantification; iv) synthesis of knowledge-based information on occurrence and ambient concentrations of these pollutants; v) providing details on various factors including meteorological parameters and source characteristics that can affect pollutant concentration in the ambient air; vi) shed light on transboundary transport in studying the fate of these pollutants; vii) modeling techniques for prediction of concentration of these pollutants; viii) finding some control technologies in treatment and reduction of these pollutants; and ix) providing important information on challenges of future climate and paving the way forward for future research.

9.2 Classification and Properties of Microplastics

9.2.1 Classification of Atmospheric Microplastics

Based on studies and synthesis found in the literature (Dris et al. 2015b; Gasperi et al. 2018; Huang et al. 2020, 2021b; Wright et al. 2020; Kabir et al. 2021; Sridharan et al. 2021; Sun et al. 2021), microplastics present in the atmosphere can be classified into the following categories, with specific abbreviations, as:

i) alkyd resin (ALK); ii) epoxy resin (EP); iii) ethylene propylene copolymer (EPC); iv) ethylene vinyl acetate copolymer (EVAC); v) styrene butadiene rubber (SBR); vi) ethylene propylene diene rubber (EPDM); vii) polyamide (PA); viii) polyacrylic acid (PAA); ix) polyacrylate (PAK); x) polyacrylonitrile (PAN); xi) polyacrylonitrile-acrylic acid (PAN-AA); xii) polyvinyl acetate (PAV); xiii) (PE); (xiv) phosphoenolpyruvic acid (PEP); xv) polyethersulfone (PES); xvi) polyethylene terephthalate (PET); xvii) poly (ethylene-co-vinyl acetate) (PEVA); xviii) phenylalanine (Phe); xix) poly (*N*-methyl acrylamide) (PMA); xx) polypropylene (PP); xxi) phenoxy resin (PR); xxii) polystyrene (PS); xxiii) polytetrafluoroethylene (teflon: PTFE); xxiv) polyurethane (PUR); xxv) poly(vinyl acetate) (PVA); xxvi) polyvinyl chloride (PVC); and xxvii) polyvinyl stearate (PVS).

Microplastics present in the atmosphere are catagorized from the perspectives of their source of generation to the atmosphere, primarily through usual degradation of large plastics by means of ultraviolet light, and/or degradation from clothing, personal health care products, and nurdles through secondary means (Gasperi et al. 2018; Smith et al. 2018; Enyoh et al. 2019; Liu et al. 2019; Evode et al. 2021; Huang et al. 2021a). The process of degradation of large plastics may occur in any of the mediums of the environment, i.e., water, air (atmosphere), or soil. It is also believed that emissions of microplastics occur during recycling of plastic wastes, as a part of waste management practice (Browne et al. 2007; Huang et al. 2013; Cai et al. 2017; Hahladakisa et al. 2018; Allen et al. 2019).

9.2.2 Characteristics of Atmospheric Microplastics

Airborne or atmospheric microplastics predominately show size ranges from 20 to 500 μm. In general, the size range of atmospheric microplastics varies over a large span from <25 to 5000 μm. Comparing the size ranges of microplastics present in aquatic and sediment environments, atmospheric microplastics are smaller than microplastics in other mediums

(Auta et al. 2017; Wang et al. 2021a; Koelmans et al. 2022). Regarding the smallest size range, it has been observed that atmospheric microplastics can vary from 5 to 100 μm, with a predominance at 20 μm. From the detection methods through Fourier transform infrared spectroscopy (FT-IR), the limit of size range occurs over 10 to 20 μm, and through the stereomicroscope, the size range detection is ~50 μm (Cai et al. 2017; Mai et al. 2018; Yukioka et al. 2020; Wang et al. 2021b). Therefore, researchers do not investigate the detection and quantification of smaller sizes of atmospheric microplastics (e.g., nanoplastics) at present. Although most of the colors of atmospheric microplastics are found as black, white, and transparent, some are also blue, brown, dark, green, gray, orange, pink, purple, red, and yellow (Prata et al. 2020; Zhang et al. 2020c; Chen et al. 2022). From reported studies, it has been seen that the shapes of airborne microplastics are fiber, film, fragment, foam, granule, and sphere, with the most common shapes of these micropollutants being fibers and fragments (Mbachu et al. 2020; Wright et al. 2020; Hanun et al. 2021; Sridharan et al. 2021). Table 9.1 shows compiled information on characteristics of various airborne microplastics.

Table 9.1 Characteristics of atmospheric microplastics in various environmental matrices.

Specific Matrix	Size (μm)	Shape	Polymer Type*	Color	References
Atmospheric fallout	50–3250	Fiber	RY, PA, PE, PP	NA	Dris et al. (2017)
Indoor	50–1650	Fiber	NA		Dris et al. (2017)
Atmospheric fallout	23–9555	Fiber, fragment, granule.	PET, PE, PES, PAN, PAA, RY, EVA, EP, ALK	Blue, black, red, transparent, brown, green.	Liu et al. (2019)
Atmospheric fallout	2–5000	Fiber	NA	Yellow, gray white-transparent, red-pink, black-gray.	Abbasi et al. (2019)
Atmospheric fallout	100–5000	Fiber, fragment.	NA	Gray	Dris et al. (2015b)
Atmospheric fallout	50–3200	Fiber	RY, Acetate, PET, PA, PU	NA	Dris et al. (2015b)
Atmospheric fallout	200–2200	Fiber, fragment, film, foam.	PET, PA, PU PE, PP, PS	NA	Cai et al. (2017)
Atmospheric fallout	50–3000	Fiber, fragment, film, foam.	PS, PE, PP, PVC, PET	White, black, red, transparent.	Allen et al. (2019)
Street dust	2–5000	Spherule, film, fiber, fragment.	NA	White, transparent, yellow-orange, red-pink, blue/green, black-gray.	Abbasi et al. (2019)
Indoor dust	17–2000	Fiber, granule.	PET, PAN, PA, PE, PP <PU, PEI, acrylic, alkyd, cellulose, rayon, protein.	NA	Liu et al. (2019)

(Continued)

Table 9.1 (Continued)

Specific Matrix	Size (μm)	Shape	Polymer Type*	Color	References
Atmospheric deposition	20–2899	Fiber, fragment, film, granule, foam.	Cellulose (69%), PAN, PET, PA, PUR, PP; nonfibrous: PS, PP, PE, PET, PUR, PVC.	NA	Wright et al. (2020)
Road dust	453–2347	Fiber, fragment, film, granule.	PE, PP, PS, PET, PAK, PVS, EPC, SBR, EPDM, PU.	Black, white, gray, yellow, green, red, brown.	Yukioka et al. (2020)

* For full expansions of these abbreviations, refer to Section 9.2.2.

9.2.3 Qualitative Assessment to Identify Microplastics

On assessing the origin of production of plastics, the trends show that the majority of the various types are manufactured based on the energy source of non-renewable fossil fuels (GESAMP 2016). The plastic-derived particles with sizes >500 μm are normally identified by visual observation qualitatively through assessing the shape and color of these particles using the stereomicroscope followed by conducting a chemical analytical method (Hidalgo-Ruz et al. 2012; Nguyen et al. 2019). It is to be noted that the method of detection for microplastics in the atmosphere is different from the above method, as the sizes of particles derived from microplastics in the atmosphere are mostly concentrated in the range below 500 μm, compared to other mediums of the environment. For example, most of the atmospheric microplastics fall in the size range from 20 to 500 μm, the details of which are provided in Section 9.2.2. The sizes of particles derived from microplastics in the atmosphere are less compared to other mediums due to the process of weathering, suspension, and resuspension of particles.

From literature reviews, identification of atmospheric microplastics can be done qualitatively with the following assessments: i) atmospheric microplastics do not have a structure similar to cellular or organic configurations; ii) to conduct proper and accurate identification process, it is preferable to remove inorganic, organic, and biofilms stuck to microplastics-derived particles to avoid the effects of artifacts; iii) in the cross-sectional view of microplastics, it is expected that fibers are even and have regular thickness along the length, and three-dimensional bending is being demonstrated; iv) in the case of usual fragmented and filmed microplastics, uniform coloring patterns with a level of transparency having more clarity are observed, and in the case of weathered microplastics-derived particles, significant internal coloring marks with damage or bleaching of color at the edges and surface are observed; v) the identification of weathered or aged microplastics can be done through irregular shapes, pitting, gouging, and scratched and torn surfaces; and vi) the color is also used as a qualitative assessment in identification of microplastics-derived particles. The colors can range from transparent to dark, and variations within white to black.

9.3 Sources of Atmospheric Microplastics

From the perspective of assessing the source of origin of the emission of microplastics-derived particles to the atmosphere, they can be divided into two groups, namely primary and secondary microplastics. Primary microplastics are generated from the basic stages of plastic manufacturing and handling processes, such as washing of milk containers, sanitizers, and many others (Browne et al. 2007). However, secondary microplastics are generated from the breaking down of large plastic materials including smaller-sized materials such as fibers from synthetic textiles and other industrial products (Zhang et al. 2020a). It should be noted that there are no natural sources of origin for emission of microplastics, with anthropogenic or man-made activities responsible for emission of all microplastics-derived particles to the atmosphere.

Overall, the sources of microplastics-derived particles to the atmosphere can be classified into major sectors like industrial, domestic, and agricultural. The activities responsible for these emissions from industrial sources include grinding and fine grinding of plastic materials and plastic garbage during waste management, recycling, and incineration processes (Dris et al. 2016; Abbasi et al. 2017; Dehghani et al. 2017; Abbasi et al. 2019). Major contributions from industrial sources come from the activities of grinding and fine grinding of plastic materials, various processes of waste incineration techniques, and landfilling (Dris et al. 2016).

The activities responsible for emission of microplastics to the atmosphere from agricultural sources include various procedures in agricultural processes such as use of plastic mulching through thin polymer films or sheets, and the use of plastics in handling organic and synthetic fertilizers in the field (Carr et al. 2016; Nizzetto et al. 2016). When incomplete removal of sludge or fertilizers from plastic materials occurs, microplastics-derived particles are retained on these surfaces (Li et al. 2018; Magni et al. 2019). In general, farmers use various types of plastic mulches of thin polymer sheets on agriculture fields during cultivation processes to enhance or curtail various properties of soil, depending on its requirements (Ekebafe et al. 2011).

The activities responsible for emission of microplastics to the atmosphere under domestic sources include use of synthetic household materials such as clothing that are made from materials of synthetic polymer types (Liu et al. 2019). These clothing materials normally contains small fibers, which can be released as microplastics-derived particles into the atmosphere during domestic activities including wearing, cleaning, drying, collection and organizing clothing, and dress materials. Domestic soft furnishings such as carpets, curtains, furniture, and many others contain fiberous materials which can enter the atmosphere during their usage and maintenance. Additionally, airborne fine particles are formed through the fragmentation process from microplastics of a fibrous nature through wind shear, abrasion, and photooxidation degradation in the presence of sunlight. Coated and packaging materials used for domestic purposes such as epoxy and alkyd resins are progressively turned into fragments of airborne microplastics after long-term exposure to ultraviolet radiation and abrasion through physical means.

Dust settled in domestic houses, on roads, and other surfaces is normally considered as a sink for atmospheric microplastics-derived particles. However, particles derived from microplastics settled on the ground surface can sometimes be re-suspended and again enter the atmosphere. Therefore, this dust can be treated as a secondary source of atmospheric microplastics. There are also several other sources, such as tire and traffic emissions that are responsible for an increase in the concentration of atmospheric microplastics debris through abrasion and weathering (Abbasi et al. 2017; Kole et al. 2017).

9.4 Measurement of Atmospheric Microplastics

Figure 9.1 depicts various steps that are normally followed for sampling preparation of atmospheric microplastics to measure microplastics-derived particles.

For sampling of microplastics-derived particles in various matrices of the atmosphere, two methods are used; i.e., passive sampling and active sampling are used in collection of these pollutants. The components of passive sampling devices include a collecting

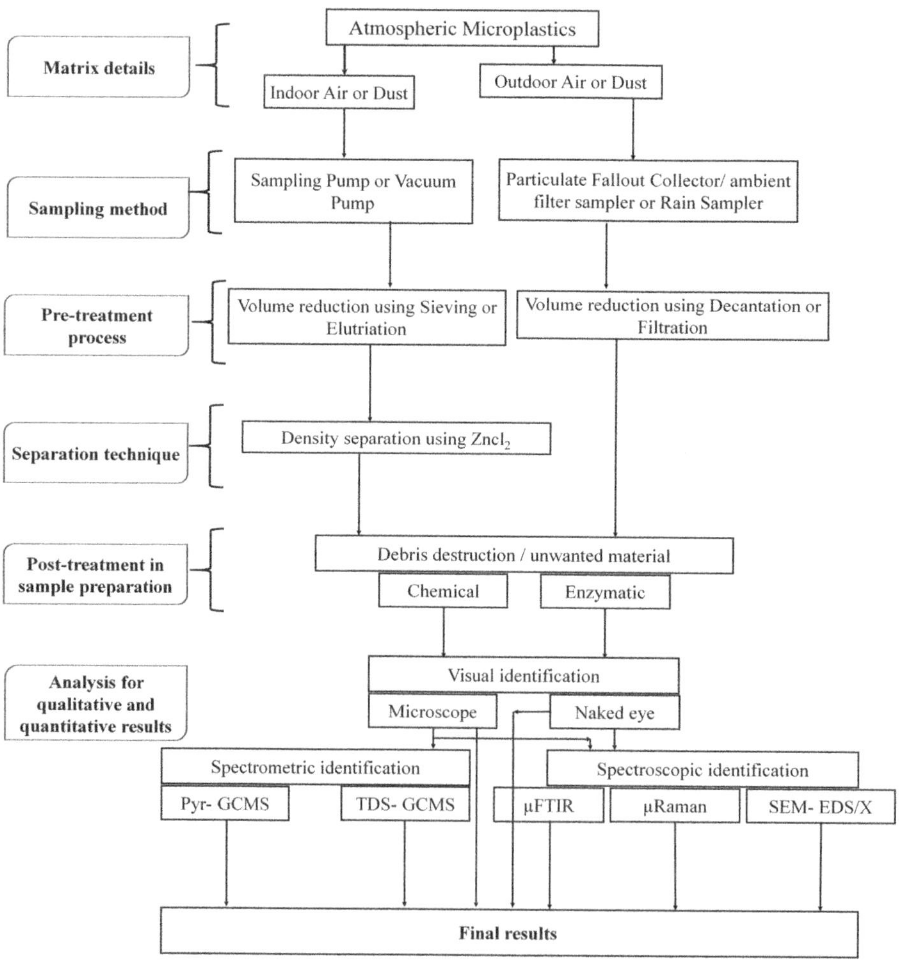

Figure 9.1 Various steps followed for sampling and sample preparation of atmospheric microplastics. The authors conceptualized the figure after Enyoh et al. (2019). *Source:* Adapted from Enyoh et al. (2019).

column or funnel, receiving tube, and terminal collection bottle (Dris et al. 2015b). The flow of air during passive sampling is maintained as equivalent to the natural speed of air. However, a higher speed of sampling velocity is maintained during active sampling through use of a suitable pumping system (Dris et al. 2017; Liu et al. 2019). Therefore, duration of sampling in active sampling is comparatively less than that of passive sampling. The active sampling method normally does not consider deposition effects, whereas the passive sampling method provides information of atmospheric microplastics deposited onto the surface of the sampler. Therefore, passive sampling should be conducted along with active sampling to achieve a complete picture of atmospheric microplastics contents. Standard brushes are normally used to collect samples of microplastics in the matrices of indoor house dust and road/street dust.

Before sending captured samples of atmospheric microplastics for analysis, in identification and quantification of polymers of microplastics, pre-treatment of atmospheric microplastics takes place through the steps of concentration, purification, and separation techniques. Screening and filtration are two widely concentration methods used. To remove biological, organic, or inorganic contaminants from captured samples, purification techniques are adopted to avoid interference by other contaminants in samples meant for analysis. For this purpose, H_2O_2 (30% volume concentration) is used to remove the above contaminants from samples. The density separation method is applied to separate the dense contaminants from light samples. For this purpose, $ZnCl_2$ (ρ 1.6–2.65 g/cm^3) is the most used agent (Dris et al. 2017; Allen et al. 2019).

Both qualitative and quantities analysis are conducted in analysis of atmospheric microplastics, which are similar to analysis of samples from other environmental mediums. Physical morphology and chemical composition/constituents characterization are conducted as parts of qualitative analysis. Properties such as shape, size, and color are analyzed through morphological analysis, and a stereomicroscope instrument is used for the same purpose (Allen et al. 2019; Liu et al. 2019). μ-FTIR and μ-Raman are used for characterization of chemical compositions of microplastics-derived particles in terms of various polymers (Zhang et al. 2020b). To obtain more accurate and precise results meant for identification of polymers and quantitative analysis of airborne microplastics, stereomicroscopy in combination with spectroscopy (μ-FTIR and μ-Raman) or SEM/EDS are used (Abbasi et al. 2019; Allen et al. 2019). Pyro-GC/MS is also used to quantify polymer degradation products to obtain an abundance of microplastics by way of an indirect approach (Liu et al. 2019).

9.5 Occurrence and Ambient Concentration of Microplastics

Because of lower density and smaller sizes, microplastics-derived particles in the atmosphere are distributed easily through wind in various matrices and these are mostly at downwind locations in greater quantity (Browne et al. 2010). Most of the reported studies are limited to specific countries including France, China, Turkey, Iran, Japan, and the UK. The deposition rate and concentration levels of microplastics-derived particles in different atmospheric matrices are presented in Table 9.2. Dris et al. (2017) conducted a study to measure polymers of microplastics-derived particles in the matrix of atmospheric fallout in Paris, France. The concentration was found in the range from 1.0 to 60.0 particle/m^3, and the deposition rate was reported as 1600 to 11,000 particle/day/m^2. Dris et al. (2017) also

Table 9.2 Deposition rates and concentration of atmospheric microplastics at various locations.

Location of Study	Matrix Type	Deposition Rate (Particle/day/m^2)	Concentration	References
Paris, France	Atmospheric fallout.	1600–11,000	1.0–60.0 particle/m^3	Dris et al. (2017)
Paris, France	Indoor.	1586–11,130	0.3–1.5 particle/m^3	Dris et al. (2017)
Shanghai, China	Atmospheric fallout.	NA	0–4.18 particle/m^3	Liu et al. (2019)
Asaluyeh County, Iran	Atmospheric fallout.	NA	1–72 particle/m^3	Abbasi et al. (2019)
Paris, France	Atmospheric fallout.	0.0–602	2.33×10^{13} particle/100 km	Dris et al. (2015a)
Dongguan, China	Atmospheric fallout.	175–313	31–47 particle/m^2	Cai et al. (2017)
Pyrenees mountains, southwest France	Atmospheric fallout.	365 ± 69	NA	Allen et al. (2019)
Asaluyeh County, Iran	Street dust.	NA	16–60 items/g	Abbasi et al. (2019)
Shanghai, China	Indoor dust.	NA	17–620 fibers/mg	Liu et al. (2019)
London, UK	Atmospheric deposition.	NA	12–925 items/m^2	Wright et al. (2020)
Karimata Strait, Japan	Road dust.	NA	0–0.8 items/m^3	Yukioka et al. (2020)

NA = Not available

quantified concentration and deposition rates in indoor dust at 0.3 to 1.5 particle/m^3, and 1586 to 11,130 particle/day/m^2, respectively. Liu et al. (2019) reported concentrations in the matrix of atmospheric fallout at Shanghai, China at 0 to 4.18 particle/m^3.

Pyr-GCMS = pyrolysis gas chromatography mass spectroscopy; TDS-GCMS = thermal desorption system gas chromatography-mass spectroscopy; μFTIR = micro Fourier transform infrared spectroscopy; μRaman = micro Raman spectroscopy; and SEM-EDS/X = scanning electron microscopy with energy dispersive X-ray spectroscopy.

Dris et al. (2015a) quantified concentration and deposition rates in the matrix of atmospheric fallout at 2.33×10^{13} particle/100 km and 0.00–602 particle/day/m^2, respectively. Abbasi et al. (2019) measured concentration levels of polymers in microplastics-derived particles in the matrices of atmospheric fallout and street dust in Asaluyeh County, Iran. The concentration levels of polymers of microplastics-derived particles were found to be 1.0 to 72.0 particle/m^3 in atmospheric fallout and 16 to 60 items/g in street dust.

Cai et al. (2017) conducted a study to measure polymers of microplastics-derived particles in the matrix of atmospheric fallout in Dongguan, China. The concentration was found in the range from 31 to 47 particle/m^2, and the deposition rate was reported as 175 to 313 particle/day/m^2. Allen et al. (2019) measured deposition rate of polymers of microplastics-derived particles in the matrix of atmospheric fallout in a remote site located in the Pyrenees mountains, southwest France. They found the deposition rate of 365 ± 69 particle/day/m^2, and such a result clearly indicated long-range transport of microplastics-derived particles from their source of origins to remote regions like mountainous areas, where hardly any microplastic pollution is observed. Liu et al. (2019) reported concentrations of microplastics-derived particles in the matrix of indoor dust in Shanghai, China and the result was found to be 17 to 620 fibers/mg.

Wright et al. (2020) observed concentrations of microplastics-derived particles in the matrix of atmospheric deposition in London, UK and found the result to be 19 to 925 items/m^2. Yukioka et al. (2020) measured concentrations of microplastics-derived particles in the matrix of road dust in Karimata Strait, Japan. The value of concentration was reported at 0 to 0.8 items/m^3. Results on occurrence of microplastics-derived particles in various matrices of the atmosphere revealed that concentrations of these pollutants vary with the characteristics of the matrix and the geographical location. Although it is difficult to draw any conclusions from the limited number of studies conducted across the world, a likely trend was observed that an indoor environment exhibits higher deposition rates of microplastics-derived particles than an outdoor environment.

9.6 Factors Affecting Pollutant Concentration

Three major factors, meteorology, climate, and anthropogenic activities, normally influence levels of microplastics-derived particles in the atmosphere. In an indoor built environment, influencing factors, such as partition, ventilation, and airflow of the rooms, play a major role in increasing the concentration and deposition of microplastics-derived particles (Prata 2018). As a result, concentration and deposition rate of microplastics-derived particles are relatively higher in indoor air than outdoor air. In addition to the above factors, the primary driving factor of microplastics in the atmosphere are source emissions, transport, dispersion, and deposition. The concentration levels of these pollutants at any location depends on proximity of sources of origin. If the sources of origin are close to the measurement site, the levels of microplastics-derived particles may be higher.

Transport of these microplastics-derived particles are driven by prevailing wind speed and wind direction in the atmosphere. Dispersion of these air pollutants depends on atmospheric turbulence or disturbance, stability condition, and height of the mixing layer or boundary layer. During deposition, microplastics-derived particles in the atmosphere fall to the ground surface. The deposition process depends mostly on precipitation, scavenging, and sedimentation. The aforementioned factors are responsible for movement of microplastics-derived particles, and these factors decide the concentration levels of these pollutants in different matrices of the atmosphere. The processes of movement of microplastics-derived particles through transport, dispersion, and deposition depend on their size, shape, and length (Zhou et al. 2017; Allen et al. 2019).

Allen et al. (2019) reported that various polymers of microplastics-derived particles varied with change in weather and climate. They observed that the levels of microplastics-derived particles in the atmosphere were higher during low precipitation periods than high precipitation periods. The study reported by Liu et al. (2019) revealed that complex landscape and densities of buildings of the study domain influenced spatial distribution and deposition of microplastics-derived particles in the atmosphere. These particles can be transformed into smaller particles, such as nanoplastics or femto plastics, depending on temperature, humidity, and solar radiation. The health effects of smaller sizes of airborne plastics will be more harmful to humans, as these smaller particles are more inhalable and can reach the deeper parts of the respiratory tract. Although fewer studies were conducted in the past on examining the influence of various factors on ambient concentration and deposition rates of microplastics-derived particles, the overall trends from the evidential studies will be matched with the hypothesis. To get more ideas and establish the facts on the role of responsible factors, comprehensive studies on dispersion modeling and the horizontal transport flux model should be conducted to simulate movement of a microplastics-derived particle.

9.7 Transport of Atmospheric Microplastics

Microplastics-derived particles are present in all components of the environment including marine, freshwater, terrestrial, and the atmosphere (Horton and Dixon 2018; Allen et al. 2019). All these components are interlinked with one other, with a varied network of pathways from source and sink for their deposition. The link between source and sink normally decides the quantity of related flux and retention of microplastics-derived particles between such environmental matrices. Atmospheric transport is known as an important vector, which can lead to deposition of microplastics-derived particles to terrestrial bodies and aquatic ecosystems (Allen et al. 2019; Zhang et al. 2019). Such tendency of transportation greatly influences the source–sink dynamics of microplastics-derived particles in various ecosystems of the environment (Bank and Hansson 2019; Windsor et al. 2019). With such interlinking or interconnection, transfer or migration of microplastics-derived particles takes place between terrestrial, atmospheric, and marine environments. Studies in the past reported that microplastics-derived particles in the atmosphere are the important source of microplastic and textile microfibers pollution in oceans (Liu et al. 2019). Although studies have reported on the role of transport of microplastics-derived particles in the atmosphere and on levels of microplastic pollution in ocean and terrestrial bodies, the processes governing transport of microplastics within air are not yet completely understood. In other words, it is difficult to assess the extent of contribution from the atmospheric fallout to aquatic and terrestrial contamination.

The findings from past studies have established that microplastics-derived particles in the atmosphere can undergo long-range transport to remote regions when emitted from high emission regions (Allen et al. 2019; Zhang et al. 2019). The presence of microplastics-derived particles in the Alps and Tibetan glaciers is being observed due to transportation of these pollutants at higher altitude in the atmosphere of source regions to the remote regions. To examine the transport pathway or trajectory of microplastics-derived particles in the atmosphere, Allen et al. (2019) used simplistic meteorological and particle settling velocity calculations and the well-known atmospheric Hybrid Single Particle Lagrangian Integrated Trajectory Model (HYSPLIT). This study showed that transport of these pollutants can occur up to 100 km from the source of emission. Although these few studies are successful in providing the facts of long-range transport of microplastics-derived particles in the atmosphere, influences of these pollutants on precipitation scavenging, atmospheric ice nucleation, and electrical charge of the particles are not completely understood. More research should be conducted with deterministic modeling exercises to fill such gaps and establish the science behind the facts of source-sink-pathway processes, details of transport parameters, and meteorological conditions responsible for such processes.

9.8 Modeling Techniques in Prediction of Fate in the Atmosphere

The measurement campaign in quantification of any air pollutant is cost-effective and requires manpower for maintenance, record-keeping, and data analysis (Behera et al. 2011, 2014). To avoid such constraints in air pollution studies, modeling techniques are becoming more familiar among researchers and policy-makers to provide a better solution for the problem of air pollution. The modeling exercise normally provides the results on occurrence of any air pollutant, both in spatial and temporal variations. Air quality models also help in evaluation of control options for their efficacy in reduction of ambient air concentration. These tools are supportive in the decision-making process by the policy-makers in implementation of control policies effectively for a region. Air quality models also forecast pollutant concentration for future years that would be helpful to researchers to find control strategies in reduction of levels of air pollutants in the future. Air quality models also help in compliance exercises of an air quality standard for any region. There are various models available as open ware or licensed versions worldwide. Dispersion models normally take into account the amount of pollutants at source level and responsible meteorological parameters, such as wind speed, wind direction, and mixing height, to predict the concentration of any pollutant at various locations during different times of interest. In addition to dispersion models, chemical transport models have the ability to consider dispersion, chemical transformation, and transportation of any pollutant followed by deposition.

Similar to dispersion, transport, and chemical transformation of atmospheric particulate matter, microplastics-derived particles in the atmosphere also have the ability to undergo dispersion, transport, and deposition. The model utility, exercise, validity, and evaluations of various models for different purposes of atmospheric particulate matter are well-established in the literature. However, modeling of microplastics-derived particles in the atmosphere is less understood and not reported in a standardized manner in literature. In the present scenario, responsible parameters required to describe adequately and characterize microplastics-derived particles in the atmosphere are unknown. Experiments should be conducted to obtain pertinent information on shape, size, and density from field samples at global, regional, and land use specific generalizations.

For microplastics-derived particles in the atmosphere, many factors required to develop modeling techniques are not available in literature. Those factors can be entrainment potential responsible for dispersion, deposition, and detention processes and potentials, and related atmospheric settling velocities with or without collation or cohesion of homogenous or heterogeneous atmospheric particles. The information on efficiency of microplastics-derived particles in precipitation scavenging are not available in the literature. There are many parametric factors that are required to develop modeling techniques meant for these pollutants. Hence, studies should be conducted to find these necessary parameters, so that development of modeling techniques will be possible and evaluation of modeling techniques can be conducted by comparing with field experiment studies. In addition to development of dispersion modeling techniques meant for microplastics-derived particles in the atmosphere, transport modeling techniques should also be developed for these pollutants to understand their atmospheric dynamics.

9.9 Control Technologies in Contaminant Treatment

The adsorption capability of microplastics-derived particles makes them prone to associate with several pollutants and carry them through transport. There are several methods available in the literature on methods to treat microplastics and to remove them from water, wastewater, and soil mediums. However, the technological methods to treat or remove microplastics from the air medium are less studied and understood in the literature. There are various technological methods invented to treat and reduce microplastics present in the water medium. These methods include the following: i) removal of microplastics using sorption and filtration methods (adsorption on green algae, removal using membrane technology, removal using advanced filtration technologies in wastewater treatment plants); ii) chemical methods to treat microplastics; and iii) biological removal and ingestion (microplastics removal by marine organisms, microplastics removal by bacteria and microplastics ingestion). These methods are based on certain mechanisms to capture or treat microplastics with various efficiency ranges, and these methods have both advantages and drawbacks.

The microplastics-derived particles in various matrices of the atmosphere are more dispersive in nature and move fast compared to other mediums of the environment. The concentration levels of microplastics-derived particles in air medium may be lower than concentration levels of atmospheric particulate matter. In general, atmospheric particulate matter can be concentrated in a specific location; however, microplastics may not be concentrated in any specific area. For example, the concentration levels of atmospheric particles may be larger in industrial and domestic regions. In the case of atmospheric microplastics, their concentration may be higher in textile industries, plastic industries, and waste disposal sites. Hence, the patterns of emission of particulate matter and microplastics are different in the atmosphere.

Atmospheric particulate matter is largely emitted from sources like industries, vehicles, domestic cooking, and biomass burning. The control of particulate matter at source normally helps in reduction of emission to the atmosphere. The control technologies used in the reduction of atmospheric particulate matter emitted from industries are electrostatic precipitators, bag filters, cyclones, venturi scrubbers, and many others. These equipments are normally installed at the source point of industry and are capable of reducing particulate matter emission by 90 to 95% from the stream of exhaust flow. To control particulate matter in residential and commercial buildings, equipment like air purifiers and special filtration provision in air-conditioning systems are used. Recently, smog tower technology has also gained popularity for deployment in open areas of residential, commercial, and industrial zones to control atmospheric particulate matter.

As there are similarities between atmospheric particulate matter and microplastics-derived particles, the above technologies should be explored from the perspectives of control of emission of these pollutants to the atmosphere. There are some variabilities between atmospheric particulate matter and microplastics-derived particles with respect to occurrence and fate. The concentration levels of microplastics-derived particles may be higher in confined areas of any related industry. To get rid of such issues in industry, the options of installation of control equipment such as bag filters, cyclones, and scrubbers should be explored and implemented. In controlling levels of concentration of microplastics-derived particles in indoor environments, air purifiers with suitable filtration options should be explored and tried. Normally, HEPA filters have the potential of removing and screening particles from the air stream. Hence, various options on use of filters and their combination for arrangement in air purifiers should be studied to provide more ways forward to find a solution for reduction of levels of microplastics-derived particles in indoor environments. In controlling levels of microplastics-derived particles near to waste disposal sites, residential areas, and commercial regions, the control option such as the smog tower concept may be explored and applied. As three mediums (water, air, soil) have many commonalities in emitting pollutants and their fate processes, the aforementioned methods applicable to control and treatment of microplastics in water medium should be explored and applied.

9.10 Challenges in Future Climate Conditions

Microplastic pollutants are being recognized as key players in climate change due to their ability to increase the release of greenhouse gases including methane and ethylene, when they are exposed to sunlight. The estimate of global carbon budget from the source of cradle to grave of plastics revealed that greenhouse gases emission would reach 1.34 and 56 Gt per year by 2030 and 2050, respectively. Such estimates comfirmed the threat from plastic pollution in changing the earth's climate. Pollutants released from plastics symbolize a huge pool of carbon in addition to the large carbon footprint inbuilt in these materials. It is noteworthy that plastic pollutants contain a large carbon footprint because of their source of origin during manufacturing from fossil fuels. As a result, the plastic cycle in all three mediums of the environment has been recognized as an unknown branch of the carbon cycle, and the plastic cycle is believed to be part of the global biogeochemical cycle. Similar to the fate of microplastics in water and soil mediums, microplastics in the air medium make a substantial contribution to climate change, as connectivity among these three components of the environment exists. These three components support the movement of microplastics in more interconnected ways. The levels of microplastics-derived particles at different heights of the atmosphere are increasing with respect to spatial and temporal variations. Such increase in levels of these pollutants have the ability to interact with sunlight and its radiations of different wavelengths at various atmospheric levels.

Because of smaller sizes and properties to suspend in air, the movement of microplastics-derived particles is believed to be faster and significantly dynamic compared to their counterparts found in other components (land/soil and water/ocean) of the environment. Such dynamic properties of microplastics-derived particles make them more likely to have impacts on climate processes. Therefore, current scientific concepts need to be developed in examining potential effects of microplastics-derived particles suspended in air on climate change. The established scientific facts on the role of various constituents of atmospheric particulate matter on climate change revealed that components like black carbon have a warming effect on the climate, whereas sulfate has a cooling effect. Microplastics-derived particles in the atmosphere can have standalone and coupled impacts on climate because of their wide ranges of shape, size, type, and color. In this way, potential capability of these microplastics-derived particles on the climate should be examined thoroughly through comprehensive research studies.

The affinity of microplastics-derived particles to undergo long-range transport followed by deposition in the oceanic bodies are larger. Hence, after deposition in oceans, these pollutants can impact the biological pump and perturb the patterns in capacity of oceans to absorb greenhouse gas of carbon dioxide. Moreover, deposited microplastics-derived particles in oceans can modify the patterns of incoming solar radiation. As a result, the chances in alteration of prevailing biogeochemical processes of greenhouse gases are higher. The issues of climate change posed by microplastics-derived particles in the atmosphere can result in compounding effects on the biosphere because of their possible associations with bioaerosols and various constituents of atmospheric particulate matter. Although there is a lack of literature to arrive at concluding remarks on specific roles of atmospheric microplastics on the radiation budget of the earth, finding warming and cooling effects of different polymers, and determination of the role of deposited microplastics-derived particles on soil and water mediums in alteration of carbon budget, the aforementioned limited studies have provided further ways forward to explore more on climate change.

9.11 Future Scope of Research

Based on literature review of the scope of research on the topic of atmospheric microplastics, several research gaps are found, which can be further studied to address all relevant issues related to these emerging air pollutants. The scope of future research in this pertinent domain is recommended, as follows:

i) Standardized operation protocols in identification and measurement of atmospheric microplastics are not available so far. The available methods are non-standardized methods, which can create disadvantages to repeatability, reproducibility, and comparability of the produced results. There factors affect sampling and detection of atmospheric microplastics, resulting in producing less accurate and precise outcomes in estimating the abundance of these pollutants. Such difficulties can cause non-detection of a large amount of microplastics with smaller sizes. Therefore, research should be conducted to develop sampling, pre-treatment, and analytical operation protocols with details of optimized quality-assurance and quality-control (QA-QC) check parameters. More accurate and precise results will always give better results in ecological risk index modeling and human health risk assessment exercises.
ii) The reported studies on atmospheric microplastics are confined to specific locations in some countries with a monitoring duration of a short period. Spatial and temporal variations of these pollutants should be developed with study locations worldwide. Validated models should be developed to predict spatial distribution of

these pollutants. Better understanding is required on assessing long-range transport (>100 km) of atmospheric microplastics in air. Mathematical and computational modeling exercises are also required to assess dispersion and long-range transport of atmospheric microplastics.

iii) The available literature are mostly limited to analysis of shape, size, color, chemical characterization, and abundance of airborne microplastics in different matrices of the atmosphere. To gain more insights into fate and transformation of these emerging pollutants, research should be conducted in establishing interactions and interrelationships between microplastics with other relevant pollutants, other ecosystems, and human exposure. Studies on identification and quantification of atmospheric pollutants adsorbed to airborne microplastics should be conducted.

iv) Detailed studies on chemical characteristics and constitutional composition should be conducted in addition to physical characteristics of atmospheric microplastics. In addition to m-FTIR, an m-Raman hyperspectral imaging technique study should used to examine chemical composition of atmospheric microplastics.

v) To conduct source identification and risk assessment studies, mass concentrations of pertinent polymers is required to obtain better results. Mass concentration results of these pollutants are sparse in the literature. Hence, studies on quantification of pertinent polymers should be conducted using the techniques of Pys-GC-MS and TDS-GC-MS.

vi) At present, atmospheric microplastics are studied with sizes specific to 50 μm. From the human health point of view, particles with size of <50 μm are subjected to inhalability and can cause several health risks. Hence, studies should be conducted to explore more advanced analytical techniques that are capable of identification and measurement of atmospheric microplastics of small sizes.

vii) Exposure risk assessment of atmospheric microplastics through the dermal route should be conducted to understand how atmospheric microplastics can penetrate into skin pores of humans to cause diseases.

viii) Modeling techniques should be developed to assess dry and wet depositions of atmospheric microplastics on terrestrial bodies.

ix) Separate studies should be conducted in developing source profiles and ambient occurrence trends to identify sources and finding their contribution to a specific polymer of atmospheric microplastics.

x) Studies should be conducted on interaction of atmospheric microplastics with cloud formation and recurrence of precipitation.

xi) Concept building on exploring interactions of atmospheric microplastics in air medium with water and soil mediums should be developed.

9.12 Conclusions

The recent trends in recognizing research in the domain of atmospheric microplastics have shown a clear way to acknowledge airborne or atmospheric microplastics as air pollutants or particulates. The presence of atmospheric microplastics is being observed in all set-ups of the environment including urban, rural, semi-urban, and remote locations, where these pollutants exhibit temporal variations while suspended in the atmosphere. The fate that atmospheric microplastics follows is similar to particulate matter after emission from the source of origin. After emission from these sources, the pollutants undergo transport in the atmosphere depending on the meteorological conditions, interact with other pollutants, and finally are deposited on terrestrial bodies through either dry or wet deposition.

At present, the scope of research on studying various perspectives of atmospheric microplastics are limited to identification of various polymers and investigating the occurrence of these pollutants in different atmospheric matrices. There is sparse literature on studying the fate of these emerging air pollutants after emission from their sources. Based on available literature and findings, it can be concluded that atmospheric microplastics have a larger potential to undergo long-range transport and these pollutants can cause severe health implications for humans through inhalation exposure. With results on occurrence of various compounds of microplastics, it has been discerned that atmospheric microplastics demonstrate a wide range of characteristics and quantities, which are influenced by patterns of anthropogenic activities, population in that domain, time of measurement, location of space, seasonal conditions, and topography of measurement sites that includes areas of vegetation and densities of buildings situated in that sampling location.

As the research scope of atmospheric microplastics are at their early stage, sampling, identification, and quantification methods of these pollutants differ substantially among inter- and intra-components of the atmospheric environment. At this stage, comparison of various methods and providing a standard protocol in identification and measurement of these pollutants is a difficult task. Despite these difficulties, two techniques (FTIR spectroscopy and SEM) are widely adopted and used by researchers to characterize atmospheric microplastics in various matrices of the atmosphere. The forms of spherules, films, fragments, fibers, and granules are being observed for these pollutants, with the most frequently identified shapes of atmospheric microplastics as fibers and fragments. Among various matrices of air components, it has been observed that concentrations of microplastics in indoor environments are higher than that of outdoor environments.

The polymers like polyamide, polyethylene, polyester, and polypropylene with fibrous shapes and varying colors comprise the largest fraction of atmospheric microplastics. It is difficult to draw conclusions on size distributions of these pollutants, as new concepts are evolving gradually. From reported results, it has been discerned that sizes of microplastics in street dust, indoor air, and outdoor air can vary with ranges from <100 to $5000\,\mu m$, from ≤ 3250 to $4850\,\mu m$, and from 2 to $9555\,\mu m$, respectively. From synthesis of the knowledge in this chapter, it can be inferred that atmospheric microplastics require more fundamental and applied research to provide a better understanding on various aspects related to fate, transport, source origin, global distribution, and human health implications through implementation of precise field sampling and analytical techniques in identification and measurement of these emerging air pollutants.

References

Abbasi, S., Keshavarzi, B., Moore, F., et al. (2017). Investigation of microrubbers, microplastics and heavy metals in street dust: a study in Bushehr City, Iran. *Environmental Earth Sciences* 76: 798.

Abbasi, S., Keshavarzi, B., Moore, F. et al. (2019). Distribution and potential health impacts of microplastics and microrubbers in air and street dusts from Asaluyeh County, Iran. *Environmental Pollution* 244: 153–164.

Alam, P. and Ahmade, K. (2013). Impact of solid waste on health and the environment. *International Journal of Sustainable Development and Green Economics (IJSDGE)* 2 (1): 165–168.

Alimi, O.S., Budarz, F.J., Hernandez, L.M., and Tufenkji, N. (2018). Microplastics and nanoplastics in aquatic environments: aggregation, deposition, and enhanced contaminant transport. *Environmental Science & Technology* 52: 1704–1724.

Allen, S., Allen, D., Phoenix, V.R. et al. (2019). Atmospheric transport and deposition of microplastics in a remote mountain catchment. *Nature Geoscience* 12 (5): 339–344.

Auta, H.S., Emenike, C.U., and Fauziah, S.H. (2017). Distribution and importance of microplastics in the marine environment: a review of the sources, fate, effects, and potential solutions. *Environment International* 102: 165–176.

Awasthi, A.K., Shivashankar, M., and Majumder, S. (2017). Plastic solid waste utilization technologies: a review. *IOP Conference Series: Materials Science and Engineering* 263 (2): 022024.

Bank, M.S. and Hansson, S.V. (2019). The plastic cycle: a novel and holistic paradigm for the anthropocene. *Environmental Science & Technology* 53: 7177–7179.

Barnes, D.K., Galgani, F., Thompson, R.C., and Barlaz, M. (2009). Accumulation and fragmentation of plastic debris in global environments. *Philosophical Transactions of the Royal Society B: Biological Sciences* 364 (1526): 1985–1998.

Behera, S.N., Sharma, M., Dikshit, O., and Shukla, S.P. (2011). GIS-based emission inventory, dispersion modeling, and assessment for source contributions of particulate matter in an urban environment. *Water, Air, & Soil Pollution* 218 (1): 423–436.

Behera, S.N., Sharma, M., Nayak, P. et al. (2014). An approach for evaluation of proposed air pollution control strategy to reduce levels of nitrogen oxides in an urban environment. *Journal of Environmental Planning and Management* 57 (4): 467–494.

Boots, B., Russell, C.W., and Green, D.S. (2019). Effects of microplastics in soil ecosystems: above and below ground. *Environmental Science & Technology* 53 (19): 11496–11506.

Browne, M.A., Galloway, T.S, and Thompson, R.C. (2007). Microplastic: an emerging contaminant of potential concern? *Integrated Environmental Assessment and Management* 3 (4): 559–561.

Browne, M.A., Galloway, T.S. and Thompson, R.C. (2010). Spatial patterns of plastic debris along estuarine shorelines. *Environmental Science & Technology* 44 (9): 3404–3409.

Cai, L., Wang, J., Peng, J. et al. (2017). Characteristic of microplastics in the atmospheric fallout from Dongguan city, China: preliminary research and first evidence. *Environmental Science and Pollution Research* 24 (32): 24928–24935.

Can-Güven, E. (2021). Microplastics as emerging atmospheric pollutants: a review and bibliometric analysis. *Air Quality, Atmosphere & Health* 14 (2): 203–215.

Carr, S.A., Liu, J. and Tesoro, A.G. (2016). Transport and fate of microplastic particles in wastewater treatment plants. *Water Research* 91: 174–182.

Chen, E.Y., Lin, K.T., Jung, C.C. et al. (2022). Characteristics and influencing factors of airborne microplastics in nail salons. *Science of the Total Environment* 806: 151472.

Chen, G., Feng, Q., and Wang, J. (2020). Mini-review of microplastics in the atmosphere and their risks to humans. *Science of the Total Environment* 703: 135504.

da Costa, J.P., Paço, A., Santos, P.S. et al. (2018). Microplastics in soils: assessment, analytics and risks. *Environmental Chemistry* 16 (1): 18–30.

Dehghani, S., Moore, F., and Akhbarizadeh, R. (2017). Microplastic pollution in deposited urban dust, Tehran Metropolis, Iran. *Environmental Science and Pollution Research* 24 (25): 20360–20371.

Dris, R., Gasperi, J., Saad, M. et al. (2016). Synthetic fibers in atmospheric fallout: a source of microplastics in the environment? *Marine Pollution Bulletin* 104(1–2): 290–293.

Dris, R., Gasperi, J., Mirande, C. et al. (2017). A first overview of textile fibers, including microplastics, in indoor and outdoor environments. *Environmental Pollution* 221: 453–458.

Dris, R., Imhof, H., Sanchez, W. et al. (2015a). Beyond the ocean: contamination of freshwater ecosystems with (micro-) plastic particles. *Environmental Chemistry* 12 (5): 539–550.

Dris, R., Gasperi, J., Rocher, V. et al. (2015b). Microplastic contamination in an urban area: a case study in Greater Paris. *Environmental Chemistry* 12 (5): 592–599.

Enyoh, C.E., Verla, A.W., Verla, E.N. et al. (2019). Airborne microplastics: a review study on method for analysis, occurrence, movement and risks. *Environmental Monitoring and Assessment* 191 (11): 1–17.

Ekebafe, L.O., Ogbeifun, D.E. and Okieimen, F.E. (2011). Polymer applications in agriculture. *Biokemistri* 23(2): 81–89.

Evode, N., Qamar, S.A., Bilal, M. et al. (2021). Plastic waste and its management strategies for environmental sustainability. *Case Studies in Chemical and Environmental Engineering* 4: 100142.

Gasperi, J., Stephanie, L.W., Rachid, D. et al. (2018). Microplastics in air: are we breathing it in? *Current Opinion in Environmental Science & Health* 1: 1–5.

GESAMP (2016). Sources, fate and effects of microplastics in the marine environment: part two of a global assessment. IMO/FAO/ UNESCO-IOC/UNIDO/WMO/IAEA/UN/ UNEP/UNDP Joint Group of Experts on the Scientific Aspects of Marine Environmental Protection: 220 p.

Gigault, J., El Hadri, H., Nguyen, B. et al. (2021). Nanoplastics are neither microplastics nor engineered nanoparticles. *Nature Nanotechnology* 16 (5): 501–507.

Hahladakisa, J.N., Costas, A.V., Roland, W.et al. (2018). An overview of chemical additives present in plastics: migration, release, fate and environmental impact during their use, disposal and recycling. *Journal of Hazardous Materials* 344: 179–199.

Hammer, J., Kraak, M.H., and Parsons, J.R. (2012). Plastics in the marine environment: the dark side of a modern gift. In: *Reviews of Environmental Contamination and Toxicology* (ed. Whitacre, D.M.), 1–44. New York: Springer.

Hanun, J.N., Hassan, F., and Jiang, J.J. (2021). Occurrence, fate, and sorption behavior of contaminants of emerging concern to microplastics: influence of the weathering/aging process. *Journal of Environmental Chemical Engineering* 9 (5): 106290.

Hartmann, N.B., Hüffer, T., Thompson, R.C. et al. (2019). Are we speaking the same language? Recommendations for a definition and categorization framework for plastic debris. *Environmental Science & Technology* 53: 1039–1047.

Hidalgo-Ruz, V., Gutow, L., Thompson, R.C. and Thiel, M.(2012). Microplastics in the marine environment: a review of the methods used for identification and quantification. *Environmental Science & Technology* 46: 3060–3075.

Horton, A.A. and Dixon, S.J. (2018). Microplastics: an introduction to environmental transport processes. *Wiley Interdisciplinary Reviews: Water* 5 (2): e1268.

Huang, D., Tao, J., Cheng, M. et al. (2021a). Microplastics and nanoplastics in the environment: macroscopic transport and effects on creatures. *Journal of Hazardous Materials* 407: 124399.

Huang, Y., He, T., Yan, M. et al. (2021b). Atmospheric transport and deposition of microplastics in a subtropical urban environment. *Journal of Hazardous Materials* 416: 126168.

Huang, Y., Qing, X., Wang, W. et al. (2020). Mini-review on current studies of airborne microplastics: analytical methods, occurrence, sources, fate and potential risk to human beings. *TrAC Trends in Analytical Chemistry* 125: 115821.

Huang, D.Y., Zhou, S.G., Hong, W. et al. (2013). Pollution characteristics of volatile organic compounds, polycyclic aromatic hydrocarbons and phthalate esters emitted from plastic wastes recycling granulation plants in Xingtan Town South China. *Atmospheric Environment* 71: 327–334.

Idumah, C.I. and Nwuzor, I.C. (2019). Novel trends in plastic waste management. *SN Applied Sciences* 1 (11): 1–14.

Junhao, C., Xining, Z., Xiaodong, G. et al. (2021). Extraction and identification methods of microplastics and nanoplastics in agricultural soil: a review. *Journal of Environmental Management* 294: 112997.

Kabir, A.E., Sekine, M., Imai, T. et al. (2021). Assessing small-scale freshwater microplastics pollution, land-use, source-to-sink conduits, and pollution risks: perspectives from Japanese rivers polluted with microplastics. *Science of the Total Environment* 768: 144655.

Koelmans, A.A., Redondo-Hasselerharm, P.E., Nor, N.H.M. et al. (2022). Risk assessment of microplastic particles. *Nature Reviews Materials* 7 (2): 1–15.

Kole, P.J., Löhr, A.J., Van Belleghem, F.G. and Ragas, A.M. (2017). Wear and tear of tyres: a stealthy source of microplastics in the environment. *International Journal of Environmental Research and Public Health* 14: 1265.

Lehner, R., Weder, C., Petri-Fink, A., and Rothen-Rutishauser, B. (2019). Emergence of nanoplastic in the environment and possible impact on human health. *Environmental Science & Technology* 53 (4): 1748–1765.

Li, J., Liu, H. and Chen, J.P. (2018a). Microplastics in freshwater systems: a review on occurrence, environmental effects, and methods for microplastics detection. *Water Research* 137: 362–374.

Liu, K., Wang, X., Fang, T. et al. (2019). Source and potential risk assessment of suspended atmospheric microplastics in Shanghai. *Science of the Total Environment* 675: 462–471.

Magni, S., Binelli, A., Pittura, L. et al. (2019). The fate of microplastics in an Italian wastewater treatment plant. *Science of the Total Environment* 652: 602–610.

Mai, L., Bao, L.J., Shi, L. et al. (2018). A review of methods for measuring microplastics in aquatic environments. *Environmental Science and Pollution Research* 25 (12): 11319–11332.

Masura, J., Baker, J., Foster, G., and Arthur, C. (2015). Laboratory methods for the analysis of microplastics in the marine environment: recommendations for quantifying synthetic particles in waters and sediments. NOAA Technical Memorandum NOS-OR&R-48.

Mbachu, O., Jenkins, G., Pratt, C., and Kaparaju, P. (2020). A new contaminant superhighway? A review of sources, measurement techniques and fate of atmospheric microplastics. *Water, Air, & Soil Pollution* 231 (2): 1–27.

Millican, J.M. and Agarwal, S. (2021). Plastic pollution: a material problem? *Macromolecules* 54 (10): 4455–4469.

Nguyen, B., Claveau-Mallet, D., Hernandez, L.M. et al. (2019). Separation and analysis of microplastics and nanoplastics in complex environmental samples. *Accounts of Chemical Research* 52(4): 858–866.

Nizzetto, L., Langaas, S. and Futter, M. (2016). Do microplastics spill on to farm soils? *Nature* 537: 488–488.

Peng, L., Fu, D., Qi, H. et al. (2020). Micro-and nano-plastics in marine environment: source, distribution and threats: a review. *Science of the Total Environment* 698: 134254.

Plastics Europe (2018). Plastics: the Facts, 2018. https://www.plasticseurope.org (accessed 20 February 2022).

Prata, J.C. (2018). Airborne microplastics: Consequences to human health? *Environmental Pollution* 234: 115–126.

Prata, J.C., Castro, J.L., da Costa, J.P. et al. (2020). The importance of contamination control in airborne fibers and microplastic sampling: experiences from indoor and outdoor air sampling in Aveiro, Portugal. *Marine Pollution Bulletin* 159: 111522.

Prata, J.C., da Costa, J.P., Duarte, A.C., and Rocha-Santos, T. (2019). Methods for sampling and detection of microplastics in water and sediment: a critical review. *TrAC Trends in Analytical Chemistry* 110: 150–159.

Rahman, A., Sarkar, A., Yadav, O.P. et al. (2021). Potential human health risks due to environmental exposure to nano-and microplastics and knowledge gaps: a scoping review. *Science of the Total Environment* 757: 143872.

Rochman, C.M., Brookson, C., Bikker, J. et al. (2019). Rethinking microplastics as a diverse contaminant suite. *Environmental Toxicology and Chemistry* 38: 703–711.

Ryan, P.G. (2015). A brief history of marine litter research. In: *Marine Anthropogenic Litter* (eds Bergmann, M., Gutow, L., and Klages, M.), 1–25. Cham: Springer.

Smith, M., Love, D.C., Rochman, C.M., and Neff, R.A. (2018). Microplastics in seafood and the implications for human health. *Current Environmental Health Reports* 5 (3): 375–386.

Sridharan, S., Kumar, M., Singh, L. et al. (2021). Microplastics as an emerging source of particulate air pollution: a critical review. *Journal of Hazardous Materials* 418: 126245.

Sun, S., Li, J., and Han, J. (2021). How human thermal plume influences near-human transport of respiratory droplets and airborne particles: a review. *Environmental Chemistry Letters* 19 (3): 1971–1982.

Thompson, R.C. (2006). Plastic debris in the marine environment: consequences and solution. In: *Marine Nature Conservation in Europe* (eds Krause, J.C., Nordheim, H., and Brager, S.), 107–115. Stralsund: Federal Agency for Nature Conservation.

Thompson, R.C., Olsen, Y., Mitchell, R.P. et al. (2004). Lost at sea: where is all the plastic? *Science* 304: 838–838.

Tourinho, P.S., Kočí, V., Loureiro, S., and Gestel, C.A. (2019). Partitioning of chemical contaminants to microplastics: sorption mechanisms, environmental distribution and effects on toxicity and bioaccumulation. *Environmental Pollution* 252: 1246–1256.

Verla, E.N., Verla, A.W., and Enyoh, C.E. (2017). Pollution assessment models of soils in Portharcourt City, Rivers State, Nigeria. *World News of Natural Sciences (WNOFNS)* 12: 1–23.

Vianello, A., Jensen, R.L., Liu, L., and Vollertsen, J. (2019). Simulating human exposure to indoor airborne microplastics using a Breathing Thermal Manikin. *Scientific Reports* 9 (1): 1–11.

Vivekanand, A.C., Mohapatra, S., and Tyagi, V.K. (2021). Microplastics in aquatic environment: challenges and perspectives. *Chemosphere* 282: 131151.

Wang, C., Zhao, J., and Xing, B. (2021). Environmental source, fate, and toxicity of microplastics. *Journal of Hazardous Materials* 407: 124357.

Wang, T., Zou, X., Li, B. et al. (2019). Preliminary study of the source apportionment and diversity of microplastics: taking floating microplastics in the South China Sea as an example. *Environmental Pollution* 245: 965–974.

Wang, Y., Huang, J., Zhu, F., and Zhou, S. (2021). Airborne microplastics: a review on the occurrence, migration and risks to humans. *Bulletin of Environmental Contamination and Toxicology* 107 (4): 657–664.

Wichai-Utcha, N. and Chavalparit, O. (2019). 3Rs Policy and plastic waste management in Thailand. *Journal of Material Cycles and Waste Management* 21 (1): 10–22.

Windsor, F.M., Tilley, R.M., Tyler, C.R. and Ormerod, S.J. (2019). Microplastic ingestion by riverine macroinvertebrates. *Science of The Total Environment* 646: 68–74.

Wright, S.L. and Kelly, F.J. (2017). Plastic and human health: a micro issue? *Environmental Science & Technology* 51 (12): 6634–6647.

Wright, S.L., Ulke, J., Font, A. et al. (2020). Atmospheric microplastic deposition in an urban environment and an evaluation of transport. *Environment International* 136: 105411.

Yousefi, M., Oskoei, V., Jonidi Jafari, A. et al. (2021). Municipal solid waste management during COVID-19 pandemic: effects and repercussions. *Environmental Science and Pollution Research* 28 (25): 32200–32209.

Yuan, W., Christie-Oleza, J.A., Xu, E.G. et al. (2022). Environmental fate of microplastics in the world's third-largest river: basin-wide investigation and microplastic community analysis. *Water Research* 210: 118002.

Yukioka, S., Tanaka, S., Nabetani, Y. et al. (2020). Occurrence and characteristics of microplastics in surface road dust in Kusatsu (Japan), Da Nang (Vietnam), and Kathmandu (Nepal). *Environmental Pollution* 256: 113447.

Zhang, Y., Gao, T., Kang, S. and Sillanpää, M. (2019). Importance of atmospheric transport for microplastics deposited in remote areas. *Environmental Pollution* 254: 112953.

Zhang, B., Yang, X., Chen, L. et al. (2020a). Microplastics in soils: a review of possible sources, analytical methods and ecological impacts. *Journal of Chemical Technology & Biotechnology* 95 (8): 2052–2068.

Zhang, Q., Xu, E.G., Li, J. et al. (2020b). A review of microplastics in table salt, drinking water, and air: direct human exposure. *Environmental Science & Technology* 54 (7): 3740–3751.

Zhang, Y., Gao, T., Kang, S. et al. (2022). Current status and future perspectives of microplastic pollution in typical cryospheric regions. *Earth-Science Reviews* 226: 103924.

Zhang, Y., Kang, S., Allen, S. et al. (2020c). Atmospheric microplastics: a review on current status and perspectives. *Earth-Science Reviews* 203: 103118.

Zhou, D., Cai, Y., and Yang, Z. (2022). Key factors controlling transport of micro-and nanoplastic in porous media and its effect on coexisting pollutants. *Environmental Pollution* 293: 118503.

Zhou, Q., Tian, C., and Luo, Y. (2017). Various forms and deposition fluxes of microplastics identified in the coastal urban atmosphere. *Chinese Science Bulletin* 62 (33): 3902–3910.

10

Modeling Microconstituents Based on Remote Sensing and GIS Techniques

Anoop Kumar Shukla[1], *Satyavati Shukla*[2], *Rao Y. Surampalli*[3], *Tian C. Zhang*[3,4], *Ying-Liang Yu*[4,5], *and Chih-Ming Kao*[4,5]

[1]*Manipal School of Architecture and Planning, Manipal Academy of Higher Education, Manipal, Karnataka, India*
[2]*Key Laboratory of Geospatial Informatics, Guilin University of Technology, Guilin, PR China*
[3]*Global Institute for Energy, Environment, and Sustainability, Lenexa, KS, USA*
[4]*Civil Engineering Department, College of Engineering, University of Nebraska, Lincoln, Omaha, NE, USA*
[5]*Institute of Environmental Engineering, National Sun Yat-Sen University, Kaohsiung, Taiwan*

10.1 Basic Components of Remote Sensing and GIS-Based Models

The primary focus of remote sensing (RS) techniques is to gather information on the subject. The various methods involved from earlier centuries to recent decades include the telescope, hot air balloon photography, unmanned aerial vehicles, observation satellites, etc.

Based on the need of a specific observer, the components of the technique vary with the devices involved, the range of electromagnetic radiations employed, etc. The devices employed may vary from cameras, hot air balloons, aircraft-mounted systems, laser systems, radar systems, aerial photogrammetry, and space satellites.

As shown in Figure 10.1, RS involves the following basic components:

a) Source of light or energy;
b) Propagation of electromagnetic radiation (EMR);
c) Interaction with the subject matter of observation;
d) Sensing system;
e) Data collection;
f) Interpretation and analysis; and
g) User information products.

Imaging sensors have evolved over the past few decades from preliminary sensors to high-resolution sensors. Lurie and Irvine (2002) strongly promulgated the unprecedented quality and quantity of geospace observations, readily available in many systems, in a very timely manner opening up wide avenues in application of RS, ranging from real-time data analysis as in the case of natural disasters, strategic surveillance, etc. to developmental

Microconstituents in the Environment: Occurrence, Fate, Removal, and Management, First Edition.
Edited by Rao Y. Surampalli, Tian C. Zhang, Chih-Ming Kao, Makarand M. Ghangrekar, Puspendu Bhunia, Manaswini Behera, and Prangya R. Rout.

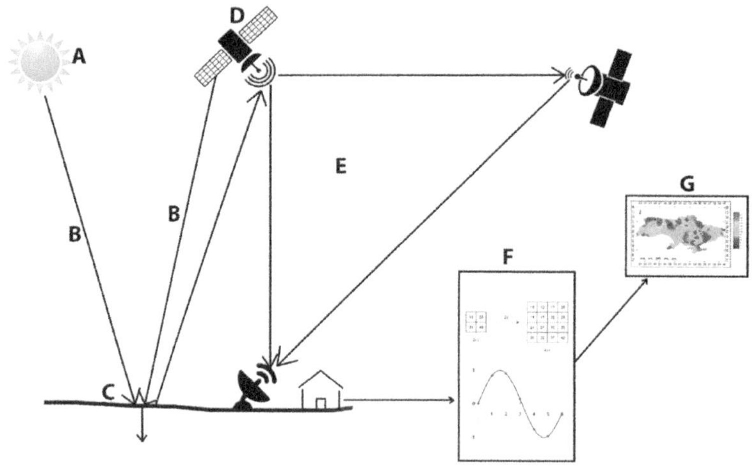

Figure 10.1 Components of remote sensing.

analysis such as in urban planning, construction monitoring, environmental impact studies, etc. The academic and commercial use benefits are second to none. The benefits derived from RS techniques have tremendous potential to overpower the initial high cost involved with the sheer variety of application possibilities.

10.1.1 Source of Light or Energy

The primary necessity of RS is to have a source of energy or illumination directed toward the subject matter of interest. The source of energy is generally in the form of EMR. The properties and behavior of these EMRs are according to wave theory basics. The two main characteristics are wavelength and frequency of radiations inversely related to each other. The range of radiations based on the wavelength and frequency are arranged in the spectrum (Table 10.1). Apart from the visible range of the spectrum, which is easily detected by the human eye, several spectra can be visible to "remote eyes" only. The most popular spectra of radiations that are particularly useful in RS are in the ultraviolet or UV.

Table 10.1 The electromagnetic spectrum used in satellite-based remote sensing.

Region	Wavelength	Property
Visible range (blue, red, green)	0.4–0.7 μm	Reflectance
Reflective Infrared range	0.7–3.0 μm	Reflectance
Thermal Infrared range	3.0–15.0 μm	Radiative Temperature
Microwave range	0.1–30 cm	Brightness Temperature (passive)
		Backscattering (active)

Source: Ray, (2012)/Mahalanobis National Crop Forecast Centre.

10.1.2 Radiation and the Atmosphere

The interaction of radiation with the atmosphere may be in the form of scattering or absorption. The phenomenon of scattering may be of Rayleigh, Mie, or a non-selective type of scattering. The absorption phenomenon is mainly due to three main components of the atmosphere, namely ozone, carbon dioxide, and water vapor. The emitted or reflected energy from the target also undergoes interaction with the atmosphere on its way to the sensor.

10.1.3 Interaction With the Subject Target

EMRs or the energy interacts with the subject matter, based on properties of the subject as well as radiation. These interactions may be in the form of absorption (A), transmission (T), or reflection (R). The proportions of these phenomena depend on the wavelength of the incident radiations and the properties of the subject matter.

From an RS point of view, reflection is the particular interaction that will be interesting to watch out for. The reflection can be of the specular type or mirror type, which originates from smooth surfaces, whereas diffuse reflection is from rough surfaces.

10.1.4 Sensing Systems

Either the emitted, dispersed, or reflected radiations from the subject are recorded or captured by the sensors. These sensors are required to be stationed on a stable platform, which is situated in ground-, aircraft-, or satellite-based sensors. Satellite-based sensors are the most expensive but also more useful based on their application and the area of study permissible.

Satellites orbit at various speeds and altitudes around the earth, offering different capabilities of earth observations. With geostationary orbit satellites, it is possible to observe a larger area constantly, making them important in real-time earth observations, weather observations, forecasting, and communication. The sun-synchronous satellite observes a particular landmass at a specific local sun time making them helpful in change detection and monitoring studies. The recording of the energy from the subject depends on various factors of the sensors such as the orbit-geostationary, sun-synchronous, near-polar, etc.

10.1.5 Data Collection

The documented inputs from the sensors are then transmitted back to the processing station for further processing of the images.

10.1.6 Interpretation and Analysis

The processed images are then interpreted based on the research scope and analyzed either in visuals, digital modes, or electronic mediums to derive information about the subject matter.

10.2 Coupling GIS With 3D Model Analysis and Visualization

The GIS data model represents how we intend to look at the world as well as giving an understanding of how the world works. GIS data gives new representation to one or more aspects of the real world.

The models created may be of the static type where the input to the model and the processed output from the model pertain to a specific study area over the same period. These studies provide general indicators, impact factors, and soft points toward the current scenario study area like soil exploration, water contamination estimates, land use/land cover (LULC) change studies, etc.

No GIS currently offers the data representation flexibility for space and time, as well as the algorithmic competence to develop process-based models internally. As a result, environmental models and GIS must be connected. The coupling can range from loose to tight. A loose coupling occurs when data is sent from one system to another. The Geographic Information System (GIS) is used to preprocess data or to create maps of input data or model outputs. Because it involves little software change, this strategy is used in the majority of applications identified in the literature. Only the file formats and accompanying input and output algorithms, which are generally model-specific, must be modified. Data management is integrated into the system through a close linkage in a tight coupling. A tight linkage will often give a single user interface for both the GIS and the model, and information transfer between the respective components will be clear (Corwin and Wagenet 1996).

However, the dynamic models represent the output for the future period from the input timelines. The output generated is supported by historical study datasets. These models are generally helpful in weather forecasting, seasonal monsoon predictions, flood prediction during storms, submergence of coastal areas due to rise in sea level, environmental impact studies for proposed developmental activities, etc. These models effectively provide a policy framework to decision-making authorities for better governance.

There are different types of 3D models. The one which gives information regarding the spatial settings is called the 3D geographic spatial object model. Another model gives information on geological settings, which is called the 3D geological temporal object model. These model creations can be analog (size replicas of real-world settings) or digital (all important data and features are digitally recorded). The spatial and temporal resolutions play an important role in the digital model.

Dynamic models are further classified as discrete models where the input details are discrete from each other and as continuous models where each variable is a continuous function of each other. A representational continuous model for local weather forecasting study has input variables such as ground altitude, temperature, pressure, natural vegetation, etc., which are all continuous functions and interwoven with each other. However, these individual discrete models may not be suitably employed in the majority of study scenarios, such as in a study of crowd behavior on beaches at vacation time. The discrete models rely on the parameter of the behavior of each individual at the study area, which is not justifiable and studied, whereas the continuous model relies on a parameter such as density of the crowd. This analogy is a clear indicator that the continuous function dynamic models provide an accurate representation of study areas.

The majority of the earlier GIS models are static with 2D space studies. However, they are increasingly approaching dynamic continuous modeling in 3D space. Since the geospace is a continuous 3D space, its representation in GIS is challenging. The majority of the study is limited to the above surface region of geospace; there are seldom attempts made in subsurface studies. In consideration of the rapid growth of urbanization and developmental decisions toward the same, the subsurface study of geospace also plays a prominent role.

10.2.1 Modeling and Simulation Approaches

Real systems are generally complicated and frequently incorporate nonlinear behaviors. To represent them in simplicity, stochastic analysis and feedback loops with varying geographical and temporal dimensions, models play a greater role. Complex system models frequently employ a blend of empirical and process-based methodologies.

Experimental models are arrived at by mathematical analysis of observed data and can usually be relevant to similar situations as that of obtained observations. For example, the universal soil loss equation (USLE), which estimates yearly soil loss from a study area, is based on observations related to topography, soil type, rainfall details, and land cover elements.

Process-based models with mathematical representation are developed over the study of physical, chemical, geological, and biological processes, as studied in hydrological and erosion models (Mitas and Mitasova 1998). Examples of several these kinds of models are as follows.

10.2.1.1 Deterministic Models

Deterministic models represent processes generally described by single input differential equations, and generating single output for the linear model and multiple output conceivable for non-linear models. A few of the numerical methods to solve equations are finite difference, finite element, path simulation, etc. The following models explain processes at multiple temporal variance levels:

- The steady-state model technique, with no temporal fluctuations, is generally employed for diagnostic purposes. With the time series input parameters, these models estimate the long-term average of spatial distributions; and
- Dynamic models are utilized for prognostic applications, such as forecasting and characterizing the spatiotemporal fluctuations during a modeled event.

10.2.1.2 Stochastic Models

Stochastic models simulate the spatiotemporal behavior of random events. Due to inputs of random components, each model run produces a distinct outcome. A single simulation produces just one conceivable result. To estimate probability distributions, many runs are used. This model enhances the characterization of geographical phenomena and combines with geostatistics for conditional simulations. Different forms can characterize the behavior of dynamic systems, such as Poisson and renewal, discrete-time, continuous-time Markov processes, transition probability matrices, Brownian processes, and diffusion.

10.2.1.3 Rule-Based Models

Rule-based models, Cellular Automata (CA), are non-linear dynamic mathematical systems based on discrete time and space that are used to simulate processes guided by local rules (Wolfram 1984). In examples of this ARIES model in diffusion model development as well as simulation of forest fires by Clarke et al. (1994) and Park and Wagner (1997), the principles of rule-based model are:

- The state of Cellular Automata (CA) is updated in disconnected time steps concerning a transition rule that is used uniformly and simultaneous to each cell at each period;
- Each cell's worth is decided by a geometrical structure of neighboring cells provided as part of the transition rule; and
- A globally complicated attitude can appear through simple local principles, making it valuable for modeling systems that are not completely understood but whose local courses are well understood.

10.2.1.4 Multi-Agent Simulation of Complex Systems

Multi-agent simulation of complex systems replicate the movement and expansion of a huge number of linked tools. A typical tool is shown as a set of rules, reactions to inputs, and is an agent in a system that generates events that effect itself and others. By using local rules, individual-based models reflect the motion and evolution of particular parameters in space and time regions. Hierarchical models are created by layering numerous collections of agents associated with activity schedules.

10.2.2 GIS Implementation

Most commercial GIS software such as GRASS, ARCGRID, ERDAS, ArcView, etc. support basic modeling within the raster subsystems. Full implementation of complicated models may necessitate enhancements to normal GIS functionalities such as temporal and 3D/4D data support, as well as meshes for finite element approach. The possibility of customization, application development tools, and availability of data templates encourage coupling of commercial GIS and modeling techniques. The use of an object-based interface allows effective implementation of GIS and capabilities to merge various layers of linkage.

10.2.2.1 Full Integration–Embedded Coupling

In full integration–embedded coupling, models are developed using programming and development tools embedded in the GIS platform such as Application Programming Interface (API), scripting tools, map algebra operations, etc. These operate as the command prompt in the GIS platform, by storing inputs and outputs in an internal database, thus eliminating data transmission. Models developed with scripting tools may be slower in terms of computing efficiency. The mobility is limited due to its reliance on a GIS within which it was created and deployed. Model creation is aided with customization and application development tools, as well as map algebra extensions. The following are the two types of full integration–embedded coupling: i) embedded coupling: hydrology and water features; and ii) implementation of map algebra: the flow of water (Shapiro and Westervelt 1992), and the simple fire spread model.

10.2.2.2 Integration Under a Common Interface–Tight Coupling

With integration under a common interface–tight coupling, the models developed are independent of GIS platforms and have their own data structures, with data interaction possibilities between model and GIS hidden from the user; albeit in some cases the datasets may be transferred. It is a single interface that connects GIS and the model. The interface frequently allows GIS integration with numerous distinct models of complicated systems with interconnected processes. This model restricts portability. Examples include SWAT, AGNPS, and ANSWERS coupled with GRASS and SWAT, IDOR3D, and BASIN-2 coupled with ArcView.

10.2.2.3 Loose Coupling

Loose coupling models are created and operated solely on the GIS platform. Model outputs are traded as GIS inputs for analysis and visualization. Examples include PAYSAGE (Hansen et al. 1996) forest and habitat change coupled with Arcview/ArcInfo, as well as SIMWE erosion and deposition coupled with GRASS (Mitas and Mitasova 1998).

10.2.2.4 Modeling Environment Linked to GIS

The modeling environment linked to GIS usually targets the construction of modular, reusable models. GIS is integrated into the modeling environment via an interface or data import/export. Typical examples of this coupling are as follows: i) spatial modeling environment (Maxwell and Costanza 1997); ii) SWARM; and iii) modular modeling system (MMS).

10.3 Emerging and Application

Remote sensing technology offers several advantages for detecting, mapping, and monitoring intruders. Remote sensing using space-borne sensors is the ultimate instrument for gathering repeated (from minutes to days) and synoptic (from local to regional coverage) measurements on the spectral behavior of various environments (land surface changes (degradation), water quality, soil and atmosphere, etc.). Integrated GIS and remote sensing have previously been used effectively to map the spread of various plant and animal species, as well as their ecosystems, landscapes, bio-climatic conditions, and invasion-facilitating variables (Skidmore 2002). Since 1972, much of the globe has had access to remote sensing (satellite) imagery. The multidate nature of satellite imagery allows for the monitoring of dynamic aspects of landscape settings, enabling for the detection of large land cover changes and the quantification of change rates (Joshi et al. 2004). Remote sensing is being used more often in environmental investigations (Kumar et al. 2013).

10.3.1 Multispectral Remote Sensing

Multispectral remote sensing is typically based on the simultaneous capture of picture data of the earth's surface at many wavelengths. As a result, we may make use of the fact that different types of surfaces reflect light of different wavelengths with varying intensities. Depending on the spatial, spectral, and radiometric resolution of the utilized sensor, different spectral behavior leads to comprehensive categorization of distinct types of land surfaces. Multispectral remote sensing entails acquiring pictures in visible, near infrared, and short-wave infrared wavelength ranges. At various wavelengths, different materials reflect and absorb differently. As a result, the spectrum reflectance characteristics found in these remotely sensed pictures can be used to distinguish between materials, but direct identification is typically not achievable. One of the most extensively used multispectral imagers is NASA's Landsat, which is used to monitor a wide range of landscape scale features. Prior to Hyperion and other aerial hyperspectral data, multispectral remote sensing data were largely used to map the feasibility of environmental consequences worldwide. Multispectral satellite data are extremely valuable for tracking temporal changes as well as ongoing monitoring of environmental damage caused by mining activity. Similarly, Synthetic Aperture Radar pictures may be used to detect changes in land use morphology caused by mining activity.

10.3.2 Hyperspectral Remote Sensing

Hyperspectral data provides a number of advantages over multispectral data, which comprises hundreds of continuous spectral bands with a limited spectrum. Because of the great spectrum resolution and reflectance spectra, particular materials may be identified directly based on their reflectance properties (Kumar et al. 2013). It enables measurements of material spectra, allowing for high-resolution identification of an area's particular mineral, rocks, soils, and plant changes over time. It has been effectively used for the identification of mine waste due to its unique ability to resolve mineral absorption properties (Swayze et al. 2000).

10.3.3 Geographic Information System (GIS)

Geographic Information Systems (GIS) are used to gather, store, analyze, transmit, and change information that is geographically related. GIS may be utilized to facilitate successful short- and long-term decision-making, socioeconomic and environmental issues, transportation, local government, and business. The use of GIS is transforming environmental planning and management. The GIS technology has greatly expanded the usefulness of remote sensing and field-based analysis. Advances in computer hardware and software, the availability of large amounts of digital data, the standardization of GIS formats and languages, the increasing interoperability of software environments, the sophistication of geoprocessing functions, and the growing use of real-time analysis and mapping on the Internet have all increased the utility and demand for GIS technology.

10.3.4 Applications

10.3.4.1 Urban Environment Management

Urbanization is a measure of the transition from traditional rural economies to modern industrial economies. It is a growing population concentration in urban units. As a result of the dramatic increase in global population over the last 50 years, most urban settlements are characterized by stock housing and water supply shortages, urban encroachment in fringe areas, inadequate sewerage, traffic congestion, pollution, poverty, and social unrest, making urban governance a difficult task to maintain a healthy urban environment. As a result, there is an urgent need to use current remote sensing technology, including both aerial and satellite-based systems, which allow us to gather a large amount of physical data quickly and repeatedly, and which, when combined with GIS, allows us to analyze the data geographically.

GIS has been frequently employed in characterization and evaluation studies that necessitate a watershed-based strategy to managing water levels and garbage in metropolitan areas. The drainage network and flow pathways of a watershed, for example, can be obtained using widely available Digital Elevation Models (DEMs). When faced with issues regarding water quality and quantity as a result of natural and man-made disasters (e.g., droughts, hazardous material spills, floods, urbanization), planning becomes critical in order to reduce their effects and guarantee optimal exploitation of existing resources.

Remote sensing may be a valuable source of data for mapping urban land use/cover and environmental monitoring. Many notable studies for environmental quality management have been conducted. Uncontrolled urbanization is to blame for many of the difficulties that our cities face today, including inferior living conditions, acute problems with drinking water, noise and air pollution, garbage disposal, traffic congestion, etc. To reduce environmental deterioration in and around cities, scientific advancement in relevant disciplines must solve the difficulties produced by increasing urbanization; only then can the advantages of progress reach the most disadvantaged. Modern remote sensing technology, which includes both aerial and satellite-based systems, allows us to collect physical data quickly and on a repetitive basis, and when combined with GIS, allows us to analyze the data spatially, offering the possibility of generating various options (modeling), thereby optimizing the entire planning process.

The dynamic character of urban environments needs examination at both the macro and micro levels. As a result, policy-makers must integrate remote sensing into urban planning and administration. The trend of employing remotely sensed data in urban studies began with first-generation satellite sensors like Landsat MSS and was accelerated by a variety of second-generation satellites including Landsat TM, ETM+, and SPOT. The introduction of a third generation of extremely high spatial resolution (5m/pixel) satellite sensors is exciting. The integrated high-resolution PAN and LISS III data may be used efficiently for urban applications. Data from IRS P-6 satellites equipped with sensors, particularly LISS IV Mono and Multispectral (MX) with 5.8m/pixel spatial resolution, is extremely valuable for intense urban mapping (Rai and Kumra 2011).

10.3.4.2 Wasteland Environment

Wetlands cover 3 to 6% of the earth's land area and provide supplies and services such as water quality maintenance, agricultural production, fisheries, leisure, floodwater retention, wildlife habitat, and soil erosion management. It is critical to identify and monitor wetlands and their associated uplands in order to avoid future loss of wetlands and protect existing wetland ecosystems for biodiversity and ecosystem services and commodities. Wetland acreage has been steadily decreasing over the last several decades owing to wetland reclamation, population pressure, water diversion, dam building, pollution, biological incursion, desertification, climate change, and other factors (Xiuwan 2002).

Remote sensing data have been used to quantify qualitative and quantitative changes in terrestrial land cover. Remote sensing (RS) data and GIS) are useful tools for monitoring wetland distribution and spatial-temporal dynamic diversity (Emadi et al. 2010). Remote sensing data from satellites has been widely used to inventory and monitor wetlands, and it may also offer information on adjacent land use and its change over time (Ghobadi et al. 2012).

The Landsat MSS, TM, and SPOT data types are commonly used for wetland characterization and spatial-temporal dynamic change (McHugh et al. 2007). Because of the satellite data's temporal revisit capacity, it is possible to monitor the wetlands seasonally or yearly. For vast geographical areas, using remote sensing data for land cover categorization is less expensive and time-consuming than aerial photography. For wetland research such as monitoring and inventory, using satellite remote sensing data is appropriate in developing nations where budgets are limited and data concerning wetland acreage, land use, and wetland losses are limited (Ozesmi and Bauer 2002).

10.3.4.3 Coastal and Marine Environment

Coastal zones in India are continually changing in shape and environment as a result of natural and human development activities. Natural processes such as wave action, erosion, changes in river courses, etc. have a long-term influence at a slower rate. However, man-made activities such as settlement, industrial operations, recreational activities, trash dumping, etc., have a much faster impact on the coastal environment. If these wetlands continue to disappear, coastal ecosystems may collapse. As a result, it is vital to track changes in the coastal zone throughout time. In reccnt years, remote sensing technology has shown to be useful in obtaining data for effective resource management, and it might therefore be applied to coastal environment monitoring and management. The high temporal resolution offered by satellite data has proved to be a significant improvement in researching the behavior of suspended sediments in coastal waters, which will aid in understanding the transportation of sediments and contaminants.

The integration of remote sensing and GIS technology provides an optimal option for understanding the spatial/temporal distribution of oil spills in the maritime environment and is regarded as the foundation of the oil spill monitoring system. The benefits of remote sensing and GIS include the ability to extract oil pollution parameters such as location and spill areas, as well as spatial and temporal information, which allows users to determine the major cause and source of oil spills and then outline risk areas in order to save the marine environment (Zhang et al. 2008). The capacity to extract oil pollution metrics such as location, size, and spill regions is one of the most significant advantages of GIS (Eljabri and Gallagher 2013). Geospatial technology-derived products aid in informed decision-making in maritime spatial planning and management.

10.4 Uncertainty in Environmental Modeling

In environmental modeling, error is inevitable, inescapable, and often invisible. The presence of multiple mistakes reveals uncertainty and, as a result, the dependability of model predictions. Uncertainty has important practical consequences. Uncertainty either casts doubt on or confirms the use of prediction outputs as a foundation for policy direction and action. Uncertainty analysis commonly assesses the uncertainty of a key decision or policy variable evaluated using a mathematical model. Uncertainty analysis in contaminant fate and transport modeling determines the reliability of the cumulative distribution of model outputs so that judgments may be made based on the likelihood that pollutant concentrations would exceed a specified regulatory threshold level.

There might be various sources of uncertainty. Uncertainty can be caused by model inaccuracies caused by the model's typical reduction of the intricacies of the real processes being depicted. Uncertainty can develop when a model fails to reflect the inherent variability of a parameter or variable in both space and time. Uncertainty also may emerge because the procedures for measuring or estimating the value are inexact.

Sensitivity analysis is used to investigate the propagation of error by models, whereas error or uncertainty analysis approaches are used to assess the sources of prediction error. Unlike sensitivity analysis, which evaluates the sensitivity of model output to small changes in an input parameter or variable, uncertainty analysis considers the inherent uncertainty in model input data and the impact this uncertainty has on model prediction. A variety of uncertainty methodologies have been developed and used for hydrological and water resource challenges in particular (Corwin 1996).

Whenever this accuracy is objectively justified then it tends to be an error, or else it remains an uncertainty. Unless this uncertainty is quantified, the reliability of the results produced remains questionable to the consumers. Various tools are available in GIS to enable the data that can be misused and manipulated without any control mechanism.

A few areas of concern are sample sizes, continuity of data captured, discreteness or separation of different parameters, overlapping, sampling biases, etc. The uncertainty of data percolates to the micro-level of study objectives if not addressed at the appropriate level of intervention mechanism. The results thus generated remain as qualitative information rather than quantitative results. There is a need to keep track of the uncertainty. These factors result in loss of data collected, resulting in falsified classification, and thus degrading the result generated.

The uncertainty in supervised classification may be overcome with accurate ground-referenced data supported with field training sites. The uncertainty of data classification varies with different sources such as class definition, the subject nature of field data collection, and pixel mixing.

The above inference may be substantiated with a common scenario. In field training sites of green cover, the class definition varies largely based on species of the plant habitat structure as each plant species has a unique signature for sensing. The accurate knowledge of the plant species is also needed for class definition by the observer added to the difficulty incompetence of the judging survey. Accounting for uncertainty is as follows: i) Probabilistic structure of Bayesian network; ii) Maximum Entropy; iii) Regression models; 4) Multivariate Analysis; v) Fuzzy Set theory; and vi) Spectral unmixing

10.5 Future of Remote Sensing and GIS Application in Pollutant Monitoring

The GIS platform is used for the continuous automated monitoring of atmospheric pollution. Its primary function is to continually and automatically monitor the contaminants in the air, obtaining continuous instantaneous information on air pollution and interpreting the data. At the same time, we can regulate the state of air pollution and meteorological conditions by continuous instantaneous monitoring when an atmospheric pollution accident occurs. It will give first-hand information for the study of the pollution occurrence, as well as a foundation for validating the air pollutant dispersion model and air quality management.

In recent years, many city governments established atmospheric environmental emergency monitoring system, which were composed primarily of an emergency command center, on-site command vehicle, and emergency environmental monitoring system, for effectively preventing, controlling, and eliminating the harm of environmental emergencies and standardizing all types of emergency work, based on establishing automatic mechanism for monitoring and forecasting air quality. The advantages of GIS were fully utilized by the atmospheric environmental emergency monitoring systems. In the case of an unintentional pollution incident, environmental protection agencies might hurry to the site to quantitatively and qualitatively monitor and assess the quantity of dangerous compounds in the atmosphere. We can determine the nature of the occurrence and the harm, and give the technical foundation for decision-making for the construction of contingency measures in a fast and accurate manner. Furthermore, the system can provide services for environmental quality assessment, prediction, warning, and enforcement in the regional environment and major projects, as well as perform background measurements for major events predicted to cause air pollution and serve as the foundation for subsequent conclusions (Rai et al. 2018).

It is often understood that the pollution concentration at the industrial settlements and high-density urban areas reach compromising levels higher than the restricted monitoring parameters. The methods of physical monitoring of study parameters in dense industrial areas are challenging due to restrictive access to the premises due to administrative burdens. Also, the temporal study for a given location demands higher human involvement to capture and record the relevant data. Data collection mechanisms (i.e., air quality monitoring stations, etc.) cannot be established at all influenced locations of a given industrial area set-up, as the influenced areas generally spread in any direction up to several kilometers. The micro-level monitoring over a larger study area is continuously challenging with the existing manual and semi-manual data collection mechanisms. New and latest technology is needed for efficient policy-making, which is supported by rapid estimates of pollution levels and in turn provides guidance for suitable intervention methods.

RS coupled with GIS offers a good advantage of spatio-temporal variations of the key study parameters. The large spectrum of study areas such as urban, semi-urban, industrial, forestry, coastal, agreeable areas, protected areas, and mining areas can be reached with these techniques.

The approach provides dynamic modeling capabilities for forecast, prevention, monitoring, and assessment-oriented actions in pollution monitoring and controls. The real extents of environmental degradation, hazards, etc. can be efficiently captured, studied, and researched with RS and GIS. Suitably required study maps at any geographical study area can be generated with the tools available with GIS techniques. Various study objectives such as pollutant concentration levels monitoring, distribution pattern, circulation mechanism aided by natural parameters, etc. can be comfortably studied. Similarly, extensive simulation patterns can be developed which can be the futuristic approach of decision-making. RS and GIS techniques can realistically be employed in tracking the origins of the pollutants, their type and nature, quantification of release, etc. A few of the inexhaustive pollutant parameters that are studied are carbon dioxide; industrial discharges like arsenic, sulphur, methane, aerosols, etc. in air contamination; lead, mercury, cadmium, and several rare earth metals in water contamination; noise levels in high traffic areas and industrial zones; and garbage and waste spills in open ocean and river basins.

RS has a wide range of applications. This holds for projecting the level of pollutants present in the environment. Environmental monitoring is defined as a systematic sampling of environmental aspects such as biotic and abiotic factors like air, water, soil, etc. to observe, study, and acquire knowledge from the process. Monitoring refers to obtaining information about current levels of harmful pollutants discharged into the environment. The main aims of monitoring the pollutants are the following: i) assess pollution interaction and patterns; ii) estimate the need for government regulations; and iii) ensure compliance with emission standards.

10.5.1 Types of Satellite-Based Environmental Monitoring

10.5.1.1 Atmosphere Monitoring

Anthropogenic activities and natural events such as forest fires and desert storms contribute to several types of pollution in the atmosphere (Somvanshi et al. 2019). Atmospheric particulate matter not only harms human health, and also affects weather and climate by influencing the earth–atmosphere system's energy budget (Zhang et al. 2021). The four basic categories of methods and procedures used for satellite-based particulate matter concentration mapping are:

1) Univariate Regression
2) CTM-based method
3) Multivariate Regression
   - Generalized linear model
   - Two-stage hierarchical model
   - Geographically and temporally weighted regression
   - Land use regression
   - Other regression methods
   - Machine learning approach
4) Semi-empirical physical approach

10.5.1.2 Air Quality Monitoring

One of the most serious environmental issues is air pollution, which is mostly a concern in urban areas. However, there is significant evidence that various types of air pollution have serious consequences for water, soil, plants, deforestation, and human health (Somvanshi et al. 2019). Weber et al. (2001) imply that the main goal of air quality monitoring is to determine concentrations of pollutants.

The air monitoring satellite provides a complete and concise idea of a large geographical area in an image, on a systematic basis due to excellent temporal resolution. The satellite's sensing system can capture many pollutants present in the air simultaneously with near real-time and continuous monitoring capabilities. Sensors used in mapping air pollution are the following: i) Landsat TM/ETM+; ii) MODIS AOT; and iii) ASTER data. Types of monitoring techniques used to map air pollutants include Aerosol, where aerosol thickness is monitored by the four following different methods: i) Ocean method; ii) Brightness method; iii) Contrast-reduction method; and iv) Dark vegetation method.

10.5.1.3 Land Use/Land Cover (LULC)

For environmental sustainability, reliable mapping and detection of Land use/Land cover (LULC) changes have become critical. Sustainable ecosystem's LULC is negatively impacted by humans (Abd El-Hamid et al. 2021). Yang and Sun (2017) imply that as a result, LULC depicts the outcome of human–environment interaction in a specific area. According to Verburg et al. (2011), the dynamics resulting from climate change processes and socio-economic factors also have an impact on LULC. Modern technologies, including RS data and GIS techniques, are presently being used extensively in the development of more effective, accurate, and time-based LULC classification algorithms and change detection findings (Lu et al. 2014).

According to Sohl and Claggett (2013), cellular- and agent-based models, as well as hybrid models based on these two model types, are the most often employed models in land-use change monitoring and analysis. Urbanization can have either beneficial or negative effects on the environment, but unmanaged urbanization always has negative consequences (El-Hamid et al. 2021).

10.5.1.4 Hazard Monitoring

Deliberate or unintentional release of dangerous compounds into the environment is an unavoidable result of human activities (Slonecker et al. 2010). According to researchers, the twentieth century's combined industrialization and population increase has led to unprecedented instances of fugitive, transitory, and ephemeral pollution releases that now threaten many plant and animal species, and may eventually challenge human life. Many hazardous waste problems are detected and remedied using a range of time-consuming and expensive monitoring and analysis procedures, such as laboratory chemical analysis. Remote Sensing (RS) is one of the technologies with a proven and developing potential to provide a non-contact and cost-effective alternative to standard sampling methods.

Information from multispectral imaging systems like the Landsat Multi Spectral Scanner (MSS) and Thematic Mapper (TM), as well as a variety of aircraft-based systems, has been used to track hazardous waste sites in terms of land use, regional risk, and the spectral characteristics of specific disposal facilities and their pollution profile (Slonecker et al. 2010). Accordingly, researchers demonstrated that spectrum differences in basic MSS data might be utilized to detect heavy metal pollution. CIR aerial photography, SPOT, and Landsat TM imagery were combined to monitor vegetative stress and recovery at a Superfund site in Michigan (Herman et al. 1994). Furthermore, other reseachers showed the importance of digital analysis of high-resolution aerial pictures, aircraft scanner data, and Landsat data for wide scope challenges like groundwater movement. Multispectral data can be utilized to track landfill gas migration and its impact on plants (Jones et al. 1994). Multispectral imaging can be used to monitor landfill development and its associated pollution profile, as shown by Kwarteng and Al-Enezi (2004).

10.5.1.5 Marine and Phytoplankton Studies

Any substances (could be algae, DOM, oils, aquatic vascular plants, thermal releases) in the surface water can remarkably change the backscattering components of the surface water. RS methods hinge on the capacity to test the changes in the spectral signature backscattered from the water and relate the measured changes by empirical or analytical models to the water-quality parameter. Types of RS techniques for water resource pollution monitoring are the following: i) snow and glacier mapping and monitoring; ii) irrigation water management; iii) flood disaster monitoring, forecasting, and management; iv) water-quality monitoring; v) watershed management; vi) groundwater prospecting; and vii) environmental impact assessment.

10.6 Identification of Microconstituents Using Remote Sensing and GIS Techniques

Some traditional analytical methods such as pyrolysis desorption-gas chromatography/mass spectrometry, Fourier transform infrared spectroscopy, and scanning electron microscopy energy-dispersive X-ray spectroscopy, have been applied for the study and investigation of microconstituents (MCs) in different environmental systems, including marine environments and watersheds (Pico et al. 2019). These techniques provide a high degree of sensitivity and reliability for MC analyses; however, the main drawback of the traditional methods is the low monitoring efficiency during sample collection and analyses for large-scale and remote sites. Moreover, in most cases, due to the limited manpower and resources, the overall samples cannot be sampled widely, resulting in low representation (Farré 2020). Park et al. (2006) applied an integrated technique using a genetic algorithm and a GIS for the design of an effective contaminant monitoring network in a large river system. Zhang et al. (2021) compared remote sensing based accumulation maps and validation against *in situ* beach sampling. In their study, they also developed a monitoring system to evaluate the pollution caused by microplastics being discharged by a river to adjacent coastlines as well as into the open ocean (Atwood et al. 2019).

Due to the rapid development of optical and sensing technologies, sensor-equipped unmanned vehicles and satellite technologies have become some of the best solutions for MC investigation, identification, and location. Davaasuren et al. (2018) identified microplastics (e.g., polyethylene, polyethylene terephthalate) in the marine environment using the dark patches in the gray scale brightness of the synthetic aperture radar graphics. Salgado-Hernanz et al. (2021) reported that many studies used satellite information and data to deal with marine microplastics using hyper- and multi-spectral optical sensors. The correspondence analysis on analyzed variables exhibited that aircrafts with high spatial resolution (<3 m) with optical sensors ($\lambda = 400$ to 2,500 nm) seem to be the most optimum combination to target marine litter, while satellites carrying Synthetic Aperture Radar (SAR) sensors ($\lambda = 3.1$ to 5.6 cm) may detect oil slicks associated to surfactants that might contain high concentration of MCs (Chaudhry and Sachdeva 2021; Salgado-Hernanz et al. 2021). Serranti et al. (2019) used the optical sensors based on hyperspectral imaging with the short-wavelength infrared range (1,000–2,500 nm) to characterize MCs in marine environments.

Remucal (2014) reported that the fates of MCs in water were often determined via the process of photolysis. Chen et al. (2020) obtained the concentration change of the photochemically produced reactive intermediates by the photochemical reaction of colored dissolved organic matter after absorbing solar radiation through the relationship between measured $\alpha 440$ and Landsat 8 reflectance. This process can be used to map reactive species production rates and concentrations as well as contaminant transformation rates (Chen et al. 2020). Through the integration of the remote sensing data with GIS technology, more appropriate management strategies for the control and management of MC pollution can be established. Mo et al. (2021) used satellite image (Landsat-8) for remote sensing analysis, and also combined the results with ArcGIS (Environmental Systems Research Institute, USA) to confirm that microplastics were the main MCs source in soils in northern China.

Compared with other sources (e.g., laundry, household dust, tire wear, personal care products), microplastics could be a significant contributor to the MCs in soil environments (Mo et al. 2021). Peng et al. (2022) used the remote sensing technology Landsat-8 (operational land imager and thermal infrared sensor) to characterize the barren land and vegetation in cities, and also input the analyzed MCs data with the QGIS system. Their results confirmed that industry and land reclamation probably were the main sources of MCs in Luoyuan Bay, China (Peng et al. 2022). Kikaki et al. (2020) applied the Landsat-8 and Sentinel-2 satellite sensors to investigate marine plastic debris by remote sensing during 2014 to 2019. Their data were processed via a QGIS system to record the spectral characteristics of floating plastic debris and determine the trajectory of MCs in the marine environment (Kikaki et al. 2020). Results indicate that the oceanic currents were important to the transport of MCs into, within, and out of the studied region (Kikaki et al. 2020). Remote sensing with GIS technology has been a cost-effective and valuable tool for the monitoring and investigation of the sources and paths of MCs in the ecosystem.

10.7 Conclusions

Today, we have Remote Sensing (RS) as one of the major tools in studying human–nature interactions in the global dimension, and it is also used as a powerful database provider of numerous data. The information is generally used for monitoring purposes and management of services. To conclude, in this chapter we put forward possible methods to integrate simulation models with GIS platforms. It is empirically proved that coupling the GIS system shows that the system is efficient and easily integrated, in comparison with more traditional methods. Additionally, we presented the coupling of 3D GIS; historically we had a 2D GIS modeling system. Recently, we have found advanced satellites to give accurate mapping. These maps are of utmost importance due to the information provided, which helps in protecting the natural settings as well as human lifestyles. These trends of 3D sensors have changed the whole viewing perceptive. Today we have point clouds and other 3D models as the tools utilizing these new sensors. Sensors have to be developed to sense built operable environments in the future. Talking about the uncertainty of RS is complexity in mapping certain services that have been tested and recognized several times in the past decade. There are researchers referring to uncertainty in a multi-faceted issue dealing with the core of spatial statistics. The uncertainty issue of spatially explicit data mapping tends to increase the RS products in GIS. In this chapter, we have presented a wide range of various terms relating to RS definitions together with their certain meanings. Moreover, with the application of RS and GIS, the distribution and fate of MCs in the vulnerable areas of the ecosystem can be effectively and efficiently characterized.

References

Abd El-Hamid, H.T., El-Alfy, M.A., and Elnaggar, A.A. (2021). Prediction of future situation of land use/cover change and modeling sensitivity to pollution in Edku Lake, Egypt based on geospatial analyses. *GeoJournal* 86 (4): 1895–1913. doi: 10.1007/s10708-020-10167-7.

Atwood, E.C., Falcieri, F.M., and Piehl, S. (2019). Coastal accumulation of microplastic particles emitted from the Po River, Northern Italy: comparing remote sensing and hydrodynamic modelling with *in situ* sample collections. *Marine Pollution Bulletin* 138: 561–574 2021.

Chaudhry, A.K. and Sachdeva, P. (2021). Microplastics' origin, distribution, and rising hazard to aquatic organisms and human health: socio-economic insinuations and management solutions. *Regional Studies in Marine Science* 48: 102018. doi: 10.1016/j.rsma.2021.102018.

Chen, Y., Hozalski, R.M., Olmanson, L.G. et al. (2020). Prediction of photochemically produced reactive intermediates in surface waters via satellite remote sensing. *Environmental Science & Technology* 54: 6671–6681. doi: 10.1021/acs.est.0c00344.

Clarke, K.C., Brass, J.A., and Riggan, P.J. (1994). A cellular automation model of wildlife propagation and extinction. *Photogrammetric Engineering and Remote Sensing* 60 (11): 1355–1367.

Corwin, D.L. (1996). GIS applications of deterministic solute transport models for regional-scale assessment of non-point source pollutants in the vadose zone. *Soil Science Society of America* 48: 69–100.

Corwin, D.L. and Wagenet, R.J. (1996). Applications of GIS to the modeling of nonpoint source in the vadose zone: a conference overview. *Journal of Environmental Quality* 25 (3): 403–411.

Davaasuren, N., Marino, A., Boardman, C. et al. (2018). Detecting microplastics pollution in world oceans using sar remote sensing. 1: School of Engineering and Innovation, The Open University, Walton Campus, MK7 6AA, UK; 2: The University of Stirling, UK; 3. The Parthenope University of Igarss, 938–941.

El-Hamid, H.T.A., Caiyong, W., and Yongting, Z. (2021). Geospatial analysis of land use driving force in coal mining area: case study in Ningdong, China. *GeoJournal* 86 (2): 605–620. doi: 10.1007/s10708-019-10078-2.

Eljabri, A. and Gallagher, C. (2013). Developing integrated remote sensing and GIS procedures for oil spill monitoring on the Libyan coast. *2nd International Conference on Environment and BioScience* 169: 17–20. doi: 10.2495/CP130111.

Emadi, M., Baghernejad, M., Pakparvar, M., and Kowsar, S.A. (2010). An approach for land suitability evaluation using geostatistics, remote sensing, and geographic information system in arid and semiarid ecosystems. *Environmental Monitoring and Assessment* 164 (1–4): 501–511. doi: 10.1007/s10661-009-0909-6.

Farré, M. (2020). Remote and *in situ* devices for the assessment of marine contaminants of emerging concern and plastic debris detection. *Current Opinion in Environmental Science & Health* 18: 79–94. doi: 10.1016/j.coesh.2020.10.002.

Ghobadi, Y., Pradhan, B., Kabiri, K. et al. (2012). Use of multi-temporal remote sensing data and GIS for wetland change monitoring and degradation. *2012 IEEE Colloquium on Humanities, Science & Engineering Research.*

Hansen, A.J., Patten, R., Degayner, E., and Marks, B.L. (1996). Simulating forest and habitat change in south-east Alaska with the landscape model PAYSAGE. *Transactions in GID 1* (2): 119–136. https://doi.org/10.1111/j.1467-9671.1996.tb00039.x.

Herman, J.D., Waltes, J.E., Ponitz, R.M., and Etzler, P. (1994). A temporal and spatial resolution remote sensing study of a Michigan superfund site. *Photogrammetric Engineering and Remote Sensing* 60 (8): 1007–1017. https://doi.org/10.3390/rs12111727.

Jones, H.K., Elgydagger, J., Survey, B.G. et al. (1994). Remote sensing to assess landfill gas migration. *Waste Management & Research* 12: 327–337.

Joshi, C., de Leeuw, J., and van Duren, I.C. (2004). Remote sensing and GIS applications for mapping and spatial modelling of invasive species. *ISPRS 2004: Proceedings of the XXth ISPRS Congress: Geo-Imagery Bridging Continents*, 669–677.

Kikaki, A., Karantzalos, K., Power, C.A., and Raitsos, D.E. (2020). Remotely sensing the source and transport of marine plastic debris in Bay Islands of Honduras (Caribbean Sea). *Remote Sensing* 12 (11): 1727. doi: 10.3390/rs12111727.

Kumar, S.S., Arivazhagan, S., and Rangarajan, N. (2013). Remote sensing and GIS applications in environmental sciences: a review. *Journal of Environmental Nanotechnology* 2 (2): 92–101. doi: 10.13074/jent.2013.06.132025.

Kwarteng, A.Y. and Al-Enezi, A. (2004). Assessment of Kuwait's Al-Qurain landfill using remotely sensed data. *Journal of Environmental Science and Health. Part A: Toxic/Hazardous Substances and Environmental Engineering* 39 (2): 351–364. doi: 10.1081/ESE-120027527.

Lu, D., Li, G., and Moran, E. (2014). Current situation and needs of change detection techniques. *International Journal of Image and Data Fusion* 5 (1): 13–38. doi: 10.1080/19479832.2013.868372

Lurie, J.B. and Irvin, E.M. Jr. (2002). Remote sensing technology and applications. *Optical Engineering* 41 (9): https://doi.org/10.1117/1.1509457.

Maxwell, T. and Costanza, R. (1997). An open geographic modeling environment. *Simulation* 68 (3): 175–185.

McHugh, O.V., McHugh, A.N., Eloundou-Enyegue, P.M., and Steenhuis, T.S. (2007). Integrated qualitative assessment of wetland hydrological and land cover changes in a data scarce dry Ethiopian highland watershed. *Land Degradation and Development* 18 (6): 643–658. doi: 10.1002/ldr.803.

Mitas, L. and Mitasova, H. (1998). Distributed soil erosion simulation for effective erosion prevention. *Water Resources Research* 34 (3). https://doi.org/10.1029/97WR03347.

Mo, X., Li, H., Lian, Y. et al. (2021). Estimation of soil microplastic input derived from plastic gauze using a simplified model. *Science of the Total Environment* 793: 148577. doi: 10.1016/j.scitotenv.2021.148577.

Ozesmi, S.L. and Bauer, M.E. (2002). Satellite remote sensing of wetlands. *Wetlands Ecology and Management* 10: 381–402. https://doi.org/10.1023/A:1020908432489.

Park, S. and Wagner, D.F. (1997). Incorporating celiular automata simulators as analyticai engines in GIS. *Transactions in GIS* 2 (3): 213–231. https://doi.org/10.1111/j.1467-9671.1997.tb00012.x.

Park, S.Y., Choi, J.H., Wang, S.Y., and Park, S.S. (2006). Design of a water quality monitoring network in a large river system using the genetic algorithm. *Ecological Modelling* 199: 289–297.

Peng, B., Hossain, K.B., Lin, Y. et al. (2022). Assessment and sources identification of microplastics, PAHs and OCPs in the Luoyuan Bay, China: based on multi-statistical analysis. *Marine Pollution Bulletin* 175: 113351. doi: 10.1016/j.marpolbul.2022.113351.

Pico, Y., Alfarhan, A., and Barcelo, D. (2019). Nano- and micro-plastic analysis: focus on their occurrence in freshwater ecosystems and remediation technologies. *Trends in Analytical Chemistry* 113: 409–425. https://doi.org/10.1016/j.trac.2018.08.022

Rai, P.K. and Kumra, V.K. (2011). Role of geoinformatics in urban planning article. *Journal of Scientific Research.* 55: 11–24. https://www.researchgate.net/publication/259572943.

Rai, P.K., Mishra, V.N., and Raju, K.N.P. (2018). Methodology and applications of remote sensing and GIS in environmental mapping and monitoring. *An International Refereed Journal (NGSI-BHU)* 64 (2): 266–276.

Ray, S.S. (2012*). Basics of remote sensing.* In: *Remote Sensing Image Analysis: Including the Spatial Domain*, pp. 1–15.

Remucal, C.K. (2014). The role of indirect photochemical degradation in the environmental fate of pesticides: a review. *Environmental Science: Processes & Impacts* 16: 628–653. doi: 10.1039/c3em00549f.

Salgado-Hernanz, P.M., Bauzà, J., Alomar, C. et al. (2021). Assessment of marine litter through remote sensing: recent approaches and future goals. *Marine Pollution Bulletin* 168.: 112347. doi: 10.1016/j.marpolbul.2021.112347.

Serranti, S., Fiore, L., Bonifazi, G. et al. (2019). Microplastics characterization by hyperspectral imaging in the SWIR range. *Proceedings SPIE 11197, SPIE Future Sensing Technologies.* 1119710. https://doi.org/10.1117/12.2542793 38. 10.1117/12.2542793.

Shapiro, M. and Westervelt, J. (1992). *R. MAPCALC: An Algebra for GIS and Image Processing.* United States Army Construction Engineering Research Laboratory Champaign.

Skidmore, A. (2002). *Environmental Modelling With GIS and Remote Sensing.* London: CRC Press. https://doi.org/10.4324/9780203302217.

Slonecker, T., Fisher, G.B., Aiello, D.P., and Haack, B. (2010). Visible and infrared remote imaging of hazardous waste: a review. *Remote Sensing* 2 (11): 2474–2508. doi: 10.3390/rs2112474.

Sohl, T.L. and Claggett, P.R. (2013). Clarity versus complexity: land-use modeling as a practical tool for decision-makers. *Journal of Environmental Management* 129: 235–243. doi: 10.1016/j.jenvman.2013.07.027.

Somvanshi, S.S., Vashisht, A., Chandra, U., and Kaushik, G. (2019). Delhi air pollution modeling using remote sensing technique. In: *Handbook of Environmental Materials Management*, 1–27. Springer International Publishing. doi: 10.1007/978-3-319-58538-3_174-1.

Swayze, G.A., Smith, K.S., Clark, R.N. et al. (2000). Using imaging spectroscopy to map acidic mine waste. *Environmental Science and Technology* 34 (1): 47–54. doi: 10.1021/es990046w.

Verburg, P.H., Neumann, K., and Nol, L. (2011). Challenges in using land use and land cover data for global change studies. *Global Change Biology* 17 (2): 974–989. doi: 10.1111/j.1365-2486.2010.02307.x.

Weber, C., Hirsch, J., and Wald, L. (2001). Air pollution mapping over a city-virtual stations and morphological indicators. *10th International Symposium Transport and Air Pollution.* https://www.researchgate.net/publication/43196159.

Wolfram, S. (1984). Universality and complexity in cellular automata. *Physica D: Nonlinear Phenomena* 10 (1–2): 1–35.

Xiuwan, C. (2002). Using remote sensing and GIS to analyse land cover change and its impacts on regional sustainable development. *International Journal of Remote Sensing* 23 (1): 107–124. doi: 10.1080/01431160010007051.

Yang, A. and Sun, G. (2017). Landsat-based land cover change in the Beijing-Tianjin-Tangshan urban agglomeration in 1990, 2000 and 2010. *ISPRS International Journal of Geo-Information* 6 (3): 59. doi: 10.3390/ijgi6030059.

Zhang, C., Luo, L., Xu, W., and Ledwith, V. (2008). Use of local Moran's I and GIS to identify pollution hotspots of Pb in urban soils of Galway, Ireland. *Science of the Total Environment* 398: 212–221.

Zhang, Y., Li, Z., Bai, K. et al. (2021). Satellite remote sensing of atmospheric particulate matter mass concentration: advances, challenges, and perspectives. *Fundamental Research* 1 (3): 240–258. doi: 10.1016/j.fmre.2021.04.007.

Part III

Various Physicochemical Treatment Techniques of Microconstituents

11

Process Feasibility and Sustainability of Struvite Crystallization from Wastewater through Electrocoagulation

Alisha Zaffar, Nageshwari Krishnamoorthy, Chinmayee Sahoo, Sivaraman Jayaraman, and Balasubramanian Paramasivan

Department of Biotechnology & Medical Engineering, National Institute of Technology Rourkela, Odisha, India

11.1 Introduction

Phosphorus (P) present in the form of organic phosphate, orthophosphate, and polyphosphate, is beneficial for plant growth and microbial activity in soil (Bakshi et al. 2020). It is mainly obtained from phosphorus rock reserves, which are being depleted at an alarming rate for the production of chemical fertilizers, due to the increase in global population and food demand. Wastewater and its sludge contain a large amount of microconstituents (MCs) in the form of phosphorus, nitrate, sulphur-related compounds, etc. that has emerged as the main cause for impairment of water quality and may also serve as a source of future pollution. Hence, it is necessary to reclaim the MCs from the source to avoid their abundance in closed waterbodies as it can lead to eutrophication, decrease in oxygen content and, in turn, affect life (Perez-Ameneiro et al. 2013). To understand nutrient loading in the surface water, specific micronutrient markers of organic origin have been tested, which can detect the attenuation pathway of the nutrients. Sucralose, atenolol, iohexol, carbamazepine, etc. were some of the organic micronutrients used, among which only atenolol was observed to have similarity with attenuation of nitrate, while none had similarity with phosphate (Badruzzaman et al. 2013). The wastewater, if separated from the source and treated properly, can benefit the environment as well as people (Bastida et al. 2018; Krishnamoorthy et al. 2021a, 2021b).

Struvite or magnesium ammonium phosphate hexahydrate (MAP) is formed by the combination of magnesium (Mg^{2+}), NH_4^+, and PO_4^{3-}, at an alkaline pH. These crystals are white, orthorhombic structured, and rod-shaped, and can possibly change to an oval shape with excess of NH_4^+ and PO_4^{3-} ions. It is poorly soluble in water (Le Corre et al. 2005; Doyle and Parsons 2002). When these ions react with magnesium, this leads to the formation of struvite due to establishment of hydrogen bonding between them (Das et al. 2017; Krishnamoorthy et al. 2020). The occurrence of these crystals was initially found as deposits

Microconstituents in the Environment: Occurrence, Fate, Removal, and Management, First Edition.
Edited by Rao Y. Surampalli, Tian C. Zhang, Chih-Ming Kao, Makarand M. Ghangrekar, Puspendu Bhunia, Manaswini Behera, and Prangya R. Rout.

on the equipment of a wastewater treatment plants affecting the hydraulic capacity and increasing maintenance costs (Bhuiyan et al. 2008). Mg^{2+}, NH_4^+, and PO_4^{3-} ions react in an equimolar ratio to form struvite crystals (Eq. 11.1).

$$Mg^{2+} + NH_4^+ + PO_4^{3-} + 6H_2O \rightarrow MgNH_4PO_46H_2O \tag{11.1}$$

Struvite is rich in nutrients like nitrogen, phosphorus, and sometimes potassium, which are the primary essential nutrients for plant growth. It is also regarded as a slow-release fertilizer, due to partial solubility in water. However, several studies have concluded its function as a superphosphate with high efficiency to be used for plant growth, as it meets nutritional demands of crops in a better way (Bastida et al. 2018; Szymańska et al. 2019; Nageshwari and Balasubramanian 2022).

According to the FAO (2017), the total fertilizer nutrient demand in the year 2015 was 184.02 million tonnes, which increased at an average rate of 1.9% every year, resulting in 201.66 million tonnes in the year 2020. Table 11.1 presents the report summary of the capacity and demands of different countries to supply different nutrients as a fertilizer and its compound annual growth rate (CAGR) (Eq. 11.2) in percentage for the year 2020 (FAO 2017).

$$CAGR = \left[\left(\frac{Endvalue}{Startvalue}\right)^{\frac{1}{No.of\ years}} - 1\right] \tag{11.2}$$

CAGR helps in assessing the mean annual growth rate of a fertilizer or its nutrients over a given period. The demand of NPK nutrients has become very high worldwide (Table 11.1). Some countries are rich in mineral rocks, which can fulfil their needs naturally, but others import such harmful chemical fertilizers. Efforts are being made to shift from chemical

Table 11.1 The availability and demand of various essential nutrients with compound annual growth rate.

	Africa	America	Asia	Europe	Oceania	Worldwide
N availability	8606	19,128	75,923	29,041	542	133,240
N demand	4302	24,564	71,476	16,504	1917	118,763
Potential N balance	4304	−5435	4447	12,537	−1375	14,477
H_3PO_4 availability	8814	8470	24,769	3762	460	46,275
H_3PO_4 demand	1610	12,418	23,328	3760	1016	42,133
Potential H_3PO_4 balance	7204	−3948	1440	2	−556	4142
K_2O availability	−100	16,011	8054	19,701	−8	43,659
K_2O demand	37,042	897	12,830	18,181	4741	393
Potential K_2O balance	−997	3181	−10,127	14,960	−401	6617
5-year CAGR for N (%)	3.78	1.77	1.52	0.78	1.50	1.54
5-year CAGR for H_3PO_4 (%)	2.8	2.6	2.1	1.6	1.3	2.19
5-year CAGR for K_2O (%)	6.76	2.05	2.56	2.52	0.05	2.44

Source: Adapted from FAO (2017).

fertilizers to eco-friendly fertilizers. On this juncture, the occurrence of struvite has proved to be a boon to the agricultural world.

Different techniques are being applied to precipitate struvite from different types of waste and wastewater rich in phosphorus and nitrogen. Many cost-intensive treatments are required to reuse the supernatant produced after struvite recovery. Among them, electrocoagulation is proving to be suitable for decentralization and large-scale applications. However, electrocoagulation is a safer, eco-friendly, and cleaner technology, which produces much less sludge and also improves the quality of the wastewater supernatant after struvite precipitation.

Electrocoagulation is an age-old technique used for treatment of wastewater through direct or indirect oxidation. It works by stabilizing the charge of the colloids with the help of an electric current, causing agglomeration and settling of the particles. Electrocoagulation is also effective in removal of organic matter, dye, heavy metal contaminants like arsenic and other contaminants, and recovery of phosphorus and nitrogen to reuse the water for other purposes. This process has been in use for a long time to treat different types of wastewaters like sewage, tannery wastewater, wastewater from textile industry, oily water, poultry wastewater, industrial wastewater, etc. The anodes used in the process were mainly iron or aluminum, because of their higher removal efficiency as compared to other metals (Kumar and Goel 2010; Mohora et al. 2012; Shim et al. 2020). Struvite precipitation through electrocoagulation (using a sacrificial magnesium electrode) can show promising results in terms of scale-up, as compared to other technologies through optimum current density, low power consumption, high current efficiency, and long-term performance resilience.

The main objective of this chapter is to elaborate the aspects of electrocoagulation and its advanced use in struvite precipitation. Electrocoagulation has been considered as an efficient way of controlling the MCs in the environment, especially wastewater. However, in the case of struvite recovery, this strategy is exploited to supply MCs of magnesium to wastewater through the sacrificial magnesium anode to recover nutrients such as phosphate and ammonium. Upon supply of power to the anode, Mg^{2+} MCs dissociate to form hydrogen bonds with the essential nutrients to form struvite. Sometimes, analogs of struvite containing MCs of potassium (K-struvite) or sodium (Na-struvite) are formed, depending on the processing conditions. This chapter provides insights into the various types of electrocoagulation employed for struvite production and the advantages over other conventional methods. Additionally, the effect of influential process parameters on struvite formation and yield has been discussed. The energy, economy, and environmental aspects of this technology have been emphasised to highlight the feasibility for scaling up and commercialization.

11.2 Struvite Crystallization Through Electrocoagulation

11.2.1 Working Principle

Electrocoagulation works on the principle of nutrients present in the form of colloids with a repulsive electrostatic force and an attractive van der Waals force between them. This latter force has certain induced or permanent dipole moments. When electricity is passed through this electrolyte, destabilization of the charges takes place, which decreases the repulsive force, and the particles easily agglomerate because of the presence of weak bonds

(Kumar and Goel 2010; Naje et al. 2017). The processes involved in electrocoagulation can be described as follows (Figure 11.1):

i) When an electric current is passed through the electrodes, the cations are released into the solution, which forms polymeric metal hydroxides that function similar to coagulants.
ii) The cations and polymeric metal hydroxides cause neutralization of the negatively charged particles present in the electrolyte.
iii) These neutralized materials combine to form flocs, which then settle down to be removed.
iv) There is also formation and emission of gases such as hydrogen and ammonia.

11.2.2 Types of Electrocoagulation

11.2.2.1 Batch Electrocoagulation

The batch electrocoagulation process is one of the most frequently used processes of electrocoagulation, where the wastewater is taken at once, without any inflow or outflow during the process. Figure 11.1 depicts the schematic and specifications of a commonly used batch electrocoagulation reactor. In this type, the main components required are a DC power supply unit, which supplies the required amount of current to the anode for ion dissipation to carry out the process of precipitation. The cathode used is usually inert like stainless steel, carbon, etc. that does not interfere with the reaction. The anode is mostly magnesium; however, other types of metals like iridium dioxide, aluminum, iron, etc. also can serve as an anode, where magnesium is added externally as salt. In the batch method, the medium is retained in the reaction vessel to allow struvite to settle before removal. Table 11.2 shows various literary works on struvite precipitation by batch electrocoagulation.

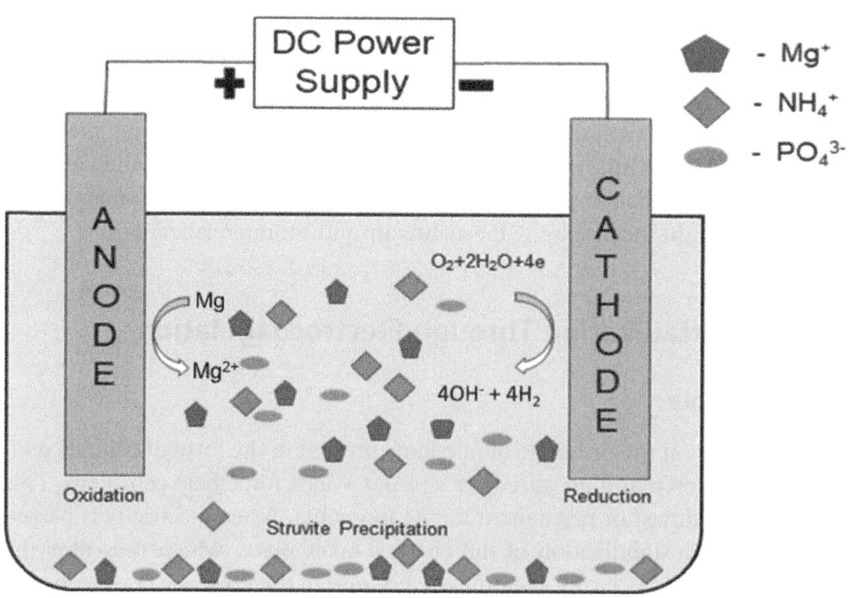

Figure 11.1 Mechanism of struvite precipitation through electrocoagulation.

Table 11.2 Various research works on utilization of batch electrocoagulation for struvite precipitation.

Wastewater Source	pH	Anode	Cathode	Surface Area (cm^2)	Inter-Electrode Distance (cm)	Voltage (V)	Current/current Density (A/ Am^{-2})	Stirring Speed (rpm)	References
Human urine	–	Magnesium	Stainless steel	19.4	5.5	1.62	4.7 mA cm^{-2}	275	Hug and Udert (2013)
			Potential air cathode	25	3	–	–	–	Kim et al. (2018)
Swine wastewater	9.5	Magnesium	Stainless steel	45	3.5		22.2 A m^{-2}	80	Lin et al. (2018)
Swine wastewater	10	Titanium with iridium oxide	Stainless steel	400	1	7	30 mA	300	Liu et al. (2011a,b)
Agricultural and municipal wastewater	7.5–9.3	Magnesium	Stainless steel	40	5	–	40–120 mA	220	Kékedy-Nagy et al., (2020)
Synthetic and authentic wastewater	7.6–9	Iron/ Aluminum	Stainless steel	70	0.5	0–65	0–5 A	125	Rajaniemi et al. (2021)
Supernatant from hydrolysis sludge	9	Magnesium sacrificial anode	Titanium	4	2	30	0–3 A	–	Zhou and Chen (2018)
Human urine	7.5–9.5	Magnesium	Stainless steel	44	–	–	0.1–0.2 A	–	Kruk et al. (2014)

11.2.2.2 Continuous Electrocoagulation

In continuous electrocoagulation, wastewater is being continuously supplied to the reaction area for producing struvite from wastewater at a particular time period. Though studies have concluded this method to be very efficient and cost-effective, the method is still in the research and experimental phase (Rajaniemi et al. 2021). Figure 11.2 shows the schematic diagram of different segments of a continuous electrocoagulation system. The basic structure contains a feeding tank, which stores and supplies the collected wastewater at particular intervals with the help of peristaltic pumps; a DC power supply unit supplying power to the system; a reaction vessel containing two or more electrodes with suitable dimensions and gaps for the reaction to occur; a magnetic stirrer for equal mixing of the reactants in the solution (Kumar and Goel 2010); and a storage tank, which stores wastewater containing struvite for a short interval to allow it to settle for easy removal process (Phalakornkule et al. 2009).

The efficiency of the system can be enhanced by increasing the number of active electrodes to increase the active electrode area to volume ratio (A/V). The flow rate has no direct influence on the process but indirectly affects other operating parameters. Mohora et al. (2012) concluded that with an increase in flow rate, the pH and temperature increased. Similarly, when the flow rate increases, the production rate decreases for the same current density. The change in specific electrical energy consumption was also negligible when flow rates were altered (Emamjomeh and Sivakumar 2009; Zodi et al. 2013). Hydraulic retention time (HRT) is an important parameter whenever a continuous flow is concerned. HRT is the amount of time the compound should remain in the functioning vessel for product formation and is calculated by dividing the volume of the vessel with the influent flowrate (Mores et al. 2016). Table 11.3 gives an overview of different operational parameters

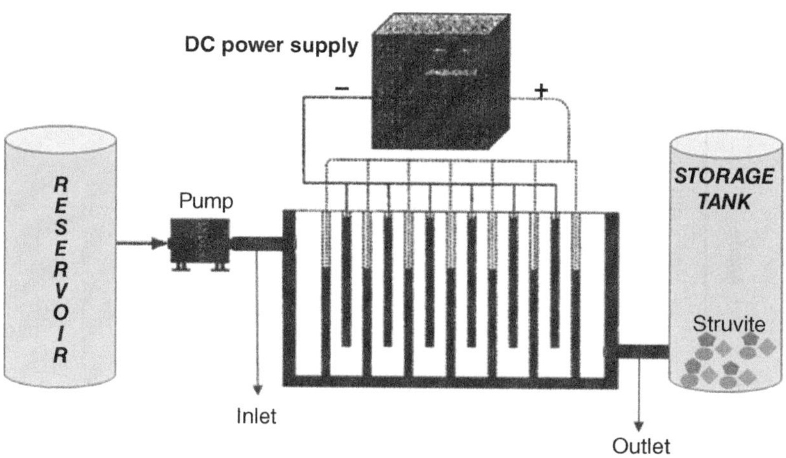

Figure 11.2 Continuous electrocoagulation process for struvite precipitation.

Table 11.3 Various research on continuous electrocoagulation for struvite precipitation.

Wastewater Source	pH	Cathode	Anode	Magnesium Source	Surface Area/ Electrode Volume (cm^2)	Inter-Electrode Distance (cm)	Current/ Voltage	Reaction Time (h)	Flow Rate ($L\,h^{-1}$)	Stirring speed (rpm)	References
Synthetic and authentic peat industry water	7.6–9	Iron	Aluminum	$MgCl_2$	307.2	5 mm	4.7 A	–	8	–	Rajaniemi et al. (2021)
Swine wastewater	–	Titanium with iridium oxide	Stainless steel	$MgCl_2$	400	1	–	4	0.73	3000	Liu et al. (2011a,b)
Swine wastewater	–	Magnesium sacrificial electrode	Carbon	Corrosion from anode	4	3.2	4 V	5	20	–	Wang et al. (2019)

Table 11.4 Dissimilarities between batch and continuous electrocoagulation methods of struvite precipitation.

Parameters	Batch Electrocoagulation	Continuous Electrocoagulation
PO_4^{3-} removal	High	Low
NH_4^+ removal	Low	High
Struvite yield	High	Low
Number of anodes and cathodes	A pair	Several pairs to increase the yield.
Electrode consumption	High	Low
Production cost	High	Low
Reaction time	Low (due to limited wastewater being treated at once).	High (due to continuous supply of wastewater).
Ease of struvite removal	High (all the struvite crystals can be recovered).	Moderate (reactor design restricts complete recovery. The leftover crystals act as seeding agent to increase the yield in consequent cycles).
Power consumption	High (due to repetition of cycles from beginning).	Low
Efficiency	Low	High and authenticated.
Wastewater treatment	Limited wastewater treated at a time.	Large volume of wastewater treated due to continuous supply.

used for struvite precipitation by continuous electrocoagulation. Table 11.4 shows the correlation and contrasts between the types of electrocoagulation.

11.2.2.3 Advantages of Electrocoagulation over Other Methods for Struvite Precipitation

Chemical and biological precipitation are the two other types of precipitation techniques used for struvite production. They include involvement of chemicals in the form of salts or biological organisms for struvite precipitation. The chemical precipitation of struvite involves dosing magnesium salts such as $MgCl_2$, $MgSO_4$, MgO, etc. In biological precipitation, struvite is produced by some selected microorganisms like *Bacillus pumilus, Brevibacterium antiquum, Myxococcus xanthus, Halobacterium salinarum*, and *Idiomarina loihiensis* (Udert et al. 2003; Chong and Thai 2014; Simoes et al. 2018). Table 11.5 states the advantages of electrocoagulation over the other two processes of struvite precipitation.

Table 11.5 Advantages of electrocoagulation over the other two methods of struvite precipitation.

	Electrocoagulation	Chemical Precipitation	Biological Precipitation
Feasibility of decentralization	High	Moderate	Low
Contamination	Low chances of contamination even if viable cells are present as they get reduced by the end of the process.	No chances of contamination.	Chances of contamination are high.
Magnesium source	Corrosion of the electrode or external addition of salt.	External addition of salt.	External addition of salt.
COD removal	Very effective (up to 86%)	Nil	Nil
pH adjustment	Self-adjustment due to release of OH^- ions due to the electrochemical reaction.	Addition of base like NaOH, etc.	Not necessary.
Struvite yield (%)	85–90	80–99	70–80
Struvite purity	High	Depends on of the quality of magnesium salt used.	Moderate
Level of maintenance	Low	Low	High
Production of unwanted compounds	Due to frequent change of electrode, rapid corrosion, active passivation and high electric charge.	Due to sludge containing chemicals.	Due to contamination. The compounds can be toxic to the microorganisms producing struvite.
Use of chemicals	Low requirement for wastewater treatment. No problem of residual chemical.	Required in large amount which can be corrosive.	Low
Influential parameters	Electrode design, pH of medium, conductivity, current density, polarity of electrodes, etc.	pH, temperature, stirring speed, PO_4^{3-}: Mg^{2+} molar ratio, etc.	Temperature, pressure, dosage, etc.

11.3 Influential Parameters Affecting Struvite Crystallization

11.3.1 pH of the Medium

The pH of the solution affects the saturation index and controls the form of existing and competing ions in the solution that will in turn determine precipitation. Struvite is formed when the solubility of all the ions exceeds the solubility product. The solubility of struvite first decreases with an increase in pH, but gradually increases with a further increase in pH. This is because at pH 9 the amount of NH_4^+ ions starts decreasing, leading to an increase in the amount of phosphate. The minimum solubility of struvite ranges between pH of 8 and 10.7 (Ryu et al. 2008; García et al. 2013).

Struvite is insoluble at higher pH, which means that the formation of struvite can also take place at low pH but is in a soluble form, and the quantity is low due to low availability of NH_4^+ ions. At a higher pH (>11), the crystal purity and size reduce due to a decrease in the agglomerative property of the crystals. Thus, the optimum pH for struvite precipitation falls between the pH of 8.9 to 9.3, especially in anaerobic digestors. pH also affects the formation of competitive crystals like magnesium carbonate, magnesium phosphate, magnesium hydroxide, etc. (Kruk et al. 2014). During the process of electrolysis, the pH increases because of the constant OH^- ions being released into the solution, which is beneficial for the commencement of crystal nucleation. However, it is harmful at later stages as it hinders the formation of struvite due to overpowering of other precipitates at higher pH. Air stripping is reported to increase the pH. In many of the studies, HCl is added constantly to maintain the pH (Kruk et al. 2014; Lin et al. 2018).

NH_3 and NH_4^+ are interconnected; hence, with an increase in pH above 8, denitrifications takes place and NH_4^+ is converted into NH_3 gas, which is volatile and thus escapes. However, the amount of resultant NH_4^+ concentration is reported to be sufficient for struvite precipitation to take place (Shim et al. 2020). For synthetic solutions, the initial pH of the solution is kept nearly neutral as during an electrochemical reaction OH^- ions are being released into the solution, which increases the pH and facilitates precipitation. Also, the soluble anode tends to increase the pH of the solution, which further facilitating struvite precipitation (Inan et al. 2013).

11.3.2 Magnesium Source and Mg^{2+}: PO_4^{3-} Molar Ratio

Struvite precipitation occurs when the molar ratio of Mg^{2+}, PO_4^{3-}, and NH_4^+ is adequate to form MAP. For this purpose, Mg^{2+} needs to be added externally to match up with the PO_4^{3-} and NH_4^+, which are present in large amounts as compared to magnesium. The most common method is to directly add the magnesium salts like $MgCl_2$, $MgSO_4$, and MgO. Due to insolubility in water, salts like $MgCO_3$ and $Mg(OH)_2$ are not used in struvite precipitation (Ben Moussa et al. 2006; Wang et al. 2010; Kim et al. 2018). Usually, struvite production is favored when the Mg:P molar ratio is 1:1 (Le Corre et al. 2009; Kabdazsku et al. 2006). However, studies have showed that when the Mg:P >1, the quantity of struvite is much higher without any effect on its quality.

When Mg^{2+} is being added to the solution with the help of an electrode, its amount is calculated by Faraday's law of electrolysis, which states that the mass of Mg^{2+} released into the solution is directly proportional to current supplied. The amount of Mg^{2+} released can be calculated using Eq. (11.3) (Kruk et al. 2014).

$$m = AIT(nF)^{-1} \tag{11.3}$$

where m is the mass of Mg^{2+} released (g); A is the atomic number of Mg (23.4 g mol^{-1}); I is current in ampere (A); T is time in seconds (s); n is the valency of Mg^{2+}; and F is Faraday's constant (96,485 C mol^{-1}).

The PO_4^{3-} removal efficiency increases with an increase in the Mg^{2+}:PO_4^{3-} ratio. Kékedy-Nagy et al. (2020) has observed that the corrosion of the Mg^{2+} anode decreases with the increase of potential, hampering the reaction. In the case of pure Mg^{2+}, the increase in electrode potential occurs at a very slow pace. However, the alloys of magnesium show an increase of potential in the initial hours of the reaction. The presence of other metals in an alloy limits the magnesium ions to form a supersaturated environment, leading to delay of nucleation and crystal growth of struvite.

Magnesium dissolution in the solution occurs due to macro-corrosion, micro-corrosion, or the chunk effect. Macro-corrosion occurs when the magnesium electrode has an anode potential higher than corrosion potential. Micro-corrosion is observed at corrosion potential where the impurities or second phases present in the magnesium anode act as the cathode for the Mg^{2+} matrix. This corrosion occurs at microscale when the electrode is not connected to any other electrode or power supply. The chunk effect is another phenomenon contributing to Mg^{2+} dissolution, as chunks from the electrode into the solution (Hug and Udert 2013).

The magnesium air battery is another source of Mg^{2+} due to its low energy density, safe to use, low toxicity, low cost, and refuellable qualities. The only problem with these batteries is the high polarization and low Coulombic efficiency. The reactions that take place inside these batteries are represented in Eqs (11.4 to 11.6) (Zhang et al. 2014).

$$\text{Anode}: Mg \rightarrow Mg^{2+} + 2e^- \tag{11.4}$$

$$\text{Cathode}: O_2 + 2H_2O + 4e^- \rightarrow 4OH^- \tag{11.5}$$

$$\text{Overall equation}: 2Mg^{2+} + O_2 + 2H_2O \rightarrow 2Mg(OH)_2 \tag{11.6}$$

11.3.3 Current Density

Current density is an important operating parameter of electrocoagulation, as it determines the corrosion of the anode and increases with an increase in current density. It also helps in determining the usage of specific energy and the consumption of the electrode, eventually contributing to determine the operating cost of the process. With an increase in current density, the purity of struvite formed increases. Current density can be calculated using (Eq. 11.7) (Emamjomeh and Sivakumar 2009; Mohora et al. 2012; Kruk et al. 2014).

$$CD = \frac{V}{RA} \tag{11.7}$$

where CD is current density (A m^{-2}); V is voltage in volt; R is external resistance in ohm; and A is area in cm^2. Current density is directly related to the amount of magnesium released into the solution according to Eq. (11.3). Hence, with an increase in the current density, current intensity increases, leading to an increase in Mg^{2+} dissolution. But this reaction will continue to increase for a only certain range, after which it will decrease. The possible reason for the decrease in dissolution of Mg^{2+} is the formation of a thin film over the anode, resulting in reduction of struvite precipitation (Lin et al. 2018).

Current density and PO_4^{3-} removal can increase with an increase in anode potential. For a cost-effective wastewater treatment, the current density should be kept low, as a high current density set-up will require more maintenance and will increase the cost of production (Hug and Udert 2013).

11.3.4 Voltage and Current Efficiency

Current efficiency is the ratio of the actual amount of Mg^{2+} dissolved and theorical value of Mg^{2+} dissolution as shown in Eq. (11.8) (Kim et al. 2018).

$$CE = \frac{M_{diss}}{M_{theo}} \times 100 \tag{11.8}$$

where M_{diss} is the mass of Mg^{2+} dissolved when a particular amount of current is passed through it; and M_{theo} is the theoretical value of Mg ion dissolution calculated by Faraday's Law Eq. (11.3). M_{diss} can be calculated from Eq. (11.9) (Kim et al. 2018; Lin et al. 2018).

$$M_{diss} = \Delta P(t) \times \frac{M_{mg}}{M_P} + m(t) - m(0) \tag{11.9}$$

where $\Delta P(t)$ is the difference in PO_4^{3-} mass (mg) before and after the struvite precipitation; M_{mg} is the molar mass of Mg (24.3 g mol^{-1}); M_p is the molar mass of P (31 g mol^{-1}); $m(t)$ is the mass of Mg (mg) at time t; and $m(0)$ is the initial mass of Mg (mg) (Lin et al. 2018).

The current efficiency is generally >100%, indicating that the dissolution of Mg^{2+} also takes place without the help of current. Increase in magnesium corrosion due to application of current is known as the negative difference effect (NDE) where anodic and cathodic reactions simultaneously occur at the anode (Zhang et al. 2014). The higher the current efficiency, the higher will be the struvite precipitation (Kim et al. 2018).

Song (2005) reported the presence of an intermediate Mg ion species during the process of dissolution. At the anode, Mg^+ is being produced, which is an intermediate species and leads to formation of Mg^{2+} on the further oxidation process along with the production of OH^- ion and hydrogen gas. Similarly, at the cathode, by redox and hydrogen evolution reactions, OH^- ions and hydrogen gas are being produced (Kékedy-Nagy et al. 2020). As both Mg^+ and Mg^{2+} are being formed during the reaction, the value of n in Eq. (11.3) may be 1 or 2. It was also observed that if the current density is higher, the value of n will be 1. The overall reactions taking place are summarized in Eqs (11.10 to 11.13) (Song 2005).

$$\text{Anode}: Mg - e^- \rightarrow Mg^+ \tag{11.10}$$

$$Mg^+ + H_2O \rightarrow Mg^{2+} + OH^- + \frac{1}{2}H_2 \tag{11.11}$$

$$\text{Cathode}: O_2 + 2H_2O + 4e^- \rightarrow 4OH^- \tag{11.12}$$

$$H_2O + 2e^- \rightarrow \frac{1}{2}H_2 + 2OH^- \tag{11.13}$$

The hydroxides produced by the cathode may lead to formation of bubbles that create turbulence in the solution and cause proper mixing of the compounds (Mores et al. 2016). Ionic conductivity of the solution also plays an important role in determining the current passing through the solution and further affects the current efficiency, the voltage applied, and the energy consumed by the system. Sometimes, salts like NaCl are added to the wastewater system to increase the efficiency of the electrolysis reaction (Wang et al. 2010; Liu et al. 2011a,b).

Deposition of $MgCO_3$ and $Mg(OH)_2$ at the surface of the Mg electrode occurs at a higher anode potential due to passivation, decreasing the current and affecting the reaction. To avoid the process of passivation, the potential should be kept above the passivation potential, which varies with the electrolyte solution and the magnesium alloy. As the presence of OH^- ions increases, the pH of the solution escalates during electrocoagulation (Hug and Udert 2013).

11.3.5 Electrode Type and Interelectrode Distance

In addition to Mg, the types of electrodes used in the struvite precipitation process can be iron, aluminum, titanium, or iridium. In the case of other metals being used, Mg salts are added externally for the formation of struvite. Mg electrode as an anode will be best suited for struvite precipitation by the method of electrocoagulation. The dissolving rate of magnesium from the anode into the solution can be calculated by Eq. (11.14) (Lin et al. 2018).

$$V_d(t) = \frac{M_{diss}(t)}{tA} \tag{11.14}$$

where $V_d(t)$ is the dissolving rate of magnesium at time t (mg cm^{-2} h^{-1}); $M_{diss}(t)$ is calculated by Eq. (11.3); t is the time duration in which the electrolysis occurs in hours; and A is the active surface area of the electrode in cm^2.

Deposition of struvite and other salts such as $MgCO_3$, $Mg(OH)_2$, newberryite, ferrous and ferric hydroxide, aluminum hydroxide, aluminum phosphate dihydrate, ferric phosphate dihydrate, etc. on the surface of the anode restricts the corrosion of anode and thus the efficiency to form struvite. It also affects the color of struvite due to an increase in thermodynamic losses. This can be prevented by the addition of chloride ions in the electrolyte to increase the current density and thus struvite precipitation (Kruk et al. 2014; Kim et al. 2018; Kékedy-Nagy et al. 2020).

Interelectrode distance plays a crucial role in maintaining the energy consumption in a reactor. The amount of energy consumed in a process triples with an increase in the gap between the electrodes. The lower the gap between the electrodes, the lower will be the energy consumption and thus the cost of the process. Additionlly, a smaller electrode gap is desirable because with an increase in gap there is an increase in the resistance, which will further delay corrosion of anode, leading to low productivity of struvite. However, the

Table 11.6 Electrode type, surface area, and interelectrode distance for struvite precipitation.

Electrode type						
Anode	**Cathode**	**Electrode Dimensions**	**Interelectrode Distance (cm)**	**Wastewater Source**	**Source of Magnesium**	**References**
Aluminum	Ti/RuO_2	–	1	Dairy industry effluent	$MgCl_2$	García et al. (2013)
Magnesium	Stainless steel	SA: 45 cm^2; T: 3 mm	3.5	Swine wastewater	Corrosion from anode	Lin et al. (2018)
Titanium plate coated with iridium oxide	Stainless steel	SA: 400 cm^2; T: 1 mm	1	Swine wastewater	$MgCl_2$	Liu et al. (2011a,b)
Magnesium	Stainless steel	SA: 80 cm^2	2	Synthetic urine	Corrosion from anode	Huang et al. (2019)

(SA = Surface area; T = Thickness)

gap should be enough to allow for proper mixing of the reactants to form the precipitate (Mohora et al. 2012; Hug and Udert 2013). Table 11.6 presents the list of electrodes being used for electrocoagulation of struvite from various wastewater sources.

11.3.6 Stirring Speed, Reaction Time, and Seeding

Struvite can be formed naturally without the help of stirring, but the size of crystals remains very small. To increase the size of the crystals at a given reaction time, stirring of the reactants is necessary for enhancing mass transfer (González-Morales et al. 2019). It is seen that during the reaction process an increase in pH occurs in the close vicinity of the electrode. Due to stirring, this pH becomes uniformly distributed throughout the bulk solution (Ben Moussa et al. 2006).

The reaction starts immediately after the current passes through the electrode, and dissolution of Mg^{2+} into the solution is observed. Most of the struvite precipitation takes place in the first 0.5 h of the experiment, after which several factors like passivation and increase of solution pH begin to affect the struvite precipitation. However, Kékedy-Nagy et al. (2020) reported that struvite production increases after a certain period due to the formation of pits, which helps to increase the active area of the Mg electrode. Hence, the reaction has to continue for several hours, to obtain a maximum yield from the reactant wastewater. Decreasing the pH may again lead to certain altercations in the reaction but increasing the reaction time to more than 1.5 h might not be very beneficial as the oxidation of NH_4^+–N also takes place, converting NH_4^+ to NH_3, and further to N_2. Increase of pH is also responsible for conversion of NH_4^+ to N_2 gas.

Seeding is done in the form of adding small crystals of struvite into the reactor, which tends to increase the adsorption of PO_4^{3+}–P and size of the struvite crystals. Decomposition of struvite and then using it for the recycling helps to increase the yield as more amounts of PO_4^{3-} and Mg^{2+} are provided to the solution, which also reduces the cost of Mg. Additionally, it is suitable for removal of NH_4^+ ions from the crystals, which would reduce the extra cost of the associated treatment. There are many methods for the decomposition purpose like acidolysis, pyrolysis, electrolysis, etc. Reusing PO_4^{3-} and Mg^{2+} after decomposition has been shown to increase the yield and purity of other struvite crystals. Also, the recovery of PO_4^{3-} increased after this process. Decomposition is influenced by pH of the solution; usually pH 6 is considered good for the decomposition process. Also, the Cl:N ratio should be around 8.5 for better performance (Wang et al. 2010; Kim et al. 2007; Liu et al. 2011a,b).

11.3.7 Presence of Competitive Ions and Purity of Struvite Crystals

The production of struvite can be influenced by the presence of various ions like K^+, Na^+, Ca^{2+}, CO_3^-, HCO_3^-, OH^-, Al^{3+}, Fe^{3+}, etc. that affect the morphology, structure, and induction time of the crystals (Di Iaconi et al. 2010; Pastor et al. 2010; Siliciano 2015). The monovalent ions like K^+ and Na^+, when present in the solution, compete with NH_4^+ ions by binding with the other two ions to form $MgNaPO_4$ (Na-struvite) and $MgKPO_4$ (K-struvite). However, it was observed that $MgNaPO_4$ is formed only when Na^+ is present in a large quantity in the solution and vice versa. Even in the case of $MgKPO_4$, production is low in the presence of excess NH_4^+ ions. However, certain studies have concluded the co-precipitation of around 30 to 40% of these ions along with struvitte (Liu et al. 2011a,b).

The presence of CO_3^{2-}, HCO_3^-, and OH^- ions corresponds to the alkalinity of the solution. These ions readily react with Mg^{2+} and NH_4^+ ions, making them unavailable for struvite formation. $MgCO_3$, $Mg(HCO_3)_2$, $Mg(OH)_2$, $(NH_4)_2CO_3$, and NH_4HCO_3 are some of the components that can possibly form in the medium. The presence of mainly Mg^{2+} compounds have more effect in the precipitation process, except for $MgCO_3$ and $Mg(OH)_2$ as they cause passivation leading to a decrease of Mg^{2+} in the solution (Siliciano 2015; Huang et al. 2017).

The calcium ions compete with Mg^{2+} to combine with PO_4^{3-} to form hydroxyapatite and other calcium phosphates. It has been concluded in various studies that when Ca^{2+}:Mg^{2+} is <0.2, the presence of calcium ions has a negligible effect on struvite precipitation; but when its ratio is >0.2, it affects the nucleation process by combining with PO_4^{3-}, making them unavailable for struvite formation. In order to avoid this, the Mg^{2+} quantity is increased, but excess of Mg^{2+} ions also results in the formation of magnesium phosphates that hinders the formation of struvite. Some studies suggest chelation of calcium ions with the help of certain chemicals like ethylenediamine tetraacetic acid (EDTA), which has a stability constant two times greater for calcium than magnesium, allowing it to reduce calcium ion concentration from wastewater without affecting phosphorus concentration (Siciliano et al. 2020). However, both the processes of chelating and increasing Mg^{2+} concentration increase the cost of the operation. Calcium ion can also be diverted to combine with carbonate ion to form $CaCO_3$, which will not only decrease the calcium ion concentration and make PO_4^{3-} readily available but also decrease the alkalinity of the solution (Ryu et al. 2008; Tao et al. 2016). Sometimes, some heavy metals also precipitate along with struvite, but their range remains under permissible limits. Cu^{2+}, Zn^{2+}, and Cd^{2+} are some of the commonly found heavy metals that precipitate along with struvite, but their concentration can easily be minimized by increasing the pH (Liu et al. 2011a, b; Huang et al. 2019).

Purity of struvite is assumed as the production of 1 mole struvite from 1 mole of NH_4^+ ion in the solution. The purity can be calculated from Eq. (11.15) (Kruk et al. 2014).

$$Struvite\ purity = [NH_4 - N]_{precipitate} \cdot [NH_4 - N]_{struvite} \tag{11.15}$$

where $[NH_4\text{–}N]_{precipitate}$ is the amount of ammonium in the precipitate (mg N g^{-1}); and $[NH_4\text{–}N]_{struvite}$ is the theoretical amount of ammonium present in struvite (57 mg N g^{-1}).

The mineral impurities present in struvite do not contain N. Molecules such as $Mg(OH)_2$, $Mg_3(PO_4)_2$, $MgHPO_4$, $MgKPO_4$, $CaHPO_4$, etc. also lack the presence of the NH_4^+ ion. Thus, even in the presence of these impurities, the struvite produced will be considered as pure (Zeng and Li 2006; Hao et al. 2008; Kruk et al. 2014). Similarly, the amount of recovered PO_4^{3-} can be determined using Eq. (11.16) (Lin et al. 2018).

$$Phosphate\ recovery = \frac{P_o - P_t}{P_o} \times 100 \tag{11.16}$$

where P_0 is the initial amount of phosphate (mg L^{-1}) present in the solution before electrolysis; and P_t is the amount of phosphate (mg L^{-1}) present at time t after the electrolysis experiment (Lin et al. 2018).

There are many methods of confirming the presence and quality of struvite such as Fourier Transform infrared spectroscopy (FTIR), X-ray diffraction (XRD), Scanning electron microscope-Energy dispersive X-ray spectroscopy (SEM-EDX), and Raman spectroscopy. Thermo-gravitometric analysis (TGA) and Differential thermal analysis (DTA) are used for assessment of the thermal stability of the crystal. The Brunauer–Emmett–Teller (BET) method helps in surface analysis of the crystal and gives information regarding the pores, micropores, and nitrogen absorption and desorption isotherm. In the case of Raman spectroscopy, a peak at 942 cm^{-1} confirms the PO_4^{3-} symmetrical stretching band in struvite (Frost et al. 2005; Heraldy et al. 2017). With the help of SEM-EDX, the atomic percentage of the crystal and its shape can be determined. Usually, in the presence of high ionic concentrations of the reactants, the kinetics increase, resulting in oval- or round-shaped crystals, but in the case of low ionic concentration, rod-shaped crystals have been observed (Manzoor et al. 2019).

11.4 Energy, Economy, and Environmental Contribution of Struvite Precipitation by Electrocoagulation

The operating cost of wastewater treatment includes the charge of electricity, labor, sludge disposal, maintenance, and equipment. However, when a product recovery is combined with the process, the overall economy becomes balanced. In the case of electrocoagulation of struvite, the only major costs will be due to the electrical energy supplied and electrodes. The electrodes can also be reused, further minimizing the cost (Inan et al. 2013).

The cost of magnesium electrode is the same as $MgCl_2$ and $MgSO_4$, but is found to be costlier than MgO. MgO is generally mixed with polyvinyl alcohol and then sent in by a screw feeder for its use; but electrochemical Mg^{2+} does not require any prior treatment and is easier to dissolve as compared to MgO. The quality of the struvite formed by an electrochemical reaction is also higher than that formed from MgO (Shen et al. 2015). Continuous supply of Mg^{2+} through electrocoagulation provides a higher struvite recovery rate as compared to processes with salt being added at the same time because continuous supply reduces the supersaturation and increases the size of the crystals, which makes them easier to filter and store (Hug and Udert 2013). Table 11.7 compares the total costs of struvite produced from different sources.

Although the electrocoagulation reaction needs input of energy, it is not very high as the oxidation of the Mg^{2+} sacrificial anode is a spontaneous process. It occurs because Mg being a reactive metal has a high electrode potential (Kékedy-Nagy et al. 2020). Equation (11.17) can be used to calculate the operational cost through the electrocoagulation method (Kuokkanen et al. 2015).

$$Operational\ cost = (a \times EEC) + (b \times EMC) + (c \times SEC) \tag{11.17}$$

where *a*, *b*, and *c* are the current market prices of electricity, electrode material, and the supporting electrolyte that are used in the process, respectively; *EMC* is electrode material consumption; *SEC* is the supporting electrolyte consumption; and *EEC* is electrical energy consumption, which can be calculated using Eq. (11.18) (Kuokkanen et al. 2015).

$$EEC = U \times I \times T(60 \times V)^{-1} \tag{11.18}$$

Table 11.7 Commercial cost of struvite produced from various methods of crystallization.

Wastewater Source	Type of Electrocoagulation	Source of Magnesium	Cost per kg of Struvite	References
Synthetic urine	Batch	$MgCl_2.6H_2O$	3.61 USD	Huang et al. (2019)
		$Mg\ (OH)_2$	6.28 USD	
		Mg sacrificial anode	3.99 USD	
Human urine	Batch and sequencing	Mg anode	4.45 EUR	Hug and Udert (2013)
		MgO	0.90 EUR	
		$MgCl_2.6H_2O$	6.80 EUR	
		$MgSO_4$	3.45 EUR	
Slurry wastewater	Batch	Mg anode	81.57 USD	Lin et al. (2018)
		MgO	31.2 USD	
		$MgCl_2.6H_2O$	98.6 USD	
Swine wastewater	Batch	Mg anode	11.34 USD	(Liu et al. 2011a)
		$MgCl_2.6H_2O$	32.4 USD	
Swine wastewater	Continuous	Electrolyser	1.45 CNY	Wang et al. (2019)
		Mg sacrificial anode	1.43 CNY	

where *U* is voltage in volts; *I* is current in ampere; *T* is the treatment time in minutes; and *V* is the volume of wastewater used in dm^3.

Though struvite production is considered economically beneficial, the environmental benefits have gained more attention in recent years leading to further research and assessment. The production of struvite helps in recovering a large amount of P from different wastewaters, reducing the reliability on the chemical phosphates. Struvite precipitation has the capability to simultaneously remove several nutrients from wastewater by forming 12 different types of analogs like K-struvite, Na-struvite, etc. (Zangarini et al. 2020).

Electrocoagulation works on almost all types of waterbodies, which when being treated chemically, can change the overall constituent of water. The amount of sludge that is being produced after chemical treatment can be very harmful which, upon inappropriated dumping, can lead to soil and groundwater pollution (Inan et al. 2013; Mores et al. 2016).

Technological advances have also led to the use of this system to attain more benefits in terms of nutrient recovery and treatment of wastewater. This system can also be combined with the ion exchange process to remove extra ions present in the wastewater. The combination of electrocoagulation with bioelectric systems (e.g., MFC and MEC) not only recovers struvite but also helps in the production of electricity without any harmful product formation, reducing the dependency on non-renewable sources of energy.

11.5 Summary and Future Perspectives

Electrocoagulation can be regarded as an efficient, more reliable, and eco-friendly technology for struvite production. The key points that can be concluded from this chapter are as follows:

- Both batch and continuous electrocoagulation have their own merits and demerits and are equally useful, but continuous electrocoagulation can be considered as a very good technology for large-scale development considering short reaction time.
- pH value from 8 to 10 is considered optimum for struvite precipitation. However, in the case of electrocoagulation, there is no need of chemical addition for pH control as the cathode releases OH^- ions.
- A higher molar ratio of Mg^{2+}:PO_4^{3-} favors the increase of yield, irrespective of the source of magnesium (salt or anode).
- Current density is an important parameter as it controls the corrosion of the anode, energy consumption, and the cost of the process.
- Although voltage and current efficiency are important, electrocoagulation is largely affected by current density, which should be >100% for dissolution of the anode.
- Mg as an anode is usually preferred over others as its dissolution will directly supply Mg^{2+} required for struvite, decreasing the cost of production. Minimum interelectrode distance is preferred as it facilitates electron supply and proper mixing of the ions.
- Stirring speed helps in distribution of ions and maintenance of pH. The reaction time depends on the type of electrocoagulation, current density, and the electrode used. Seeding can increase the overall PO_4^{3-} recovery and yield.

- K^+, Na^+, Ca^{2+}, CO_3^-, HCO_3^-, OH^-, Al^{3+}, Fe^{3+}, etc. are the usual competitive ions and can be easily controlled in other ways such as pH adjustments. The heavy metals deposited along with struvite are very limited compared to the commercial fertilizers and found to be below the permissible limits.
- Most struvite analogs lack ammonium, still promising the purity and quality of struvite produced.

The research gaps that must be addressed to overcome some of the limitations of this process as well as increase the efficiency and cost benefits are:

- A reliable method to avoid passivation or cleaning of the anode should be researched to increase the life of the anode for better production of struvite and reduction of process cost.
- Foaming can be another issue, which can be solved by adding anti-foaming agents without compromising the quality and quantity of struvite produced. However, this might further add to the production costs.
- Using renewable energy for electrocoagulation can be another way to reduce the operational cost.
- Collecting the released gases like ammonia, hydrogen, etc. generated during the struvite formation process can further add advantages and a new source of income to the overall process.

Acknowledgments

The authors thank the Department of Biotechnology and Medical Engineering of the National Institute of Technology Rourkela for providing the necessary research facilities. The authors greatly acknowledge the Science and Engineering Research Board (SERB) of the Department of Science and Technology (DST) and Government of India (GoI) for sponsoring the research through Technology Development Programme [File No. DST/TDT/Agro43/2020].

References

Badruzzaman, M., Oppenheimer, J.A., and Jacangelo, J.G. (2013). Impact of environmental conditions on the suitability of microconstituents as markers for determining nutrient loading from reclaimed water. *Water Research* 47 (16): 6198–6210. doi: 10.1016/j.watres.2013.07.029.

Bakshi, A., Verma, A.K., and Dash, A.K. (2020). Electrocoagulation for removal of phosphate from aqueous solution: statistical modeling and techno-economic study. *Journal of Cleaner Production* 246: 118988. doi: 10.1016/j.jclepro.2019.118988.

Bastida, F., Jehmlich, N., Martínez-Navarro, J. et al. (2019). The effects of struvite and sewage sludge on plant yield and the microbial community of a semiarid Mediterranean soil. *Geoderma* 337: 1051–1057. https://doi.org/10.1016/j.geoderma.2018.10.046.

Ben Moussa, S., Maurin, G., Gabrielli, C., and Ben Amor, M. (2006). Electrochemical precipitation of struvite. *Electrochemical and Solid-State Letters* 9 (6): C97. doi: 10.1149/1.2189222.

Bhuiyan, M.I.H., Mavinic, D., and Koch, F. (2008). Thermal decomposition of struvite and its phase transition. *Chemosphere* 70 (8): 1347–1356. doi: 10.1016/j.chemosphere.2007.09.056.

Chong, N.M. and Thai, Q.M. (2014). Optimization and kinetics of nutrient removal from wastewater by chemical precipitation of struvite. *Desalination and Water Treatment* 54 (12): 3422–3431. doi: 10.1080/19443994.2014.908416.

Das, P., Gupta, G., Velu, V. et al. (2017). Formation of struvite urinary stones and approaches towards the inhibition: a review. *Biomedicine & Pharmacotherapy* 96: 361–370. doi: 10.1016/j.biopha.2017.10.015.

Di Iaconi, C., Pagano, M., Ramadori, R., and Lopez, A. (2010). Nitrogen recovery from a stabilized municipal landfill leachate. *Bioresource Technology* 101 (6): 1732–1736. doi: 10.1016/j.biortech.2009.10.013.

Doyle, J.D. and Parsons, S.A. (2002). Struvite formation, control and recovery. *Water Research* 36 (16): 3925–3940. doi: 10.1016/s0043-1354(02)00126-4.

Emamjomeh, M.M. and Sivakumar, M. (2009). Fluoride removal by a continuous flow electrocoagulation reactor. *Journal of Environmental Management* 90 (2): 1204–1212. doi: 10.1016/j.jenvman.2008.06.001.

Food and Agricultural Organization (FAO) (2017). Agriculture Organization of the United Nations. In: *World Fertilizer Trends and Outlook to 2020. Rome.* https://www.fao.org/3/i6895e/i6895e.pdf (accessed 1 May 2021).

Frost, R.L., Weier, M.L., Martens, W.N. et al. (2005). Raman spectroscopy of newberyite, hannayite and struvite. *Spectrochimica Acta Part A: Molecular and Biomolecular Spectroscopy* 62 (1–3): 181–188. doi: 10.1016/j.saa.2004.12.024.

García, F.P., Hernández, J.C., Reyes Cruz, V.E. et al. (2013). Recovery and characterization of struvite from sediment and sludge resulting from the process of acid whey electrocoagulation. *Asian Journal of Chemistry* 25 (14): 8005–8009. doi: 10.14233/ajchem.2013.14933.

González-Morales, C., Camargo-Valero, M.A., Molina-Pérez, F.J., and Fernández, B. (2019). Effect of the stirring speed on the struvite formation using the centrate from a WWTP. *Revista Facultad de Ingeniería Universidad de Antioquia* 92: 42–50.

Hao, X.D., Wang, C.C., Lan, L., and van Loosdrecht, M.C.M. (2008). Struvite formation, analytical methods and effects of pH and Ca2+. *Water Science and Technology* 58 (8): 1687–1692. doi: 10.2166/wst.2008.557.

Heraldy, E., Rahmawati, F., Heriyanto, and Putra, D.P. (2017). Preparation of struvite from desalination waste. *Journal of Environmental Chemical Engineering* 5 (2): 1666–1675. doi: 10.1016/j.jece.2017.03.005.

Huang, H., Li, B., Li, J. et al. (2019). Influence of process parameters on the heavy metal (Zn2+, Cu2+ and Cr3+) content of struvite obtained from synthetic swine wastewater. *Environmental Pollution* 245: 658–665. doi: 10.1016/j.envpol.2018.11.046.

Huang, H., Zhang, D., Zhao, Z. et al. (2017). Comparison investigation on phosphate recovery from sludge anaerobic supernatant using the electrocoagulation process and chemical precipitation. *Journal of Cleaner Production* 141: 429–438. doi: 10.1016/j.jclepro.2016.09.127.

Hug, A. and Udert, K.M. (2013). Struvite precipitation from urine with electrochemical magnesium dosage. *Water Research* 47 (1): 289–299. doi: 10.1016/j.watres.2012.09.036.

Inan, H. and Alaydın, E. (2013). Phosphate and nitrogen removal by iron produced in electrocoagulation reactor. *Desalination and Water Treatment* 52 (7–9): 1396–1403. doi: 10.1080/19443994.2013.787950.

Kabdazsli, I., Parsons, S.A., and Tünay, O. (2006). Effect of major ions on induction time of struvite precipitation. *Croatica Chemica Acta* 79 (2): 243–251.

Kékedy-Nagy, L., Abolhassani, M., Perez Bakovic, S.I. et al. (2020). Electroless production of fertilizer (struvite) and hydrogen from synthetic agricultural wastewaters. *Journal of the American Chemical Society* 142 (44): 18844–18858. doi: 10.1021/jacs.0c07916.

Kim, D., Ryu, H.D., Kim, M.S. et al. (2007). Enhancing struvite precipitation potential for ammonia nitrogen removal in municipal landfill leachate. *Journal of Hazardous Materials* 146 (1–2): 81–85.

Kim, J.H., An, B.M., Lim, D.H., and Park, J.Y. (2018). Electricity production and phosphorous recovery as struvite from synthetic wastewater using magnesium-air fuel cell electrocoagulation. *Water Research* 132: 200–210. doi: 10.1016/j.watres.2018.01.003.

Krishnamoorthy, N., Arunachalam, T., and Paramasivan, B. (2021). A comparative study of phosphorus recovery as struvite from cow and human urine. *Materials Today: Proceedings 47: 391–395.*.

Krishnamoorthy, N., Dey, B., Arunachalam, T., and Paramasivan, B. (2020). Effect of storage on physicochemical characteristics of urine for phosphate and ammonium recovery as struvite. *International Biodeterioration & Biodegradation* 153: 105053.

Krishnamoorthy, N., Dey, B., Unpaprom, Y. et al. (2021a). Engineering principles and process designs for phosphorus recovery as struvite: a comprehensive review. *Journal of Environmental Chemical Engineering* 9 (5): 105579. https://doi.org/10.1016/j.jece.2021.105579)

Krishnamoorthy, N., Zaffar, A., Arunachalam, T. et al. (2021b). Municipal wastewater as a potential resource for nutrient recovery as struvite. In: *Urban Mining for Waste Management and Resource Recovery*, 187–215. CRC Press.

Kruk, D.J., Elektorowicz, M., and Oleszkiewicz, J.A. (2014). Struvite precipitation and phosphorus removal using magnesium sacrificial anode. *Chemosphere* 101: 28–33. doi: 10.1016/j.chemosphere.2013.12.036.

Kumar, N.S. and Goel, S. (2010). Factors influencing arsenic and nitrate removal from drinking water in a continuous flow electrocoagulation (EC) process. *Journal of Hazardous Materials* 173 (1–3): 528–533. doi: 10.1016/j.jhazmat.2009.08.117.

Kuokkanen, V., Kuokkanen, T., Rämö, J. et al. (2015). Removal of phosphate from wastewaters for further utilization using electrocoagulation with hybrid electrodes: techno-economic studies. *Journal of Water Process Engineering* 8: e50–e57. doi: 10.1016/j.jwpe.2014.11.008.

Le Corre, K.S., Valsami-Jones, E., Hobbs, P., and Parsons, S.A. (2005). Impact of calcium on struvite crystal size, shape and purity. *Journal of Crystal Growth* 283 (3–4): 514–522. doi: 10.1016/j.jcrysgro.2005.06.012.

Le Corre, K.S., Valsami-Jones, E., Hobbs, P., and Parsons, S.A. (2009). Phosphorus recovery from wastewater by struvite crystallization: a review. *Critical Reviews in Environmental Science and Technology* 39 (6): 433–477. doi: 10.1080/10643380701640573.

Lin, X., Han, Z., Yu, H. et al. (2018). Struvite precipitation from biogas digestion slurry using a two-chamber electrolysis cell with a magnesium anode. *Journal of Cleaner Production* 174: 1598–1607. doi: 10.1016/j.jclepro.2017.10.224.

Liu, Y., Kwag, J.H., Kim, J.H., and Ra, C. (2011a). Recovery of nitrogen and phosphorus by struvite crystallization from swine wastewater. *Desalination* 277 (1–3): 364–369. doi: 10.1016/j.desal.2011.04.056.

Liu, Y., Kumar, S., Kwag, J. et al. (2011b). Recycle of electrolytically dissolved struvite as an alternative to enhance phosphate and nitrogen recovery from swine wastewater. *Journal of Hazardous Materials* 195: 175–181. doi: 10.1016/j.jhazmat.2011.08.022.

Manzoor, M.A., Mujeeburahiman, M., Duwal, S.R., and Rekha, P. (2019). Investigation on growth and morphology of *in vitro* generated struvite crystals. *Biocatalysis and Agricultural Biotechnology* 17: 566–570. doi: 10.1016/j.bcab.2019.01.023.

Mohora, E., Rončević, S., Dalmacija, B. et al. (2012). Removal of natural organic matter and arsenic from water by electrocoagulation/flotation continuous flow reactor. *Journal of Hazardous Materials* 235–236: 257–264. doi: 10.1016/j.jhazmat.2012.07.056.

Mores, R., Treichel, H., Zakrzevski, C.A. et al. (2016). Remove of phosphorous and turbidity of swine wastewater using electrocoagulation under continuous flow. *Separation and Purification Technology* 171: 112–117. doi: 10.1016/j.seppur.2016.07.016.

Nageshwari, K. and Balasubramanian, P. (2022). Evolution of struvite research and the way forward in resource recovery of phosphates through scientometric analysis. *Journal of Cleaner Production* 357: 131737.

Naje, A.S., Chelliapan, S., Zakaria, Z. et al. (2017). A review of electrocoagulation technology for the treatment of textile wastewater. *Reviews in Chemical Engineering* 33 (3). doi: 10.1515/revce-2016-0019.

Pastor, L., Mangin, D., Ferrer, J., and Seco, A. (2010). Struvite formation from the supernatants of an anaerobic digestion pilot plant. *Bioresource Technology* 101 (1): 118–125. doi: 10.1016/j.biortech.2009.08.002.

Perez-Ameneiro, M., Vecino, X., Vega, L. et al. (2013). Elimination of micronutrients from winery wastewater using entrapped grape marc in alginate beads. *CyTA – Journal of Food* 12 (1): 73–79. doi: 10.1080/19476337.2013.797923.

Phalakornkule, C., Polgumhang, S., and Tongdaung, W. (2009). Performance of an electrocoagulation process in treating direct dye: batch and continuous upflow processes. *International Journal of Chemical and Molecular Engineering* 3 (9): 499–504.

Rajaniemi, K., Hu, T., Nurmesniemi, E.T. et al. (2021). Phosphate and ammonium removal from water through electrochemical and chemical precipitation of struvite. *Processes* 9 (1): 150. doi: 10.3390/pr9010150.

Ryu, H.D., Kim, D., and Lee, S.I. (2008). Application of struvite precipitation in treating ammonium nitrogen from semiconductor wastewater. *Journal of Hazardous Materials* 156 (1–3): 163–169. doi: 10.1016/j.jhazmat.2007.12.010.

Shen, Y., Ye, Z.L., Ye, X. et al. (2015). Phosphorus recovery from swine wastewater by struvite precipitation: compositions and heavy metals in the precipitates. *Desalination and Water Treatment* 57 (22): 10361–10369. doi: 10.1080/19443994.2015.1035342.

Shim, S., Reza, A., Kim, S. et al. (2020). Simultaneous removal of pollutants and recovery of nutrients from high-strength swine wastewater using a novel integrated treatment process. *Animals* 10 (5): 835. doi: 10.3390/ani10050835.

Siciliano, A. (2015). Assessment of fertilizer potential of the struvite produced from the treatment of methanogenic landfill leachate using low-cost reagents. *Environmental Science and Pollution Research* 23 (6): 5949–5959. doi: 10.1007/s11356-015-5846-z.

Siciliano, A., Limonti, C., Curcio, G.M., and Molinari, R. (2020). Advances in struvite precipitation technologies for nutrients removal and recovery from aqueous waste and wastewater. *Sustainability* 12 (18): 7538. doi: 10.3390/su12187538.

Simoes, F., Vale, P., Stephenson, T., and Soares, A. (2018). The role of pH on the biological struvite production in digested sludge dewatering liquors. *Scientific Reports* 8 (1). doi: 10.1038/s41598-018-25431-7.

Song, G. (2005). Recent progress in corrosion and protection of magnesium alloys. *Advanced Engineering Materials* 7 (7): 563–586. doi: 10.1002/adem.200500013.

Szymańska, M., Szara, E., Wąs, A. et al. (2019). Struvite: an innovative fertilizer from anaerobic digestate produced in a bio-refinery. *Energies* 12 (2): 296. doi: 10.3390/en12020296.

Tao, W., Fattah, K.P., and Huchzermeier, M.P. (2016). Struvite recovery from anaerobically digested dairy manure: a review of application potential and hindrances. *Journal of Environmental Management* 169: 46–57. doi: 10.1016/j.jenvman.2015.12.006.

Udert, K., Larsen, T., and Gujer, W. (2003). Biologically induced precipitation in urine-collecting systems. *Water Supply* 3 (3): 71–78. doi: 10.2166/ws.2003.0010.

Wang, J., Burken, J G., and Zhang, X.J. (2006). Effect of seeding materials and mixing strength on struvite precipitation. *Water Environment Research* 78 (2): 125–132. https://doi.org/10.2175/106143005x89580.

Wang, C.C., Hao, X.D., Guo, G.S., and van Loosdrecht, M. (2010). Formation of pure struvite at neutral pH by electrochemical deposition. *Chemical Engineering Journal* 159 (1–3): 280–283. doi: 10.1016/j.cej.2010.02.026.

Wang, F., Fu, R., Lv, H. et al. (2019). Phosphate recovery from swine wastewater by a struvite precipitation electrolyzer. *Scientific Reports* 9 (1). doi: 10.1038/s41598-019-45085-3.

Zangarini, S., Pepè Sciarria, T., Tambone, F., and Adani, F. (2020). Phosphorus removal from livestock effluents: recent technologies and new perspectives on low-cost strategies. *Environmental Science and Pollution Research* 27 (6): 5730–5743. doi: 10.1007/s11356-019-07542-4.

Zeng, L. and Li, X. (2006). Nutrient removal from anaerobically digested cattle manure by struvite precipitation. *Journal of Environmental Engineering and Science* 5 (4): 285–294. doi: 10.1139/s05-027.

Zhang, T., Tao, Z., and Chen, J. (2014). Magnesium–air batteries: from principle to application. *Materials Horizons* 1 (2): 196–206. doi: 10.1039/c3mh00059a.

Zhou, X. and Chen, Y. (2018). An integrated process for struvite electrochemical precipitation and ammonia oxidation of sludge alkaline hydrolysis supernatant. *Environmental Science and Pollution Research* 26 (3): 2435–2444. doi: 10.1007/s11356-018-3667-6.

Zodi, S., Merzouk, B., Potier, O. et al. (2013). Direct red 81 dye removal by a continuous flow electrocoagulation/flotation reactor. *Separation and Purification Technology* 108: 215–222. doi: 10.1016/j.seppur.2013.01.052.

12

Adsorption of Microconstituents

Challa Mallikarjuna, Rajat Pundlik, Rajesh Roshan Dash, and Puspendu Bhunia

School of Infrastructure, Indian Institute of Technology, Bhubaneswar, Odisha, India

12.1 Introduction

Many trace microconstituents (MCs) like pharmaceuticals, endocrine disrupting compounds, and personal care products are present in wastewater, which are not easily removed by conventional wastewater treatment systems. There are numerous pathways through which these MCs are identified in the aquatic ecosystems. The sources of these compounds are groundwater, surface water, wastewater effluent, and industrial effluent. The existing monitoring method does not evaluate the presence of these compounds in treated wastewater. These compounds may have adverse effects on humans and aquatic ecosystems with the application of treated water for reuse. Most conventional treatment units (e.g., flocculation, coagulation, filtration, sedimentation) are inefficient in removing these compounds. The secondary treatment processes (e.g., activated sludge processes) and tertiary treatment processes (e.g., filtration) have lower removal efficiencies for most of the MCs of concern. The main reasons for the lower removal efficiency of such compounds are the slow rate of degradation and their low concentrations (ranging from ng/L to μg/L). As a result, advanced tertiary treatment processes are extensively studied for the removal of MCs. These processes include advanced oxidation process (AOP), membrane filtration, and adsorption, which have the potential to tackle the complex dynamic aspect of MCs. Advanced oxidation processes such as ozone-based processes, Fenton oxidation, or UV/chlorine oxidation have provided the prospect of treating emerging MCs. Nevertheless, these technologies are energy-intensive and may produce more toxicity with by-products formation, which needs special attention.

Sorption is a process by which the materials like MCs are easily either adsorbed in or on another material. Sorption involves a variety of mechanisms including absorption, adsorption, ion exchange, surface complexation, and surface precipitation as a non-mechanistic approach by which surfaces remove the contaminants from the solution. In this chapter, most of the studies work on the principle of all these mechanisms, but the term "adsorption"

Microconstituents in the Environment: Occurrence, Fate, Removal, and Management, First Edition.
Edited by Rao Y. Surampalli, Tian C. Zhang, Chih-Ming Kao, Makarand M. Ghangrekar, Puspendu Bhunia, Manaswini Behera, and Prangya R. Rout.

is used instead of "sorption" due to lack of data to justify the occurance of all these mechanisms. Adsorption is one of the most widely used processes for MCs removal from an aqueous solution. As a sustainable approach, adsorption involves cost-effective, energy-efficient, and easy-to-design/operating processes and has been used for water purification and removal of various constituents such as dye, heavy metals, and MCs. Conventional adsorbents such as activated carbons (ACs) are commonly used to remove specific MCs due to their special characteristics (i.e., high surface area, porous texture), which enhance the adsorption mechanism. These adsorbents are produced in large quantities because of their ability to remove not only various MCs but also organic and inorganic matter. On the other hand, emerging or non-conventional adsorbents are low-cost alternatives that utilize naturally occurring substances and waste from agricultural and industrial sectors. Non-conventional adsorbents are studied to produce adsorbents of renewable sources and reduce the load on conventional adsorbents. The characteristics of non-conventional adsorbents are improved with surface modification through physical and chemical methods. In recent years, a number of studies have been carried out with these modified non-conventional adsorbents for the removal of specific MCs. Majorly, these emerging adsorbents constitute easily available solid waste in agricultural and industrial sectors.

This chapter describes the fundamentals of adsorption and the interactive behavior of adsorbents and adsorbates for the removal of MCs. Moreover, the multi-component preference analysis is discussed to understand the robust mechanism in a real wastewater scenario. It also briefly elaborates on various aspects such as conventional adsorbents, non-conventional adsorbents with their surface modification, disposal methods of adsorbents, and advantage and disadvantages of adsorption.

12.2 Adsorption Mechanism

Adsorption is the process that is widely used for the removal of specific contaminants in the aqueous environment (e.g., water and wastewater) or gas–phases interface owing to the surface forces. The term adsorption was introduced by Kayser (1881) to differentiate surface deposition from intermolecular interaction. Adsorption is a surface phenomenon wherein a foreign material accumulates at a solid surface from liquid or gas surroundings. The material accumulation occurs on the interface of two different phases, which can be liquid–liquid, gas–liquid, gas–solid, or liquid–solid. The material adsorbed on the surface is called "adsorbate," and the surface on which the adsorbate is adsorbed is called "adsorbent." In the case of MCs, several adsorption forces such as hydrophobic interaction, π–π bond, London or van der Waals forces, electrostatic forces, and hydrogen bonding act individually or simultaneously during the adsorption processes, and the associated mechanisms may be elucidated with kinetic and isotherm studies. The majority of interactions between the adsorbent and adsorbate are attractive or repulsive, depending on the chemical characteristics of the material and MCs involved. The single-layer adsorption is the capacity of the adsorbent to accommodate a single or monolayer of species of adsorbate on its surface. However, when adsorbate concentration is high, additional layers may be formed on the monolayer with intermolecular attraction forces, resulting in multilayer adsorption. The functional groups play an important role in determining the adsorption forces of a particular adsorbate. For example, the aromatic compounds of π-donor with π-acceptor provide effective adsorption potential with π–π bond for the removal of aromatic functional contaminants. The heterogeneity of the surface morphology of the adsorbent significantly changes the adsorption mechanism with the variation of adsorption equilibrium and kinetics. It provides numerous ways to bind adsorbate, depending on the heat of adsorption and mass transfer mechanism. The adsorption considers mass transfer of adsorbate on the adsorbent with the following three-step process: i) film diffusion (external diffusion), in which adsorbate from aqueous solution adsorbs on the surface of the adsorbent; ii) pore diffusion (intraparticle diffusion), in which adsorbate travels into the crevices of the adsorbent; and iii) surface reaction, in which chemical reaction of adsorbate occurs on the internal surface of the adsorbent. Each step provides resistance, which impacts the adsorption rate and dominant resistance from any one of the steps, resulting in the rate-controlling step. Figure 12.1 shows the mass transfer mechanism and basic terms of adsorption.

There are two types of adsorption: *physisorption* and *chemisorption* based on the nature of the force of attraction applied at the interface. Physisorption is the process in which forces of attraction between adsorbent and adsorbate are physical in nature. These interactions are attached with van der Waals forces, which are weak in nature and result in a reversible reaction. In contrast to physisorption, chemisorption has strong forces of attraction at the interface of adsorbent and adsorbate with limited sites of interaction, resulting in monolayer and irreversible reaction. At times, both chemisorption and physisorption can occur, one followed by the other or simultaneously. Nevertheless, the adsorption step acts as rate-limiting in chemisorption with high activation energy, whereas physisorption requires less activation energy and favors adsorption at low temperatures. Thermodynamic properties such as adsorption enthalpy are lower in physisorption and higher in chemisorption.

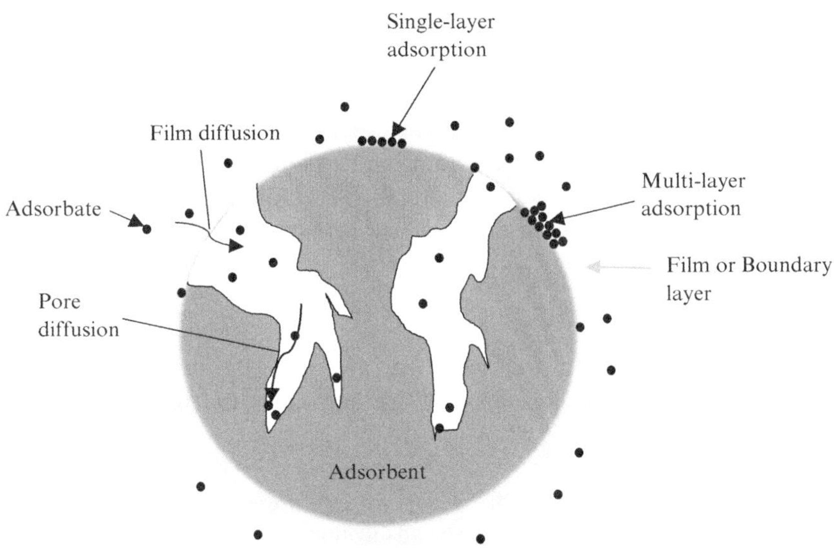

Figure 12.1 Mass transfer mechanism with basic adsorption terms.

12.3 Adsorption Isotherms and Kinetics

12.3.1 Adsorption Isotherms

The adsorption isotherm is the fundamental tool to predict the interactive behavior of adsorbents and adsorbates. It is evaluated with equilibrium data obtained from the batch adsorption studies. There are various sets of isotherm models (i.e., Langmuir, Freundlich, Dubinin–Radushkevich, Temkin), which are analyzed by fitting the equilibrium data. The isotherm constants of these models help to calculate the adsorption capacities and mass transfer characteristics of batch adsorption studies. The validation of models is done by comparing the R^2 value and perfectly describing the adsorption process with calculated and experimental adsorption capacities (Ngulube et al. 2018). Among single solute adsorption isotherms, Langmuir, Freundlich, and Dubinin–Radushkevich, are widely used for their best fitting result and simple mathematical expressions. The Redlich–Peterson isotherm is commonly used in the category of three parameters isotherm and the Brunauer–Emmett–Teller isotherm among multilayer physisorption isotherms. These most popular models used in the study of micronutrients are discussed as follows.

12.3.1.1 Langmuir Isotherm

The Langmuir equation is primarily assumed for the interaction of gases on the solid surface. The Langmuir adsorption is valid for a single layer as a fixed number of active sites are available for the adsorbate (Sullivan et al. 2010). The non-linear form of the Langmuir isotherm is described in Eq. (12.1).

$$q_e = \frac{q_m b C_e}{1 + b C_e} \tag{12.1}$$

where C_e is the equilibrium concentration (mg/L); q_m is the maximum adsorption capacity (mg/g) for a complete monolayer coverage; b is the Langmuir isotherm constant (L/mg); and q_e is the amount adsorbed at equilibrium. The separation factor R_L is employed for the Langmuir isotherms. R_L shows the different conditions of adsorption, whether favorable, unfavorable, or linear, depending on its value. $R_L = 0$ means irreversible adsorption; $R_L = 1.0$ shows linear adsorption; $0 < R_L < 1.0$ indicates favorable adsorption, while $R_L > 1$ shows unfavorable adsorption. The R_L equation (Hall et al. 1966 is expressed as Eq. (12.2).

$$R_L = \frac{1}{\left(1 + b C_o\right)} \tag{12.2}$$

where b is the Langmuir isotherm constant (L/mg); and C_o is the initial adsorbate concentration (mg/L).

12.3.1.2 Freundlich Isotherm

The Freundlich isotherm is mostly used to describe adsorption characteristics of heterogeneous surfaces and is best fitted for the liquid phase adsorbate. As compared to Langmuir, it lacks the linearity range at low concentration and does not determine the saturation value of the adsorbent. The nonlinear form and linear form of the Freundlich equation are expressed as Eqs (12.3) and (12.4), respectively.

$$q_e = K_F C_e^n \tag{12.3}$$

$$\log q_e = n \log C_e + \log K_F \tag{12.4}$$

where C_e is the equilibrium concentration (mg/L); K_F is the Freundlich constant ((mg/g)/(mg/L)^n); n is the Freundlich intensity (dimensionless); and q_e is the amount adsorbed at equilibrium. The study of the isotherm becomes linear, favorable, or unfavorable, depending on the n value. The Freundlich intensity gives the magnitude of the driving force of adsorption or the surface heterogeneity, when $n = 1$ means linear adsorption, while $n < 1$ shows the favorable condition, and $n > 1$ shows the unfavorable condition.

12.3.1.3 Dubinin–Radushkevich Isotherm

The Dubinin–Radushkevich equation is derived with the effect of the micro pore-filling mechanism of adsorbate and estimates adsorption energy or mean free energy. It is assumed the heterogeneous surface is based on the adsorption potential theory. The linear form of the Dubinin–Radushkevich is expressed as Eqs (12.5) and (12.6).

$$\ln q_e = -K_{DR}\varepsilon^2 + \ln q_{DR} \tag{12.5}$$

$$\varepsilon = RT \ln\left(1 + \frac{1}{C_e}\right) \tag{12.6}$$

where K_{DR} is a constant related to the sorption energy (mol^2/kJ^2); ε is the Polanyi potential; q_{DR} is adsorption capacity (mg/g); q_e is the amount adsorbed at equilibrium; R is the universal gas constant (8.314 J/mol/K); T is the absolute temperature (K); and C_e is the equilibrium concentration (mg/L). The mean free energy helps to distinguish between the physical and chemical adsorption process using a constant related to sorption energy expressed in Eq. (12.5).The mean adsorption energy for the adsorbate is expressed as

$$E = \frac{1}{\sqrt{2K_{DR}}} \tag{12.7}$$

where E is the mean adsorption energy (kJ mol).

12.3.1.4 Redlich–Peterson Isotherm

The Redlich–Peterson isotherm is a three parameter isotherm, which considers the hybrid features of the Langmuir and Freundlich isotherms. It is applied to a wide range of adsorbate concentrations considering homogeneous or heterogeneous systems. This empirical isotherm uses linear dependence on concentrations in the numerator and exponential derivative in the denomination, allowing it to function as a Freundlich isotherm at high concentrations and a Langmuir isotherm at lower concentrations by following Henry's law. The non-linear form of this isotherm is given as Eq. (12.8).

$$q_e = \frac{K_{RP} C_e}{1 + a_{RP} C_e^{\beta}} \tag{12.8}$$

where K_{RP} (L/g) and a_{RP} $(mg/L)^{-\beta}$ are the Redlich–Peterson constants; C_e is the equilibrium concentration (mg/L); and β is an exponent whose value must lie between 0 and 1. In Eq. (12.8), if β is equal to 0, then it becomes a linear isotherm, whereas if β is equal to 1, it becomes a Langmuir isotherm, and if $\beta = 1$ and K_{RP} and $a_{RP} \gg 1$, then it becomes a Freundlich isotherm.

12.3.1.5 Brunauer–Emmett–Teller (BET) Isotherm

The BET isotherm is employed for multilayer adsorption systems of gas–solid equilibrium. It is a theoretical equation which gives effective surface area and adsorption energy characteristics. The isotherm related to liquid–solid interface is expressed as

$$q_e = \frac{q_s C_{BET} C_e}{(C_s - C_e)\left[1 + (C_{BET} - 1)(C_e / C_s)\right]} \tag{12.9}$$

where C_{BET} is the BET adsorption isotherm (L/mg); C_s is the adsorbate monolayer saturation concentration (mg/L); C_e is the equilibrium concentration (mg/L); q_s is the theoretical isotherm saturation capacity (mg/g); and q_e is the equilibrium adsorption capacity (mg/g) (Ngulube et al. 2018).

12.3.2 Adsorption Kinetics

Adsorption kinetics determine the correlation of various parameters and can be used to find the rate limiting step in the adsorption processes. Basically, the purposes of the kinetics study are to obtain an adsorption rate with minimal complexity and to identify the favorable operating conditions with a type of mechanism involved in the adsorbent performance. Most common kinetics models include the pseudo-first-order equation, the pseudo-second-order equation, Elovich equation, and intraparticle diffusion model, which are discussed below.

12.3.2.1 Pseudo-First-Order Equation

In 1918, Langmuir presented a first-order rate equation to analyze the kinetics of sorption of oxalic acid and malonic acid onto charcoal. Pseudo-first-order models are generally best fitted for the initial period of the first reaction step. The non-linear and linear form of the equations are shown in Eqs (12.10) to (12.12).

$$q_t = q_e\left(1 - e^{-k_1 t}\right) \tag{12.10}$$

$$\log(q_e - q_t) = -\frac{k_1}{2.303} t + \log(q_e) \tag{12.11}$$

$$\ln(q_e - q_t) = -k_1 t + \ln(q_e) \tag{12.12}$$

where k_1 is the rate constant for the first-order pseudo equation; and q_e and q_t are the amounts of adsorbate uptake per mass of adsorbent at equilibrium and at any time t (min), respectively (Langmuir 1918).

12.3.2.2 Pseudo-Second-Order Equation

The analysis of sorption kinetics of solid–liquid adsorption can be done with a pseudo-second-order equation. The non-linear and linear form of the equations are expressed as Eqs (12.13) and (12.14).

$$q_t = \frac{q_e^2 k_2 t}{1 + k_2 q_e t} \tag{12.13}$$

$$\frac{t}{q_t} = \left(\frac{1}{q_e}\right) t + \frac{1}{k_2 q_e^2} \tag{12.14}$$

where k_2 is the rate constant of the second-order pseudo equation; and q_e and q_t are the amounts of adsorbate uptake per mass of adsorbent at equilibrium and at any time t (min), respectively. This model is based on the assumption that the rate-limiting step may be chemisorption and has shown better correlation with the experimental data (Ho and McKay 1999).

12.3.2.3 Elovich Model

The Elovich model is applied to the system with heterogeneous adsorbing surfaces and is used majorly for chemisorption mechanism. The Elovich model is mathematically expressed as Eq. (12.15).

$$\frac{dq_t}{dt} = \alpha \exp\left(-\beta q_t\right) \tag{12.15}$$

Applying boundary conditions $q_t = 0$ at $t = 0$, the model can be expressed as Eq. (12.16).

$$q_t = \frac{1}{\beta} \ln\left(1 + \alpha\beta t\right) \tag{12.16}$$

Assuming $\alpha\beta \gg 1$, the Elovich equation in linear form is expressed as Eq. (12.17).

$$q_t = \frac{1}{\beta} \ln\left(t\right) + \frac{1}{\beta} \ln\left(\alpha\beta\right) \tag{12.17}$$

where α is the initial rate constant (mg/g/min); and β is the desorption constant during any one experiment (mg/g).

12.3.2.4 Intraparticle Diffusion Model

Weber and Morris (1964) proposed the Intraparticle Diffusion model model to consider the controlled step with intraparticle diffusion for analysis of adsorption kinetics, as shown in Eq. (12.18).

$$q = k_{id}\sqrt{t} + B \tag{12.18}$$

where k_{id} is the intraparticle diffusion constant (mg = g/min$^{0.5}$); and B is the initial adsorption (mg/g).

12.4 Factors Affecting Adsorption Processes

12.4.1 Surface Area

Surface area is a structural property of adsorbent analyzed with the BET method in m^2/g. The surface area plays a major role in increasing the efficiency of adsorbents with more adsorbing sites. The adsorbent is activated with different thermal and chemical treatment methods for modified surface area. The surface area of ACs ranges from 600 to 2000 m^2/g. As the adsorption process is a surface phenomenon, an direct increase in specific BET surfaces would increase the adsorption capacity of the adsorbent. Thus, clays with high surfaces are used for the adsorption of different ionic species. Even zeolites are also used due to their distinct features of the different cavity structures in the ion-exchange process (Crini 2006).

12.4.2 Contact Time

Contact time plays a crucial role in determining the adsorption kinetics and drawing valid conclusions. The adsorption rate of initial contact time is faster depending on the affinity of adsorbate to the adsorbent, and adsorption studies kept for a certain period may not have reached true equilibrium. Mailler et al. (2016) observed that the microconstituents (MCs) such as ciprofloxacin would have a removal of 29, 40, 62, 66, and 71% at 5, 10, 30, 45, and 60 min contact time, respectively, and diclofenac would have a removal of 19, 23, 32, 36, and 39%, respectively. All these cases did not reach true equilibrium, but the difference in removal efficiency was <10%. Additionally, the initial contact time of 5 min for adsorption of MCs (e.g., atrazine, norfloxacin, ofloxacin, sulfamethoxazole) resulted in more than half of the removal efficiency achieved in 60 min. However, higher molecular weight MCs (e.g., erythromycin or roxithromycin) showed slower adsorption kinetics with a removal efficiency of 3 to 7% in an initial contact time of 5 min. Thus, the adsorption kinetics studies usually are carried out first to find the optimal contact time, which are then employed in tertiary treatment for efficient adsorption of MCs.

12.4.3 Nature and Initial Concentration of Adsorbate

The initial concentration mostly increases the adsorption capacity under equilibrium conditions due to the higher accessibility of adsorbate on porous structures with a more solid–liquid interface prospect. At lower initial concentrations, the adsorbate attaches to higher energy sites of macropores, whereas an increase in the initial concentration leads to an increase in the adsorbate–adsorbent interaction, resulting in occupying energy sites of micropores. The adsorption capacity of the few MCs (e.g., sulfamethoxazole, carbamazepine, ketoprofen) increased from 0.5 to 19.181 mg/g, with the initial concentration variation of 1 to 50 mg/L (Turk Sekulic et al. 2019).

12.4.4 pH

pH is an important factor for the adsorption of some MCs as it changes the surface charge of the adsorbent and also ionizes some of the compounds present in the solution. In some cases, the cationic species of MCs bond with adsorbent at $pH > pH_{pzc}$, as its surface becomes negatively charged. On the other hand, $pH < pH_{pzc}$ favors anionic adsorption with a positively charged adsorbent (Turk Sekulic et al. 2019). Likewise, clay mineral's sorption capacity varies highly with the basic dye and acidic dye, which shows the influence of pH in the adsorption process. Again, siliceous material efficiency drops with an increase in the pH beyond 8 (Crini 2006).

12.4.5 Nature and Dose of Adsorbent

The increase of adsorbent dose leads to an increased rate of adsorption potential with more active sites. The higher surface area of the adsorbent increases the accessibility of the adsorbate on higher numbers of the active sites available for the adsorption. The adsorbent not only requires a high surface area but also should be a porous material for good adsorption. The pores of adsorbents are classified into three categories: micropores (~0.8 – 2 nm); mesopores (2–5 nm); and macropores (≥50 nm), depending on their width. Most of the AC materials are of micropores in nature, while for different applications like catalytic support, a higher mesopores or macropores to micropores ratio is a viable option. The higher dose of powered AC shows higher removal of MCs; for example, 80% in carbamazepine, 60% in diclofenac, 95% in propranolol, and 80% in sulfamethanoxazole (Mailler et al. 2016).

12.4.6 Interfering Substance

In real applications of adsorption studies, the major factor which affects the removal efficiency of the adsorbents are the interfering substances. The competitive nature of MCs for the adsorption sites under such conditions reduces the adsorption capacity of each MC in multi-component analysis. As a result, the other MCs become the interfering agent to infringe on the adsorption mechanism. The adsorbent characteristics become selective in nature to a particular group of MCs with interfering substance, depending on site blockages, ion charges, and interaction between different interfering substances.

12.5 Multi-Component Preference Analysis

Several studies are conducted with a single component of MCs being uptaken by modified adsorbents for achieving higher adsorption capacity, although in a real scenario, MCs form a group of multi-components combining together in the wastewater. Under such conditions, these components can react differently with kinetic transport, adsorbent selectivity, and bed usage. Moreover, interaction among these components also plays a crucial role to determine the nature of the adsorption process. The interaction between the components can be found by formulating the relationship between the adsorption capacity parameter of participating component ($q_{c,i}$) individually with respect to the adsorption capacity of the participating component in the multi-component solution ($q_{c,m}$). The relationships can be classified as (Olu-Owolabi et al. 2010; Singh and Balomajumder 2016, 2017):

- *Synergic interaction* ($q_{c,i} < q_{c,m}$): when the adsorption capacity of a particular component ($q_{c,i}$) is less than the adsorption capacity of the same present in the multi-component solution ($q_{c,m}$);

- *Antagonistic interaction* ($q_{c,i} > q_{c,m}$): when the adsorption capacity of a particular component ($q_{c,i}$) is more than the adsorption capacity of the same present in the multi-component solution ($q_{c,m}$);
- *Non-interaction* ($q_{c,i} > q_{c,m}$): when the adsorption capacity of a particular component remains unaffected with the participation of the other components in analysis.

The robust mechanism of the removal of MCs studies is majorly synthesized through transition metal–clay combination adsorbent and nanoparticles adsorbent. However, very few studies of multi-component analysis exist on MCs. The dominant mechanism in such multi-component preference analysis are hydrophobicity and metal interactions at the solid–liquid interface in the case of adsorption of caffeine. The best result of these analyses preferred adsorption bed assembly with serially layered bed structures of different modified adsorbents (Cabrera-Lafaurie et al. 2015).

The surface characteristic of a porous structure also plays a major role in preparing inorganic–organic adsorbents with transition metals. The porous structure consists of mesopores and macropores, which provide more surface area for the metal modified adsorbent to adsorb more MCs. The nanoparticles are synthesized with enhanced surface morphology for the removal of the multi-components. The eco-friendly adsorbent such as synthesized silica nanoparticles shows better results for the removal of three different cationic dyes than the multi-component analysis (Mahmoodi and Maghsoodi 2014). Hydrogel was used in the adsorption of multi-type pollutants originated from non-point source pollution. Hydrogel's highly porous structure incorporates different targeted functional groups, which provide effective competitive sorption capacities in multi-component analysis. However, some studies also evaluated the AC or agriculture waste-derived ash in the multi-component system. Rice husk ash adsorbent was used for simultaneous removal of seven different pesticides (alachlor, metolachlor, chlorpyriphos, fipronil, α-endosulfan, β-endosulfan, p,p′-DDT) in a pesticides mixture with a removal efficiency of 90% (Saha et al. 2014).

12.6 Conventional and Emerging Adsorbents

12.6.1 Conventional Adsorbents

ACs, inorganic materials (i.e., activated alumina, silica gel, zeolites, molecular sieves but not zeolites), and ion-exchange resins are included in the conventional type of adsorbents. However, the generic adsorbents domain was dominated by only four commercially available conventional adsorbents (i.e., ACs, zeolites, silica gel, activated alumina) (Crini et al. 2019). The classification of the conventional adsorbents used for the treatment of water/wastewater is shown in Figure 12.2.

12.6.2 Commercial Activated Carbons

The demand for ACs is increasing in everyday life due to their extensive use in different fields like purification and separation. Moreover, they are also widely used in different industrial processes such as medicinal use, gas storage, pollutant/odor removal, gas separation, catalysis, pharmaceuticals, as electrode materials in electrochemical devices, and also in wastewater treatment. The high surface areas and large pore volumes that are

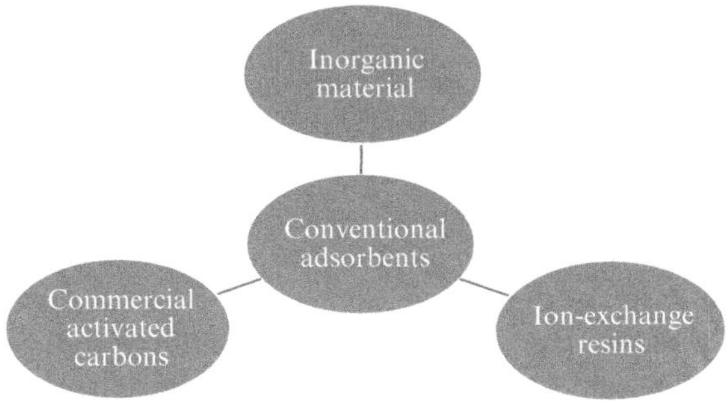

Figure 12.2 Classification of conventional adsorbents.

generated by the heterogeneous porous structures make the ACs perfect adsorbents (Sayğılı and Güzel 2016). Moreover, the chemical nature/properties can be easily modified with a simple chemical treatment (Crini et al. 2019). A growing interest has developed in the remediation of water/wastewater effluents using granular or powdered ACs, due to the strict guidelines imposed on drinking water standards (Streat et al. 1995). Different types of natural or synthetic materials including the lignocellulosic material precursors are widely used in the production of ACs (Sayğılı and Güzel 2016). Naturally available materials like coconut shell, wood, or coal are generally used in the manufacture of commercial ACs with precise surface properties. The surface area and pore-opening (microporous structure) of typical ACs are 600–2000 m^2 g and 1.5 nm (average), respectively (Streat et al. 1995). A good-quality output can result from an effectively designed adsorption system with commercial ACs. Batch type adsorption processes are generally preferred for the usual ACs, where the AC is added to the container containing the polluted effluent. The packed bed column containing the AC has been used with the aqueous solution being fed continuously (Crini et al. 2019).

Secondary commercial ACs will work as adsorbents for the treatment of heavily polluted wastewater such as textile effluents (characterized by color, chemical oxygen demand, total organic carbon) and also for reducing the organic load on secondary and/or tertiary treatment. The commercial ACs can also work as an effective adsorbent for a broad spectrum of compounds like pesticides, aromatic and phenolic derivatives, pharmaceutics, volatile organic compounds, hydrocarbons and surfactants, minerals, metals, and dyes. Moreover, the bacteria retained on the macroporous ACs can degrade part of the adsorbed organic matter, which helps in the *in-situ* sorbent regeneration (Radovic et al. 2000).

The treatment with commercial ACs results in additional benefits like generating no by-products, which are common in the chemical oxidation treatment systems. Though the ACs have been used for water/wastewater treatment for a long time, still the development is being pursued due to the continuous demand for very clean water. Both fundamental- and application-oriented research have been actively conducted in various aspects, like the following: i) possible use of agricultural and industrial wastes, water bottles made of polyethylene terephthalate, scrap tires as new precursors for the ACs; ii) production of AC cloth, and nanotubes as a novel class of adsorbent materials; and iii) understanding of the activation, sorption, and regeneration mechanisms in the case of microwave techniques and the processes that do not require heat (Crini et al. 2019).

12.6.3 Inorganic Material

Clay, silica, and zeolite are inorganic materials that are currently being used in the removal of heavy metals such as Cr(IV). Alumina and silica are composed of minerals of clay, which also include a bound water molecule (Rosales-Landeros et al. 2013). Due to the presence of a large surface area, clays are able to absorb cationic, anionic, and neutral metal species. Moreover, these species are able to participate in cationic and anionic exchange processes. The properties of an adsorbent, such as sorption capacities, cation and anion exchange properties, and binding energies were found to vary widely (Mohan and Pittman 2006). Clays show the chemical properties of colloids due to their extremely fine particle nature. Because of the high specific surface areas, chemical and mechanical stability, layered structure, and high cation exchange capacity (CEC), clays are identified as excellent adsorbent substances (Sharma et al. 2009). Clays are considered as a host material and contain a layered structure; they are also classified based on the layered structure differences. Currently, different types of clays used for the adsorption studies include smectites (montmorillonite, saponite), mica (illite), kaolinite, serpentine, pylophyllite (talc), vermiculite, and sepiolite (Shichi and Takagi 2000). The net negative charge present over the clays provides the opportunity for the adsorption of positively charged ions. Moreover, the high surface area and high porosity of clays also favor the sorption properties positively. Out of the different types of clays, the montmorillonite type exhibits the larger surface area and the higher CEC and is available in the market at 20 times lower price than AC (Crini 2006).

The abundant availability of silicon and oxygen in nature has made silica (SiO_2) a very common mineral. However, this mineral can also be produced in the laboratory, which is a polymer of silicic acid, consisting of inter-linked tetrahedral SiO_4, and has SiO_2 stoichiometry. The silica tetrachloride or substituted chlorosilane/orthosilicate is the basis for the formation of porous and granular types of silica gel. Additionally, this sol–gel method is a popular versatile process for the production of silica. Hydrolysis of sodium silicate or alkoxysilane with the presence of acid or base will yield the Si-$(OH)_4$ entities, which is known as the sol–gel process. Moreover, the process also results in the solution on controlled condensation, which has an important role in adsorption and ion exchange processes due to the presence of an active silica surface with a large specific surface area (Rosales-Landeros et al. 2013).

The structural framework of linked tetrahedral with four oxygen atoms surrounded by a cation defines the crystalline morphology of zeolite minerals (Coombs et al. 1997). Natural zeolites minerals have a uniform pore size, which is of hydrated crystalline aluminosilicate and contain the 3D network of AlO_4 and SiO_4 that are tetrahedrally linked to one another by the sharing of oxygens (Rosales-Landeros et al. 2013). Out of more than 30 natural zeolites minerals, only 7 (mordenite, clinoptilolite, chabazite, erionite, ferrierite, phillipsite, analcime) are available at a sufficient volume and purity, and can be considered for utilization (Mohan and Pittman 2006). Reversible hydration, ion exchange, and uniform pore size are major prime characteristics of zeolites. Moreover, these zeolites can be modified through the ion exchange process for the adsorption of various metal ions. Due to the isomorphic substitution of cations in the crystal lattice, the zeolites have a net negative structural charge that has made them the selective and favorite ion exchange mineral for certain cations (Faghihian and Bowman 2005). The most abundantly available among the natural zeolite minerals is clinoptilolite, which belongs to the heulandite group and has an open reticular structure formed by open channels of 8–10 membered rings. Moreover, the structure also contains Na(I), K(I), Ca(II), and Mg(II) as exchangeable ions (Rosales-Landeros et al. 2013).

12.6.4 Ion-Exchange Resins

The ion exchange resins (IERs) are a class of polymeric materials that are commonly utilized in water treatment, and are cross-linked polymers based on acrylic or styrene monomers. The IERs contain amine functional groups, which act as the anion exchanger for the water treatment. Also, IERs are functionalized with the quaternary amines for sustaining the exchange capacity at a wider range of pH. However, in some special cases, other types of amines are also used (Parvazinia et al. 2018). The ion exchange process has advantages like recovery of valuable metals, selectivity, lesser volume of sludge production, and generation of effluent as per strict guidelines (Tehrani et al. 2013), and polymeric resins are generally utilized in the ion exchange systems (Rengaraj et al. 2001). The IERs can effectively remove the heavy metal compounds from the lower concentration of influents (Rengaraj and Moon 2002). The IERs have successfully reduced the pollution strength of effluent before releasing it into the atmosphere; for example, their implementation with the rinsing water or polluting metals from pickling baths have proved efficient in the removal of minerals and organic substances with the inclusion of numerous types of dye molecules. The IERs system also has benefits such as no loss of adsorbent during regeneration, solvent recovery after use, and elimination of contaminants (soluble) to trace levels. In spite of these benefits of IERs, as commercial activated carbons they are also costly, an important factor to be considered before implementing them into the system (Crini et al. 2019).

Rengaraj et al. (2001) have used the IRN77 and SKN1 cation exchange resins for the chromium from water to compare their ion-exchange behavior and observed >95% chromium removal under optimal conditions. The adsorption capacity (k) for chromium calculated from the Freundlich adsorption isotherm was found to be 35.38 and 46.34 mg/g for IRN77 and SKN1 resins, respectively, and the cation exchange resins have followed the first-order reversible kinetics. The ion exchange resins investigated in this study showed reversible uptake of chromium and so have good application potential for the removal/recovery of chromium from aqueous solutions. Tiravanti et al. (1997) have used the ion exchange process for the pretreatment of tannery wastewater and have observed a considerable decrease in the use of chemical compounds and sludge production reduced by 80% with respect to the traditional treatment. Jørgensen (1979) has examined the applicability of cellulose ion exchangers for water and wastewater treatment. Three cellulose ion exchangers were examined, and it was found that they are able to remove proteins, azo-dyes, DBS, humic acid, chromate, and heavy metal ions. The selectivity of the ion exchangers for removal of these compounds is very high, which explains their ability to remove the examined compounds with high efficiency and capacity, although the three cellulose ion exchangers have only a small capacity.

12.6.5 Emerging/Non-Conventional Adsorbents

Though carbon has gained wide recognition as an effective adsorbent for wastewater treatment, due to the huge costs involved in the carbon-adsorption process, continuous research is ongoing to find effective alternatives for replacing activated carbon. ACs are highly structured and robust materials, which are produced from relatively expensive feed input materials. Moreover, larger quantities of AC are needed for the new challenges and developments in wastewater treatment. Commercial ACs are also costly because their utilization requires extensive regeneration and reactivation process. Hence, there is a need to

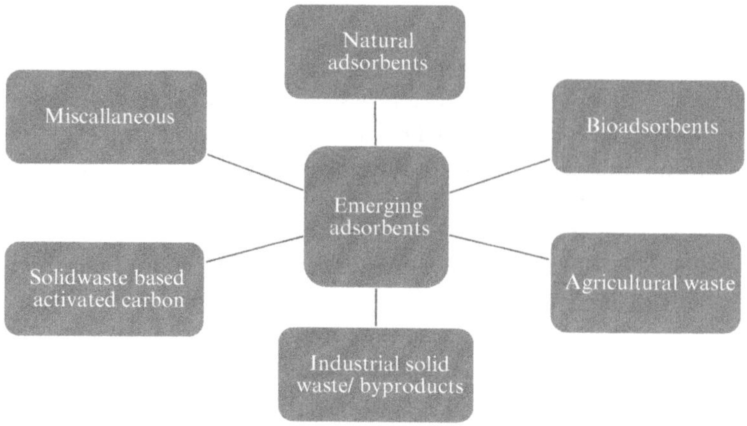

Figure 12.3 Classification of emerging adsorbents.

consider alternative input source materials for producing the ACs, which have the same effective adsorptive capacities as those commercially available (Streat et al. 1995). Moreover, the substitute should be easily available, economically viable, and also able to regenerate chemically and with good quantities of adsorbate recovery (Mohan and Singh 2002). Hence there is a need for identifying more effective and tailor-made substances for satisfying the demands stated (Mohan and Singh 2002).

The schematic layout for the non-conventional adsorbents is shown in Figure 12.3, which includes the ACs produced from agricultural solid waste, industrial by-products, natural materials (e.g., clays), industrial by-products (e.g., red mud), bio-sorbents (e.g., chitosan), and a few miscellaneous adsorbents like alginates (Crini et al. 2019). The non-conventional adsorbents that are alternatives to the commercial ACs majorly include biological products. They also include by-products such as bio-sorbents (including biomasses of industrial and agricultural origin) and green adsorbents (from forest industries). These non-conventional adsorbents have been studied and used in remediation processes as effective and inexpensive materials. They also included substances like algae, bacteria, fungi and yeasts, bark, sawdust, peat, natural products (e.g., cotton, flax, hemp), polysaccharides (e.g., starch, cellulose, chitosan, alginates), industrial by-products (e.g., red mud, sludge), as well as plants and innovative nanomaterial substances (Crini et al. 2019).

12.6.5.1 Natural Adsorbents

Generally, natural materials are available in abundant quantities; also, some of the waste materials generated from industrial or agricultural activities can serve as an effective and economical adsorbent. Moreover, because of their low cost, they can be disposed of without any regeneration after their successful utilization. Due to their cost-effectiveness and abundant availability, these agricultural by-products become an appropriate source of input for the production of ACs. The carbonaceous materials like bark, coal, lignite, coconut shells, wood, dead biomass, seaweed, pecan shell, and peat are widely used for the production of commercial ACs (Chuah et al. 2005).

Natural clay materials have been familiar to humans from the early days of civilization. Due to their low cost, a wide range of availability on most continents, higher sorption capacities, and effective ion-exchange nature, clay materials stood as a major adsorbent component (Shichi and Takagi 2000). For instance, the interactions between the clay minerals and clay are extensively studied, and the results have shown a strong affinity toward heteroatomic cationic and anionic dyes. Moreover, with an adsorption capacity of 360.5 mg of dye/g, the bentonite clay stood as an effective adsorbent for the removal of basic dyes (Crini 2006). Among the various natural siliceous adsorbents, silica beads have received strong attention because of their hydrophilic surface due to the presence of silanol groups (Krysztafkiewicz et al. 2002). Because of their porous texture, high surface area, and mechanical stability, these siliceous sorbents are also identified as attractive for effluent remediation applications (Crini 2006). Zeolites are high porous aluminosilicates with various cavity structures that also have a negatively charged lattice. The strong and attractive adsorbent behavior of the zeolites is due to their high ion-exchange capacity, high specific surface areas, and availability at lower prices. Moreover, they have shown effective sorption capacities for remediating the effluents with trace quantities of pollutants (e.g., heavy metals, phenols) (Crini 2006).

12.6.5.2 Agricultural Wastes

Agricultural by-products have received much attention from researchers due to their renewable, low-cost, and environmentally-friendly nature (Sayğılı and Güzel 2016). Also, they have been proved to be effective and cost-effective sorbents (Zahar 2020). A large number of products like coffee grounds, cocoa shell, waste tea, coffee husks, pomegranate shell, mahogany sawdust, rice husk, bael shell, grape industrial processing waste, tomato processing solid waste, etc., are successfully utilized in the generation of low-cost ACs (Sayğılı and Güzel 2016). The following forestry-based agricultural, industrial by-products like tea waste, coffee, hazelnut shells, peanut hull, sawdust, barks, palm kernel husk, coconut husk, peanut skins, cellulosic and lignocellulosic wastes, hemp-based products, cotton, and modified cotton, corncobs, rice hulls, apple wastes, wool fibers, olive cake, almond shells, cactus leaves, banana, orange peels, sugar beet pulp, palm fruit bunch, maize leaf, and other different substances are also used as effective adsorbents (Vandenbossche et al. 2015). The olive cake or olive pomace (solid residue generated during the extraction of olive oil from olives) has been successfully used in the adsorption of a limited number of heavy metal ions. However, the ACs prepared from olive pomace are utilized in the adsorption of methylene blue, RR22, and Azo-dye orange (II) (Akar et al. 2009).

12.6.5.3 Industrial By-Product (Industrial Solid Wastes)

Considerable attention has been paid in recent times for developing effective and low-cost adsorbents from industrial by-products, since it will be beneficial to the development of a sustainable society by using industrial wastes to develop inexpensive alternative adsorbents (Huang et al. 2008). Various studies have been published on the generation of ACs from several industrial wastes like PET bottles, waste tires, refuse-derived fuel, by-products of lactic acid fermentation processes, garbage, sewage sludges, waste newspaper, waste carbon slurries, and blast furnace slag (Crini 2006).

Steel slag, a waste product from the steel manufacturing industry, has shown effective results in the removal of heavy metals like Zn, Cu, Pb Cd, Ni, Cr, and phosphorus. Steel slag contains both iron oxides and calcium hydroxides, which favors the elimination of pollutants from the aqueous solutions. The iron oxide provides the sites for the adsorption of anions, whereas calcium hydroxide helps in raising the solution pH for the precipitation of heavy metals (Zahar 2020). Red mud (RM) is also another industrial waste produced during bauxite leaching by the Bayer method (Liu et al. 2007), that contains aluminum oxide, iron oxide, silica, titanium oxides, and hydroxides (Huang et al. 2008). RM has been successfully used in the removal of fluoride, hexavalent chromium, and dyes from aqueous solution (Liu et al. 2007) and has also been proved effective in the removal of arsenate, toxic heavy metals, dyes, and phenol (Huang et al. 2008). Being produced from a fertilizer plant during the dihydrate wet phosphoric acid process, phosphogypsum is a non-conventional adsorbent and has been effectively used in the removal of lead from aqueous solution (Cesur and Balkaya 2007).

12.6.5.4 Solid Waste-Based Activated Carbons

To generate carbon in the charred form (under controlled conditions), most of the naturally available vegetable substances can be utilized. Also, if abundantly available material such as organic agricultural solid waste is used for this purpose, there would be no limitation of feed material. Straw is one such kind of effective, cheap, and readily available material for the generation of ACs. Various other agricultural solid wastes like inter alia and waste apple pulp are also widely used in the production of ACs (Streat et al. 1995). Sawdust and bark, which are agricultural solid wastes and waste by-products generated from the forestry industry, also have been iused for the adsorption process (Garg et al. 2004), and the sawdust and bark (rich in polyphenol) have been proved as an effective adsorbent for the removal of dye materials (Palma et al. 2003; Crini 2006). Due to the presence of its high tannin content, which contains the polyhydroxy polyphenol groups, bark works as an effective adsorbent (Crini 2006). In recent times, the tree fern, which is an agricultural waste material, has been actively studied for the removal of pollutants; also, it has a complex structure with lignin and cellulose as major components (Ho et al. 2005). Chemical bonding and ion exchange are the major sorption mechanisms observed. Tree fern has a maximum sorption capacity of 408 mg/g achieved for the basic red 13 (Ho et al. 2005; Crini 2006). The other low-cost and readily available agricultural by-products like date pits, pith, corncob, barley husk, wheat straw, wood chips, and orange peel have been tested successfully for the removal of dye materials (Crini 2006). Recently, biochar is emerging as an effective adsorbent for treatment of wastewater. Biochar is principally derived from organic matter (biomass) when heated at high temperatures in the absence or limited quantities of oxygen. Different feedstocks such as algae, crop residues, forests biomass, and animal wastes are used in the pyrolysis, microwave pyrolysis, torrefaction, hydrothermal carbonization/liquifaction, and gasification processes for production of biochar by thermos-chemical heating (Srivatsav et al. 2020).

12.6.5.5 Bio-Sorbents

The use of biological substances for the removal of pollutants from aqueous solutions is known as biosorption, and is regarded as a novel, competitive, effective, and economical approach to the remediation of effluents. Biological substances like chitin, chitosan, peat, yeasts, and fungi or bacterial biomass have been tested successfully as chelating and complexing adsorbents for the removal of pollutants such as dyes. These bio-sorbents are more selective when compared with the ion-exchange resins and commercial ACs and can reduce the targeted pollutants (like dyes) to the ppb level (Crini 2006). Biopolymers like chitin and chitosan are widely recognized bio-sorbents for the removal of pollutants to ppm or ppb levels. These are renewable and biodegradable and are abundantly available in nature. However, chitin and chitosan are commercially produced from 0crustaceans (crab, krill, crayfish) (Teng et al. 2001). Peat is another bio-sorbent that has porous and complex soil material; also, the peat contains lignin, cellulose, fulvic, and humic acid. This lignin and humic acid contain polar functional groups, like alcohols, aldehydes, ketones, carboxylic acids, phenolic hydroxides, and ethers, which can favor the formation of chemical bonding. Due to these factors, peat is able to remove the pollutants effectively, such as dyes from the aqueous solutions (Crini 2006).

Starch is one of the richest sources of carbohydrate present in plants as an energy storage material. Starch has characteristics like hydrophilicity, biodegradability, polyfunctionality, high chemical reactivity, and adsorption capacities; also it can yield the cyclic derivative and cyclodextrin (Crini 2006). These cyclodextrins can produce the inclusion compounds with different aromatic molecules. Due to their hydrophobic cavity nature, the cyclodextrins can capture the pollutants (Crini and Morcellet 2002). Starches and cyclodextrins can produce water-insoluble cross-linked networks by reaction between the hydroxyl groups of the chains with a coupling agent. These cross-linked starches have a tremendous swelling capacity in water, which helps in the faster diffusion process of pollutants. The maximum adsorption capacities of 249, 88, and 77.4 mg/g were observed for the removal of acid blue 25 using starch-based, cross-linked cyclodextrin and chitosan/cyclodextrin mixed adsorbents, respectively (Crini 2006).

12.6.5.6 Miscellaneous Adsorbents

Cotton is an abundantly available and naturally occurring organic substrate, which has excellent physical and chemical properties with reference to stability, water absorbency, and removal of pollutants like dye (Crini 2006). For the removal of acid blue 25, acid yellow 99, and reactive yellow 23, using the treated cotton in a continuous system, Bouzaida and Rammah (2002) have achieved 589, 448, and 302 mg/g of removal from the aqueous solution. The Cucurbituril (CB) is a hexameric macrocyclic substance that is self-assembled by the acid-catalyzed condensation reaction of glycoluril and formaldehyde. The hydrophobic interior sites of the CB are the favorable locations for holding the hydrocarbon molecules and also produce stable complexes with different protonated alkyl and arylamines (Lee et al. 2003). Moreover, the CB has been implemented successfully for the remediation of pollutants like colorant aqueous solutions or volatile organic compounds from the air (Márquez et al. 2004). The CB was found success in the removal of dye compounds from water (Chen et al. 2015).

Hydrogels of polysaccharide-related compounds have gained attention due to their special characteristics like biocompatibility, biodegradability, non-toxic, and high swelling nature (Paulino et al. 2011). Hydrogels possess a 3D hydrophilic structure, which has the ability to absorb and retain higher volumes of water (Dorkoosh et al. 2002). At present, hydrogels are produced with various characteristics, like containing magnetite, for remediating the effluents with the metal ions (Paulino et al. 20110). However, different types of polysaccharide-based hydrogels and chitosan with and without a magnetic nature have shown effective results in remediating the effluents from the battery manufacturing industry, which produce huge amounts of metal ions such as Pb(II), Cd(II), and Cu(II) (Paulino et al. 2008). The presence of oxygen and nitrogen atoms in the hydrogel, which are prepared by the chemical cross-linking of either polyacrylic acid or chitosan, has a key role in the adsorption of metal ions (specifically cations). Moreover, the polymer hydrogels, which are produced by the simultaneous cross-linking of acrylic acid and chitosan monomers, have also shown excellent adsorption capabilities with the metals ions (Vilela et al. 2019).

12.7 Desirable Properties and Surface Modification of Adsorbents

12.7.1 Desorption/Regeneration Studies

The adsorbents that are used in the remediation of effluents have the option of regeneration, disposal, or recycling. Desorption is one the most important process for recovering the valuable metals from the adsorbents and simultaneous regeneration of the adsorbent for reuse (Bishnoi 2005). The regeneration of adsorbent can be carried out with acids, alkalis, or other chemicals. pH has a major role in the successful regeneration of the adsorbent. The desorption studies on building materials and agricultural wastes have not been attempted, since these materials are available at larger volumes. The regeneration studies on these materials have not yielded successful results. Desorption studies require a practical adsorbent, which can produce both recoveries of valuables and good regeneration capacity (Yadav et al. 2018).

The removal mechanism utilized and the mechanical strength of the adsorbent are the major points on which the success of the desorption depends. Generally, most of the biosorbents use the ion-exchange process for the removal of highly toxic compounds, hence mild to strongly acidic conditions are sufficient for the desorption of these substances (Younas et al. 2021). Moreover, utilization of acids for the desorption process is cost-effective since a majority of the industrial waste are acidic solutions. However, acidic conditions should not affect the efficiency of adsorbent adversely (Bishnoi 2005).

The effectiveness of an adsorbent depends on the ease of regeneration and recycling, along with the major aspect of its adsorption capacity (Bishnoi 2005). However, most of the studies mainly concentrate on adsorption capacity only, not focusing on its reusability (Vijayaraghavan and Yun 2008). The choice of adsorbent is considerably affected by its recycling aspect. The recycling of adsorbents can make the process cost-effective and also reduces the burden of supplying the input material constantly (Younas et al. 2021). For the past few years, the regeneration and discarding aspects of waste adsorbents have been in discussion to assess their recyclability and also to minimize the negative impact of waste adsorbents (Baig et al. 2015). The effective rejuvenation of adsorbent will be decided by a major principle, i.e., whether the desorbing adsorbent can reach its original/initial adsorbing capabilities or not. The reproducibility studies of composite materials have shown that even after six desorption stages, the regeneration efficacy remains constant. About 90% of the tested conventional adsorbents do not have regeneration data. Hence, future studies have to focus on the selection of high-performing absorbents, which can be regenerated directly for the adsorption-desorption sequence without losing their adsorption capabilities (Younas et al. 2021).

Combustion or dissolution in strong acids/bases is the only method used for the removal of the adsorbate. The destructive recovery is economically feasible, only in the case of low-cost biomass being utilized as a bio-sorbent for the precious metal recovery. However, non-destructive desorption from the stacked bio-sorbent has gained major attention so far. In the majority of scenarios for desorption of the adsorbent, dilute inorganic acids or bases are preferred due to their effectiveness. However, they may create structural harm to the adsorbent, which can eventually reduce the capacity of the adsorbent after regeneration. Since pH has a major role in the adsorption process, controlling the pH of the desorption solution can have a positive effect on both adsorbent regeneration and contaminant recovery (Park et al. 2010). Different chemicals such as HCl, H_2SO_4, NaOH, $NaNO_3$, ethylenediamine tetraacetic acid (EDTA), HNO_3, NaCl, CH_3COONa, and $NaHCO_3$ are tested for the regeneration of the adsorbents (Younas et al. 2021). From the different studies, it was observed that some chemicals like EDTA have effectively regenerated the adsorbents without any harm. The adsorbents (e.g., crab shells, some large-scale algae industrial wastes, agricultural wastes) are found stable without immobilizing their macrostructure when they are in strong acidic or alkaline environments (Vijayaraghavan and Balasubramanian 2015).

Recently, Rout et al. (2015) has conducted a desorption and reusability study on phosphate-loaded adsorbent (dolochar), in order to regenerate the adsorbent for subsequent reuse. The authors have mentioned that the adsorption-desorption cycle was repeated up to three times and observed that both the adsorption and desorption capacity of the adsorbent were maintained well above 80% of the original capacity in the third cycle. They have also mentioned that "despite the high rate of desorption, the process is not encouraged in this study primarily due to the following two factors: i) dolochar is copiously generated industrial waste and utilizing it as an adsorbent will contribute positively toward the disposal issue; and ii) applicability of phosphate enriched dolochar in agricultural fields for serving the purpose of phosphate availability to plants and enhanced hydraulic conductivity."

12.7.2 Column Studies

The batch adsorption data are suitable for small-scale treatment systems but are not favorable for large-scale treatment systems due to the overestimation of sorption capacities. The data obtained from the fixed-bed column studies are more realistic in nature, as they resemble the flow conditions of full-scale constructed filters. For predicting the column bed functional life span, which depends on the column breakthrough, continuous flow experiments are necessary (Mateus and Pinho 2010). Moreover, for designing the commercial/industrial fixed-bed adsorbents and to know the practical application of adsorbents, it is necessary to have column studies data (Yadav et al. 2018). The flow-through adsorbent packed bed has similar characteristics (flow conditions) to large-scale treatment systems; hence, the adsorption process of the packed bed systems becomes the basis for various important applications (Rout et al. 2017). For designing and improving the operation of the large-scale adsorption systems, the fixed bed continuous type of activities plays a vital role when compared with the batch type systems. Moreover, these types of continuous systems have a greater role in predicting the breakthrough curves, which can be useful in determining the functional lifespan of adsorbent beds (Mateus and Pinho 2010). The successful design of the fixed-bed column includes the estimation of time-concentration profile through breakthrough curves (Rout et al. 2014). The breakthrough curves of the column bed can be generated by plotting effluent (C_t) to influent (C_0) adsorbate concentration vs. time (t). By using the area under the breakthrough curve (for unsymmetric), q_{total} and t_s can be estimated using Eqs (12.19) and (12.20) (Rout et al. 2017).

$$q_{total} = Q\int_{t=0}^{t=t_e} C_r dt \tag{12.19}$$

$$t_s = \frac{1}{C_0}\int_{t=t_e}^{t=0} C_r dt \tag{12.20}$$

where Q is the volumetric flow rate (L/h^{-1}); C_r = (C_0−C_t) is the adsorbate removal (mg/L^{-1}) concentration; q_{total} is the total adsorbate adsorbed in the column; t_s is the stoichiometric time; t_e is exhaustion time (h); t_t is the total time (h); and for a symmetrical breakthrough curve, t_s is the time at which $C_t/C_0 = 0.5$. The following parameters are also important while performing the column studies, such as equilibrium adsorbate uptake capacity of the column (q_{eq}), the total amount of adsorbate entering the column (P_{total}), percentage removal of adsorbate (R), mass transfer zone (MTZ), the time required for the MTZ to move the length of its own height up/down the column (t_z), the rate of movement of MTZ (U_z), empty bed contact time (EBCT), adsorbent usage rate (U_r), time equivalent to the usable capacity (t_u), the length of bed used up to the breakthrough time (L_b), and length of the unused bed (LUB) (Khan et al. 2015). For predicting the breakthrough curves from the continuous type of adsorption studies, numerous kinetics models were developed. However, the Thomas, Clark, MDR, and BDST models are some of the most frequently used models for predicting the dynamic nature of the adsorption process over packed bed columns. Along with predicting the breakthrough curves accurately, these models are also helpful in generating the data for scaling up the packed column adsorption systems (Rout et al. 2017).

Recently, Rout et al. (2017) have studied the packed columns under dynamic conditions for phosphate removal from aqueous solutions, using dolochar as an adsorbent. Thomas modified dose-response (MDR) and Clark models were applied to simulate the experimental breakthrough curves following nonlinear regression analysis. Also, the bed depth service time (BDST) model was engaged to scale up the adsorption process by means of a linear regression approach. They have observed that statistical modeling proposed quadratic models for breakthrough and exhaustion time with correlation coefficient (R^2) values of 0.99 in both cases. In another study, the same authors have analyzed the phosphate adsorption potential and behavior of a mixture of Ground Burnt Patties (a solid waste generated from cooking fuel used in earthen stoves) and Red Soil (a natural substance) in a fixed-bed column and have observed that both the breakthrough and exhaustion time increased with an increase in bed depth and a decrease in flow rate and influent concentration. The authors have identified that the values of R^2 and the Sum of the Squares of Error (SSE) revealed the Modified Dose-Response model as the best-fitted model to the experimental data (Rout et al. 2014).

12.7.2.1 Surface Modification of Adsorbents

Although results obtained from the conventional and non-conventional adsorbents studies are found effective in remediating the pollutants, the activation of the adsorbent prior to the application may reduce the higher dosage consumptions (Liu et al. 2007). Various methods currently being used by different researchers for the preparation of adsorbents and the surface modifications are mentioned in Table 12.1. The adsorption characteristics can be affected

Table 12.1 Preparation of adsorbents and the surface modification methods.

S. No	Adsorbent	Adsorbate	Adsorbent Preparation/Surface Modification Processes	References
1	Sugar beet bagasse (SBB).	Cr (VI).	• Crushed SBB rinsed in deionized (DI) water and dried at 70 °C for 24 h. • 50 g of SBB powder was mixed with 50 mL of H_3PO_4 (60%) and refluxed at 100 °C for 2 h. After cooling, it was rinsed with deionized water several times, dried at 85 °C for 24 h, and converted to the pellets. • Pellets were pyrolyzed in a cylindrical stainless-steel reactor in a tubular electrical furnace under nitrogen (N_2) gas flow (0.5 L/min) at 450 °C. • The obtained AC_{SBB} was rinsed with DI water under N_2 gas to a neutral pH and dried at 105 °C for 12 h. Finally, the AC_{SBB} was crushed and sieved by 100 mesh for further usages.	(Ghorbani et al. 2020).
2	Olive pomace (OP).	Sulfadiazine.	• The washed OP dried at 103 °C, was ground, and sieved. • Carbonized at 450 °C for 2 h under N_2 gas in a reactor made of stainless steel; and • Impregnated with KOH in a mass ratio of 1:1, 1:2, and 1:4 g OP/g KOH and the suspension was stirred in a water bath at 60 °C for 12 h, and then dried in an oven at 103 °C for 24 h. • The mixture was activated under N_2 flow at different temperatures of 560, 700, and 840 °C for 2 h. • The excess KOH in AC samples was removed with 1 M HCl. Then the obtained AC samples were washed and dried at 103 °C for 24 h.	(Aslan and Şirazi 2020)
3	Tomato processing solid waste (TW).	Methylene blue.	• TW was washed with water to remove impurities and subsequently dried naturally. • Crushed and ground to pass through 80-mesh sieves prior to its chemical activation. $ZnCl_2$ of purity 99.9% was used as a chemical activator in the AC production.	(Saygılı and Güzel 2016)

(Continued)

Table 12.1 (Continued)

S. No	Adsorbent	Adsorbate	Adsorbent Preparation/Surface Modification Processes	References
4	Oil palm fiber (OPF).	Methylene blue.	• OPF of 1–2 mm size is carbonized at 700 °C under purified nitrogen (99.995%) using a tubular reactor placed in a microwave (at 2.45 GHz frequency). • Produced char was soaked in KOH with an impregnation (char: KOH) ratio of 1:0.5 (wt%).	(Foo and Hameed 2011)
5	Red mud (RM).	Phosphate.	• The solid red mud (RM) samples were mixed at room temperature for 24 h with 2 M HCl or 2 M HNO_3 solution, respectively, at a liquid/solid ratio of 20 ml/g, referred to as RM–HCl and RM–HNO_3, respectively. After acid treatment, the residue was washed with distilled water and dried overnight at 100 °C. Acid thermal treatment was carried out by placing the acid pretreated RM samples in a muffle furnace for calcination at 700 °C in air for 5 h (referred to as RM–HCl-700 and RM HNO_3-700).	(Huang et al. 2008)
6	Red mud (RM).	Phosphate.	• The heat-treated RM samples were ground in a mortar and sieved through a 100 mesh sieve. The acid-heat activation of red mud was performed in a glass reactor fitted with a stirrer and a thermometer. The samples were loaded into the glass reactor, and the ratio of liquid to solid was 20 ml/g. The effect of hydrochloric acid concentration on activation was tested by acidified raw red mud with a specified amount of a known concentration (0.01, 0.05, 0.25, 0.5, 1 mol/L HCl) at a specific temperature. The effect of temperature on acid-heat activation was carried out at 25, 40, 60, 80, 100 °C for 2 h, and the effect of heating time on acid–heat activation was carried out for 0.5, 1, 2, and 3 h. At the end of the experiment, the mixture was centrifuged for 10 min at 7000 r/min. The residue was washed with distilled water and dried at 100 °C overnight.	(Liu et al. 2007)
7	Bagasse pitch.	Cadmium and zinc.	• The activated carbon was prepared by treating one part of raw bagasse (cut into small pieces) with two parts (by weight) of concentrated sulfuric acid and kept in an oven maintained at 150–165 °C for a period of 24 h. The carbonized material was washed well with double distilled water to remove the free acid and dried at 105–110 °C for 24 h. The dried material was subjected to thermal activation at 800–850 °C for 30 min. For batch studies, the adsorbent material was powdered and sieved to the desired particle size (<325 mesh).	(Mohan and Singh 2002)

considerably by the material modifications (Rosales-Landeros et al. 2013); for instance, materials like clay and sepiolite have shown improvements in their adsorption capacities when their surfaces are modified with the quaternary amines. Compared to initial/non-modified compounds, NaOH treated kaolinite and acid-treated bentonite have yielded higher adsorption capacities (Crini 2006). Clays such as bentonite, stevensite, kaolinite, illite, palygorskite, and sepiolite, which are modified with the cationic surfactants, have been used in the removal of chromate from aqueous solutions (Rosales-Landeros et al. 2013).

The surface of the adsorbent can be modified in two ways, i.e., by physical treatment or by chemical treatment. Physical treatment includes thermal or hydrothermal modification; whereas, in chemical treatment, the chemical characteristics of the adsorbent are affected. Simple immobilization, electrostatic interaction, and hydrogen bond formation are the most preferred methods for achieving chemically modified surfaces. For example, silica compounds, which are functionalized with carboxylic acids, sulfonic acids, and quaternary ammonium groups, are proved efficient in removing heavy metals (Rosales-Landeros et al. 2013). Also, the silica surface modified with the silane coupling agents (amino functional groups) and silica beads can enhance the interaction with dyes and effectively remove the acid dyes of colored effluents (Krysztafkiewicz et al. 2002; Crini 2006).

Some adsorbents like zeolites carry a negative charge on their surface, making them constraints to the adsorption of a few cations only and debarring them from removing anions. The negative charge present over this kind of material can be changed to positive by the surface modification technique. Also, some of the metals like chromium exist as cations, anions, or non-ionized species. Hence, it is important to make the adsorbent a favorable one for the removal of cations, anions, or non-ionized/neutral pollutants. Zeolites, which are treated with cationic surfactants, are effective in the removal of non-polar hydrophobic organic contaminants due to the presence of fractional organic carbon on their surface (Rosales-Landeros et al. 2013). Similarly, some adsorbents with very lower adsorption capabilities, such as raw clinoptilolite, can be improved by chemical modification with quaternary amines. These chemical modifications are also found effective for chitosan in terms of improving removal performance and selectivity toward dyes (Crini 2006). Sun and Yang (2003) have modified the peat-resin by mixing it with polyvinyl alcohol and formaldehyde, which has resulted in the macro-reticular porous structure with improved physical properties. Moreover, the modified peat resulted in 400 and 350 mg/g removal (maximum) of basic violet 14 and basic green 4, respectively. Similarly, various surface modification methods followed for adsorbent regeneration and activation are mentioned in Table 12.1.

12.8 Disposal Methods of Adsorbents and Concentrate

Improper handling of the waste adsorbents, which contain the adsorbed pollutants, may create serious issues in the neighboring environment. Hence, the methods such as landfills, mixing them with animal manure, and incorporating them into building materials are generally used for the safe disposal of these waste adsorbents (Sullivan et al. 2010). Stabilization/solidification is also considered as a suitable method for the treatment of waste adsorbents. Moreover, this method is widely applied for disposing of toxic waste, as it can convert these toxic waste materials into less-/non-toxic compounds by chemical, physical, or thermal techniques. These techniques are generally used as pre-landfill waste remediation methods for the safe disposal of hazardous wastes. These methodologies involve mixing the waste, either in the form of sludge, liquid, or solid, into a cementitious binder system (Chen et al. 2009). Dumping into a small sand covered brick-lined pit is a common practice used for the disposal of sorptive media. These pits are generally about 1 m^3 in size and are rarely sealed. Sometimes, adsorbent wastes containing arsenic (As) is mixed with cow dung, which can reduce soluble arsenic species to gaseous arsine (AsH_3). Also, utilization of these As-based wastes into the cement blocks and cement plinths is a common practice in the urban locations of Bangladesh and India (Sullivan et al. 2010).

12.9 Advantages and Disadvantages of Adsorption

12.9.1 Advantages

- Due to the benefits of low cost and higher efficiency, adsorption is preferred among the various methods. Also, the technique does not generate any by-products from secondary pollutants (Aslan and Şirazi 2020).
- The process is simple and easily applicable to various remediation systems. Moreover, the technology has benefits such as the availability of a wide range of commercial adsorbents, working efficiently with fast kinetics, yielding high-quality treated effluent, and highly effective for the removal/separation of various kinds of contaminants (e.g., refractory molecule) (Crini and Lichtfouse 2019).
- The literature studies have shown that the process is efficient with respect to the treatment of effluents containing the emerging contaminants (Grassi et al. 2012):
  - The process has shown good removal efficiencies without generation of any by-products that are more toxic than the initial pollutants.
  - The adsorption process can be easily implemented with integrated treatment facilities, which are devoid of the issues like space constraints, waste disposal, and capital/operation cost. Such types of systems (at optimal operating conditions) will produce higher efficiencies with minimal cost.
- Some of the easily available and low-cost materials (like zeolites) are in wide use for various applications, despite their lesser adsorption capabilities compared with the other commercial clay materials. Also, a similar trend has been observed with natural materials like peat, a widely available bio-sorbent that can effectively remove the different kinds of pollutants (Crini 2006).
- The biopolymer materials such as chitosan have become an alternative to other biomaterials due to the following special aspects, like chemical stability, high reactivity, excellent chelation behavior, and high selectivity for the contaminants. Moreover, the compound has drawn wide attention because of its lower price (compared to ACs). Furthermore, possessing higher concentration of amino and hydroxy functional groups has made the chitosan attractive for the removal of various compounds like phenolic substances, dyes, and metal ions (Guibal 2004).

12.9.2 Disadvantages

- Along with the stated benefits, the adsorption process has negative aspects like the chemical requirement for frequent regeneration, fouling, and adsorbent corrosion of plants, as well as disposal of the exhaust air (Younas et al. 2021).
- Requires the success of the process, depending on the following aspects like safe disposal of the spent/saturated adsorbent, maintenance of the pH and temperature to controlled conditions, and selection of adsorbent since the efficiency of the process varies with adsorbent (Bhatnagar and Sillanpää 2011).
- Along with the higher cost and non-selectivity of the ACs, it also has the following disadvantages (Crini et al. 2019):
  - Along with high adsorption capacities, the AC is facing issues related to the disposal of spent ACs, rapid saturation, and regeneration.
  - The widespread use of ACs is restricted since the regeneration is expensive and not straightforward. The regeneration process has resulted in the loss of adsorbent. Moreover, it is not economical for small and medium-sized enterprises to use ACs due to their high cost.
- The regeneration of ACs causes the gradual loss of the material, which eventually affects the economic viability of the process (Streat et al. 1995).
- Though some adsorbents like zeolite gave promising results at the laboratory scale, the real field application of these kinds of materials is unknown. Moreover, this type of materials needs artificial support when they are utilized in column studies (Calzaferri et al. 2000).
- There are many limitations when raw materials like peat are used as adsorbents (Couillard 1994):
  - Natural peat has lesser mechanical strength;
  - Higher affinity toward the water;
  - Low chemical stability;
  - Shrinking and/or swelling tendency; and
  - Releases fulvic acid.

12.10 Conclusions

This chapter includes various aspects like adsorption mechanisms, adsorption isotherms, and kinetic factors affecting adsorption processes, multi-component preference analysis, conventional and non-conventional adsorbent, surface modification of adsorbents, disposal methods of adsorbents, as well as advantages and disadvantages of adsorption. Mass transfer mechanisms and types of adsorption (physisorption and chemisorption) are thoroughly discussed in the adsorption mechanisms section of this chapter. Information related to both isotherms (Freundlich, Dubinin–Radushkevich, Redlich–Peterson, and BET isotherms) and kinetics models (Pseudo-first-order, Pseudo-second-order, Elovich, Intraparticle diffusion models) are discussed briefly. The major factors affecting the adsorption process such are surface area, contact time, pH, temperature, nature, and dose of adsorbate, nature and dose of adsorbent, and interfering substances are also included in the present study.

The multi-component preference analysis aspect included in this discussion has also been supported by recently published data. The conventional and non-conventional (emerging) adsorbents are classified appropriately and are supported with suitable data. The aspects related to the desorption and regeneration of used adsorbent as well as the necessity of the surface modification of the adsorbents are discussed thoroughly in the chapter. The importance of column studies in the adsorption process and handling/disposal of used adsorpbents are explained briefly. The column studies are supported with the latest literature data. Finally, the aspects of advantages and disadvantages of the adsorptions process are discussed at the end of the chapter with suitable examples.

References

Akar, T., Tosun, I., Kaynak, Z. et al. (2009). An attractive agro-industrial by-product in environmental cleanup: dye biosorption potential of untreated olive pomace. *Journal of Hazardous Materials* 166 (2–3): 1217–1225.

Aslan, S. and Şirazi, M. (2020). Adsorption of sulfonamide antibiotic onto activated carbon prepared from an agro-industrial by-product as low-cost adsorbent: equilibrium, thermodynamic, and kinetic studies. *Water, Air, & Soil Pollution* 231 (5): 1–20.

Baig, U., Rao, R.A.K., Khan, A.A. et al. (2015). Removal of carcinogenic hexavalent chromium from aqueous solutions using newly synthesized and characterized polypyrrole–titanium(IV)phosphate nanocomposite. *Chemical Engineering Journal* 280: 494–504.

Bhatnagar, A. and Sillanpää, M. (2011). A review of emerging adsorbents for nitrate removal from water. *Chemical Engineering Journal* 168 (2): 493–504.

Bishnoi, N.R. (2005). Fungus. An alternative for bioremediation of heavy metal containing wastewater: a review. *Journal of Scientific & Industrial Research* 64: 93–100.

Bouzaida, I. and Rammah, M.B. (2002). Adsorption of acid dyes on treated cotton in a continuous system. *Materials Science and Engineering: C* 21 (1–2): 151–155.

Cabrera-Lafaurie, W.A., Román, F.R., and Hernández-Maldonado, A.J. (2015). Single and multi-component adsorption of salicylic acid, clofibric acid, carbamazepine and caffeine from water onto transition metal modified and partially calcined inorganic-organic pillared clay fixed beds. *Journal of Hazardous Materials* 282: 174–182.

Calzaferri, G., Brühwiler, D., Megelski, S. et al. (2000). Playing with dye molecules at the inner and outer surface of zeolite L. *Solid State Sciences* 2 (4): 421–447.

Cesur, H. and Balkaya, N. (2007). Zinc removal from aqueous solution using an industrial by-product phosphogypsum. *Chemical Engineering Journal* 131 (1–3): 203–208.

Chen, C., Liang, B., Ogino, A. et al. (2009). Oxygen functionalization of multiwall carbon nanotubes by microwave-excited surface-wave plasma treatment. *Journal of Physical Chemistry C* 113 (18): 7659–7665.

Chen, R., Qiao, H., Liu, Y. et al. (2015). Adsorption of methylene blue from an aqueous solution using a cucurbituril polymer. *Environmental Progress & Sustainable Energy* 34 (2): 512–519.

Chuah, T.G., Jumasiah, A., Azni, I. et al. (2005). Rice husk as a potentially low-cost biosorbent for heavy metal and dye removal: an overview. *Desalination* 175 (3): 305–316.

Coombs, D.S., Alberti, A., Armbruster, T. et al. (1997). Recommended nomenclature for zeolite minerals: report of the subcommittee on zeolites of the International Mineralogical Association, Commission on new Minerals and Mineral names. *Canadian Mineralogist* 35 (6): 1571–1606.

Couillard, D. (1994). The use of peat in wastewater treatment. *Water Research* 28 (6): 1261–1274.

Crini, G. (2006). Non-conventional low-cost adsorbents for dye removal: a review. *Bioresource Technology* 97 (9): 1061–1085.

Crini, G. and Lichtfouse, E. (2019). Advantages and disadvantages of techniques used for wastewater treatment. *Environmental Chemistry Letters* 17 (1): 145–155.

Crini, G., Lichtfouse, E., Wilson, L.D., and Morin-Crini, N. (2019). Conventional and non-conventional adsorbents for wastewater treatment. *Environmental Chemistry Letters* 17 (1): 195–213.

Crini, G. and Morcellet, M. (2002). Synthesis and applications of adsorbents containing cyclodextrins. *Journal of Separation Science* 25 (13): 789–813.

Dorkoosh, F.A., Verhoef, J.C., Borchard, G. et al. (2002). Intestinal absorption of human insulin in pigs using delivery systems based on superporous hydrogel polymers. *International Journal of Pharmaceutics* 247 (1–2): 47–55.

Faghihian, H. and Bowman, R.S. (2005). Adsorption of chromate by clinoptilolite exchanged with various metal cations. *Water Research* 39 (6): 1099–1104.

Foo, K.Y. and Hameed, B.H. (2011). Microwave-assisted preparation of oil palm fiber activated carbon for methylene blue adsorption. *Chemical Engineering Journal* 166 (2): 792–795.

Garg, V.K., Amita, M., Kumar, R., and Gupta, R. (2004). Basic dye (methylene blue) removal from simulated wastewater by adsorption using Indian Rosewood sawdust: a timber industry waste. *Dyes and Pigments* 63 (3): 243–250.

Ghorbani, F., Kamari, S., Zamani, S. et al. (2020). Optimization and modeling of aqueous Cr(VI) adsorption onto activated carbon prepared from sugar beet bagasse agricultural waste by application of response surface methodology. *Surfaces and Interfaces* 18: 100444.

Grassi, M., Kaykioglu, G., Belgiorno, V., and Lofrano, G. (2012). Removal of emerging contaminants from water and wastewater by adsorption process. *Journal of the Serbian Chemical Society* 79 (4): 15–37.

Guibal, E. (2004). Interactions of metal ions with chitosan-based sorbents: a review. *Separation and Purification Technology* 38 (1): 43–74.

Hall, K.R., Eagleton, L.C., Acrivos, A., and Vermeulen, T. (1966). Pore-and solid-diffusion kinetics in fixed-bed adsorption under constant-pattern conditions. *Industrial & Engineering Chemistry Fundamentals* 5 (2): 212–223.

Ho, Y.S., Chiang, T.H., and Hsueh, Y.M. (2005). Removal of basic dye from aqueous solution using tree fern as a biosorbent. *Process Biochemistry* 40 (1): 119–124.

Ho, Y.S. and McKay, G. (1999). Pseudo-second order model for sorption processes. *Process Biochemistry* 34 (5): 451–465.

Huang, W., Wang, S., Zhu, Z. et al. (2008). Phosphate removal from wastewater using red mud. *Journal of Hazardous Materials* 158 (1): 35–42.

Jørgensen, S.E. (1979). Examination of the applicability of cellulose ion exchangers for water and waste water treatment. *Water Research* 13 (12): 1239–1247.

Kayser, H. (1881). Ueber die Verdichtung von Gasen an Oberflächen in ihrer Abhängigkeit von Druck und Temperatur. *Annals of Physics* 248 (1): 526–537.

Khan, M.M.R., Rahman, M.W., Mozumder, M.S.I. et al. (2015). Performance of a submerged adsorption column compared with conventional fixed-bed adsorption. *New Publication: Balaban* 57 (21): 9705–9717.

Krysztafkiewicz, A., Binkowski, S., and Jesionowski, T. (2002). Adsorption of dyes on a silica surface. *Applied Surface Science* 199 (1–4): 31–39.

Langmuir, I. (1918). The adsorption of gases on plane surfaces of glass, mica and platinum. *Journal of the American Chemical Society* 40 (9): 1361–1403.

Lee, J.W., Samal, S., Selvapalam, N. et al. (2003). Cucurbituril homologues and derivatives: new opportunities in supramolecular chemistry. *Accounts of Chemical Research* 36 (8): 621–630.

Liu, C., Li, Y., Luan, Z. et al. (2007). Adsorption removal of phosphate from aqueous solution by active red mud. *Journal of Environmental Sciences* 19 (10): 1166–1170.

Mahmoodi, N.M. and Maghsoodi, A. (2014). Kinetics and isotherm of cationic dye removal from multicomponent system using the synthesized silica nanoparticle. *New Publication: Balaban* 54 (2): 562–571.

Mailler, R., Gasperi, J., Coquet, Y. et al. (2016). Removal of emerging micropollutants from wastewater by activated carbon adsorption: experimental study of different activated carbons and factors influencing the adsorption of micropollutants in wastewater. *Journal of Environmental Chemical Engineering* 4 (1): 1102–1109.

Márquez, C., Hudgins, R.R., and Nau, W.M. (2004). Mechanism of host–guest complexation by cucurbituril. *Journal of the American Chemical Society* 126 (18): 5806–5816.

Mateus, D.M.R. and Pinho, H.J.O. (2010). Phosphorus removal by expanded clay: six years of pilot-scale constructed wetlands experience. *Water Environment Research* 82 (2): 128–137.

Mohan, D. and Pittman, C.U. (2006). Activated carbons and low cost adsorbents for remediation of tri- and hexavalent chromium from water. *Journal of Hazardous Materials* 137 (2): 762–811.

Mohan, D. and Singh, K.P. (2002). Single- and multi-component adsorption of cadmium and zinc using activated carbon derived from bagasse: an agricultural waste. *Water Research* 36 (9): 2304–2318.

Ngulube, T., Gumbo, J.R., Masindi, V., and Maity, A. (2018). Calcined magnesite as an adsorbent for cationic and anionic dyes: characterization, adsorption parameters, isotherms and kinetics study. *Heliyon* 4 (10): e00838.

Olu-Owolabi, B.I., Popoola, D.B., and Unuabonah, E.I. (2010). Removal of Cu2+ and Cd2+ from aqueous solution by bentonite clay modified with binary mixture of goethite and humic acid. *Water, Air, and Soil Pollution* 211 (1–4): 459–474.

Palma, G., Freer, J., and Baeza, J. (2003). Removal of metal ions by modified *Pinus radiata* bark and tannins from water solutions. *Water Research* 37 (20): 4974–4980.

Park, D., Yun, Y.S., and Park, J.M. (2010). The past, present, and future trends of biosorption. *Biotechnology and Bioprocess Engineering* 15 (1): 86–102.

Parvazinia, M., Garcia, S., and Maroto-Valer, M. (2018). CO_2 capture by ion exchange resins as amine functionalised adsorbents. *Chemical Engineering Journal* 331: 335–342.

Paulino, A.T., Belfiore, L.A., Kubota, L.T. et al. (2011). Effect of magnetite on the adsorption behavior of Pb(II), Cd(II), and Cu(II) in chitosan-based hydrogels. *Desalination* 275 (1–3): 187–196.

Paulino, A.T., Santos, L.B., and Nozaki, J. (2008). Removal of Pb2+, Cu2+, and Fe3+ from battery manufacture wastewater by chitosan produced from silkworm chrysalides as a low-cost adsorbent. *Reactive and Functional Polymers* 2 (68): 634–642.

Radovic, L.R., Moreno-Castilla, C., and Rivera-Utrilla, J. (2000). Carbon materials as adsorbents in aqueous solutions. *Chemistry and Physics of Carbon* 27: 227–405.

Rengaraj, S. and Moon, S.-H. (2002). Kinetics of adsorption of Co(II) removal from water and wastewater by ion exchange resins. *Water Research* 36 (7): 1783–1793.

Rengaraj, S., Yeon, K.-H., and Moon, S.-H. (2001). Removal of chromium from water and wastewater by ion exchange resins. *Journal of Hazardous Materials* 87 (1–3): 273–287.

Rosales-Landeros, C., Barrera-Díaz, C.E., Bilyeu, B. et al. (2013). A review on Cr(VI) adsorption using inorganic materials. *American Journal of Analytical Chemistry* 04 (07): 8–16.

Rout, P.R., Bhunia, P., and Dash, R.R. (2015). Effective utilization of a sponge iron industry by-product for phosphate removal from aqueous solution: a statistical and kinetic modelling approach. *Journal of the Taiwan Institute of Chemical Engineers* 46: 98–108.

Rout, P.R., Bhunia, P., and Dash, R.R. (2017). Evaluation of kinetic and statistical models for predicting breakthrough curves of phosphate removal using dolochar-packed columns. *Journal of Water Process Engineering* 17 (April): 168–180.

Rout, P.R., Dash, R.R., and Bhunia, P. (2014). Modelling and packed bed column studies on adsorptive removal of phosphate from aqueous solutions by a mixture of ground burnt patties and red soil. *Advances in Environmental Research* 3 (3): 231–251.

Saha, A., Gajbhiye, V.T., Gupta, S. et al. (2014). Simultaneous removal of pesticides from water by rice husk ash: batch and column studies. *Water Environment Research* 86 (11): 2176–2185.

Sayğılı, H. and Güzel, F. (2016). High surface area mesoporous activated carbon from tomato processing solid waste by zinc chloride activation: process optimization, characterization and dyes adsorption. *Journal of Cleaner Production* 113: 995–1004.

Sharma, Y.C., Srivastava, V., Singh, V.K. et al. (2009). Nano-adsorbents for the removal of metallic pollutants from water and wastewater. 30 (6): 583–609. https://doi.org/10.1080/09593330902838080.

Shichi, T. and Takagi, K. (2000). Clay minerals as photochemical reaction fields. *Journal of Photochemistry and Photobiology C: Photochemistry Reviews* 1 (2): 113–130.

Singh, N. and Balomajumder, C. (2016). Simultaneous removal of phenol and cyanide from aqueous solution by adsorption onto surface modified activated carbon prepared from coconut shell. *Journal of Water Process Engineering* 9: 233–245.

Singh, N. and Balomajumder, C. (2017). Equilibrium isotherm and kinetic studies for the simultaneous removal of phenol and cyanide by use of *S. odorifera* (MTCC 5700) immobilized on coconut shell activated carbon. *Applied Water Science* 7 (6): 3241–3255.

Srivatsav, P., Bhargav, B.S., Shanmugasundaram, V., et al.(2020). Biochar as an eco-friendly and economical adsorbent for the removal of colorants (Dyes) from aqueous environment: a review. *Water (Switzerland)* 12 (12): 1–27.

Streat, M., Patrick, J.W., and Perez, M.J.C. (1995). Sorption of phenol and para-chlorophenol from water using conventional and novel activated carbons. *Water Research* 29 (2): 467–472.

Sullivan, C., Tyrer, M., Cheeseman, C.R., and Graham, N.J.D. (2010). Disposal of water treatment wastes containing arsenic: a review. *Science of the Total Environment* 408 (8): 1770–1778.

Sun, Q. and Yang, L. (2003). The adsorption of basic dyes from aqueous solution on modified peat–resin particle. *Water Research* P37 (7): 1535–1544.

Tehrani, M.S., Azar, P.A., Namin, P.E., and Dehaghi, S.M. (2013). Removal of lead ions from wastewater using functionalized multiwalled carbon nanotubes with Tris(2-Aminoethyl) Amine. *Journal of Environmental Protection* 4 (6): 529–536.

Teng, W.L., Khor, E., Tan, T.K. et al. (2001). Concurrent production of chitin from shrimp shells and fungi. *Carbohydrate Research* 332 (3): 305–316.

Tiravanti, G., Petruzzelli, D., and Passino, R. (1997). Pretreatment of tannery wastewaters by an ion exchange process for Cr(III) removal and recovery. *Water Science and Technology* 36 (2–3): 197–207.

Turk Sekulic, M., Boskovic, N., Milanovic, M. et al. (2019). An insight into the adsorption of three emerging pharmaceutical contaminants on multifunctional carbonous adsorbent: mechanisms, modelling and metal coadsorption. *Journal of Molecular Liquids* 84: 372–382.

Vandenbossche, M., Jimenez, M., Casetta, M., and Traisnel, M. (2015). Remediation of heavy metals by biomolecules: a review. 45 (15): 1644–1704. http://dx.doi.org/10.1080/10643389.2014.966425.

Vijayaraghavan, K. and Balasubramanian, R. (2015). Is biosorption suitable for decontamination of metal-bearing wastewaters? A critical review on the state-of-the-art of biosorption processes and future directions. *Journal of Environmental Management* 160: 283–296.

Vijayaraghavan, K. and Yun, Y.S. (2008). Bacterial biosorbents and biosorption. *Biotechnology Advances* 26 (3): 266–291.

Vilela, P.B., Matias, C.A., Dalalibera, A. et al. (2019). Polyacrylic acid-based and chitosan-based hydrogels for adsorption of cadmium: equilibrium isotherm, kinetic and thermodynamic studies. *Journal of Environmental Chemical Engineering* 7 (5): 103327.

Weber Jr, W.J. and Morris, J.C. (1964). Equilibria and capacities for adsorption on carbon. *Journal of the Sanitary Engineering Division* 90 (3): 79–108.

Yadav, K.K., Gupta, N., Kumar, V. et al. (2018). A review of emerging adsorbents and current demand for defluoridation of water: bright future in water sustainability. *Environment International* 111: 80–108.

Younas, F., Mustafa, A., Farooqi, Z.U.R. et al. (2021). Current and emerging adsorbent technologies for wastewater treatment: trends, limitations, and environmental implications. *Water* 13 (2): 215.

Zahar, M.M. (2020). Treatment of Acid Mine Drainage (AMD) using industrial by-product: sorption behavior of steel slag for metal-rich mine water. *Global NEST Journal* 22 (2): 258–267.

13

Ion Exchange Process for Removal of Microconstituents From Water and Wastewater

Muhammad Kashif Shahid[1], H.N.P. Dayarathne[2], Bandita Mainali[3], Jun Wei Lim[4,5], and Younggyun Choi[6]

[1]*Research Institute of Environment & Biosystem, Chungnam National University, Daejeon, Republic of Korea*
[2]*School of Engineering and Mathematical Sciences, La Trobe University, Bendigo, Australia*
[3]*School of Engineering, Faculty of Science and Engineering, Macquarie University, Sydney, Australia*
[4]*HICoE-Centre for Biofuel and Biochemical Research, Institute of Self-Sustainable Building, Department of Fundamental and Applied Sciences, Universiti Teknologi PETRONAS, Seri Iskandar, Perak Darul Ridzuan, Malaysia*
[5]*Department of Biotechnology, Saveetha School of Engineering, Saveetha Institute of Medical and Technical Sciences, Chennai, India*
[6]*Department of Environmental & IT Engineering, Chungnam National University, Daejeon, Republic of Korea*

13.1 Introduction

The ion exchange process has achieved significant attention in water treatment systems. This process includes a reversible exchange of ions that occurs between the solid and liquid phases of ion exchangers (Dixit et al. 2021). In this process, the targeted ions from the liquid phase are substituted by the mobile ions from the ion exchangers. Several ion exchange resins contain diverse polymers and different ligands attached with polymers (Altarawneh et al. 2020). Due to the insoluble resin phases, the ion exchange resins have great importance. Followed by interactions with ionic solutions, the resin can be easily separated through the filtration process. These materials are also feasible for the continuous operations including chromatographic and column separations. Their insolubility makes them eco-friendly since the series of processes including packing, regeneration, and repacking enables their use for several years (Chandrasekara and Pashley 2015). These materials have been applied in water and wastewater treatment, removal of contaminants of emerging concerns, and MCs from water, water softening, sensors, hydrometallurgy, biological phase separations, and chromatography (Shahid et al. 2021a).

Water and wastewater treatment plants (WWTPs) are progressively designed to remove the micropollutants and contaminants from water originating from different sources (Shahid et al. 2017, Shahid et al. 2021b). For instance, natural organic matter (NOM) causes a serious issue in water supplies because the disinfection practices result in the transformation of natural compounds into toxic by-products (Zheng et al. 2016). The micropollutants (e.g., pesticides, synthetic and natural hormones, pharmaceutically active products) are

Microconstituents in the Environment: Occurrence, Fate, Removal, and Management, First Edition.
Edited by Rao Y. Surampalli, Tian C. Zhang, Chih-Ming Kao, Makarand M. Ghangrekar, Puspendu Bhunia, Manaswini Behera, and Prangya R. Rout.

known to be of serious concern for environmental and human health (Shahid et al. 2021a). For instance, the estrogenic substances including 17α-ethinylestradiol and 17β-estradiol (with ng/L concentration) can lead to endocrine disrupting impacts in fish species including estuarine flounder, minnow, and trout (Liu 2011). The perfluorinated compounds (PFCs) widely applied in paper, textile, and oil industries are discharged to the environment mainly from production processes such as the coating stage. These MCs may be found everywhere in the environment. A study reported 200 ng/L concentration of perfluorooctane sulfonate in surface water (McLachlan et al. 2007). The quantity, properties, and effects of MCs differ significantly based on water sources and the biogeochemical cycles of their environments.

The possible hazardous effects of MCs toward aquatic ecosystem and human life cannot be ignored (Shahid and Choi 2020; Kashif et al. 2022). Hence, an effective removal technology is essential to eliminate these substances from the aqueous environment. Conventional wastewater treatment processes include chlorination, activated sludge system, advanced oxidation processes, adsorption, and bioremediation (Shahid et al. 2019; Chhuon et al. 2020; Dayarathne et al. 2021). Recently, ion exchange resins received attention for their promising treatment of MCs in water and wastewater (Johnson et al. 2016). This chapter discusses the properties of different types of ion exchange resins and their applications for removal of MCs from aqueous solution. The efficiency of ion exchangers significantly depends on several factors including solution pH, type of resin, dosage, ionic strength of the solution, and physiochemical properties of the adsorbate.

13.2 Properties of Different Ion Exchange Resin

The ion exchange resins are high-molecular weight materials, generally solid, inorganic or organic compounds, and remain insoluble in water and several other solvents. The ion exchange resins have the ability to exchange their active ions (cations or anions) with the ions being attracted from the neighboring electrolyte (Eqs 3.1–3.2). Chemically, these substances are polybases, polyacids, or both (amphoteric ion exchangers, also known as polyampholyte). Ion exchange resins are categorized as cation and anion exchangers. The cation exchangers carry positively charged exchangeable mobile ions, whereas anion exchangers have negatively charged exchangeable ions. A few ion exchangers synthesized via modification of several kinds of substances, mainly natural ones, in addition to exchanging ions, display distinctive sorption properties (Kumar et al. 2017). For instance, giant reed is a naturally existing composite of cellulose, hemicelluloses, and lignin, which is inexpensive and readily available. The modification of giant reed with quaternary amine functionalities indicated high removal efficiency for perchlorates (Baidas et al. 2011). Similarly, modified chitosan and wheat straw were found effective to remove MCs such as perchlorates in water (Xie et al. 2010). Generally, the cation exchangers have functionalities including $-COOH$, $-SO_3H$, and $-OH$. The cation exchanging resins with $-SO_3H$ functional groups are classified as strong cation exchangers. The cation exchanger dissociates to discharge hydrogen ions that assists other positively charged ions to take up the ion exchange vicinities. The $-COOH$ and $-PO_3^{2-}$ functionalities constitute the weak cation exchange resins, being less chemically resistant.

$$2Na^{+}R^{-} + Ca^{2+} \rightleftharpoons Ca^{2+}R_2^{-} + 2Na^{+} \; [\text{cation exchange reaction}] \tag{13.1}$$

$$2R^+OH^- + H_2SO_4 \rightleftharpoons R_2^+SO_4^{2-} + 2H_2O \ \left[\text{anion exchange reaction}\right] \tag{13.2}$$

where R indicates the cation exchange resin and anion exchange resin in Eqs (13.1) and (13.2), respectively.

Anion exchange resins are made up of a polymeric matrix to which different functionalities are linked. The quaternary ammonium group is found in strong basic anion exchangers, whereas tertiary amino groups constitute the weakest basic anion exchanger. The primary and secondary amino groups are also characteristics of weak basic anion exchangers. The commercially available exchangers comprise of either $-N^+(CH_3)_3$ groups (known as type 1 resins) or $-N^+(CH_3)_2C_2H_4OH$ groups (known as type 2 resins). The ion exchange resins are also categorized based on material, structure, functionalities, and their synthesis methods (Figure 13.1). These materials are also categorized based on origin such as synthetic, semi-synthetic, and natural ion exchangers (Al-Asheh and Aidan 2021).

The properties of the ion exchange resins primarily depend on the physico-chemical nature of their functionalities such as their dissociation properties, selective bonding for certain species or ions, charge, or degree of ionization (Chandrasekara and Pashley 2015). It is noteworthy that the characteristics of an ion exchange resin are also influenced by the polymeric support and the degree of cross-linking. Generally, two types of interactions can be found among the functional groups and the target species including ion substitution

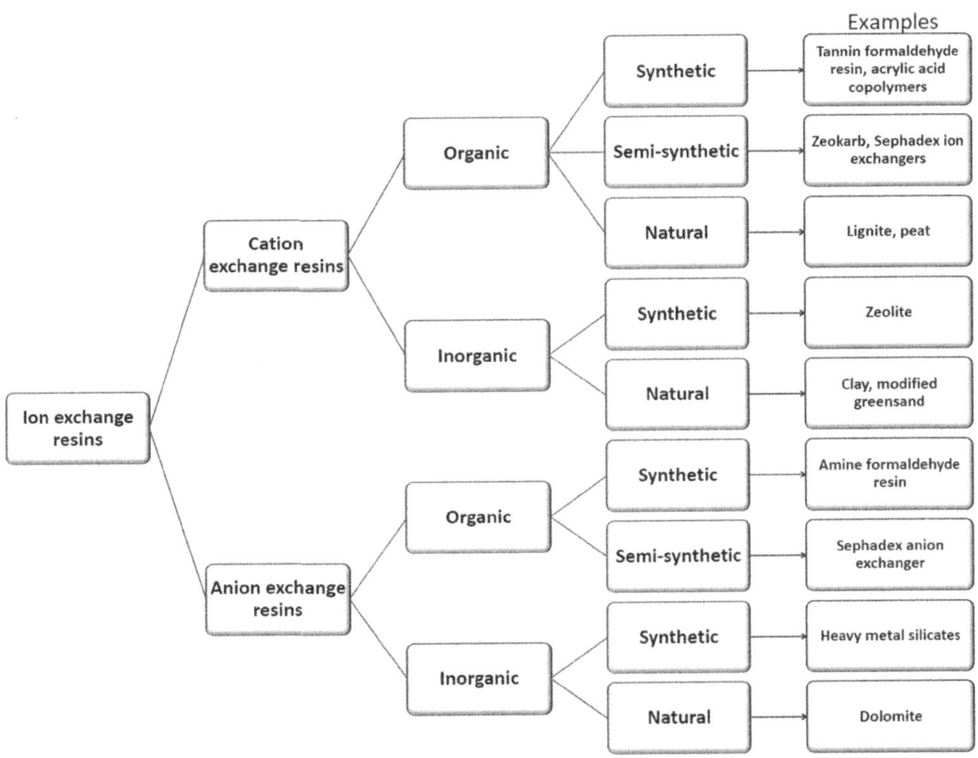

Figure 13.1 A common grouping of ion exchange resins based on their sources.

coordination bonding. An ideal ion exchanger is expected to have physical and chemical stability, hydrophilicity, adequate ion exchange capacity, reasonably high speed of ion exchange, adequate particle size and effective surface, and cost-effectiveness (Levchuk et al. 2018).

The ion exchange resins are generally produced by addition and condensation polymerization reactions. Addition polymerization involves the repetitive monomers addition which possess double or triple bonds to synthesize polymers (Li et al. 2021). Condensation polymerization is a phenomenon where repetitive condensation reactions occur among two diverse bi-functional or tri-functional monomers (Li et al. 2021). An example of condensation polymerization is the synthesis of terpolymer cation-exchange resin using 0.1 mol p-hydroxybenzaldehyde, 0.2 mol formaldehyde, and 0.1 mol biuret (Patle and Gurnule 2016). The reaction is carried out in an HCl medium under oil bath ($126 \pm 2°C$) for 5 h. The production of anion exchange resin from divinylbenzene and 2-methyl-5-vinylpyridine might be an example of addition polymerization (Vorotyntsev et al. 2018). Herein, polymerization was carried out in a thermostatic water bath (80°C; 400 rpm; 24 h). The condensation polymerization occurs through a step-growth pattern while an addition polymerization happens in chain-growth patterns. A single monomer is required in addition polymerization, whereas a minimum of two monomers are required for the condensation polymerization.

13.3 Functionalities of Polymeric Resins

A range of functional groups utilizing oxygen, sulfur, nitrogen, or phosphorus as an active atom are found in ion exchange resins (Table 13.1). Followed by the ionization of a functional group, an exchange of ions happens between the mobile ions of resin and the target ions in the bulk solution. The widely applied polymeric matrixes in ion exchange resins are divinylbenzene (DVB), polystyrene (PS), polyacrylic divinylbenzene (PA-DVB), and polystyrene cross-linked with divinylbenzene (PS-DVB) (Nesterenko and Nesterenko 2021). The nitrogen-based resins need a primary protonation to trigger an anion exchange property. The secondary and tertiary amino groups, quaternary ammonium groups, guanidine groups, and picolylamine groups are the most frequently found nitrogen-carrying functionalities in ion exchange resins (Wołowicz and Hubicki 2012). The protonation characteristics of the amino groups enable the synthesis of ion exchange resins with a range of physico-chemical characteristics. Amino groups are distinguished based on the number of H-atoms replaced by aryl or alkyl groups. The replacement of a single H-atom in the amino group is classified as the primary amine. Once two or three hydrogen atoms are replaced, secondary or tertiary amines are produced, respectively. The protonation of the amine group by a fourth aryl or alkyl group results in the formation of the quaternary ammonium group. Due to the greater dissociation strength, the quaternary ammonium groups have the ability to treat a solution containing a broad range of anionic complexes (Jinhua et al. 2010). Most of the commercially applied strong anion exchange resins contain quaternary ammonium groups (Li et al. 2022). The ion exchange resins having secondary and tertiary amino groups are weak anionic resins with approximate pKa values of 10.7 and 9.7, respectively. Resins with a mixture of tertiary amine and quaternary ammonium groups are

Table 13.1 Classification of widely applied commercial resins based on functionalities.

Functionality	Resin Name	Type
Quaternary ammonium functionality	Amberlite IRA-458	$-N^+(CH_3)_3$/nsa
	Purolite A850DL	$-N^+R_3/Cl^-$
	Dowex Minix	$-N^+$((CH2)3CH3)3/ns
	Dowex MAS-1	$-N^+R_3$/ns
	Indep. NOTREN	$-N^+R_3/Cl^-$
	Purolite A850DL	$-N^+R_3/Cl^-$
	Purolite A500/2788	$-N^+(CH_3)_3$/ns
	Purolite A300OH	$-N(CH_3)_2(C_2H_5OH)/OH^-$
	Purofine PFA300	$-N(CH_3)_2(C_2H_5OH)/Cl^-$
Combined amino functionality	Amberlite IRA 96	$-NR_2$/FB
	Amberlite IRA-67	$-NR_2$/FB
	Purolite A172	Complex amine/FB
	Purolite A100/2412	$-N^+R_2$; $-N^+R_3$/ns
	Lewatit MP64	$-N^+R_2$; $-N^+R_3/Cl^-$
	TOKEM AN-25	Pyridine/ns
	Resinex HM-WBP-79 C	$-N^+R_2$; $-N^+R_3$/FB;Cl^-
Picolylamine functionality	Dowex XUS 43578	–N(picolyl)$_2$/ns
	Dowex M4195	–N(picolyl)$_2$/ns
	Dowex XUS 43605	$-OH(C_3H_6)N$–(picolyl)/ns
	Lewatit MDS TP 220	–N(picolyl)$_2$/SO_4^{2-}
	Purolite S960	–N(picolyl)$_2$/FB; SO_4^{2-}
Sulfonic functionality	BioRad AG 50W-X8	$-S(=O)_2-OH$/ns
	Dowex G-25	$-S(=O)_2-OH/H^+$
	Ionac CFP-110	$-S(=O)_2-ONa/Na^+$
	Amberlite 252	$-S(=O)_2-ONa/Na^+$
	Lewatit S1468	$-S(=O)_2-ONa/Na^+$
	Dowex XVS	$-S(=O)_2-OH$/ns
	Xi'an LST D001	$-S(=O)_2-OH$/ns
Thiourea functionality	Dowex XUS 43600	$-CH_2SC=NHNH_2$/ns
	Lewatit TP-214	$-CH_2NHC=SNH_2$/ns
	Purolite S-920	$-CH_2SC=NHNH_2$/H+
Thiol functionality	Ambersep GT74	$-SH/H^+$
	Chelite S	SNa/Na^+
	Duolite GT73	$-SH/H^+$
	Smopex 234	$-C(=O)OC_2H_4SH$/ns

(Continued)

Table 13.1 (Continued)

Functionality	Resin Name	Type
Phosphonic functionality	Bio-Rex 63	$-P(=O)(OH)_2$/ns
	Cationite FV	$-P(=O)(OH)_2$/ns
	Cationite PM-52	$-P(OH)_2$/ns
	Cationite KF-3	$-P(=O)(OH)_2$/ns
	Duolite C-65	$OP(=O)(OH)_2$/ns
	Duolite ES-463	$-P(=O)(OH)_2$/ns
	Eichrom Actinide Resin	DEHMPA/ns
	Nalcite X-219	$-P(=O)(OH)_2$/ns
Phosphinic functionality	Diaion CRP200	$-P(=O)OHR$/ns
	Lewatit TP 272	DTMPPA/ns
Phosphoric functionality	Lewatit VP OC 1026	D2EHPA/ns
	Eichrom LN Resin	D2EHPA/ns
Phosphine functionality	Eichrom TRU Resin	CMPO in TBP/ns
	Eichrom RE Resin	CMPO in TBP/ns
Glucamine functionality	Amberlite IRA 743	Glucamine/ns
	Diaion CRB03	Glucamine/ns
	Purolite S-108	Glucamine/ns
Oxime functionality	Eichrom Nickel Resin	DMG/ns
	Purolite 910	$-C(=NOH)NH_2$/FB
	Sumichelate MC900	Amidoxime/ns
Aminophosphonic functionality	Amberlite RC747UPS	$-CH(NH_2)-P(=O)(OH)_2$/ns
	Amberlite IRC747	$-NH-CH_2-P(=O)(OH)_2$/ns
	Amberlite IRC747	$-NH-CH_2-P(=O)(ONa)_2/Na^+$
	Lewatit MDS TP 260	$-NH-CH_2-P(=O)(ONa)_2/Na^+$
	Purolite S940	$-CH(NH_2)P(=O)(ONa)_2/Na^+$
	Xi'an LST LSC-500	$-NH-CH_2-P(=O)(OH)_2$/ns
	Serdolit Chelite P	$-P(=O)(OH)_2$/ns
Iminoacetic functionality	Amberlite IRC748	$-N(CH_2C(=O)ONa)_2/Na^+$
	BioRad Chellex-100	$-N(CH_2C(=O)OH)_2$/ns
	Dowex A-1	$-CH_2N(CH_2C(=O)ONa)_2/Na^+$
	Diaion CR10	$-CH_2N(CH_2C(=O)OH)_2$/ns
	Diaion CR11	$-N(CH_2C(=O)ONa)_2/Na^+$
	KT-2 (ANKB-50)	$-CH_2N(CH_2C(=O)OH)_2$/ns
	Muromac A-1	$-N(CH_2C(=O)OH)_2$/ns
	Serdolit Chelite CHE	$-N(CH_2C(=O)OH)_2$/ns
	Wofatit CM-50	$-CH_2N(CH_2C(=O)OH)_2$/ns
	Varion CH	$-CH_2N(CH_2C(=O)OH)_2$/ns
	Xi'an LST LSC-100	$-N(CH_2C(=O)OH)_2$/ns

Table 13.1 (Continued)

Functionality	Resin Name	Type
Carboxylic functionalities	IMAC HP333	–COOH/ns
	Dowex MAC-3	$-COOH/H^+$
	Lewatit S 8528	$-COOH/H^+$
	Purolite C106	$-COOH/H^+$
	Wofatit CA-20	$-COONa/Na^+$

CMPO = octylphenyl-*N,N*-di-isobutyl carbamoylphosphine oxide; DMG = dimethylglyoxim; D2EHPA = di(2-ethylhexyl) phosphoric acid; DTMPPA = di(2,4,4trimethylpentyl) phosphinic acid; DEHMPA = di(2-ethylhexyl) methanediphosphonic acid; TBP = tri-n-butyl phosphate; ns = not specified.
Source: Adapted from Silva et al., 2018.

known to have higher selectivity for targeted ions. The combination of weak and strong anion exchanger groups has been applied in large-scale water and wastewater treatment plants.

The ion exchange resins with sulfonic groups are widely applied in metal separation processes. This group of ion exchangers exhibits the robust adsorption capacity among the cation exchangers and has been applied in the treatment of industrial wastewater (Moosavirad et al. 2015). A recent study examined the interactions between metaldehyde (pesticide) and the sulfonic functional group of an ion exchange resin. It is found that the addition of high concentrations of sulfonic groups on an appropriate adsorption media can enhance the efficiency of metaldehyde removal (Altarawneh et al. 2020). The ion exchange resins with a thiol functional group have been used for treatment of wastewater containing arsenic (Duarte et al. 2021) and mercury (Wang et al. 2021). Thiourea has a strong affinity for the metal ions and has been commercially integrated into ion exchange resin (Riley et al. 2021). The thiourea functionalized resins have been applied for removal of iron, copper, lead, and zinc from sewage sludge leachates (Bezzina et al. 2019).

The phosphorus-based resins are generally applied for pre-concentration of transition metals, uranium, lanthanides, and thorium. These ion exchangers comprise of either phosphinic, phosphonic, or phosphate functionality (Hérès et al. 2018). Contrary to the sulfur- or nitrogen-based groups, a limited number of phosphorus-based resins are commercially available. The phosphonic group has high, stable, and distinct chelating associations with polyvalent metal ions owing to four potential donor atoms with ligand potential. The ion exchange resins having phosphorous groups are rare and usually synthesized merely at lab-scale as solvent impregnated resins. The most effective phosphorous-based resins include di(2,4,4-trimethylpentyl)-phosphinic acid, di(2-ethylhexyl) phosphoric acid, tri-isobutyl-phosphine sulfide, tri-n-octylphosphine oxide, tri-n-butylphosphate, and O-methyldihexyl-phosphine oxide O'-hexyl-2-ethyl phosphoric acid (Silva et al. 2018). The ion exchange resins with immobilized phosphinic groups appeared to be effective in removing divalent ions such as Cu^{2+}, Zn^{2+}, Co^{2+}, and Cd^{2+}. While phosphorous-based resins display high performance for metal separation technologies, the higher production costs limit the large-scale production of these resins.

Oxygen atoms are generally found in resins as a part of hydroxyl and carboxyl functionalities (Silva et al. 2018), though very few ion-exchange resins are based solely on oxygen-bearing functionalities as ion exchangers. Resins with the carboxylic functional groups are a widely known type of weak cation exchange resin. These resins have been applied in water softening, and metal recovery during water and wastewater treatment (Mukhamediev et al. 2021). Apart from carboxylic functionalities, oxygen is generally found as an active ligand site (in combination with nitrogen, sulfur, and phosphorus) in various ion exchange resins such as the oxime, glucamine, and iminoacetic functional groups. The oxime groups exhibit greater selectivity for trivalent metals, and lab-scale investigations highlight significant desorption rates using organic acids (Zhang et al. 2016). Glucamine resins has been applied for the removal of arsenic, mercury, and boron (Cyganowski et al. 2021). Resins with iminoacetic groups are widely applied in metal recovery from water and wastewater treatments preferring the adsorption of base metals. They can be easily regenerated with strong acids.

13.4 Ion Exchange Mechanism

The ion exchange rate significantly depends on the chemical interactions involved in the ion exchange material such as pellets, beads, dispersions, membranes, etc. The kinetic selectivity of activities occurring inside the material phase plays an important role in this regard. The reaction rates immensely depend on the interior structure of ion exchange materials (Kumar and Jain 2013). The widely known ion exchange process contains a water-swollen ion exchanger and an aqueous environment. Along the lines of any heterogeneous system, an ion exchange is achieved through ions transfer to and from the interphase boundaries. This phenomenon involves the chemical interactions itself, diffusion into the material, and diffusion within the neighboring solution. Apart from two main phases, a thin layer (or film) of solution onto the surface of the exchanger must be considered distinctly. The characteristics of the thin film varies from the characteristics of the neighboring bulk solution. The production of this layer cannot be ignored and severe agitation can only decrease the thickness of the interphase layer.

The schematic of mass transfer during the ion exchange process is depicted in Figure 13.2. It can be seen that "Y" ions diffuse from the bulk solution via a thin layer into ion exchange material, and "X" ions diffuse out of the ion exchange material crossing the thin film into the bulk solution. This interphase diffusion of counterions is known as ion exchange. The counterions are replaceable ions found in ion exchangers and tend to move freely within the system; however, their motion needs to compensate for respective counter motion of other ions with similar change satisfying the law of electroneutrality (Dong et al. 2021). The term "counterion" can be applied in a wider prospective, either for the ions inside the exchanger or outside the exchanger in bulk solution (Pidoux et al. 2022). All the ions with a charge contrary to the functionality can be referred as counterions. The term "co-existing ion" is applied to the ions with the charge similar to the functional group. "Ionic form" is the term defining which counterion is present in the ion exchange.

Considering the law of electroneutrality, all the counterions that leave the ion exchange resin will be substituted by an equal number of other counterions (Pidoux et al. 2022). As

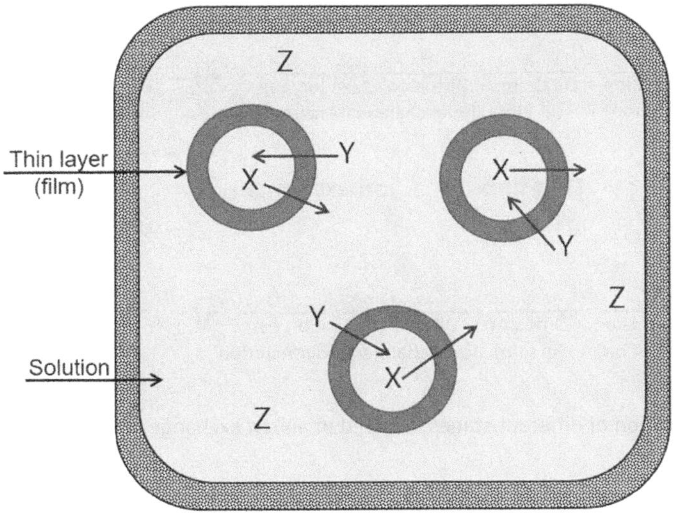

Figure 13.2 A schematic illustration of mass transfer taking place in ion exchange. "X" indicates the counter ion attached with ion exchange material. "Y" and "Z" show the counter ion and co-existing ion in the solution, respectively. During ion exchange process, Y moves into the ion exchanger substituting X.

the counterion passes the interphase bridge, it results in generation of an electric potential between two phases. This potential needs to be counterbalanced through transportation of another counterion in an opposing way (ion exchange) or by the transport of a co-existing ion in the similar direction. In various operations, specifically in low concentrated solution, Donnan exclusion (Gao et al. 2022) hinders the co-existing ions to enter the exchanger throughout the entire process. Under these circumstances, the co-existing ions do not take part in the reaction and only slightly influence the reaction rate. Figure 13.3 briefly highlights the steps involved in the ion exchange process. The entire process can be divided in ten steps that can be described as follows:

1) In an aqueous solution, the dissociation of dissolved complexes occurs.
2) First ion diffuses from the solution toward the thin layer, known as the interphase film. This step can be influenced via agitation to enhance the ion transfer.
3) The first ion diffuses out of the interface film. Agitation can be applied to decrease thickness of film. Herein, transport can only occur due to the mobility of the ion.
4) Owing to the concentration gradient, the first ion diffuses inside the ion exchanger.
5) The first ion attaches with the functional group of the exchanger.
6) The second ion dissociates from the functional group of the ion exchanger material following the law of electroneutrality.
7) The second counterion diffuses from the bulk of an ion exchange material to the surface.
8) The counterion diffuses through the thin interface layer toward bulk solution.
9) The counterion diffuses into the bulk solution.
10) The counterion attaches with respective charge species in the solution phase.

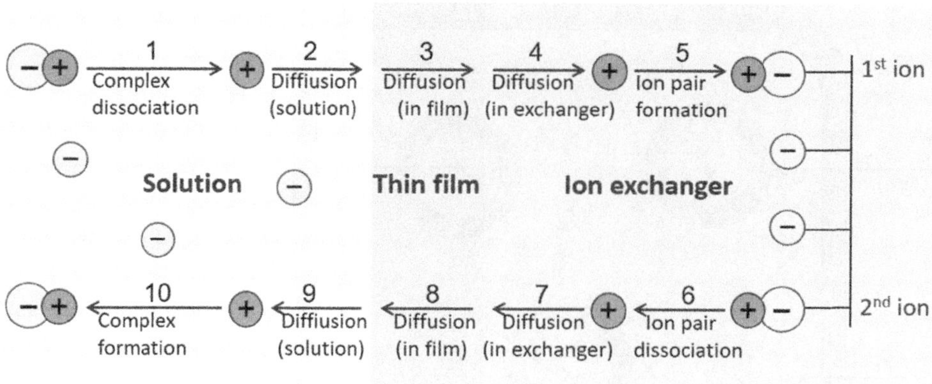

Figure 13.3 A schematic illustration of different stages involved in an ion exchange process.

13.5 Ion Exchange Kinetics

The ion exchange process only happens if an ion is able to stoichiometrically substitute another ion and an adequate flux is available for ion exchange (Bonin et al. 2021). The diffusion of the slowest ion significantly limits the mass transport kinetics. Although a simple illustration of the ion exchange process is depicted in Figure 13.3, a relatively complex process may occur during real ion exchange. Principally, the highly mobile ions diffuse at a fast rate; however, ion diffusion is always affected by the charge balance throughout the material. The charge balance can speed up the slower ions and slow down the faster ions. The mechanism involved in an ion exchange process describes potential rate-determining steps. Based on Figure 13.3, the scientific illustration can be presented as:

- *Mass transfer (diffusion) toward solution or in other outer phase* (Figure 13.3, stages 2 and 9) is an inevitable stage during ion associations. This is a recognized process as defined by conventional diffusion equations. It can be simply supported through hydrodynamic turbulences (e.g., stirring) and so is not thought of as a potential restrictive stage for ion exchange.
- *Mass transfer via interface layer (film)* covering an ion exchanger is signified by stages 3 and 8 in Figure 13.3. The interface layer is a solution region of definite thickness without convection. The mass transfer inside the thin layer depends on the rate constant known as the diffusion coefficient. The thickness of the layer can be minimized by agitation; however, zero thickness cannot be achieved.
- *Mass transfer inside the ion exchanger* (Figure 13.3, stages 4 and 7) signif icantly depends on the physico-chemical properties of an ion exchanger and cannot be altered unless the material properties are changed.
- *The association and dissociation reactions for ion pair* (Figure 13.3, stages 5 and 6) are the main chemical reactions, which can influence the overall rate of reaction of ion exchange.
- *The dissociation and formation of complexes in bulk solution* (Figure 13.3, stages 1 and 10) is not assumed as a unit of an ion exchange process. However, the formation of complex formation might be a bottleneck for the entire process.

Based on the aforementioned illustrations, it can be summarized that the majority of ion exchange processes exclusively follow the diffusion phenomenon. The processes are not controlled by real chemical reactions but by diffusion of counterions (Dong et al. 2021). In most cases, two major rate determining steps are the mass transport (diffusion) inside the ion exchange material or mass transfer (diffusion) through the interface layer (film). In both circumstances, mass transfer is described merely by system properties. The rate of these stages cannot be changed by external workout without changing the system itself. Diffusion into the ion exchange material is known as particle diffusion, and the diffusion by interface layer is known as film diffusion. In simple circumstances, the ion exchange rate is estimated through the slowest process out of them. The leading mechanism can be anticipated applying approximate criterion (Eqs 13.3 and 13.4).

$$\frac{Q\cdot \bar{D}\cdot \delta}{C\cdot D\cdot r0}\left(5+2\alpha_{\mathrm{A}}^{\mathrm{B}}\right) \ll 1 \text{ Particle diffusion control} \tag{13.3}$$

$$\frac{Q\cdot \bar{D}\cdot \delta}{C\cdot D\cdot r0}\left(5+2_{\mathrm{A}}^{\mathrm{B}}\right) \gg 1 \text{ Film diffusion control} \tag{13.4}$$

where *Q*, *C*, and *D* signify the concentration of fixed group, solute concentration, and the interdiffusion coefficient in the ion exchange, respectively. The interdiffusion coefficient inside an ion exchanger is signified by *D* with over-line mark, whereas δ, $r0$, and $\alpha^{B}{}_{A}$ indicated the film thickness, bead radius, and separation factor, respectively.

As per criterion, all factors that tend to enhance the interdiffusion rate inside the beads and to decrease the rate in the layer, support the film diffusion control and vice versa. Therefore, the film diffusion control can succeed in the system characterized by distinct properties including low degree of cross-linking, higher concentration of exchange localities in the material, dilute solution, small particle size, and ineffective agitation.

13.6 Application of Ion Exchange for Treatment of Microconstituents

The microconstituents (MCs), also known as micropollutants, are defined as the natural and synthetic substances, containing elements and organic and inorganic compounds, identified within the environment and water for which regular evaluation of the possible effect on environment and human life is a prudent procedure. Several organic micropollutants, including industrial chemicals, pharmaceutical and personal care products, pesticides, and endocrine disruptors, enter surface waters through effluents from WWTPs and stormwater runoff. The traces of these MCs have been identified in treated drinking water worldwide (Shahid 2021a). Besides trace concentrations of MCs in drinking water, there is growing concern over the long-standing impacts on life due to its exposure to a blend of MCs. Ion exchange resins are widely applied in treatment of MCs found in aqueous environments. Several studies reported successful removal of heavy metals from water streams using cationic exchange resins (Abo-Farha et al. 2009). Table 13.2 details the properties and effectiveness of some commercial ion exchange resins for removal of MCs from water.

Table 13.2 Summary of some ion exchange resins, their properties, and ion exchange capacity.

Resin	Matrix	Functionality	Microconstituents	Uptake Capacity (μmol/g)	References
A520E	Polystyrene	Quaternary ammonium.	Perfluorooctanoate.	211	(Gagliano et al. 2020)
IRA 400	Polystyrene	Quaternary ammonium.	Perfluorooctanesulfonic acid.	570	(Deng et al. 2010)
IRA 958	Polyacrylic	Quaternary ammonium.	Perfluorooctanesulfonic acid.	5251	(Deng et al. 2010)
IRA410	Polystyrene	Benzyldimethyl (2-hydroxyethyl) ammonium.	Perfluorooctanesulfonic acid.	787	(Deng et al. 2010)
MN100	Polystyrene	Tertiary amine.	17α-Ethinylestradiol.	2000	(Liu 2011)
MN100	Polystyrene	Tertiary amine.	17β-Estradiol.	1250	(Liu 2011)
MN100	Polystyrene	Tertiary amine.	Trimegestone.	769	(Liu 2011)
MN100	Polystyrene	Tertiary amine.	Progesterone.	1111	(Liu 2011)
MN100	Polystyrene	Tertiary amine.	Gestodene.	1429	(Liu 2011)
MN100	Polystyrene	Tertiary amine.	Estrone.	667	(Liu 2011)
A530E	Polystyrene	Quaternary ammonium.	Perfluorotridecanoic acid.	3016	(Liu 2011)
A530E	Polystyrene	Quaternary ammonium.	Perfluoroundecanoic acid.	3894	(Liu 2011)
A530E	Polystyrene	Quaternary ammonium.	Tricosafluorododecanoic acid.	5868	(Liu 2011)
A532E	Polystyrene	Bifunctional Quaternary amines.	Tricosafluorododecanoic acid.	7403	(Liu 2011)
A532E	Polystyrene	Bifunctional Quaternary amines.	Perfluoroheptanoic acid.	2837	(Liu 2011)
IRA67	Polyacrylic divinylbenzene.	Combined amino functional group.	Perfluorooctanoate.	2820	(Du et al. 2015)
AI400	Polystyrene divinylbenzene.	Quaternary ammonium.	Perfluorooctanoate.	2920	(Yu et al. 2009)

Lalmi et al. (2018) achieved over 90% removal of Pb^{2+} from contaminated water employing a cation exchange resin (Purolitre C100 E) with a sulfonic acid group. The highest Pb^{2+} removal efficiency was obtained under acidic pH conditions (2.4–4.8). A significant rise in lead removal is observed with increasing resin bed height; however, an increased flowrate caused decrease in removal rate (Lalmi et al. 2018). The applied resin is efficiently regenerated using 2 molar $Ca(NO_3)_2$. Zawierucha et al. (2022) achieved over 98% removal efficiency for arsenate when applying permethylated β-cyclodextrin impregnated ion exchange resin in static and continuous mode of operations. A study examined the efficiency of different ion exchange resins for mercury recovery from mine wastewater, and achieved the highest removal efficiency (≈100% removal rate) with Dowex XZS-1 that is a strong cation resin with sulfonic acid functional group (Monteagudo and Ortiz 2000). The resin is successfully regenerated with 0.5 molar HCl (Monteagudo and Ortiz 2000). A study utilized polymeric sub-micron ion exchange resin to remove copper, nickel, zinc, and lead from the surface water and wastewater (Murray and Örmeci 2019). The study reported the nickel (II) removal using a wide range of resins and found different complexation (Figure 13.4) based on the basicity level of the solution and the functionalities of resin (Wołowicz and Wawrzkiewicz 2021).

Zeolite-based materials have been used as cation exchangers for many heavy metal ions (Figueiredo and Quintelas 2014). A few studies also reported the tailored zeolite materials for the treatment of anionic MCs such as phosphates (Choi et al. 2012), iodides (Warchoł et al. 2006), nitrates (Onyango et al. 2010), arsenate, and chromate (Yusof and Malek 2009). A magnetically enhanced anion and cation exchanger has been employed for the removal of dissolved organic matter in natural water treatment (Apell and Boyer 2010). Tang et al. (2021) used a hybrid magnetic ion exchange resin followed by ozonation for the treatment of MCs (Tang et al. 2021). The magnetic ion exchange resin prior to ozonation reduced the generation of nitrate–nitrogen and ammonia–nitrogen effectively, since the ion exchanger adsorbed nitrite nitrogen and organic nitrogen. Naim et al. (2022) conducted a study on the removal of 13 different dyes from textile wastewater using macroreticular strong base resins and weak base resins (Naim et al. 2022). Both resins were found to be effective for removal of dyes from wastewater; however, the removal efficiency was dependent on the number, molecular size, configuration, structure, kind of solubilizing groups, and resin type (Naim et al. 2022).

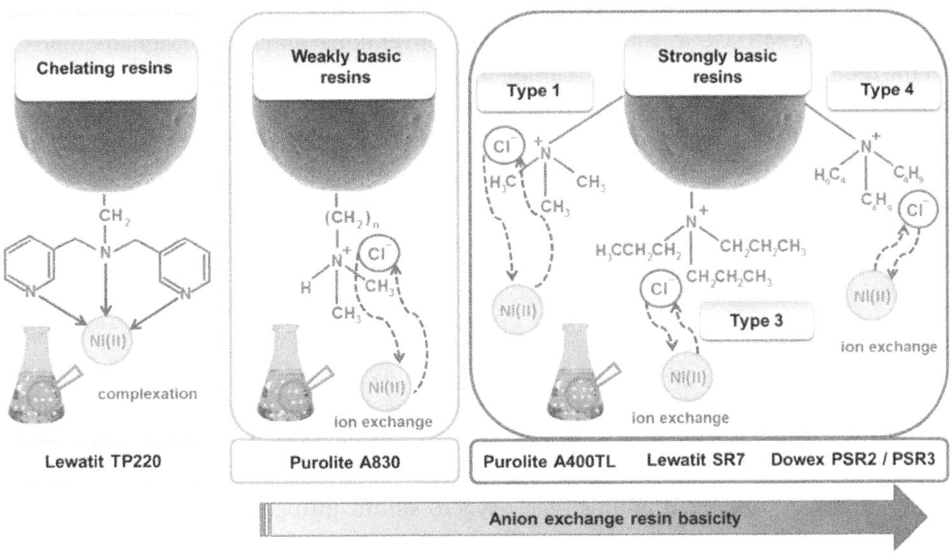

Figure 13.4 Mechanism involved in removal of nickel (II) ion with different resins at different values of basicity. *Source:* Anna Wołowicz & Wawrzkiewicz, 2021/MDPI/CC BY 4.0.

Anion exchangers have been used for the treatment of NOM in water. Ion exchange appears to be more effective than that of coagulation to remove the charged organic species from water. A study compared the effectiveness of several sorbent medias such as anion exchange resins, carbonaceous resins, metal oxides, and activated carbon to remove humic substances from water, which obtained leading removal efficiency for the ion exchange process (Fettig 1999; Dixit et al. 2021). The exchange process has been applied successfully to remove per- and poly-fluoroalkyl substances (PFAS) from the aqueous environment. Owing to their ubiquitous existence in the environment and potential toxicity, PFAS receive considerable attention (Domingo and Nadal 2017). The promising properties of anion exchange resin such as pore structure, polymer composition, and functionalities (e.g., tertiary amine, quaternary amine, etc.) are supposed to effect the PFAS removal rate (Dixit et al. 2021). It is found that the hydrophobic and electrostatic interactions contribute to removal of PFAS substances using ion exchangers and the PFAS removal efficiency can be altered in the presence of co-existing ions (Dixit et al. 2021). The existence of electrostatic interactions in the ion exchange process has been confirmed through the influence of varying solution pH upon PFAS removal (Gagliano et al. 2020). The surface charge of ion exchange material can be altered through protonation of surface functionalities with altering pH that directly influences the PFAS removal efficiency. Another study on removal of perfluorinated ether acids with Purolite A860 resin has confirmed the equivalent exchange of chloride ions (in milli-equivalents) with negligible effect of hydrophobic adsorption (Dixit et al. 2020).

Johnson et al. (2016) achieved the successful removal of NOM using nanoscale resin (single-walled carbon nanotubes functionalized with a strong base anion exchange resin having quaternary ammonium functional groups) (Johnson et al. 2016). Allpike et al. (2005) operated a full-scale potable water treatment plant with a combination of MIEX process (magnetic ion exchange) and coagulation (MIEX-C), and compared the process efficiency with conventional enhanced coagulation processes (Allpike et al. 2005). Results exhibited the greater performance of MIEX-C as compared with the conventional enhanced coagulation process for the removal of dissolved organic carbon in each of the eight separate apparent molecular weight fractions recognized. Though the conventional process preferred to remove the fractions with highest apparent molecular weight, and those showing the highest aromatic character, MIEX-C significantly removed the dissolved organic carbon across all apparent molecular weight fractions and did not seem to differentiate as strongly based on differences in apparent molecular weight or aromatic character (Allpike et al. 2005).

13.7 Summary

This chapter details a brief review of the ion exchange resins for removal of microconstituents (MCs) from the aqueous environment. The properties of different types of ion-exchange resins are deliberated. The ion exchange resins are discussed based on the main acting atoms in functional groups (i.e., oxygen, sulfur, nitrogen, phosphorous) and ion exchange properties. Special attention is given to the mechanism involved in an ion exchange process that is the mass transfer (diffusion) via particle and film control diffusion. Finally, the applications of different ion exchange resins in water and wastewater treatment are discussed.

Acknowledgment

This work was supported by Brain Pool Fellowship Program through the National Research Foundation of Korea (NRF) funded by the Ministry of Science and ICT (Grant No.: 2019H1D3A1A02071191).

References

Abo-Farha, S.A., Abdel-Aal, A.Y., Ashour, I.A., and Garamon, S.E. (2009). Removal of some heavy metal cations by synthetic resin purolite C100. *Journal of Hazardous Materials* 169 (1): 190–194.

Al-Asheh, S. and Aidan, A. (2021). A comprehensive method of ion exchange resins regeneration and its optimization for water treatment. In: *Promising Techniques for Wastewater Treatment and Water Quality Assessment*, 1–458. IntechOpen.

Allpike, B.P., Heitz, A., Joll, C.A. et al. (2005). Size exclusion chromatography to characterize DOC removal in drinking water treatment. *Environmental Science & Technology* 39 (7): 2334–2342.

Altarawneh, M., Waters, D., Goh, B.-M. et al. (2020). Adsorptive interactions between metaldehyde and sulfonic functional group in ion exchange resin. *Journal of Molecular Liquids* 313: 113555.

Apell, J.N. and Boyer, T.H. (2010). Combined ion exchange treatment for removal of dissolved organic matter and hardness. *Water Research* 44 (8): 2419–2430.

Baidas, S., Gao, B., and Meng, X. (2011). Perchlorate removal by quaternary amine modified reed. *Journal of Hazardous Materials* 189 (1): 54–61.

Bezzina, J.P., Ruder, L.R., Dawson, R., and Ogden, M.D. (2019). Ion exchange removal of Cu(II), Fe(II), Pb(II) and Zn(II) from acid extracted sewage sludge: resin screening in weak acid media. *Water Research* 158: 257–267.

Bonin, L., Deduytsche, D., Wolthers, M. et al. (2021). Boron extraction using selective ion exchange resins enables effective magnesium recovery from lithium rich brines with minimal lithium loss. *Separation and Purification Technology* 275: 119177.

Chandrasekara, N.P.G.N. and Pashley, R.M. (2015). Study of a new process for the efficient regeneration of ion exchange resins. *Desalination* 357: 131–139.

Chhuon, R., Shahid, M.K., Kim, S., and Choi, Y. (2020). Mill scale as a ballasted flocculant for enhancing the settleability of activated sludge. *Journal of Environmental Chemical Engineering* 8 (5): 104237.

Choi, J.-W., Lee, S.-Y., Lee, S.-H. et al. (2012). Comparison of surface-modified adsorbents for phosphate removal in water. *Water, Air, & Soil Pollution* 223 (6): 2881–2890.

Cyganowski, P., Şen, F., Altıok, E. et al. (2021). Surface-activated chelating resins containing *N*-methyl-*D*-glucamine functional groups for desalination of geothermal water aimed for removal of boron and arsenic. *Solvent Extraction and Ion Exchange* 39 (5–6): 584–603.

Dayarathne, H.N.P., Angove, M.J., Aryal, R. et al. (2021). Removal of natural organic matter from source water: review on coagulants, dual coagulation, alternative coagulants, and mechanisms. *Journal of Water Process Engineering* 40: 101820.

Deng, S., Yu, Q., Huang, J., and Yu, G. (2010). Removal of perfluorooctane sulfonate from wastewater by anion exchange resins: effects of resin properties and solution chemistry. *Water Research* 44 (18): 5188–5195.

Dixit, F., Barbeau, B., Mostafavi, S.G., and Mohseni, M. (2020). Efficient removal of GenX (HFPO-DA) and other perfluorinated ether acids from drinking and recycled waters using anion exchange resins. *Journal of Hazardous Materials* 384: 121261.

Dixit, F., Dutta, R., Barbeau, B. et al. (2021). PFAS removal by ion exchange resins: a review. *Chemosphere* 272: 129777.

Domingo, J.L. and Nadal, M. (2017). Per- and polyfluoroalkyl substances (PFASs) in food and human dietary intake: a review of the recent scientific literature. *Journal of Agricultural and Food Chemistry* 65 (3): 533–543.

Dong, H., Wu, Z., Liu, M.J., and Tarpeh, W.A. (2021). The role of intraparticle diffusion path length during electro-assisted regeneration of ion exchange resins: implications for selective adsorbent design and reverse osmosis pretreatment. *Chemical Engineering Journal* 407: 127821.

Du, Z., Deng, S., Chen, Y. et al. (2015). Removal of perfluorinated carboxylates from washing wastewater of perfluorooctanesulfonyl fluoride using activated carbons and resins. *Journal of Hazardous Materials* 286: 136–143.

Duarte, G., Teixeira, M.C., Olusegun, S.J., and Ciminelli, V.S.T. (2021). Mechanisms of arsenic removal from simulated surface water based on As(III) retention on thiol chelating resins. *Environmental Nanotechnology, Monitoring & Management* 16: 100532.

Fettig, J. (1999). Removal of humic substances by adsorption/ion exchange. *Water Science and Technology* 40 (9): 173–182.

Figueiredo, H. and Quintelas, C. (2014). Tailored zeolites for the removal of metal oxyanions: overcoming intrinsic limitations of zeolites. *Journal of Hazardous Materials* 274: 287–299.

Gagliano, E., Sgroi, M., Falciglia, P.P. et al. (2020). Removal of poly- and perfluoroalkyl substances (PFAS) from water by adsorption: role of PFAS chain length, effect of organic matter and challenges in adsorbent regeneration. *Water Research* 171: 115381.

Gao, K.W., Yu, X., Darling, R.M. et al. (2022). Increased Donnan exclusion in charged polymer networks at high salt concentrations. *Soft Matter* 18 (2): 282–292.

Hérès, X., Blet, V., Di Natale, P. et al. (2018). Selective extraction of rare earth elements from phosphoric acid by ion exchange resins. *Metals* 8 (9): 682.

Jinhua, W., Xiang, Z., Bing, Z. et al. (2010). Rapid adsorption of Cr (VI) on modified halloysite nanotubes. *Desalination* 259 (1–3): 22–28.

Johnson, B.R., Eldred, T.B., Nguyen, A.T. et al. (2016). High-capacity and rapid removal of refractory NOM using nanoscale anion exchange resin. *ACS Applied Materials & Interfaces* 8 (28): 18540–18549.

Kashif, A., Rehman, R., Fuwad, A. et al. (2022). Current advances in the classification, production, properties and applications of microbial biosurfactants: a critical review. *Advances in Colloid and Interface Science* 306: 102718.

Kumar, P., Pournara, A., Kim, K.-H. et al. (2017). Metal-organic frameworks: challenges and opportunities for ion-exchange/sorption applications. *Progress in Materials Science* 86: 25–74.

Kumar, S. and Jain, S. (2013). History, introduction, and kinetics of ion exchange materials. *Journal of Chemistry* 2013: 957647.

Lalmi, A., Bouhidel, K.-E., Sahraoui, B., and el Houda Anfif, C. (2018). Removal of lead from polluted waters using ion exchange resin with $Ca(NO_3)_2$ for elution. *Hydrometallurgy* 178: 287–293.

Levchuk, I., Rueda Márquez, J.J., and Sillanpää, M. (2018). Removal of natural organic matter (NOM) from water by ion exchange: a review. *Chemosphere* 192: 90–104.

Li, F., Xu, J., Wang, Y. et al. (2021). Pressure-induced polymerization: addition and condensation reactions. *Molecules* 26 (24) 7581.

Li, S., Du, S., Xie, N. et al. (2022). Imidazole-functionalized multiquaternary side-chain polyethersulfone anion-exchange membrane for fuel cell applications. *ACS Applied Energy Materials 5 (8): 10023–10033.*

Liu, D. (2011). Removal of emerging contaminants from aqueous solutions by using polymeric resins. Temple University. doi:10.34944/dspace/1740. https://scholarshare.temple.edu/handle/20.500.12613/1758.

McLachlan, M.S., Holmström, K.E., Reth, M., and Berger, U. (2007). Riverine discharge of perfluorinated carboxylates from the European continent. *Environmental Science & Technology* 41 (21): 72607265.

Monteagudo, J.M. and Ortiz, M.J. (2000). Removal of inorganic mercury from mine wastewater by ion exchange. *Journal of Chemical Technology & Biotechnology* 75 (9): 767–772.

Moosavirad, S.M., Sarikhani, R., Shahsavani, E., and Mohammadi, S.Z. (2015). Removal of some heavy metals from inorganic industrial wastewaters by ion exchange method. *Journal of Water Chemistry and Technology* 37 (4): 191–199.

Mukhamediev, M.G., Bekchanov, D.Z., Juraev, M.M. et al. (2021). A new sulfonic acid cation-exchange resin based on polyvinyl chloride and its evaluation in water softening. *Russian Journal of Applied Chemistry* 94 (12): 1594–1601.

Murray, A. and Örmeci, B. (2019). Use of polymeric sub-micron ion-exchange resins for removal of lead, copper, zinc, and nickel from natural waters. *Journal of Environmental Sciences* 75: 247–254.

Naim, M.M., Al-harby, N.F., El Batouti, M., and Elewa, M.M. (2022). Macro-reticular ion exchange resins for recovery of direct dyes from spent dyeing and soaping liquors. *Molecules* 27 (5): 1593.

Nesterenko, P.N. and Nesterenko, E.P. (2021). Hydrophobicity of polymer based anion-exchange columns for ion chromatography. *Heliyon* 7 (6): e07290.

Onyango, M.S., Masukume, M., Ochieng, A., and Otieno, F. (2010). Functionalised natural zeolite and its potential for treating drinking water containing excess amount of nitrate. *Water SA* 36 (5): 655–662.

Patle, D.B. and Gurnule, W.B. (2016). An eco-friendly synthesis, characterization, morphology and ion exchange properties of terpolymer resin derived from p-hydroxybenzaldehyde. *Arabian Journal of Chemistry* 9: S648–S658.

Pidoux, L., Shorney-Darby, H., Vaudevire, E. et al. (2022). Impact of resin loading on ion exchange equilibrium for removal of organic matter and inorganic ions. *Journal of Hazardous Materials* 431: 128530.

Riley, A.L., Porter, C.P., and Ogden, M.D. (2021). Selective recovery of copper from a synthetic metalliferous waste stream using the thiourea-functionalized ion exchange resin puromet MTS9140. *Eng* 2 (4): 512–530.

Shahid, M.K. and Choi, Y. (2020). Characterization and application of magnetite particles, synthesized by reverse coprecipitation method in open air from mill scale. *Journal of Magnetism and Magnetic Materials* 495 (August 2019): 165823.

Shahid, M.K., Kashif, A., Fuwad, A., and Choi, Y. (2021a). Current advances in treatment technologies for removal of emerging contaminants from water: a critical review. *Coordination Chemistry Reviews* 442: 213993.

Shahid, M.K., Kashif, A., Pathak, P. et al. (2021b). Water reclamation, recycle, and reuse. In: *Clean Energy and Resource Recovery*, 2nd edn. 39–49. Elsevier Inc.

Shahid, M.K., Phearom, S., Choi, Y.-G.-G. et al. (2019). Adsorption of arsenic (V) on magnetite-enriched particles separated from the mill scale. *Environmental Earth Sciences* 78 (3): 65.

Shahid, M.K., Pyo, M., and Choi, Y. (2017). Carbonate scale reduction in reverse osmosis membrane by CO_2 in wastewater reclamation. *Membrane Water Treatment* 8 (2): 125–136.

Silva, R.A., Hawboldt, K., and Zhang, Y. (2018). Application of resins with functional groups in the separation of metal ions/species: a review. *Mineral Processing and Extractive Metallurgy Review* 39 (6): 395–413.

Tang, Y., Chen, Z., Wen, Q. et al. (2021). Evaluation of a hybrid process of magnetic ion-exchange resin treatment followed by ozonation in secondary effluent organic matter removal. *Science of the Total Environment* 754: 142361.

Vorotyntsev, A.V., Petukhov, A.N., Makarov, D.A. et al. (2018). Synthesis, properties and mechanism of the ion exchange resins based on 2-methyl-5-vinylpyridine and divinylbenzene in the catalytic disproportionation of trichlorosilane. *Applied Catalysis B: Environmental* 224: 621–633.

Wang, G., Guo, K., Wang, B. et al. (2021). Mercury adsorption on thiol-modified porous boron nitride: a combined experimental and theoretical investigation. *Industrial & Engineering Chemistry Research* 60 (35): 12984–12998.

Warchoł, J., Misaelides, P., Petrus, R., and Zamboulis, D. (2006). Preparation and application of organo-modified zeolitic material in the removal of chromates and iodides. *Journal of Hazardous Materials* 137 (3): 1410–1416.

Wołowicz, A. and Hubicki, Z. (2012). The use of the chelating resin of a new generation Lewatit MonoPlus TP-220 with the bis-picolylamine functional groups in the removal of selected metal ions from acidic solutions. *Chemical Engineering Journal* 197: 493–508.

Wołowicz, A. and Wawrzkiewicz, M. (2021). Screening of ion exchange resins for hazardous Ni(II) removal from aqueous solutions: kinetic and equilibrium batch adsorption method. *Processes* 9 (2): 285.

Xie, Y., Li, S., Wang, F., and Liu, G. (2010). Removal of perchlorate from aqueous solution using protonated cross-linked chitosan. *Chemical Engineering Journal* 156 (1): 56–63.

Yu, Q., Zhang, R., Deng, S. et al. (2009). Sorption of perfluorooctane sulfonate and perfluorooctanoate on activated carbons and resin: kinetic and isotherm study. *Water Research* 43 (4): 1150–1158.

Yusof, A.M. and Malek, N.A.N.N. (2009). Removal of Cr(VI) and As(V) from aqueous solutions by HDTMA-modified zeolite Y. *Journal of Hazardous Materials* 162 (2): 1019–1024.

Zawierucha, I., Nowik-Zajac, A., Girek, T. et al. (2022). Arsenic(V) removal from water by resin impregnated with cyclodextrin ligand. *Processes* 10 (2): 253.

Zhang, Y., Liu, Q., and Li, L. (2016). Removal of iron from sythetic copper leach solution using a hydroxy-oxime chelating resin. *Hydrometallurgy* 164: 154–158.

Zheng, Q., Yang, X., Deng, W. et al. (2016). Characterization of natural organic matter in water for optimizing water treatment and minimizing disinfection by-product formation. *Journal of Environmental Sciences (China)* 42: 1–5.

14

Membrane-Based Separation Technologies for Removal of Microconstituents

Sanket Dey Chowdhury[1], *Rao Y. Surampalli*[2], *and Puspendu Bhunia*[1]

[1]*School of Infrastructure, Indian Institute of Technology Bhubaneswar, Odisha, India*
[2]*Global Institute for Energy, Environment, and Sustainability, Lenexa, KS, USA*

14.1 Introduction

Currently, the occurrence of microconstituents (MCs), including pharmaceuticals and personal care products (PPCPs), pesticides, surfactants, hormones, per- and polyfluoroalkyl substances (PFASs), flame retardants, and plasticizers, in water and wastewater has become a major concern for authorities (Shojaee Nasirabadi et al. 2016; García et al. 2020). Although these contaminants exist in trivial concentrations (μg/L to ng/L) in the aquatic matrices (Coppens et al. 2015), their presence in drinking water makes them a potential threat to human health because these MCs are biologically active and bio-persistent in nature (Verlicchi et al. 2012; Simazaki et al. 2015). These contaminants remain physiologically and pharmacologically active, thereby having adverse impacts on humans at such low concentrations (Simazaki et al. 2015). For instance, inflammatory reaction in the human liver has been observed due to their exposure to diclofenac (Li et al. 2014). This is why these MCs are also termed as emerging contaminants (ECs) (Yang et al. 2011; Jiang et al. 2013).

The aforementioned MCs can enter the environment by several pathways. Personal hygiene products, industrial waste and wastewater, wastewater from the agricultural fields, domestic wastewater, pharmaceutical waste, hospital wastewater, and wastewater from firefighting training grounds, military bases, and airports have been the predominant contributors of MCs to water and wastewater (Zhang et al. 2008; Deblonde et al. 2011; Rivera-Utrilla et al. 2011; Stuart et al. 2012; Nasseri et al. 2018; Vo et al. 2020).

Predominantly, MCs are removed through the secondary and tertiary treatment methods (García et al. 2020; Vo et al. 2020). The conventional water and wastewater treatment methods such as coagulation, bio-oxidation, adsorption, and sedimentation are highly pollutant specific and thus become futile in removing diversified ECs (Jones et al. 2005; Conkle et al. 2008; Garcia et al. 2010; Zhang et al. 2014). In fact, the bio-persistent nature of MCs leads to the inadequate performance of conventional wastewater

Microconstituents in the Environment: Occurrence, Fate, Removal, and Management, First Edition.
Edited by Rao Y. Surampalli, Tian C. Zhang, Chih-Ming Kao, Makarand M. Ghangrekar, Puspendu Bhunia, Manaswini Behera, and Prangya R. Rout.

treatment plants (CWWTPs) and drinking water treatment plants (DWTPs) in eliminating MCs from aquatic matrices. As a consequence, the effluents discharged from CWWTPs and DWTPs embrace a significant pharmaceutical load, degrading the quality of the receiving ecosystem (Manaia et al. 2018; Cacace et al. 2019). Moreover, various advanced oxidation processes (AOPs) such as ozonation, ultra-violet (UV) photolysis, Fenton process, sonolysis, anodic oxidation, and photocatalysis have been explored as the tertiary treatment step to eradicate MCs (Klamerth et al. 2010; Naddeo et al. 2010; Kanakaraju et al. 2018; Miklos et al. 2018; Tufail et al. 2020). However, ineffectiveness in removing the wide spectrum of ECs, formation of toxic by-products, and huge monetary investment associated with the AOPs encourage researchers to explore other innovative advanced treatment methods for eliminating ECs from aqueous solutions (Hai et al. 2007; Dewil et al. 2017).

Among various water and wastewater treatment methods, membrane-based separation technologies (MBSTs) have drawn the attention of researchers because of their following advantages (Judd and Jefferson 2003; Javier Benitez et al. 2011; Zhang et al. 2012; Mohammad et al. 2015; Shojaee Nasirabadi et al. 2016; Krzeminski et al. 2017; Dharupaneedi et al. 2019):

a) MBSTs do not necessitate the change of phase of the contaminants during separation;
b) Steady-state condition prevails for a longer duration due to negligible accumulation;
c) MBSTs eliminate the need for the external addition of chemicals;
d) These are energy-efficient processes;
e) MBSTs produce high-quality permeates;
f) These processes can be carried out at moderate temperatures;
g) Maintenance cost is minimal; and
h) They also eliminate the possibility of toxic by-product formation.

Even though membrane technology is relatively new, it has already been widely employed for the separation of MCs from water and wastewater. Particularly, the pressure-driven membrane processes such as microfiltration (MF), ultrafiltration (UF), nanofiltration (NF), reverse osmosis (RO), and forward osmosis (FO) have been employed on an industrial scale (Cui and Muralidhara 2010). Among the aforementioned pressure-driven membrane processes, UF, NF, and RO processes are capable of removing MCs to a greater extent from the aquatic environment and produce high-quality permeates during the course of treating the industrial wastewater for reuse (Javier Benitez et al. 2011).

In this chapter, a comprehensive overview on the application of the MBSTs, especially MF, UF, NF, RO, and FO processes, for removing various MCs from aqueous solutions has been presented. Moreover, a detailed discussion on the classification of the available MBSTs, types of membrane materials used during the aforementioned pressure-driven membrane processes, and their pore characteristics has also been incorporated. Furthermore, the fundamental principle involved behind the removal of the MCs using membrane separation technology has also been highlighted in this chapter. Additionally, a brief discussion on membrane fouling has been included. This chapter also focuses on the future perspectives regarding the removal of MCs using MBSTs. Finally, a conclusion has been drawn, highlighting the content of the entire chapter.

14.2 Classification of Available MBSTs

Depending upon the driving force, causing the permeant to flow through the membrane, the available MBSTs, employed for eradicating the MCs from the aqueous matrices, can be classified into the following groups (Kárászová et al. 2020; Matsuura 2020):

a) Pressure difference across the membrane is the driving force:
   - UF
   - MF
   - NF
   - FO
   - RO
   - Membrane degasification (MDG)
   - Pervaporation (PV)
b) Temperature difference across the membrane is the driving force:
   - Membrane distillation (MD)
c) Concentration gradient across the membrane is the driving force:
   - Dialysis (D)
   - Membrane extraction (ME)
d) Electrical potential difference across the membrane is the driving force:
   - Electrodialysis (ED)
   - Electrodeionization (EDI)

The basic information regarding the aforementioned MBSTs has been compiled in Table 14.1. However, in this chapter, the application of the frequently used MBSTs for removing MCs from the aquatic environment has been emphasized. Thus, this chapter comprises detailed discussions on the UF, MF, NF, FO, and RO processes.

14.3 Classification of Membranes and Membrane Materials and Their Properties

14.3.1 Classification of Membranes

The membranes can be classified based upon the following four broader criteria:

a) ***Based on the size of the pore-openings***
   Based on the size of the pore-openings, membranes can be classified into three types: (Matsuura 2020; Metcalf and Eddy 2003).
   i) *Macroporous membrane*
      In the macroporous membrane, the size of the pore-openings is >50 nm. This is commonly used in the MF process (Crespo and Brazinha 2015; Dhangar and Kumar 2020).
   ii) *Mesoporous membrane*
      The size of the pore-openings is kept between 2 and 50 nm in mesoporous membranes. The application of the mesoporous membrane is commonly found in UF processes (Acero et al. 2010; Urtiaga et al. 2013).

Table 14.1 MBSTs employed in removing MCs from water and wastewater.

Mbsps	Driving Force	Type of Membrane	Pore Size	Flux Through Membrane		Function	References
				Water	Microconstituents		
MF	TMP difference 1–3 bar.	Porous, symmetric, asymmetric.	0.05–10 μm	Yes	No	Filtration.	Crespo and Brazinha (2015); Dhangar and Kumar (2020); Kárászová et al. (2020); Van der Bruggen and Luis (2015)
UF	TMP difference 2–5 bar.	Microporous, asymmetric.	0.001–0.005 μm	Yes	No	Protein and pathogen removal.	Acero et al. (2010); Urtiaga et al. (2013)
NF	TMP difference 5–15 bar.	Porous, TFC, charged.	<2 nm	Yes	No	Desalination.	Radjenović et al. (2008); Van der Bruggen et al. (1999); Yangali-Quintanilla et al. (2010)
RO	TMP difference 15–75 bar.	Asymmetric, TFC, charged.	0.5 nm	Yes	No		Radjenović et al. (2008); Urtiaga et al. (2013); Wang et al. (2009)
FO	Osmotic pressure.						D'Haese et al. (2013)
MDG	TMP difference (vacuum on permeate side).	Porous, symmetric, asymmetric.	0.05–0.1 μm	No	No	Degasification.	Kárászová et al. (2020)

PV	Transmembrane fugacity difference.	Non-porous, hydrophilic.	4 Å	Yes	Yes		Desalination.	Halakoo and Feng (2020); Wu et al. (2018)
MD	Temperature gradient.	Highly porous, symmetric.	100 Å–1 µm	Yes	No			Kárászová et al. (2020); Matsuura (2020)
D	Concentration gradient.	Semipermeable film.	5–10 nm	–	–		Deionization.	Matsuura (2020)
ME		Microporous film supported liquid membrane.					Acid separation (lactic acid, valeric acid, etc.).	
ED	Electrical potential difference.	Swollen gel, charged.	10–100 Å	No	No		Desalination.	Bazinet et al. (1998); Kárászová et al. (2020)
EDI		Symmetric.	<0.2 nm					Kárászová et al. (2020)

Note: MF = Microfiltration; UF = Ultrafiltration; NF = Nanofiltration; RO = Reverse osmosis; FO = Forward osmosis: MDG = Membrane degasification; PV = Pervaporation; MD = Membrane distillation; D = Dialysis: ME = Membrane extraction; ED = Electrodialysis; EDI = Electrodeionization; TFC = Thin film composite. Fugacity is the measure of chemical potential in the form of adjusted pressure.

Source: Adapted from Kárászová et al. (2020).

iii) *Microporous membrane*

In a microporous membrane, the size of the pore-openings is kept below 2 nm. These membranes are widely used in NF, FO, and RO processes (Radjenović et al. 2008; D'Haese et al. 2013). Moreover, owing to their very small pore-openings, these membranes are also termed as dense membranes (Metcalf and Eddy 2003). In fact, the microporous membranes can further be classified into three types:

1) such as supermicroporous membrane (size of pore-openings = 0.7–2 nm);
2) ultramicroporous membrane (size of pore-openings = 0.35–0.70 nm); and
3) ultraporous membrane (size of pore-openings <0.35 nm) (Nath 2017).

b) *Based on symmetry of structure*

Depending upon symmetry of structure, membranes can also be classified into two types (Metcalf and Eddy 2003):

i) *Symmetric membrane*

The cross-section of this membrane is uniform and is made of the same material (Ersahin et al. 2012).

ii) *Asymmetric membrane*

The asymmetric membrane consists of a thin skin having a thickness of 0.1 to 10 μm on a porous supporting layer (Nath 2017). The cross-section of this membrane is not uniform. Thin film composite is a type of asymmetric membrane, which is particularly developed for NF and RO applications (Ersahin et al. 2012). In thin film composite membranes, a thin skin of dense polymeric layer is developed on a microporous support film (Nath 2017).

c) *Based on geometry*

Based on geometry, the membranes can be classified into the following four types such as flat sheet or plate and frame, tubular, spiral wound, and hollow fiber membranes (Metcalf and Eddy 2003; Nath 2017). These membranes necessitate supports for their installation. Thus, they can also be termed as membrane modules (Metcalf and Eddy 2003).

i) *Flat sheet or plate and frame membrane*

In this type of membrane, the plate is supported by the frame, and the membrane contains a series of flat sheets (Figure 14.1a). The plates often act as the phonographic records in that flat surfaces contain small grooves through which the permeate flows. This type of membrane is widely used in RO, MF, and UF processes (Kárászová et al. 2020).

ii) *Tubular membrane*

The tubular membrane resembles a cylinder with negligible thickness (Figure 14.1b). In tubular membrane modules, feed is normally applied from outside of the membrane tube, and permeate is collected inside the module through vacuum pumping (Figure 14.2). These membranes are widely used in RO and UF processes (Nath 2017).

iii) *Spiral wound membrane*

The spiral wound membrane is a modification of the tubular membrane. In this membrane, a flat sheet is folded with respect to a central support, and the perme-

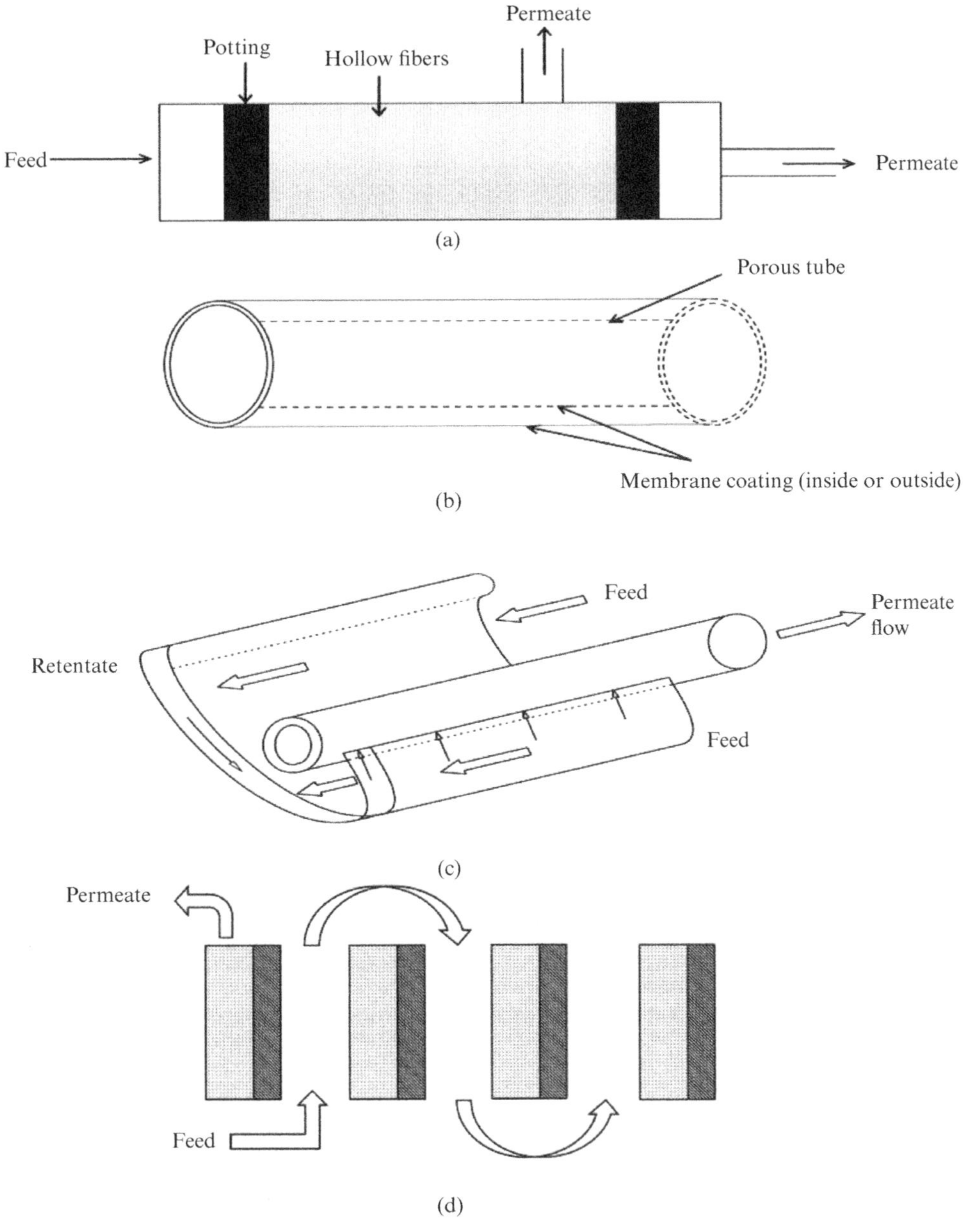

Figure 14.1 Classification of membranes based on geometry: (a) flat sheet or plate and frame; (b) tubular; (c) spiral wound; and (d) hollow fiber.

ate spacer is placed between two membranes (Figure 14.1c). The feed is allowed to have a spiral flow through the membrane module. These membranes are commonly used in RO, UF, and MF processes (Kárászová et al. 2020; Metcalf and Eddy 2003; Nath 2017).

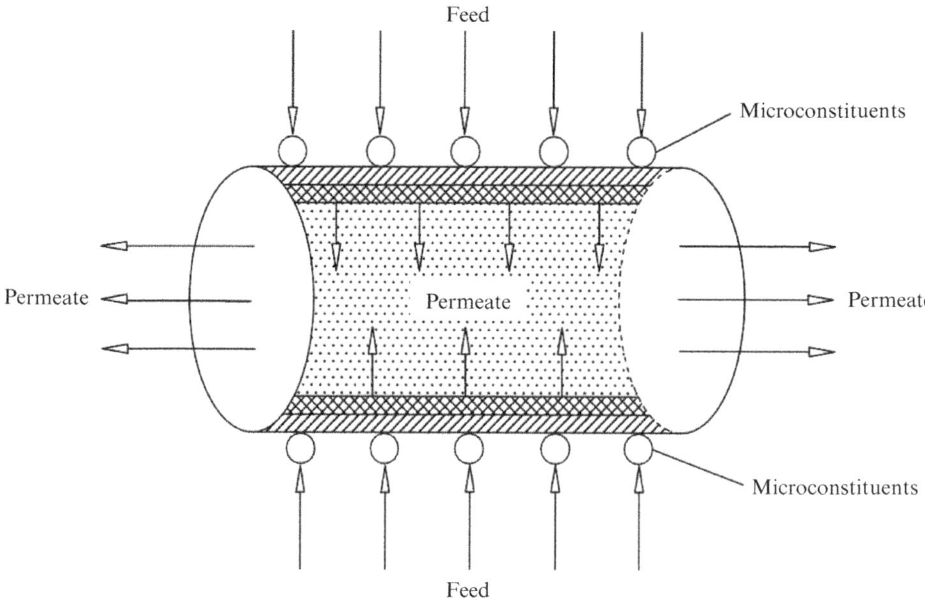

Figure 14.2 Outside-in flow mode of operation of the membrane modules.

iv) Hollow fiber membrane

The hollow fibre membrane is also a modification of the tubular membrane. It consists of fibers with an opening diameter of 25 to 250 μm and wall thickness of 5 to 50 μm (Figure 14.1d). Usually, the hollow fiber membrane is an asymmetric membrane. Due to the wall thickness, the MCs gain an additional narrow path to flow through the membrane module. Hollow fiber membranes are extensively used in RO and UF processes (Nath 2017).

d) *Based on material of preparation*

Based on the materials used for preparing the membranes, membranes can be classified into two types.

i) Organic membrane

Organic membranes are mainly prepared from various synthetic or naturally occurring polymers/co-polymers. Generally, the organic membranes cannot withstand high temperatures (>200°C) (Metcalf and Eddy 2003). A brief discussion on the various types of polymers widely used in membrane preparation has been incorporated under Section 14.3.2.

ii) Inorganic membrane

Inorganic membranes, also known as ceramic membranes, can be effectively employed at elevated temperatures. The metal membranes can survive up to temperatures ranging between 500 and 800°C. In fact, many ceramic membranes have the potential to survive beyond 1000°C temperature (Kárászová et al. 2020).

Various inorganic materials such as aluminum oxide, silicon oxides, zinc oxide, and titanium oxides have been explored as the inorganic membrane materials due to their high stability, pollutant selectivity, and permeability at elevated temperatures (Nath 2017). Moreover, these materials also resist the corrosive liquids at high temperatures (Metcalf and Eddy 2003). Recently, zeolite has attracted researchers' attention as an inorganic membrane material because it is capable of molecular sieving, selective adsorption of the contaminants, ion exchange, and catalysis (Nath 2017). Meanwhile, the ceramic membranes are generally asymmetric membrane containing a microporous top layer and mesoporous intermediate layer. These membranes can be extensively employed in MF and UF processes (Metcalf and Eddy 2003; Nath 2017).

14.3.2 Classification and Properties of Membrane Materials

As mentioned earlier, the membranes are mostly prepared from the synthetic polymers/co-polymers and various inorganic (ceramic or mineral) materials (Ersahin et al. 2012; Kárászová et al. 2020). In this section, a brief discussion on the various types of polymers used in membrane preparation and their characteristics has been incorporated.

14.3.2.1 Membrane Classification

Basically, the membranes made of a large variety of polymers can be broadly classified into three groups:

i) *Rubbery polymer*
The polymer operating above the glass transition temperature (T_g) is known as a rubbery polymer. It behaves like a viscous liquid (Pabby et al. 2008). In the rubbery polymer, the polymeric molecules move like the liquid molecules. However, these polymers ensure high organic flux and low water flux through them. Polydimethyl siloxane, silicon rubber, and polyvinylammonium thiocyanate are some common examples of rubbery polymers (Nath 2017).

ii) *Glassy polymer*
The polymer operating below T_g is termed a glassy polymer. Unlike rubbery polymers, these are either crystalline or amorphous in nature (Nath 2017). Moreover, attributing to the obstruction in the motion of the polymeric molecules, the polymer structure becomes heterogeneous. Cellulose acetate (CA), polycarbonate (PC), and polysulphone (PS) are the most extensively used varieties of glassy polymer (Nath 2017).

iii) *Ion exchange polymer*
Basically, the ion exchange polymer membranes are the ion exchanging resins prepared in the form of a membrane. These membranes contain surface charges. A membrane with positive surface charge allows the anionic pollutants to pass through it. However, the membrane with negative surface charge permits the cationic pollutants to pass through it. The former membrane is known as an anionic membrane, whereas the latter is known as a cationic membrane (Nath 2017).

In the following subsections, some common variants of the polymers used for preparing the membrane, including their salient properties, have been highlighted.

14.3.2.1.1 Cellulose Derivatives

Cellulose derivatives are the pivotal polymers for membrane applications, which occur naturally in the environment. The potential source of cellulose and its derivatives is cotton liners containing the highest percentage of alpha cellulose (Nath 2017). Cellulose does not dissolve in water owing to its homogeneous structure and the presence of hydrogen bonds (Ersahin et al. 2012). Cellulose nitrate (CN) (inorganic) and CA (organic) are the most explored derivatives of cellulose as membrane materials (Narbaitz et al. 2013; Nath 2017). CA is cheaper and less inflammable than CN. The main advantage of the CA membrane is that it ensures high flux attributed to the existence of a wide range of pore-openings. Additionally, cellulose triacetate (CTA) is another cellulose derivative showing high hydrophilicity, thereby cutting down the frequency of membrane fouling (Nath 2017). However, the cellulose derivatives impart a less critical T_g value (68.6°C), limiting their application at high temperatures (Nath 2017). CN is widely used in MF processes whereas both CA and CTA are extensively employed in NF, RO, MF, and UF processes (Javier Benitez et al. 2011; Shojaee Nasirabadi et al. 2016).

14.3.2.1.2 Aromatic Polyamides

Aromatic polyamides are also known as aramides. Usually these are crystalline polymers, imparting better resistance to various solvents and higher thermal stability than their aliphatic counterparts (Nath 2017). These are identified by the presence of amide bonds in their molecular structure. The polyamide (PA) membranes outcompete the CA membranes in terms of tolerating pH fluctuations; however, the PA membranes are surpassed by the CA membranes in terms of chlorine resistance (Yangali–Quintanilla et al. 2010). This could be because the aromatic rings present in the PA membranes are substituted by chlorine, destroying the intermolecular hydrogen bonds of PAs (Metcalf and Eddy 2003). Polyamide hydrazide and polyetheramide hydrazide have enough potential to be utilized in NF and RO processes (Dolar et al. 2011).

14.3.2.1.3 Polysulphone

Polysulphone (PS) is prepared by condensing dichlorodiphenyl sulphone and bisphenol A. PSs and polyethersulphones (PESs) are immobile, rigid, resistant to creep, and impart high dimensional stability (Nath 2017). These materials can withstand high temperatures (up to 125 and 75°C for PES and PS, respectively) and impart high tolerance to pH variations and fair resistance to chlorine (Yangali-Quintanilla et al. 2010; Ersahin et al. 2012). The sulphone groups consume the electrons present in the adjacent aromatic groups, thereby restricting thermal oxidation of the membrane. Moreover, owing to the presence of aromatic rings, PS and PES can survive intensive energy irradiations. Furthermore, attributed to the existence of diversified pore-openings, PS and PES can be potentially used in RO, MF, and UF processes (Vergili 2013; Shojaee Nasirabadi et al. 2016; Kárászová et al. 2020).

14.3.2.1.4 Polyimides

Polyimides are the most extensively investigated variety of polymer as the RO membrane material is polyimide (Nath 2017). Polyimides are prepared by condensing the polyamic acid obtained as a by-product of the reaction between dianhydrides and diamines. Polyimides are widely used in RO and UF processes (Metcalf and Eddy 2003).

14.3.2.1.5 Polytetrafluoroethylene

Polytetrafluoroethylene, also known as Teflon, can be identified as polyethylene with fluorine substituting for all the hydrogen atoms present. The monomer of Teflon, tetrafluoroethylene, is prepared from decarboxylation of sodium perfluoro-propionate or dehalogenation of dichloro tetrafluoroethane (Nath 2017). Teflon is highly crystalline in nature and consists of straight molecular chains, thereby possessing high strength and better thermal resistivity (Pabby et al. 2008). Teflon is commonly used in MF processes (Nath 2017).

14.3.2.1.6 Polycarbonates

Polycarbonate (PC) is a type of polyester, portraying toughness and ductility below T_g. Attributed to its thermoplastic nature, PC can be converted to different shapes, including sheets and films (Nath 2017).

14.3.2.1.7 Polypropylene

Polypropylene, polymer of propylene, is inert and hydrophobic in nature and can survive at elevated temperatures. It is widely used in MF processes (Nath 2017).

14.3.2.2 Cutting-Edge Membranes

Apart from the above-mentioned conventional materials, currently, researchers have emphasized the development and application of cutting-edge membranes containing advanced materials, e.g., nanocomposites, graphene, etc. (Table 14.2). Basically, the addition of the nanocomposites modifies the membrane properties including porosity, roughness, and hydrophilicity. Moreover, such membrane modification also results in the development of various unique membrane properties, e.g., photocatalytic activity (TiO_2-graphene oxide (GO) nanocomposite membranes), facilitating abiotic removal of the ECs from the aquatic matrix (de Oliveira et al. 2022). Among various nanocomposite membranes, the TiO_2-GO membrane has attracted major attention because it improves the hydrophilicity of the polymeric membranes, exhibits antimicrobial properties, encompasses the degradation of the pollutants present in the feed, and cuts down membrane fouling to a great extent (de Oliveira et al. 2022). Apart from TiO_2-GO membranes, the PVDF $ZnO/Ag_2CO_3/Ag_2O$ membrane has also been employed for removing ECs, especially ibuprofen (Rosman et al. 2021) (Table 14.2). However, the use of graphene-based membranes, such as GO, improves selectivity of the pollutants, permeate flux, and antifouling properties, thereby elevating the overall treatment efficacy of the process (Januário et al. 2021). For example, Lou et al. (2020) have employed cross-linked GO self-supporting membranes for removing ibuprofen and obtained 44.9% removal of the same (Table 14.2).

Usually, the membranes used in MF are homogeneous in nature and are referred to as isotropic or symmetric membranes, whereas the membranes used in UF are asymmetric, containing a thin active layer with smaller pores and a support layer with larger pores. Moreover, the asymmetric membrane used in UF is mostly made of a single polymer. Hence, it can also be termed as an integral-asymmetric membrane (Kárászová et al. 2020). Meanwhile, hollow fiber or tubular membrane modules are mostly employed in MF and UF processes (Shojaee Nasirabadi et al. 2016).

On the other hand, membranes used in RO, FO, and RO, also known as dense membranes, do not have distinct pores and are asymmetric composite membranes with a porous support layer of PS or PES (Shojaee Nasirabadi et al. 2016; Nath 2017; Kárászová et al. 2020). These dense membranes ensure high water flux. As already mentioned, PA and CA have been extensively used as dense membrane materials (Kárászová et al. 2020). However, the membrane materials used in NF are comparatively less dense as compared to those used in RO. Hence, NF can also be termed as the loose RO process (Metcalf and Eddy 2003). Moreover, in NF, the membrane materials contain slight charges. Furthermore, the key difference between the NF and RO membrane is that the RO membrane restricts the monovalent ions to pass through it. However, the NF membrane allows the monovalent ions to pass through, but inhibits the permeation of the bivalent ions (Metcalf and Eddy 2003). Generally, the spiral wound membrane modules have been predominantly employed in NF and RO applications (Kárászová et al. 2020; Metcalf and Eddy 2003; Shojaee Nasirabadi et al. 2016). The application of different membrane materials in various MBSTs to eradicate the ECs from water and wastewater is portrayed in Table 14.2.

14.4 Fundamental Principles and Hydraulics of Microconstituents Removal via Different MBSTs

In this section, the fundamentals as well as the hydraulics of MCs removal via MF, UF, NF, FO, and RO processes are described (Zhang et al. 2012).

14.4.1 Fundamental Principles

As observed in Table 14.1, the size of the membrane pore-openings is significantly higher in UF and MF than that in NF, FO, and RO processes. Thus, in the case of UF and MF, the separation of MCs from water and wastewater (also known as feed) fed to the membrane is accompanied by straining, sieving, or size exclusion (Figure 14.3a). However, in the case of NF, RO, and FO, a thin water layer is adsorbed onto the surface of the dense membrane (Figure 14.3b), facilitating the rejection of the micropollutants through diffusion from the feed (Matsuura 2020; Metcalf and Eddy 2003; Nath 2017).

In the case of NF, RO, and FO, apart from their removal by the adsorbed water layer, the MCs are removed via diffusion, hydrophobic interactions (for neutral ECs), and electrostatic interactions (for positive or negatively charged ECs) between the pollutants and membrane surface (Radjenović et al. 2008) as well as physical sieving (for both neutral and charged ECs) on the membrane pores (Coday et al. 2014). Basically, the neutral and hydrophilic ECs are removed through physical sieving by the membrane, whereas the hydrophobic and neutral ECs are removed via adsorption by the membrane (Coday et al. 2014). However, the charged ECs are removed via electrostatic interaction with the charged membrane materials used during NF, RO, and FO applications (Kárászová et al. 2020). The charged micropollutants are also removed via the sieving mechanism during NF, RO, and FO processes (Khanzada et al. 2020).

However, since the pore-openings of the membranes used in UF and MF are large, physical sieving or size exclusion is the governing EC removal mechanism in these

Table 14.2 Application of various MBSTs for removing MCs from water and wastewater.

Compounds	Filtration Type	Membrane Material	Feed Concentration	Source	Rejection (%)	References
Pharmaceuticals and personal care products (PPCPs)						
	NF	Aromatic PA.	5–18 μg/L	Surface water.	96	Yangali-Quintanilla et al. (2010)
		TFC membranes with a cross-linked aromatic PA top layer.	2–30 μg/L	Surface water.	100	Verliefde et al. (2007)
		Made of thin film PA.	–	–	88	Acero et al. (2010)
		PES	0.025–0.1 μg/L	Influent of a DWTP.	10–92	Vergili (2013)
		CA	26.44 mg/L	Synthetic wastewater.	54.3–59.1	Rana et al. (2012)
		CA membrane developed by incorporating charged surface modifying macromolecules.	140–4706 ng/L	Influent of a WTP.	27–80.4	Narbaitz et al. (2013)
		Thin Film Nanocomposite Dually Charged (polydopamine and quaternate chitosan).	–	Synthetic water.	76–89	Javier Benitez et al. (2011)
		0.4 Modified montmorillonite.	–	–	85–97	Bojnourd and Pakizeh (2018)
	RO	–	259, 302 ng/L	Saline ground water.	90.3	Snyder et al. (2007)
		CTA-TFC	–	Synthetic wastewater.	25–95	Fujioka et al. (2015)

(Continued)

Table 14.2 (Continued)

Compounds	Filtration Type	Membrane Material	Feed Concentration	Source	Rejection (%)	References
		PA-TFC		Synthetic water in line with the quality of North Bay Regional Water Treatment Plant, California, US.	87–99	Huang et al. (2011)
				Anaerobic riverbank filtrate.	75–99	Albergamo et al. (2019)
		PVDF $ZnO/Ag_2CO_3/Ag_2O$.	20 mg/L	Synthetic wastewater.	35.27	Rosman et al. (2021)
	UF	PES	381 ng/L	Effluent from a WWTP.	69	Acero et al. (2010)
		PES, Regenerated Cellulose, and Cattle intestine.	–	Synthetic wastewater spiked with micropollutants.	70–80	Burba et al. (2005)
		Alumina/TiO_2.		Wastewater effluent.	35–62	Garcia–Ivars et al. (2017)
		Cross-linked GO self-supporting membranes.	–	Synthetic wastewater.	44.9	Lou et al. (2020)
	FO	CTA	250 μg/L	10 mM NaCl.	82–83	Jin et al. (2012)
		TFC			> 95	
Diclofenac	NF	TFC with aromatic PA coated with an ultrathin polyimide.	–	–	38	Yoon et al. (2007)

		TFC membranes with a cross-linked aromatic PA top layer.	2–5 μg/L	Surface water.	100	Verliefde et al. (2007)
		Made of thin film polyamide.	–	–	92	Acero et al. (2010)
		–	115, 156 ng/L	Effluent from a WWTP.	56–98	Kimura et al. (2009)
		Thin Film Nanocomposite.	–	–	98–99	Paseta et al. (2019)
	RO	PA-TFC		Ground Water, NE Spain.	45–98%	Radjenović et al. (2008)
		CTA-TFC		Synthetic wastewater.	25–95	Fujioka et al. (2015)
		PA-TFC		Tap water.	73–99	Hofs et al. (2013)
	UF	PES	94 ng/L	Effluent from a WWTP.	71	Acero et al. (2010)
		TFC with sulfonated PES coated with an ultrathin polyimide.	2–<150 ng/L	Surface water.	9	Yoon et al. (2007)
	FO	CTA	100 μg/L	Synthetic wastewater.	92.9	Kong et al. (2015)
Caffeine	NF	Aromatic PA.	5–18 μg/L	Synthetic wastewater.	91	Yangali–Quintanilla et al. (2010)
		TFC with aromatic PA coated with an ultrathin polyimide.	–	–	29	Yoon et al. (2007)

(Continued)

Table 14.2 (Continued)

Compounds	Filtration Type	Membrane Material	Feed Concentration	Source	Rejection (%)	References
		TFC	–	Lake Ontario water, membrane bioreactor effluent, and laboratory-grade water (Milli-Q).	0–100	Comerton et al. (2008)
	RO	PA-TFC	–	Synthetic water in line with the quality of North Bay Regional Water Treatment Plant, California, US.	87–99	Huang et al. (2011)
		–	196–311 ng/L	Saline ground water.	83.3	Snyder et al. (2007)
	UF	TFC with sulfonated PES coated with an ultrathin polyimide.	2–<150 ng/L	Surface water.	1	Yoon et al. (2007)
		PES	–	Secondary wastewater effluent.	12	Acero et al. (2017)
Ketoprofen	NF	Aromatic PA.	5–18 μg/L	Surface water.	94	Yangali–Quintanilla et al. (2010)
		TFC membranes with a cross-linked aromatic PA top layer.	2–50 μg/L	Surface water.	100	Verliefde et al. (2007)
		TFC	–	–	87	Acero et al. (2010)
	UF	PES	608 ng/L	Effluent from a WWTP.	58	

	RO	PA-TFC	–	Ground Water, NE Spain.	45–98	Radjenović et al. (2008)
		CTA-TFC	–	Synthetic wastewater.	25–95	Fujioka et al. (2015)
Acetaminophen	NF	Aromatic PA.	5–18 μg/L	Surface water.	77	Yangali–Quintanilla et al. (2010)
		PA-TFC	–	Ground Water, NE Spain.	30–98	Radjenović et al. (2008)
				Lake Ontario water, membrane bioreactor effluent, and laboratory-grade water (Milli-Q).	100	Comerton et al. (2008)
	RO			Ground Water, NE Spain.	45–98	Radjenović et al. (2008)
				Lake Ontario water, membrane bioreactor effluent, and laboratory-grade water (Milli-Q).	82–100	Comerton et al. (2008)
				Synthetic water in line with the quality of North Bay Regional Water Treatment Plant, California, US.	87–99	Huang et al. (2011)

(Continued)

Table 14.2 (Continued)

Compounds	Filtration Type	Membrane Material	Feed Concentration	Source	Rejection (%)	References
	UF	PES		Secondary wastewater effluent.	17	Acero et al. (2017)
		Alumina/TiO_2.		Synthetic wastewater.	20–45	Garcia–Ivars et al. (2017)
		CA/MWNT-TiO_2			24	Zaib et al. (2013)
Naproxen	NF	TFC with aromatic PA coated with an ultrathin polyimide.	2–<150 ng/L	Surface water.	9	Yoon et al. (2007)
		CA	–	–	74–84	Javier Benitez et al. (2011)
	RO	PA-TFC		Tap water.	73–99	Hofs et al. (2013)
		–	91–118 ng/L	Saline groundwater.	78.8	Snyder et al. (2007)
			24 ng/L	Effluent from a WWTP.	12.5	
	UF	PES	384 ng/L		47	Acero et al. (2010)
			–	Ground water.	75	Javier Benitez et al. (2011)
	FO	CTA	100 μg/L	Synthetic wastewater.	97.9	Kong et al. (2015)
		PA-TFC	250 μg/L	10 mM NaCl.	> 95	Jin et al. (2012)
Ciprofloxacin	UF	PES	229 ng/L	Effluent from a WWTP.	65	Acero et al. (2010)
	NF	PA	–	–	99–99.9	Dolar et al. (2011)

			10 mg/L	Milli-Q water, tap water, model water, real pharmaceutical wastewater.	94–99.9	
Trimethoprim	UF	TFC with sulfonated PES coated with an ultrathin polyimide.	2–<150 ng/L	Surface water.	22	Yoon et al. (2007)
		–	138 ng/L	Effluent from a WWTP.	18.1	Snyder et al. (2007)
		PES	521 ng/L		74	Acero et al. (2010)
	NF	TFC with aromatic PA coated with an ultrathin polyimide.	–	–	56	Yoon et al. (2007)
		Thin film PA.	–		86	Acero et al. (2010)
		PA	10 mg/L	Milli-Q water, tap water, model water, real pharmaceutical wastewater.	97.9–99.9	Dolar et al. (2011)
	RO	–	265–278 ng/L	Saline groundwater.	90.6	Snyder et al. (2007)
			10 mg/L	Synthetic wastewater.	98.2–100	Košutić et al. (2007)
		Thin film nanocomposite.	–	Tap water.	61–99	Hofs et al. (2013)
Sulfamethoxazole	NF	Aromatic PA.	5–18 μg/L	Surface water.	95	Yangali-Quintanilla et al. (2010)
		TFC with aromatic PA coated with an ultrathin polyimide.	–	–	Snyder et al. (2007)	Yoon et al. (2007)

(Continued)

Table 14.2 (Continued)

Compounds	Filtration Type	Membrane Material	Feed Concentration	Source	Rejection (%)	References
		PA-TFC		Ground Water, NE Spain.	30–98	Radjenović et al. (2008)
				Synthetic water.	19.6–95	Guo et al. (2017)
		Thin film nanocomposite (Embedded zeolite nanoparticles).		–	84–95	Dong et al. (2016)
	RO	PA-TFC		Ground Water, NE Spain.	45–98	Radjenović et al. (2008)
				Lake Ontario water, membrane bioreactor effluent, and laboratory-grade water (Milli-Q).	82–100	Comerton et al. (2008)
		CTA-TFC		Synthetic wastewater.	25–95	Fujioka et al. (2015)
		PA-TFC		Anaerobic riverbank filtrate.	75–99	Albergamo et al. (2019)
	UF	PES, regenerated cellulose, and cattle intestine.		Synthetic wastewater.	26–50	Burba et al. (2005)
		PES		Secondary wastewater effluent.	94	Acero et al. (2017)
		Alumina/TiO_2/GO.		Synthetic wastewater.	47	Chu et al. (2017)
	FO	TiO_2 modified membrane.			90–99	Huang et al. (2015)

		CTA	200 µg/L	Antibiotic model synthetic wastewater.	> 95	Liu et al. (2015)
			5 µM	–	90	Heo et al. (2013)
Carbamazepine	NF	Aromatic PA.	5–18 µg/L	Surface water.	90	Yangali–Quintanilla et al. (2010)
		TFC membranes with a cross-linked aromatic PA top layer.	2–25 µg/L	Surface water.	96	Verliefde et al. (2007)
		–	116–136 ng/L	Effluent from a WWTP.	20–99	Kimura et al. (2009)
		CA	26.25 mg/L	Synthetic wastewater.	60.1–60.2	Rana et al. (2012)
		PA-TFC	–	Ground Water, NE Spain.	30–98	Radjenović et al. (2008)
		Thin film nanocomposite dually charged (polydopamine and quaternate chitosan).		Synthetic water.	76–89	Javier Benitez et al. (2011)
	RO	PA-TFC		Laboratory grade water (Milli-Q).	57–91	Kimura et al. (2004)
				Synthetic water.	30–67	Baransi–Karkaby et al. (2019)
		TFC (glycidyl methacrylate modified)			59–100	
	UF	PES/N-doped CNTs.		Drinking water and wastewater sources.	88.97	Wanda et al. (2017)
		PES/silica/germanium dioxide nanoparticles		Surface water.	90	Wanda et al. (2018)

(Continued)

Table 14.2 (Continued)

Compounds	Filtration Type	Membrane Material	Feed Concentration	Source	Rejection (%)	References
		CA/MWNT-TiO_2.		Synthetic wastewater.	80	Zaib et al. (2013)
		PVDF			13	Chen et al. (2019)
	FO	CTA	5 μM		82.6	Xie et al. (2012)
			100 μg/L		94.4	Kong et al. (2015)
			5 μM		83	Heo et al. (2013)
Atenolol	NF	TFC membranes with a cross-linked aromatic PA top layer.	2–50 μg/L	Surface water.	91	Verliefde et al. (2007)
		Thin film PA.	–	–	76	Acero et al. (2010)
	RO	CTA-TFC		Synthetic wastewater.	25–95	Fujioka et al. (2015)
				Tap water.	73–99	Hofs et al. (2013)
	UF	PES	1435 ng/L	Effluent from a WWTP.	11	Acero et al. (2010)
Clofibric acid	NF	TFC membranes.	2–100 μg/L	Surface water.	100	Verliefde et al. (2007)
		–	32–80 ng/L	Effluent from a WWTP.	56–90	Kimura et al. (2009)
	RO				78–100	
Bisphenol A	NF	Aromatic PA.	5–18 μg/L	Surface water.	89	Yangali–Quintanilla et al. (2010)
	RO	TFC (glycidyl methacrylate modified).	–	Synthetic water.	59–100	Baransi–Karkaby et al. (2019)

		PA-TFC		Lake Ontario water, membrane bioreactor effluent, and laboratory-grade water (Milli-Q).	82–100	Comerton et al. (2008)
		CA-TFC		Laboratory grade water (Milli-Q).	0–85	Kimura et al. (2004)
	UF	PVDF		Synthetic secondary effluent.	64–76	Hu et al. (2014)
		PES/SWCNT		Synthetic wastewater.	80	Kaminska et al. (2015)
		TiO_2 and ZrO_2.		Secondary wastewater effluent.	48–100	Zielińska et al. (2017)
Methylparaben	NF	PA-TFC		Synthetic water.	19.6–95	Guo et al. (2017)
		TFC with Fe.		Synthetic wastewater.	33–98	
Ethylparaben		PA-TFC		Synthetic water.	19.6–95	
		TFC with Fe.		Synthetic wastewater.	33–98	
Propylparaben		PA-TFC		Synthetic water.	19.6–95	
		TFC with Fe.		Synthetic wastewater.	33–98	
Hormones						
Estradiol	NF	TFC with aromatic PA coated with an ultrathin polyimide.	–	–	39	Yoon et al. (2007)
	UF	Cellulose/PP.			25–45	Neale and Schäfer (2012)

(Continued)

Table 14.2 (Continued)

Compounds	Filtration Type	Membrane Material	Feed Concentration	Source	Rejection (%)	References
				Synthetic wastewater.	12	Imbrogno et al. (2017)
		Alumina		Surface water.	50	Fan et al. (2014)
	FO	CTA	290 ng/L	Synthetic wastewater.	> 96	Cartinella et al. (2006)
Testosterone	UF	Cellulose/PP.	–		4–26	Neale and Schäfer (2012)
		TFC with sulfonated PES coated with an ultrathin polyimide.	2–<150 ng/L	Surface water.	Yoon et al. (2007)	Yoon et al. (2007)
		–	81 ng/L	Effluent from a WWTP.	71.6	Snyder et al. (2007)
	NF	TFC	–	–	62	Yoon et al. (2007)
Estrone	RO	PA-TFC		Lake Ontario water, membrane bioreactor effluent, and laboratory-grade water (Milli-Q).	82–100	Comerton et al. (2008)
	UF	–	98 ng/L	Effluent from a WWTP.	90.8	Snyder et al. (2007)
		Cellulose/PP.	–	Synthetic wastewater.	11–42	Neale and Schäfer (2012)
		PVDF		Synthetic secondary effluent.	28–46	Hu et al. (2014)

	NF	PA-TFC		Lake Ontario water, membrane bioreactor effluent, and laboratory-grade water (Milli-Q).	0–100	Comerton et al. (2008)
		Aromatic PA.	5–18 µg/L	Surface water.	93	Yangali–Quintanilla et al. (2010)
	FO	CTA	330 ng/L	Synthetic wastewater.	> 96	Zaib et al. (2013)
Progesterone	RO	PA-TFC	–	Synthetic water in line with the quality of North Bay Regional Water Treatment Plant, California, US.	87–99	Huang et al. (2011)
		–	250–285 ng/L	Saline groundwater.	< 91.2	Snyder et al. (2007)
	UF	Cellulose/PP.	–	Synthetic wastewater.	35–65	Neale and Schäfer (2012)
	NF	TFC with aromatic PA coated with an ultrathin polyimide.	–	–	66	Yoon et al. (2007)
Estriol	RO	–	128 ng/L	Saline groundwater.	80.5	Snyder et al. (2007)
		PA-TFC	–	Lake Ontario water, membrane bioreactor effluent, and laboratory-grade water (Milli-Q).	82–100	Comerton et al. (2008)

(Continued)

Table 14.2 (Continued)

Compounds	Filtration Type	Membrane Material	Feed Concentration	Source	Rejection (%)	References
	NF				0–100	
	UF	PVDF		Synthetic secondary effluent.	10–17	Hu et al. (2014)
		Cellulose/PP.		Synthetic feed.	12	Imbrogno et al. (2017)
Pesticides						
Atrazine	RO	CTA-TFC	–	Synthetic wastewater.	25–95	Fujioka et al. (2015)
		PA-TFC		Anaerobic riverbank filtrate.	75–99	Albergamo et al. (2019)
		Thin film PA.	2–4 mg/L	Groundwater, Slavonia, Croatia.	95.9	Kosutic et al. (2005)
	NF	Thin film nanocomposite (Oleic acid modified silica nanocomposite).	–	–	90–98	Rakhshan and Pakizeh (2015)
		Thin film PA.	2–4 mg/L	Groundwater, Slavonia, Croatia.	85.2	Kosutic et al. (2005)
		PA-TFC	20 mg/L	Groundwater, Denmark.	95	Fini et al. (2019)
	UF	PES	–	Secondary wastewater effluent.	9	Acero et al. (2017)
	FO	CTA	5 μM	Synthetic wastewater.	48.7	Xie et al. (2012)

					49	Heo et al. (2013)
		Aquaporin	1 mg/L		>98	Madsen et al. (2015)
		CTA			70	
Dichlorvos	NF	Thin film PA.	2–4 mg/L	Groundwater, Slavonia, Croatia.	62	Kosutic et al. (2005)
	RO				94.7	
Triadimefon	NF				99.8	
	RO				78.3	
Diazinon	NF				93.9	
Dimethoate		PA-TFC	2 mg/L	Groundwater, Denmark.	85	Fini et al. (2019)
Furan		ZrO_2 nanoparticles.	–	Wastewater.	89	Donato and Drioli (2021)
Phenylurea (diuron and isoproturon).	RO	TFC-PA	2 mg/L	Agricultural effluent.	95	Jatoi et al. (2021)
Flumequine	UF	–	–	Secondary wastewater effluent.	65	Acero et al. (2017)
Isoproturon					70	
Ketorolac					99	
Diuron	RO	CTA-TFC	–	Synthetic wastewater.	25–95	Fujioka et al. (2015)
		PA-TFC		Anaerobic riverbank filtrate.	75–99	Albergamo et al. (2019)
2,6-dichlorobenzamide	FO	Aquaporin	1 mg/L	Synthetic wastewater.	>98	Madsen et al. (2015)

(Continued)

Table 14.2 (Continued)

Compounds	Filtration Type	Membrane Material	Feed Concentration	Source	Rejection (%)	References
		CTA			56	
	RO	PA-TFC	–	Anaerobic riverbank filtrate.	75–99	Albergamo et al. (2019)
		PA-TFC		Groundwater, Denmark.	95	Fini et al. (2019)
	NF				Acero et al. (2010)	
Surfactants						
Anionic surfactants (belong to ABS group).	UF	PES	15–600 mg/L	Synthetic wastewater.	71–91	Kowalska (2008)
		PS			55–81	
Disponil	RO	Thin film PA top active layer.	11649 mg/L	–	13	Baudequin et al. (2014)
Tegotens			3390 mg/L		41	
Simulsol			1948 mg/L		67	
Fluorinated surfactant.			825 mg/L		92	
Per- and polyfluoroalkyl substances (PFASs)						
PFBA	Membrane separation.	CA	400 ng/L	Milli-Q water.	5.6	Jin et al. (2021)
		AF			0	
PFHxA		CA			8.9	
		AF			28.3	

	RO	PA-TFC	–	Synthetic wastewater.	96–99	Soriano et al. (2019)
	NF				82–96	
PFOA	UF	PVDF		Lake water.	39	Pramanik et al. (2017)
		TiO_2/ZrO_2		Synthetic wastewater.	80	Tsai et al. (2010)
	Membrane separation.	CA	400 ng/L	Milli-Q water.	37.2	Jin et al. (2021)
		AF			77.1	
PFDA		CA			76	
		AF			97.4	
PFBS		CA			8.2	
		AF			0	
PFHxS		CA			29.3	
		AF			69	
PFOS	UF	PVDF	–	Lake water.	25	Pramanik et al. (2017)
		TiO_2/ZrO_2		Synthetic wastewater.	84	Tsai et al. (2010)
	Membrane separation.	CA	400 ng/L	Milli-Q water.	73.2	Jin et al. (2021)
		AF			97.1	
PFDS		CA			96	
		AF			98.1	

Note: PA = Polyamide; PES = Polyethersulphone; CA = Cellulose acetate; CTA = Cellulose triacetate; DWTP = Drinking water treatment plant; WWTP = Wastewater treatment plant; MWNT = Multi-walled nanotube; TiO_2= Titanium dioxide; GO = Graphene oxide; SWCNT = Single-walled carbon nanotube; ZrO_2= Zirconium dioxide; PP = Polypropylene; PS = Polysulphone; AF = Amyloid-Fibril.

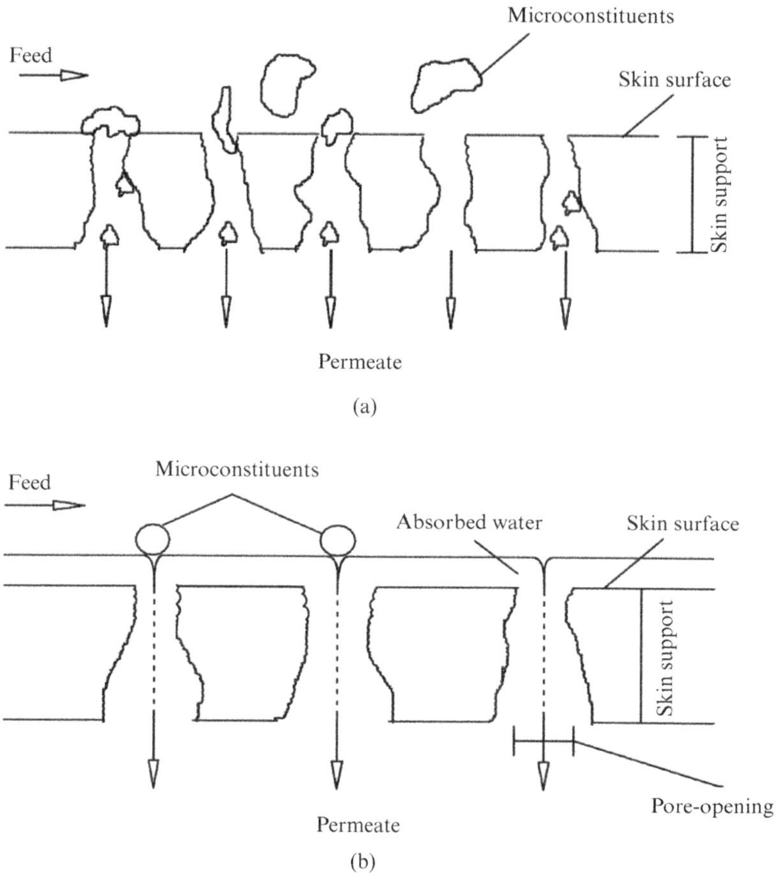

Figure 14.3 Fundamental principles involved in different MBSTs: removal of the pollutants by (a) sieving; and (b) the thin water layer adsorbed onto the dense membrane surface.

processes (Hu et al. 2014). Apart from size exclusion, hydrophobic interaction (Burba et al. 2005) and adsorption (Javier Benitez et al. 2011) have been the major EC removal pathways during UF and MF processes. However, unlike NF, RO, and FO, the membranes used in UF and MF do not contain any charge. Hence, electrostatic interaction between the ECs and the membrane surface does not occur during these processes (Nath 2017).

Meanwhile, in all the aforementioned processes, the difference in hydrostatic pressure across the membrane (also known as transmembrane pressure (TMP)) encompasses the separation of the micropollutants (Shojaee Nasirabadi et al. 2016). It has to be noted that only MF and UF can be termed as the filtration processes if filtration is defined depending upon sieving, or straining, or size exclusion. On the other hand, in NF, FO, and RO, the predominant EC removal mechanism is diffusion. Thus, these processes can be termed as the membrane separation processes (Nath 2017).

14.4.2 Hydraulics of Microconstituents Removal

In the case of UF and MF, the feed containing the MCs is pressurized to flow through the membrane with the help of a pump. Here, the flow is placed from the feed side to the permeate side (Dharupaneedi et al. 2019). The micropollutants present in the feed solution are filtered or sieved by the membrane pores, and clear permeate is obtained on the other side of the membrane (Shojaee Nasirabadi et al. 2016; Nath 2017). In UF and MF, the feed (water or wastewater) can be applied in the two following ways: i) parallel to the membrane surface (also known as cross-flow filtration) (Figure 14.4a); and ii) perpendicular to the membrane surface (also known as dead-end filtration) (Figure 14.4b). Both these two modes of membrane operations are briefly discussed later.

The hydraulics of MCs removal via FO and RO processes have been schematically represented in Figures 14.5a and 14.5c, respectively. As already mentioned, FO and RO processes are driven by the difference in osmotic pressure across the membrane (Metcalf and Eddy 2003). When two solutions, spiked with different concentrations of the MCs, are separated by a dense membrane (membrane with very less pore-opening size), due to the potential difference, water from the relatively low concentration side (high potential) flows toward the high concentration side to dilute the solution. This phenomenon is known as the FO process (Metcalf and Eddy 2003) (Figure 14.5a). After a certain time, the concentration of both the solutions, resting opposite to the membrane, becomes equal. At that time, the pressure difference balances the potential difference. This condition is termed as the equilibrium condition (Figure 14.5b). At the equilibrium condition, no net flow takes place through the membrane, and the balancing pressure is termed as osmotic pressure (Nath 2017). If a pressure greater than the osmotic pressure is applied externally from the opposite direction, the flow through the membrane takes place from the high concentration side to the low concentration side (Figure 14.5c). This phenomenon is referred to as the RO process (Metcalf and Eddy 2003). Hence, in the RO process, the feed embraces a high concentration of MCs. During the course of passage through the membrane, the MCs are rejected by the thin film of water adsorbed onto the membrane surface (Matsuura 2020). Hence, RO ensures better removal of the ECs than that of FO.

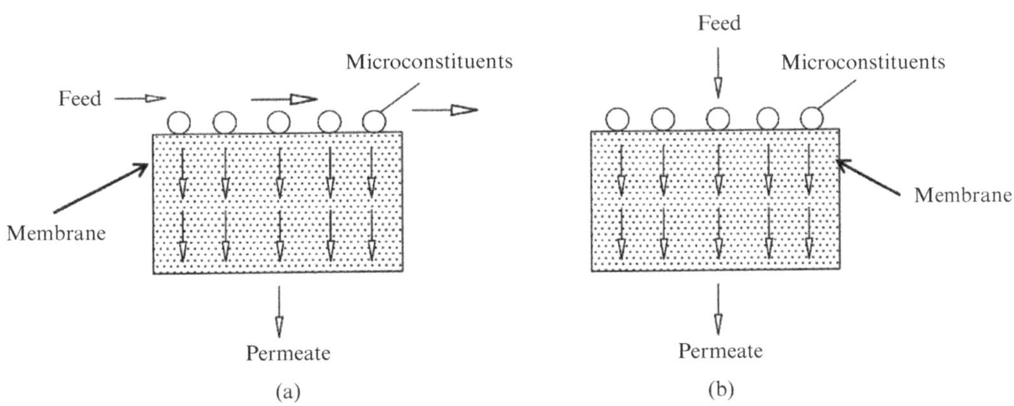

Figure 14.4 Modes of operation of the MBSTs depending upon the direction of feed flow: (a) cross-flow filtration; and (b) dead-end filtration.

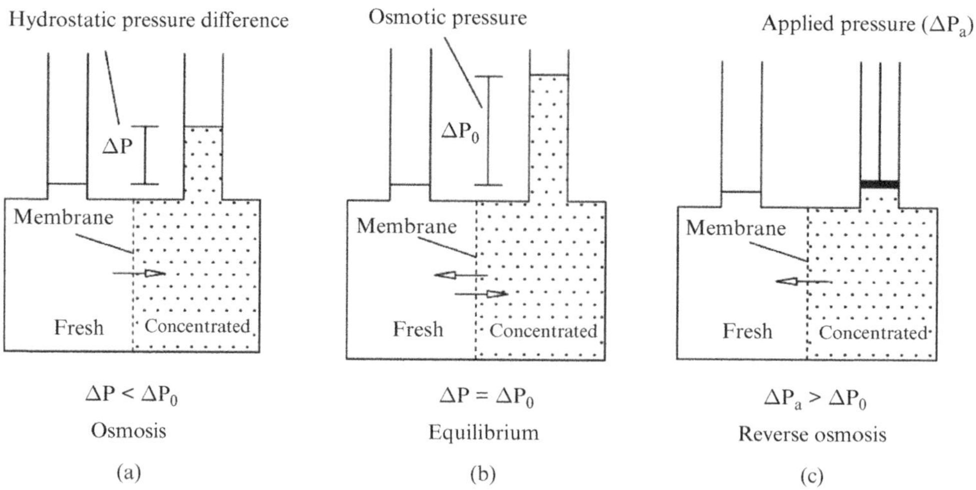

Figure 14.5 Schematic representation of osmotic flow: (a) osmosis; (b) osmotic equilibrium; and (c) reverse osmosis. *Source:* Adapted from Zhang et al. 2012..

14.4.2.1 Modes of Operation

The modes of operation in MBSTs can be broadly classified into the two following groups:

a) ***Depending upon the direction of feed flow***
   Depending upon the direction of flow of feed, the membrane filtration or separation processes can be classified into the two following types:
   i) *Cross-flow filtration*
      In cross-flow filtration, the feed flows parallel to the membrane surface. During the course of passage of feed over the membrane, permeate becomes separated (Figure 14.4a), and the pollutants deposited on the membrane surface are washed away to some extent by the feed flow (Metcalf and Eddy 2003). Hence, in cross-flow filtration, the frequency of clogging is somewhat less. However, some of the pollutant particles may become deposited on the membrane surface, resulting in clogging of membrane pores after a certain time. Chemically enhanced backwashing (CEB) can be used to clean the membrane pores. However, in CEB, there is always a possibility of permanent damage to the membrane (Metcalf and Eddy 2003). Cross-flow filtration is commonly adopted in NF and RO applications, where dead-end filtration is not feasible owing to the frequent clogging of the membrane (Nath 2017).
   ii) *Dead-end filtration*
      In dead-end filtration, the feed solution flows perpendicular to the membrane surface, which means the entire feed solution passes through the membrane where the micropollutants are filtered, and clean permeate is obtained on the other side of the membrane (Shojaee Nasirabadi et al. 2016; Nath 2017) (Figure 14.4b). However, as time progresses, more pollutants accumulate on the membrane surface, resulting in a reduction in the permeate flux. Therefore, the frequency of clogging is much higher in dead-end filtration than in cross-flow filtration. In dead-end filtration, clean in place (CIP) can be a good way of cleaning the membranes (Metcalf and

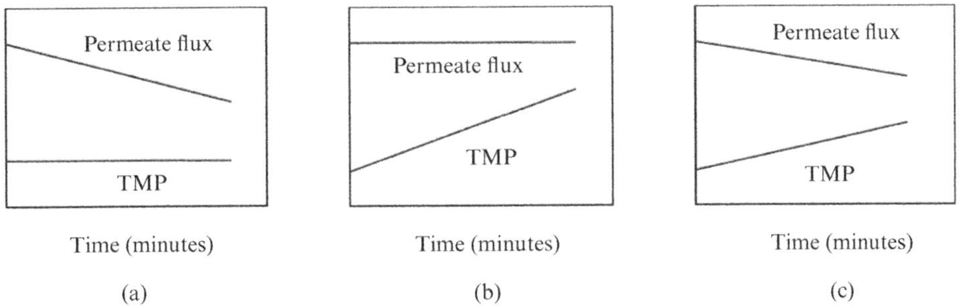

Figure 14.6 Modes of operation of the MBSTs depending upon permeate flux and TMP: (a) constant TMP mode; (b) constant permeate flux mode; and (c) hybrid mode.

Eddy 2003). In the CIP technique, chemicals are added to the feed solution; when it comes in contact with the membrane, the pollutant particles, clogged in the membrane pores, are dissolved, thereby enhancing the permeate flow (Metcalf and Eddy 2003). This is mainly adopted in MF and UF processes (Nath 2017).

b) ***Depending upon the permeate flux and TMP***

With respect to the permeate flux and TMP, the membrane separation processes can be operated in the three following ways:

i) *Constant TMP*

In this mode, the TMP is kept constant and thus the permeate flux keeps on reducing with time due to the deposition of the pollutants inside the membrane pores (Figure 14.6a). This is not an economical option as the obtained throughput volume is less (Matsuura 2020; Metcalf and Eddy 2003).

ii) *Constant permeate flux*

Here, the aim is to obtain a constant permeate flux with time. Thus, to counteract the impact of the pollutant deposition on the membrane surface, the TMP has to be increased with time (Figure 14.6b). However, when the critical TMP is reached, CEB has to be performed to obtain the constant permeate flux, causing possible permanent damage to the membranes (Metcalf and Eddy 2003).

iii) *Hybrid mode*

To overcome the hurdles associated with both the aforementioned modes of membrane operation, both the permeate flux and TMP are allowed to vary with time. Hence, this is termed as a hybrid mode of operation (Bourgeous et al. 1999; Metcalf and Eddy 2003) (Figure 14.6c).

14.4.2.2 Definitions of Some Frequently Used Terms in MBSTs

In this section, some commonly used terms in MBSTs are described (Matsuura 2020; Metcalf and Eddy 2003; Nath 2017).

a) ***Flux***

The transfer rate of the solvents in terms of mass or volume through a unit area of the membrane surface per unit time is known as flux. It is expressed in $m^3/m^2/h$ or $kg/m^2/h$.

b) *Permeate*

The clean throughput obtained after the pollutants become separated from the feed through the membrane is termed as permeate. In other words, permeate is the part of feed that passes through the membrane.

c) *Retentate or concentrate*

Retentate is the part of feed that does not pass through the membrane. Hence, feed is the summation of permeate and retentate.

d) *Molecular weight cut-off (MWCO)*

The molecular weight of the smallest pollutant that is being rejected by the membrane with around 90% efficiency is termed as molecular weight cut-off (MWCO). It is generally expressed in Daltons.

e) *TMP*

TMP is the difference in hydrostatic pressure between feed side and permeate side.

f) *Recovery rate (r)*

Volume of the feed water that can be recovered on the permeate side is known as the recovery rate (r). It is expressed in %, as in Eq. (14.1).

$$r = \left(\frac{Q_p}{Q_f} \times 100 \right) \% \tag{14.1}$$

where Q_p and Q_f denote permeate flow rate and feed flow rate, respectively.

g) *Rejection efficiency (R)*

The rejection efficiency is the difference in concentration of a particular pollutant (here, MCs) between the feed side and permeate side, expressed as a percentage of its concentration on the feed side, as shown in Eq. (14.2).

$$R = \left(\frac{C_f - C_p}{C_f} \times 100 \right) \% \tag{14.2}$$

where C_f and C_p denote the concentration of the MC on the feed side and permeate side, respectively.

The following section is a brief discussion on the potential of various MBSTs to eradicate the MCs from aqueous matrices.

14.5 Application of the MBSTs for Removing Microconstituents From Aqueous Matrices

The application of various MBSTs for removing different MCs, especially PPCPs, hormones, pesticides, surfactants, and PFASs existing in the water and wastewater matrices, has been represented in Table 14.2. From the data listed in the table, it is clearly understood that the MBSTs have remarkable potential to remove various MCs from aqueous matrices. For example, Verliefde et al. (2007) found that ibuprofen and diclofenac were completely removed from the surface water sample via an NF process when thin film composite (TFC) membranes with a PA active layer were applied. However, NF is not as effective in removing hormones as it is in eliminating PPCPs from the aqueous environment. For instance, Yoon et al. (2007) examined the performance of TFC membranes for removing testosterone via an NF process. They found that only 62% testosterone was removed. However, the RO process using a TFC membrane was able to eradicate testosterone by up to 82 to 100% from the samples collected from Lake Ontario, membrane bioreactor effluent, and laboratory grade Milli-Q water (Comerton et al. 2008). Even though the EC removal mechanisms are the same in both NF and RO processes, the membrane pore-openings in RO membranes are finer than those in NF membranes (Kárászová et al. 2020). This could be the possible reason behind the aforementioned results. On the other hand, the pore-openings in UF membranes are larger than those in NF and RO membranes, and the removal of the ECs is triggered by sieving or straining in the UF process (Shojaee Nasirabadi et al. 2016; Kárászová et al. 2020). Thus, UF permits the micropollutants with a size relatively smaller than its pore-openings to pass through the membrane, cutting down the efficacy of the UF process to some extent. For instance, atrazine was removed by up to 95.9 and 90 to 98% via RO (Kosutic et al. 2005) and NF (Rakhshan and Pakizeh 2015), respectively, but only 9% via a PES UF membrane (Acero et al. 2017) (Table 14.2).

FO also captures commendable removal of various MCs from the aqueous matrices. For instance, Zaib et al. (2013) have investigated the potential of CTA membranes to remove estrone hormone from synthetic wastewater in an FO process. They reported that the FO process facilitated >96% removal of estrone. Since estrone is neutral, its removal via FO was governed by the diffusion through the water layer adsorbed onto the dense membrane surface (Nath 2017). Here, the electrostatic interaction between estrone molecules and the surface of the CTA membrane did not occur. Furthermore, Jin et al. (2021) have tried to evaluate the efficacy of amyloid-fibril (AF)-based membranes in removing various PFASs (Table 14.2) from Milli-Q water. They found that AF membranes assured better removal of the PFASs than that of CA membranes (Table 14.2). The supremacy of AF membranes over CA membranes could be attributed to the synergy between the amyloidal properties of the membrane and the molecular traits of the PFASs (Jin et al. 2021). Furthermore, perfluorooctane sulfonic acid (PFOS) and perfluorooctanoic acid (PFOA) were also removed through the attractive electrostatic interaction with the AF membrane surface (Jin et al. 2021).

14.6 Membrane Fouling

Membrane fouling can be defined as the deposition and accumulation of various constituents, also known as foulants, on the membrane surface or inside the membrane pores (Li and Chen 2010). As a consequence, the permeate flux is reduced, and also the performance of the membrane falls drastically (Metcalf and Eddy 2003).

14.6.1 Classification of Membrane Fouling

Depending upon the type of foulants, membrane fouling can be predominantly classified into the following four types.

14.6.1.1 Particulate or Colloidal Fouling

When the suspended solids or the colloidal particles become attached to the membrane surface or deposited inside the membrane pores, the resulting membrane fouling is known as particulate or colloidal fouling (Ye and Chen 2005). Colloidal or particulate fouling is engendered by the inorganic and non-biological particles such as clay, silt, etc. (Metcalf and Eddy 2003). This type of fouling is common during MF and UF processes. Air scrubbing and back flushing can be the possible remedies for particulate or colloidal fouling (Li and Chen 2010). Moreover, appropriate upstream filtration and coagulation can effectively reduce the risk of particulate fouling (Metcalf and Eddy 2003).

14.6.1.2 Biological or Microbial Fouling

Biological or microbial fouling is also known as membrane biofouling (Li and Chen 2010). It occurs when algae, plants, microbes, and other biological means grow on the membrane surface and inside the membrane pores (Flemming 2002). Basically, the biofoulants, especially the microbes, excrete extracellular polymeric substance (EPS) (Hodgson et al. 1993) and form a slimy layer over the membrane surface, also known as biofilm, changing the surface properties of the membrane and cutting down the permeate flux (Li and Chen 2010). Biofouling can be controlled to some extent by the following: i) eliminating the biodegradable compounds from the feed; ii) applying pure chemical dosage; and iii) disinfecting the feed by chlorine, ozone, or UV radiation (Li and Chen 2010).

14.6.1.3 Scaling or Precipitation Fouling

Scaling or precipitation fouling is also termed inorganic fouling. It occurs when the dissolved constituents such as crystallized salts (e.g., calcium sulphate, calcium carbonate, barium sulphate, etc.), oxides (silica), or hydroxides present in the feed are dislodged and accumulate on the membrane surface or inside the membrane pores. This is a common occurrence during NF and RO processes (Li and Chen 2010). Softening can be done to restrict the scaling of the membrane (Metcalf and Eddy 2003).

14.6.1.4 Organic Fouling

Organic fouling occurs due to the deposition of various carbon-based organic materials such as protein, carbohydrates, polysaccharides, etc. onto the membrane surface or inside the membrane pores (McMurry 2000). Pretreatment of the feed or proper selection of the membrane materials to restrict the adsorption of organic foulants to the membrane can be possible ways to prevent organic fouling (Nath 2017).

14.6.2 Mechanisms of Membrane Fouling

There are broadly two ways in which the fouling can be initiated during the course of membrane operations (Metcalf and Eddy 2003):

a) ***Membrane fouling due to the build-up of solids***

   Basically, the solvent flux through the membrane can be restricted by the solids, deposited on the membrane surface or inside the membrane pores, in the three following ways:

   i) *Pore narrowing*

      Pore narrowing is the deposition of the solids, present in the feed, to the interior surface of the membrane pores, reducing the permeate flux through the membrane pores. As the name suggests, pore narrowing is caused by the deposition of the solid particles having sizes smaller than the membrane pore-openings (Figure 14.7a). Backwashing can be done to reduce pore narrowing and regain the permeate flux (Metcalf and Eddy 2003).

ii) *Pore plugging*

Pore plugging occurs when the solid particles having sizes equal to the membrane pore-openings become stuck in the membrane pores during the passage of feed through the membrane (Figure 14.7b). As a result, the membrane pores become blocked causing membrane fouling. Pore plugging can also be reduced to some extent by backwashing (Metcalf and Eddy 2003).

iii) *Gel or cake formation*

When the solid particles present in the feed are deposited and accumulate on the membrane surface, as shown in Figure 14.7c, the solvent flux through the membrane pores becomes prohibited. In this case, a large amount of solids accumulate on the membrane surface and form a gel or cake layer (Li and Chen 2010; Metcalf and Eddy 2003; Nath 2017).

b) ***Permanent damage of the membranes***

Permanent damage of the membranes takes place if the feed accommodates acids, bases, free chlorine, and free oxygen molecules. Even if the pH of the feed is extremely low or extremely high, permanent damage of the membranes can be observed (Metcalf and Eddy 2003).

14.6.3 Control of Membrane Fouling

Membrane fouling can be controlled to some extent by adapting the following two preventive measures:

1) Fouling due to build-up of solids can be reduced by implementing integrated membrane filtration techniques. In other words, instead of taking the feed directly for NF or RO, it can be passed through MF, UF, NF, and RO membranes in sequence. As a result, the larger particles will be eliminated beforehand, and the frequency of clogging of the NF

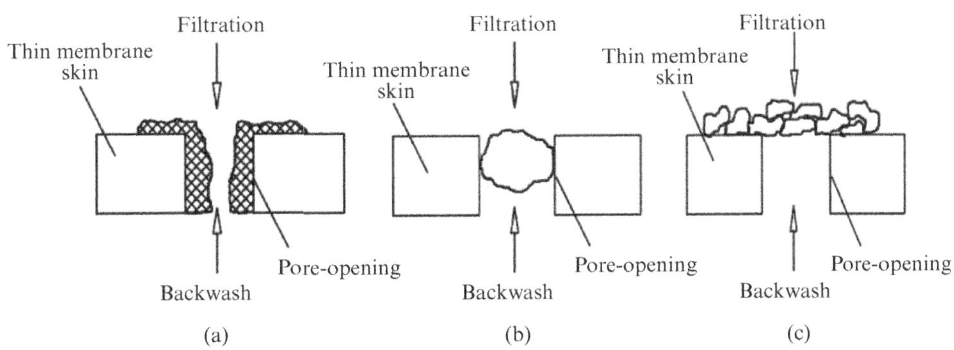

Figure 14.7 Mechanisms of membrane fouling via build-up of solids. (a) pore narrowing; (b) pore plugging; and (c) gel or cake formation.

and RO membranes can be delayed to some extent (Li and Chen 2010; Metcalf and Eddy 2003); and

2) Permanent damage of the membrane can be controlled by changing the feed water characteristics, i.e., by controlling the previously-mentioned substances that may cause permanent damage to the membrane. However, the extent of damage depends upon the nature of the membrane materials. Hence, appropriate selection of the membrane materials is of paramount importance to fight against membrane fouling (Metcalf and Eddy 2003).

14.7 Future Perspectives

Even though the MBSTs have been extensively explored for eradicating various MCs from the aqueous matrices, there are a few concerns associated with this technology, which warrant further research:

- The techno-economical aspects of the MBSTs have not yet been adequately explored. Moreover, the cost comparisons between the various MBSTs and the other available MC removal technologies have to be performed in detail to evaluate the feasibility of using the MBSTs to remove the MCs from water and wastewater.
- It is clear from the literature that very limited study has been reported on application of the cutting-edge membranes using nanocomposite, graphene, etc. for eradicating ECs from the aquatic matrices. Further study is of utmost importance to further improve the treatment efficacy of the MBSTs.
- Most of the studies on the removal of MCs using MBSTs have been limited to the lab-scale and performed using synthetic wastewater spiked with a single MC. In reality, however, the scenario is completely different. Moreover, the complexity of real wastewater differs from that of synthetic wastewater to a great extent. Hence, future research works should concentrate on the implementation of the MBSTs for eradicating the MCs from real wastewater in field-scale applications.

14.8 Conclusions

In this chapter, a comprehensive overview on the application of various MBSTs for removing various microconstituents (MCs) from aqueous matrices has been presented. From the extensive review of the existing literature, it has been observed that UF, NF, RO, and FO processes have shown commendable potential in eradicating MCs from water and wastewater, whereas the application of MF has now become obsolete. Usually, UF facilitates the removal of MCs through sieving or size exclusion, whereas in NF, RO, and FO, the adsorbed water layer on the dense membrane surface captures the separation of the MCs. Additionally, the dense membranes used in RO, NF, and FO accommodate slight charges, ensuring enhanced removal of the charged pollutants via electrostatic interactions. Usually, the cross-flow filtration mode is more suitable to NF, RO, and FO than dead-end filtration because the frequency of clogging of the membrane pores is less in cross-flow filtration mode. Meanwhile, the hybrid mode of operation (i.e., both the TMP and the permeate flux are allowed to vary with time) has been the most feasible mode of membrane operation. Usually, the ceramic or inorganic membranes have been found to be better than the organic or polymeric membranes in terms of thermal stability, pollutant selectivity, and resisting the corrosive liquids at elevated temperature. However, aromatic PA and cellulose derivatives have been vigorously explored as the membrane materials. Even though the MBSTs are extensively implemented for separating the ECs from aqueous matrices, there is always a chance that the membrane pores will become clogged, known as membrane fouling, due to the deposition of different types of foulants. However, by proper selection of membrane materials, by giving appropriate pre-treatment to the feed solution, or by changing the feed water characteristics, the risk of occurrence of membrane fouling can be avoided to some extent. Nevertheless, issues regarding the practicability of the MBSTs such as economic feasibility and field-scale application for remediating the ECs from water and wastewater necessitate further attention.

References

Acero, J.L., Benitez, F.J., Leal, A.I. et al. (2010). Membrane filtration technologies applied to municipal secondary effluents for potential reuse. *Journal of Hazardous Materials* 177 (1–3): 390–398.

Acero, J.L., Benitez, F.J., Real, F.J., and Teva, F. (2017). Removal of emerging contaminants from secondary effluents by micellar-enhanced ultrafiltration. *Separation and Purification Technology* 181: 123–131.

Albergamo, V., Blankert, B., Cornelissen, E.R. et al. (2019). Removal of polar organic micropollutants by pilot-scale reverse osmosis drinking water treatment. *Water Research* 148: 535–545.

Baransi-Karkaby, K., Bass, M., and Freger, V. (2019). *In situ* modification of reverse osmosis membrane elements for enhanced removal of multiple micropollutants. *Membranes* 9 (2): 28.

Baudequin, C., Mai, Z., Rakib, M. et al. (2014). Removal of fluorinated surfactants by reverse osmosis role of surfactants in membrane fouling. *Journal of Membrane Science* 458: 111–119.

Bazinet, L.F.L.L., Lamarche, F., and Ippersiel, D. (1998). Bipolar-membrane electrodialysis: applications of electrodialysis in the food industry. *Trends in Food Science & Technology* 9 (3): 107–113.

Bojnourd, F.M. and Pakizeh, M. (2018). Preparation and characterization of a nanoclay/PVA/PSf nanocomposite membrane for removal of pharmaceuticals from water. *Applied Clay Science* 162: 326–338.

Bourgeous, K.N., Tchobanoglous, G., and Darby, J. (1999). Performance evaluation of the Koch Ultrafiltration (UF) membrane system for wastewater reclamation. Center for Environment [i.e., Environmental] and Water Resources Engineering, Department of Civil and Environmental Engineering, University of California, Davis.

Burba, P., Geltenpoth, H., and Nolte, J. (2005). Ultrafiltration behavior of selected pharmaceuticals on natural and synthetic membranes in the presence of humic-rich hydrocolloids. *Analytical and Bioanalytical Chemistry* 382 (8): 1934–1941.

Cacace, D., Fatta-Kassinos, D., Manaia, C.M. et al. (2019). Antibiotic resistance genes in treated wastewater and in the receiving water bodies: a pan-European survey of urban settings. *Water Research* 162: 320–330.

Cartinella, J.L., Cath, T.Y., Flynn, M.T. et al. (2006). Removal of natural steroid hormones from wastewater using membrane contactor processes. *Environmental Science & Technology* 40 (23): 7381–7386.

Chen, Y., Xu, W., Zhu, H. et al. (2019). Effect of turbidity on micropollutant removal and membrane fouling by MIEX/ultrafiltration hybrid process. *Chemosphere* 216: 488–498.

Chu, K.H., Fathizadeh, M., Yu, M. et al. (2017). Evaluation of removal mechanisms in a graphene oxide-coated ceramic ultrafiltration membrane for retention of natural organic matter, pharmaceuticals, and inorganic salts. *ACS Applied Materials & Interfaces* 9 (46): 40369–40377.

Coday, B.D., Yaffe, B.G., Xu, P., and Cath, T.Y. (2014). Rejection of trace organic compounds by forward osmosis membranes: a literature review. *Environmental Science & Technology* 48 (7): 3612–3624.

Comerton, A.M., Andrews, R.C., Bagley, D.M., and Hao, C. (2008). The rejection of endocrine disrupting and pharmaceutically active compounds by NF and RO membranes as a function of compound and water matrix properties. *Journal of Membrane Science* 313: 323–335.

Conkle, J.L., White, J.R., and Metcalfe, C.D. (2008). Reduction of pharmaceutically active compounds by a lagoon wetland wastewater treatment system in Southeast Louisiana. *Chemosphere* 73: 1741–1748.

Coppens, L.J., van Gils, J.A., Ter Laak, T.L. et al. (2015). Towards spatially smart abatement of human pharmaceuticals in surface waters: defining impact of sewage treatment plants on susceptible functions. *Water Research* 81: 356–365.

Crespo, J.G. and Brazinha, C. (2015). 1: Fundamentals of pervaporation. In: *Pervaporation, Vapour Permeation and Membrane Distillation* (eds. Basile, A., Figoli, A., and Khayet, M.), 3–17. Oxford, UK: Woodhead Publishing.

Cui, Z.F. and Muralidhara, H.S. (2010). *Membrane Technology: A Practical Guide to Membrane Technology and Applications in Food and Bioprocessing*. Elsevier.

D'Haese, A., Le-Clech, P., Van Nevel, S. et al. (2013). Trace organic solutes in closed-loop forward osmosis applications: influence of membrane fouling and modeling of solute build-up. *Water Research* 47 (14): 5232–5244.

de Oliveira, C.P.M., Farah, I.F., Koch, K. et al. (2022). TiO_2-Graphene oxide nanocomposite membranes: a review. *Separation and Purification Technology* 280: 119836.

Deblonde, T., Cossu-Leguille, C., and Hartemann, P. (2011). Emerging pollutants in wastewater: a review of the literature. *International Journal of Hygiene and Environmental Health* 214 (6): 442–448.

Dewil, R., Mantzavinos, D., Poulios, I., and Rodrigo, M.A. (2017). New perspectives for advanced oxidation processes. *Journal of Environmental Management* 195: 93–99.

Dhangar, K. and Kumar, M. (2020). Tricks and tracks in removal of emerging contaminants from the wastewater through hybrid treatment systems: a review. *Science of the Total Environment* 738: 140320.

Dharupaneedi, S.P., Nataraj, S.K., Nadagouda, M. et al. (2019). Membrane-based separation of potential emerging pollutants. *Separation and Purification Technology* 210: 850–866.

Dolar, D., Vuković, A., Ašperger, D., and Košutić, K. (2011). Effect of water matrices on removal of veterinary pharmaceuticals by nanofiltration and reverse osmosis membranes. *Journal of Environmental Sciences* 23 (8): 1299–1307.

Donato, L. and Drioli, E. (2021). Imprinted membranes for sustainable separation processes. *Frontiers of Chemical Science and Engineering* 1–18.

Dong, L.X., Huang, X.C., Wang, Z. et al. (2016). A thin-film nanocomposite nanofiltration membrane prepared on a support with *in situ* embedded zeolite nanoparticles. *Separation and Purification Technology* 166: 230–239.

Ersahin, M.E., Ozgun, H., Dereli, R.K. et al. (2012). A review on dynamic membrane filtration: materials, applications and future perspectives. *Bioresource Technology* 122: 196–206.

Fan, X., Tao, Y., Wang, L. et al. (2014). Performance of an integrated process combining ozonation with ceramic membrane ultra-filtration for advanced treatment of drinking water. *Desalination* 335 (1): 47–54.

Fini, M.N., Madsen, H.T., and Muff, J. (2019). The effect of water matrix, feed concentration and recovery on the rejection of pesticides using NF/RO membranes in water treatment. *Separation and Purification Technology* 215: 521–527.

Flemming, H.C. (2002). Biofouling in water systems: cases, causes and countermeasures. *Applied Microbiology and Biotechnology* 59 (6): 629–640.

Fujioka, T., Khan, S.J., McDonald, J.A., and Nghiem, L.D. (2015). Rejection of trace organic chemicals by a hollow fibre cellulose triacetate reverse osmosis membrane. *Desalination* 368: 69–75.

García, J., García-Galán, M.J., Day, J.W. et al. (2020). A review of emerging organic contaminants (EOCs), antibiotic resistant bacteria (ARB), and antibiotic resistance genes (ARGs) in the environment: increasing removal with wetlands and reducing environmental impacts. *Bioresource Technology* 307: 123228.

Garcia, J., Rousseau, D.P., Morato, J. et al. (2010). Contaminant removal processes in subsurface–flow constructed wetlands: a review. *Critical Reviews in Environmental Science and Technology* 40 (7): 561–661.

Garcia-Ivars, J., Durá-María, J., Moscardó-Carreño, C. et al. (2017). Rejection of trace pharmaceutically active compounds present in municipal wastewaters using ceramic fine ultrafiltration membranes: effect of feed solution pH and fouling phenomena. *Separation and Purification Technology* 175: 58–71.

Guo, H., Yao, Z., Yang, Z. et al. (2017). A one–step rapid assembly of thin film coating using green coordination complexes for enhanced removal of trace organic contaminants by membranes. *Environmental Science & Technology* 51 (21): 12638–12643.

Hai, F.I., Yamamoto, K., and Fukushi, K. (2007). Hybrid treatment systems for dye wastewater. *Critical Reviews in Environmental Science and Technology* 37 (4): 315–377.

Halakoo, E. and Feng, X. (2020). Layer-by-layer assembly of polyethyleneimine/graphene oxide membranes for desalination of high-salinity water via pervaporation. *Separation and Purification Technology* 234: 116077.

Heo, J., Boateng, L.K., Flora, J.R. et al. (2013). Comparison of flux behavior and synthetic organic compound removal by forward osmosis and reverse osmosis membranes. *Journal of Membrane Science* 443: 69–82.

Hodgson, P.H., Leslie, G.L., Fane, A.G. et al. (1993). Cake resistance and solute rejection in bacterial microfiltration: the role of the extracellular matrix. *Journal of Membrane Science* 79 (1): 35–53.

Hofs, B., Schurer, R., Harmsen, D.J.H. et al. (2013). Characterization and performance of a commercial thin film nanocomposite seawater reverse osmosis membrane and comparison with a thin film composite. *Journal of Membrane Science* 446: 68–78.

Hu, Z., Si, X., Zhang, Z., and Wen, X. (2014). Enhanced EDCs removal by membrane fouling during the UF process. *Desalination* 336: 18–23.

Huang, H., Cho, H., Schwab, K., and Jacangelo, J.G. (2011). Effects of feedwater pretreatment on the removal of organic microconstituents by a low fouling reverse osmosis membrane. *Desalination* 281: 446–454.

Huang, M., Chen, Y., Huang, C.H. et al. (2015). Rejection and adsorption of trace pharmaceuticals by coating a forward osmosis membrane with TiO_2. *Chemical Engineering Journal* 279: 904–911.

Imbrogno, A., Biscarat, J., and Iris Schafer, A. (2017). Estradiol uptake in a combined magnetic ion exchange–ultrafiltration (MIEX–UF) process during water treatment. *Current Pharmaceutical Design* 23 (2): 328–337.

Januário, E.F.D., Vidovix, T.B., Beluci, N.D.C.L. et al. (2021). Advanced graphene oxide-based membranes as a potential alternative for dyes removal: a review. *Science of the Total Environment* 789: 147957.

Jatoi, A.S., Hashmi, Z., Adriyani, R. et al. (2021). Recent trends and future challenges of pesticide removal techniques: comprehensive review. *Journal of Environmental Chemical Engineering* 9 (4): 105571–105588.

Javier Benitez, F., Acero, J.L., Real, F.J. et al. (2011). Ultrafiltration and nanofiltration membranes applied to the removal of the pharmaceuticals amoxicillin, naproxen, metoprolol and phenacetin from water. *Journal of Chemical Technology & Biotechnology* 86 (6): 858–866.

Jiang, J.Q., Zhou, Z., and Sharma, V.K. (2013). Occurrence, transportation, monitoring and treatment of emerging micro-pollutants in waste water: a review from global views. *Microchemical Journal* 110: 292–300.

Jin, T., Peydayesh, M., Joerss, H. et al. (2021). Amyloid-fibril based membranes for PFAS removal from water. *Environmental Science: Water Research & Technology* 7 (10): 1873–1884.

Jin, X., Shan, J., Wang, C. et al. (2012). Rejection of pharmaceuticals by forward osmosis membranes. *Journal of Hazardous Materials* 227: 55–61.

Jones, O.A.H., Voulvoulis, N., and Lester, J.N. (2005). Human pharmaceuticals in wastewater treatment processes. *Critical Reviews in Environmental Science and Technology* 35 (4): 401–427.

Judd, S. and Jefferson, B. (eds.) (2003). *Membranes for Industrial Wastewater Recovery and Re-use*. Elsevier.

Kaminska, G., Bohdziewicz, J., Calvo, J.I. et al. (2015). Fabrication and characterization of polyethersulfone nanocomposite membranes for the removal of endocrine disrupting micropollutants from wastewater: mechanisms and performance. *Journal of Membrane Science* 493: 66–79.

Kanakaraju, D., Glass, B.D., and Oelgemöller, M. (2018). Advanced oxidation process-mediated removal of pharmaceuticals from water: a review. *Journal of Environmental Management* 219: 189–207.

Kárászová, M., Bourassi, M., and Gaálová, J. (2020). Membrane removal of emerging contaminants from water: which kind of membranes should we use? *Membranes* 10 (11): 305.

Khanzada, N.K., Farid, M.U., Kharraz, J.A. et al. (2020). Removal of organic micropollutants using advanced membrane-based water and wastewater treatment: a review. *Journal of Membrane Science* 598: 117672.

Kimura, K., Iwase, T., Kita, S., and Watanabe, Y. (2009). Influence of residual organic macromolecules produced in biological wastewater treatment processes on removal of pharmaceuticals by NF/RO membranes. *Water Research* 43 (15): 3751–3758.

Kimura, K., Toshima, S., Amy, G., and Watanabe, Y. (2004). Rejection of neutral endocrine disrupting compounds (EDCs) and pharmaceutical active compounds (PhACs) by RO membranes. *Journal of Membrane Science* 245 (1–2): 71–78.

Klamerth, N., Rizzo, L., Malato, S. et al. (2010). Degradation of fifteen emerging contaminants at $\mu g\ L^{-1}$ initial concentrations by mild solar photo-Fenton in MWTP effluents. *Water Research* 44: 545–554.

Kong, F.X., Yang, H.W., Wu, Y.Q. et al. (2015). Rejection of pharmaceuticals during forward osmosis and prediction by using the solution-diffusion model. *Journal of Membrane Science* 476: 410–420.

Košutić, K., Furač, L., Sipos, L., and Kunst, B. (2005). Removal of arsenic and pesticides from drinking water by nanofiltration membranes. *Separation and Purification Technology* 42 (2): 137–144.

Košutić, K., Dolar, D., Ašperger, D., and Kunst, B. (2007). Removal of antibiotics from a model wastewater by RO/NF membranes. *Separation and Purification Technology* 53 (3): 244–249.

Kowalska, I. (2008). Surfactant removal from water solutions by means of ultrafiltration and ion exchange. *Desalination* 221 (1–3): 351–357.

Krzeminski, P., Leverette, L., Malamis, S., and Katsou, E. (2017). Membrane bioreactors: review on recent developments in energy reduction, fouling control, novel configurations, LCA and market prospects. *Journal of Membrane Science* 527: 207–227.

Li, H. and Chen, V. (2010). Membrane fouling and cleaning in food and bioprocessing. In: *Membrane Technology*, 213–254. Butterworth-Heinemann. doi: https://doi.org/10.1016/B978-1-85617-632-3.00010-0.

Li, Y., Zhu, G., Ng, W.J., and Tan, S.K. (2014). A review on removing pharmaceutical contaminants from wastewater by constructed wetlands: design, performance and mechanism. *Science of the Total Environment* 468: 908–932.

Liu, P., Zhang, H., Feng, Y. et al. (2015). Integrating electrochemical oxidation into forward osmosis process for removal of trace antibiotics in wastewater. *Journal of Hazardous Materials* 296: 248–255.

Lou, Y., Tan, F.J., Zeng, R. et al. (2020). Preparation of cross-linked graphene oxide on polyethersulfone membrane for pharmaceuticals and personal care products removal. *Polymers* 12 (9): 1921.

Madsen, H.T., Bajraktari, N., Hélix-Nielsen, C. et al. (2015). Use of biomimetic forward osmosis membrane for trace organics removal. *Journal of Membrane Science* 476: 469–474.

Manaia, C.M., Rocha, J., Scaccia, N. et al. (2018). Antibiotic resistance in wastewater treatment plants: tackling the black box. *Environment International* 115: 312–324.

Matsuura, T. (2020). *Synthetic Membranes and Membrane Separation Processes*. CRC Press.

McMurry, J. (2000). *Organic Chemistry*. Pacific Grove. Albany. Belmont. Boston. Cincinati. Johannesburg. London. Madrid. Melbourne. Mexico City. New York. Scottsdale. Tokyo. Toronto: Brooks/Cole, a Division of Thomson Learning.

Miklos, D.B., Remy, C., Jekel, M. et al. (2018). Evaluation of advanced oxidation processes for water and wastewater treatment: a critical review. *Water Research* 139: 118–131.

Mohammad, A.W., Teow, Y.H., Ang, W.L. et al. (2015). Nanofiltration membranes review: recent advances and future prospects. *Desalination* 356: 226–254.

Naddeo, V., Belgiorno, V., Kassinos, D. et al. (2010). Ultrasonic degradation, mineralization and detoxification of diclofenac in water: optimization of operating parameters. *Ultrasonics Sonochemistry* 17 (1): 179–185.

Narbaitz, R.M., Rana, D., Dang, H.T. et al. (2013). Pharmaceutical and personal care products removal from drinking water by modified cellulose acetate membrane: field testing. *Chemical Engineering Journal* 225: 848–856.

Nasseri, S., Ebrahimi, S., Abtahi, M., and Saeedi, R. (2018). Synthesis and characterization of polysulfone/graphene oxide nano-composite membranes for removal of bisphenol A from water. *Journal of Environmental Management* 205: 174–182.

Nath, K. (2017). *Membrane Separation Processes*. PHI Learning Pvt. Ltd.

Neale, P.A. and Schäfer, A.I. (2012). Quantification of solute–solute interactions in steroidal hormone removal by ultrafiltration membranes. *Separation and Purification Technology* 90: 31–38.

Pabby, A.K., Rizvi, S.S., and Requena, A.M.S. (2008). *Handbook of Membrane Separations: Chemical, Pharmaceutical, Food, and Biotechnological Applications. CRC Press.* doi: https://doi.org/10.1201/9781420009484.

Paseta, L., Antorán, D., Coronas, J., and Téllez, C. (2019). 110th anniversary: polyamide/metal-organic framework bilayered thin film composite membranes for the removal of pharmaceutical compounds from water. *Industrial & Engineering Chemistry Research* 58 (10): 4222–4230.

Pramanik, B.K., Pramanik, S.K., Sarker, D.C., and Suja, F. (2017). Removal of emerging perfluorooctanoic acid and perfluorooctane sulfonate contaminants from lake water. *Environmental Technology* 38 (15): 1937–1942.

Radjenović, J., Petrović, M., Ventura, F., and Barceló, D. (2008). Rejection of pharmaceuticals in nanofiltration and reverse osmosis membrane drinking water treatment. *Water Research* 42 (14): 3601–3610.

Rakhshan, N. and Pakizeh, M. (2015). Removal of triazines from water using a novel OA modified SiO_2/PA/PSf nanocomposite membrane. *Separation and Purification Technology* 147: 245–256.

Rana, D., Scheier, B., Narbaitz, R.M. et al. (2012). Comparison of cellulose acetate (CA) membrane and novel CA membranes containing surface modifying macromolecules to remove pharmaceutical and personal care product micropollutants from drinking water. *Journal of Membrane Science* 409: 346–354.

Rivera–Utrilla, J., Sánchez-Polo, M., Gómez–Serrano, V. et al. (2011). Activated carbon modifications to enhance its water treatment applications: an overview. *Journal of Hazardous Materials* 187 (1–3): 1–23.

Rosman, N., Salleh, W.N.W., Razali, N.A.M. et al. (2021). Ibuprofen removal through photocatalytic filtration using antifouling PVDF-ZnO/Ag_2CO_3/Ag_2O nanocomposite membrane. *Materials Today: Proceedings* 42: 69–74.

Shojaee Nasirabadi, P., Saljoughi, E., and Mousavi, S.M. (2016). Membrane processes used for removal of pharmaceuticals, hormones, endocrine disruptors and their metabolites from wastewaters: a review. *Desalination and Water Treatment* 57 (51): 24146–24175.

Simazaki, D., Kubota, R., Suzuki, T. et al. (2015). Occurrence of selected pharmaceuticals at drinking water purification plants in Japan and implications for human health. *Water Research* 76: 187–200.

Snyder, S.A., Adham, S., Redding, A.M. et al. (2007). Role of membranes and activated carbon in the removal of endocrine disruptors and pharmaceuticals. *Desalination* 202 (1–3): 156–181.

Soriano, A., Gorri, D., and Urtiaga, A. (2019). Selection of high flux membrane for the effective removal of short-chain perfluorocarboxylic acids. *Industrial & Engineering Chemistry Research* 58 (8): 3329–3338.

Stuart, M., Lapworth, D., Crane, E., and Hart, A. (2012). Review of risk from potential emerging contaminants in UK groundwater. *Science of the Total Environment* 416: 1–21.

Tsai, Y.T., Yu-Chen Lin, A., Weng, Y.H., and Li, K.C. (2010). Treatment of perfluorinated chemicals by electro-microfiltration. *Environmental Science & Technology* 44 (20): 7914–7920.

Tufail, A., Price, W.E., and Hai, F.I. (2020). A critical review on advanced oxidation processes for the removal of trace organic contaminants: a voyage from individual to integrated processes. *Chemosphere* 260: 127460.

Urtiaga, A.M., Pérez, G., Ibáñez, R., and Ortiz, I. (2013). Removal of pharmaceuticals from a WWTP secondary effluent by ultrafiltration/reverse osmosis followed by electrochemical oxidation of the RO concentrate. *Desalination* 331: 26–34.

Van der Bruggen, B. and Luis, P. (2015). Pervaporation. In: *Progress in Filtration and Separation* (ed. Tarleton, S.), 101–154. Academic Press.

Van der Bruggen, B., Schaep, J., Wilms, D., and Van de Casteele, C. (1999). Influence of molecular size, polarity and charge on the retention of organic molecules by nanofiltration. *Journal of Membrane Science* 156 (1): 29–41.

Vergili, I. (2013). Application of nanofiltration for the removal of carbamazepine, diclofenac and ibuprofen from drinking water sources. *Journal of Environmental Management* 127: 177–187.

Verlicchi, P., Al Aukidy, M., and Zambello, E. (2012). Occurrence of pharmaceutical compounds in urban wastewater: removal, mass load and environmental risk after a secondary treatment: a review. *Science of the Total Environment* 429: 123–155.

Verliefde, A.R., Heijman, S.G.J., Cornelissen, E.R. et al. (2007). Influence of electrostatic interactions on the rejection with NF and assessment of the removal efficiency during NF/GAC treatment of pharmaceutically active compounds in surface water. *Water Research* 41 (15): 3227–3240.

Vo, H.N.P., Ngo, H.H., Guo, W. et al. (2020). Poly-and perfluoroalkyl substances in water and wastewater: a comprehensive review from sources to remediation. *Journal of Water Process Engineering* 36: 101393.

Wanda, E.M.M., Mamba, B.B., and Msagati, T.A.M. (2017). Nitrogen-doped carbon nanotubes/polyethersulfone blend membranes for removing emerging micropollutants. *CLEAN: Soil, Air, Water* 45 (4): 1500889.

Wanda, E.M.M., Nyoni, H., Mamba, B.B., and Msagati, T.A.M. (2018). Application of silica and germanium dioxide nanoparticles/polyethersulfone blend membranes for removal of emerging micropollutants from water. *Physics and Chemistry of the Earth* 108: 28–47.

Wang, L., Albasi, C., Faucet-Marquis, V. et al. (2009). Cyclophosphamide removal from water by nanofiltration and reverse osmosis membrane. *Water Research* 43 (17): 4115–4122.

Wu, Y., Fu, X., Tian, G. et al. (2018). Pervaporation of phenol wastewater with PEBA–PU blend membrane. *Desalination and Water Treatment* 102: 101–109.

Xie, M., Nghiem, L.D., Price, W.E., and Elimelech, M. (2012). Comparison of the removal of hydrophobic trace organic contaminants by forward osmosis and reverse osmosis. *Water Research* 46 (8): 2683–2692.

Yang, Y., Fu, J., Peng, H. et al. (2011). Occurrence and phase distribution of selected pharmaceuticals in the Yangtze Estuary and its coastal zone. *Journal of Hazardous Materials* 190 (1–3): 588–596.

Yangali–Quintanilla, V., Maeng, S.K., Fujioka, T. et al. (2010). Proposing nanofiltration as acceptable barrier for organic contaminants in water reuse. *Journal of Membrane Science* 362 (1–2): 334–345.

Ye, Y. and Chen, V. (2005). Reversibility of heterogeneous deposits formed from yeast and proteins during microfiltration. *Journal of Membrane Science* 265 (1–2): 20–28.

Yoon, Y., Westerhoff, P., Snyder, S.A. et al. (2007). Removal of endocrine disrupting compounds and pharmaceuticals by nanofiltration and ultrafiltration membranes. *Desalination* 202 (1–3): 16–23.

Zaib, Q., Mansoor, B., and Ahmad, F. (2013). Photo–regenerable multi-walled carbon nanotube membranes for the removal of pharmaceutical micropollutants from water. *Environmental Science: Processes & Impacts* 15: 1582–1589.

Zhang, D., Gersberg, R.M., Ng, W.J., and Tan, S.K. (2014). Removal of pharmaceuticals and personal care products in aquatic plant-based systems: a review. *Environmental Pollution* 184: 620–639.

Zhang, T.C., Surampalli, R.Y., Vigneswaran, S. et al. (2012). *Membrane Technology and Environmental Applications*. Reston, Virginia: ASCE.

Zhang, Y., Geißen, S.U., and Gal, C. (2008). Carbamazepine and diclofenac: removal in wastewater treatment plants and occurrence in water bodies. *Chemosphere* 73 (8): 1151–1161.

Zielińska, M., Cydzik-Kwiatkowska, A., Bułkowska, K. et al. (2017). Treatment of bisphenol A-containing effluents from aerobic granular sludge reactors with the use of microfiltration and ultrafiltration ceramic membranes. *Water, Air, & Soil Pollution* 228 (8): 1–9.

15

Advanced Oxidation Processes for Microconstituents Removal in Aquatic Environments

Sanket Dey Chowdhury[1], *Rao Y. Surampalli*[2], *and Puspendu Bhunia*[1]

[1]*School of Infrastructure, Indian Institute of Technology Bhubaneswar, Odisha, India*
[2]*Global Institute for Energy, Environment, and Sustainability, Lenexa, KS, USA*

15.1 Introduction

Currently, the existence of microconstituents (MC) in aqueous matrices has become a burning issue owing to their ability to pose potential risks to humans and aquatic life (Zhang et al. 2014). Basically, MCs embrace a group of diversified synthetic chemicals hidden in the environment for a long duration, but the recent technological and analytical advancements can detect their presence in the water and wastewater samples, even in parts per trillion (ng/L) (Richardson and Ternes 2011; Sangion and Gramatica 2016; Cerreta et al. 2020). Hence, these MCs are termed as emerging contaminants (ECs) or contaminants of emerging concern (CEC) (Garcia et al. 2020; Yang et al. 2011; Jiang et al. 2013). Meanwhile, the MCs comprise the endocrine disruptor compounds (EDCs). Several groups of chemicals fall under the EDCs, namely pharmaceuticals and personal care products (PPCPs), pesticides, surfactants, per- and polyfluoroalkyl substances (PFASs), flame retardants, hormones, and plasticizers (Cerreta et al. 2020).

Minimal exposure to these MCs endangers acute as well as chronic toxicity to humans and aquatic life (Carlsson et al. 2006; Fent et al. 2006; Corcoran et al. 2010; Bayen 2012; Biel-Maeso et al. 2018). For instance, feminization of male fishes and fish-liver impairments have been reported due to exposure to the pharmaceuticals (Corcoran et al. 2010). Furthermore, diclofenac was reported to have inflammatory impacts to human livers (Li et al. 2014). MCs are mostly non-biodegradable and persistent in nature, making the conventional biological and chemical treatment methods futile in terms of their removal. Hence, the performance of conventional wastewater treatment plants (WWTPs) becomes trivial in eradicating such contaminants (Manaia et al. 2018; Cacace et al. 2019). As a consequence, the treated effluent discharged from the WWTPs remains spiked with MCs (both as parent compounds and their metabolites), deteriorating the quality of the receiving aquatic matrices (Mirzaei et al. 2018; Xiang et al. 2018). Hence, there is a need to explore advanced technologies to efficiently eradicate MCs from the environment.

Microconstituents in the Environment: Occurrence, Fate, Removal, and Management, First Edition.
Edited by Rao Y. Surampalli, Tian C. Zhang, Chih-Ming Kao, Makarand M. Ghangrekar, Puspendu Bhunia, Manaswini Behera, and Prangya R. Rout.

Among various advanced treatment methods employed so far, advanced oxidation processes (AOPs) have shown promise in remediating MCs from water and wastewater (Ganiyu et al. 2018; Liu et al. 2019; Miklos et al. 2018; Wols et al. 2013). AOPs are employed with the objective to oxidize the complex organic pollutants present in the aquatic matrices, which are resistant to microbial degradation (Metcalf and Eddy 2003). It is not necessary for the contaminants to be completely oxidized. Sometimes, partial oxidation may serve the purpose because the by-products may possess higher biodegradability and reduced toxicity than the parent compounds (Kanakaraju et al. 2018). The oxidation of the pollutants can be classified into the following four types, depending upon the extent of degradation: i) primary degradation (structural change in the parent compound); ii) acceptable degradation (partial oxidation of the parent compounds leading to the formation of less toxic by-products); iii) ultimate degradation (complete mineralization of the parent compound to carbon dioxide (CO_2), water, and ammonia); and iv) unacceptable degradation (the by-products possess higher toxicity than the parent compounds) (Rice 1996).

In AOPs, the *in-situ* generation of various reactive oxidizing species (ROS) such as hydroxyl radicals (OH•) and superoxide anion radicals ($O_2^{\bullet -}$), with minimal selectivity during the course of AOP, facilitates the enhanced mineralization of the bio-persistent, toxic, and recalcitrant MCs under favorable conditions (Dalrymple et al. 2007; Dewil et al. 2017). However, the inefficacy of AOPs in removing a wide spectrum of MCs, formation of toxic by-products, and huge monetary investment are a few of the limitations associated with individual AOPs (Hai et al. 2007; Dewil et al. 2017). Moreover, the persistent nature of MCs and opaqueness of the wastewater matrix also hinder the performance of individual AOPs (Liu et al. 2019; Yang and Pignatello 2017; Zhang et al. 2011). These concerns related to individual AOPs have encouraged researchers to shift their focus on implementing integrated AOPs for MCs removal (Ribeiro et al. 2015; Huang et al. 2018; Miklos et al. 2018). For example, highly turbid wastewater restricts the penetration of UV radiation, making UV-based AOPs ineffective in eradicating MCs (Autin et al. 2013). Moreover, carbamazepine, highly persistent in nature, showed reluctance to be removed through individual AOPs (Dai et al. 2012; Deng et al. 2013).

However, integrated AOPs not only ensure enhanced degradation of the recalcitrant contaminants, but also cut down the time required to achieve complete mineralization of the parent compounds (Tufail et al. 2020). For instance, Moreira et al. (2016) found that photocatalysis standalone accomplished only 26% removal of diclofenac in one and half hours, whereas photocatalysis in combination with ozonation ensured complete removal of diclofenac in 10 minutes. This could be because the incorporation of ozone (O_3) during photocatalysis resulted in more OH• radical formation, promoting better removal of diclofenac in quick succession (Tufail et al. 2020).

In this chapter, the potential of individual as well as combined or integrated AOPs in removing PPCPs, pesticides, surfactants, and PFASs from the aqueous matrices has been comprehensively documented. Moreover, the classification and the fundamental mechanisms of various individual as well as combined AOPs have been elaborated on in this chapter. Additionally, future perspectives regarding the treatment of MCs employing the AOPs have been suggested and finally, a conclusion has been drawn highlighting the main themes in this chapter.

15.2 Classification of AOPs

Depending upon the oxidizing agents used, the AOPs employed to eradicate MCs from water and wastewater can be predominantly classified into the following six categories (Tufail et al. 2020):

1) *Ultra-violet (UV) photolysis*: In this method, the MCs absorb UV radiation and become degraded (Carlson et al. 2015; Duca et al. 2017).
2) *Photocatalysis*: In this case, various photocatalysts are used, which are excited by exposure to sunlight (solar photocatalysis) or UV irradiation (UV photocatalysis) and facilitate the degradation of MCs (Tufail et al. 2020). Among different photocatalysts, titanium dioxide (TiO_2) has been widely used because it is chemically and thermally stable and non-toxic in nature (Metcalf and Eddy 2003; Kanakaraju et al. 2018).
3) *Ozonation*: Here, ozone (O_3) is used as the oxidizing agent (Metcalf and Eddy 2003).
4) *Sonolysis*: In the case of sonolysis of MCs, an ultrasonic or longitudinal wave with a frequency range of 40 to 1000 kHz has been propagated in the aqueous medium, generating cavitation bubbles (Wu et al. 2013). However, it is a chemical-free AOP (Tufail et al. 2020).
5) *Anodic oxidation (AO)*: This is carried out in an electrolytic cell in the presence of an anode and electrolytic solution (Feng et al. 2013). Basically, this method necessitates direct energy input for degrading the contaminants (Madhavan et al. 2010; Naddeo et al. 2010).
6) *Fenton process*: In this process, the OH• radicals have been generated by the Fenton reagent (hydrogen peroxide (H_2O_2) and ferrous (Fe^{2+}) salts) under highly acidic conditions with a pH of ~3 (Rozas et al. 2010; Sun et al. 2016).

Apart from the abovementioned individual AOPs, researchers have also focused on the implementation of different AOPs in combination to achieve better removal of MCs (Huang et al. 2018; Miklos et al. 2018). Some of the popular examples of the integrated AOPs, employed to eradicate the MCs, are as follows:

a) *UV radiation-based integrated AOPs*: In this case, UV photolysis is integrated with one or more individual AOPs, as mentioned above, to escalate the generation of the ROS (Tufail et al. 2020). From the available literature, it has been observed that the following combinations have gained the major attention of researchers for eradicating MCs: UV/ozonation, UV/H_2O_2, UV/ozonation/H_2O_2, UV/sonolysis, UV/sonolysis/ozonation, UV/Fenton process, UV photocatalysis/ozonation, UV/persulfate (PS), and UV/chlorine (Cl_2) (Joseph et al. 2009; Cheng et al. 2016).
b) *Ozonation-based integrated AOPs*: Here, ozonation is coupled with H_2O_2, Cl_2, or PS for degrading the MCs (Homem and Santos 2011; Ribeiro et al. 2015). The common examples of ozonation-based integrated AOPs are the peroxone process (O_3/H_2O_2) and ozonation/SO_4^{2-} (De Witte et al. 2009; Cong et al. 2015; Yang et al. 2015).
c) *Fenton-based integrated AOPs*: In this case, the Fenton reagent is coupled with O_3, H_2O_2, or PS. This is also known as the heterogeneous Fenton process (Tufail et al. 2020). Some common examples of Fenton-based integrated AOPs are the heterogeneous Fenton process, Fenton process/sonolysis, and photo-Fenton process (Fenton process/UV- or solar-irradiation) (Li et al. 2015; Kanakaraju et al. 2018).

d) *Electrochemical-based integrated AOPs*: These combine the AOP process with other AOPs such as sonolysis, UV photocatalysis, and Fenton process. The major combinations of electrochemical-based integrated AOPs are electro-Fenton process, sono-electro-Fenton process, and photo-electro-Fenton process (mainly exposed to UV irradiation) (Oturan et al. 2008; El-Ghenymy et al. 2013; Feng et al. 2013).

15.3 Fundamentals of Different AOPs

As already discussed, in AOPs, the *in-situ* ROS triggers the degradation of parent MCs to less-, non-, or even more toxic and simpler by-products (Metcalf and Eddy 2003; Kanakaraju et al. 2018). Basically, these ROS are free radicals embracing at least one unpaired electron (Andreozzi et al. 1999). OH•, O_2^-, hydroperoxyl radicals (O_2H•), and alkoxyl radicals (RO•) are some common examples of ROS. Among all these ROS, OH• has gained major attention owing to its non-selective nature, high reactivity, and capability to oxidize a wide spectrum of organic micropollutants (Legrini et al. 1993). Moreover, the reaction between the organic MCs and OH• radicals occurs in the three following pathways: i) direct electron transfer (Eq. 15.1); ii) radical-radical interaction forming peroxyl radicals (Eq. 15.2): and iii) hydrogen abstraction (Eq. 15.3) (Kanakaraju et al. 2018).

$$RX + OH^{\cdot} \rightarrow RX^{\cdot+} + OH^- \tag{15.1}$$

$$R + O_2 \rightarrow RO_2 \tag{15.2}$$

$$RH + OH \rightarrow R + H_2O \tag{15.3}$$

In this section, the fundamentals of individual as well as combined AOPs for removing different MCs (e.g., PPCPs, PFASs, surfactants, pesticides) have been emphasized.

15.4 Fundamentals of Individual AOPs

15.4.1 Fundamentals of Microconstituents Degradation by Ozonation Process

O_3, a strong and highly selective oxidizing agent, initiates the degradation of the MCs in the two following ways: i) directly attacking the specific sites of the MCs; and ii) generating OH• radicals that will engender the change in the structure of the contaminants, i.e., indirect degradation of the MCs (Tufail et al. 2020). The MCs, possessing a high density of the electrons or embracing phosphorus (P), sulphur (S), nitrogen (N), oxygen (O), and double bond are very susceptible to degradation by ozonation (Kuang et al. 2013; Zhao et al. 2017). Under basic conditions, O_3 reacts with OH^- ions in the water and produces OH• radicals (Miklos et al. 2018). The indirect degradation of the MCs by the OH• radicals is a second-order reaction. However, the OH• radical possesses a higher degradation rate than that of O_3. From the available literature, it is clearly evident that ozonation ensures a commendable removal (92–100%) of the ECs (Jin et al. 2012; Gomes et al. 2017). O_3 dosage, characteristics of the MCs, and pH of the aqueous medium are the key factors governing the efficiency of the ozonation process in degrading the ECs (Von Gunten 2003).

Even though ozonation portrays remarkable degradation of the MCs, the formation of toxic by-products has been a major concern associated with this method. However, the implementation of ozonation in combination with the Fenton process, UV irradiation, or H_2O_2 may be able to suppress toxic by-product formation by immediately initiating OH• radicals (Tufail et al. 2020).

15.4.2 Fundamentals of Microconstituents Degradation by UV-Irradiation

This is also termed as UV photolysis, in which the MCs are degraded after absorbing UV radiation (Carlson et al. 2015; Chianese et al. 2017; Duca et al. 2017). Basically, low-, medium-, or high-pressure mercury (Hg) lamps are used as the source of UV radiation, which emit light in the wavelength of 254 nm (Tufail et al. 2020). Recently, light-emitting diode (LED) lamps have gained the attention of researchers because LED lamps eliminate the presence of Hg, sharply emit the spectrum wavelength, and offer flexibility and compact size (Sirtori et al. 2010; Autin et al. 2013; Song et al. 2016). Direct UV radiation captures the photosensitive moieties in the MCs, thereby oxidizing them, whereas with indirect UV photolysis of the pollutants, oxidants like O_3 and H_2O_2 are added externally, which will generate the ROS that will further degrade the ECs. Like ozonation, indirect UV photolysis is more efficient than direct photolysis in terms of degrading the ECs (Kanakaraju et al. 2018). The literature suggests that UV photolysis secures 47 to 99% removal of the ECs (Tufail et al. 2020). However, the efficacy of UV photolysis depends on the following factors: intensity of UV lamps, turbidity of the aqueous medium, coefficient of molar absorption, pH of the solution, and the chemical structure of the pollutants (Yuan et al. 2009; Baeza and Knappe 2011; Luo et al. 2018).

15.4.3 Fundamentals of Microconstituents Degradation by Photocatalysis

In photocatalytic degradation of MCs, an electron, present in the valence band of the photocatalyst, is photo-excited due to the incidence of UV or solar radiation and moves to the conduction band of the photocatalyst. As a result, a hole (h^+) is generated in the valence band of the photocatalyst (Eq. 15.4) (Bamba et al. 2008; Martinez et al. 2013). The electron, on reaching the conduction band, reacts with O_2 and forms $O_2^{\bullet -}$ radicals (Eq. 15.5) and h^+ reacts with OH^- ions to form OH• radicals (Eq. 15.6) (Alvarez et al. 2016; Zhao et al. 2018; Dong et al. 2019). These OH• and $O_2^{\bullet -}$ radicals trigger photocatalytic degradation of the MCs.

$$\boldsymbol{TiO_2 + h\nu \rightarrow TiO_2(photo - excited) + (e^- + h^+)} \tag{15.4}$$

$$\boldsymbol{e^- + O_2 \rightarrow O_2^{\cdot -}} \tag{15.5}$$

$$\boldsymbol{h^+ + OH^- \rightarrow OH^{\cdot}} \tag{15.6}$$

where e^- denotes the electron. The basic mechanism of photo-excitation of the photocatalyst is schematically represented in Figure 15.1. When UV radiation (wavelength <400 nm) is used to excite the photocatalyst, the subsequent degradation of the ECs is termed UV photocatalysis, whereas the photocatalytic degradation of the ECs in the presence of visible light or solar radiation (400 nm < wavelength < 700 nm) is termed solar photocatalysis (Kanakaraju et al. 2018; Miklos et al. 2018). The efficiency of photocatalytic degradation of

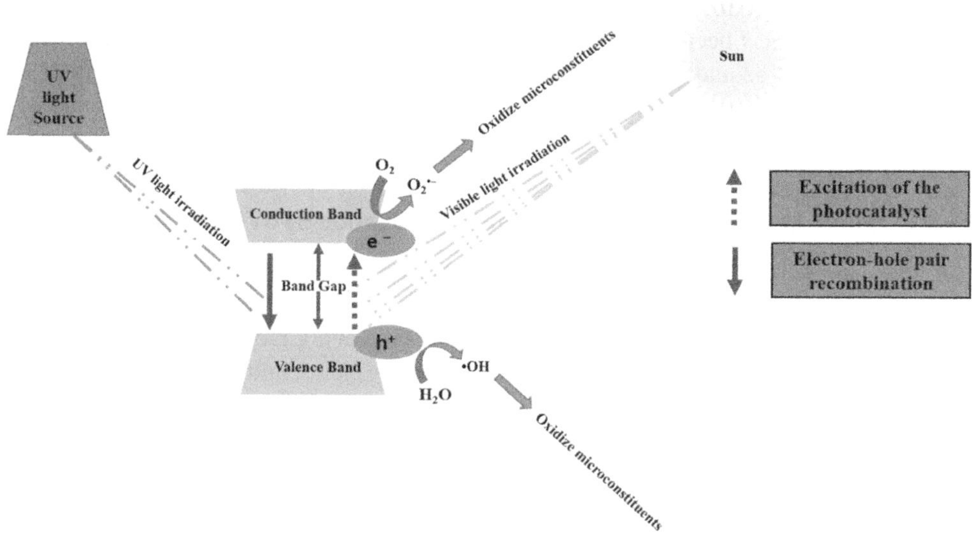

Figure 15.1 Schematic representation of photo-excitation of the photocatalyst.

the ECs depends on the nature of the photocatalyst (i.e., electron-hole pair recombination time), intensity of incident light, and the type of target pollutant (Tufail et al. 2020). As already mentioned, the most commonly used photocatalyst is TiO_2 owing to its enlarged surface area, cost-effectiveness, and thermal and chemical stability (Metcalf and Eddy 2003; Miklos et al. 2018). However, the quick electron-hole pair recombination has been a major concern associated with TiO_2 because it reduces the radical formation and subsequently results in declination in the photodegradation efficiency. To overcome this, platinum (Pt), palladium (Pd), gold (Au), zinc oxide (ZnO), tungsten oxide (WO_3), and tungsten sulphide (WS_2) can be effectively utilized as photocatalysts for degrading MCs (Wang et al. 2018). Most importantly, the energy of the incident solar or UV radiation should be greater or equal to the band-gap energy of the photocatalysts to make them photo-excited (Metcalf and Eddy 2003). However, the post-treatment separation of the exhausted photocatalysts from the treated effluent has always been a major hurdle associated with this process.

Electrochemical oxidation (EO) has become a popular AOP for remediating the micropollutants because it is a chemical-free process and generates ROS using electricity. Hence, this process does not produce any secondary sludge (Madhavan et al. 2010; Ganiyu et al. 2018; Garcia-Segura et al. 2018). EO can be classified into the two following types: i) electrochemical degradation (via anodic oxidation (AO)); and ii) electrochemical separation (electrocoagulation and electrodialysis) (Sires and Brillas 2012). EO, especially AO of the MCs, takes place in accordance with the three following pathways: i) the electrons are directly transferred to the anode surface; ii) water molecules are oxidized at the anode surface, leading to the formation of ROS, which in turn highlights mediated oxidation of the ECs; and iii) the ROS can be produced from the ions of the electrolyte used, which in turn capture electrochemical degradation of MCs. The direct charge transfer at the electrode surface is termed as direct oxidation, whereas the latter two cases indicate the indirect oxidation of the ECs (Feng et al. 2013). In the case of EO, an anode possessing high affinity toward O_2 generation is used because it will readily react and degrade the MCs (Tufail et al. 2020). Ruthenium dioxide (RuO_2), lead oxide (PbO_2), graphite, Pt, tin dioxide (SnO_2), iridium dioxide (IrO_2), and boron-doped diamond (BDD) are some common examples of anode materials (Feng et al. 2013; Moreira et al. 2017). However, BDD anodes are extensively employed in AO because they are corrosion-resistant, cost-effective, possess inert surfaces, and facilitate *in-situ* generation of OH• radicals (Svorc et al. 2017; Garcia-Espinoza et al. 2018; Kanakaraju et al. 2018). Moreover, the *in-situ* generation of OH• radicals cuts down further electrolyte requirement, thereby making BDD an eco-friendly anode material for degrading MCs (Chaplin 2014).

15.4.4 Fundamentals of Microconstituents Degradation by Electrochemical Oxidation (EO) or Anodic Oxidation (AO) and Sonolysis

Microconstituents degradation via sonolysis is initiated by propagating the longitudinal or ultrasonic wave having a frequency ranging from 40 to 1000 kHz in aqueous medium (Miklos et al. 2018). As a consequence, cavitation bubbles are formed (Tufail et al. 2020). High intensity of these acoustic wave-induced cavity bubbles in the aquatic medium leads to the implosion of the bubbles, producing hot spots with pressure and temperature of around 1000 bar and 5000 K, respectively (Wu et al. 2013; Miklos et al. 2018). As a result of water pyrolysis, highly reactive OH• radicals are formed, which degrade the micropollutants present in the aqueous medium (Guyer and Ince 2011). The reactions involved in sonolysis of the ECs are represented by Eqs (15.7) to (15.9).

$$H_2O + US\,wave \rightarrow OH^{\cdot} + H^{\cdot} \tag{15.7}$$

$$2OH^{\cdot} \rightarrow H_2O_2 \tag{15.8}$$

$$2OH^{\cdot} \rightarrow H_2O + O \tag{15.9}$$

where *US wave* denotes ultrasound or ultrasonic wave. Sonolysis completely eliminates the chemical requirement. Meanwhile, sonolysis proficiently eradicates the volatile contaminants such as phenol (Sunartio et al. 2007; Chand et al. 2009; Torres-Palma et al. 2010). However, the OH• radicals can also help in degrading the non-volatile pollutants (Tufail et al. 2020). The input power and frequency of the US wave are the major factors that govern the MC degradation efficacy of sonolysis (Ameta et al. 2018; Ince 2018; Kıdak and Doğan, 2018). The literature suggests that sonolysis ensures 80 to 100% removal of the ECs (Tufail et al. 2020). However, the high energy input and reduced electrical efficiency are the prime concerns related to sonolysis (Goel et al. 2004). In fact, cavitation bubbles formed due to the propagation of the US wave do not break down entirely to form OH• radicals, cutting down the efficiency of this process (Pang et al. 2011; Rayaroth et al. 2016).

15.4.5 Fundamentals of Microconstituents Degradation by the Fenton Process

In the Fenton process, iron salts (Fe (II)) and H_2O_2 (also known as the Fenton reagent) react with each other under acidic ambiance (pH ~3) and produce OH• radicals (Eq. 15.10) (Sun et al. 2016). Basically, at higher pH, ferric ions (Fe (III)) scavenge OH• radicals and precipitate in the form of ferric hydroxide ($Fe(OH)_3$), thereby reducing the efficiency of this process (Wadley and Waite 2004; Clarizia et al. 2017). Moreover, the generation of $Fe(OH)_3$ sludge, at high pH, involves additional sludge handling, escalating the cost of the Fenton process (Tufail et al. 2020). Hence, the pH is maintained in the acidic range during the Fenton process. Here, Fe^{2+} ions act as the catalysts and can be regenerated by incorporating additional H_2O_2 (Eq. 15.11) (Bautista et al. 2008). In light of the aforementioned discussion, it can be stated that the Fe^{2+} and H_2O_2 dosage ratio, and the pH of the aqueous media are the predominant factors affecting the efficacy of the Fenton process. Ultimately, the cost-effectiveness and post-treatment separation of Fe^{2+} through magnetic means trigger the acceptability of the Fenton process, even in full-scale industrial applications (Bae et al. 2015).

$$Fe^{2+} + H_2O_2 \rightarrow Fe^{3+} + OH^- + OH^{\cdot} \tag{15.10}$$

$$Fe^{3+} + H_2O_2 \rightarrow Fe^{2+} + H^+ + HO_2^{\cdot} \tag{15.11}$$

15.5 Fundamentals of Integrated AOPs

Various limitations associated with the individual AOPs, discussed in earlier subsections, encourage researchers to explore different combinations of AOPs for better removal of the MCs (Kanakaraju et al. 2018; Miklos et al. 2018; Tufail et al. 2020). The fundamentals of different combinations of AOPs are briefly discussed in the coming subsections.

15.6 Fundamentals of UV-Irradiation-Based Integrated AOPs

As discussed earlier, in UV irradiation-based integrated AOPs, UV photolysis is combined with other AOPs (Cheng et al. 2016).

15.6.1 UV/H_2O_2

The major advantage of incorporating H_2O_2 in the UV photolysis of the MCs is that UV irradiation helps in generating two OH• radicals from one molecule of H_2O_2 (Eq. 15.12). However, these *in-situ* OH• radicals may again recombine to form H_2O_2 or remain as free OH• radicals to carry out degradation of the contaminants (Tufail et al. 2020). But the photolysis of H_2O_2 takes place if the wavelength of the UV irradiation is <254 nm. Thus, UV-C-irradiation, with a wavelength <254 nm, is employed to serve this purpose. Moreover, the opaqueness of the aqueous medium governs the penetration of UV radiation and thereby affects the MCs degradation efficiency of the UV/H_2O_2 process (Buxton et al. 1988; Nicole et al. 1990). Hence, the integration of UV irradiation with H_2O_2 facilitates *in-situ* generation of the OH• radicals due to the photolysis of H_2O_2. These OH• radicals can efficiently degrade a wide spectrum of MCs (Keen et al. 2016; Miklos et al. 2018).

$$H_2O_2 + UV\ radiation \rightarrow 2OH^{\cdot} \tag{15.12}$$

15.6.2 UV Photocatalysis/Ozonation

UV photocatalysis alone is capable of degrading a wide range of MCs, but the major concerns associated with this process are the following: i) it necessitates high energy input; and ii) the mineralization of the MCs takes place at a sluggish rate (Agustina et al. 2005; Mozia and Morawski 2012). Basically, the integration of ozonation and UV photocatalysis triggers OH• radical formation through a series of processes such as photocatalytic degradation, direct ozonation, and generation of OH• radicals due to the interaction between O_3 and UV radiation in the presence of O_2 (Fatta-Kassinos et al. 2011). Another advantage of incorporating O_3 during UV photocatalysis is that O_3 acts as the electron scavenger, thereby cutting down the possibilities of electron-hole pair recombinations, leading to enhancement of ECs degradation efficacy (Moreira et al. 2016; Tufail et al. 2020).

15.6.3 UV/Fenton Process

The fundamentals of the Fenton process have already been discussed. As seen in Eq. (15.10), the ferrous (Fe (II)) salts and H_2O_2 react with each other and produce ferric ions (Fe (III)) under acidic conditions. Integration of UV irradiation and the Fenton process is termed the photo-Fenton process (Kanakaraju et al. 2018). However, if the Fenton process is carried out in the presence of solar radiation, it is also referred to as the photo-Fenton process (Miklos et al. 2018). The photo-Fenton process is proved to be superior to the Fenton process because the former process is capable of producing more OH• radicals than the latter process, either via photo-induced reduction of Fe^{3+} ions to Fe^{2+} ions or by direct interaction between UV radiation and H_2O_2. Moreover, the Fe^{2+} ions produced further react with H_2O_2 to produce more OH• radicals (Clarizia et al. 2017; Mirzaei et al. 2017). Hence, compared to the standalone Fenton process, the photo-Fenton process can effectively degrade a wide spectrum of MCs through a series of pathways: Fenton process, UV photolysis, reaction between UV and H_2O_2, and photo-induced reduction of Fe^{3+} ions to Fe^{2+} ions (Pérez-Estrada et al. 2005; Ameta et al. 2018).

15.6.4 UV/Persulfate (PS) or Permonosulfate (PMS)

$SO_4^{\bullet 2-}$ radical-based AOPs in part enhance redox potential, prolong radical life, and at the same time, negotiate the radical scavenging to a great extent (Sharma et al. 2015). Basically, $SO_4^{\bullet 2-}$ radicals are generated when PS or PMS ions are activated through the application of transition metals, heat, US wave, and UV irradiation (Tufail et al. 2020). Among the aforementioned activation methods, UV irradiation has gained major attention (Yang et al. 2019). Generally, on activation, the peroxide bond present in PS or PMS is broken and as a result, $SO_4^{\bullet 2-}$ radicals are formed. Likewise, on breakage of the peroxide bond in H_2O_2, OH• radicals are formed. Moreover, Yang et al. (2019) reported that the generation of $SO_4^{\bullet 2-}$ radicals from PS or PMS necessitates lesser energy input than that in the case of OH• radical formation from H_2O_2 via UV irradiation, because the peroxide bond energy in PS or PMS (140 kJ/mol) is less than that in H_2O_2 (213.3 kJ/mol). Moreover, they also noticed that UV photolysis of one molecule of PMS results in the formation of one $SO_4^{\bullet 2-}$ radical, whereas two $SO_4^{\bullet 2-}$ radicals are generated due to UV photolysis of one molecule of PS. Apart from the $SO_4^{\bullet 2-}$ radicals, OH• radicals are also formed on activation of PS or PMS via UV irradiation, resulting in better degradation and mineralization of the MCs (Tufail et al. 2020).

15.6.5 UV/Cl_2

The integration of UV irradiation and free Cl_2 results in the formation of various ROS such as chloride radicals ($Cl^{\bullet}$), hypochlorite radicals ($OCl^{\bullet}$), and OH• radicals (Cerreta et al. 2020). $Cl^{\bullet}$ radicals are pollutant-specific and can effectively degrade the MCs containing electron-donating functional groups (Lu et al. 2018). The integrated UV/Cl_2 process can efficiently remove PPCPs at low pH (Cerreta et al. 2020; Gao et al. 2020). However, the formation of chlorinated by-products has been a major issue associated with this process (Tufail et al. 2020).

15.7 Fundamentals of Ozonation-Based Integrated AOPs

As discussed earlier, in ozonation-based integrated AOPs, ozonation is coupled with the other AOPs (Kanakaraju et al. 2018).

15.7.1 Ozonation/H_2O_2

The combination of O_3 and H_2O_2 is also termed the peroxone process (De Witte et al. 2009; Kanakaraju et al. 2018; Miklos et al. 2018). Integration of O_3 and H_2O_2 not only triggers OH• radicals formation, but also cuts down the time required to degrade the ECs. The type of aquatic matrix and reaction conditions determine the extent of O_3 dosage required. Typically, in a peroxone process, the O_3 dosage varies between 1 and 20 mg/L for degrading the ECs (Lin et al. 2009; Katsoyiannis et al. 2011).

15.7.2 Ozonation/PS or PMS

Apart from all the PS/PMS-activation processes mentioned earlier, O_3 can also be applied to activate PS or PMS (Cheng et al. 2016). Moreover, the application of O_3 triggers the generation of OH• radicals along with the production of $SO_4^{\bullet 2-}$ radicals by activating PS or PMS and thereby improving the ECs degradation potential of this AOP (Deniere et al. 2018).

15.8 Fundamentals of Fenton Process-Based Integrated AOPs

Here, the Fenton reagent is employed in combination with the other AOPs to improve the MCs degradation efficiency (Miklos et al. 2018; Wang et al. 2019; Tufail et al. 2020).

15.8.1 Heterogeneous Fenton Process

The fundamentals of the heterogeneous Fenton process are similar to those of the Fenton process (Eq. 15.10). However, in the heterogeneous Fenton process, the mixture of ferrous salts and H_2O_2 is employed in combination with O_3 or PS (Amr and Aziz 2012; Qin et al. 2018; Wang et al. 2019). Especially, in the O_3/Fenton process, the iron salt helps in activating O_3 by transferring an electron from metal to O_3 and forms an $O_3^{\cdot}$ radical which further produces OH• radicals (Li et al. 2015). Moreover, the Fenton process has to be carried out in acidic pH, and neutralization has to done before its disposal (Tufail et al. 2020). The implementation of the heterogeneous Fenton process (based on iron oxide) can entirely negotiate this limitation associated with Fenton process standalone (Ikhlaq et al. 2018). Hence, the heterogeneous Fenton process has been mostly preferred over the classical Fenton process to substantially degrade and mineralize the MCs.

15.8.2 Photo-Fenton Process

As discussed earlier, the photo-Fenton process implies the combination of the Fenton process with either UV radiation or solar radiation for degrading a wide spectrum of ECs (Kanakaraju et al. 2018; Sirés et al. 2007). The fundamentals of the photo-Fenton process have already been discussed in the Section 15.3.2.1.3.

15.8.3 Sono-Fenton Process

Sonolysis captivates the breakdown of the cavitation bubbles (Wu et al. 2013; Miklos et al. 2018). Sonolysis is suitable for volatile ECs, but the degradation of non-volatile ECs is initiated by OH• radicals, generated via pyrolytic degradation of water molecules (Tufail et al. 2020). The literature has shown that the integration of sonolysis and the Fenton process has resulted in substantial eradication of MCs with an efficiency higher than those of the Fenton process or sonolysis individually (Chand et al. 2009; Pang et al. 2011). In the Fenton process, iron salts (Fe^{2+} ions) react with H_2O_2 and produce Fe^{3+} ions and OH• radicals (Tufail et al. 2020). However, if H_2O_2 is present in sufficient concentrations, Fe^{3+} ions further react with H_2O_2 and form O_2H• complex, reducing the availability of free OH• radicals, which further cuts down the efficiency of the Fenton process. In the sono-Fenton process, the presence of a US wave propels the breakdown of O_2H• complex and releases free OH• radicals, enhancing the MCs degradation efficiency of the process (Pang et al. 2011; Tufail et al. 2020). Furthermore, the Fenton process can only be employed under acidic conditions, whereas sono-Fenton can potentially degrade the MCs over a wider pH range (Huang et al. 2012).

15.9 Fundamentals of Electrochemical-Based Integrated AOPs

From extensive literature, it has been observed that three electrochemical-based integrated AOPs such as electro-Fenton process, sono-electro-Fenton process, and photo-electro-Fenton process have gained major attention of researchers (Oturan et al. 2008; Feng et al. 2013). The fundamentals of the aforementioned integrated AOPs are discussed below.

15.9.1 Electro-Fenton Process

The electro-Fenton process is the combination of the AO and Fenton processes. However, this is an indirect process that facilitates the generation of OH• radicals within the system

(Tufail et al. 2020). Here, H_2O_2 is produced electrochemically through O_2 reduction, whereas the ferrous iron salts are either added externally to the system or regenerated from Fe^{3+} ions within the reaction media. The OH• radicals produced trigger the degradation and mineralization of the MCs. Moreover, owing to the *in-situ* generation of H_2O_2, the concerns related to its storage, transportation, or handling can be easily eliminated in this process (Sirés et al. 2014). However, the *in-situ* generation of OH• radicals in the electro-Fenton process is governed by the pH of the media, anode material, and O_2 stability (Nidheesh and Gandhimathi 2012; Sirés et al. 2014).

15.9.2 Sono-Electro-Fenton Process

In sonolysis, propagation of a US wave results in the formation and breakdown of cavity bubbles (Miklos et al. 2018). If the cavity bubbles are formed in the proximity of the anode, mass transfer to the anode is escalated, and anode fouling also is delayed, resulting in O_2 transfer to the anode surface, which in turn boosts H_2O_2 formation and triggers the Fenton process (Oturan et al. 2008). Thus, when sonolysis is coupled with the electro-Fenton process, the removal of ECs is heightened.

15.9.3 Photo-Electro-Fenton Process

The photo-electro-Fenton process is the combination of UV photolysis and the electro-Fenton process. Here, the presence of UV radiation helps to produce the OH• radicals from Fe^{3+} ions, thereby eliminating the OH• radical scavenging effect imparted by $Fe(OH)_3$ (Miklos et al. 2018). Basically, in the photo-electro-Fenton process, the degradation and mineralization of the MCs have been achieved through photochemical (photo-Fenton process) as well as electrochemical (electro-Fenton process) pathways (Brillas et al. 2007). Current intensity plays a pivotal role in the photo-electro-Fenton process. It has been found that the formation of OH• radicals in the system is directly proportional to the applied current intensity. Therefore, increase in the applied current intensity results in better removal of MCs (El-Ghenymy et al. 2012, 2013).

15.10 Application of Individual/Integrated AOPs For Microconstituents Removal

In last few years, AOPs have been extensively employed for removing the MCs from water and wastewater matrices. The performance of various individual as well as integrated AOPs for the degradation of ECs, especially PPCPS, pesticides, surfactants, and PFASs, is thoroughly discussed below.

15.10.1 PPCP Removal

The potential of various AOPs, individually or in combination with other AOPs, for removing a wide spectrum of PPCPs from the aquatic matrices, is represented in Table 15.1. From the data in the table, it is clearly evident that the AOPs have remarkable potential to

Table 15.1 Application of AOPs for removing some frequently-detected species of different MCs.

Microconstituents	AOPs	Experimental Conditions	Removal Efficiency (%)		References
			Individual AOP	Integrated AOP	
Pharmaceuticals and Personal Care Products (PPCPs)					
Paracetamol	UV/H_2O_2	$C_0 = 3.02\,\mu g/L$; $H_2O_2 = 0.17\,g/L$; UV lamp (254 nm, 17 W); pH = 2–7; duration = 120 min.	20 (UV)	100	Andreozzi et al. (2003a)
	UV/PS	$C_0 = 50\,mg/L$; UV lamp (254 nm); PS = 50 mg/L; pH = 6; duration = 90 min.	7 (UV)	80	Moussavi et al. (2017)
	$UV/US/H_2O_2$	$C_0 = 20\,mg/L$; US bath (20 kHz, 200 W); $H_2O_2 = 900\,mg/L$; UV lamp (253.7 nm; 13 W); pH = 3.9; duration = 120 min.	30 (UV)	100	Ghafoori et al. (2015)
	O_3/H_2O_2	$C_0 = 5.3\,mg/L$; $H_2O_2 = 51\,mg/L$; $O_3 = 36\,mg/L\text{-min}$; T = 24°C; pH = 5.5; duration = 2 min.	20 (O_3)	90	Andreozzi et al. (2003a)
	Fenton oxidation	$C_0 = (100–500)\,mg/L$; pH = 3; T = 25°C; H_2O_2: Fe (II) = 2:1; duration = 60 min.	70.37	–	Van et al. (2020)
		$C_0 = (100–500)\,mg/L$; pH = 3; T = 25 °C; H_2O_2: Fe-S = 1:2; duration = 60 min.	80.77		
	O_3/H_2O_2	$C_0 = 0.117\,\mu g/L$; $O_3 = 2.5\,mg/L$; $H_2O_2 = 0.5\,mg/L$; duration = 24 min.	> 99	–	Edrees et al. (2017)
	Electro-Fenton process	$C_0 = 5\,mM$; pH = 2.	72 (FP)	74	
	Electrolysis	$C_0 = 96\,mg/L$; pH = 5; recirculated vitreous carbon electrode; duration = 240 min.	90	–	

(Continued)

Table 15.1 (Continued)

Microconstituents	AOPs	Experimental Conditions	Removal Efficiency (%)		References
			Individual AOP	Integrated AOP	
Clofibric acid	UV/H_2O_2	$C_0 = 214\,mg/L$; $H_2O_2 = 34\,g/L$; UV lamp (17 W; 254 nm); pH = 5; duration = 60 min.	40 (UV)	100	Andreozzi et al. (2003b)
	O_3/H_2O_2	$C_0 = 2\,\mu g/L$; $H_2O_2 = 0.5\,mg/L$; $O_3 = 1\,mg/L$-min; duration = 10 min.	8 (O_3)	92	Zwiener and Frimmel (2000)
		$C_0 = 321\,mg/L$; $H_2O_2 = 34\,g/L$; $O_3 = 36\,mg/L$-min; pH = 5; duration = 5 min.	8 (O_3)	100	Andreozzi et al. (2003b)
	Photo-electro-Fenton process	$C_0 = 89\,mg/L$; Fe(II) = 0.202 g/L; UV-A (360 nm, 56 W); $Na_2SO_4 = 7\,g/L$; pH = 3; T = 35ºC; current = 150 mA/cm^2; duration = 360 min.	43 (AO)	100, 97 (electro-Fenton).	Brillas et al. (2007)
Carbamazepine	UV/H_2O_2	$C_0 = 20\,\mu g/L$; low pressure Hg lamp (254 nm); $H_2O_2 = 0.17\,g/L$; pH = 5; duration = 4 min.	8 (UV)	100	Vogna et al. (2004a)
		$C_0 = 0.26\,mg/L$; low pressure Hg lamp (254 nm, 1000 W); $H_2O_2 = 10\,mg/L$.	25 (UV)	60	Gogate and Pandit (2004)
	UV/Cl_2	$C_0 = 10\,\mu g/L$; $Cl_2 = 5\,mg/L$; UV lamp (253.7 nm; 4 W); pH = 7; duration = 15 min.	5 (UV)	98	Wang et al. (2016a)
	O_3/H_2O_2	$C_0 = 2000\,mg/L$; $H_2O_2 = 0.17\,g/L$; $O_3 = 0.5\,mg/L$-min; pH = 7; T = 25ºC; duration = 30 min.	80 (O_3)	90	Liu et al. (2019)
	$UV\text{-}TiO_2$	UV-A lamp (365 nm, 3 × 8 W); $TiO_2 = 1.5\,g/L$; duration = 90 min.	–	99	Carabin et al. (2015)

Ibuprofen	UV-TiO_2	Medium pressure Hg lamp (125 W); ultra-pure water matrix.	–	92	Candido et al. (2016)
		$C_0 = 60$ mg/L; UV-LED (382 nm; 4 × 10 W); duration = 30 min.		100	Jallouli et al. (2018)
	UV/Cl_2	$C_0 = 100$ μg/L; $Cl_2 = 7.1$ mg/L; UV fluence rate = 1.08 mW/cm^2; duration = 20 min.	8 (UV)	100	Xiang et al. (2016)
	O_3/H_2O_2	$C_0 = 2$ μg/L; $H_2O_2 = 0.5$ mg/L; $O_3 = 1$ mg/L-min; duration = 10 min.	12 (O_3)	90	Zwiener and Frimmel (2000)
	Photo-electro-Fenton process	$C_0 = 41$ mg/L; Fe(II) = 0.202 g/L; UV (365 nm, 6 W); $Na_2SO_4 = 7$ g/L; pH = 3; T = 25ºC; current = 33.3 A/cm^2; duration = 360 min.	22 (AO)	98	Wang et al. (2016b)
		$C_0 = 41$ mg/L; Fe(II) = 0.202 g/L; UV-A (360 nm, 56 W); $Na_2SO_4 = 7$ g/L; pH = 3; T = 35 ºC; current = 150 mA/cm^2; duration = 36 min.	28 (AO)	92	Skoumal et al. (2009)
Diclofenac	UV/H_2O_2	$C_0 = 151$ μg/L; low pressure lamp (254 nm, 17 W); $H_2O_2 = 34$ g/L; pH = 7.	43 (UV)	100	Vogna et al. (2004b)
	UV-TiO_2/ O_3	$C_0 = 0.1$ mM; UV lamp (313 nm); $TiO_2 = 1.5$ g/L; $O_3 = 10$ mg/L-min; pH = 7; duration = 120 min.	50 (UV)		Aguinaco et al. (2012)
		$C_0 = 32$ mg/L; UV lamp (382 nm, 20 W); $TiO_2 = 1.5$ g/L; $O_3 = 10.5$ mg/L-min; duration = 90 min.	56 (UV)		Moreira et al. (2016)
	O_3/H_2O_2	$C_0 = 2$ μg/L; $H_2O_2 = 0.5$ mg/L; $O_3 = 1$ mg/L-min; duration = 10 min.	96.8 (O_3)	98	Zwiener and Frimmel (2000)

(Continued)

Table 15.1 (Continued)

Microconstituents	AOPs	Experimental Conditions	Removal Efficiency (%)		References
			Individual AOP	Integrated AOP	
Bisphenol A	UV/H_2O_2	$C_0 = 3\,\mu g/L$; $H_2O_2 = 6.8\,g/L$; pH = 6–8; medium and low pressure UV lamp (254 nm).	6 (UV)	94	Rosenfeldt and Linden (2004)
		$C_0 = 60\,\mu g/L$; $H_2O_2 = 0–50\,mg/L$; pH = 4.3–5.3; low pressure UV lamp (254 nm, 5 W).	6 (UV)	100	Chen et al. (2006)
	$US/UV\text{-}TiO_2$	$C_0 = 100\,\mu g/L$; US bath (20 kHz; 400 kW); solar lamp (350–400 nm, 30 W); $TiO_2 = 0.5–4\,g/L$; duration = 120 min.	4 (UV)		Torres-Palma et al. (2008)
		$C_0 = 5\,mg/L$; $TiO_2 = 0.05\,g/L$; UV lamp (300–800 nm); US bath (300 kHz; 80 W); duration = 360 min.	35 (UV)		
	US/Fenton process	$C_0 = 20\,mg/L$; US bath (40 kHz, 100 W); $H_2O_2 = 200\,mM$; pH = 3–9; duration = 50 min.	40 (FP)	90	Huang et al. (2012)
Amoxicillin	UV/H_2O_2	$C_0 = 1.827\,\mu g/L$; $H_2O_2 = 102\,mg/L$; pH = 5.5; low pressure UV lamp (254 nm).	10 (UV)	100	Andreozzi et al. (2003a)
	O_3/Fenton process	$C_0 = 100\,mg/L$; Fe (II) = 242 mg/L; pH = 3; T = 25ºC; $H_2O_2 = 34\,mg/L$; duration = 2 min.	10 (FP)	65	Li et al. (2015)
	Electro-Fenton process	$C_0 = 36\,mg/L$; Fe (II) = 0.202 g/L; $Na_2SO_4 = 7\,g/L$; pH = 5.3; T = 35ºC; current = 20.83 mA/cm^2; duration = 360 min.	72 (FP)	100	Sopaj et al. (2015)
Penicillin	$UV/O_3/H_2O_2$	$C_0 = 2.5\,mg/L$; $O_3 = 177\,mg/L\text{-}min$; $H_2O_2 = 0.03\,\mu g/L$; pH = 7; T = 25ºC; UV (254 nm); duration = 70 min.	80 (UV)	93	Keen et al. (2012)
	O_3/H_2O_2	$C_0 = 2.5\,mg/L$; $O_3 = 49.3\,mg/L\text{-}min$; $H_2O_2 = 0.034\,mg/L$; pH = 7; T = 25°C; duration = 70 min.	56 (O_3)	89	

Tetracycline	US/Fenton process	C_0 = 100 mg/L; US bath (20 kHz, 100 W); H_2O_2 = 1 mM; pH = 3–7; duration = 60 min.	56 (FP)	92	Hou et al. (2016)
Trimethoprim	Photo-electro-Fenton process	C_0 = 20 mg/L; Fe (II) = 0.202 g/L; UV-A (350 nm, 6 W); Na_2SO_4 = 7 g/L; pH = 3–4.5; T = 20 °C; current = 150 mA/cm^2; duration = 420 min.	21 (AO)	73	Moreira et al. (2014)
	Electro-Fenton process	C_0 = 20 mg/L; Fe (II) = 0.202 g/L; H_2O_2 = 15 mg/L; Na_2SO_4 = 7 g/L; pH = 3–4.5; T = 20°C; current = 150 mA/cm^2; duration = 420 min.		57	
Sulfamethoxazole	UV/Cl_2	C_0 = 50,000 mg/L; Cl_2 = 5 mg/L; UV lamp (253.7 nm; 11 W); pH = 7; duration = 30 min.	40 (UV)	100	Kwon et al. (2018)
	O_3/H_2O_2	C_0 = 40 mg/L; H_2O_2 = 0.02 mg/L; O_3 = 170 mg/L-min; pH = 4–6; T = 24°C; duration = 10 min.	7 (O_3)		Lin et al. (2009)
	Electro-Fenton process	C_0 = 52.7 mg/L; Fe(II) = 0.08 g/L; Na_2SO_4 = 7 g/L; pH = 3; T = 25°C; current = 7.5 mA/cm^2; duration = 600 min.	82 (FP)	96	Dirany et al. (2010)
Atenolol	Photo-electro-Fenton process	C_0 = 0.12 mg/L; Fe (II) = 0.202 g/L; UV-A (273 nm, 15 W); Na_2SO_4 = 7 g/L; pH = 3; T = 35°C; H_2O_2 = 15 mg/L; current = 0.5 A/cm^2; duration = 360 min.	18 (FP)	100	Isarain-Chávez et al. (2011)
	Electro-Fenton process	C_0 = 158 mg/L; Fe(II) = 0.202 g/L; Na_2SO_4 = 14 g/L; pH = 3; T = 35°C; current = 55 mA/cm^2; duration = 480 min.	58 (FP)	97	
Pesticides					
Malathion or Karbofos	UV/H_2O_2	C_0 = 10–100 mg/L; Hg vapor lamp, pH = 3–12.	–	1	Chenna et al. (2016)
	Fenton process	C_0 = 33 mg/L; [Malathion]:[H_2O_2] = 1:100; [H_2O_2]:[Fe(II)] = 40:1; pH = 3; duration = 100 min.		1	Zhang and Pagilla (2010)
	Microwave-electrodeless UV/Fenton process	C_0 = 24.53 mg/L; pH = 5; Fe (II) = 0.8 mM; H_2O_2 = 0–300 mM; T = 2°C; UV (254, 313; 365, and 405 nm); duration = 120 min.		100	Cheng et al. (2015)

(Continued)

Table 15.1 (Continued)

Microconstituents	AOPs	Experimental Conditions	Removal Efficiency (%)		References
			Individual AOP	Integrated AOP	
Carbofuran	Sono-Fenton process	C_0 = 20 mg/L; pH = 3; Fe(II) = 0.8 mM; H_2O_2 = 0–300 mM; T = 25°C; duration = 120 min (sonolysis) and 30 min (sono-Fenton process).	> 40 (FP)	> 99	Ma and Sung (2010)
Chlorpyrifos	EO	$[COD]_{C0}$ = 450 mg/L; pH = 2; Niobium (Nb)/PbO_2 electrodes; T = 70°C; current = 50 mA/cm^2; duration = 600 min.	76 (as COD)	–	Samet et al. (2010)
	Sonolysis/EO	C_0 = 900 mg/L; pH = 7; T = 15–35ºC; stainless steel electrode.	72.8 (EO)	93.3	Ren et al. (2019)
	UV/H_2O_2	C_0 = 200 µg/L; pH = 7; high pressure Hg vapor lamp.	97 (UV)	97	Utzig et al. (2019)
	Solar photocatalysis	C_0 = 2.74 mg/L; TiO_2 = 15.72 mg/L; synthetic solar radiation by RPR; pH = 7; duration = 62.5 min.	–	84.01	Amiri et al. (2018)
	UV/Fenton process	C_0 = 805.56 mg/L; pH = 3; $[H_2O_2]$:[COD] = 2:1; $[H_2O_2]$:[Fe(II)] = 25:1; duration = 60 min.		64.8	Affam et al. (2014)
	Sonolysis	C_0 = 1 mg/L; pH = 9; T = 25°C; US bath (130 kHz, 500 W).	98.96	–	Agarwal et al. (2016)
Diazinon	UV photocatalysis	C_0 = 20 mg/L; pH = 7; UV lamp (125 kW); ZnO + TiO_2 = 0.5 g/L; duration = 120 min.	–	87.26	Jonidi-Jafari et al. (2015)
		C_0 = 20 mg/L; pH = 7; UV lamp (125 kW); Cu doped ZnO = 0.5 g/L; duration = 120 min.		96.79	Shirzad-Siboni et al. (2017)
	Sonolysis/ UV photocatalysis	C_0 = 30 mg/L; pH = 5.5; iron doped TiO_2 = 0.4 g/L; T = 110°C; duration = 100 min.	19 (sonolysis), 36 (UV)	85	Tabasideh et al. (2017)
	Sono-Fenton process	C_0 = 50 mg/L; US probe (20 kHz); pH = 3; Fe(II) = 20 mg/L; H_2O_2 = 150 mg/L; T = 25°C; duration = 60 min.	–	98.3	Wang and Shih (2015)

Acetamiprid	UV/H_2O_2	$C_0 = 100\,\mu g/L$; pH = 2.7; $H_2O_2 = 25\,mg/L$; duration = 20 min.	75 (UV)	100	Carra et al. (2016)
	UV/PS	$C_0 = 100\,\mu g/L$; PS = 50 mg/L; UV (254 nm); duration = 5 min.	40 (UV)		
	Photo-Fenton process	$C_0 = 100\,\mu g/L$; pH = 2.7; Fe(II) = 1 mg/L; $H_2O_2 = 25\,mg/L$; duration = 10 min.	55 (UV)		
Atrazine	O_3/H_2O_2	$C_0 = 0.215\,mg/L$; $H_2O_2 = 0.34\,mg/L$; $O_3 = 2\,mg/L$ min; pH = 8; T = 15ºC; duration = 30 min.	20 (O_3)	80	Yang et al. (2015)
		$C_0 = 2000\,mg/L$; $H_2O_2 = 170\,mg/L$; $O_3 = 0.5\,mg/L$-min; pH = 7; T = 25°C; duration = 30 min.	8 (O_3)	18	Liu et al. (2019)
	O_3/PMS	$C_0 = 0.215$ mg/L; PMS = 6.14 mg/L; $O_3 = 2$ mg/L-min; pH = 7; T = 15°C; duration = 30 min.	20 (O_3)	100	Yang et al. (2015)
	Photo-electro-Fenton process	$C_0 = 20\,mg/L$; Fe(II) = 0.404 g/L; UV-A (360 nm, 34 W); $Na_2SO_4 = 7\,g/L$; pH = 3; T = 25°C; current = 100 mA/cm^2; duration = 60 min.	38 (FP)	95	Garza-Campos et al. (2014)
	UV photocatalysis/ electro-Fenton process	$C_0 = 20\,mg/L$; Fe(II) = 0.40 g/L; $TiO_2 = 0.3\,g/L$; UV-A (360 nm, 34 W); $Na_2SO_4 = 7\,g/L$; pH = 3; T = 25°C; current = 100 mA/cm^2; duration = 60 min.		100	
	Electro-Fenton process	$C_0 = 20\,mg/L$; Fe(II) = 0.404 g/L; $Na_2SO_4 = 7\,g/L$; pH = 6.7; T = 25°C; current = 1 A/cm^2; duration = 180 min.	80 (FP)	97	Oturan et al. (2012)

(Continued)

Table 15.1 (Continued)

Microconstituents	AOPs	Experimental Conditions	Removal Efficiency (%)		References
			Individual AOP	Integrated AOP	
2, 4-D	Electro-Fenton process	C_0 = 217 mg/L; Fe(II) = 0.404 g/L; Na_2SO_4 = 7 g/L; pH = 3; T = 35°C; current = 100 mA/cm^2; duration = 360 min.	48 (FP)	83	Brillas et al. (2007)
	Photo-electro-Fenton process	C_0 = 217 mg/L; Fe (II) = 0.404 g/L; Na_2SO_4 = 7 g/L; pH = 3; T = 35°C; current = 100 mA/cm^2; UV-A (360 nm, 6 W); duration = 240 min.	30 (FP)	70	
	Sono-electro-Fenton process	C_0 = 217 mg/L; Fe(II) = 0.04 g/L; US bath (2 kHz, 60 W); Na_2SO_4 = 7 g/L; pH = 3; T = 35°C; current = 100 mA/cm^2; H_2O_2 = 15 mg/L; duration = 120 min.	78 (FP)	100	Oturan et al. (2008)
Surfactants					
LASs	$UV/O_3/H_2O_2$	$[COD]_{C0}$ = 300–350 mg/L; H_2O_2 = 100 mM; O_3 = 0.27 g/h; pH = 7; UV-A (365 nm, 15 W); industrial wastewater; duration = 180 min.	5 (O_3)	58	Zangeneh et al. (2014)
	O_3/H_2O_2	$[COD]_{C0}$ = 300–350 mg/L; H_2O_2 = 100 mM; O_3 = 0.27 g/h; pH = 7; industrial wastewater; duration = 180 min.	30 (H_2O_2)	53	
	UV/O_3	$[COD]_{C0}$ = 300–350 mg/L; O_3 = 0.27 g/h; pH = 7; UV-A (365 nm, 15 W); industrial wastewater; duration = 180 min.	5 (O_3)	51	
	UV/H_2O_2	$[COD]_{C0}$ = 300–350 mg/L; H_2O_2 = 100 mM; pH = 7; UV-A (36 nm, 15 W); industrial wastewater; duration = 180 min.	32 (UV)	49	
	O_3	C_0 = 4.63–5.30 mg/L; O_3 = 15.9–16.7 mg/L; UASB effluent (domestic wastewater); duration = 30–50 min.	67–90	–	Gasi et al. (1991)

LAS and PL	UV/O_3	$C_0 = 84$ mg/L; $O_3 = 15.6$ mg/L; high pressure UV lamp; T = 10–40°C; laundry wastewater.	–	88.57	Matsuo et al. (1997)
LAS (C_{12}) and ABS (C_{12})	Fenton oxidation	$C_0 = 10$ mg/L; pH = 3; $FeSO_4 = 90$ mg/L; $H_2O_2 = 60$ mg/L; duration = 50 min.	> 95	–	Lin and Leu (1999)
LAS mixture (C_{10}-C_{13})	UV/H_2O_2	$C_0 = 2500$ mg/L; low pressure UV lamp (254 nm, 8 W); $[H_2O_2]$:[surfactant] = 15:1; pH = 2.2–3.	–	97	Sanz et al. (2003)
APEs	Fenton oxidation	$C_0 = 250–2750$ mg/L; T = 22°C; $H_2O_2 > 750$ mg/L; duration = 30 min	88	–	Kitis et al. (1999)
SDS	UV/H_2O_2	$C_0 = 200$ mg/L; UV-A (254 nm); $[H_2O_2]$:[surfactant] = 2:1; municipal wastewater; duration = 7 min.	–	81	Mondal et al. (2019)
	$UV/O_3/H_2O_2$	$C_0 = 2.5$ mg/L; $O_3 = 10$ mg/L; pH = 7; $H_2O_2 = 10$ mM; duration = 55 min.		96	Arslan et al. (2018)
Anionic surfactants	EO	$C_0 = 119.2$ mg/L; real carwash wastewater; current = 50 mA; BDD anode and carbon-felt cathode; $Na_2SO_4 = 0.05$ M; pH = 3; duration = 240 min.	96 (EO)	–	Ganiyu et al. (2018)
	EO/H_2O_2			100	
	Electro-Fenton process	$C_0 = 119.2$ mg/L; real carwash wastewater; current = 50 mA; BDD anode and carbon-felt cathode; $Na_2SO_4 = 0.05$ M; pH = 3; duration = 120 min.	–		
Per- and polyfluoroalkyl substances (PFASs)					
PFOA	UV/H_2O_2	$C_0 = 29.6$ mM; duration = 72 h.	13.3 (UV in 24 h)	26.5	Hori et al. (2004)
	PS	$C_0 = 29.6$ mM; duration = 240 min.	100	–	Hori et al. (2005)
	$UV\text{-}TiO_2$	$C_0 = 50\ \mu M$; UV lamp (254 nm, 400 W); duration = 12 h.	–	18.6	Chen et al. (2016)
	$Pb\text{-}TiO_2/UV$			99.9	
	UV photolysis	$C_0 = 50–67.3\ \mu M$; low pressure UV lamp (254 nm, 20 W); duration = 240 min.	98.6	–	Giri et al. (2014)
PFOS	UV photolysis	$C_0 = 40\ \mu M$; low pressure Hg lamp (254 nm, 32 W); duration = 240 h.	27.2		Yamamoto et al. (2007)

(Continued)

Table 15.1 (Continued)

Microconstituents	AOPs	Experimental Conditions	Removal Efficiency (%)		References
			Individual AOP	Integrated AOP	
PFCA	Fe^{3+}/Fe^{2+} O_2 via photo-induced complexation	$C_0 = 67.3\,\mu M$; Xe-Hg lamp (220–460 nm); duration = 24 h.	–	64.5	Hori et al. (2007)
PFHxS	Iron-induced decomposition	$C_0 = 5.93\,\mu M$; T = 350°C; duration = 360 min.	99	–	Hori et al. (2008)
18 PFASs (12 PFCAs, 3 PFSAs, FOSA, 6:2 FTSA and 8:2 FTSA)	O_3/PS/catalyst	$(C_0)_{18\ PFASs} = 20{,}000\,\mu g/mL$; catalyst = 2.5 g/L; pH = 6–7; duration = 180 min; T = 22°C; O_3 = 300 mg/h; [PFAS]:[ammonium persulfate] = 1:50.	22 (O_3)	47 (O_3/PS/catalyst), 24 (O_3/PS), 26 (O_3/catalyst).	Franke et al. (2019)
FtOH (TFHFESE)	UV/H_2O_2	C_0 = 10 mg/L; UV lamp (244 nm, 3 mW/cm^2); H_2O_2 = 25 mM; duration = 45 min.	73.7 (UV)	97.2	Barisci and Suri (2021)
	O_3/H_2O_2	C_0 = 10 mg/L; O_3 flow rate = 75 L/h; H_2O_2 = 5–100 mM; duration = 120 min.	76 (O_3)	Presence of H_2O_2 reduces the performance.	

Note: C_0 = Initial concentration of the MC; T = Temperature; FP = Fenton process; EO = Electrochemical oxidation; COD = Chemical oxygen demand; RPR = Raceway pond reactor; LASs = Linear alkylbenzene sulfonates; PL = Potassium laurate; ABS = Acrylonitrile butadiene styrene; APEs = Alkylphenol ethoxylates; SDS = Sodium dodecyl sulfate; PFOA = Perfluorooctanoic acid; PFOS = Perfluorooctane sulfonic acid; PFCA = Perfluoroalkyl carboxylic acid; PFHxS = Perfluorohexane sulfonic acid; PFSAs = Perfluoroalkylated sulfonate acid; FOSA = Perfluorooctane sulfonamide; FTSA = Fluorotelomer sulfonic acid; FtOH = Fluorotelomer alcohol; TFHFESE = 2-(1,1,2-trifluoro-2-hepta fluoropyloxy ethylsulfonyl)-ethanol; 2,4-D 2,4-dichlorophenoxyacetic acid.

eradicate the PPCPs, especially when implemented as integrated AOPs. For instance, UV irradiation (having a wavelength of 254 nm) alone ensured 20% removal of paracetamol in 120 minutes, whereas UV irradiation in combination with H_2O_2 oxidant ensured complete degradation of paracetamol within two hours (Andreozzi et al. 2003a). As discussed earlier, incorporation of H_2O_2 enhances OH• radical formation via UV photolysis of H_2O_2 (Tufail et al. 2020). This could be the reason for the better removal of paracetamol in the UV/H_2O_2 process. Liu et al. (2019) investigated the performances of the ozonation and peroxone processes for removing carbamazepine. O_3, being a very strong oxidant, captivated 80% removal of carbamazepine, although it is a recalcitrant compound. Integration of O_3 and H_2O_2 further improved the efficiency of removing carbamazepine (up to 90%) by triggering OH• radical formation. Moreover, the peroxone process also lessened the reaction time. Additionally, carbamazepine, containing electron-withdrawing functional group ($-CONH_2$), was found to be reluctant toward direct UV photolysis (5%), but addition of Cl_2 with UV photolysis improved the degradation of carbamazepine (98%) to a great extent (Wang et al. 2016a) (Table 15.1). As mentioned earlier, the combination of Cl_2 with UV irradiation improves the removal of PPCPs by producing various ROS such as Cl•, ClO•, and OH• radicals, which could be the possible reason behind the better performance of the UV/Cl_2 process (Cerreta et al. 2020). A similar observation was also reported for diclofenac (Vogna et al. 2004b).

15.10.2 Pesticide Removal

Most pesticides, hidden in the environment, are very persistent in nature (Khan and Pathak 2020). From Table 15.1, it is observed that UV-based, electrochemical-based, and Fenton process-based AOPs have been predominantly employed for removing pesticides. For example, atrazine was removed by up to 80% via a Fenton process, and an electro-Fenton process ensured 97% removal of atrazine (Oturan et al. 2012) (Table 15.1). The electro-Fenton process facilitates *in-situ* generation of OH• radicals, thereby improving the degradation of atrazine (Sirés et al. 2014). Also the electro-Fenton process negotiates the sensitivity toward pH fluctuations, which has been a major concern with the classical Fenton process. Oturan et al. (2012) observed that the electro-Fenton process effectively removed 97% of atrazine at pH 6.7 (Table 15.1).

A similar trend has also been observed for 2,4-dichlorophenoxyacetic acid (Brillas et al. 2007; Oturan et al. 2008). Furthermore, the combination of the Fenton process or electro-Fenton process with sonolysis can be a wise choice for removing the pesticides. For illustration, the Fenton process in integration with sonolysis achieved commendable removal of carbofuran (>99%) (Ma and Sung 2010) and diazinon (98.3%) (Wang and Shih 2015) (Table 15.1). Basically, in a sono-electro-Fenton process, cavity bubbles near the anode improve the mass transfer to the anode and cut down anode fouling, thereby improving OH• radical formation (Oturan et al. 2008). However, the removal of pesticides via ozonation was trivial. For example, Liu et al. (2019) found that only 8% atrazine was removed through ozonation. This could be because O_3 is highly pollutant specific (Tufail et al. 2020). Furthermore, acetamiprid contains both electron-donating and electron-withdrawing ($-CH_3$, $-Cl$) functional groups in its structure, making it slightly recalcitrant toward UV photolysis; however, UV irradiation in combination with other AOPs improves the degradation of acetamiprid due to the presence of OH• radicals. For example, Carra et al. (2016)

reported that the removal of acetamiprid via UV radiation was 40% in 5 minutes, whereas the addition of PS ensured complete degradation of acetamiprid within 5 minutes.

15.10.3 Surfactant Removal

Based on the limited literature, surfactant removal efficiency of various AOPs has been represented in Table 15.1. For example, a mixture of linear alkylbenzene sulfonates (LASs), when exposed to the combination of UV irradiation and H_2O_2, was removed by up to 97% (Sanz et al. 2003). As discussed before, UV irradiation in combination with H_2O_2 triggers OH• radical formation, ensuring high removal of LASs mixture (Tufail et al. 2020). Mondal et al. (2019) explored the potential of UV/H_2O_2 for removing sodium dodecyl sulfate (SDS). They found that SDS was removed by up to 81%. Furthermore, another group of researchers integrated ozonation with H_2O_2 for removing SDS (Arslan et al. 2018). They reported 96% removal of SDS. Again, Ganiyu et al. (2018) proved that the electro-Fenton process is more efficient in removing anionic surfactants present in real carwash wastewater as compared to AO and AO/H_2O_2 (Table 15.1). Hence, from the data collected in Table 15.1, it can be said that electrochemical-based, UV-based, and ozonation-based AOPs have been emphasized by researchers for removing surfactants from aqueous matrices.

15.10.4 PFAS Removal

Similar to the surfactants, in the case of PFASs, extensive work has not been reported on the application of AOPs. Hence, based on the limited literature, the efficacy of different AOPs in removing some of the commonly-detected PFASs in water and wastewater has been listed in Table 15.1. Usually, the long-chain PFASs (number of carbon atoms ≥7; Vo et al. (2020)) are very persistent in nature. For example, only 13.3% perfluorooctanoic acid (PFOA) was removed via UV irradiation. Even the integration of H_2O_2 oxidant with UV irradiation failed to magnify the removal of PFOA (Hori et al. 2004). The UV/H_2O_2 process encompassed only 26.5% removal of PFOA in 72 hours, signifying high resistence of PFOA toward the UV/H_2O_2 process (Hori et al. 2004). However, the removal of fluorotelomer alcohol (FtOH), 2-(1,1,2-trifluoro-2-hepta fluoropyloxy-ethylsulfonyl)-ethanol (TFHFESE), was reported to be 97.2% via the UV/H_2O_2 process (Barisci and Suri 2021). Additionally, they also reported that the degradation of TFHFESE was higher via ozonation (76%), but the integration of H_2O_2 with O_3 lessened the removal of the same. This could be attributed to the scavenging of OH• radicals by H_2O_2, when present in excess of aqueous O_3 (Yao et al. 2016; Turkay et al. 2017).

15.11 Future Perspectives

Regarding MCs from water and wastewater, there are some concerns that need further attention:

- The application of different individual and combined AOPs has been mostly reported for removing various MCs from deionized water. Thus, the effect of the water matrix on the degradation of MCs has not been explored. This needs to be addressed because the existence of various chemical species in water matrices affects the rate of degradation of ECs (Heeb et al. 2014). Moreover, to understand the actual potential of AOPs to remove ECs, the effects of organic matter as well as inorganic ions on the removal of MC need to be evaluated.

- The MCs with high persistence result in partial or incomplete degradation or mineralization. As a consequence, some intermediate by-products may be produced. However, very little evidence regarding the degradation pathway of the ECs via AOPs is available. This area needs further investigation.
- Moreover, the toxicity analysis of the AOP-treated effluent on the receiving ecosystems has not been vigorously performed. Sometimes, owing to incomplete degradation of persistent ECs, by-products with higher toxicity than their parent compounds may be produced. Hence, the toxicity analysis of the AOP-treated effluent is of utmost importance.
- Particularly, in photocatalysis, once the process is over, the ultimate fate of the photocatalyst remains a major concern regarding this particular AOP, which requires further investigations.
- Mostly, the application of AOPs for eradicating MCs has been limited to the lab-scale only. Due to high capital cost, the AOPs have not been implemented on the field-scale. Even though few studies regarding the cost analysis of different AOPs are there, more detailed cost analysis of various individual as well as combined AOPs have to be emphasized to enhance their acceptability.
- Currently, an awareness among researchers has been observed to modify the photocatalysts (e.g., TiO_2) with sulphur (S), nitrogen (N), carbon (C), multivalent cations (e.g., Fe^{2+}, Cr^{3+}, Mn^{2+}, etc.), and noble metals [e.g., gold (Au), platinum (Pt), silver (Ag), etc.] in order to increase the photocatalytic efficiency through reducing the electron-hole pair (EHP) recombination and reducing the external energy requirement (from UV range to visible range) to overcome the band-gap energy, also known as the red-shift of the absorption band of the photocatalysts, by multiple excitation (Prashanth et al. 2021) (Figure 15.2). Basically, through doping, the excitation of the composite can be done with lesser energy or longer wavelength, especially for the metallic cations because for

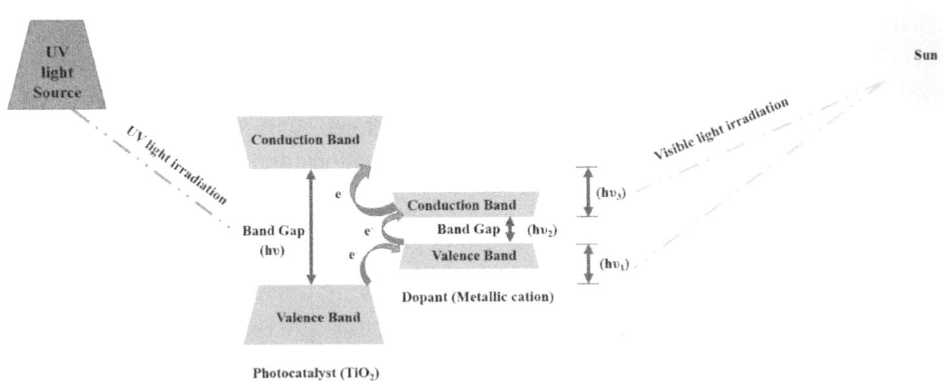

Figure 15.2 Schematic representation of the doping of TiO_2 photocatalyst with metallic cation dopants. ($hυ$ = external energy, $υ_1$, $υ_2$, $υ_3 \ll υ$, and $hυ_1 + hυ_2 + hυ_3 = hυ$).

the metals, the conduction band and valence band are very close to each other. However, when the photocatalysts are doped with noble metals, the noble metals help to absorb the electrons, thereby helping in radical formation through reducing the EHP recombination which, in turn, increases the photocatalytic efficiency (Prashanth et al. 2021). Sometimes the photocatalyst, especially TiO_2, is coupled with other semiconductors (e.g., cadmium sulphide (CdS)), resulting in the red-shift of the absorption band of the composite. Meanwhile, both the valence band and the conduction band of the semiconductor dopant or the photocatalyst should be at higher levels than the respective valence band and the conduction band of the other materials (Prashanth et al. 2021). Vigorous exploration regarding the application of the doped TiO_2 for removing the ECs is of utmost importance to enhance the cost-effectiveness of the photocatalysis process.

- Finally, instead of randomly opting for the AOPs or a combination of AOPs for removing the ECs, further research should be carried out on how the characteristics of MCs govern their removal through different AOPs. In other words, an in-depth knowledge regarding the best possible AOP or combination of different AOPs for a particular type of MCs will help in reducing the resource requirements (both temporal and monetary).

15.12 Conclusions

In this chapter, a comprehensive review on the classification, fundamentals, and application of different AOPs, individual or in combination with other AOPs, have been represented. The AOPs facilitate *in-situ* generation of the ROS such as OH• radicals and O_2^- radicals that carry out the degradation of MCs. However, individual AOPs possess some limitations that can be overcome by implementing the integrated AOPs. For instance, the Fenton process is highly sensitive to pH fluctuations. It has to be performed in acidic pH (around 3). Otherwise, $Fe(OH)_3$ will be produced at high pH, reducing the treatment efficiency of the Fenton process by scavenging free OH• radicals. However, electro-Fenton, sono-Fenton, or O_3/Fenton processes negotiate such limitations of the classic Fenton process. Furthermore, integrated AOPs ensure better mineralization or degradation and also makes their degradation faster. For example, compared to ozonation alone, application of O_3/H_2O_2 magnifies the degradation of the MCs and also speeds up their degradation reaction due to the higher formation of the OH• radicals in the latter process. However, the toxicity of the AOP-treated effluent due to incomplete mineralization of the persistent MCs has to be vigorously explored. Furthermore, the economic feasibility of the individual as well as integrated AOPs has to be evaluated to strengthen the acceptability of the AOPs as the MC remediation alternatives.

References

Affam, A.C., Chaudhuri, M., Kutty, S.R.M., and Muda, K. (2014). UV Fenton and sequencing batch reactor treatment of chlorpyrifos, cypermethrin and chlorothalonil pesticide wastewater. *International Biodeterioration & Biodegradation* 93: 195–201.

Agarwal, S., Tyagi, I., Gupta, V.K. et al. (2016). Degradation of azinphos-methyl and chlorpyrifos from aqueous solutions by ultrasound treatment. *Journal of Molecular Liquids* 221: 1237–1242.

Aguinaco, A., Beltrán, F.J., García-Araya, J.F., and Oropesa, A. (2012). Photocatalytic ozonation to remove the pharmaceutical diclofenac from water: influence of variables. *Chemical Engineering Journal* 189: 275–282.

Agustina, T.E., Ang, H.M., and Vareek, V.K. (2005). A review of synergistic effect of photocatalysis and ozonation on wastewater treatment. *Journal of Photochemistry and Photobiology C: Photochemistry Reviews* 6 (4): 264–273.

Alvarez-Corena, J.R., Bergendahl, J.A., and Hart, F.L. (2016). Advanced oxidation of five contaminants in water by UV/TiO_2: reaction kinetics and byproducts identification. *Journal of Environmental Management* 181: 544–551.

Ameta, R., Chohadia, A.K., Jain, A., and Punjabi, P.B. (2018). Fenton and photo-Fenton processes. In: *Advanced Oxidation Processes for Waste Water Treatment (eds. Ameta, S.C. and Ameta, R.)*, 49–87. Academic Press.

Amiri, H., Nabizadeh, R., Martinez, S.S. et al. (2018). Response surface methodology modeling to improve degradation of *Chlorpyrifos* in agriculture runoff using TiO_2 solar photocatalytic in a raceway pond reactor. *Ecotoxicology and Environmental Safety* 147: 919–925.

Amr, S.S.A. and Aziz, H.A. (2012). New treatment of stabilized leachate by ozone/Fenton in the advanced oxidation process. *Waste Management* 32 (9): 1693–1698.

Andreozzi, R., Caprio, V., Insola, A., and Marotta, R. (1999). Advanced oxidation processes (AOP) for water purification and recovery. *Catalysis Today* 53 (1): 51–59.

Andreozzi, R., Caprio, V., Marotta, R., and Radovnikovic, A. (2003b). Ozonation and H_2O_2/UV treatment of clofibric acid in water: a kinetic investigation. *Journal of Hazardous Materials* 103 (3): 233–246.

Andreozzi, R., Caprio, V., Marotta, R., and Vogna, D. (2003a). Paracetamol oxidation from aqueous solutions by means of ozonation and H_2O_2/UV system. *Water Research* 37 (5): 993–1004.

Arslan, A., Topkaya, E., Bingöl, D., and Veli, S. (2018). Removal of anionic surfactant sodium dodecyl sulfate from aqueous solutions by O_3/UV/H_2O_2 advanced oxidation process: process optimization with response surface methodology approach. *Sustainable Environment Research* 28 (2): 65–71.

Autin, O., Romelot, C., Rust, L. et al. (2013). Evaluation of a UV-light emitting diodes unit for the removal of micropollutants in water for low energy advanced oxidation processes. *Chemosphere* 92 (6): 745–751.

Bae, I.K., Ham, H.M., Jeong, M.H. et al. (2015). Simultaneous determination of 15 phenolic compounds and caffeine in teas and mate using RP-HPLC/UV detection: method development and optimization of extraction process. *Food Chemistry* 172: 469–475.

Baeza, C. and Knappe, D.R. (2011). Transformation kinetics of biochemically active compounds in low-pressure UV photolysis and UV/H_2O_2 advanced oxidation processes. *Water Research* 45 (15): 4531–4543.

Bamba, D., Atheba, P., Robert, D. et al. (2008). Photocatalytic degradation of the diuron pesticide. *Environmental Chemistry Letters* 6 (3): 163–167.

Barisci, S. and Suri, R. (2021). Removal of polyfluorinated telomer alcohol by Advanced Oxidation Processes (AOPs) in different water matrices and evaluation of degradation mechanisms. *Journal of Water Process Engineering* 39: 101745.

Bautista, P., Mohedano, A.F., Casas, J.A. et al. (2008). An overview of the application of Fenton oxidation to industrial wastewaters treatment. *Journal of Chemical Technology & Biotechnology: International Research in Process, Environmental & Clean Technology* 83 (10): 1323–1338.

Bayen, S. (2012). Occurrence, bioavailability and toxic effects of trace metals and organic contaminants in mangrove ecosystems: a review. *Environment International* 48: 84–101.

Biel-Maeso, M., Baena-Nogueras, R.M., Corada-Fernández, C., and Lara-Martín, P.A. (2018). Occurrence, distribution and environmental risk of pharmaceutically active compounds (PhACs) in coastal and ocean waters from the Gulf of Cadiz (SW Spain). *Science of the Total Environment* 612: 649–659.

Brillas, E., Baños, M.Á., Skoumal, M. et al. (2007). Degradation of the herbicide 2, 4-DP by anodic oxidation, electro-Fenton and photoelectro-Fenton using platinum and boron-doped diamond anodes. *Chemosphere* 68 (2): 199–209.

Buxton, G.V., Greenstock, C.L., Helman, W.P., and Ross, A.B. (1988). Critical review of rate constants for reactions of hydrated electrons, hydrogen atoms and hydroxyl radicals $OH/{\cdot}O^-$ in aqueous solution. *Journal of Physical and Chemical Reference Data* 17 (2): 513–886.

Cacace, D., Fatta-Kassinos, D., Manaia, C.M. et al. (2019). Antibiotic resistance genes in treated wastewater and in the receiving water bodies: a pan-European survey of urban settings. *Water Research* 162: 320–330.

Candido, J.P., Andrade, S.J., Fonseca, A.L. et al. (2016). Ibuprofen removal by heterogeneous photocatalysis and ecotoxicological evaluation of the treated solutions. *Environmental Science and Pollution Research* 23 (19): 19911–19920.

Carabin, A., Drogui, P., and Robert, D. (2015). Photo-degradation of carbamazepine using TiO_2 suspended photocatalysts. *Journal of the Taiwan Institute of Chemical Engineers* 54: 109–117.

Carlson, J.C., Stefan, M.I., Parnis, J.M., and Metcalfe, C.D. (2015). Direct UV photolysis of selected pharmaceuticals, personal care products and endocrine disruptors in aqueous solution. *Water Research* 84: 350–361.

Carlsson, C., Johansson, A.K., Alvan, G. et al. (2006). Are pharmaceuticals potent environmental pollutants? Part I: Environmental risk assessments of selected active pharmaceutical ingredients. *Science of the Total Environment* 364 (1–3): 67–87.

Carra, I., Sánchez Pérez, J.A., Malato, S. et al. (2016). Performance of different advanced oxidation processes for tertiary wastewater treatment to remove the pesticide acetamiprid. *Journal of Chemical Technology & Biotechnology* 91 (1): 72–81.

Cerreta, G., Roccamante, M.A., Plaza-Bolaños, P. et al. (2020). Advanced treatment of urban wastewater by UV-C/free chlorine process: micro-pollutants removal and effect of UV-C radiation on trihalomethanes formation. *Water Research* 169: 115220.

Chand, R., Ince, N.H., Gogate, P.R., and Bremner, D.H. (2009). Phenol degradation using 20, 300 and 520 kHz ultrasonic reactors with hydrogen peroxide, ozone and zero valent metals. *Separation and Purification Technology* 67 (1): 103–109.

Chaplin, B.P. (2014). Critical review of electrochemical advanced oxidation processes for water treatment applications. *Environmental Science: Processes & Impacts* 16 (6): 1182–1203.

Chen, M.J., Lo, S.L., Lee, Y.C. et al. (2016). Decomposition of perfluorooctanoic acid by ultraviolet light irradiation with Pb-modified titanium dioxide. *Journal of Hazardous Materials* 303: 111–118.

Chen, P.J., Linden, K.G., Hinton, D.E. et al. (2006). Biological assessment of bisphenol A degradation in water following direct photolysis and UV advanced oxidation. *Chemosphere* 65 (7): 1094–1102.

Cheng, G., Lin, J., Lu, J. et al. (2015). Advanced treatment of pesticide-containing wastewater using Fenton reagent enhanced by microwave electrodeless ultraviolet. *BioMed Research International* 2015: Article ID 205903, 8 pages, doi: http://dx.doi.org/10.1155/2015/205903.

Cheng, X., Guo, H., Liu, H. et al. (2016). Performance and mechanism on degradation of estriol using O_3/PS process. *Ozone: Science & Engineering* 38 (5): 358–366.

Chenna, M., Messaoudi, K., Drouiche, N., and Lounici, H. (2016). Study and modeling of the organophosphorus compound degradation by photolysis of hydrogen peroxide in aqueous media by using experimental response surface design. *Journal of Industrial and Engineering Chemistry* 33: 307–315.

Chianese, S., Iovino, P., Leone, V. et al. (2017). Photodegradation of diclofenac sodium salt in water solution: effect of HA, NO_3^- and TiO_2 on photolysis performance. *Water, Air, & Soil Pollution* 228 (8): 1–9.

Clarizia, L., Russo, D., Di Somma, I. et al. (2017). Homogeneous photo-Fenton processes at near neutral pH: a review. *Applied Catalysis B: Environmental* 209: 358–371.

Cong, J., Wen, G., Huang, T. et al. (2015). Study on enhanced ozonation degradation of para-chlorobenzoic acid by peroxymonosulfate in aqueous solution. *Chemical Engineering Journal* 264: 399–403.

Corcoran, J., Winter, M.J., and Tyler, C.R. (2010). Pharmaceuticals in the aquatic environment: a critical review of the evidence for health effects in fish. *Critical Reviews in Toxicology* 40 (4): 287–304.

Dai, C.M., Zhou, X.F., Zhang, Y.L. et al. (2012). Comparative study of the degradation of carbamazepine in water by advanced oxidation processes. *Environmental Technology* 33 (10): 1101–1109.

Dalrymple, O.K., Yeh, D.H., and Trotz, M.A. (2007). Removing pharmaceuticals and endocrine-disrupting compounds from wastewater by photocatalysis. *Journal of Chemical Technology & Biotechnology: International Research in Process, Environmental & Clean Technology* 82 (2): 121–134.

De Witte, B., Dewulf, J., Demeestere, K., and Van Langenhove, H. (2009). Ozonation and advanced oxidation by the peroxone process of ciprofloxacin in water. *Journal of Hazardous Materials* 161 (2–3): 701–708.

Deng, J., Shao, Y., Gao, N. et al. (2013). Degradation of the antiepileptic drug carbamazepine upon different UV-based advanced oxidation processes in water. *Chemical Engineering Journal* 222: 150–158.

Deniere, E., Van Hulle, S., Van Langenhove, H., and Demeestere, K. (2018). Advanced oxidation of pharmaceuticals by the ozone-activated peroxymonosulfate process: the role of different oxidative species. *Journal of Hazardous Materials* 360: 204–213.

Dewil, R., Mantzavinos, D., Poulios, I., and Rodrigo, M.A. (2017). New perspectives for advanced oxidation processes. *Journal of Environmental Management* 195: 93–99.

Dirany, A., Sirés, I., Oturan, N., and Oturan, M.A. (2010). Electrochemical abatement of the antibiotic sulfamethoxazole from water. *Chemosphere* 81 (5): 594–602.

Dong, W., Wang, D., Wang, H. et al. (2019). Facile synthesis of In_2S_3/UiO-66 composite with enhanced adsorption performance and photocatalytic activity for the removal of tetracycline under visible light irradiation. *Journal of Colloid and Interface Science* 535: 444–457.

Duca, C., Imoberdorf, G., and Mohseni, M. (2017). Effects of inorganics on the degradation of micropollutants with vacuum UV (VUV) advanced oxidation. *Journal of Environmental Science and Health, Part A* 52 (6): 524–532.

Edrees, W.H.A., Abdullah, Q.Y.M., Al-Kaf, A.G., and Naji, K.M. (2017). A review on comparative study between the physicochemical and biological processes for paracetamol degradation. *Universal Journal of Pharmaceutical Research* 2: 12.

El-Ghenymy, A., Cabot, P.L., Centellas, F. et al. (2013). Mineralization of sulfanilamide by electro-Fenton and solar photoelectro-Fenton in a pre-pilot plant with a Pt/air-diffusion cell. *Chemosphere* 91 (9): 1324–1331.

El-Ghenymy, A., Garrido, J.A., Centellas, F. et al. (2012). Electro-Fenton and photoelectro-Fenton degradation of sulfanilic acid using a boron-doped diamond anode and an air diffusion cathode. *The Journal of Physical Chemistry A* 116 (13): 3404–3412.

Fatta-Kassinos, D., Vasquez, M.I., and Kümmerer, K. (2011). Transformation products of pharmaceuticals in surface waters and wastewater formed during photolysis and advanced oxidation processes-degradation, elucidation of byproducts and assessment of their biological potency. *Chemosphere* 85 (5): 693–709.

Feng, L., van Hullebusch, E.D., Rodrigo, M.A. et al. (2013). Removal of residual anti-inflammatory and analgesic pharmaceuticals from aqueous systems by electrochemical advanced oxidation processes. A review. *Chemical Engineering Journal* 228: 944–964.

Fent, K., Weston, A.A., and Caminada, D. (2006). Ecotoxicology of human pharmaceuticals. *Aquatic Toxicology* 76 (2): 122–159.

Franke, V., Schäfers, M.D., Lindberg, J.J., and Ahrens, L. (2019). Removal of per- and polyfluoroalkyl substances (PFASs) from tap water using heterogeneously catalyzed ozonation. *Environmental Science: Water Research & Technology* 5 (11): 1887–1896.

Ganiyu, S.O., Dos Santos, E.V., de Araújo Costa, E.C.T., and Martínez-Huitle, C.A. (2018). Electrochemical advanced oxidation processes (EAOPs) as alternative treatment techniques for carwash wastewater reclamation. *Chemosphere* 211: 998–1006.

Gao, Y.Q., Gao, N.Y., Chen, J.X. et al. (2020). Oxidation of β-blocker atenolol by a combination of UV light and chlorine: kinetics, degradation pathways and toxicity assessment. *Separation and Purification Technology* 231: 115927.

García, J., García-Galán, M.J., Day, J.W. et al. (2020). A review of emerging organic contaminants (EOCs), antibiotic resistant bacteria (ARB), and antibiotic resistance genes (ARGs) in the environment: increasing removal with wetlands and reducing environmental impacts. *Bioresource Technology* 307: 123228.

García-Espinoza, J.D., Mijaylova-Nacheva, P., and Avilés-Flores, M. (2018). Electrochemical carbamazepine degradation: effect of the generated active chlorine, transformation pathways and toxicity. *Chemosphere* 192: 142–151.

Garcia-Segura, S., Ocon, J.D., and Chong, M.N. (2018). Electrochemical oxidation remediation of real wastewater effluents: a review. *Process Safety and Environmental Protection* 113: 48–67.

Garza-Campos, B.R., Guzmán-Mar, J.L., Reyes, L.H. et al. (2014). Coupling of solar photoelectro-Fenton with a BDD anode and solar heterogeneous photocatalysis for the mineralization of the herbicide atrazine. *Chemosphere* 97: 26–33.

Gasi, T.M.T., Amaral, L.A.V., Pacheco, C.E.M. et al. (1991). Ozone application for the improvement of UASB reactor effluent. I: Physical-chemical and biological appraisal. *Ozone: Science & Engineering* 13 (2): 179–193.

Ghafoori, S., Mowla, A., Jahani, R. et al. (2015). Sonophotolytic degradation of synthetic pharmaceutical wastewater: statistical experimental design and modeling. *Journal of Environmental Management* 150: 128–137.

Giri, R.R., Ozaki, H., Guo, X. et al. (2014). Oxidative-reductive photodecomposition of perfluorooctanoic acid in water. *International Journal of Environmental Science and Technology* 11 (5): 1277–1284.

Goel, M., Hongqiang, H., Mujumdar, A.S., and Ray, M.B. (2004). Sonochemical decomposition of volatile and non-volatile organic compounds: a comparative study. *Water Research* 38 (19): 4247–4261.

Gogate, P.R., and Pandit, A. (2004). A review of imperative technologies for wastewater treatment. II: Hybrid methods. *Advances in Environmental Research* 8 (3–4): 553–597.

Gomes, J., Costa, R., Quinta-Ferreira, R.M., and Martins, R.C. (2017). Application of ozonation for pharmaceuticals and personal care products removal from water. *Science of the Total Environment* 586: 265–283.

Güyer, G.T. and Ince, N.H. (2011). Degradation of diclofenac in water by homogeneous and heterogeneous sonolysis. *Ultrasonics Sonochemistry* 18 (1): 114–119.

Hai, F.I., Yamamoto, K., and Fukushi, K. (2007). Hybrid treatment systems for dye wastewater. *Critical Reviews in Environmental Science and Technology* 37 (4): 315–377.

Heeb, M.B., Criquet, J., Zimmermann-Steffens, S.G., and Von Gunten, U. (2014). Oxidative treatment of bromide-containing waters: formation of bromine and its reactions with inorganic and organic compounds: a critical review. *Water Research* 48: 15–42.

Homem, V. and Santos, L. (2011). Degradation and removal methods of antibiotics from aqueous matrices: a review. *Journal of Environmental Management* 92 (10): 2304–2347.

Hori, H., Hayakawa, E., Einaga, H. et al. (2004). Decomposition of environmentally persistent perfluorooctanoic acid in water by photochemical approaches. *Environmental Science & Technology* 38 (22): 6118–6124.

Hori, H., Nagaoka, Y., Sano, T., and Kutsuna, S. (2008). Iron-induced decomposition of perfluorohexanesulfonate in sub- and supercritical water. *Chemosphere* 70 (5): 800–806.

Hori, H., Yamamoto, A., Hayakawa, E. et al. (2005). Efficient decomposition of environmentally persistent perfluorocarboxylic acids by use of persulfate as a photochemical oxidant. *Environmental Science & Technology* 39 (7): 2383–2388.

Hori, H., Yamamoto, A., Koike, K. et al. (2007). Photochemical decomposition of environmentally persistent short-chain perfluorocarboxylic acids in water mediated by iron (II)/(III) redox reactions. *Chemosphere* 68 (3): 572–578.

Hou, L., Wang, L., Royer, S., and Zhang, H. (2016). Ultrasound-assisted heterogeneous Fenton-like degradation of tetracycline over a magnetite catalyst. *Journal of Hazardous Materials* 302: 458–467.

Huang, R., Fang, Z., Yan, X., and Cheng, W. (2012). Heterogeneous sono-Fenton catalytic degradation of bisphenol A by Fe_3O_4 magnetic nanoparticles under neutral condition. *Chemical Engineering Journal* 197: 242–249.

Huang, W., Bianco, A., Brigante, M., and Mailhot, G. (2018). UVA-UVB activation of hydrogen peroxide and persulfate for advanced oxidation processes: efficiency, mechanism and effect of various water constituents. *Journal of Hazardous Materials* 347: 279–287.

Ikhlaq, A., Kazmi, M., Tufail, A. et al. (2018). Application of peanut shell ash as a low-cost support for Fenton-like catalytic removal of methylene blue in wastewater. *Desalination and Water Treatment* 111: 338–344.

Ince, N.H. (2018). Ultrasound-assisted advanced oxidation processes for water decontamination. *Ultrasonics Sonochemistry* 40: 97–103.

Isarain-Chávez, E., Rodríguez, R.M., Cabot, P.L. et al. (2011). Degradation of pharmaceutical beta-blockers by electrochemical advanced oxidation processes using a flow plant with a solar compound parabolic collector. *Water Research* 45 (14): 4119–4130.

Jallouli, N., Pastrana-Martínez, L.M., Ribeiro, A.R. et al. (2018). Heterogeneous photocatalytic degradation of ibuprofen in ultrapure water, municipal and pharmaceutical industry wastewaters using a TiO_2/UV-LED system. *Chemical Engineering Journal* 334: 976–984.

Jiang, J.Q., Zhou, Z., and Sharma, V.K. (2013). Occurrence, transportation, monitoring and treatment of emerging micro-pollutants in waste water: a review from global views. *Microchemical Journal* 110: 292–300.

Jin, X., Peldszus, S., and Huck, P.M. (2012). Reaction kinetics of selected micropollutants in ozonation and advanced oxidation processes. *Water Research* 46 (19): 6519–6530.

Jonidi-Jafari, A., Shirzad-Siboni, M., Yang, J.K. et al. (2015). Photocatalytic degradation of diazinon with illuminated ZnO-TiO_2 composite. *Journal of the Taiwan Institute of Chemical Engineers* 50: 100–107.

Joseph, C.G., Puma, G.L., Bono, A., and Krishnaiah, D. (2009). Sonophotocatalysis in advanced oxidation process: a short review. *Ultrasonics Sonochemistry* 16 (5): 583–589.

Kanakaraju, D., Glass, B.D., and Oelgemöller, M. (2018). Advanced oxidation process-mediated removal of pharmaceuticals from water: a review. *Journal of Environmental Management* 219: 189–207.

Katsoyiannis, I.A., Canonica, S., and von Gunten, U. (2011). Efficiency and energy requirements for the transformation of organic micropollutants by ozone, O_3/H_2O_2 and UV/H_2O_2. *Water Research* 45 (13): 3811–3822.

Keen, O.S., Baik, S., Linden, K.G. et al. (2012). Enhanced biodegradation of carbamazepine after UV/H_2O_2 advanced oxidation. *Environmental Science & Technology* 46 (11): 6222–6227.

Keen, O.S., Love, N.G., Aga, D.S., and Linden, K.G. (2016). Biodegradability of iopromide products after UV/H_2O_2 advanced oxidation. *Chemosphere* 144: 989–994.

Khan, S.H. and Pathak, B. (2020). Zinc oxide based photocatalytic degradation of persistent pesticides: a comprehensive review. *Environmental Nanotechnology, Monitoring & Management* 13: 100290.

Kıdak, R. and Doğan, Ş. (2018). Medium-high frequency ultrasound and ozone based advanced oxidation for amoxicillin removal in water. *Ultrasonics Sonochemistry* 40: 131–139.

Kitis, M., Adams, C.D., and Daigger, G.T. (1999). The effects of Fenton's reagent pretreatment on the biodegradability of nonionic surfactants. *Water Research* 33 (11): 2561–2568.

Kuang, J., Huang, J., Wang, B. et al. (2013). Ozonation of trimethoprim in aqueous solution: identification of reaction products and their toxicity. *Water Research* 47 (8): 2863–2872.

Kwon, M., Yoon, Y., Kim, S. et al. (2018). Removal of sulfamethoxazole, ibuprofen and nitrobenzene by UV and UV/chlorine processes: a comparative evaluation of 275 nm LED-UV and 254 nm LP-UV. *Science of the Total Environment* 637: 1351–1357.

Legrini, O., Oliveros, E., and Braun, A.M. (1993). Photochemical processes for water treatment. *Chemical Reviews* 93 (2): 671–698.

Li, M., Zeng, Z., Li, Y. et al. (2015). Treatment of amoxicillin by O_3/Fenton process in a rotating packed bed. *Journal of Environmental Management* 150: 404–411.

Li, Y., Zhu, G., Ng, W.J., and Tan, S.K. (2014). A review on removing pharmaceutical contaminants from wastewater by constructed wetlands: design, performance and mechanism. *Science of the Total Environment* 468: 908–932.

Lin, A.Y.C., Lin, C.F., Chiou, J.M., and Hong, P.A. (2009). O_3 and O_3/H_2O_2 treatment of sulfonamide and macrolide antibiotics in wastewater. *Journal of Hazardous Materials* 171 (1–3): 452–458.

Lin, S.H. and Leu, H.G. (1999). Operating characteristics and kinetic studies of surfactant wastewater treatment by Fenton oxidation. *Water Research* 33 (7): 1735–1741.

Liu, Z., Hosseinzadeh, S., Wardenier, N. et al. (2019). Combining ozone with UV and H_2O_2 for the degradation of micropollutants from different origins: lab-scale analysis and optimization. *Environmental Technology* 40 (28): 3773–3782.

Lu, X., Shao, Y., Gao, N. et al. (2018). Investigation of clofibric acid removal by UV/persulfate and UV/chlorine processes: kinetics and formation of disinfection byproducts during subsequent chlor(am)ination. *Chemical Engineering Journal* 331: 364–371.

Luo, S., Wei, Z., Spinney, R. et al. (2018). UV direct photolysis of sulfamethoxazole and ibuprofen: an experimental and modelling study. *Journal of Hazardous Materials* 343: 132–139.

Ma, Y.S. and Sung, C.F. (2010). Investigation of carbofuran decomposition by a combination of ultrasound and Fenton process. *Journal of Environmental Engineering and Management* 20 (4): 213–219.

Madhavan, J., Grieser, F., and Ashokkumar, M. (2010). Combined advanced oxidation processes for the synergistic degradation of ibuprofen in aqueous environments. *Journal of Hazardous Materials* 178 (1–3): 202–208.

Manaia, C.M., Rocha, J., Scaccia, N. et al. (2018). Antibiotic resistance in wastewater treatment plants: tackling the black box. *Environment International* 115: 312–324.

Martínez, C., Vilariño, S., Fernández, M.I. et al. (2013). Mechanism of degradation of ketoprofen by heterogeneous photocatalysis in aqueous solution. *Applied Catalysis B: Environmental* 142: 633–646.

Matsuo, T., Nishi, T., Matsuda, M., and Izumida, T. (1997). Compatibility of the ultraviolet light-ozone system for laundry waste water treatment in nuclear power plants. *Nuclear Technology* 119 (2): 149–157.

Metcalf, L. and Eddy, H.P. (2003). *Wastewater Engineering, Treatment and Reuse, 4th edn*. New York: Tata McGraw-Hill.

Miklos, D.B., Remy, C., Jekel, M. et al. (2018). Evaluation of advanced oxidation processes for water and wastewater treatment: a critical review. *Water Research* 139: 118–131.

Mirzaei, A., Chen, Z., Haghighat, F., and Yerushalmi, L. (2017). Removal of pharmaceuticals from water by homo/heterogonous Fenton-type processes: a review. *Chemosphere* 174: 665–688.

Mirzaei, R., Yunesian, M., Nasseri, S. et al. (2018). Occurrence and fate of most prescribed antibiotics in different water environments of Tehran, Iran. *Science of the Total Environment* 619: 446–459.

Mondal, B., Adak, A., and Datta, P. (2019). Degradation of anionic surfactant in municipal wastewater by UV-H_2O_2: process optimization using response surface methodology. *Journal of Photochemistry and Photobiology A: Chemistry* 375: 237–243.

Moreira, F.C., Boaventura, R.A., Brillas, E., and Vilar, V.J. (2017). Electrochemical advanced oxidation processes: a review on their application to synthetic and real wastewaters. *Applied Catalysis B: Environmental* 202: 217–261.

Moreira, F.C., Garcia-Segura, S., Boaventura, R.A. et al. (2014). Degradation of the antibiotic trimethoprim by electrochemical advanced oxidation processes using a carbon-PTFE air-diffusion cathode and a boron-doped diamond or platinum anode. *Applied Catalysis B: Environmental* 160: 492–505.

Moreira, N.F., Sousa, J.M., Macedo, G. et al. (2016). Photocatalytic ozonation of urban wastewater and surface water using immobilized TiO_2 with LEDs: micropollutants, antibiotic resistance genes and estrogenic activity. *Water Research* 94: 10–22.

Moussavi, G., Momeninejad, H., Shekoohiyan, S., and Baratpour, P. (2017). Oxidation of Acetaminophen in the contaminated water using $UVC/S_2O_8^{2-}$ process in a cylindrical photoreactor: efficiency and kinetics of degradation and mineralization. *Separation and Purification Technology* 181: 132–138.

Mozia, S. and Morawski, A.W. (2012). The performance of a hybrid photocatalysis-MD system for the treatment of tap water contaminated with ibuprofen. *Catalysis Today* 193 (1): 213–220.

Naddeo, V., Belgiorno, V., Kassinos, D. et al. (2010). Ultrasonic degradation, mineralization and detoxification of diclofenac in water: optimization of operating parameters. *Ultrasonics Sonochemistry* 17 (1): 179–185.

Nicole, I., De Laat, J., Dore, M. et al. (1990). Utilisation du rayonnement ultraviolet dans le traitement des eaux: mesure du flux photonique par actinometrie chimique au peroxyde d'hydrogene. *Water Research* 24 (2): 157–168.

Nidheesh, P.V. and Gandhimathi, R. (2012). Trends in electro-Fenton process for water and wastewater treatment: an overview. *Desalination* 299: 1–15.

Oturan, M.A., Sirés, I., Oturan, N. et al. (2008). Sonoelectro-Fenton process: a novel hybrid technique for the destruction of organic pollutants in water. *Journal of Electroanalytical Chemistry* 624 (1–2): 329–332.

Oturan, N., Brillas, E., and Oturan, M.A. (2012). Unprecedented total mineralization of atrazine and cyanuric acid by anodic oxidation and electro-Fenton with a boron-doped diamond anode. *Environmental Chemistry Letters* 10 (2): 165–170.

Pang, Y.L., Abdullah, A.Z., and Bhatia, S. (2011). Review on sonochemical methods in the presence of catalysts and chemical additives for treatment of organic pollutants in wastewater. *Desalination* 277 (1–3): 1–14.

Pérez-Estrada, L.A., Malato, S., Gernjak, W. et al. (2005). Photo-Fenton degradation of diclofenac: identification of main intermediates and degradation pathway. *Environmental Science & Technology* 39 (21): 8300–8306.

Prashanth, V., Jayasree, P., Rajput, P., and Remya, N. (2021). Solar photocatalysis and its application for emerging contaminant removal from wastewater. In: *Advanced Oxidation Processes for Effluent Treatment Plants* (ed. Shah, M.P.), 69–85. Elsevier.

Qin, Y., Luo, S., Geng, S. et al. (2018). Degradation and mineralization of aniline by O_3/Fenton process enhanced using high-gravity technology. *Chinese Journal of Chemical Engineering* 26 (7): 1444–1450.

Rayaroth, M.P., Aravind, U.K., and Aravindakumar, C.T. (2016). Degradation of pharmaceuticals by ultrasound-based advanced oxidation process. *Environmental Chemistry Letters* 14 (3): 259–290.

Ren, Q., Yin, C., Chen, Z. et al. (2019). Efficient sonoelectrochemical decomposition of chlorpyrifos in aqueous solution. *Microchemical Journal* 145: 146–153.

Ribeiro, A.R., Nunes, O.C., Pereira, M.F., and Silva, A.M. (2015). An overview on the advanced oxidation processes applied for the treatment of water pollutants defined in the recently launched Directive 2013/39/EU. *Environment International* 75: 33–51.

Rice, R.G. (1996). Applications of ozone for industrial wastewater treatment: a review. *Ozone: Science & Engineering* 18 (6): 477–515.

Richardson, S.D. and Ternes, T.A. (2011). Water analysis: emerging contaminants and current issues. *Analytical Chemistry* 83 (12): 4614–4648.

Rosenfeldt, E.J. and Linden, K.G. (2004). Degradation of endocrine disrupting chemicals bisphenol A, ethinyl estradiol, and estradiol during UV photolysis and advanced oxidation processes. *Environmental Science & Technology* 38 (20): 5476–5483.

Rozas, O., Contreras, D., Mondaca, M.A. et al. (2010). Experimental design of Fenton and photo-Fenton reactions for the treatment of ampicillin solutions. *Journal of Hazardous Materials* 177 (1–3): 1025–1030.

Samet, Y., Agengui, L., and Abdelhédi, R. (2010). Anodic oxidation of chlorpyrifos in aqueous solution at lead dioxide electrodes. *Journal of Electroanalytical Chemistry* 650 (1): 152–158.

Sangion, A. and Gramatica, P. (2016). Hazard of pharmaceuticals for aquatic environment: prioritization by structural approaches and prediction of ecotoxicity. *Environment International* 95: 131–143.

Sanz, J., Lombrana, J.I., De Luis, A.M., and Varona, F. (2003). UV/H_2O_2 chemical oxidation for high loaded effluents: a degradation kinetic study of LAS surfactant wastewaters. *Environmental Technology* 24 (7): 903–911.

Sharma, J., Mishra, I.M., and Kumar, V. (2015). Degradation and mineralization of Bisphenol A (BPA) in aqueous solution using advanced oxidation processes: UV/H_2O_2 and UV/$S_2O_8^{2-}$ oxidation systems. *Journal of Environmental Management* 156: 266–275.

Shirzad-Siboni, M., Jonidi-Jafari, A., Farzadkia, M. et al. (2017). Enhancement of photocatalytic activity of Cu-doped ZnO nanorods for the degradation of an insecticide: kinetics and reaction pathways. *Journal of Environmental Management* 186: 1–11.

Sirés, I. and Brillas, E. (2012). Remediation of water pollution caused by pharmaceutical residues based on electrochemical separation and degradation technologies: a review. *Environment International* 40: 212–229.

Sirés, I., Brillas, E., Oturan, M.A. et al. (2014). Electrochemical advanced oxidation processes: today and tomorrow: a review. *Environmental Science and Pollution Research* 21 (14): 8336–8367.

Sirés, I., Centellas, F., Garrido, J.A. et al. (2007). Mineralization of clofibric acid by electrochemical advanced oxidation processes using a boron-doped diamond anode and Fe2+ and UVA light as catalysts. *Applied Catalysis B: Environmental* 72 (3–4): 373–381.

Sirtori, C., Agüera, A., Gernjak, W., and Malato, S. (2010). Effect of water-matrix composition on Trimethoprim solar photodegradation kinetics and pathways. *Water Research* 44 (9): 2735–2744.

Skoumal, M., Rodríguez, R.M., Cabot, P.L. et al. (2009). Electro-Fenton, UVA photoelectro-Fenton and solar photoelectro-Fenton degradation of the drug ibuprofen in acid aqueous medium using platinum and boron-doped diamond anodes. *Electrochimica Acta* 54 (7): 2077–2085.

Song, K., Mohseni, M., and Taghipour, F. (2016). Application of ultraviolet light-emitting diodes (UV-LEDs) for water disinfection: a review. *Water Research* 94: 341–349.

Sopaj, F., Rodrigo, M.A., Oturan, N. et al. (2015). Influence of the anode materials on the electrochemical oxidation efficiency. Application to oxidative degradation of the pharmaceutical amoxicillin. *Chemical Engineering Journal* 262: 286–294.

Sun, M., Xu, D., Ji, Y. et al. (2016). Using Fenton oxidation to simultaneously remove different estrogens from cow manure. *International Journal of Environmental Research and Public Health* 13 (9): 917.

Sunartio, D., Ashokkumar, M., and Grieser, F. (2007). Study of the coalescence of acoustic bubbles as a function of frequency, power, and water-soluble additives. *Journal of the American Chemical Society* 129 (18): 6031–6036.

Švorc, Ľ., Borovska, K., Cinková, K. et al. (2017). Advanced electrochemical platform for determination of cytostatic drug flutamide in various matrices using a boron-doped diamond electrode. *Electrochimica Acta* 251: 621–630.

Tabasideh, S., Maleki, A., Shahmoradi, B. et al. (2017). Sonophotocatalytic degradation of diazinon in aqueous solution using iron-doped TiO_2 nanoparticles. *Separation and Purification Technology* 189: 186–192.

Torres-Palma, R.A., Nieto, J.I., Combet, E. et al. (2008). Influence of TiO_2 concentration on the synergistic effect between photocatalysis and high-frequency ultrasound for organic pollutant mineralization in water. *Applied Catalysis B: Environmental* 80 (1–2): 168–175.

Torres-Palma, R.A., Nieto, J.I., Combet, E. et al. (2010). An innovative ultrasound, Fe^{2+} and TiO_2 photoassisted process for bisphenol a mineralization. *Water Research* 44 (7): 2245–2252.

Tufail, A., Price, W.E., and Hai, F.I. (2020). A critical review on advanced oxidation processes for the removal of trace organic contaminants: a voyage from individual to integrated processes. *Chemosphere* 260: 127460.

Turkay, O., Ersoy, Z.G., and Barışçı, S. (2017). The application of an electro-peroxone process in water and wastewater treatment. *Journal of the Electrochemical Society* 164 (6): E94.

Utzig, L.M., Lima, R.M., Gomes, M.F. et al. (2019). Ecotoxicity response of chlorpyrifos in *Aedes aegypti* larvae and *Lactuca sativa* seeds after UV/H_2O_2 and UVC oxidation. *Ecotoxicology and Environmental Safety* 169: 449–456.

Van, H.T., Nguyen, L.H., Hoang, T.K. et al. (2020). Heterogeneous Fenton oxidation of paracetamol in aqueous solution using iron slag as a catalyst: degradation mechanisms and kinetics. *Environmental Technology & Innovation* 18: 100670.

Vo, H.N.P., Ngo, H.H., Guo, W. et al. (2020). Poly- and perfluoroalkyl substances in water and wastewater: a comprehensive review from sources to remediation. *Journal of Water Process Engineering* 36: 101393.

Vogna, D., Marotta, R., Andreozzi, R. et al. (2004a). Kinetic and chemical assessment of the UV/H_2O_2 treatment of antiepileptic drug carbamazepine. *Chemosphere* 54 (4): 497–505.

Vogna, D., Marotta, R., Napolitano, A. et al. (2004b). Advanced oxidation of the pharmaceutical drug diclofenac with UV/H_2O_2 and ozone. *Water Research* 38 (2): 414–422.

Von Gunten, U. (2003). Ozonation of drinking water. Part I: Oxidation kinetics and product formation. *Water Research* 37 (7): 1443–1467.

Wadley, S. and Waite, T.D. (2004). Fenton processes applied to water and wastewater treatment. In: *Advanced Oxidation Processes for Water and Wastewater Treatment* (ed. Parsons, S.), vol. 4, 111–136. Cornwall, UK: IWA Publishing.

Wang, C. and Shih, Y. (2015). Degradation and detoxification of diazinon by sono-Fenton and sono-Fenton-like processes. *Separation and Purification Technology* 140: 6–12.

Wang, S., Wu, J., Lu, X. et al. (2019). Removal of acetaminophen in the Fe^{2+}/persulfate system: kinetic model and degradation pathways. *Chemical Engineering Journal* 358: 1091–1100.

Wang, W.L., Wu, Q.Y., Huang, N. et al. (2016a). Synergistic effect between UV and chlorine (UV/chlorine) on the degradation of carbamazepine: influence factors and radical species. *Water Research* 98: 190–198.

Wang, W.L., Wu, Q.Y., Huang, N. et al. (2018). Potential risks from UV/H_2O_2 oxidation and UV photocatalysis: a review of toxic, assimilable, and sensory-unpleasant transformation products. *Water Research* 141: 109–125.

Wang, Y., Shen, C., Li, L. et al. (2016b). Electrocatalytic degradation of ibuprofen in aqueous solution by a cobalt-doped modified lead dioxide electrode: influencing factors and energy demand. *RSC Advances* 6 (36): 30598–30610.

Wols, B.A., Hofman-Caris, C.H.M., Harmsen, D.J.H., and Beerendonk, E.F. (2013). Degradation of 40 selected pharmaceuticals by UV/H_2O_2. *Water Research* 47 (15): 5876–5888.

Wu, T.Y., Guo, N., Teh, C.Y., and Hay, J.X.W. (2013). Theory and fundamentals of ultrasound. In: *Advances in Ultrasound Technology for Environmental Remediation* (eds Wu, Y.T., Guo, N., Teh, C.Y. et al.), 5–12. Dordrecht: Springer.

Xiang, J., Wu, M., Lei, J. et al. (2018). The fate and risk assessment of psychiatric pharmaceuticals from psychiatric hospital effluent. *Ecotoxicology and Environmental Safety* 150: 289–296.

Xiang, Y., Fang, J., and Shang, C. (2016). Kinetics and pathways of ibuprofen degradation by the UV/chlorine advanced oxidation process. *Water Research* 90: 301–308.

Yamamoto, T., Noma, Y., Sakai, S.I., and Shibata, Y. (2007). Photodegradation of perfluorooctane sulfonate by UV irradiation in water and alkaline 2–propanol. *Environmental Science & Technology* 41 (16): 5660–5665.

Yang, Q., Ma, Y., Chen, F. et al. (2019). Recent advances in photo-activated sulfate radical-advanced oxidation process (SR-AOP) for refractory organic pollutants removal in water. *Chemical Engineering Journal* 378: 122149.

Yang, Y., Fu, J., Peng, H. et al. (2011). Occurrence and phase distribution of selected pharmaceuticals in the Yangtze Estuary and its coastal zone. *Journal of Hazardous Materials* 190 (1–3): 588–596.

Yang, Y., Jiang, J., Lu, X. et al. (2015). Production of sulfate radical and hydroxyl radical by reaction of ozone with peroxymonosulfate: a novel advanced oxidation process. *Environmental Science & Technology* 49 (12): 7330–7339.

Yang, Y. and Pignatello, J.J. (2017). Participation of the halogens in photochemical reactions in natural and treated waters. *Molecules* 22 (10): 1684.

Yao, W., Wang, X., Yang, H. et al. (2016). Removal of pharmaceuticals from secondary effluents by an electro-peroxone process. *Water Research* 88: 826–835.

Yuan, F., Hu, C., Hu, X. et al. (2009). Degradation of selected pharmaceuticals in aqueous solution with UV and UV/H_2O_2. *Water Research* 43 (6): 1766–1774.

Zangeneh, H., Zinatizadeh, A.A.L., and Feizy, M. (2014). A comparative study on the performance of different advanced oxidation processes (UV/O_3/H_2O_2) treating linear alkyl benzene (LAB) production plant's wastewater. *Journal of Industrial and Engineering Chemistry* 20 (4): 1453–1461.

Zhang, D., Gersberg, R.M., Ng, W.J., and Tan, S.K. (2014). Removal of pharmaceuticals and personal care products in aquatic plant-based systems: a review. *Environmental Pollution* 184: 620–639.

Zhang, N., Liu, G., Liu, H. et al. (2011). Diclofenac photodegradation under simulated sunlight: effect of different forms of nitrogen and kinetics. *Journal of Hazardous Materials* 192 (1): 411–418.

Zhang, Y. and Pagilla, K. (2010). Treatment of malathion pesticide wastewater with nanofiltration and photo-Fenton oxidation. *Desalination* 263 (1–3): 36–44.

Zhao, X., Du, P., Cai, Z. et al. (2018). Photocatalysis of bisphenol A by an easy-settling titania/titanate composite: effects of water chemistry factors, degradation pathway and theoretical calculation. *Environmental Pollution* 232: 580–590.

Zhao, Y., Kuang, J., Zhang, S. et al. (2017). Ozonation of indomethacin: kinetics, mechanisms and toxicity. *Journal of Hazardous Materials* 323: 460–470.

Zwiener, C.F.F.H. and Frimmel, F.H. (2000). Oxidative treatment of pharmaceuticals in water. *Water Research* 34 (6): 1881–1885.

Part IV

Various Physico-Chemical Treatment Techniques of Microconstituents

16

Aerobic Biological Treatment of Microconstituents

Hung-Hsiang Chen[1], *Thi-Manh Nguyen*[1], *Ku-Fan Chen*[1], *Chih-Ming Kao*[2], *Rao Y. Surampalli*[3], *and Tian C. Zhang*[4]

[1]*Department of Civil Engineering, National Chi Nan University, Puli, Nantou County, Taiwan*
[2]*Institute of Environmental Engineering, National Sun Yat-sen University, Kaohsiung, Taiwan*
[3]*Global Institute for Energy, Environment and Sustainability, Lenexa, KS, USA*
[4]*Civil & Environmental Engineering, College of Engineering, University of Nebraska, Lincoln, Omaha, NE, USA*

16.1 Introduction

Modern technology brings a high quality of life to humans, which not only improves the convenience of life, but also greatly prolongs the human lifespan. Fot example, the invention of antibiotics has freed humans from some diseases and so increased their quality of life and lifespan. The emergence of plastic has replaced some heavier raw materials, such as metal and wood, etc., making life more convenient. However, many of these materials and drugs are shown as microconstitutes (MCs) once discharged into the environment, and many of them cause harm to the environment and humans. Contaminants of emerging contaminants (CECs) are mainly synthetic organic chemicals recently discovered in the environment that may damage organisms at environmentally relevant concentrations (Ahmed et al. 2015). CECs mainly include pharmaceutical organic pollutants, pharmaceuticals and personal care products (PPCP), endocrine disrupting compounds (EDC), surfactants, polycyclic aromatic hydrocarbons (PAH), perfluorinated substances, microplastics (MPs), pesticides, and nanomaterials (Naidu et al. 2016; Tijani et al. 2016). Due to their hydrophobic properties, CECs can bioaccumulate in lipid-rich tissues of organisms, causing injury to the endocrine systems of humans and animals (Rodriguez-Narvaez et al. 2017).

Endocrine disrupting chemicals (EDCs) have been linked to endometriosis and prostate, testicular, and breast cancer (Zlatnik 2016; Rehman et al. 2018; Street et al. 2018). PPCPs, including analgesics, antibiotics, stimulant drugs, lipid modifiers, diuretic preservatives, sunscreens, antimicrobials, and cosmetics, are associated with reduced reproductive health and increased antimicrobial resistance, and increase the burden on ecosystems and are a threat to the health of lifeforms (Ahmed et al. 2017; Sharma et al. 2019). MPs are tiny plastic particles with an upper size limit of <5 mm in diameter (Wu et al. 2021). Common types of MP polymers found in aquatic environments are

Microconstituents in the Environment: Occurrence, Fate, Removal, and Management, First Edition.
Edited by Rao Y. Surampalli, Tian C. Zhang, Chih-Ming Kao, Makarand M. Ghangrekar, Puspendu Bhunia, Manaswini Behera, and Prangya R. Rout.

polyethylene (PE), polypropylene (PP), polystyrene (PS), polyvinyl chloride (PVC), polyacrylamide (PAM), and polyethylene terephthalate diol ester (PET) (Eerkes-Medrano et al. 2015; Di and Wang 2018). Previous studies have shown that MPs can act as carriers of harmful substances, including persistent organic pollutants (POPs) and heavy metals (Brennecke et al. 2016; Rodrigues et al. 2019). The pollution sources mentioned above all contain MCs. The existence of MCs in wastewater is a serious problem, and their removal has also become an issue that has been receiving attention recently.

This chapter overviews aerobic biological processes in wastewater treatment applications for removal of MCs. After briefly introducing the high- and low-rate aerobic biological treatment systems, this chapter describes the performance of different aerobic biological systems/processes for treatment of MCs, followed by discussions about biodegradation of some selected MCs. Finally, the current challenges and prospects in aerobic biological treatment of MCs are outlined. This chapter potentially provides insight into the development of next-generation technologies for removal of MCs in aquatic environments.

16.2 Aerobic Biological Systems/Processes

The secondary treatment of wastewater is mostly carried out by biological treatment, which can be divided into aerobic and anaerobic decomposition. These two terms are directly related to the type of microorganisms that are involved in the degradation of organic impurities in wastewater and the operating conditions of the bioreactor (Mittal 2011). Aerobic biological treatment utilizes the oxidative decomposition and transformation functions of microorganisms (e.g., bacteria, algae, fungi) for rapid biodegradation of organic matter in the presence of free dissolved oxygen (Eq. 16.1) (Metcalf et al. 2014; Samer 2015). Traditionally, there are two major categories of aerobic biological treatment systems, high-rate systems and low-rate systems. Most of centralized WWTPs use high-rate systems, while low-rate systems are more popular in rural areas (Metcalf et al. 2014; Ngo et al. 2016; Surampalli et al. 2018).

$$\text{Organic matter} + O_2 \xrightarrow{\text{Aerobic bacteria}} \text{New cell} + \text{Energy} + H_2O + CO_2 + \text{by product} \quad (16.1)$$

16.2.1 High-Rate Systems

High-rate systems can be classified as suspended growth and attached growth. Table 16.1 shows the comparison of suspended and attached growth processes. Sometimes, a combination of suspended and attached growth biological treatment is used.

16.2.1.1 Suspended Growth Processes

Suspended growth processes are biological treatment methods in which the inflowing wastewater or sludge is fully in contact with the microorganisms by aeration or mechanical stirring when the microorganisms are suspended in the aeration tank to carry out biological reactions. The activated sludge process (ASP) is the most widely used aerobic biochemical treatment technology, which is mainly composed of aeration tanks, sedimentation tanks,

Table 16.1 Comparison of suspended and attached growth processes.

	Suspended Growth Processes	Attached Growth Processes
The way of microbial growth.	Use stirring or aeration to suspend microorganisms so they can grow in the tank.	Makes microorganisms to form biofilms and attach to the medium.
Space buffer.	Due to directly facing the influent water, there is no buffer space, and so is sensitive to toxicity.	The buffer space can be increased by the presence of the medium, and the tolerance value of toxicity is higher.
Sedimentation tank.	It is necessary to set up, and the operation of returning sludge.	May or may not need to be set up.
Current status of use and design.	It is more widely used.	It still needs continuous research and development.
Sludge volume.	More	Less

aeration systems, and sludge return systems (Figure 16.1). After pre-treatment, the wastewater flows into the aeration tank and contacts with the activated sludge growing in suspension, and then the sludge is separated from the treated water in the sedimentation tank. To separate the solids retention time from the hydraulic retention time, a portion of the separated sludge is continuously returned to the aeration tank. To operate the system at steady state conditions, a portion of the sludge is discarded from either the final clarifiers or aeration basins, and the wasted sludge needs to be treated in a sludge treatment system. Depending on the feeding, aeration, and operational methods or reactor types, activated sludge systems have many different alternatives, such as conventional plug-flow, completely-stirred tank reactors, tapered aeration, step aeration, extended aeration, oxidation ditch method, sequencing batch reactor (SBR), deep aeration, pure oxygen, contact-stabilization, and membrane bioreactor process (MBR).

The suspended growth activated sludge processes may be used to remove nutrients (e.g., N and P). The nitrogen and phosphorus in urban wastewater mainly come from domestic

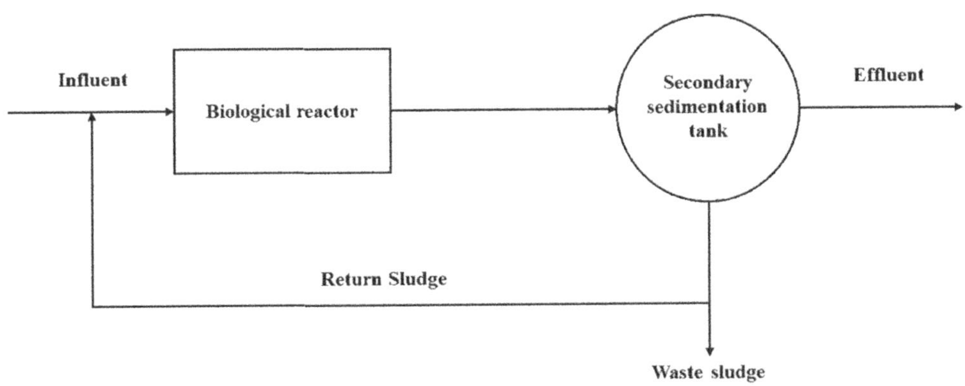

Figure 16.1 Schematic of activated sludge process.

wastewater and partly from industrial wastewater. These nutrients seriously affect the normal function of the urban water environment. The main hazards of nitrogen and phosphorus is to cause receiving water eutrophication, affect the quality of the water resource, increase the cost of water treatment, and cause toxicity to humans and other organisms. Many alternative processes are available for nutrient removal in addition to removal of organic matter, such as the anoxic/aerobic (A/O) process, step-feed A/O process, intermittent aeration (oxidation ditch), SBRs, post-anoxic denitrification with addition of external carbon (e.g., CH_3OH), simultaneous and cyclic nitrification and denitrification (SNdN and CNdN), modified Ludzack–Ettinger (MLE), 4-stage Bardenpho, Bio-denitro™, and Orbal™ (Metcalf 2014). Figure 16.2 shows an Anaerobic/Anoxic/Oxic (A/A/O) process.

The aerobic biological process is the most useful method for the removal of pollutants and nutrients from wastewater because of its cost-effectiveness and high treatment efficiency. To date, conventional ASPs are the most common way to treat municipal and industrial wastewater. However, treating the wastewater by ASPs shows some disadvantages, such as high sludge production and limited removal efficiency for many organic and inorganic pollutants (Zhao et al. 2019). To overcome the drawbacks of ASPs, membrane bioreactors (MBRs) have been considered as a potential treatment technology to remove microbial and chemical pollutants in wastewater due to their high removal efficiencies, especially for suspended solids, nutrients, fecal coliforms, pathogens, and organic micropollutants (Luo et al. 2014; Tran et al. 2016; Le et al. 2018).

16.2.1.2 Attached Growth Processes

Attached growth processes are wastewater treatment methods that simulate the natural purification of the soil in nature. In attached growth biological treatment processes, free-form microorganisms are attached to certain carriers, such as synthetic materials, fibers, natural materials, pebbles, etc., to form a film-like biological sludge (biofilm). When the biofilm comes into contact with organic pollutants, the pollutants are absorbed by microorganisms and become substrates for microbial growth. Many different types of biofilm reactors have been used for wastewater treatment, such as trickling filters, rotating disc contactors, submerged beds with biofilm particles immersed in the liquid (up-flow or down-flow), fluidized beds, moving beds, membrane reactors, and sequencing batch reactors (Metcalf et al. 2014). It should be noted that because the biofilm is usually relatively thick (e.g., >100 to 200 μm), the biofilm processes often involve both aerobic and anaerobic respiration mechanisms.

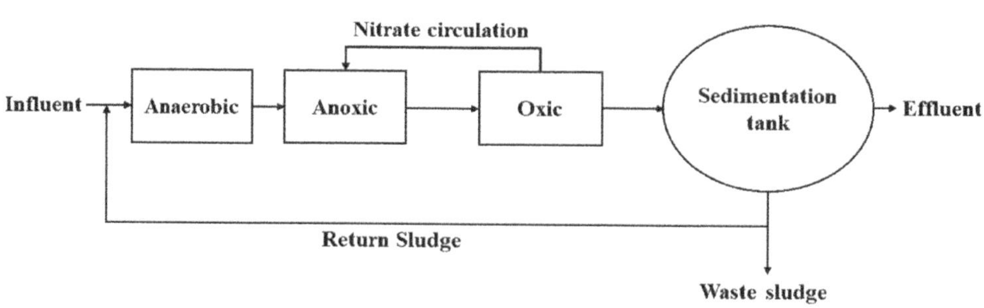

Figure 16.2 Schematic of the Anaerobic/Anoxic/Oxic (A/A/O) process.

16.2.2 Low-Rate Systems

Low-rate systems usually belong to decentralized wastewater treatment systems, such as stabilization ponds, anaerobic/aerated lagoons, land treatment systems, wetland systems, vermifiltration/vermicomposting, and different structural BMPs (Best Management Practices) (Surampalli et al. 2018). For example, constructed wetlands (CWs) have become very popular over the last few years for the removal of CECs from domestic, industrial, agricultural wastewaters, leachate, urban runoff, and contaminated groundwater, mainly due to their cost-effectiveness, easy construction, and environmental friendliness (Chowdhury et al. 2021). Vermifiltration (VF) of the wastewater (Shokoohi et al. 2020) and vermicomposting (VC) of sewage sludge (Huang et al. 2018) are two economical and environmentally-friendly technologies that have gained attention for CECs treatment (Chowdhury et al. 2021).

16.3 Removal of CECs By Different Aerobic/Anoxic Treatment Processes

Table 16.2 shows the degradation of some CECs with aerobic biological processes. This section describes some studies on how different aerobic/anoxic treatment processes remove CECs.

Table 16.2 Degradation of CECs with aerobic biological systems/processes.

Treatment	Contaminant	Initial Conc. (μg/L)	Removal (%)	References
Wastewater treatment plants.	Estradiol-3-glucuronide Tetracycline.	0.05 0.006–0.05	100 66–90	Lin et al. 2009; Chen et al. 2018
Membrane bioreactor.	Androsterone and Naproxen.	1.2 50	98 86–89	Tambosi et al. 2010; Monsalvo et al. 2014
Batch reactors.	LAS and SAS[a] ibuprofen.	2000 1000	86.2±9.4 14.9	Motteran et al. 2022
Sequencing batch reactors.	17β-estradiol and 17α-ethinyl estradiol.	500	> 95	Racz et al. 2012
Activated sludge process.	Cephalexin.	2	96	Costanzo et al. 2005
Constructed wetlands.	Many kinds.	0.1–110	20–99	Chowdhury et al. 2021
Aerobic granular sludge cultivated in a SBR.	Tetracycline.	300	92	Wang et al. 2021

(Continued)

Table 16.2 (Continued)

Treatment	Contaminant	Initial Conc. (μg/L)	Removal (%)	References
Vermifiltration.	Ciprofloxacin. SMZ Tetracycline.	–	86–98 54–73 87–96	Shokoohi et al. 2020
Aerobic granular bioreactor.	2-fluorophenol.	12,400	100	Ramos et al. 2017
Trickling filters and biofilm reactors.	PPCPs.	–	< 70	Kasprzyk-Hordern et al. 2010
Moving bed biofilm reactor.	PPCPs and others.		80–94 for PPCPs; 0–28 for others.	Zupanc et al. 2013
High-rate algae ponds.	12 PPCPs and 26 intermediates.		40–60	García-Galán et al. 2021

[a]LAS = linear alkylbenzene sulfonates; SAS = secondary alkyl sulfonate.

16.3.1 ASPs

Aerobic biological treatments have been widely used to remediate contamination in wastewater. In general, ASPs have an efficacy of 30–50% for pharmaceuticals and 40% for some personal care products (Prasad et al. 2019). Ben et al. (2018) investigated the occurrence, removal, and risk of 42 CECs (30 PPCPs + 12 EDCs) in 14 WWTPs distributed across China and found that different MP categories showed similar distributions among the WWTPs studied. Of all the secondary treatment processes studied, the A/A/O process combined with a moving-bed biofilm reactor achieved the highest CEC removal. The dominant CECs in the influent, effluent, and excess sludge were phenolic estrogenic compounds (PEs), macrolides, and fluoroquinolones. Tetracyclines, bezafibrate, caffeine, steroid estrogens, and PEs showed high and stable removal efficiencies, whereas other CECs showed varied removal efficiencies. Diclofenac, an anti-inflammatory drug, is often found in the wastewater. Elshikh et al. (2022) used ASP to treat diclofenac in wastewater. The adsorption efficiency of diclofenacc in wastewater at an initial concentration of 1 mg/L was 80%. The adsorption of diclofenac on the sludge sample was time dependent. Additionally, pH affected diclofenac adsorption, and the maximum removal was observed at the acidic pH in the medium.

16.3.2 Removal of CECs By Different Aerobic/Anoxic Treatment Processes

Suarez et al. (2010) used the conventional activated sludge reactor to operate under nitrifying (aerobic) and denitrifying (anoxic) conditions for over 1.5 years and evaluated the reactor removal efficiency for 16 PPCPs. The results show that 11 of the PPCPs (17β-estradiol, estrone, ethinylestradiol, ibuprofen, galaxolide, tonalide, ccelestolide, fluoxetine, roxithromycin, erythromycin, naproxen) have the removal efficiency of >85% with the aerobic reactor. Although it did not have a better removal effect than the aerobic reactor, the anoxic reactor still has a good effect on the removal of galaxolide, tonalide, celestolide, and natural estrogens (>70%) (Suarez et al. 2010). According to the above examples, both aerobic and anoxic treatment have a good performance for the treatment of CECs.

In general, the organic nitrogen in wastewater is converted into ammonia gas through bacterial decomposition and hydrolysis, and then oxidized to nitrite and nitrate by autotrophic bacteria via the so-called nitrification process. In the anoxic environment, organic carbon is used as an electron donor to convert nitrate into nitrogen via the so-called denitrification process. Sun et al. (2019) evaluated the removal of ibuprofen and triclosan simultaneously with a solid-phase denitrification (SPD) system. After 602 days, the results show that the removal efficiency of ibuprofen and triclosan are 79.69 ± 6.35% and 65.96 ± 7.62%, respectively, under stable influent conditions of 50 μg/L^{-1}. In this system heterotrophic denitrifying bacteria and ammonia oxidizing bacteria have important functions for the biodegradation of ibuprofen and triclosan (Sun et al. 2019).

16.3.3 MBR and Membranes Technology

Among biological treatment methods, MBR is one of the most effective methods for removing CECs, as MBR has a smaller pore size (0.01–5 μm) for filtration, thereby preventing the passage of CECs (Meng et al. 2017). Lares et al. (2018) reported the use of MBR in a WWTP in Finland to remove 60.0% of CEC from conventional activated sludge effluent with a total CEC removal rate of 98.3%. A variety of emerging contaminants, their occurrence in aquatic environments, their health effects, and their removal by membrane processes has been summarized (Zhang et al. 2012).

16.3.4 ASPs and/or Trickling Filters

Kasprzyk-Hordern et al. (2010) determined the fate of 55 PPCPs, EDCs, and illicit drugs in two contrasting WWTPs (ASPs and trickling filters). The impact of treated wastewater effluent on the quality of receiving waters was also assessed. PPCPs were found to be present at high loads reaching 10 kg/d in raw sewage, and their removal efficiency was found to be strongly dependent on the technology implemented in the WWTPs. In general, the WWTP with trickling filters resulted in, on average, <70% removal of all 55 PPCPs studied, while that with ACPs gave a much higher removal efficiency of >85%. The treated wastewater effluents were the main contributors to PPCPs concentrations (up to 3 kg of PPCPs/d) in the rivers studied. Because the WWTP effluents were also major contributors to river flow (dilution factor for the studied rivers did not exceed 23 times), the effect of WWTP effluent on the quality of river water is significant and cannot be underestimated.

16.3.5 Lagoons and Constructed Wetlands

Matamoros et al. (2015) used two pilot high-rate algal ponds (HRAPs) to remove 26 CECs from urban wastewater. The removal efficiency of the CECs ranged from negligible removal to more than 90%, depending on the compound. CECs removal efficiencies were enhanced during the warm season. Biodegradation and photodegradation are the most important removal pathways, whereas volatilization and sorption were solely achieved for hydrophobic compounds (log $K_{ow} > 4$) with moderately high Henry's law constant values (11–12 $Pa/m^3 mol$) such as musk fragrances. As a green, low-cost, and more sustainable technology, HRAPs can be a feasible alternative to ASPs in terms of overall wastewater treatment, and will yield similar removal efficiency of PPCPs with and without primary treatment (García-Galán et al. 2021). Saptarshi et al. (2019) reported that fungal reactors can effectively treat selective PPCPs with 95–100% efficacy and microalgal systems can treat a wide array of contaminants, even at concentrations as high as 24 μg/L. They are considered better for treatment of high-load PPCPs.

Guerra et al. (2019) reported that removal of Triclosan (TCS) (an antimicrobial agent used in many PPCPs and cleaning products) from lagoons as well as secondary and advanced treatment facilities were much higher than primary treatment facilities ($p < 0.001$). TCS removal was strongly correlated with organic nitrogen removal and was higher during summer compared with winter. Aerobically digested solids contained the lowest levels (median 555 ng/g), while anaerobically digested primary solids contained the highest levels of TCS (median 22,700 ng/g). Thus, TCS is consistently present in wastewater and biosolids at relatively high concentrations and that removal from wastewater and levels in biosolids are strongly influenced by the wastewater and solids treatment types. Bakare and Adeyinka (2022) reported that the concentration of TCS ranged from 1.906 to 73.462 μg/L, from 1.732 to 6.980 μg/L, and from 0.138 to 2.455 μg/kg in influent, effluent, and sludge samples, respectively.

Chowdhury et al. (2021) reviewed the potential of the CWs and the earthworm-based treatment technologies (VF and VC) to remove CECs. They reported that photolytic degradation, phytoremediation, sorption, and microbial degradation govern the removal of CECs in the CWs, and the predominant EC removal pathway in a particular CW mainly depends on the configuration of the CWs. They reported that VF and VC have some additional advantages over CWs, such as natural aeration and CEC uptake by the earthworms. Compared with the conventional high-rate methods, CWs and VF reduce the level of toxicity imposed by the CECs to a higher extent, and help to reduce the bioavailability of CECs. However, one of the major disadvantages for CW and VF technologies is the requirement of a large footprint.

16.3.6 Mixed Technologies

New technologies have been developed worldwide to treat CECs in combined systems. Materials such as biochar (BC), hydrochar (HC), and digestate (DG) have been used together with biological wastewater treatment technologies for CEC removal (Loffredo 2022). Malamis et al. (2009) have examined the removal of Cr(III) by adopting a system combining a membrane bioreactor (MBR) together with low-cost mineral additives (zeolite, bentonite, vermiculite). Ultrafiltration membranes without added minerals could remove significant amounts of Cr(III) from the final effluent with removal efficiencies between 43.2 and 69.2%. With the addition of 10 g/L zeolite, vermiculite, and bentonite, the removal efficiencies were 81.5, 95.2, and 99.5%, respectively. Thus, effective removal of chromium was achieved by the MBR system combined with low-cost minerals. The integration of VF or the CW technologies with other technologies, such as solar-driven photocatalysis (which is economical compared with UV photocatalysis) or microbial fuel cells, has been proposed (Chowdhury et al. 2021).

16.4 Aerobic Biodegradation of Selected CECs

16.4.1 Hormones and Their Conjugates

The primary and secondary treatment of conventional WWTPs may or may not be efficient for hormone removal (Table 16.3), with removal efficiency of 17ß-estradiol (E_2) and estrone (E_1) being 85–99 and 25–80%, respectively (Khanal et al. 2006). Although nitrification and other advanced treatment processes (e.g., activated carbon adsorption, chemical oxidation, membrane filtration) can remove hormones more efficiently, their costs are relatively high (Vader et al. 2000; Alum et al. 2004; Westerhoff et al. 2005; Khanal et al. 2006). The maximum concentrations of hormones in 12 municipal WWTP effluent (in ng/L) were 12 for E_1, 4 for E_2, 6.1 for testosterone, and 4.5 for androstenedione (Kolodziej et al. 2003). Biodegradation of hormones are inhibited by the presence of antibiotics such as sulfamethazine, tylosin, and chlorotetracycline (Hemmings and Hartel 2006; Khanal et al. 2006). Yu et al. (2022) reported that most progestogens were effectively eliminated in the aerobic unit, with biodegradation being the primary removal pathway.

It should be noted that a majority of excreted hormones are in conjugated forms (e.g., 96% for swine and 46% for cattle) (Lange et al. 2002; D'Ascenzo et al. 2003; Shore and Shemesh 2003; Shrestha 2011). Up to now, most of the studies on estrogen conjugates have focused on WWTPs (e.g., Reddy et al. 2005). Deconjugation efficiencies for estrogen glucuronides in WWTPs are usually very high, but can vary over a wide range (Shrestha 2011). Yost et al. (2013) reported that hormones and their conjugates in the lagoon solid phase had significantly higher concentrations than that in the lagoon liquid (slurry) phase (Table 16.2). Dissipation of conjugated hormones is commonly accepted (Yost et al. 2014). Usually, glucuronides can easily transform into their free estrogen forms, but sulfate conjugates are more recalcitrant, persisting longer in septic tanks and manure storage ponds (Hutchins et al. 2007; Shrestha 2011; Goeppert et al. 2015).

Table 16.3 Summary of concentrations of hormones and their conjugates at different phases from source to the environment of selected studies.[a]

Sampling Site or Material	Detected Compounds and Concentration (ng/L)	Remarks	References
WWTPs	Influent: E3 = 100; E2 = 1.9; EE2 = 19; E1 = 100; 2OH-E1 = 13; 16OH-E1 = 125; E1-3G = 3.1; E1-3S = 3.9 Effluent: E3 = 0.5; EE2 = 5; E1 = 5; E1-3G = 2.5; E1-3S = 2.9	Simple volume = 100 mL for influent and 250 mL for effluent.	Gentili et al. (2002)

(Continued)

Table 16.3 (Continued)

Sampling Site or Material	Detected Compounds and Concentration (ng/L)	Remarks	References
• WWTPs; • Tamagawa River; • Lake Kasumigaura.	• E1-3S = 2.2 and 0.3; E2-3S = 1.0 and LDL; E1 = 34 and 2.5; β-E2 = 2.5 and 0.3 • E1-3S = 0.3–0.8; E2-3S = 0.2–0.8; E1 = 3.4–6.6; β-E2 = 0.8–1.0 • E1-3S = 0.3–0.8; E2-3S = LDL-0.4; E1 = 0.2–0.8; β-E2 = LDL	• 2 WWWTPs; • Three samples from Tamagawa River; • Samples from 5 different locations.	Isobe et al. (2003)
WWTPs	Influent: E1-3S = 34.1(4.7); E2-3S = 3.2(7.6); E1-3G = 0.4(15.5); E3-3G = 0.3(13.3); E1 = 24.2; E2 = 7.6 Effluent: E1-3S = 0.29(0.3); E2-3S = LDL; E1-3G = LDL; E3-3G = LDL; E1 = 0.71; E2 = 0.26	Data = Average (Standard Deviation). Measured concentrations of steroid sulfates in the influent to a WWTPs were ~100 times greater than that of the respective steroid glucuronides.	Reddy et al. (2005)
Lagoons handling swine, poultry, and cattle CAFO wastes.	E1-3S = 2–91; E2-3S = 8–44; E2α-3S = 141–182; E2-17S = 72–84; Total free estrogen levels (E1, E2, E3, 17α-estradiol) in whole lagoon samples of swine PL = 1000–21,000; poultry PL = 1800–4000; dairy SL = 370–550, and beef SL = 22–24	In dairy waste lagoons, 57% of the total estrogens were detected as conjugates; in poultry lagoons, ~95% estrogens were conjugates. PL = primary lagoon, SL = secondary lagoon.	Hutchins et al. (2007)
WWTPs secondary effluent.	E1 = 5.9–60.0; 2-ClE1 = ND-4.0; 4-ClE1 = ND-14.5; 2,4-diClE1 = ND-9.8	Disinfection by hypochlorite. Three samples were analyzed between May and December of 2005.	Nakamura et al. (2007)
WWTPs final effluent.	E3-3G = 0.5–2.7; E1-3G = 5.2–18; E2-3G = ND-2.0; EE2-3G = ND-1.9; E1-3S = 0.5–1.8; E2-3S = ND-0.88 EE2-3S = 0.11–0.42; E3 = 2.9–11; monoClE3 = ND-0.26; E = 10–24; E2 = 3.1–9.0; EE2 = 3.0–7.0; diClE3 = ND-9.8; monoClE1 = ND-1.8; monoBrEE2 = ND-0.85; monoBrE2 = 21–49; monoClEE2 = ND-5.6; monoClE2 = ND-0.96; diClE1 = ND-0.38; diClEE2 = ND-0.48; diClE2 = 6.7–22; diBrEE2 = ND-9.9; diBrE2 = ND-7.6	ND = no peak was found. Measurements of 8 grab and/or 24 h composite samples of 4 days between March and May. Free estrogen concentrations were consistent with other WWTPs; sulfate conjugates were lower than previous studies; and halogenated forms were high.	Griffith et al. (2014)

[a] E3 = Estriol; E2 = 17ß-estradiol; EE2 = 17α-ethinylestradiol; E1 = Estrone; E1-3G = Estriol-3-glucuronide; E1-3S = Estrone-3-sulfate; E3-3G = Estriol-3-glucuronide; EE2-3G = Ethinylestradiol-3-glucuronide; EE2-3S = Ethinylestradiol-3-sulfate; CAFO = Concentrated animal feeding operations.

16.4.2 Nanoparticles (NPs) and Nanomaterials (NMs)

Increasing use of inorganic nanoparticles in consumer products and industrial processes has resulted in greater presence of NPs/NMs in wastewater. NMs made of biodegradable materials will naturally be the best candidates for biotransformation. For example, C60 and C70 fullerenes were reported to be metabolized by wood decay fungi after 12 weeks (Filley et al. 2005). Biotransformation or biodegradation may induce transformation of the NMs or their alteration products as reported. Designing and applying biodegradable NMs as remedial agents in groundwater and wastewater management systems can be a challenge (Brar et al. 2015).

Under typical conditions, >90% of the NMs may attach to biomass and then be removed in the WWTP (Westerhoff et al. 2013). Chen and Bergendahl (2021) reported that after going through wastewater treatment processes, a significant inorganic NP mass remains in the effluent, ranging from 10^3 to 10^6 particles/mL, with NP sizes ranging from ~20 to 120 nm. Some species of plants and fungi can accumulate (and thus remove) NPs in the environment (e.g., water, air, soil) (Sánchez et al. 2011). Westerhoff et al. (2013) reported that some metallic NMs (e.g., silver-, zinc-, copper-based NPs) may dissolve, while some others (e.g., fullerenes) can biodegrade in wastewater and subsequently be adsorbed to settable biomass and then be removed from the wastewater. However, we need to be careful while handling sludge with absorbed NMs/NPs.

The probable role of biofilms in the removal of NMs in WWTP has been considered. Peulen and Wilkinson (2011) reported that NPs diffused into biofilms, and the diffusion coefficients decreased exponentially with the square of the radius of NPs. Sheng and Liu (2011) reported that extracellular polymers on the biofilm and biofilm diversity were important in controlling the antimicrobial effects of Ag-NPs. Grün et al. (2018) reported that Ag-NPs led to an altered biofilm community composition due to the displacement of putatively AgNP-sensitive bacterial taxa *Actinobacteria*, *Chloroflexi*, and *Cyanobacteria* by taxa known for their enhanced adaptability toward metal stress, such as *Acidobacteria*, *Sphingomonadales*, and *Comamonadaceae*. These results cause serious concerns with respect to the broad application of AgNPs and their potentially adverse impact on ecosystems.

Ladner et al. (2012) reported that commercially available functionalized nano-oxides (positively and negatively charged) can pass through a range of microfiltration and ultrafiltration membranes. While negatively charged NPs were less well rejected, >99% positively charged NMs were rejected by negatively charged membranes. Westerhoff et al. (2011) reported that microfiltration clarification was more than 10% efficient in removing TiO_2 than the full-scale WWTPs that uses gravity secondary settling. Membranes were used to recover nanoscale iron oxides and titanium dioxide (Brar et al. 2015). The immobilization of NPs on a solid support has been used to remove toxic metal ions (Mn^{2+}, Cu^{2+}, Ni^{2+}, $Co^{2+\cdot}$ Ag+) present in waste water (Zhang et al. 2009, 2012; Zhang 2011).

16.4.3 Microplastics

Chen et al. (2022b) reviewed various methods for separation and degradation of micro- and nano-plastics from urban waters. Ebrahimbabaie et al. (2022) reviewed biodegradation of micro- and nano-plastics, which often occurs in multiple phases. They concluded that the associated mechanisms are often complex, and relevant pathways have not been clearly delineated. Many factors can affect the biodegradation of micro- and nano-plastics, such as abiotic factors (e.g., hydrophilicity, crystallinity, chemical composition), biotic factors (e.g., enzymatic activity, metabolic activity), and environmental factors (e.g., humidity, light, temperature) (Ebrahimbabaie et al. 2022; Chen et al. 2022b). While microorganisms (e.g., bacteria, fungi, algae) and invertebrates (e.g., insects, snails) enable the mineralization of diverse micro- and nano-plastics, the biodegradation usually takes months or years. Song et al. (2020) studied the interaction between microplastics (PP, PE, PET, PVC) and microalgae (*Chlorella* sp. L38 and *Phaeodactylum tricornutum* MASCC-0025) and found an obvious inhibition effect of microplastics on *Phaeodactylum tricornutum* MASCC-0025 growth with an inhibition ratio up to 21.1%. By contrast, *Chlorella* sp. L38 presented strong adaptive capacity to microplastics. The toxic effect might be explained by the possible leaching of additives of four tested microplastics. They concluded that microalgae have a potential to be used as an alternative bio-solution for microplastics treatment. Chen et al. (2022a) reported that both traditional petroleum-based and emerging biodegradable microplastics promoted sedimentary N_2O production by affecting the microbes associated with nitrification and denitrification processes. This work implicates that the emerging biodegradable microplastics represented by polylactic acid may have a greater potential to enhance estuarine N_2O emissions and accelerate global climate change.

16.5 Challenges and Future Perspectives

Microconstituents (MCs) cause a huge impact on the environment (Boxall Alistair et al. 2012; Bickley et al. 2017; Dong et al. 2020; Hwang et al. 2020). Biological treatments use different biological organisms or biological processes to eliminate MCs. Biological treatment is extensively used because it is more cost-effective than chemical or physical treatments (Samer 2015). This method utilizes nematodes, bacteria, or other organisms to decompose organic waste (Mani et al. 2020). These treatments typically take place at the secondary or tertiary stage of treatment and are designed to significantly remove contaminants through biodegradation.

However, it has been reported that some non-biodegradable organic micropollutants cannot be adequately removed using biological treatment processes. WWTPs are often seen as a defense against contaminants entering the environment; however, they are also one of the most serious sources of CEC contamination. Studies have shown that the removal efficiency of CECs in traditional wastewater treatment plants can reach 97%. The main concern lies with the production of metabolites that might be more toxic than parent molecules or significant amounts of CECs being transported into wasted sludge (Lajayer et al. 2022).

Several studies showed that CECs inhibited EPS secretion in sludge and reduced the EPS protein, humic acid, and fatty acid content. Some CECs also show varying degrees of inhibition of nitrification and denitrification reactions, resulting in reduced wastewater treatment efficiency in bioreactors and increased production of residual sludge (Zhang et al. 2020). Kruglova et al. (2014) indicated that carbamazepine shows no biodegradation in laboratory-scale sequencing batch reactors (SBR), and the concentration of the substance is reduced to the background level each time by rinsing with sewage. Ben et al. (2018) reported that sulfamethoxazole, ofloxacin, ciprofloxacin, clarithromycin, erythromycin, estrone, and bisphenol A in the effluent, as well as β-estradiol 3-sulfate in the excess sludge, could pose high risks. Since most CECs are retained in wastewater sludge, they may also cause damage to the terrestrial environment (Gherghel et al. 2019).

Due to the non-biodegradability, toxicity, and structural complexity of CECs, the current treatment technologies used in WWTPs cannot effectively eliminate CECs in water at low levels (Sheng et al. 2016; Alvarino et al. 2018). In this case, different treatment methods are combined to improve the removal of CECs, such as biological treatment methods combined with advanced oxidation processes (e.g., membrane bioreactors + ozonation + constructed wetlands + UV irradiation, etc.) and membranes with biological treatment separation processes (e.g., membrane bioreactors + reverse osmosis, etc.) (Dhangar and Kumar 2020; Rathi et al. 2021). While many remediation methods exist to remove CECs from water, there is still considerable scientific interest in the effects of fungi in wastewater bioremediation. Some studies describe the ability of white-rot fungi (WRF) to treat EDC. Zhang et al. (2012) designed a novel plate bioreactor to inoculate white-rot fungi into the reactor and conduct carbamazepine removal experiments. Carbamazepine is difficult to biodegrade under either aerobic or anaerobic conditions in WWTPs, and its removal rate is mostly <10%. Long-term continuous operation (~100 days) effectively eliminates carbamazepine (60–80%) with sufficient nutrient supply. However, the disadvantage of using fungal reactors to remove pollutants is that the residual chemical oxygen demand (COD) concentration in the treated effluent is relatively high. Although these methods show good potential for removing CECs, they are not economical in terms of energy consumption and high cost participation (Dhangar and Kumar 2020). Therefore, there are still many challenges in improving the drawbacks of aerobic biological treatment systems for removing CECs.

16.6 Conclusions

Removing MCs from wastewater to achieve reuse of the wastewater is important. Aerobic biological treatment is one of most commonly used technology for mitigating MCs in wastewater. The use of microorganisms to treat pollutants is also a bio-friendly alternative. Although many studies point to aerobic treatment as a useful way to remove CECs in wastewater, there are still limitations of biological treatment to remove MCs completely. MCs are frequently detected in the effluent of conventional wastewater treatment plant using biological treatment. Therefore, it is necessary to either improve the efficacy of existed WWTPs or apply newly developed biological technologies to remove MCs in wastewater. There should be more in-depth research in the future to solve the problem of MCs. More practical and cost-effective technologies are needed for the treatment of MCs to achieve sustainable use of water resource.

References

Ahmed, M.B., Zhou, J.L., Ngo, H.H., and Guo, W. (2015). Adsorptive removal of antibiotics from water and wastewater: progress and challenges. *Science of the Total Environment* 532: 112–126.

Ahmed, M.B., Zhou, J.L., Ngo, H.H. et al. (2017). Progress in the biological and chemical treatment technologies for emerging contaminant removal from wastewater: a critical review. *Journal of Hazardous Materials* 323: 274–298.

Alum, A., Yoon, Y., Westerhoff, P., and Abaszadegan, M. (2004). Oxidation of BPA, E_2 and EE_2 and by0product estrogenicity. *Environmental Toxicology* 19: 257–264.

Alvarino, T., Suarez, S., Lema, J., and Omil, F. (2018). Understanding the sorption and biotransformation of organic micropollutants in innovative biological wastewater treatment technologies. *Science of the Total Environment* 615: 297–306.

Bakare, B.F. and Adeyinka, G.C. (2022). Occurrence and fate of triclosan and triclocarban in selected wastewater systems across Durban Metropolis, KwaZulu-Natal, South Africa. *International Journal of Environmental Research and Public Health* 19 (11): 6769.

Ben, W.W., Zhu, B., Yuan, X.J. et al. (2018). Occurrence, removal and risk of organic micropollutants in wastewater treatment plants across China: comparison of wastewater treatment processes. *Water Research* 130: 38–46.

Bickley, L.K., van Aerle, R., Brown, A.R. et al. (2017). Bioavailability and kidney responses to diclofenac in the fathead minnow (*Pimephales promelas*). *Environmental Science & Technology* 51 (3): 1764–1774.

Boxall Alistair, B.A., Rudd Murray, A., Brooks Bryan, W. et al.(2012). Pharmaceuticals and personal care products in the environment: what are the big questions? *Environmental Health Perspectives* 120 (9): 1221–1229.

Brar, S.K., Zhang, T.C., Verma, M. et al. (2015). *Nanomaterials in the Environment*. Reston, VA: ASCE.

Brennecke, D., Duarte, B., Paiva, F. et al. (2016). Microplastics as vector for heavy metal contamination from the marine environment. *Estuarine, Coastal and Shelf Science* 178: 189–195.

Chen, C., Pan, J.G., Xiao, S.X. et al. (2022a). Microplastics alter nitrous oxide production and pathways through affecting microbiome in estuarine sediments. *Water Research* 221 (2022): 118733.

Chen, Q., Li, Z., and Hua, X. (2018). Fate of estrogens in a pilot-scale step-feed anoxic/oxic wastewater treatment system controlling by nitrogen and phosphorus removal. *Environmental Science and Pollution Research* 25 (13): 12981–12991.

Chen, Y.D. and Bergendahl, J. (2021). Identification and quantification of a wide variety of inorganic nanoparticles in municipal wastewater. *Journal of Hazardous, Toxic, and Radioactive Waste* 25 (4): 04021030.

Chen, Z.J., Liu, X.Q., Wei, W. et al. (2022b). Removal of microplastics and nanoplastics from urban waters: separation and degradation. *Water Research* 221 (2022): 118820.

Chowdhury, S.D., Surampalli, R.Y., and Bhunia, P. (2021). Potential of the constructed wetlands and the earthworm-based treatment technologies to remove the emerging contaminants: a review. *Journal of Hazardous, Toxic, and Radioactive Waste* 26 (2): 04021066.

Costanzo, S.D., Murby, J., and Bates, J. (2005). Ecosystem response to antibiotics entering the aquatic environment. *Marine Pollution Bulletin* 51 (1): 218–223.

D'Ascenzo, D., Di Corcia, A., Gentili, A. et al. (2003). Fate of natural estrogen conjugates in municipal sewage transport and treatment facilities. *The Science of the Total Environment* 302: 199–209.

Dhangar, K. and Kumar, M. (2020). Tricks and tracks in removal of emerging contaminants from the wastewater through hybrid treatment systems: a review. *Science of the Total Environment* 738: 140320.

Di, M. and Wang, J. (2018). Microplastics in surface waters and sediments of the Three Gorges Reservoir, China. *Science of the Total Environment* 616–617: 1620–1627.

Dong, C.D., Chen, C.W., Chen, Y.C. et al. (2020). Polystyrene microplastic particles: *in vitro* pulmonary toxicity assessment. *Journal of Hazardous Materials* 385: 121575.

Ebrahimbabaie, P., Yousefi, K., and Pichtel, J. (2022). Photocatalytic and biological technologies for elimination of microplastics in water: current status. *The Science of the Total Environment* 806 (2): 150603.

Eerkes-Medrano, D., Thompson, R.C., and Aldridge, D.C. (2015). Microplastics in freshwater systems: a review of the emerging threats, identification of knowledge gaps and prioritisation of research needs. *Water Research* 75: 63–82.

Elshikh, M.S., Hussien, D., Alkhattaf, F.S. et al. (2022). Diclofenac removal from the wastewater using activated sludge and analysis of multidrug resistant bacteria from the sludge. *Environmental Research* 208: 112723.

Filley, T.R., Ahn, M., Held, B.W., and Blanchette, R.A. (2005). Investigations of fungal mediated C60–C70 fullerene decomposition. Preprints of *Extended Abstracts Presented at the ACS National Meeting, American Chemical Society, Division of Environmental Chemistry* 45 (1): 446–450.

García-Galán, M.J., Arashiro, L., Santos, L.H.M.L. et al. (2021). Fate of priority pharmaceuticals and their main metabolites and transformation products in microalgae-based wastewater treatment systems. *Journal of Hazardous Materials* 390: 121771.

Gentili, A., Perret, D., Marchese, S. et al. (2002). Analysis of free estrogens and their conjugates in sewage and river waters by solid-phase extraction then liquid chromatography-electrospray-tandem mass spectrometry. *Chromatographia* 56 (1): 25–32.

Gherghel, A., Teodosiu, C., and De Gisi, S. (2019). A review on wastewater sludge valorisation and its challenges in the context of circular economy. *Journal of Cleaner Production* 228: 244–263.

Goeppert, N., Dror, I., and Berkowitz, B. (2015). Fate and transport of free and conjugated estrogens during soil passage. *Environmental Pollution* 206: 80–87.

Griffith, D.R., Soule, M.C.K., Matsufuji, H. et al. (2014). Measuring free, conjugated, and halogenated estrogens in secondary treated wastewater effluent. *Environmental Science & Technology* 48: 2569–2578.

Grün, A.Y., App, C.B., Breidenbach, A. et al. (2018). Effects of low dose silver nanoparticle treatment on the structure and community composition of bacterial freshwater biofilms. *PLoS One* 13 (6): e0199132.

Guerra, P., Teslic, S., Shah, A. et al. (2019). Occurrence and removal of triclosan in Canadian wastewater systems. *Environmental Science and Pollution Research.* 26 (31): 31873–31886.

Hemmings, S.N.J. and Hartel, P.G. (2006). Mineralization of hormones in breeder and broiler litters at different water potentials and temperatures. *Journal of Environmental Quality* 35: 701–706.

Huang, K., Xia, H., Wu, Y. et al. (2018). Effects of earthworms on the fate of tetracycline and fluoroquinolone resistance genes of sewage sludge during vermicomposting. *Bioresource Technology* 259: 32–39.

Hutchins, S., White, M.V., Hudson, F.M., and Fine, D. (2007). Analysis of lagoon samples from different concentrated animal feeding operations for estrogens and estrogen conjugates. *Environmental Science & Technology* 41 (3): 738–744.

Hwang, J., Choi, D., Han, S. et al. (2020). Potential toxicity of polystyrene microplastic particles. *Scientific Reports* 10 (1): 7391.

Isobe, T., Shiraishi, H., Yasuda, M. et al. (2003). Determination of estrogens and their conjugates in water using solid-phase extraction followed by liquid chromatography-tandem mass spectrometry. *Journal of Chromatography* A 984 (2): 195–202.

Kasprzyk-Hordern, B., Dinsdale, R.M., and Guwy, A.J. (2010). The removal of pharmaceuticals, personal care products, endocrine disruptors and illicit drugs during wastewater treatment and its impact on the quality of receiving waters. *Water Research* 44 (6): 2076–2084.

Khanal, S.K., Xie, B., Thompson, M.L. et al. (2006). Fate, transport, and biodegradation of natural estrogens in the environment and engineered systems. *Environmental Science & Technology* 40 (21): 6537–6546.

Kolodziej, E.P., Gray, J.L., and Sedlak, D.L. (2003). Quantification of steroid hormones with pheromonal properties in municipal wastewater effluent. *Environmental Toxicology and Chemistry* 22: 2622–2629.

Kruglova, A., Ahlgren, P., Korhonen, N. et al. (2014). Biodegradation of ibuprofen, diclofenac and carbamazepine in nitrifying activated sludge under 12°C temperature conditions. *Science of the Total Environment* 499: 394–401.

Ladner, D.A., Steele, M., Weir, A. et al. (2012). Functionalized nanoparticle interactions with polymeric membranes. *Journal of Hazardous Materials* 211 (4): 288–295.

Lajayer, B.A., Nobaharan, K., Moghiseh, E. et al. (2022). Chapter 8: Removal of pharmaceuticals and personal care products from water and wastewater through biological processes: an overview. In: *Hazardous and Trace Materials in Soil and Plants: Sources, Effects and Management* (eds. Naeem, M., Aftab, T., Ansari, A. et al.), 87–96. Academic Press.

Lange, I., Daxenberger, A., Schiffer, B. et al. (2002). Sex hormomes originating from different livestock production systems: fate and potential disrupting activity in the environment. *Analytica Chimica Acta* 473 (1–2): 27–37.

Lares, M., Ncibi, M.C., Sillanpää, M., and Sillanpää, M. (2018). Occurrence, identification and removal of microplastic particles and fibers in conventional activated sludge process and advanced MBR technology. *Water Research* 133: 236–246.

Le, T.-H., Ng, C., Tran, N.H. et al. (2018). Removal of antibiotic residues, antibiotic resistant bacteria and antibiotic resistance genes in municipal wastewater by membrane bioreactor systems. *Water Research* 145: 498–508.

Lin, A.Y.-C., Yu, T.-H., and Lateef, S.K. (2009). Removal of pharmaceuticals in secondary wastewater treatment processes in Taiwan. *Journal of Hazardous Materials* 167 (1): 1163–1169.

Loffredo, E. (2022). Recent advances on innovative materials from biowaste recycling for the removal of environmental estrogens from water and soil. *Materials (Basel)* 15 (5): 1894.

Luo, Y., Guo, W., Ngo, H.H. et al.(2014). A review on the occurrence of micropollutants in the aquatic environment and their fate and removal during wastewater treatment. *Science of the Total Environment* 473–474: 619–641.

Malamis, S., Katsou, E., Chazilias, D., and Loizidou, M. (2009). Investigation of Cr(III) removal from wastewater with the use of MBR combined with low-cost additives. *Journal of Membrane Science* 333 (1): 12–19.

Mani, S., Chowdhary, P., and Zainith, S. (2020). Microbes mediated approaches for environmental waste management. In: *Microorganisms for Sustainable Environment and Health* (eds Chowdhary, P.Raj, A., Verma, D., and Akhter, Y.), 17–36. Elsevier.

Matamoros, V., Gutierrez, R., Ferrer, I. et al. (2015). Capability of microalgae-based wastewater treatment systems to remove emerging organic contaminants: a pilot-scale study. *Journal of Hazardous Materials* 288: 34–42.

Meng, F., Zhang, S., Oh, Y. et al.(2017). Fouling in membrane bioreactors: an updated review. *Water Research* 114: 151–180.

Metcalf, E., Abu-Orf, M., Bowden, G. et al.and AECOM. (2014). Wastewater Engineering: Treatment and Resource Recovery, 5th edn. McGraw-Hill Education.
Mittal, A. (2011). Biological wastewater treatment. *Water Today* 1: 32–44.
Monsalvo, V.M., McDonald, J.A., Khan, S.J., and Le-Clech, P. (2014). Removal of trace organics by anaerobic membrane bioreactors. *Water Research* 49: 103–112.
Motteran, F., Varesche, M.B.A.V., and Lara-Martin, P.A. (2022). Assessment of the aerobic and anaerobic biodegradation of contaminants of emerging concern in sludge using batch reactors. *Environmental Science and Pollution Research.* doi: 10.1007/s11356-022-21829-1.
Naidu, R., Arias Espana, V.A., Liu, Y., and Jit, J. (2016). Emerging contaminants in the environment: risk-based analysis for better management. *Chemosphere* 154: 350–357.
Nakamura, H., Kuruto-Niwa, R., Uchida, M., and Terao, Y. (2007). Formation of chlorinated estrones via hypochlorous disinfection of wastewater effluent containing estrone. *Chemosphere* 66 (8): 1441–1448.
Ngo, H.H., Guo, W.S., Surampalli, R.Y., and Zhang, T.C. (2016). *Green Technologies for Sustainable Water Management.* Reston, VA: ASCE.
Peulen, T.O. and Wilkinson, K.J. (2011). Diffusion of nanoparticles in a biofilm. *Environmental Science & Technology* 45 (8): 3367–3373.
Prasad, M.N.V., Vithanage, M., and Kapley, A. (2019). *Pharmaceuticals and Personal Care Products Waste Management and Treatment Technology: Emerging Contaminants and Micro Pollutants.* Butterworth-Heinemann.
Racz, L., Muller, J.G., and Goel, R.K. (2012). Fate of selected estrogens in two laboratory-scale sequencing batch reactors fed with different organic carbon sources under varying solids retention times. *Bioresource Technology* 110: 35–42.
Ramos, C., Amorim, C.L., Mesquita, D.P. et al. (2017). Simultaneous partial nitrification and 2-fluorophenol biodegradation with aerobic granular biomass: reactor performance and microbial communities. *Bioresource Technology* 238: 232–240.
Rathi, B.S., Kumar, P.S., and Show, P.-L. (2021). A review on effective removal of emerging contaminants from aquatic systems: current trends and scope for further research. *Journal of Hazardous Materials* 409: 124413.
Reddy, S., Iden, C.R., and Brownawell, B.J. (2005). Analysis of steroid conjugates in sewage influent and effluent by liquid chromatography-tandem mass spectrometry. *Analytical Chemistry* 77: 7032–7038.
Rehman, S., Usman, Z., Rehman, S. et al. (2018). Endocrine disrupting chemicals and impact on male reproductive health. *Translational Andrology and Urology* 7 (3): 490.
Rodrigues, J.P., Duarte, A.C., Santos-Echeandía, J., and Rocha-Santos, T. (2019). Significance of interactions between microplastics and POPs in the marine environment: a critical overview. *TrAC Trends in Analytical Chemistry* 111: 252–260.
Rodriguez-Narvaez, O.M., Peralta-Hernandez, J.M., Goonetilleke, A., and Bandala, E.R. (2017). Treatment technologies for emerging contaminants in water: a review. *Chemical Engineering Journal* 323: 361–380.
Samer, M. (2015). Biological and chemical wastewater treatment processes. *Wastewater Treatment Engineering* 150.
Sánchez, A., Recillas, S., Font, X. et al. (2011). Ecotoxicity of, and remediation with, engineered inorganic nanoparticles in the environment. *Trends in Analytical Chemistry* 30 (3): 507–516.
Saptarshi, D., Bano, F., and Malik, A. (2019). Pharmaceuticals and personal care product (PPCP) contamination: a global discharge inventory. In: *Pharmaceuticals and Personal Care*

Products: Waste Management and Treatment Technology: Emerging Contaminants and Micro Pollutants (eds. Prasad, M.N.V, Vithanage, M., and Kapley, A.), Chapter 1. Butterworth-Heinemann.

Sharma, B.M., Bečanová, J., Scheringer, M. et al.(2019). Health and ecological risk assessment of emerging contaminants (pharmaceuticals, personal care products, and artificial sweeteners) in surface and groundwater (drinking water) in the Ganges River Basin, India. *Science of the Total Environment* 646: 1459–1467.

Sheng, C., Nnanna, A.G.A., Liu, Y., and Vargo, J.D. (2016). Removal of trace pharmaceuticals from water using coagulation and powdered activated carbon as pretreatment to ultrafiltration membrane system. *Science of the Total Environment* 550: 1075–1083.

Sheng, Z.Y. and Liu, Y. (2011). Effects of silver nanoparticles on wastewater biofilms. *Water Research* 45 (18): 6039–6050.

Shokoohi, R., Ghobadi, N., Godini, K. et al. (2020). Antibiotic detection in a hospital wastewater and comparison of their removal rate by activated sludge and earthworm-based vermifilteration: environmental risk assessment. *Process Safety and Environmental Protection* 134: 169–177.

Shore, L.S. and Shemesh, M. (2003). Naturally produced steroid hormones and their release into the environment. *Pure and Applied Chemistry* 75: 1859–1871.

Shrestha, S.L. (2011). *Fate and Transformation of a Conjugated Natural Hormone 17β-Estradiol-3-Glucuronide in Soil–Water Systems*. PhD Dissertation, North Dakota State University, Fargo, North Dakota, USA.

Song, C.F., Liu, Z.Z., Wang, C.L. et al. (2020). Different interaction performance between microplastics and microalgae: the bio-elimination potential of *Chlorella* sp. L38 and *Phaeodactylum tricornutum* MASCC-0025. *The Science of the Total Environment* 723: 138146. https://www.sciencedirect.com/science/article/abs/pii/S0048969720316594.

Street, M.E., Angelini, S., Bernasconi, S. et al. (2018). Current knowledge on endocrine disrupting chemicals (EDCs) from animal biology to humans, from pregnancy to adulthood: highlights from a national Italian meeting. *International Journal of Molecular Sciences* 19 (6): 1647.

Suarez, S., Lema, J.M., and Omil, F. (2010). Removal of pharmaceutical and personal care products (PPCPs) under nitrifying and denitrifying conditions. *Water Research* 44 (10): 3214–3224.

Sun, H., Wang, T., Yang, Z. et al. (2019). Simultaneous removal of nitrogen and pharmaceutical and personal care products from the effluent of waste water treatment plants using aerated solid-phase denitrification system. *Bioresource Technology* 287: 121389.

Surampalli, R.Y., Zhang, T.C., Brar, S.K. et al. (2018). *Handbook of Environmental Engineering*. New York: McGraw Hill.

Tambosi, J.L., de Sena, R.F., Favier, M. et al. (2010). Removal of pharmaceutical compounds in membrane bioreactors (MBR) applying submerged membranes. *Desalination* 261 (1): 148–156.

Tijani, J.O., Fatoba, O.O., Babajide, O.O., and Petrik, L.F. (2016). Pharmaceuticals, endocrine disruptors, personal care products, nanomaterials and perfluorinated pollutants: a review. *Environmental Chemistry Letters* 14 (1): 27–49.

Tran, N.H., Chen, H., Reinhard, M. et al. (2016). Occurrence and removal of multiple classes of antibiotics and antimicrobial agents in biological wastewater treatment processes. *Water Research* 104: 461–472.

Vader, J.S., Finkel, C.G., Sperling, F.M.G.M. et al. (2000). Degradation of ethinyl estradiol by nitrifying activated sludge. *Chemosphere* 41: 1239–1243.

Wang, X., Li, J., Zhang, X. et al. (2021). The performance of aerobic granular sludge for simulated swine wastewater treatment and the removal mechanism of tetracycline. *Journal of Hazardous Materials* 408: 124762.

Westerhoff, P., Song, G., Hristovski, K., and Kiser, A. (2011). Occurrence and removal of titanium at full-scale wastewater treatment plants: implications for TiO_2 nanomaterials. *Journal of Environmental Monitoring* 13 (5): 1195.

Westerhoff, P., Yoon, Y., Snyder, S., and Wert, E. (2005). Fate of endocrine-disruptor, pharmaceutical, and personal care product chemicals during simulated drinking water treatment processes. *Environmental Science & Technology* 39: 6649–6663.

Westerhoff, P.K., Kiser, M.A., and Hristovski, K. (2013). Nanomaterial removal and transformation during biological wastewater treatment. *Environmental Engineering Science* 30 (3): 109–117.

Wu, M., Tang, W., Wu, S. et al. (2021). Fate and effects of microplastics in wastewater treatment processes. *Science of the Total Environment* 757: 143902.

Yost, E.E., Meyer, M.T., Dietze, J.E. et al. (2013). Comprehensive assessment of hormones, phytoestrogens, and estrogenic activity in an anaerobic swine waste lagoon. *Environmental Science & Technology* 47: 13781–13790.

Yost, E.E., Meyer, M.T., Dietze, J.E. et al. (2014). Transport of steroid hormones, phytoestrogens, and estrogenic activity across a swine lagoon/sprayfield system. *Environmental Science & Technology* 48: 11600–11609.

Yu, Q.M., Yang, X.D., Zhao, F.Z. et al. (2022). Occurrence and removal of progestogens from wastewater treatment plants in China: spatiotemporal variation and process comparison. *Water Research* 211: 118038.

Zhang, D. (2011). Removal of heavy metal ions from water using nano-barium strontium titanate immobilized on the silica gel G as an adsorbent. *Applied Mechanics and Materials* 71–78: 3500–3503.

Zhang, T.C., Surampalli, R., Lai, K.C.K. et al. (eds.) (2009). *Nanotechnologies for Water Environment Applications*. Reston, VA: ASCE, 630 p.

Zhang, T.C., Surampalli, R.Y., Vigneswaran, S., et al. (eds.) (2012). *Membrane Technology and Environmental Applications*. Reston, VA: ASCE.

Zhang, Y.-T., Wei, W., Huang, Q.-S. et al. (2020). Insights into the microbial response of anaerobic granular sludge during long-term exposure to polyethylene terephthalate microplastics. *Water Research* 179: 115898.

Zhao, Y., Liu, D., Huang, W. et al. (2019). Insights into biofilm carriers for biological wastewater treatment processes: current state-of-the-art, challenges, and opportunities. *Bioresource Technology* 288: 121619.

Zlatnik, M.G. (2016). Endocrine-disrupting chemicals and reproductive health. *Journal of Midwifery & Women's Health* 61 (4): 442–455.

Zupanc, M., Kosjek, T., Petkoysek, M. et al. (2013). Removal of pharmaceuticals from wastewater by biologicalprocesses, hydrodynamic cavitation and UV treatment. *Ultrasonics Sonochemistry* 20 (4): 1104–1112.

17

Anaerobic Biological Treatment of Microconstituents

Thi-Manh Nguyen[1], Hung-Hsiang Chen[1], Ku-Fan Chen[1], Chih-Ming Kao[2], Rao Y. Surampalli[3], and Tian C. Zhang[4]

[1]*Department of Civil Engineering, National Chi Nan University, Puli, Nantou, Taiwan*
[2]*Institute of Environmental Engineering, National Sun Yat-sen University, Kaohsiung, Taiwan*
[3]*Global Institute for Energy, Environment and Sustainability, Lenexa, KS, USA*
[4]*Civil & Environmental Engineering, College or Engineering, University of Nebraska, Lincoln, Omaha, NE, USA*

17.1 Introduction

In the eighteenth century, when organic garbage was left to decompose in large piles, according to Van Helmont, flammable gas was produced (Abbasi et al. 2012). In the next century, combustible gas methane was discovered by John Dalton and Humphrey Davy (Tietjen 1975). The early 1970s saw a substantial increase in the price of fossil fuels, which encouraged the use of anaerobic digestion (AD) systems to produce biogas while reducing industrial effluent pollution simultaneously (Abbasi et al. 2012; Angelidaki et al. 2003). These innovations allowed AD to be applied to significantly less concentrated industrial wastewaters instead of highly concentrated (in volatile organics content, VOC) manure slurry or sewage sludge.

Anaerobic biotechnology, or AD, is a biological method for converting organic matter into its most oxidized and reduced states (CO_2 and CH_4, respectively) using anaerobic microorganisms in the absence of dissolved oxygen. Carbon dioxide (25–50%) and methane (50–75%) are the two major gaseous products of the AD process. Other minor by-products are also produced, including nitrogen, hydrogen, ammonia, and hydrogen sulfide (usually <1% of the total gas volume) (Angelidaki et al. 2003; Kougias and Angelidaki 2018). The resulting product is commonly termed as biogas. This AD process is often applied on an industrial scale for waste and wastewater treatment.

Since the 2010s, researchers have recognized that many contaminants previously considered to be recalcitrant under anaerobic conditions can be biotransformed in the AD process due to the diversity of anaerobic microbial metabolism. As a natural extension of this recognition, considerable research has been conducted in the past 10 years on using AD to treat microconstituents (MCs). To this end, significant information has been accumulated on removal of a wide array of MCs in different AD systems [e.g.,

Microconstituents in the Environment: Occurrence, Fate, Removal, and Management, First Edition.
Edited by Rao Y. Surampalli, Tian C. Zhang, Chih-Ming Kao, Makarand M. Ghangrekar, Puspendu Bhunia, Manaswini Behera, and Prangya R. Rout.

anaerobic membrane reactors (AnMBRs), upflow anaerobic sludge blanket (UASB), anaerobic contact reactor (ACR)]. Several critical reviews have been published to evaluate the fate and removal of different MCs in different AD systems (Bhandari et al. 2009; Surampalli et al. 2018; Oberoi et al. 2022; Zhang et al. 2021a). However, in-depth analyses of removal efficiencies, pathways, and the associated biodegradation mechanisms for different classes of MCs in different AD systems under some critical influencing factors are still lacking.

Herein, we present a non-exhaustive yet synthesized overview of recent advances in using AD processes for MCs treatment. This chapter first introduces a few representative AD systems widely used for wastewater treatment and the current status of AD technology in general, followed by a presentation about the general mechanisms of pollutant removal in AD processes. The chapter then describes recent advances in AD processes for MCs removal, focusing on removal efficiencies of AD processes or combinations of AD with other treatment methods, advancements in configurations, design and operation of AD processes, as well as brief descriptions on removal mechanisms for different MCs and critical influencing factors (e.g., the inhibitory effects of MCs on the microbial community). The chapter also presents the challenges and future perspectives of AD technology, followed by conclusions.

17.2 Types of AD Reactors and Current Status of AD Technology

AD processes have long been used to treat wastewater and produce fuel. According to Liu (1989), the solids retention time (SRT) is an important consideration in reactor design. A low hydraulic retention time (HRT) is a requirement for system economics, whereas a lengthy SRT is necessary to maintain optimum bioconversion efficiency. SRT and HRT are equivalent in a typical complete mixed anaerobic reactor without sludge recycling. Different types of reactors have been created to decrease HRT while increasing SRT. These types of reactors are divided into attached growth reactors and suspended growth reactors (Angelidaki et al. 2003; Ghangrekar and Behera 2014; Metcalf and Eddy/AECOM 2014; Aslanzadeh 2014; Najafpour and Ebrahimi 2016). The next section introduces some AD reactors, followed by a brief description of the current status of the AD technology.

17.2.1 Suspended Growth Process

The suspended growth processes involve using microorganisms that grow in liquid form to treat high intensity industrial wastewater and organic sludge. There are various configurations available, including the ACR, fluidized bed feed, complete-mix digestion, upflow anaerobic sludge blanket (UASB), two-stage bed leachate filter, and membrane solids separation. Generally, anaerobic suspension growth is operated in an airtight reactor (Hassan et al. 2013; Aslanzadeh 2014). Especially, suspended growth reactors can increase SRT by separating and recycling solids from the reactor output (contact process) or forming a sludge layer at the bottom of the reactor UASB. ACR and UASB, shown in Figures 17.1 and 17.2, are briefly introduced below.

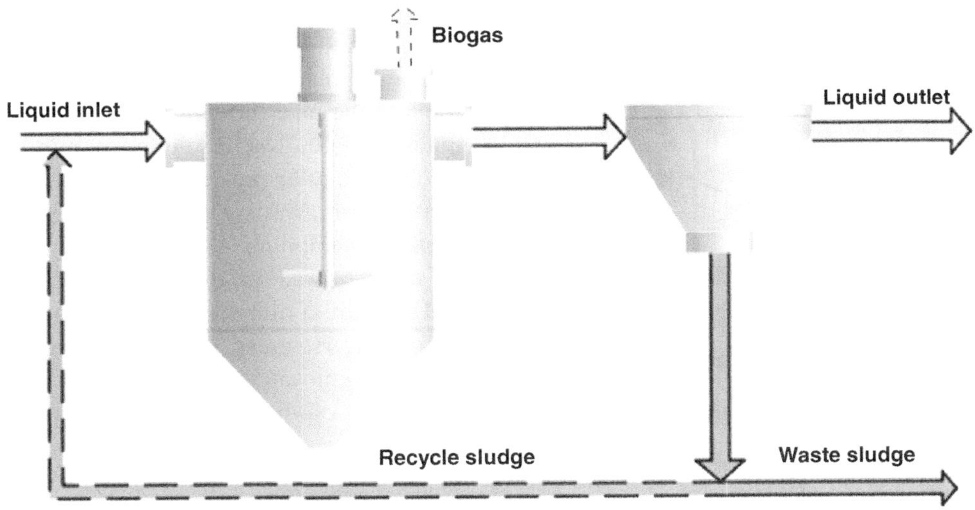

Figure 17.1 An anaerobic contact reactor.

17.2.1.1 Anaerobic Contact Reactor (ACR)

Figure 17.1 shows an ACR. ACP produces high biomass concentration, improved efficiency, and reduced reactor sizes. Anaerobic sludge recycled to the reactor after settling in a settling tank provides for more contact between biomass and unprocessed wastewater (Nähle 1991; Hassan et al. 2013). Decomposition takes place in closed tanks with circulating sludge. Sludge and wastewater are thoroughly mixed in the tank. After decomposition, the mixture is transferred to a settling (or flotation) tank to separate the sludge and liquid in an external clarifier or vessel, and the solids are then recycled back into the reactor tank (Hassan et al. 2013). The amount of residual sludge is usually very little due to the slow growth of microorganisms. When granulation is challenging or the concentration of undesirable components such as suspended solids in wastewater is higher than expected, ACR systems are especially successful. ACP systems are successful at retaining flocculent, such as non-granular sludge, allowing maintenance of the proper anaerobic biomass inventory levels.

17.2.1.2 Upflow Anaerobic Sludge Blanket (UASB) Reactor

The idea of an unpacked high-rate reactor known as the UASB reactor was created in the 1970s (Hassan et al. 2013). It is the most commonly employed high-rate anaerobic system for treating domestic and industrial wastewater globally. Anaerobic sludge in the UASB reactor has good settling performance and does not need vigorous mechanical agitation. A high enough upflow velocity and gas production contribute to the agitation in the reactor to achieve well mixing. By regulating the upflow velocity, biomass is held in suspension and retained as a granular matrix or blanket (Figure 17.2). The sludge blanket in the lower part of the reactor allows wastewater to flow upward through it, and a three-phase

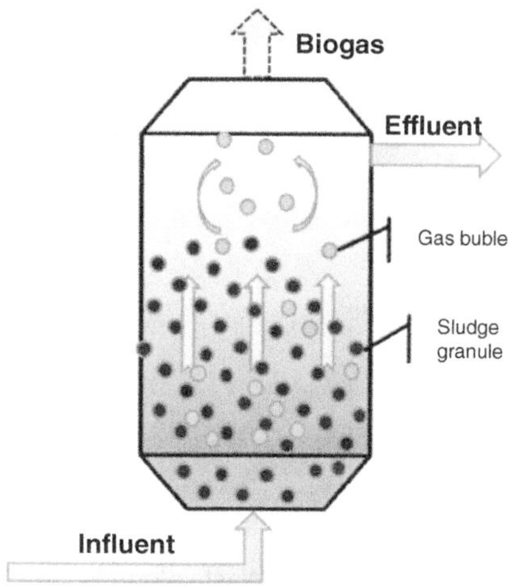

Figure 17.2 A UASB reactor.

separation system is fabricated in the reactor's upper part to separate the produced gases, microorganisms, and treated water. The produced gases flow into the pipe through the absorption tank containing 5–10% NaOH solution for purification. The sludge after separating from the air bubbles settles again. Wastewater follows the serrated overflow membrane leading to the next treatment stage. Nowadays, research and applications of USAB are continuously growing in the following four diverse areas: i) energy recovery/production; ii) removal of specific pollutants (e.g., MCs); iii) UASB-based hybrid systems for different treatments; and iv) characterization of microbial community and population dynamics of granular sludge (Collivignarelli et al. 2021).

17.2.2 Attached Growth Process

An attractive solution for on-site wastewater treatment are anaerobic attached growth techniques. In attached anaerobic reactors, the bacteria are first only weakly connected to a surface (reversible absorption), but after a few hours they become securely attached (irreversibly absorbed) (Marshall et al. 1971), and then bacterial activity is increased due to binding. According to Switzenbaum and Eimstad (1987), increasing levels of inorganic elements were discovered as loading rates rose, accounting for 40 to 50% of biofilms. The substrate conversion efficiency will be greatly decreased if the substrate cannot reach the bacteria inside the biofilm, and diffusion into the biofilm may be rate-limited. When evaluating the alkalinity or acidity of bulk liquids, an accumulation of alkalis or acids that can hinder microbial activity in biofilms cannot be found. Due to the distinctions in solubility and chemical reactivity between CH_4 and CO_2, these two substances constitute two entirely different circumstances. In contrast to CH_4, which is difficult to dissolve and can create air bubbles in biofilms, CO_2 is readily soluble and its build-up will cause the pH to decrease. The support medium in the attached growth reactor retains biomass and so improves wastewater quality by reducing the total suspended solids concentration of the wastewater. The amount of volume the reactor needs to contain these bacteria in order to get a specified conversion efficiency decreases as they grow attached to create a biofilm in the reactor and considerably boost SRT as compared to HRT.

In upflow fixed-bed biofilm reactors (Figure 17.3), solids from the effluent of an anaerobic contact process were retained using packing media. A biofilm of closely packed bacteria develops on the surface of the packaging material (Hassan et al. 2013), which can be fixed or suspended in reactors. The main function of microorganisms in the attached-growth process is to primarily remove soluble organic matter measured as biochemical oxygen demand (BOD) or chemical oxygen demand (COD) from wastewater as it passes through the medium. BOD gradually decreases with the passage of wastewater through the media, increasing cell mass as a result. The substrate diffuses into the biofilm, where the bacteria break it down. Small levels of oxygen do not cause the biofilms to disintegrate since they are extremely stable. Organic matter is removed by attached microorganisms by absorbing its soluble and suspended components. However, Liu (1989) point out that the lengthy warm-up period and potential for the packaging to become clogged with inert suspended particulates or chemical precipitates (metal sulfides, calcium carbonate, etc.) are the main drawbacks of this system. Exopolysaccharides released by acidogenic and hydrolytic bacteria are the primary cause of blockage (Liu 1989).

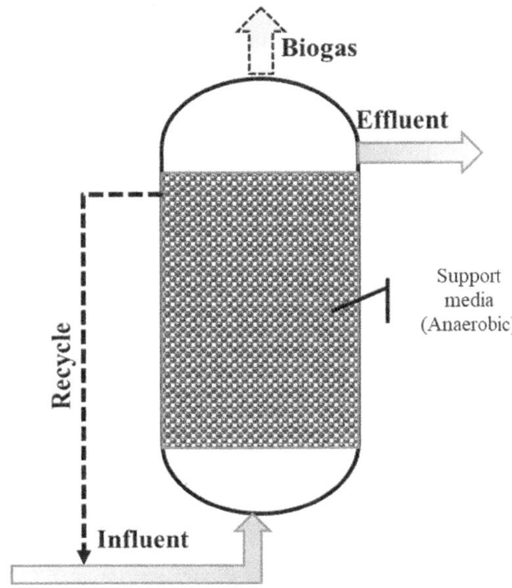

Figure 17.3 An upflow fixed-bed anaerobic reactor.

Under anaerobic conditions, attached-growth systems typically like a type of filter (Loupasaki and Diamadopoulos 2013). Because they produce less solid residue than other varieties and have relatively short HRT, these reactors are becoming increasingly common. Although many different kinds of support materials have been researched, polymeric materials appear to be the most common. According to experimental findings, the porosity and roughness of the media's surface have a large impact on how well anaerobic filters work. Due to the increased retention of attached biofilm in the reactor's open-pored surface roughness, its COD removal effectiveness was ~78% (Loupasaki and Diamadopoulos 2013). Due to the larger growth of suspended biomass, a reactor with a higher media porosity also showed high removal efficiency of ~77%. In contrast, the efficiency is as low as 57% in environments with smooth surfaces and low porosity (Show and Tay 1999; Loupasaki and Diamadopoulos 2013).

The inflow wastewater to an anaerobic filter can flow downward hydraulically (Figure 17.4). Soluble organic matter is eliminated as it comes into contact with the anaerobic biofilm. Low concentrations of suspended particles may also be eliminated by retaining them in the interstices of the medium and letting them finally decompose. Secondary sedimentation is not necessary for anaerobic filters, but sediments must be periodically removed from the filter due to biofilm sloughing occurring in them less frequently than in trickling filters (de Lemos Chernicharo 2007). The influent can also enter anaerobic filters from the bottom, move upward, and exit from the top while the medium is still saturated (void spaces occupied by liquid).

17.2.3 AnMBRs

In an AnMBR, an anaerobic bioreactor (e.g., a UASB or ACR) is incorporated with a membrane module that may be placed either outside or inside of the anaerobic reactor (i.e., either external/side-stream or submerged/immersed with the digester). Because the

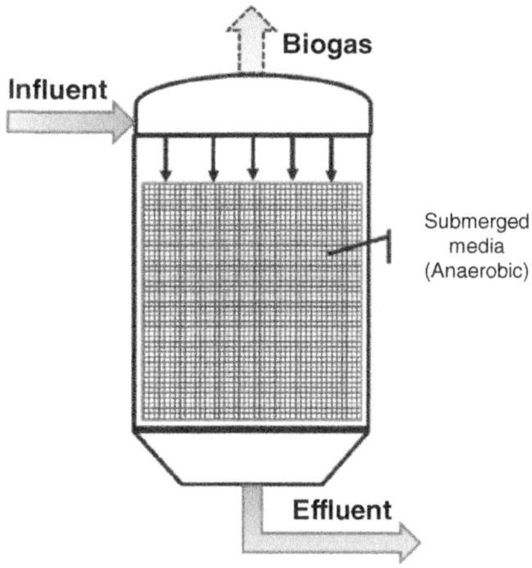

Figure 17.4 A downflow anaerobic filter with submerged medium.

membrane separates anaerobic biomass completely from the effluent of the anaerobic reactor, an AnMBR decouples HRT from SRT, significantly extending its applicability for treating various strengths of wastewater (Cheng et al. 2018a; Ji et al. 2020; Lim et al. 2020; Oberoi et al. 2022; Zhang et al. 2021a). Obviously, AnMBR can belong to either suspended- or attached-growth or both processes, depending on the type of anaerobic reactor. The membrane materials and different configurations are well documented in the literature (Stuckey 2012; Zhang et al. 2012; Abdelrahman et al. 2021; Lu et al. 2021; Oberoi et al. 2022). Microfiltration (MF) and ultrafiltrate (UF) are used widely as membrane modules in AnMBRs (Oberoi et al. 2022). Recently, many new AnMBR configurations (Oberoi et al. 2022) or integration of AnMBR with physico-chemical treatment technologies (Rosli et al. 2021; Ebrahimbabaie et al. 2022) have been developed for removal of MCs/other recalcitrant pollutants, such as anaerobic osmotic membrane bioreactor (AnOMBR), anaerobic fluidized-bed membrane bioreactor (AnFMBR), anaerobic electrochemical membrane reactor (AnEMBR), microaeration-based anaerobic membrane bioreactor systems, and anaerobic dynamic membrane bioreactor (AnDMBR). Regardless of membrane configuration, ACR-based AnMBRs are most popular due to ACR's simplicity (e.g., ease of operation and maintenance).

17.2.4 Current Status of AD Technology

Nowadays, AD is one of the main technologies for industrial wastewater treatment. It is now possible to treat almost all types of biodegradable wastewater using many types of high-speed anaerobic tanks because of the advancements in AD technology. In comparison to aerobic processes, anaerobic processes not only use less energy to operate, but also produce energy in the form of biogas, a clean fuel, which are the reasons for AD having gained additional significance in recent years (Abbasi et al. 2012).

AD facilities are expanding rapidly worldwide. China is currently the largest producer of AD, with an estimated 2500 industrial plants treating wastewater (mostly for high-intensity wastewater) (Kougias and Angelidaki 2018). Building and expanding AD facilities are becoming more popular in Europe. However, AD facilities still lack automatic control, and this process is mostly based on human-made experimentation. As a result, methane digesters can become overloaded or operation can lose control, which reduces biogas capacity.

It is well known that in an AD system wastewater is digested anaerobically in four stages: hydrolysis, acidogenesis, acetogenesis, and methanogenesis (see Section 17.3 for details). Nevertheless, the understanding of AD processes has greatly expanded in recent years. One of the most recent technological and methodological advancements, for instance, is the development of novel applications or solutions to AD problems, especially the considerable decrease in the price and time needed for AD bacteria sequencing (Kougias and Angelidaki 2018).

17.3 Mechanisms of Pollutant Removal in AD Processes

As a complex process, methane is produced in AD by the synchronous and interconnected biological processes of hydrolysis, acidogenesis, acetogenesis, and methanogenesis, which transform the initial organic component into biogas (Aslanzadeh 2014) as shown in the general reaction in Eq. (17.1).

$$Organic + H_2O \xrightarrow{Microorganism} Biomass + CH_4 + CO_2 + NH_3 \quad (17.1)$$

Initially, extracellular enzymes in fermentative bacteria hydrolyze organic polymers into soluble oligomers and monomers. Then the bacterial species go on to utilize the soluble products to create short-chain fatty acids. By further oxidizing short-chain fatty acids, acetogenic bacteria create acetate, formate, hydrogen, and carbon dioxide. Finally, archaea use the latter chemicals to produce methane.

In an AD process, the rate limiting step is typically the phase that results in process failure when kinetic stress is applied (Aslanzadeh 2014). In a continuous culture, kinetic stress is defined as the application of an SRT that is gradually reduced until it falls below the limiting value, which will cause the microbe to wash out (Pavlostathis and Giraldo-Gomez 1991). The hydrolysis step is generally reported as the rate-limiting stage for complex organic substrate (Fernandes et al. 2009; Izumi et al. 2010; Ma et al. 2011), either dangerous by-products (complex heterocyclic compounds), or unpleasant volatile fatty acids (VFAs) (Neves et al. 2006; Lu et al. 2008), whereas the rate-limiting stage for substrates that are easily biodegradable is methanogenesis (Lu et al. 2008).

The physiology, nutritional requirements, growth kinetics, and environmental sensitivity of the microorganisms driving the degrading reactions in each of these phases vary greatly (Adekunle and Okolie 2015). For instance, the two types of microorganisms, one producing acid and one producing methane, often find it challenging to maintain a balance, which results in an unstable reaction and low methane yield. The characteristics and associated mechanisms of each stage in AD processes are briefly described below.

17.3.1 The Hydrolysis Stage

In the hydrolysis stage, extracellular enzymes are made by the hydrolytic bacteria hydrolyze organic polymers (e.g., lipids, polysaccharides, proteins, fats, nucleic acids, etc.) into substances that can be used as sources of energy and cell carbon, such as monosaccharides, amino acids, and other simple organic compounds. Several strict anaerobes including *bacteroides, clostridia*, or *streptococci* can perform this phase (Christy et al. 2014). The organic molecules are too large for microorganisms to directly consume and utilize as a source of nourishment, thus some microbes release extracellular enzymes that enable them to degrade organic materials into smaller ones, which the microbes may then use as a source of energy and nutrients. Several bacteria have particular functions; for instance, they release enzymes that break down proteins or carbohydrates. Glycolytic microorganisms break down various sugars, while proteolytic microorganisms break down proteins (Schnürer and Jarvis 2010). The type of substrate has a significant impact on how quickly the substrate decomposes during the hydrolysis stage. The breakdown of proteins typically occurs more quickly than the transformation of cellulose and hemicellulose (Schnürer and Jarvis 2010). The rate of hydrolysis is also decreased by low temperature or low pH. However, studies have shown that maintaining a neutral pH (6.5–7.5) promotes the rate of hydrolysis (Veeken et al. 2000).

17.3.2 The Acidogenesis Stage

The monomers produced during the hydrolytic stage are broken down into short-chain organic acids such as butyric acids, propanoic acids, acetic acids, alcohols, hydrogen, and carbon dioxide by a variety of obligatory anaerobic bacteria. The amount of hydrogen produced during this step as an intermediate product has an impact on the kind of end product that is created during the fermentation process. For instance, fewer reduced compounds would be present if the hydrogen partial pressure was too high. Typically, at this stage, simple sugars and long-chain fatty acids are converted to shorter-chain organic acids (propionic acid, valeric acid, acetic acid), alcohols (ethanol), ketones, a little hydrogen, and carbon dioxide (Gerardi 2003). This stage is commonly known as the fermentation stage. The mechanism of the acidogenesis stage of fatty acids and glycerine (which is the product of lipid hydrolysis) is relatively complex. Glycerine is degraded to a number of intermediate products such as 1,3-propanediol to form the end products, propionate and acetate (Qatibi et al. 1991). VFAs are also products of this stage. However, VFAs are the substrate for the acetogenesis stage and VFAs can inhibit the methanogens stage and thus affect the whole process.

17.3.3 The Acetogenesis Stage

The other microorganisms working in the acetogenesis stage use the products produced in the acidogenic phase as substrates (Aslanzadeh 2014). Substances such as VFAs and alcohols that methanogenic bacteria cannot directly convert to methane are oxidized into acetate, hydrogen, and carbon dioxide, as shown in Eqs (17.2) to (17.4) (Murphy and Thanasit 2013):

$$CH_3CH_2OH(ethanol) + H_2O \rightarrow CH_3COO^- + H^+ + 2H_2 \tag{17.2}$$

$$CH_3CH_2COO^-(propionic) + 3H_2O \rightarrow CH_3COO^- + HCO_3^- + H^+ + 3H_2 \tag{17.3}$$

$$CH_3(CH_2)_2COO^-(butyric) + 2H_2O \rightarrow 2CH_3COO^- + H^+ + 2H_2 \tag{17.4}$$

The partial pressure of hydrogen gas in the AD has a significant impact on how much material is decomposed at this stage. Long-chain fatty acid degradation is halted by the accumulation of VFAs if the partial pressure of hydrogen gas increases. It will therefore be sufficient to prevent the development of bacteria that create acid, acetate, and methane (Chowdhury 2005). By contrast, the symbiotic interactions between hydrogen-producing bacteria (acetogenic bacteria) and hydrogen-consuming microorganisms (methane-producing microorganisms and sulfate reducing bacteria), often termed interspecies H_2 transfer, are able to maintain low partial pressure of hydrogen (between 10^{-4} and 10^{-6} atm) to enable the proper progress of methane generation (Schink 1997; Schnürer and Jarvis 2010; Deublein and Steinhauser 2011; Chandra et al. 2012). Additionally, a group of bacteria that are homoacetogenic from hydrogen and carbon dioxide (autotrophs) or from organic materials can contribute to the occurrence of another pathway of acetogenesis (heterotrophs). However, in environments with low hydrogen concentrations, methanogens have a larger affinity for hydrogen than homoacetogen; therefore, the amount of acetic acid produced through this pathway is very minor (Pereira 2003).

17.3.4 The Methanogenesis Stage

The methanogenesis stage is the last stage of AD, which uses a mixture of the following pathways to create the desired result, biogas with the valuable component methane gas. Each pathway relates to a distinct substrate group and assortment of bacteria that produce methane (in the totality of methane-producing organisms) (Gerardi 2003).

Pathway 1:

$$CO_2 + 4H_2 \rightarrow CH_4 + 2H_2O \tag{17.5}$$

Hydrogen and carbonic substrates are used by methanogen bacteria (*Hydrogenotrophic Methanogen*) to create CH_4. This method produces <30% of the methane.

Pathway 2:

$$CH_3COOH \rightarrow CO_2 + CH_4 \tag{17.6}$$

$$4CO + 2H_2O \rightarrow CH_4 + 3CO_2 \tag{17.7}$$

Acetate is converted to methane and carbonic acid by *Acetotrophic Methanogen* bacteria. This pathway accounts for ~70% of methane production. However, the energy released from this pathway is relatively small. The resulting carbon dioxide is reduced to methane by Pathway 1. Carbon monoxide (CO) is just a substrate for a small subset of methane-producing bacteria.

Pathway 3:

$$CH_3OH + H_2 \rightarrow CH_4 + H_2O \tag{17.8}$$

$$4(CH_3)_3 - N + 6H_2O \rightarrow 9CH_4 + 3CO_2 + 4NH_3 \tag{17.9}$$

Substrates with methyl groups ($-CH_3$) are degraded by *Methylotrophic Methanogen* bacteria. This route generates a very small quantity of methane. Methanogenic bacteria work under anaerobic conditions to produce methane and carbon dioxide from intermediate products during the methanogenic stage (Aslanzadeh 2014). Being the slowest biological reaction in the process, methanogenesis is a crucial step in the overall AD process (Murphy and Thanasit 2013).

Figure 17.5 summaries the whole biochemical process of AD. The division of anaerobic fermentation into four stages is only theoretical and conventional. In fact, all the chemical processes of all four stages work at the same time and in sync with each other. The products produced in stage 1 are used up immediately in stages 2 and 3, and all products produced in stage 3 will be used up by methanogenic bacteria in stage 4 to create biogas. After the

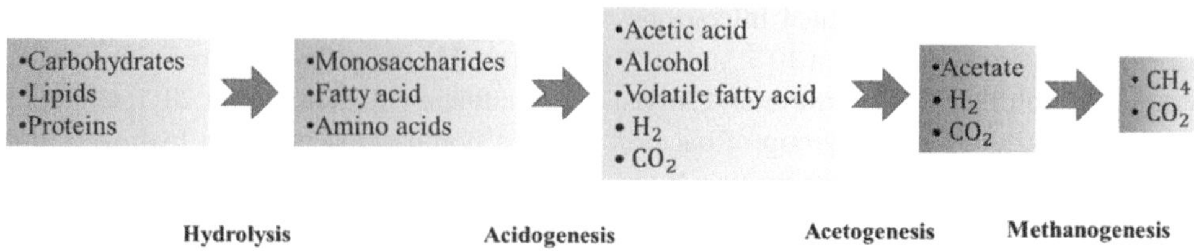

Figure 17.5 The main steps of the anaerobic degradation process.

entire system has reached equilibrium, the pH of the medium will be neutral. If the system immediately becomes unbalanced, the end product, methane gas, will become less or none. For example, when too much acid is produced in the second stage, the acetate bacteria and methanogenic bacteria cannot consume it in time, and the pH of the system will drop sharply. Low pH will strongly affect the whole process, especially stage 4, and can even destroy all neutral bacteria (Eastman and Ferguson 1981).

17.4 AD Technology for Treatment of MCs

Although MCs concentration are in the range of ng/L to μg/L in the aquatic environment or wastewater (Venegas et al. 2021; Choi et al. 2022), they can adversely affect environmental systems and human health due to their persistence, bioaccumulation, and biological activity (Bhandari et al. 2009; Luo et al. 2014; Venegas et al. 2021). Treating MCs with AD is necessary as in most cases some MCs may be concentrated in biosolids after secondary (biological) treatment of wastewater (e.g., MPs in WAS) (He et al. 2021), and the presence of contaminants in some biosolids has raised concerns about expanding pollution and other negative effects. In other cases, MCs in the natural water or wastewater is in relatively high concentrations. For example, sulfamethoxazole (SMX) is a typical antibiotic widely used in human and animal medicine to prevent disease; its maximum concentration was reported as 20 μg/L worldwide, but as 230 μg/L in Asia (Aus der Beek et al. 2016). Another reason is that some MCs can be removed more efficiently in AD processes. For example, Motteran et al. (2022) reported that artificial sweeteners were not removed under aerobic conditions, but were steadily decreased in the anaerobic reactors.

To give a non-exhaustive yet synthesized overview on using AD to remove MCs is a formidable task, as the topic would involve many different AD processes and MCs [e.g., pharmaceuticals and personal care products (PPCPs), pesticides, hormones, industrial chemicals, nanomaterials/nanoparticles (NMs/NPs), microplastics] (Bhandari et al. 2009; Luo et al. 2014; Lim et al. 2019). Accordingly, in this section, we try to organize the information on key characteristics of selected AD systems for MCs removal and on biodegradation of selected MCs in AD processes.

17.4.1 Key Characteristics of Selected AD Systems for MCs Removal

17.4.1.1 Reactor Configurations and Combinations of Different Methods

Table 17.1 shows the removal efficiency and performance of selected AD systems for MCs treatment. In the past, it was commonly accepted that AD had a limited capacity to

Table 17.1 MC removal by and key characteristics of selected AD processes/systems.

Reactor/System	MC Removal (%)	Operating Conditions/ Other Results/Comments
AnMBR (ACR or UASB or Stainless reactor + submerged or side-stream membrane).	**PPCPS**[1]: ACT: 18.0–76.5; ATN: 12–85.9; AMI: 47.0– 99.6; AMOX: 73.2; AMPI: 34.6–98.9; AMT: 47.0; CAFF: 23.0–92.8; CBZ: 0.3–34.0; CEFO: 79.4; CEFT: 46.7; CIP: 20.0–76.4; DIAZ: 61.6; DCF: <1.0–15.0; DIL: 10.0; DIPH: 93.0; ENA: 36.6; ERY: 40.0–88.0; Etodolac: 15.0–49.0; GMF: 12.2–30.0; IBU: <1.0–28.0; KET: 24.0–27.2; METF: >99.0; OME: 99.0; NPX: 52.0–74.7; PRI: 1.8–54.0; SDZ: 22.4–46.8; SMX: 69.0–99.6; SMZ: 20.6–43.7; TMP: 35.4–97.5; VER: >99.0. **MPs**;[16] 94 (87 for fiber and 100 for particles) and CBZ (49.8–70.0).	[1]**Operation conditions**: HRT = 6–120 h; SRT = 300–700 d; T: 35°C; and pH: 7–7.5; PPCPs initial conc. = 1.1 µg/L–215 mg/L; Yield (L/g_{COD} removed): biogas = 0.04–0.6 and CH_4 = 0.15–1.7; COD removal (%): 78.0–98.0. For,[16] Pore size: 0.03 µm; T: 30°C; HRT: 6 h; UASB + AnMBR, where up to 58% decrease in methanogenic activity was observed at the exposure of 50 PP-MPs/gTS.
Novel AnMBR (e.g., AnO-, AnF-, AnE-, MA-n- and AnD-MBR).	**PPCPS:**[2] ATN: 99.6–100; CAFF: 96; CBZ: 96.4; CIP: 100; *Cytostatic drugs: 95.6; DCF: 78.2–80.0; NPX: 96.4; SMX: 93.0–100; TMP: 95.0–100.	**Important factors**: draw solution (NaCl, $MgCl_2$, Na_2SO_4) and conc. for AnOMBR; GAC, PAC, Zeolite-Fluidized media for AnFMBR; electrical potential and energy for AnEMBR; micro-aeration rate for MA-AnMBR.
Batch reactor.	**In suspended growth systems**:[3] LAS and SAS: 0 within 72 d; ACE: 0–12.3; ASP: 76.1–78.9; CYC: 11.1–30.8; DIF: 43.2–60.6; IBU: 25.2–30.7; NAX: 57.7–81.1; SUC: 25.7–52.6; can reduce sweeteners. **In soil reactors**:[4] nadolol, sulfamethizole, SMX, sulfamethopyridazone, CBZ, IBU, DCF, hydrochlorothiazide, and GMF) and four artificial sweeteners (ACE, saccharin, CYC, SUC) are relatively persistent under anaerobic conditions (e.g., 18–24% of nadolol is lost after 1 month reaction; $t_{1/2}$ = 95–103/d). **In batch reactors seeded with microbes from deep sea sediments**: E2 not removed and lag phase can be year long;[8] removal of DON and ZEN (= >90% in 20/d) with mycotoxins (AFB1, FBs, OTA, DON, ZEN, T-2 toxin ergot alkaloid) being added in the reactor.	**Test procedure**:[3] Organization for Economic Co-operationand Development (OCDE); inoculation sludge (100 mg/L) from UASB; MS initial conc. = 1–2 mg/L with ethanol being co-substrate in the anaerobic reactors; test solution = 100 mL at 35°C. SUC and CBZ may be used as potential tracers for soil and groundwater contamination.[4]

(Continued)

Table 17.1 (Contnued)

Reactor/System	MC Removal (%)	Operating Conditions/ Other Results/Comments
CWs (HSSFCW, FWSCW + or their combinations).	18 selected MCs (PPCPs): 39–98 in HSSFCW;[5] Others PPCPs: 0–99, but for CMZ = −316 to −44 and for TMP −293.[6]	The environmental risk posed by residual MCs was not fully reduced in most of the cases; design info.[6] MCs may be removed due to anaerobic/aerobic biodegradation, plant uptake, photolytic degradation, and sorption.[5,6]
Anaerobic-aerobic biofilm	No reduction of CBZ by either anaerobic or aerobic biofilms over a dosing period of 51 d followed by an additional 23 d of observation. CAFF was reduced by 11.09% in the upflow anaerobic packed bed biofilm reactor (UAnPBBR) and by 91.90% in the aerobic trickling filter biofilm reactor (TF). E1 and E2 had minimal reduction in the UAnPBBR but 99.67% reduction in the TF, while EE2 was reduced by 1.62% in the AnPBBR and 20.36% in the TF. On average, reduction was 3-log for *E. coli* (CN-13) and 1-log for F+ specific coliphage (MS2).[7]	The system comprised an upflow anaerobic packed bed biofilm reactor (UAnPBBR) + a trickling filter (TF). HRT = 1.21 d, and organic loading rate = 2.0 kg $BOD_5/m^3/d$ for combined volume of the UAnPBBR and TF.
UASB	Inefficient in bisphenol A and DCF removal; partial removal of bezafibrate and TMP; increased the concentration of nonylphenol; SMX removal increased with HRT;[9] can remove 500–1226 mg/L phenol effectively (>99% at HRT = 24 h) with co-substrates such as glucose, VFA, and dilute molasses;[12,13] removal of E1 and EE2 >90, heavy metals (Ca^{2+}, Mg^{2+}, Cu^{2+}, Pb^{2+}, Zn^{2+}, Fe^{2+} (= 65–100)), total nitrogen (= 55–85), total phosphorus (61.8–65.1), sulfate (= 100), TCE > 85, SMX (6.2–85.8), TMP (6.2–91.1), CBZ (38.2–48.9), LC50 of *D. rerio* (0–30) and *D. dubia* (10–25%).[15] The UASB can tolerate up to a concentration of 18 PP-MPs/gTS, while a further elevated concentration of 50 PP-MPs/gTS caused a remarkable inhibition (58%) of CH_4 production.[16]	[9]Three UASB used: vol = 7.2, 16.8, and 14.2 m^3; HRT = 5.4, 8.7, and 11.0 h; SRT = 32, 47, and 57 h; the UASBs were fed with wastewater after primary sedimentation.[13] HRT affects gas production (at HRT = 12 h, the methane yield was 0.308 L/g_{COD} removed).

Table 17.1 (Contnued)

Reactor/System	MC Removal (%)	Operating Conditions/ Other Results/Comments
UASB + SABF + HSSFCW.	**UASB:**[10] 13 MCs (i.e., lipid regulator, nervous stimulant, anti-inflammatory and endocrine disrupters) = 51–58; mass reduction of total MCs = 62.1 nervous stimulant = 61.3; anti-inflammatory = 99.1; endocrine disruptors = 32.5.	Organic loading rates: 1, 2, 3, and 4 kg $COD/m^3/d$ for d. 1–163, 164–300, 301–378; 379–420, respectively. HSSFCW optimized and maintained the final effluent quality for most MCs, reducing concentrations and ecological risks.
UASB + DHS.	The removal of total coliform (TC), fecal coliform (FC), and fecal streptococci (FS) was very low. Most of them were removed by the follow-up DHS.[11]	HRT = 6.0 h for UASB and 3.2 h for DHS.
UASB + TF.	Removed 89–95% of triclosan, 15 PAHs, estrogens (E1, E2, E3, EE2) and 8 PBDE congeners; removed 92% (33% in aqueous phase) PAHs; removed 86% (43% in aqueous phase) PBDEs, but primary treatment accounted for 90% PBDE removal; UASB removal for E1 = ~10% and E3 = ~20%; observed an increment in the aqueous concentration of some PBDE congeners (BDE 47, 100, 183) after primary treatment.[14]	Data were from 3 field-scale WWTPs. Populations served (in millions): ASP = 1.6; UASB + TF = 1.1; and WSP = 0.015; total removal 95–99% for WSPs, 89–95% for UASB + TF, and 79–94% for ASP. The partitioning of the chemicals (specifically PAHs, triclosan and PBDEs) onto the suspended solids indicates that sorption plays an important role in their removal in these systems. The effluent concentrations of triclosan, some estrogens, PAHs and BDE 209 were above European environmental quality standards (EQS).
WAPs (= anaerobic pond + facultative pond).	Removed 95–99% of triclosan, 15 PAHs, estrogens (E1, E2, E3, EE2) and 8 PBDE congeners; removed 96% (64% in aqueous phase) PAHs; removed 95% (94% in aqueous phase) PBDEs; removed 99% E1 and 98% E3.[14]	
Landfill (lysimeter).	**MPs:**[17] after 854 d treatment, removal of HDBP = 13 and oxo-degradable plastics = 27.	Each lysimeter is a cylinder (1.9 m height, 0.5 m diameter) built with acrylic.

(Continued)

Table 17.1 (Contnued)

Reactor/System	MC Removal (%)	Operating Conditions/ Other Results/Comments
Anammox reactors (batch, CSTR, UASB).	The following pollutants have inhibition effects on anammox processes: antibiotics, aromatics, azoles, surfactants, microplastics, organic solvents, humic substances, biodegradable organic matter, or metals and metallic nanoparticles.[18]	Anammox bacteria live as aggregates, enhancing their tolerance to adverse environments and increasing protection to toxic loads. However, several MCs can disrupt this protection and cause irreversible toxicity toward anammox cells.

References: 1 = Oberoi et al. (2022); 2 = supporting material in Oberoi et al. (2022); 3 = Motteran et al. (2022); 4 = Biel-Maeso et al. (2019); 5 = Ilyas et al. (2021); 6 = Chowdhury et al. (2021); 7 = Lin et al. (2020); 8 = Fernández et al. (2017); 9 = Queiroz et al. (2012); 10 = de Oliveira et al. (2020); 11 = Tawfik et al. (2015); 12 = Veeresh et al. (2005); 13 = Ke et al. (2004); 14 = Komolafe et al. (2021); 15 = Collivignarelli et al. (2021); 16 = Pittura et al. (2021); 17 = Xochitl et al. (2021); 18 = Madeira and de Araújo (2021).
MSs: Analgesic (ACT = Acetaminophen); anticonvulsant (CBZ = Carbamazepine, CMZ = Chlomethiazole; PRI = Primidone); antibiotics (SMX = Sulfamethoxazole, TMP = Trimethoprim, CIP = Ciprofloxacin, AMOX = Amoxicillin, AMPI = Ampicillin, ERY = Erythromycin); antidepressant (AMI = Amitriptyline); anti-inflammatory (NPX = Naproxen, IBU = Ibuprofen, KET = Ketoprofen, DCF = Diclofenac); β-blocker (ATN = Atenolol); *Cytostatic drugs = cyclophosphamide, azathioprine, methotrexate, doxorubicin, epirubicin, flutamide, mitotane, and tamoxifen; lipid regulator (GMF = gemfibrozil); and stimulant (CAFF = caffeine). ACE = acesulfame; SUC = sucralose; ASP = aspartame; CYC = cyclamate; TCE = Trichloroethylene. MPs = Microplastics. Hormones: E1 = Estrone, E2 = 17β-estradiol, EE2 = 17α-ethynylestradiol. PAHs = Polycyclic aromatic hydrocarbons. PBDE = Polybrominated diphenyl ether. Mycotoxins: (FBs: fumonisins (FB1 + FB2 + FB3), AFB1 = aflatoxin B1, DON = deoxynivalenol, ZEN = zearalenone, OTA = ochratoxin A).
Novel AnMBR: AnO-, AnF-, AnE-, MA-n-, and AnD-MBR = Anaerobic osmotic-, Anaerobic fluidized-, Anaerobic electrochemical-, Microaeration enhanced anaerobic-, and Anaerobic dynamic-membrane bioreactor.
Treatment systems: ASP = activated sludge process; TF = Trickling filter; CWs = Constructed Wetlands (including HSSFCE = Horizontal subsurface flow constructed wetland; FWSCW = Free water surface constructed wetland; VFCW = Vertical flow constructed wetland); UASB = upflow anaerobic sludge blanket reactor; WAP = Waste stabilization pond; SABF = Submerged aerated biological filters; DHS = Downflow aerobic hanging sponge system.

eliminate MCs through biotransformation (Stasinakis 2012). Table 17.1 indicates that AD (or its combination with other processes) can remove many MCs efficiently; the combination of the AD with other processes would affect the treatment efficiency significant. The use of sludge pretreatment methods does not seem to enhance MCs removal, whereas encouraging results have been reported when AD was combined with other treatment methods (Table 17.1). The UASB reactor may be crucial in the removal of MCs; thus, it is widely utilized in wastewater treatment technologies in developing countries (Do Nascimento et al. 2021; Komolafe et al. 2021; Rodrigues-Silva et al. 2022). However, for some MCs, UASB alone may not be efficient enough. Table 17.1 shows that AnMBRs represent a potential method for the effective treatment of MCs in industrial and municipal wastewater (Lim et al. 2019; Ji et al. 2020). MBR can significantly increase the rate of MCs removal and biogas production when compared to conventional AD processes, although MCs are still present in the final biological material. However, these alterations might not be technically and economically feasible, depending on the socio-economic conditions in specific regions of the aforementioned developing countries (Do Nascimento et al. 2021). Recent research has attempted to apply physical retention mechanisms in systems intended to enhance the removal of MCs during conventional anaerobic processing (Venegas et al. 2021).

These methods include selective membrane separation downstream using nanofiltration (NF) or membrane distillation, the addition of ingredients that improve absorption, including activated carbon, support materials (attached growth reactors), or highly selective membranes (AnMBRs) (Harb et al. 2019; Lim et al. 2019), and the hybrid anaerobic and aerobic processes (Harb et al. 2019). Activated carbon adsorption offers a partial solution despite its outstanding MCs removal efficiency since it separates MCs from one phase and concentrates them in another. Since total removal of MCs has not been accomplished, activated carbon saturated with MCs is a feasible alternate option for MCs treatment, but its reuse and final disposal still requires additional research.

17.4.1.2 Removal of Different MCs and Associated Mechanisms

Considerable research has been mainly focused on specific groups of compounds (e.g., PPCPs, endocrine-disrupting chemicals (EDCs), heavy metals, PAHs), while there are fewer or no data for others (brominated flame retardants, organotins, NMs/NPs, MPs) (Bhandari et al. 2009; Surampalli et al. 2018). Table 17.1 indicates that AD alone or AD-based hybrid systems can remove most of PPCPs, hormones, heavy metals, PAHs, PBDEs, pesticides, nutrients (N and P), and MPs, but not much for pathogens/microbial cells, surfactants, and NPs/NMs (Surampalli et al. 2018).

According to previous research, the two main methods for removing MCs in closed reactors are biotransformation and adsorption (Liu and Wong 2013; Venegas et al. 2021), which are also the mechanisms for removing MCs in natural systems (e.g., CWs, ponds), in addition to plant uptake and photolytic degradation (Chowdhury et al. 2021; Ilyas et al. 2021). Usually, acidogenesis, acetogens, and methanogens are significant contributors to the anaerobic biotransformation of MCs. Hydrolysis appears to play a minimal role. Since they catalyze biochemical reactions, enzymes are essential to all biological processes. It should be noted that reductive dehalogenation and co-metabolism are the two other mechanisms that often occur in site remediation (or other aqueous remediation systems) of hazardous pollutants, many of which belong to MCs (even though their concentrations may be much higher) (USEPA 1986; Sharma and Reddy 2004; Metcalf and Eddy/AECOM 2014). For example, diclofenac (DCF) could be degraded by reductive dechlorination followed by decarboxylation of phenylacetate carboxylic acid group (catalyzed by decarboxylase) under anaerobic conditions (DCF-II) (Gonzalez-Gil et al. 2019; Granatto et al. 2020).

The primary determinants of biotransformation are the physico-chemical characteristics of the MCs, process temperature, SRT, and organic loading rate (Gonzalez-Gil et al. 2018). The nature of the microbial community and metabolism are additional factors affecting the elimination of various MCs in AD. For the majority of MCs, anaerobic systems are typically less effective than aerobic systems, particularly when running at a short HRT (<10 hours) (Falås et al. 2016; Do Nascimento et al. 2021). The duration of the compound–biomass interaction is an important element in the anaerobic biotransformation of MCs, i.e., longer contact time enhances the AD process (Harb et al. 2019). Using a longer HRT and/or reduced organic loading rate are the simplest ways to lengthen the exposure period.

MCs (e.g., mycotoxins, antibiotics, MPs, NMs/NPs) could positively or negatively affect the AD system. For example, some NPs may stimulate AD processes and methane yield. Wolfson et al. (2018) reported that NPX is possibly O-demethylated to 6-O-desmethylnaproxen (catalyzed by O-methyltransferases) via a syntrophic relationship between homoacetogens, acetate oxidizers, and hydrogenotrophic methanogens (NPX-IB). Zhang et al. (2021) reported that aflatoxin B1(AFB1) (one of the most concerned mycotoxins) was poorly degraded in AD and would cause inhibitions in the short term. Oxytetracycline, norfloxacin (the two representatives of antibiotics), and MPs would inhibit methane production of the AD system. Usually, MCs adverse effects on AD performance include the following: i) decreased COD removal and methane yield; ii) inhibition of specific microbial groups or metabolic reactions or stimulation of some microbial products (e.g., extracellular polymeric substances, EPS); and iii) change in population dynamics and structure of the microbial community (Cheng et al. 2021; Oberoi et al. 2022; Do et al. 2022). For example, antibiotics could inhibit the VFAs degrading bacteria and shift acetoclastic methanogens to hydrogenotrophic methanogens that could lead to a lower yield of methane (Cheng et al. 2018b). Currently, sufficient information is not available on which factor plays a major role in the negative effect on AD.

17.4.2 Biodegradation of Selected MCs in AD Processes

Table 17.2 shows biodegradation of four classes of selected MCs in AD systems, as significant information has been accumulated on biodegradation of PPCPs, many EDCs, PAHs, pesticides, antimicrobials/antibiotics, etc. The table shows that AD processes can remove hormones and their conjugates fairly well, but may not reduce plasticizers, surfactants, and fire retardants in a significant way. Thus, combining AD with other processes to treat wastewater contaminated with MCs would be a better direction. Additionally, information on treatment/management of MPs and NPs/NMs has been accumulated recently, which is introduced below.

17.4.2.1 MPs

The forms of MPs include fragment, pellets, fibers, film, and foam, with fiber being the one most enriched in WAS (62.5–78.5%) and also more harmful than other forms to the ecosystem (He et al. 2021; Miri et al. 2022). MPs can be classified into polyethylene (PE), acylate and polypropylene (PP), High Density Polyethylene (HDPE), Low Density Polyethylene (LDPE), Polyvinyl Chloride (PVC), Polyethylene Terephthalate (PET), Polybutylene Terephthalate (PBT), Polyamide (PA6), Polystyrene (PS), Polycarbonates (PC), Polyoxymethylene (POM), Polyphenylenesulfide (PPS), etc. The six types of most frequently detected MPs are PS, PE, PVC, PET, PES, and polyamide (PA6) (Zhang et al. 2021b).

In WWPTs, 72–99.4% of MPs in wastewater are mainly removed by transferring them to the solid phase (e.g., WAS) (Gatidou et al. 2019; He et al. 2021), which definitely would affect the follow-up treatment (e.g., in AD processes) and final disposal of the sludge (land

Table 17.2 Biodegradation of selected MCs in AD processes.

MCs Class	Removal (%) of MCs in AD Processes
Plasticizers	DEHP (95.3) and DBP (94.7): removed in mesophilic anaerobic reactor;[1] composting removed 58% of the DEHP in the raw sludge, and 34% of that in the anaerobically digested sludge during 85 days stabilization in compost bins;[2] characterization and genome analysis of phthalate esters-degrading strains were reported.[3–6]
Surfactant	In AD, the degradation of surfactants is less likely to occur (e.g., in full-scale anaerobic digesters: 18–35);[7] LAS removal rate decreased from 77.2 to 55.3 when the organic loading rate increased;[8] AO-R_{12} and AO-R_{14} inhibited biogas production by ~90%. AO-cocoamido was biodegraded by 60.8%.[9] Surfactant can enhance the efficiency of hydrolysis, affect the transformation of short-chain fatty acids, change the activity of methanogens, and promote the solubility of contaminants in sludge, such as organic contaminants and heavy metals.[10]
BFRs and Plastic additives.	Most of BFRs are associated with suspended solids; BDR-15 was removed via reductive debromination in a biofilm plug-flow reactor;[11] anaerobic microbes that debrominate PBDEs and their associated synergistic partnerships with non-dehalogenating microbes were reviewed;[12] the strains mediating organohalide respiration of both PBDEs and chloroethenes were reported;[13] the use and abundance of plastic additives, their health-related risks, the microorganisms capable of degrading them, the associated mechanism of biodegradation, and current innovations capable of improving the efficiency of bioremediation were reviewed.[14]
Hormones and conjugates.	Anaerobic swine waste lagoon:[1] Lagoon slurry (= liquid phase, including hormones in both dissolved and suspended slurry phase) (conc. and removal): E1 = 10,200 (30%); E2β = 237 (72%); E2α = 793 (53%); E3 = 325 (46%); E1-3S = 0.62 (77%); E2β-17S = 0.019 (466%); not detected for: E1-3G, E2β-3S, E2β-17G, E3-3S, E3-17S, E3-3G. Lagoon sludge (sampled 0.3 m below the sludge/slurry interface) (conc. and removal): E1 = 206,000 (109%); E2β = 12,400 (59%); E2α = 8940 (86%); E3 = 9830 (93%); E1-3S = 45 (115%); E2β-17S = 0.019(466%); not detected for: E1-3G, E2β-3S, E2β-17S, E2β-17G, E3-3S, E3-17S, E3-3G. 17α-E2: more obstinate under anaerobic condition compared to aerobic condition; Fe(III) and DOM could synergistically facilitate biodegradation of E2 in anaerobic aqueous environment; E1 was able to convert to 17α-E2 under aerobic and anaerobic, or methanogenic nitrate and sulfate-reducing conditions by wastewater irrigation practices.[16]

References: 1 = Chang et al. (2005); 2 = Marttinen et al. (2003); 3 = Feng et al. (2018); 4 = Wright et al. (2020); 5 = Tiwari et al. (2022); 6 = Chiu et al. (2022). 7 = Bhandari et al. (2009); 8 = Faria et al. (2019). 9 = Ríos et al. (2017); 10 = He et al. (2019); 11 = Rayne et al. (2003); 12 = Zhao et al. (2019); 13 = Zhao et al. (2021); 14 = Lumio et al. (2021); 15 = Yost et al. (2013); Yost et al. (2014); 16 = Yu et al. (2019).
MCs: DBP = Dibutyl phthalate; DEHP = Di(2-ethylhexyl)phthalate; LAS = Linear alkyl sulfonate; AO-R_{12} = Lauramine oxide; AO-R_{14} = Myristamine oxide; AO-cocoamido = Cocamidopropylamine oxide; BFRs = Brominated fire retardants; BDR-15 = 4,4'-dibromodiphenyl ether; PBDEs = polybrominated diphenyl ethers; Hormones: E1 = Estrone; E2 = 17ß-estradiol; E3 = Estriol; EE2 = 17α-ethinylestradiol; E1-3G = Estriol-3-glucuronide; E1-3S = Estrone-3-sulfate; E3-3G = Estriol-3-glucuronide; EE2-3G = Ethinylestradiol-3-glucuronide; EE2-3S = Ethinylestradiol-3-sulfate; 17α-E2 = 17α-Estradiol.
Source: Adapted from Bhandari et al., 2009; Surampalli et al., 2018.

application). Several papers reported that the current information on occurrence, effect, and fate of MPs in anaerobic digestion of WAS (He et al. 2021; Pittura et al. 2021). He et al. (2021) reported that different MPs would affect WAS anaerobic digestion in a different way. For example, PVC, PE, PBS, and PET may stimulate the solubilization step of the WAS, but may have an inhibitive effect on hydrolysis, acidogenic, and methanogenic steps. PA6, PS (5 and 80 nm), and PES may have no (or some inhibition) effect on the solubilization, hydrolysis, and acidogenic steps of WAS, but an inhibitive effect on the methanogenic step. However, all MPs affect microbial community structure, with some species being reduced (e.g., 4.3–15.2% for hydrolysis, 0.6–63.2% for acidogenic, and 10.0–26.0% for methanogenic), depending on the types and concentrations of the MPs. Zhang et al. (2021b) summarized that exposures of AD functional bacteria to PE, PVC, and PET could induce the generation of reactive oxygen species (ROS) causing reduction of cell viability and restrained hydrolysis, acidification, and methanogenesis. PVC was found to decrease both ammonia and nitrite removal efficiencies, while PBS had a mild inhibitory effect only at 0.5 g/L. Reduced amounts of amino acids and increased polysaccharide in the sludge exposed to 0.5 g/L PVC or PBS led to a decrease in the hydrophobicity and looser sludge structure. PBS stimulated the anammox bacteria *Ca. Brocadia* and the relative nitrogen metabolism modules, while PVC reduced the abundance of the species (Madeira and de Araújo 2021). The presence of MPs can change the effects of other pollutants on anaerobic digestion of WAS as MPs are a sink of hydrophobic organic contaminants that could be transportation vectors of contaminants. More information on biodegradation of MPs can be found in the literature (Jiménez et al. 2022; Miri et al. 2022; Temporiti et al. 2022).

17.4.2.2 NMs/NPs

NPs have been used to enhance the overall performance of AD processes, such as the following: i) zero-valent iron (ZVI) NPs; ii) metallic and metal oxides NPs; iii) carbon-based NPs; and iv) multi-compound NPs (Dehhaghi et al. 2019; Zhu et al. 2021; Jadhav et al. 2022). For example, ZVI NPs could stabilize the AD process and stimulate the growth of beneficial microorganisms to the AD process and the enzymes involved. Metallic and metal oxides NPs (e.g., Ni, Co, Fe, Fe_3O_4, Fe_2O_3, MgO, Al_2O_3) can enhance biogas production and methane content; however, NPs such as ZnO and TiO_2 may have an inhibitory effect on the biogas production process when used at higher concentrations. Generally, carbon nanotubes (CNTs) could noticeably enrich electroactive bacteria such as *Caloramator sp.* or *Geobacter sp.* along with other bacteria or methanogenic archaea (*Methanosarcina*, *Methanosaeta*, etc.). Single-wall carbon nanotubes (SWCNTs) could facilitate direct interspecies electron transfer (DIET) during the AD process, while multi-wall carbon nanotube (MWCNT) could enhance biogas and methane production rates by transferring electrons toward methanogens (Ambuchi et al. 2017).

NPs can facilitate hydrolysis processes by providing a large surface area to volume ratio for microorganisms to attach to the active sites of organic molecules (Hsieh et al. 2016). Alternatively, NPs can also function as electron donors/acceptors and cofactors of important enzymes involved in various bioprocesses enhancing their yields (Dehhaghi et al. 2019). The NPs are observed to stick on the surface of anaerobic sludge due to their negative surface charges, then these nanoparticles are easier to penetrate through cell membranes, breakdown complex organics, produce more biomass for microorganisms, and thus facilitate metabolism in the following phases of AD (Zhu et al. 2021). The impact of NPs on AD highly depends on their released free ions as well as their characteristics (sizes, shapes, surface area, surface charge), dispersion degree, and stability (Nel et al. 2006; Zhu et al. 2021).

However, NPs/NMs can exhibit inhibitory effects if used at beyond-optimal concentrations. Zhu et al. (2021) summarized tolerate concentrations of selected metallic NPs on the methane-production stage and overall AD process. Most of the iron-based nano-additives had the highest reported tolerant concentrations up to 20 mg/L, whilst the appearance of nano-ZnO and Cu-based nano-additives provoked adverse impacts on the methane-production process at very low concentrations (<1 mM) (Zhu et al. 2021). Cell membrane disruption and generation of reactive oxygen species have been proposed as the main toxicity mechanisms of some NPs such as ZVI NPs. The toxicity of NPs depends on their type, exposure time, and concentration in a given reaction (Wang and Chen 2016). More importantly, the large-scale application of NPs could trigger some concerns, such as environmental and health-related complications (Gottschalk et al. 2013). Thus, optimizing the dose of the NPs of interest in AD must be understood.

17.5 Challenges and Future Perspectives

AD has long been widely used because of its special ability not only to treat industrial wastewater but also to produce methane that provides a renewable energy source for global demand (McCarty 2001). To further expand AD's applications for MCs removal and management, considerable research is still needed to address the following issues:

- Feasibility/reliability of AD processes for MCs treatment: We need to know for different MCs from different sources, what AD process works better with higher removal efficiency, lower odor generation, and easier management of refractory organic matter decomposition.
- Sustainable AD methods or AD-based combinations to remove MCs more efficiently: MCs are difficult components to remove completely from wastewater with the AD method. Considering removal efficiency and sustainability principles, more research is needed to design robust and cost-effective AD treatment processes. Moreover, the optimum application scenarios and strategies to selecting AD processes or combinations of different treatment systems need to be evaluated further.
- Better understanding of biotransformation processes and other removal phenomena occurred in AD process, as well as on identification of (bio)transformation products. MCs are usually at low concentrations, and the complex matrices may interfere with detection. The potential impact of combined contaminants on AD performance are not clear.

To take NPs/NMs as an example, large-scale applications of using NPs such as ZVI NPs in the AD process and biogas production have not yet been considered in detail. Future research should focus on the following: i) economic considerations (e.g., reducing the price of NPs); ii) system improvement (e.g., various attributes of NPs when used as additives in AD processes, including facilitating mixing and pumping operations, enriching the population and diversity of beneficial microorganisms in AD systems, improving biogas release, and inducing the production and activity of AD-related enzymes); iii) environmental issues (the fate of NPs in AD processes and minimizing adverse environmental/health consequences); and iv) effective guidelines that encompass environmental, economic, and technical (e.g., preventing inhibitory effects) aspects of NPs applications).

17.6 Conclusions

AD is a successful way to manage wastewater because of its safety, cost savings, and energy recovery. Usually, the suspended growth AD process is used to treat high-strength industrial wastewater and/or organic sludge, while the attached growth AD process uses media for microbes to grow. AD has demonstrated its potential as a technology for MCs removal as AD, or its combination with other processes. can remove many MCs very efficiently. The use of sludge pretreatment methods may not enhance MCs removal, whereas encouraging results have been reported when AD was combined with other treatment methods. UASB may be crucial in removal of many MCs but, for some MCs, may not be efficient enough. AnMBRs represent a potential method for effective treatment of MCs in industrial and municipal wastewater, but the associated cost is an issue. HRT/SRT, organic volumetric loading rate, temperature, and pH are the most important operational parameters affecting AD performance. Sludge characteristics (type of sludge, adaptation on the compound), physico-chemical properties of MCs, and metabolism would also affect MCs biodegradation. To further expand AD's applications for MCs removal and management, considerable research is still needed.

References

Abbasi, T., Tauseef, S.M., and Abbasi, S.A. (2012). A brief history of anaerobic digestion and "Biogas." In: *Biogas Energy* (eds. Abbasi, T., Tauseef, S.M., and Abbasi, S.A.), 11–23. New York: Springer.

Abdelrahman, A.M., Ozgun, H., Dereli, R.K. et al. (2021). Anaerobic membrane bioreactors for sludge digestion: current status and future perspectives. *Critical Reviews in Environmental Science and Technology* 51: 2119–2157.

Adekunle, K.F. and Okolie, J.A. (2015). A review of biochemical process of anaerobic digestion. *Advances in Bioscience and Biotechnology* 6 (3): 205–212.

Ambuchi, J.J., Zhang, Z., Shan, L. et al. (2017). Response of anaerobic granular sludge to iron oxide nanoparticles and multi-wall carbon nanotubes during beet sugar industrial wastewater treatment. *Water Research* 117: 87–94.

Angelidaki, I., Ellegaard, L., and Ahring, B.K. (2003). Applications of the anaerobic digestion process. In: *Biomethanation II* (eds. Ahring, B.K. and and Angelidaki, I.), 1–33. Berlin, Heidelberg: Springer.

Aslanzadeh, S. (2014). *Pretratment of Cellulosic Waste and High-Rate Biogas Production.* Doctoral Thesis on Resource Recovery. Boråz: University of Borås, 1–50.

Aus der Beek, T., Weber, F.A., Bergmann, A. et al. (2016). Pharmaceuticals in the environment: global occurrences and perspectives. *Environmental Toxicology and Chemistry* 35: 823–835.

Bhandari, A., Surampalli, R., Adams, C.D. et al. (2009). *Contaminants of Emerging Environmental Concern*. Reston, VA: ASCE.

Biel-Maeso, M., González-González, C., Lara-Martín, P.A., and Corada-Fernández, C. (2019). Sorption and degradation of contaminants of emerging concern in soils under aerobic and anaerobic conditions. *Science of the Total Environment* 666: 662–671.

Chandra, R., Takeuchi, H., and Hasegawa, T. (2012). Methane production from lignocellulosic agricultural crop wastes: a review in context to second generation of biofuel production. *Renewable and Sustainable Energy Reviews* 16 (3): 1462–1476.

Chang, B.V., Liao, G.S., and Yuan, S.Y. (2005). Anaerobic degradation of di-n-butyl phthalate and di-(2-ethylhexyl) phthalate in sludge. *Bulletin of Environmental Contamination and Toxicology* 75 (4): 775–782.

Cheng, D., Ngo, H.H., Guo, W. et al. (2021). Improving sulfonamide antibiotics removal from swine wastewater by supplying a new pomelo peel derived biochar in an anaerobic membrane bioreactor. *Bioresource Technology* 319: 124160.

Cheng, D.L., Ngo, H.H., Guo, W.S. et al. (2018a). Anaerobic membrane bioreactors for antibiotic wastewater treatment: performance and membrane fouling issues. *Bioresource Technology* 267: 714–724.

Cheng, D.L., Ngo, H.H., Guo, W.S. et al. (2018b). Problematic effects of antibiotics on anaerobic treatment of swine wastewater. *Bioresource Technology* 263: 642–653.

Chiu, K.K., Bashir, S.T., Abdel-Hamid, A.M. et al. (2022). Isolation of DiNP-degrading microbes from the mouse colon and the influence DiNP exposure has on the microbiota, intestinal integrity, and immune status of the colon. *Toxics* 10 (2): 75.

Choi, S., Yoom, H., Son, H. et al. (2022). Removal efficiency of organic micropollutants in successive wastewater treatment steps in a full-scale wastewater treatment plant: bench-scale application of tertiary treatment processes to improve removal of organic micropollutants persisting after secondary treatment. *Chemosphere* 288: 132629.

Chowdhury, N. (2005). *Microbial production of hydrogen under mesophilic conditions*. Electronic Theses and Dissertations. University of Windsor, Canada. https://scholar.uwindsor.ca/etd/2722

Chowdhury, S.D., Surampalli, R.Y., and Bhunia, P. (2021). Potential of the constructed wetlands and the earthworm-based treatment technologies to remove the emerging contaminants: a review. *Journal of Hazardous, Toxic, and Radioactive Waste* 26 (2): 04021066.

Christy, P.M., Gopinath, L.R., and Divya, D. (2014). A review on anaerobic decomposition and enhancement of biogas production through enzymes and microorganisms. *Renewable and Sustainable Energy Reviews* 34: 167–173.

Collivignarelli, M.C., Abbà, A., Caccamo, F.M. et al. (2021). Applications of up-flow anaerobic sludge blanket (UASB) and characteristics of its microbial community: a review of bibliometric trend and recent findings. *International Journal of Environmental Research and Public Health* 18 (19): 10326.

de Lemos Chernicharo, C.A. (2007). Anaerobic reactors. In: *Biological Wastewater Treatment Series*, vol. 4, 59–60 London: IWA Publishing.

de Oliveira, J.F., Fia, R., Rodrigues, F.N. et al. (2020). Quantification, removal and potential ecological risk of emerging contaminants in different organic loads of swine wastewater treated by integrated biological reactors. *Chemosphere* 260: 127516.

Dehhaghi, M., Tabatabaei, M., Aghbashlo, M. et al. (2019). A state-of-the-art review on the application of nanomaterials for enhancing biogas production. *Journal of Environmental Management* 251: 109597.

Deublein, D. and Steinhauser, A. (2011). *Biogas from Waste and Renewable Resources*. Weinheim, Germany: Wiley-VCH. https://doi.org/10.1002/9783527632794.

Do Nascimento, J.G.D.S., Silva, E.V.A., Dos Santos, A.B. et al. (2021). Microaeration improves the removal/biotransformation of organic micropollutants in anaerobic wastewater treatment systems. *Environmental Research* 198: 111313.

Do, T.M., Choi, D., Oh, S., and Stuckey, D.C. (2022). Anaerobic membrane bioreactor performance with varying feed concentrations of ciprofloxacin. *The Science of the Total Environment* 803: 150108.

Eastman, J.A. and Ferguson, J.F. (1981). Solubilization of particulate organic carbon during the acid phase of anaerobic digestion. *Water Pollution Control Federation* 53: 3. https://www.osti.gov/biblio/7152148-solubilization-particulate-organic-carbon-during-acid-phase-anaerobic-digestion.

Ebrahimbabaie, P., Yousefi, K., and Pichtel, J. (2022). Photocatalytic and biological technologies for elimination of microplastics in water: current status. *The Science of the Total Environment* 806 (2): 150603.

Falås, P., Wick, A., Castronovo, S. et al. (2016). Tracing the limits of organic micropollutant removal in biological wastewater treatment. *Water Research* 95: 240–249. https://doi.org/10.1016/j.watres.2016.03.009.

Faria, C.V., Delforno, T.P., Okada, D.Y., and Varesche, M.B.A. (2019). Evaluation of anionic surfactant removal by anaerobic degradation of commercial laundry wastewater and domestic sewage. *Environmental Technology* 40 (8): 988–996.

Feng, L., Liu, H., Cheng, D. et al. (2018). Characterization and genome analysis of a phthalate esters-degrading strain *Sphingobium yanoikuyae* SHJ. *BioMed Research International* 2018: 3917054.

Fernandes, T.V., Klaasse Bos, G.J., Zeeman, G. et al. (2009). Effects of thermo-chemical pre-treatment on anaerobic biodegradability and hydrolysis of lignocellulosic biomass. *Bioresource Technology* 100 (9): 2575–2579.

Fernández, L., Louvado, A., Esteves, V.I. et al. (2017). Biodegradation of 17β-estradiol by bacteria isolated from deep sea sediments in aerobic and anaerobic media. *Journal of Hazardous Materials* 323 (A): 359–366.

Gatidou, G., Arvaniti, O.S., and Stasinakis, A.S. (2019). Review on the occurrence and fate of microplastics in sewage treatment plants. *Journal of Hazardous Materials* 367: 504–512.

GerardiṛGwrarG, M.H. (2003). Fermentation. In: *The Microbiology of Anaerobic Digesters* (ed. Gerardi, M.H.), 43–50. John Wiley & Sons, Ltd. https://doi.org/10.1002/0471468967.ch6.

Ghangrekar, M.M. and Behera, M. (2014). 3.5-suspended growth treatment processes. In: *Comprehensive Water Quality and Purification* (ed. Ahuja, S.), 74–89. Waltham: Elsevier. https://doi.org/10.1016/B978-0-12-382182-9.00087-6.

Gonzalez-Gil, L., Krah, D., Ghattas, A.-K. et al. (2019). Biotransformation of organic micropollutants by anaerobic sludge enzymes. *Water Research* 152: 202–214. https://doi.org/10.1016/j.watres.2018.12.064.

Gonzalez-Gil, L., Mauricio-Iglesias, M., Serrano, D. et al. (2018). Role of methanogenesis on the biotransformation of organic micropollutants during anaerobic digestion. *Science of the Total Environment* 622–623: 459–466.

Gottschalk, F., Sun, T., and Nowack, B. (2013). Environmental concentrations of engineered nanomaterials: review of modeling and analytical studies. *Environmental Pollution* 181: 287–300.

Granatto, C.F., Grosseli, G.M., Sakamoto, I.K. et al. (2020). Methanogenic potential of diclofenac and ibuprofen in sanitary sewage using metabolic cosubstrates. *The Science of the Total Environment* 742: 140530.

Harb, M., Lou, E., Smith, A.L., and Stadler, L.B. (2019). Perspectives on the fate of micropollutants in mainstream anaerobic wastewater treatment. *Energy Biotechnology: Environmental Biotechnology* 57: 94–100.

Hassan, S., Zwain, H., and Dahlan, I. (2013). Development of anaerobic reactor for industrial wastewater treatment: an overview, present stage and future prospects. *Journal of Advanced Scientific Research* 4: 07–12.

He, Q., Xu, P., Zhang, C. et al. (2019). Influence of surfactants on anaerobic digestion of waste activated sludge: acid and methane production and pollution removal. *Critical Reviews in Biotechnology* 39 (5): 746–757.

He, Z.W., Yang, W.J., Ren, Y.X. et al. (2021). Occurrence, effect, and fate of residual microplastics in anaerobic digestion of waste activated sludge: a state-of-the-art review. *Bioresource Technology* 331: 125035.

Hsieh, P.H., Lai, Y.C., Chen, K.Y., and Hung, C.H. (2016). Explore the possible effect of TiO_2 and magnetic hematite nanoparticle addition on biohydrogen production by *Clostridium pasteurianum* based on gene expression measurements. *International Journal of Hydrogen Energy* 41: 21685–21691.

Ilyas, H., Masih, I., and van Hullebusch, E.D. (2021). The anaerobic biodegradation of emerging organic contaminants by horizontal subsurface flow constructed wetlands. *Water Science and Technology* 83 (11): 2809–2828.

Izumi, K., Okishio, Y., Nagao, N. et al. (2010). Effects of particle size on anaerobic digestion of food waste. *International Biodeterioration & Biodegradation* 64 (7): 601–608. https://doi.org/10.1016/j.ibiod.2010.06.013.

Jadhav, P., Khalid, Z.B., Zularisam, A.W. et al. (2022). The role of iron-based nanoparticles (Fe-NPs) on methanogenesis in anaerobic digestion (AD) performance. *Environmental Research* 204 (B): 112043.

Ji, J., Kakade, A., Yu, Z. et al. (2020). Anaerobic membrane bioreactors for treatment of emerging contaminants: a review. *Journal of Environmental Management* 270: 110913. https://doi.org/10.1016/j.jenvman.2020.110913.

Jiménez, D.J., Öztürk, B., Wei, R. et al. (2022). Merging plastics, microbes, and enzymes: highlights from an international workshop. *Applied and Environmental Microbiology* 88 (14): e0072122.

Ke, S.Z., Shi, Z., Zhang, T., and Fang, H.H. (2004). Degradation of phenol in an upflow anaerobic sludge blanket (UASB) reactor at ambient temperature. *Journal of Environmental Sciences (China)* 16 (3): 525–528.

Komolafe, O., Mrozik, W., Dolfing, J. et al. (2021). Occurrence and removal of micropollutants in full-scale aerobic, anaerobic and facultative wastewater treatment plants in Brazil. *Journal of Environmental Management* 287: 112286. https://doi.org/10.1016/j.jenvman.2021.112286.

Kougias, P.G. and Angelidaki, I. (2018). Biogas and its opportunities: a review. *Frontiers of Environmental Science & Engineering* 12 (3): 14.

Lim, M., Ahmad, R., Guo, J. et al. (2019). Removals of micropollutants in staged anaerobic fluidized bed membrane bioreactor for low-strength wastewater treatment. *Process Safety and Environmental Protection* 127: 162–170. https://doi.org/10.1016/j.psep.2019.05.004.

Lim, M., Patureau, D., Heran, M. et al. (2020). Removal of organic micropollutants in anaerobic membrane bioreactors in wastewater treatment: critical review. *Environmental Science: Water Research & Technology* 6: 1230–1243.

Lin, S., Rong, K., Lamichhane, K.M. et al. (2020). Anaerobic-aerobic biofilm-based digestion of chemical contaminants of emerging concern (CEC) and pathogen indicator organisms in synthetic wastewater. *Bioresource Technology* 299: 122554.

Liu, J.-L. and Wong, M.-H. (2013). Pharmaceuticals and personal care products (PPCPs): a review on environmental contamination in China. *Environment International* 59: 208–224. https://doi.org/10.1016/j.envint.2013.06.012.

Liu, Y.B. (1989). Modeling for anaerobic fixed-bed biofilm reactors. Urbana, IL: University of Illinois. https://www.osti.gov/biblio/5536999-modeling-anaerobic-fixed-bed-biofilm-reactors.

Loupasaki, E. and Diamadopoulos, E. (2013). Attached growth systems for wastewater treatment in small and rural communities: a review. *Journal of Chemical Technology and Biotechnology* 88: 190–204. https://doi.org/10.1002/jctb.3967.

Lu, J., Gavala, H.N., Skiadas, I.V. et al. (2008). Improving anaerobic sewage sludge digestion by implementation of a hyper-thermophilic prehydrolysis step. *Journal of Environmental Management* 88 (4): 881–889.

Lu, X., Zheng, C., Zhen, G. et al. (2021). Roles of colloidal particles and soluble biopolymers in long-term performance and fouling behaviors of submerged anaerobic membrane bioreactor treating methanolic wastewater. *Journal of Cleaner Production* 290: 125816.

Lumio, R.T., Tan, M.A., and Magpantay, H.D. (2021). Biotechnology-based microbial degradation of plastic additives. *Biotechnology* 11 (7): 350.

Luo, Y., Guo, W., Ngo, H.H. et al. (2014). A review on the occurrence of micropollutants in the aquatic environment and their fate and removal during wastewater treatment. *Science of the Total Environment* 473–474: 619–641.

Ma, J., Duong, T.H., Smits, M. et al. (2011). Enhanced biomethanation of kitchen waste by different pre-treatments. *Bioresource Technology* 102 (2): 592–599.

Madeira, C.L. and de Araújo, J.C. (2021). Inhibition of anammox activity by municipal and industrial wastewater pollutants: a review. *Science of the Total Environment* 799: 149449.

Marshall, K.C., Stout, R., and Mitchell, R. (1971). Mechanism of the initial events in the sorption of marine bacteria to surfaces. *Microbiology* 68 (3): 337–348.

Marttinen, S.K., Hänninen, K., and Rintala, J.A. (2003). Removal of DEHP in composting and aeration of sewage sludge. *Chemosphere* 54: 265–272.

McCarty, P.L. (2001). The development of anaerobic treatment and its future. *Water Science and Technology: A Journal of the International Association on Water Pollution Research* 44 (8): 149–156.

Metcalf & Eddy/AECOM. (2014). *Wastewater Engineering: Treatment and Resource Recovery*, 5th edn. McGraw-Hill, Inc.

Miri, S., Saini, R., Davoodi, S.M. et al. (2022). Biodegradation of microplastics: better late than never. *Chemosphere* 286 (1): 131670.

Motteran, F., Varesche, M.B.A.V., and Lara-Martin, P.A. (2022). Assessment of the aerobic and anaerobic biodegradation of contaminants of emerging concern in sludge using batch reactors. *Environmental Science and Pollution Research* 29 (56): 84946–84961. https://doi.org/10.1007/s11356-022-21829-1.

Murphy, J.D. and Thanasit, T. (2013). Chapter 5: Fundamental science and engineering of the anaerobic digestion process for biogas production. In: *The Biogas Handbook* (eds. Wellinger, A., Murphy, J., and Baxter, D.), 104–130. Woodhead Publishing.

Nähle, C. (1991). The contact process for the anaerobic treatment of wastewater: technology, design and experiences. *Water Science and Technology* 24 (8): 179–191.

Najafpour, G. and Ebrahimi, A. (2016). Biological treatment processes: suspended growth vs. attached growth. *Iranica Journal of Energy and Environment* 7: 114–123.

Nel, A., Xia, T., Mädler, L., and Li, N. (2006). Toxic potential of materials at the nanolevel. *Science* 311: 622–627.

Neves, L., Ribeiro, R., Oliveira, R., and Alves, M.M. (2006). Enhancement of methane production from barley waste. *Biomass & Bioenergy* 30 (6): 599–603.

Oberoi, A.S., Surendra, K.C., Wu, D. et al. (2022). Anaerobic membrane bioreactors for pharmaceutical-laden wastewater treatment: a critical review. *Bioresource Technology* 361: 127667. https://doi.org/10.1016/j.biortech.2022.127667.

Pavlostathis, S.G. and Giraldo-Gomez, E. (1991). Kinetics of anaerobic treatment: a critical review. *Critical Reviews in Environmental Control* 21 (5–6): 411–490.

Pereira, M.A. (2003). Anaerobic biodegradation of long chain fatty acids: biomethanisation of biomass-associated LCFA as a challenge for the anaerobic treatment of effluents with high lipid/LCFA content. Doctoral thesis. University of Minho, Portugal. Retrieved from http://repositorium.sdum.uminho.pt.

Pittura, L., Foglia, A., Akyol, Ç. et al. (2021). Microplastics in real wastewater treatment schemes: comparative assessment and relevant inhibition effects on anaerobic processes. *Chemosphere* 262: 128415.

Qatibi, A.-I., Bories, A., and Garcia, J.-L. (1991). Sulfate reduction and anaerobic glycerol degradation by a mixed microbial culture. *Current Microbiology* 22: 47–52.

Queiroz, F.B., Brandt, E.M.F., Aquino, S.F. et al. (2012). Occurrence of pharmaceuticals and endocrine disruptors in raw sewage and their behavior in UASB reactors operated at different hydraulic retention times. *Water Science and Technology* 66 (12): 2562–2569.

Rayne, S., Ikonomou, M.G., and Whale, M.D. (2003). Anaerobic microbial and photochemical degradation of 4,4'-dibromodiphenyl ether. *Water Research* 37 (3): 551–560.

Ríos, F., Lechuga, M., Fernández-Arteaga, A. et al. (2017). Anaerobic digestion of amine-oxide-based surfactants: biodegradation kinetics and inhibitory effects. *Biodegradation* 28 (4): 303–312.

Rodrigues-Silva, F., Masceno, G.P., Panicio, P.P. et al. (2022). Removal of micropollutants by UASB reactor and post-treatment by Fenton and photo-Fenton: matrix effect and toxicity responses. *Environmental Research* 212: 113396.

Rosli, F.A., Ahmad, H., Jumbri, K. et al. (2021). Efficient removal of pharmaceuticals from water using graphene nanoplatelets as adsorbent. *Royal Society Open Science* 8 (1): 201076.

Schink, B. (1997). Energetics of syntrophic cooperation in methanogenic degradation. *Microbiology and Molecular Biology Reviews: MMBR* 61 (2): 262–280.

Schnürer, A. and Jarvis, Ä. (2010). Microbiological handbook for biogas plants. *Swedish Gas Centre Report* 2009.

Sharma, H.D. and Reddy, K.R. (2004). *Geoenvironmental Engineering: Site Remediation, Waste Containment, and Emerging Waste Management Technologies*. Wiley.

Show, K.-Y. and Tay, J.-H. (1999). Influence of support media on biomass growth and retention in anaerobic filters. *Water Research* 33 (6): 1471–1481.

Stasinakis, A.S. (2012). Review on the fate of emerging contaminants during sludge anaerobic digestion. *Bioresource Technology* 121: 432–440.

Stuckey, D.C. (2012). Recent developments in anaerobic membrane reactors. *Bioresource Technology* 122: 137–148.

Surampalli, R.Y., Zhang, T.C., Brar, S.K. et al. (2018). *Handbook of Environmental Engineering*. New York: McGraw Hill.

Switzenbaum, M.S. and Eimstad, R.B. (1987). Analysis of anaerobic biofilms. *Environmental Technology Letters* 8 (1–12): 21–32.

Tawfik, A., El-Zamel, T., Herrawy, A., and El-Taweel, G. (2015). Fate of parasites and pathogenic bacteria in an anaerobic hybrid reactor followed by downflow hanging sponge system treating domestic wastewater. *Environmental Science and Pollution Research* 22: 12235–12245.

Temporiti, M.E.E., Nicola, L., Nielsen, E., and Tosi, S. (2022). Fungal enzymes involved in plastics biodegradation. *Microorganisms* 10 (6): 1180.

Tietjen, C. (1975). From biodung to biogas: historical review of European experience. In: *Energy, Agriculture, and Waste Management: Proceedings of the 1975 Cornell Agricultural Waste Management Conference*. (ed. Jewell, W.J.), 207–260. Ann Arbor: Ann Arbor Science.

Tiwari, N., Bansal, M., Santhiya, D., and Sharma, J.G. (2022). Insights into microbial diversity on plastisphere by multi-omics. *Archives of Microbiology* 204 (4): 216.

USEPA. (1986). *Mcrobial Decomposition of Chlorinated Aromatic Compounds*. EPA/600/2-86/090, Hazardous Waste Engineering Research Laboratory, Office of Research and Development, USEPA, Cincinnati, OH 45268.

Veeken, A., Kalyuzhnyi, S., Scharff, H., and Hamelers, H.V.M. (2000). Effect of pH and VFA on hydrolysis of organic solid waste. *Journal of Environmental Engineering* 126: 12, 126.

Veeresh, G.S., Kumar, P., and Mehrotra, I. (2005). Treatment of phenol and cresols in upflow anaerobic sludge blanket (UASB) process: a review. *Water Research* 39 (1): 154–170.

Venegas, M., Leiva, A.M., Reyes-Contreras, C. et al. (2021). Presence and fate of micropollutants during anaerobic digestion of sewage and their implications for the circular economy: a short review. *Journal of Environmental Chemical Engineering* 9 (1): 104931.

Wang, D. and Chen, Y. (2016). Critical review of the influences of nanoparticles on biological wastewater treatment and sludge digestion. *Critical Reviews in Biotechnology* 36: 816–828.

Wolfson, S.J., Porter, A.W., Campbell, J.K., and Young, L.Y. (2018). Naproxen is transformed via acetogenesis and syntrophic acetate oxidation by a methanogenic wastewater consortium. *Microbial Ecology* 76: 362–371.

Wright, R.J., Bosch, R., Gibson, M.I., and Christie-Oleza, J.A. (2020). Plasticizer degradation by marine bacterial isolates: a proteogenomic and metabolomic characterization. *Environmental Science & Technology* 54 (4): 2244–2256.

Xochitl, Q.P., María Del Consuelo, H.B., María Del Consuelo, M.S. et al. (2021). Degradation of plastics in simulated landfill conditions. *Polymers (Basel)* 13 (7): 1014.

Yost, E.E., Meyer, M.T., Dietze, J.E. et al. (2013). Comprehensive assessment of hormones, phytoestrogens, and estrogenic activity in an anaerobic swine waste lagoon. *Environmental Science & Technology* 47: 13781–13790.

Yost, E.E., Meyer, M.T., Dietze, J.E. et al. (2014). Transport of steroid hormones, phytoestrogens, and estrogenic activity across a swine lagoon/sprayfield system. *Environmental Science & Technology* 48: 11600–11609.

Yu, W., Du, B., Yang, L. et al. (2019). Occurrence, sorption, and transformation of free and conjugated natural steroid estrogens in the environment. *Environmental Science and Pollution Research (International)* 26 (10): 9443–9468.

Zhang, S., Lei, Z., Dzakpasu, M. et al. (2021a). Removal of trace organic contaminants in municipal wastewater by anaerobic membrane bioreactor: efficiencies, fates and impact factors. *Journal of Water Process Engineering* 40: 101953.

Zhang, T.C., Surampalli, R.Y., Vigneswaran, S. et al. (2012). *Membrane Technology and Environmental Applications*. Reston, VA: ASCE.

Zhang, S., Lei, Z., Dzakpasu, M. et al. (2021a). Removal of trace organic contaminants in municipal wastewater by anaerobic membrane bioreactor: efficiencies, fates and impact factors. *Journal of Water Process Engineering* 40: 101953.

Zhang, Y.C., Wu, D., Su, Y.L., and Xie, B. (2021b). Occurrence, influence and removal strategies of mycotoxins, antibiotics and microplastics in anaerobic digestion treating food waste and co-digestive biosolids: a critical review. *Bioresource Technology* 330: 124987.

Zhao, S., Rogers, M.J., Cao, L. et al. (2021). Identification of reductive dehalogenases that mediate complete debromination of penta- and tetrabrominated diphenyl ethers in *Dehalococcoides* spp. *Applied and Environmental Microbiology* 87 (17): e0060221.

Zhao, S., Rogers, M.J., Ding, C., and He, J.Z. (2019). Reductive debromination of polybrominated diphenyl ethers: microbes, processes and dehalogenases. *Frontiers in Microbiology* 9: 1292.

Zhu, X., Blanco, E., Bhatti, M., and Borrion, A. (2021). Impact of metallic nanoparticles on anaerobic digestion: a systematic review. *Science of the Total Environment* 757: 143747.

18

Bio-Electrochemical Systems for Micropollutant Removal

Rishabh Raj[1], *Sovik Das*[2], *Manaswini Behera*[3], *and Makarand M. Ghangrekar*[1,2]

[1]*School of Environmental Science and Engineering, Indian Institute of Technology Kharagpur, West Bengal, India*
[2]*Department of Civil Engineering, Indian Institute of Technology Kharagpur, West Bengal, India*
[3]*School of Infrastructure, Indian Institute of Technology Bhubaneswar, Odisha, India*

18.1 The Concept of Bio-Electrochemical Systems

Energy consumption worldwide is increasing at an alarming rate and has led to the overexploitation of fossil fuels in order to meet escalating energy demands (Rahimnejad et al. 2011). Thus, the impetus is being laid on manufacturing energy-efficient appliances and overhauling existing systems to reduce power consumption. In this era of an energy crisis, a similar paradigm shift is also taking place in the domain of wastewater treatment and allied disciplines, where the recent inclination is on the development of more economic wastewater treatment techniques with lower energy footprint. Prevailing conventional biological treatment methods, such as activated sludge process, biological contactors, and more modern technologies, like advanced oxidation process, membrane technologies, electrochemical oxidation, etc., are all energy-intensive processes. Their operational power requirements constitute a sizeable share in the overall cost of the treatment. Dwindling fossil reserves and inflating electricity costs have instigated researchers to probe for more cost-effective and energy-efficient wastewater treatment alternatives.

The internal chemical energy of 7.6 kJ/L is estimated to be stored in the bonds of organic ingredients of domestic wastewater, which is considered as a misplaced resource and can be exploited to recover useful energy by employing innovative technologies (Heidrich et al. 2011). Bio-electrochemical systems (BESs) are such innovative, efficient, and eco-friendly technologies applied for the abatement of contaminated water, in which chemical energy of organic contaminants present in wastewater is converted into bioelectricity or valuable chemical products, like H_2, H_2O_2, methane, ethanol, acetate, etc., through the application of electroactive microorganisms (Pant et al. 2012; Logan et al. 2019; Das et al. 2020f). The unique capability of BESs to treat wastewater with concomitant energy or chemical production makes them a lucrative alternative for sustainable wastewater management. The

Microconstituents in the Environment: Occurrence, Fate, Removal, and Management, First Edition.
Edited by Rao Y. Surampalli, Tian C. Zhang, Chih-Ming Kao, Makarand M. Ghangrekar, Puspendu Bhunia, Manaswini Behera, and Prangya R. Rout.

BESs can be further categorized into the microbial fuel cell (MFC), the microbial electrolysis cell (MEC), the microbial electrosynthesis (MES), and the microbial desalination cell (MDC), depending on the valuable recovered and their respective mechanism of operation.

A typical MFC is analogous to a fuel cell except for the reactant fuel, which is wastewater in the case of MFC and H_2, and alcohol, etc. in the case of fuel cells (Jung et al. 2020). In its most generalized form, MFC consists of two distinct chambers housing anode (anaerobic environment) and cathode (aerobic environment) separated by a cation-selective membrane referred to as the separator. The electroactive microorganisms oxidize organic substrate in the anodic chamber, releasing electrons, protons, and CO_2. Protons migrate to the cathodic chamber through separators and the electrons released at the anode are further transferred to a terminal electron acceptor in the cathodic chamber through a conductive material, thereby generating electricity.

Usually, the two following categories of electroactive microorganism are employed in a typical BES, namely: i) exoelectrogens, which act as anodic bio-catalysts in performing anaerobic oxidation of organic matter by donating electrons to the electrode; and ii) electrotrophs, which are cathodic bio-catalysts responsible for carrying out the reduction of CO_2/HCO_3^- by accepting electrons from the electrodes (Zou and He 2018; Das et al. 2020e; Nath et al. 2021). The MDC is an extension of MFC, where the reactor is designed to desalinate salt water in addition to chemical oxygen demand (COD) removal and generation of bioelectricity, whereas MEC is a modified form of MFC in which hydrogen is produced by electro-hydrogenesis of biodegradable organic contaminants in a sealed anaerobic cathodic chamber (Call and Logan 2008; Logan et al. 2015; Saeed et al. 2015).

In 1911, Professor Potter (1911) pioneered research on MFC with his experiments on electricity generation by the decomposition of organic compounds via *Escherichia coli*, trail-blazing the idea of utilizing microbes to produce electricity. However, research on MFC took a significant technological leap forward in the mid-1960s, when the current producing ability of microbes was combined with fuel cells to devise the first MFC (Berk and Canfield 1964; van Hees 1965). Thereafter different systems for harnessing value-added products from wastewater have been developed and investigated. Nevertheless, MFC, MEC, and MDC are among the most established technologies representing the BESs.

The mechanism of treatment and removal efficiency of pollutants in BESs are broadly governed by reactor design, electrode material, electrolyte pH, operating temperature, microbial inoculum, substrate, mediators, type of separator, and the characteristics of wastewater being treated (Gul et al. 2021). Different types of wastewaters, including petrochemical effluent, metal-containing waste streams, pharmaceutical wastewater, and other contaminated water, have been subjected to bio-electrochemical treatment and have yielded a substantial reduction in COD of the influent wastewater along with the simultaneous recovery of bioelectricity, hydrogen, biofuels, nutrients, etc. (Nancharaiah et al. 2015; Priyadarshini et al. 2021; Wilberforce et al. 2021).

Removal of recalcitrant micropollutants has been the main focus of BESs of late. Significant removal for xenobiotic compounds like oxyfluorfen, sulphonamide, surfactant, heavy metals, etc., has been achieved using MFCs (Zhang et al. 2018a; Cheng et al. 2020; Sathe et al. 2020; Chakraborty et al. 2020b). Furthermore, composite systems have also been fabricated by coupling MFC with other treatment technologies, like electro-Fenton and constructed wetlands for creating a synergetic effect leading to the enhanced removal of persistent and bio-refractory compounds, such as antibiotics, betablockers, heavy metals, etc. (Monteil et al. 2019; Wang et al. 2020a; Wen et al. 2021; Sathe et al. 2022). Evidently, the dual advantage of BESs, i.e., contaminant abatement and energy production, showcase them as a promising waste-to-energy/chemical conversion platform with the potential to become a part of the circular bioeconomy (Jung et al. 2020).

18.2 Bio-Electrochemical Systems: Materials and Configurations

The BESs are distinguished based on their desired end product; however, the core components of bio-electrochemical reactors are identical irrespective of the system used. The three fundamental components of a typical MFC are an anodic chamber, cathodic chamber, and a proton exchange membrane (PEM). A similar reactor set-up is assembled in MEC, with the only difference being the external voltage supply, which is provided to facilitate hydrogen production or any other valuables depending upon the type of cathode catalysts employed. Another type of BES, i.e., MDC, contains an additional desalination chamber other than the two regular electrode housing compartments, which is formed by inserting anion exchange membrane on the anode side and cation exchange membrane toward the cathode side of the MDC. These BESs comprise of many functional components which are discussed below.

18.2.1 Electrodes

Electrodes are an indispensable part of any BES and are essential for carrying out electrochemical reactions in the feed wastewater to achieve the requisite level of contaminant removal. Additionally, electrodes are vital for sustaining the microbial population as they provide the surface area required for the formation of electroactive biofilm to complete the metabolic pathway and to produce energy or chemical products. Both electrodes, i.e., anode and cathode, are encased in separate compartments of the reactor known as the anodic chamber and the cathodic chamber, respectively.

The anodic chamber is the essential component and the working chamber of an MFC, wherein the anode is placed, and oxidation of the organic matter occurs in an anaerobic environment. In an MFC, the anode functions as a media by providing a site for the growth of a microbial biofilm, which is critical for the effective biodegradation of organic substrates with the simultaneous recovery of bioelectricity. Moreover, the anode also facilitates the conductance of electrons released by exoelectrogenic microorganisms, while degrading organic substrate to the cathode via an external circuit connecting the two electrodes, generating bioelectricity in the process. In MECs, external energy is required to be supplied to drive electrons from the anode to the cathode as the reduction potential of proton to H_2 at the cathode is relatively lower than the generated cathode potential, which inhibits spontaneous hydrogen evolution reaction at the cathode.

Selection of a suitable anode is imperative to attain the desired result during the treatment of wastewater through MFC, as the growth of biofilm and flow of electrons is directly affected by the nature of anode material used in an MFC. An ideal anode in MFC must possess high conductivity to facilitate electron transfer and should be bio-compatible for nurturing microbial growth and biofilm attachment (Hindatu et al. 2017). At the same time, the anode should also have a large surface area for better biofilm anchorage and must be chemically stable in different wastewater streams and operating conditions (Hindatu et al. 2017). Low fabrication cost, higher conductivity, and resistance to bio-fouling are also desired properties for an anode material for the economic operation of an MFC. Anode fabricated from materials like carbon cloth, carbon fiber, carbon felt, tungsten carbide, graphite granules, etc. are frequently employed in BESs owing to their low cost and relatively superior conductivity.

The cathode also has a pivotal role in bioelectricity production and product recovery through BESs. The cathode accepts electrons released from the anode through an external circuit and receives protons from the anodic chamber through PEM in an MFC. Furthermore, from the cathode, electrons are conducted to a terminal electron acceptor like oxygen in the aerobic cathodic chamber of an MFC to produce bioelectricity. In an anaerobic cathodic chamber like the one used in MEC, a proton acts as a terminal electron acceptor and gets reduced to form hydrogen. Additionally, the removal of recalcitrant compounds and metals as well as CO_2 reduction can also be achieved via a cathodic reduction reaction (Jain and He 2018).

Cathode electrode material should display high conductivity and electrocatalytic activity and must also be chemically stable in a stringent operating environment. Materials such as platinum, activated carbon, graphene, carbon black/graphite hybrid, magnetic cathodes, etc., have been employed as cathode material in different BESs (Li et al. 2016; Carlo et al. 2017; Zhou et al. 2019; Liang et al. 2020). To escalate the rate of reduction reaction at cathode metal catalysts, ferricyanide and noble metals such as platinum, can be used; however, this incurs extra expense and hence usually avoided (Mani et al. 2020). Different cathode catalysts employed in BESs with the relative enhancement in the valuable recovery are listed in Table 18.1.

Table 18.1 Different cathode catalysts used in BESs.

BES/Cathode Material	Cathode Catalyst	Power/Product Generation	Relative Improvement After Catalyst Addition	References
MFC:				
Carbon black, activated carbon, and polytetrafluorethylene mix	Fe–N–C	36.9 W/m^3	1.25	Santoro et al. (2019)
Blended carbon black, activated carbon, and polytetrafluorethylene.	Fe–AAPyr–GNS	235 μW/cm^3	2.24	Kodali et al. (2018)
Carbon paper	CNT/PPy	113.5 mW/m^2	1.64	Ghasemi et al. (2016)

Table 18.1 (Continued)

BES/Cathode Material	Cathode Catalyst	Power/Product Generation	Relative Improvement After Catalyst Addition	References
Carbon paper	Pt	122.85 mW/m^2	1.77	Ghasemi et al. (2016)
Carbon black on steel mesh	N–P/ chitosan	1603.6 mW/m^2	5	Liang et al. (2019)
Carbon black.	CuZn	75.1 mW/m^2	64	Das et al. (2020b)
Carbon black supported on titanium mesh	Fe–NSC	2068 mW/m^2	1.55	Liang et al. (2021)
MEC:				
Stainless steel	Ni–Co–P	Hydrogen (4.2 m^3/m^3)	1.45	Chaurasia et al. (2020)
Nickel foam	$NiMoO_4$	Hydrogen (0.12 L/L)	2.63	Jayabalan et al. (2021)
Carbon felt	Ni–Pd	Hydrogen peroxide (233 mg/L)	3	Gupta et al. (2020a)
Carbon paper	CeO_2–rGO	Hydrogen (5.05 m^3/m^3)	1.42	Pophali et al. (2020)

Fe–AAPyr–GNS = Iron aminoantipyrine and graphene nanosheet derived catalyst; CNT/PPy = Carbon nanotube polypyrrole nanocomposite; N–P/chitosan = Nitrogen and phosphorus dual-doped chitosan derived carbon catalyst; CuZn = copper and zinc bimetallic catalyst; Fe–NSC = Iron and sulphur co-doped nitrogen-enriched hydrothermal carbon catalyst; Ni–Co–P = Ni, cobalt and phosphorous catalyst; $NiMoO_4$ = Nickel molybdate nanocatalyst; Ni–Pd = Nickel and palladium catalyst; Ni/Co = Nickel and cobalt catalyst; CeO_2–rGO = Cerium oxide (CeO_2) and reduced graphene oxide

Moreover, the use of biocathodes has made it possible to circumvent the need of adding costly metal catalysts by employing electroactive microorganisms as biocatalysts for carrying out redox reactions (Jafary et al. 2015). Biocathodes have been readily applied in MECs to generate hydrogen, mineralize complex organic dyes, and reduce heavy metals (Huang et al. 2018; Arulmani et al. 2021). More recently, researchers have been exploring the application of biocathodes in MFCs and have reported enhanced bioelectricity generation along with improved COD removal efficiency and abatement of heavy metals, like cadmium and nickel (Yadav et al. 2020; Singh and Kaushik 2021). The surface properties of a biocathode play a crucial role in cathode–microbe interaction and determine the formation of a biofilm over the cathode surface (Guo et al. 2015). Preferably, metals and carbon-based electrodes are employed as biocathodes owing to their higher specific surface area favoring biofilm adhesion and better conductivity (Rusli et al. 2019).

18.2.2 Separators

Separators are typically semipermeable polymers or ceramic membranes used for partitioning the compartments of BESs and facilitate the movement of ions to the adjacent chamber. Besides acting as a physical barrier between the compartments, the separator facilitates the exchange of protons from the anodic chamber to cathodic chamber in BES and are often referred to as PEMs, whereas in MDC, anion and cation exchange membranes are coupled together to exchange ions between the two compartments and the desalination chamber. Moreover, separators prevent the passage of air from the aerobic cathodic chamber to the anaerobic anodic chamber and also avert the crossover of the substrate, which otherwise could negatively impact the performance of BES (Stratford et al. 2014; Sadhasivam et al. 2017). The selection of suitable separator material is imperative in deciding the removal efficiency of contaminants as well as for bioelectricity and biofuel recovery from BES. An ideal separator material must be efficient in conducting ions, have low internal resistance, should not allow inter-flow of substrate and oxygen, and should possess high mechanical and chemical stability (Rahimnejad et al. 2014).

Separators are broadly classified as the following: i) anion exchange membranes; ii) cation exchange membranes; iii) bipolar membranes; iv) porous membranes; and v) PEM (Shabani et al. 2020). Cation exchange membranes, such as sulfonated polyether ether ketone (SPEEK), Nafion 117, CMI-7000, etc., are proficient in transferring protons and are widely used in MFCs. Anion exchange membranes have positively charged groups attached to their polymer matrix, allowing movement of anions across the membranes and are primarily employed in MDCs (Jadhav et al. 2014). However, bipolar membranes are formed by clubbing anion and cation exchange membrane together, which are capable of transferring both protons and hydroxide ions efficiently.

Traditionally, Nafion is the most prevalent PEM used in MFCs due to its excellent proton conductivity, with SPEEK, Flemion, Hyflon, etc. being the other polymer membrane alternatives (Shabani et al. 2020). Nevertheless, Nafion membranes are costly, prone to biofouling, and have relatively lower energy recovery (Chia et al. 2020; Shabani et al. 2020). Many investigations have been performed to find suitable and cost-effective Nafion alternatives for cutting down the fabrication cost of MFCs. Table 18.2 illustrates different materials used as PEMs in MFCs and their relative performance with Nafion 117 membrane. Sirajudeen et al. (2021) utilized biopolymer composite as PEM in MFC, which displayed superior performance in terms of COD removal and power generation when compared to Nafion membrane-based MFC. The MFC with biopolymer composite attained a COD removal efficiency of 72.7% and power density of 601 mW/m^2, while the MFC with Nafion as PEM could only attain 64.5% and 520 mW/m^2 of COD removal efficiency and power density, respectively (Sirajudeen et al. 2021). The elevated performance of the MFC was due to the greater proton conductivity and lesser oxygen diffusion of the biopolymer PEM (Sirajudeen et al. 2021).

Porous sulphonated biochar has exhibited comparable performance with Nafion, when employed as PEM in MFC; however, the power harvested per unit cost of sulphonated biochar PEM was 26 times more than that using Nafion as PEM, thus presenting a more economical alternative to expensive conventional separators (Chakraborty et al. 2020a). Furthermore, ceramic polymers are inexpensive material and have been incorporated as PEM in MFC for bioelectricity generation. For instance, Das et al. (2020a) tested goethite and

Table 18.2 Different PEMs used in MFCs and their relative performance with respect to conventional Nafion membrane.

PEM	Power Generation (mW/m^2)	COD Removal (%)	CE (%)	Power Generation Relative to Nafion-117	COD Removal Relative to Nafion-117	CE Relative to Nafion-117	References
SPEEK/SGO	1028	–	51	2*	–	2*	Ayyaru and Ahn (2021)
SPEEK/PES	6.665	68	26.3	1.83	1.13	1.77	Lim et al. (2012)
PES/SPES	58.726	86	22	1.29	1.20	1.22	Zinadini et al. (2017)
PHA/PHB	601	72.7	65.7	1.16	1.13	1.01	Sirajudeen et al. (2021)
SBC	41.08	81	6.36	0.70	0.92	0.86	Chakraborty et al. (2020a)
S-OBPI	110	–	–	1.47	–	–	Singha et al. (2016)
PVDF/Nafion	4.9	78	12.1	1.15	0.92	1.23	Shahgaldi et al. (2014)
ACCS/clay	3.7**	81.05	13	1.94	1.12	1.4	Neethu et al. (2019)
G-5	112.81	87.7	23.74	1.05	1.22	1.09	Das et al. (2020a)

* = data compared with Nafion 115; ** = data unit is W/m^3; PES = polyethersulfone; SPES = sulfonated polyethersulfone; PHA = polyhydroxyalkanoate; PHB = poly-(R)-3-hydroxybutyrate; SGO = sulfonated graphene oxide; SBC = sulfonated biochar; S-OBPI = sulfonated oxy-polybenzimidazole; PVDF = polyvinylidene fluoride; ACCS = activated carbon derived from coconut shell; G-5 = 5% Goethite supplemented natural clay ceramic

natural clay-based PEM in MFC for wastewater treatment, which was five folds cheaper in comparison to Nafion 117 and yielded 22% higher COD removal. Nonetheless, such PEMs are still at the developmental or lab-scale stage and require extensive investigation before they can replace expensive polymer membrane in MFCs, which remains a major hurdle in advancing BESs as a field-scale technology.

18.3 Different Types of Bio-Electrochemical Systems

A BES integrates microorganisms and electrochemical principles for wastewater treatment and has the distinct ability to produce bioelectricity or biofuel, which sets it apart from the other contemporary wastewater treatment technologies. Depending on the nature of the end product recovered through BESs, it can be categorized as an MFC for bioelectricity generation, MEC for hydrogen production, MDC for desalination of brackish water along with treatment of wastewater in the anodic chamber for organic matter removal, MES for the synthesis of valuable chemicals like organic acids, alcohols, etc., and microbial carbon-capture cell (MCC) for CO_2 sequestration (Jafary et al. 2015; Saeed et al. 2015; Das et al. 2019a, 2020f; Dessì et al. 2020). However, MFC, MEC, and MDC are the most established technologies representing BESs and are described in the following section.

18.3.1 Microbial Fuel Cell

A fuel cell is referred to as an MFC that utilizes wastewater as feedstock and electroactive microorganisms as biocatalyst in the anaerobic anodic chamber for degrading organic matter present in the wastewater along with the concomitant generation of bioelectricity. In its simplest form, a dual compartment MFC consists of a reactor bifurcated into two parts, namely anodic chamber and cathodic chamber (Figure 18.1). Semipermeable PEM is used to physically separate the anodic chamber from the cathodic chamber; however, at the same time, PEM keeps both the electrodes chemically and ionically conjugated by facilitating the migration of protons from the anodic chamber to the cathodic chamber. Moreover, PEM prevents intermixing of electrolytes of the two regions as well as preventing the diffusion of oxygen from the aerobic cathodic chamber into the anaerobic anodic chamber, which could otherwise restrict bioelectricity generation of the MFC (Rahimnejad et al. 2015).

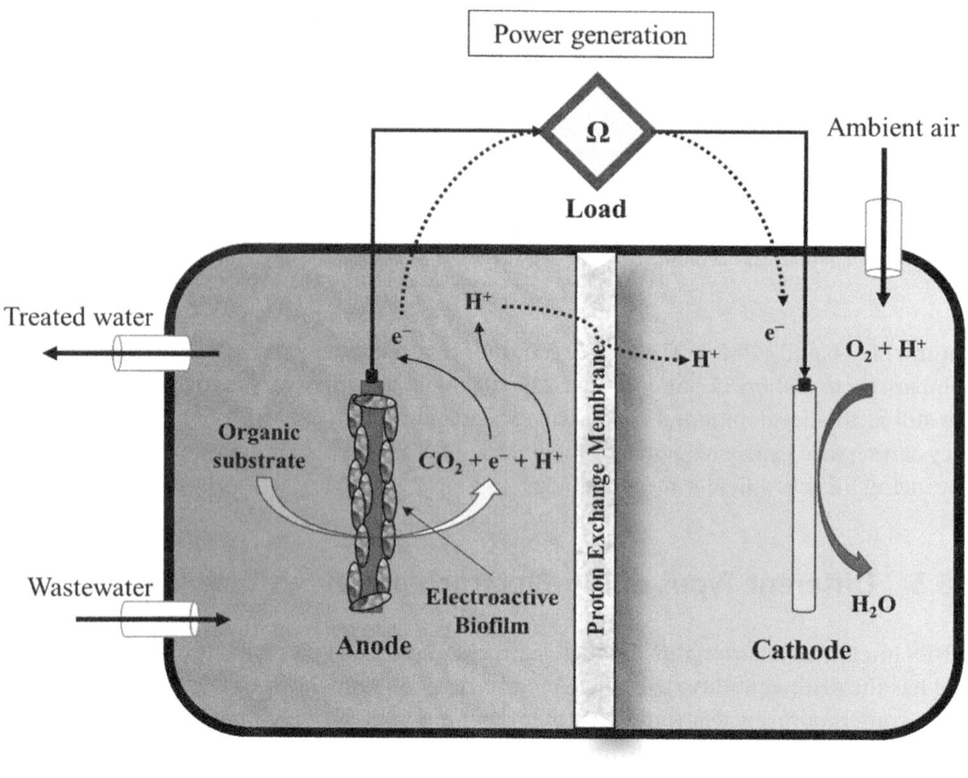

Figure 18.1 Schematic diagram of a microbial fuel cell.

Before the steady-state performance of a BES, ample time is provided for the proliferation of microbial mass over the anode surface to form an electroactive layer referred as the biofilm. This biofilm consists of a diverse exoelectrogenic microbial population capable of transferring electrons to the anode under anaerobic conditions (Logan 2009). The biodegradation of organic substrate occurs in the anodic chamber of an MFC, resulting in the release of electrons, protons, and CO_2 (Eq. 18.1). The electrons are shuttled to the anode by exoelectrogenic bacteria from where they are conveyed to the cathode through an external circuit, thereby producing electricity (Li et al. 2018).

Furthermore, electrons from the cathode are used to produce water by reducing oxygen through oxygen reduction reaction (Eq. 18.2), with oxygen functioning as the terminal electron acceptor. Protons migrate to the cathodic chamber via PEM that maintains the overall electroneutrality of the system. Furthermore, based on the mechanism of electron transfer, MFCs are classified as MFC operated with mediators, like humic acids, methylene blue, etc., and mediator-less MFC, where the electrons transfer is assisted through electroactive microorganisms, such as bacteria of *Geobacter* and *Shewanella* genera (Sund et al. 2007; Chaturvedi and Verma 2016; Zheng et al. 2020).

$$\underset{\text{Organic matter}}{C_6H_{12}O_6 + 6H_2O} \xrightarrow[\text{Anaerobic environment}]{\text{Electro active microorganisms}} 6CO_2 + 24\,e^- + 24H^+ \tag{18.1}$$

$$6O_2 + 24H^+ + 24e^- \rightarrow 12H_2O \tag{18.2}$$

Different factors may affect the contaminant removal efficiency and energy production in an MFC, such as the substrate concentration, pH of the electrolyte, operating temperature, nature of the electrode material used, salinity, etc. However, the development of biofilm and effectiveness of electroactive microorganisms in the transportation of electrons are the most critical parameters governing the performance of an MFC (Das et al. 2019b; Ghorai et al. 2021). Also, internal resistance offered by PEM is also a major limitation of MFC, resulting in a diminished power and current density (Shreeram et al. 2018). Nonetheless, high material cost and ordinary power generation drastically reduce the cost-effectiveness of these systems, restraining their field-scale application (Zheng et al. 2020).

18.3.2 Microbial Electrolysis Cell

An MEC is the modified version of MFC, wherein the cathode also operates under anaerobic conditions in addition to the anode (Figure 18.2). The valuables recovered from MEC can be H_2, methane, H_2O_2, and organic acids depending on the redox mechanism targeted in the MEC (Hua et al. 2019). The organic substrate is oxidized by exoelectrogenic bacteria colonizing on the anode resulting in the release of electrons at the anode, CO_2, and protons in the solution (Yu et al. 2018). Electrons and protons are conveyed to the cathode through an external circuit and PEM, respectively. In the cathodic chamber, electrons and protons interact to produce hydrogen (Eq. 18.3), whereas in a single-chamber MEC, protons combine with CO_2 to form methane (Das et al. 2020e). However, spontaneous flow of electrons from anode to cathode does not occur in MEC due to higher cathode potential in comparison to the anode potential, and thus, an external voltage of greater than +0.2 V vs.

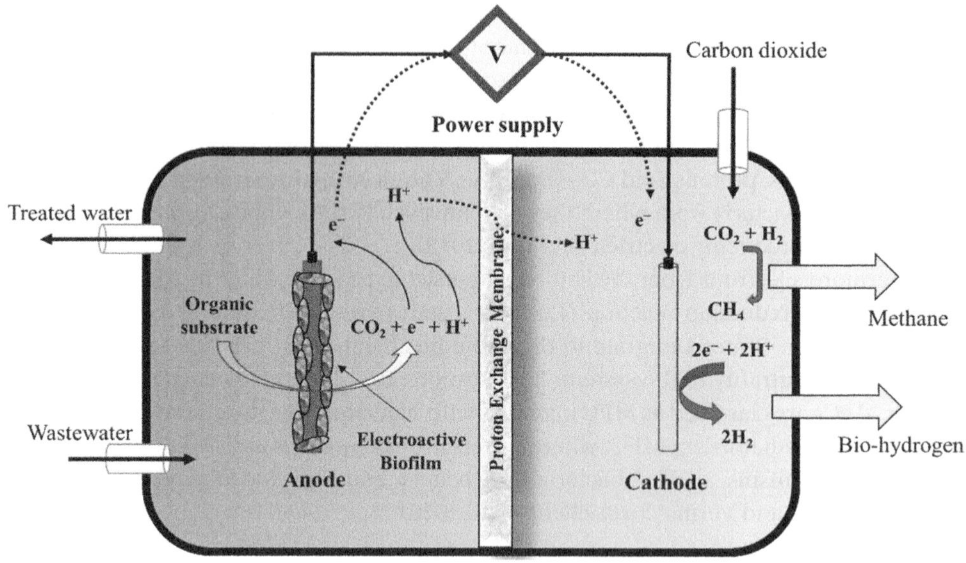

Figure 18.2 Schematic diagram of a microbial electrolysis cell.

standard hydrogen electrode (SHE) is applied to initiate electron flow between the electrodes (Zikmund et al. 2018; Das et al. 2019b).

$$2H^+ + 2e^- \rightarrow H_2 \quad (18.3)$$

Diverse types of wastewaters and organic compounds, including sewage, winery wastewater, sugar industry effluent, glycerol, acetic acid, etc., have been successfully applied as feedstock in MEC for H_2 production (Cusick et al. 2011; Heidrich et al. 2013a; Kadier et al. 2014). Furthermore, MECs have also been pertinent in abating complex xenobiotic compounds and heavy metals, such as antibiotics, dyes, lead, nickel, etc. (Chakraborty et al. 2020b). Despite versatility in treatment, scale-up applications of MECs are discouraged because of low energy efficiency and sluggish H_2 production rate. Increasing externally applied voltage can augment the rate of H_2 formation in MECs, though higher applied voltage may inhibit the growth of electrogenic bacteria residing on the surface of the bioanode, which would be catastrophic for the efficient function of MEC (Lim et al. 2017). Nonetheless, the added advantage of H_2 production while treating contaminated water makes MEC an compelling prospect for becoming part of a sustainable wastewater remediation platform in the near future.

18.3.3 Microbial Desalination Cell

The current freshwater demand is approximated to be around 4.6 trillion/m^3 per year and is expected to grow in the near future (Boretti and Rosa 2019). Depleting freshwater resources and continuously increasing water demand have created a global water crisis. Many water stressed regions with limited freshwater supply have been using desalinated sea water to meet their water requirements (Maxwell 2016). However, the existing desalination technologies, such as deionization, reverse osmosis, and thermal processes like a multistage flash evaporator, multiple effect desalination, etc., consume a significant amount of electricity, making the cost of the desalination process exorbitant.

The MDCs offer more sustainable solutions to desalinate sea water owing to their substantial energy savings. For instance, MDCs can conserve up to 4 and 196 kWh/m^3 of energy compared to reverse osmosis and thermal technologies, respectively (Tawalbeh et al. 2020). An MFC is modified into MDC by attaching an additional central desalination chamber between the electrode housing compartments for brackish water desalination with simultaneous electricity production (Figure 18.3). The desalination chamber is constructed by placing the anion exchange membrane (AEM) toward the anode and the cation exchange membrane (CEM) on the side of the cathode. Principally, the organic constituents of wastewater are degraded by exoelectrogens residing in the anodic chamber of MDC to produce electrons, which are subsequently transferred to the cathode through an external circuit. This leads to the aggregation of protons in the anolyte triggering outward movement of anions, like Cl^- and SO_4^-, from the salt water present in the desalination chamber via AEM (Sevda and Abu-Reesh 2019). The transported electrons in the cathodic chamber further reduce terminal electron acceptors like O_2 and instigate migration of cations like Na^+ and Ca^+ passaging through the CEM into the cathodic chamber (Morvarid et al. 2017). All these processes culminate in the production of bioelectricity and desalinated water.

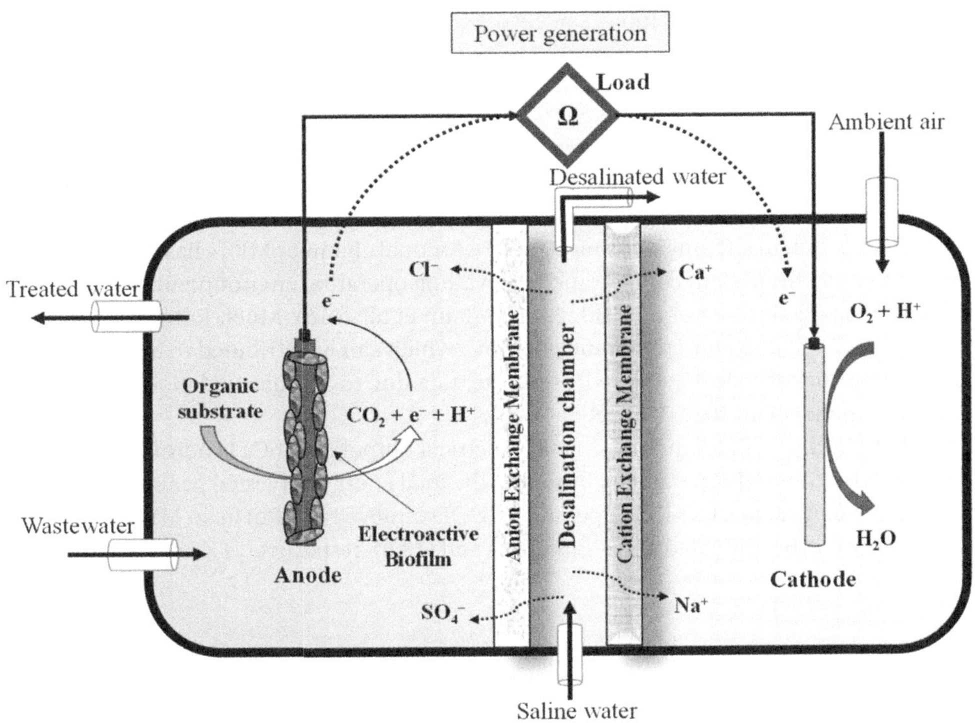

Figure 18.3 Schematic diagram of microbial desalination cell.

Although MDCs have delivered promising results, the nominal production rate of electricity and salt-free water impedes the progress of this neoteric technology toward larger applications. The significant drawbacks encountered during the operation of MDCs are membrane biofouling, pH fluctuation, and the higher internal resistance of separators, which are responsible for poor efficiency in terms of power generation and slow desalination rate (Tawalbeh et al. 2020). Comprehensive and large-scale investigations are required to overcome these roadblocks so that MDC technology can realize its full potential.

18.4 Performance Assessment of Bio-Electrochemical Systems

The performance of a BES is assessed primarily on the extent to which contaminants are removed from the influent wastewater and by the efficiency at which energy, biofuels, or valuable chemicals are produced during the treatment process. In MFC, Coulombic efficiency (CE) is often used as an indicator for gauging the system's overall efficacy in terms of substrate removal and bioelectricity generation. The CE can be defined as the ratio of actual electric current produced during the operation of MFC to the theoretical current expected to be produced based on the COD consumption. Calculation of CE can be done using Eq. (18.4).

$$CE_{MFC} = \frac{M\int_{0}^{t} Idt}{Fbv\Delta COD} \times 100 \tag{18.4}$$

where M denotes the molecular weight of oxygen; I is the current generated; F stands for Faraday's constant = 96,485 C/mol; v represents the working volume of the anodic chamber of MFC; b is the number of electrons exchanged per mole of oxygen = 4; and ΔCOD is the difference in COD concentration of influent and effluent after time t (Bhowmick et al. 2019).

A wide variation in CE ranging from 20 to 89% for dual chamber MFCs has been reported depending upon the system configuration, prevailing operating environment, and the type of substrate applied (Chen et al. 2013, 2021; Tursun et al. 2016). Much lower CEs of 25 to 52% have been obtained for membraneless MFC, which can be attributed to higher oxygen flux from the air cathode leading to loss of substrate due to aerobic oxidation (Pinto et al. 2011; Pugazhendi et al. 2020; Yang et al. 2020).

Another effective tool for quantifying the electrical output of MFCs is normalized energy recovery (NER). The NER parameter evaluates the total energy harvested per unit volume of wastewater being treated (NERv) or per unit of COD removed (NERc) in an MFC for a given duration and can be estimated using Eqs (18.5) and (18.6), respectively (Xiao et al. 2014).

$$NERv = \frac{P \times t}{v} \tag{18.5}$$

$$NERs = \frac{P \times t}{COD \times v} \tag{18.6}$$

where P is the power generated during an MFC operation. The NERv and NERs are reported in kWh/m^3 and kWh/kg COD, respectively, which are widely used units in wastewater treatment in MFCs and assist in easier comparison of energy efficiency. Therefore, most of the recent investigations prefer NER for assessing the performance of MFC. Many documented literature on MFC have reported an NERv value $<1.5\,kWh/m^3$ and NERs of 1 kWh/kg COD (Ge et al. 2014).

In MECs, the reactor performance is generally assessed through the parameter CE, representing the quantity of hydrogen gas produced per unit of electrical current supplied. CE can be calculated using Eq. (18.7).

$$CE_{MEC} = \frac{PV_H \int_0^t Idt}{2FRT} \times 100 \tag{18.7}$$

where P is pressure in atm; V_H is the volume of hydrogen gas produced; R stands for ideal gas constant = 0.082 L/atm/K/mol; T is the temperature in K; I is the measured current in MEC; F is Faraday's constant = 96,485 C/mol; dt represents the reaction time of the experiment; and factor 2 is taken to convert moles of the electron to moles of hydrogen (Son et al. 2021). However, some researchers also like to provide the H_2 or methane gas yield of MEC with respect to COD consumed, as per Eq. (18.8).

$$Y = \frac{V_{gas}}{v\,\Delta COD} \tag{18.8}$$

where Y is gas yield in m^3/kg COD; V_{gas} is the total volume of H_2 or methane produced in m^3; v is the working volume of the reactor; and ΔCOD is the amount of COD consumed during the treatment process in kg (Tenca et al. 2013).

In the case of MDCs, desalination efficiency is also estimated in addition to CE to analyze the system's overall performance. Calculation of CE in MDCs is the same as that in an MFC and can be estimated using Eq. (18.4). The desalination efficiency (η_{DE}) of an MDC can be computed through Eq. (18.9).

$$\eta_{DE} = \frac{C_i - C_f}{C_i} \times 100 \tag{18.9}$$

where C_i is the initial salt concentration of the influent brackish water; and C_f is the final salt concentration after treatment (Bejjanki et al. 2021).

The removal efficiency of contaminants and valuable product recovery in BES is affected by many biological and physiochemical parameters, such as electroactive biofilm development on the electrode surface, nature of substrate used, type of electrode material, characteristics of the separator used, pH of the electrolyte, temperature, and so on. The growth of a healthy and mature electroactive biofilm is a prerequisite for the stable performance of any BES. Improved biofilm formation over the anode surface in MFCs has resulted in higher power production and CE. In this regard, Zhang et al. (2017) developed a hybrid biofilm by inserting a multiwalled carbon nanotube into the anodic biofilm, which resulted in 58.8% improvement in power density from 0.71 to $1.13\,mW/m^2$ and 84% enhancement in CE with an attained value of 44.9%. The carbon nanotube matrix acted as an electron transfer tunnel that amplified the transfer of electrons leading to the ameliorated performance of the MFC (Zhang et al. 2017).

The selection of suitable material for electrode fabrication is instrumental to the biofilm genesis and adherence over the electrode surface. In addition to being conducive to microbial colonization, the electrodes in BES should possess high conductivity to facilitate electron flow with minimum resistance. Metals and carbon-based materials are the two most frequently used electrodes in BESs owing to their easy availability, straightforward construction, and relatively higher service life. Many modifications have been applied to boost the physiochemical characteristics of the electrode to suit the requirement of BESs. For instance, coating the carbon cloth anode of an MFC with tungsten carbide led to an augmented CE of 83.2%, which was 1.71 times higher than the CE of 48.6% obtained from an MFC operated with plain carbon cloth (Liu et al. 2020). The tungsten carbide coating enhanced the electron conductivity of the anode, and nanoparticles provided a greater number of sites for electrocatalytic redox reactions in the MFC (Liu et al. 2020). Iron(II) molybdate ($FeMoO_4$) nanoparticles, when coated over graphite plate electrodes, improved their biocompatibility toward exoelectrogenic bacteria that helped in microbial proliferation on the anode surface of the MFC and achieved a higher CE of 21.3% due to a healthier biofilm growth, when compared to the CE of 15.3% for an MFC operated with plain graphite electrodes (Mohamed et al. 2020).

Membrane separators are another vital component of dual chamber BESs that is imperative in determining the CE of the system. The internal resistance of the PEM is considered as one of the primary reasons for lower electricity generation in MFC. In this regard, alkali pre-treatment of a clayware ceramic membrane increased the CE of the MFC from 6 to 10.2% (Das et al. 2020c). The alkali pre-treated membrane was found to have higher proton conductance and greater resistance to oxygen diffusion from cathodic chamber to anodic chamber, thus attaining a better CE than the untreated PEM (Das et al. 2020c). Though polymeric membranes, like Nafion and SPEEK, are frequently employed as PEM owing to their chemical stability and high conductivity, they are expensive and prone to biofouling. Alternate inexpensive materials, such as biopolymer composites and biochar-derived membranes, have produced comparable CE at a much lower production cost (Chakraborty et al. 2020a; Sirajudeen et al. 2021).

Other factors like temperature, pH of electrolyte, and anolyte additives can also affect the efficiency of BESs. Jadhav and Ghangrekar (2009) found that when the ambient temperature is decreased from 22–35°C to 8–22°C, a substantial decrease in COD removal efficiency of an MFC from 90 to 59% was observed due to lower activity of bacteria and inhibition of methanogens. However, this lead to the improvement in CE of MFC from 1.5 to 5% owing to reduction in parasitic loss of charge because of bacterial inactivity at a lower temperature (Jadhav and Ghangrekar 2009). Furthermore, the addition of 0.2 g/L of sodium citrate in anolyte reduced the charge transfer resistance in the MFC, thereby leading to an elevated power density of 784.05 mW/m^2 (1.58 times higher) and CE of 56.17% (1.64 times better) (Chen et al. 2021). Nevertheless, mediocre CE and energy efficiency of BESs remains a major roadblock in elevating these technologies for scaled-up applications.

18.5 Pollutant Removal in Bio-Electrochemical Systems

18.5.1 Treatment of Different Wastewaters in Bio-Electrochemical Systems

Wastewaters containing organic contaminants and nitrogen, phosphorous, metals, dyes, volatile organic compounds, emerging contaminants, etc., have been effectively remediated using BESs (Table 18.3) (Mu et al. 2009; Nimje et al. 2012; Nancharaiah et al. 2015, 2016; Zhang et al. 2018b; Ahmad et al. 2021). Anaerobic oxidation by electroactive microorganisms is the principal mechanism by which readily degradable organic matter is removed in BESs. Putrescible organic contaminants are utilized by electroactive bacteria as a substrate for carrying out metabolic activities and for reproduction, meanwhile breaking down the pollutants into simpler organic compounds. Evidently, BESs are more adapted to treat wastewater amenable to biodegradation, such as domestic sewage.

Min and Logan (2004) first investigated domestic wastewater for bioelectricity production in carbon electrode-based MFC and attained 79% COD removal with initial COD of 1000 mg/L. Moreover, a dip in the power density from 72 to 43 mW/m^2 was also observed when hydraulic retention time (HRT) was increased from 1.1 to 4 h, suggesting power output is a function of HRT (Min and Logan 2004). Modification in reactor configuration has allowed higher current density even for low strength wastewater. For instance, single chamber exchangeable carbon cloth equipped air cathode MFC with a graphite felt anode was able to attain 80 to 95% COD removal efficiency and produced a maximum power density of 149 and 754 mW/m^2 for domestic and synthetic wastewater, respectively, both having a relatively lower influent COD of 100 mg/L (Yu et al. 2012).

The lower power density for domestic wastewater can be attributed to poor ionic strength and the presence of complex organic ingredients, which are not natural substrates for exoelectrogens (Yu et al. 2012). Concurring results have been obtained while treating similar wastewater with higher biodegradability as feedstock in MFCs. Brewery wastewater, having an average COD of 3197 mg/L when subjected to MFC treatment employing carbon fiber cloth electrodes, demonstrated 94% of COD removal, whereas MFC with platinum coated carbon cloth electrodes was able to reduce COD by 92.31% from dairy industry effluent having an initial COD of 6000 mg/L (Lu et al. 2017; Choudhury et al. 2021).

Much lower COD removal efficiency has been reported for MFC while treating complex wastewater originating from industries. For example, a mere 17% decrease in COD was observed for winery wastewater having an average influent COD of 13,245 mg/L after treating it in MFC fabricated with carbon felt electrode (Penteado et al. 2016). Likewise, for distillery wastewater with a COD concentration of 3200 mg/L, only 63.8% of COD was removed during post-treatment in MFC using graphite plate electrode (Samsudeen et al. 2015). The inefficiency in COD removal by MFC is due to the presence of recalcitrant organic compounds in winery and distillery wastewater that could not be effectively degraded by microbes residing in the anodic chamber of the MFC.

Wastewater containing biorefractory contaminants, such as pharmaceutical wastewater, petrochemical wastewater, etc., are particularly difficult to treat, especially with conventional biological methods due to their high toxicity, fluctuating composition, and low

Table 18.3 Treatment of different types of wastewaters using BES.

BES	Electrodes	Contaminant/ Wastewater	Initial Concentration/ COD (mg/L)	Removal Efficiency (%)	Valuable Recovery	References
MFC	A: Carbon veil. C: Carbon veil.	Electroplating industry wastewater.	1500	87	Bioelectricity (Power density of 6.2 W/m^3)	Karuppiah et al. (2021)
MFC	A: Carbon felt. C: Carbon cloth.	High-salinity mustard tuber wastewater.	520.25	87	Bioelectricity (Power density of 5.4 W/m^3)	Zhang et al. (2020)
MFC	A: Carbon brush. C: Carbon cloth.	Cu^{2+}	12.5	98.3	Bioelectricity (Power density of 10.2 W/m^3)	Wu et al. (2018)
MFC	A: Carbon cloth. C: Carbon cloth.	Norfloxacin.	4	65.5	Bioelectricity (Power density of 700 W/m^2)	Ondon et al. (2020)
MEC	A: CeO_2-rGO-PC. C: CeO_2-rGO-PC.	Thermal power plant wastewater.	13,500	95	Hydrogen (Yield of 5.05 m^3/m^3)	Pophali et al. (2020)
MEC	A: Carbon brush. C: Carbon cloth.	Coagulated swine wastewater.	12,500	98	Hydrogen (Yield of 57 L/m^3)	Han et al. (2021)
MEC	A: Graphite felt. C: Carbon cloth/ platinum.	Domestic wastewater.	104.4	94.1	Hydrogen (Yield of 0.92 m^3/m^3)	Jwa et al. (2019)
MEC	A: MoS_2-GO NF. C: MoS_2-GO NF.	Alizarin yellow R dye.	100	92.44	Hydrogen (Yield of 0.183 m^3/m^3)	Hou et al. (2021)
MDC	A: Plain graphite plate. C: Plain graphite plate.	Dairy wastewater.	2560	80.2	Bioelectricity (Power density of 44.1 mW/m^2)	Bejjanki et al. (2021)

Table 18.3 (Continued)

BES	Electrodes	Contaminant/ Wastewater	Initial Concentration/ COD (mg/L)	Removal Efficiency (%)	Valuable Recovery	References
MDC	A: Carbon graphite sheet. C: Carbon graphite sheet.	Yeast wastewater.	9280	90.3	Bioelectricity (Power density of 3.44 mW/m^2)	Goren and Okten (2021)

BES = bio-electrochemical system; MFC = microbial fuel cell; MEC = microbial electrolysis cell; MDC = microbial desalination chamber; A = anode; C = cathode; COD = chemical oxygen demand; CeO_2-rGO-PC = carbon film dispersed with cerium oxide (CeO_2) and reduced graphene oxide (rGO) electrode; MoS_2-GO NF = Molybdenum disulphide, graphene oxide and nickel foam composite electrode

biodegradability, which has an inhibitory effect on microbes involved in the treatment process (Ismail and Habeeb 2017; Tian et al. 2020). However, BESs have exhibited relatively high COD removal for such complex wastewater streams, mainly because of the superior biocatalytic activity of electroactive microorganisms (Ahmad et al. 2021). Rashid et al. (2021) treated real pharmaceutical wastewater with an initial COD concentration of 5460 mg/L using graphite electrode-based membrane-less MFC and obtained a COD removal of 80.55%.

Furthermore, the COD removal efficiency of synthetic sewage having an initial COD of 300 mg/L in an MFC utilizing graphite plate electrodes increased from 90.67 to 96.28%, when HRT was extended from 5 to 8 h (Chang et al. 2018). The increased duration of HRT facilitates prolonged substrate exposure to microbes leading to the elimination of soluble organic matter, consequently augmenting COD removal in the MFC (Chang et al. 2018). Furthermore, the COD removal efficiency of MFC tends to decrease with an increase in initial COD concentration of the wastewater. A similar trend was demonstrated during the treatment of petrochemical wastewater by an MFC incorporating polyacrylonitrile coated carbon felt electrodes, where COD removal efficiency drastically reduced from 82 to 40%, when the initial influent COD was elevated from 5000 to 45,000 mg/L (Sarmin et al. 2019). However, power density soared from 150 to 850 mW/m^2 when the initial influent COD was varied from 5000 to 45,000 mg/L owing to the higher quantity of substrate degradation releasing a large number of electrons in the anodic chamber of the MFC (Sarmin et al. 2019).

The MECs are an extension of MFCs modified to produce valuable chemicals, like hydrogen and methane, which are regarded as bio-fuels with different industrial applications. Performance of MECs is considered to be better in terms of CE and yield of valuables due to the operation of the cathodic chamber in an anaerobic environment, which minimizes the oxygen infusion in the bio-electrochemical reactor, thus leading to higher CE in comparison to MFCs (Cusick et al. 2010). However, lesser oxygen leakage into the anodic chamber of MECs is also responsible for relatively lower COD removal efficiency in respect to MFCs (Cusick et al. 2010). For example, COD removal efficiency of synthetic wastewater fed to a single-chamber MEC, having a carbon cloth anode and a nickel foam cathode with an initial COD concentration of 1120 mg/L, was 76% at an applied cell voltage of 0.9 V along with a CE of 154% (Hou et al. 2015). The COD removal efficiency was 18.3% lower in comparison to the MFC treating dairy wastewater of a similar nature, whereas CE attained in MEC was 126% higher than that in MFC (Choudhury et al. 2021).

Moreover, MECs have displayed relatively better heavy metal remediation than MFCs, as affirmed by the investigation carried out by Qin et al. (2012), in which MEC employing a carbon felt anode and a stainless steel cathode attained 99% removal of Ni^{2+}. The Ni^{2+} removal was three times higher than that achieved by MFC, mainly due to the development of higher cathode potential in MEC that induced electrodeposition of Ni^{2+} ions (Qin et al. 2012). In the case of zinc (Zn^{2+}), the removal efficiency of 99% was manifested by MEC operating under a controlled cathode potential of −1.38 V vs. SHE with a carbon felt wrapped graphite rod anode and a titanium cathode from an initial Zn^{2+} concentration of 800 mg/L (Modin et al. 2017), whereas Zn^{2+} removal attained by plain carbon felt electrode-based MFC was 96% for a much lower initial Zn^{2+} concentration of 131 mg/L (Lim et al. 2021). Evidently, MEC presents a more efficient alternative for remediating waste streams containing heavy metals.

The BESs have also demonstrated great promise in the remediation of xenobiotic compounds, such as dyes, antibiotics, chlorinated organic compounds, etc., and are much more sustainable than their contemporary energy-intensive counterparts, like electrochemical advanced oxidation, advanced oxidation, and electrocoagulation (Ahmad et al. 2020; Chakraborty et al. 2020b; Raj et al. 2021). Sulfamethoxazole (SMX) is an antibiotic commonly used to treat bacterial infections and is widely reported to be present in wastewater and natural water bodies, which can have detrimental environmental effects (Nagulapally et al. 2009). Carbon cloth electrodes-based MFC and MEC were investigated for removing SMX with an initial concentration of 30 mg/L in the test sample and yielded a comparable removal efficiency of 85.1 and 92.53%, respectively (Xue et al. 2019a, 2019b). Microbial metabolism of electroactive microorganisms was identified as the dominant pathway of SMX degradation in both MFC and MEC. However, a substantial drop in removal efficiency of SMX was observed when synthetic swine wastewater containing a mixture of pharmaceutical compounds in addition to SMX was used as feedstock for MFC.

The carbon electrode-based MFC was able to partially eliminate SMX achieving a maximum removal efficiency of 59.4% from an initial concentration of 100 μg/L (Cheng et al. 2021). The deterioration in the removal efficiency of SMX may be due to the presence of other complex pharmaceutical compounds, which could inhibit the metabolic activity of anodic microbes of the MFC. In another experiment, a single chamber MEC having carbon coated anode and platinum coated cathode was able to remediate 99% of a mixture of psychoactive drugs containing paroxetine, venlafaxine, and o-desmethylvenlafaxine, each having an initial concentration of 750 ng/L (Akagunduz et al. 2021). Clearly, persisted bioelectrochemical simulation of microbes in BESs can augment the metabolic activity of electroactive bacteria that leads to the better removal of even so-called recalcitrant xenobiotic compounds, thereby reinforcing BESs as wholesome technology for sustainable wastewater remediation.

Furthermore, the advent of MDC has made it possible to reduce the salinity of brackish water in addition to subduing the organic load with concomitant bioelectricity generation.

Bejjanki et al. (2021) treated dairy wastewater with an initial COD of 2560 and 3621 mg/L of salinity by graphite plate electrode-based MDC employing *Oscillatoria princeps* algae as the biocatalyst and this system was able to achieve 80.2 and 65.8% of COD removal efficiency and desalination efficiency, respectively. In a separate experiment, carbon electrode fitted MDC yielded 95.5% COD removal efficiency while treating yeast wastewater with an initial COD of 9280 mg/L with concurrent power generation of 13.4 mW/m^3 (Goren and Okten 2021). The above results advocate the use of MDCs for a more energy-efficient wastewater treatment and desalination processing. Nevertheless, MDCs are still in the developmental stage, and more comprehensive investigations are required to assess the techno-economic feasibility of the technology for becoming part of the mainstream wastewater management systems.

18.5.2 Micropollutant Remediation

Micropollutants are usually complex organic compounds that are not a natural substrate for electroactive microorganisms, thus these are difficult to break down by microbial action. Nevertheless, facile modifications in BES reactors and integration with other technologies have shown efficacy in removing recalcitrant contaminants (Table 18.4). Lin et al. (2021) was able to anaerobically biodegrade the feed containing 10 mg/L tetrabromobisphenol A by 97% in carbon electrode fitted MFC. Also, the degradation intermediate of tetrabromobisphenol A (i.e., bisphenol A and bromine (Br^-)), was detected at much lower concentrations suggesting better mineralization of tetrabromobisphenol A in the MFC system (Lin et al. 2021). Interestingly, in another investigation, power density rose by 18% to 40 W/m^3 when 20 mg/L of Sulfamethoxazole was added to the anolyte of a dual chamber MFC (Wu et al. 2020). The final removal efficiency of Sulfamethoxazole was 98%, implying SMX acted as a favorable substrate for enhancing the activity of anodic microorganisms (Wu et al. 2020).

Constructed wetlands (CW) are well known for attenuating pharmaceuticals, personal care products, and other micropollutants. However, they require a very large area for handling effluent even in relatively small quantities. Integrating CW and BES seems to be a realizable alternative to circumnavigate the drawbacks of both the systems, when operated as a standalone technology. For instance, an up-flow MFC-coupled CW was able to remove 95.7% of ibuprofen and 91.3% of bisphenol A from the effluent, with both having an initial concentration of 10 mg/L (Li et al. 2019). Supplementing MFC with UV radiation can also augment the degradation of biorefractory pollutants as observed during photo-assisted MFC degradation of ethinylestradiol. The removal efficiency for 1 mg/L of ethinylestradiol spiked from 70 to 89.82% in the photo-assisted MFC for a similar experimentation period (Sogani et al. 2021).

Another neoteric technology of bio-electro-Fenton (BEF) is attracting increasing attention owing to its proficiency in removing persistent trace contaminants. Such systems promote *in situ* cathodic generation of H_2O_2, which is further catalyzed using Fenton's reaction to produce highly potent hydroxyl radicals (·OH). Recalcitrant compounds are attacked and oxidized by ·OH, leading to their rapid removal. Soltani et al. (2021) reported 99% removal for tetracycline with an initial concentration of 10 mg/L in a BEF system along with a maximum power density of 141.60 mW/m^2. Faster degradation of pollutants can be achieved in the BEF

Table 18.4 Removal of different micropollutants through bio-electrochemical systems.

Type of BES	Micropollutant	Initial Concentration (mg/L)	Removal Efficiency (%)	References
MFC-EF	Mesotrione	50	99	Zhao and Zhang (2021)
PMFC	SDS	10	96	Sathe et al. (2020)
FMFC	Acetaminophen	300	100	Shabani et al. (2021)
MFC	Oxytetracycline	1	95	Chen et al. (2018)
MFC	Neomycin	100	88	Catal et al. (2018)
MFC	Ibuprofen	50	43.6	Aguilera Flores et al. (2021)
SMFC	Copper	172*	57	Liu et al. (2021)
MFC-EF	Pyraclostrobin	30	100	Zhao and Kong (2018)
MFC-CW	Sulfamethoxazole, Tetracycline.	100	99.4, 97.8	Wen et al. (2021)
MDC	Boron	5	79	Goren and Okten (2021)
MEC	Bisphenol A	10	95.4	Amin et al. (2021)
MEC	Antimony	25	98.4	Arulmani et al. (2021)
MEC	Sulfamethoxazole	30	92.5	Xue et al. (2019b)
MEC-F	Aniline	4460	97.1	Li et al. (2017)
MEC-F	Diclofenac, Ibuprofen	0.5	100	Zou et al. (2020)

*Unit is in mg/kg; MFC = microbial fuel cell; MFC-EF = electro Fenton-assisted MFC; PMFC = photo-assisted MFC; FMFC = fungal MFC; SMFC = sediment MFC; MFC-CW = constructed wetland integrated MFC; MDC = microbial desalination cell; MEC = microbial electrolysis cell; MEC-F = Fenton assisted MEC; SDS = sodium dodecyl sulphate.

system operated under the MEC mode, as the externally applied voltage boosts H_2O_2 production. Concurring results were obtained in the investigation carried out by Yang et al. (2021), where metoprolol degradation efficiency increased from 29% at no external voltage to 65% at an applied cell voltage of +0.2V. However, an additional cost of power supply is incurred in a BEF-MEC integrated assembly. Clearly, BESs have shown potential in remediating micropollutants efficiently; however, further research is required to amend the flaws in this technology to make it more robust and sustainable for tackling the issue of the presence of persistent pollutants in wastewater.

18.6 Scale-Up of BES

Different lab-scale investigations and research works have been performed worldwide, which ascertained the applicability of BESs as robust wastewater treatment and bioenergy recovery technology. The BESs are now being examined further on a larger magnitude to engineer a working prototype, which is technically sound as well as economically viable. Moreover, the scale-up investigations of BESs will also aid in understanding the operational and practical complications, which may not be encountered during laboratory testing. Pilot-scale experiments on MFCs have been performed since the late 2000s; however, most of the initial pilot-scale investigations were under-designed in terms of electrode specific area and spacing, leading to underwhelming power production (Logan 2010). Later, scale-up MFCs were fabricated by connecting smaller MFC units in series or parallel by maintaining the relative aspect ratio of the electrode area to the reactor working volume. Ghadge and Ghangrekar (2015) parallelly connected a multiple carbon felt electrode-based MFC fabricated with a clayware separator, which had a working volume of 26 L. Moreover, 78.8% COD removal efficiency and 47.28 mA of current generation were achieved while treating synthetic wastewater with the initial COD of 510 mg/L (Ghadge and Ghangrekar 2015).

To further improve the power generation of MFC, during an investigation, 560 miniaturized terracotta cylinder MFCs with carbon-based electrodes were stacked together and fed with neat human urine (Gajda et al. 2018b). The stacked MFC had a much higher electrode surface area-to-liquid volume ratio yielding a power output of 250 mW and an average power density of 20.4 mW/m^3, which was 2.9 times higher than the power density achieved by a larger terracotta MFC (Gajda et al. 2018b). The relatively poor power generation efficiency of a large size assembled MFC set-up was also highlighted during the investigation carried out by Das et al. (2020d), wherein 6 MFCs of 120 L capacity each were compiled together to treat real sanitary sewage with a varying influent COD in the range of 976 to 8220 mg/L, and a COD removal efficiency of 87.29% was obtained along with maximum generated power of 61 mW. In another interesting investigation, a 1000 L MFC assembled by stacking 50 smaller MFC units, each having granular activated carbon as electrodes, was operated in a municipal wastewater treatment plant in China (Liang et al. 2018). For a varying influent COD of 60 to 400 mg/L, the COD removal efficiency of the field-scale MFC was observed to be in the range of 70 to 90% over a working period of 1 year, whereas the power density fluctuated from 7 to 60 W/m^3 (Liang et al. 2018). Another 1000 L MFC was fabricated by tethering 64 smaller units together and fed with real municipal wastewater with an average COD of 400 mg/L in an upscaled investigation. The stretched MFC attained COD removal efficiency in the range of 85 to 95%, while generating a maximum power density of 0.060 kWh/m^3 (Gurav et al. 2021).

Pilot-scale MECs have also demonstrated their potential as an efficient hydrogen producing wastewater remediation technology. For instance, a 175 L MEC with stainless steel wire mesh cathode and graphite felt anode using primary treated domestic wastewater with an initial COD of 347 mg/L as feedstock, attained an average COD removal efficiency of 63.5%, while concurrently producing 0.8 L/d of 93% pure H_2 gas (Cotterill et al. 2017). The COD removal efficiency in a BES is directly proportional to the duration for which feedstock remains in the system, and generally, an HRT ranging from 18 to 36 h is adopted in pilot-scale investigations for achieving satisfactory treatment of wastewater (Cusick et al. 2011; Heidrich et al. 2013b; Ieropoulos et al. 2016; Das et al. 2020d). The large HRT requirement of BESs severely decreases their organic loading rates, thus reducing their chances to compete with existing biological treatment units, like the activated sludge process. In an attempt to optimize the HRT of an MEC treating return sludge liquor with a COD of 4535 mg/L in terms of the energy cost of treatment, Leicester et al. (2020) varied

HRT from 0.015 to 18 days and reported an optimized HRT of 0.5 days with a COD removal efficiency and hydrogen production of 55% and 0.0111 m^3/m^3 of treated return sludge liquor, respectively, at an energy consumption of 0.59 kJ/g of COD removed.

The MDCs are relatively newer BESs with minimal data available on the operation of up-scaled versions of MDC. In one such investigation, a 1 L up-flow stacked MDC equipped with carbon cloth electrodes yielded long-term desalination efficiency in the range of 87.2 to 96% and a maximum power density of 32.91 W/m^3 during the treatment of brine solution with the salinity of 25 g/L in the form of NaCl, thereby affirming the possible scale-up potential of MDCs (Wang et al. 2020b). It is evident that different large-scale BESs have illustrated great promise in concurrent wastewater treatment and bioenergy recovery. However, such systems have drawbacks, like insignificant energy recovery efficiency, relatively milder COD removal than existing wastewater treatment technologies, high fabrication cost, and requirement of low biological loading rates. An added effort is required for overcoming limitations associated with pilot-scale BESs before applying these systems for field-scale applications and subsequent commercialization.

18.7 Challenges and Future Outlook

Although BESs have the distinctive ability to produce valuable chemical by-products, biofuels, and bioelectricity while concomitantly treating wastewater, innovative technologies, such as MFC, MEC, MDC, etc., have failed to pave their way into a real wastewater treatment facility. One of the main hurdles in the up-scaling of BESs is the poor energy or product generation rate and the high fabrication cost of the reactors. The power produced in MFCs is insufficient for most practical applications due to the loss associated with the higher internal resistance of reactor components and poor ionic conductivity of wastewater. The low power output efficiency of MFCs is further aggravated when the size of the reactor is increased due to higher internal resistance offered by dimensionally larger reactor components and transportation losses (Logan 2010). Researchers have attempted to design smaller stackable MFCs units, enhancing the electrode to feedstock volume and improving power output. However, manufacturing readily stackable MFC units at a reasonable price is a challenge in itself, owing to the complications in the reactor architecture.

A significant fraction of reactor fabrication cost is contributed by electrodes and PEM, accounting for approximately 60% of the total cost (Ge and He 2016). The use of carbon-based electrodes, such as carbon cloth, carbon sheet, graphite rod, graphite plate, and carbon fiber, dramatically reduces the initial capital investment; however, they are the culprit of inefficient power generation and subdued wastewater treatment efficacy because of poor physico-chemical stability, low specific surface area, and poor conductivity and hydrophobicity of carbon materials that impede the electron transfer mechanism (Yaqoob et al. 2020). Metal electrodes are durable and have excellent conductivity, though they are not widely employed in BESs due to their high cost, susceptibility to corrosion, and non-biocompatibility.

The build up of salt deposits on the cathode and surface fouling by microbial metabolic secretions further deteriorates the operational efficiency of bio-electrochemical reactors (Hindatu et al. 2017; Gajda et al. 2018a). Many efforts have been made to develop novel electrode materials and to modify existing electrodes, which would augment the performance of BESs. For example, MFC employing a chemically reduced graphene oxide electrode demonstrated superior performance compared to carbon electrode-based MFCs, whereas titanium dioxide nanoclusters modified carbon felt electrode in an MFC amplified the microbial growth leading to improved bioelectricity production (Pareek et al. 2019; Duarte and Kwon 2020). Researchers have also investigated the applicability of waste material like medicine wrappers to function as inexpensive electrode material for MFC; however, the electrochemical efficiency of such materials was found to be below par (Noori et al. 2018).

Membrane separators are another vital component of dual chamber BESs, which are imperative in producing bioelectricity and valuable by-products. Polymeric membranes such as Nafion 117 are outstanding proton conductors; nonetheless, they are expensive and are also prone to biofouling. Many affordable materials, such as ceramics, biochar, and activated carbon, have been used as PEMs in MFCs. However, the power generation was relatively lower than conventional PEMs due to lower proton conductivity. The focus has been shifted toward developing membrane-less BESs to reduce the cost and complications involved in a dual chamber system. Membrane-less reactors often suffer from inferior product recovery and bio-electricity generation rates resulting from electrolyte intermixing and oxygen diffusion from cathodic to anodic chambers. Besides, BESs are also marred by a mediocre quality of treated effluent and sensitivity toward toxic contaminants, as the electroactive microorganisms are susceptible to hazardous environments.

Furthermore, a thorough apprehension of reaction kinetics, substrate utilization, and electrochemical performance is critical for the efficient and dependable designing of BESs for commercial platforms. Underlying principles governing the reaction kinetics in BESs can then be exploited to develop suitable models for the precise estimation of the bioelectrochemical responses of a system. One of the most fundamental kinetic expressions used in bioelectrochemistry is the Nernst-Monod model, which relates the electron shuttle's kinetic response to the potential of an anodic biofilm (Torres et al. 2008). This model has been extensively applied and modified for investigating the effect of competitive electron acceptors, pH, external poised potential, and varying substrate concentration on bioanode performance (Lusk et al. 2018; Su et al. 2018; Suzuki et al. 2018; Kubannek et al. 2022). However, these models have their own limitations; thus, finding an adaptive mathematical model that can produce an accurate response during a parallel occurring biological and electrochemical process is still an unresolved challenge.

Another conceptual solution for economizing the BES production and operation without compromising the performance is by nurturing suspended growth instead of attached biofilms. Suspended growth culture that can perform exogenic electron transfer will avoid the need for expensive electrodes and diminish potential losses. Such systems would be ideal for handling more significant loading rates by reducing clogging related issues and will have the potential to supply cheap, sustainable energy from wastewater. Also, the modeling of suspended growth bioelectrochemical reactors will not be as complex as the existing attached growth systems due to the relatively higher homogeneity. Production of H_2 gas has already been achieved in a suspended growth environment bioreactor, though at a rate slower than the attached growth reactor (Chiranjeevi et al. 2014). Nevertheless, considerable progress is required in this direction before a working model of suspended growth BES can be materialized.

Integrating BESs with other wastewater treatment methods like constructed wetlands, up-flow anaerobic sludge blanket reactors, coagulation, advanced oxidation processes, etc., can alleviate the quality of treated effluent (Elmaadawy et al. 2020; Gupta et al. 2020b). Hence, impetus should be made on designing BESs that are easy to integrate with other contemporary wastewater remediating technologies, which will help to merge BES with existing wastewater treatment infrastructure, rendering them an added viability. Moreover, comprehensive redesigning of BES architecture is required to make it robust in handling feedstock of varying nature and concentration. Novel reactor configuration should be envisioned such that it decreases the fabrication and maintenance cost, and facilitates the combining of individual units to form larger prototypes, as well as suppressing the internal resistance of the entire system for producing plausible energy. Furthermore, endeavors for ameliorating the long-term stability of BESs are required to make the technology economically viable to drive large-scale commercial applications.

18.8 Summary

Environmental pollution, water crisis, fossil fuel depletion, etc., are imminent problems looming large over society. The BESs offer a much more energy-efficient and eco-friendly method for wastewater remediation with concomitant value-added product recovery. In addition to the treatment of contaminated water, BESs are capable of producing bioelectricity, biofuels, and useful chemicals that can be used further to bring down the overall footprint and cost of these innovative technologies. Besides, the upscaled prototypes of BESs have also been effectively employed as decentralized wastewater management systems. The unique ability of chemical and energy recovery with concomitant wastewater treatment sets BESs apart from other contemporary technologies, rendering them as a potential prospect for becoming viable waste to wealth conversion platforms. However, poor power/product recovery efficiency has been a significant obstacle in the progress of BES to become a more streamlined wastewater treatment technology. The poor power generation of MFCs is mainly due to the higher internal resistance offered by PEM and other reactor components. Moreover, the extravagant fabrication cost of reactor severely hampers the economic feasibility of the BESs. Further investigations are required to address these drawbacks and conceptualize a novel architecture for BESs suitable for stacking into larger units for commercial applications. Also, emphasis should be given to improve the techno-economic feasibility of BESs, helping them to become a sustainable alternative to existing wastewater management systems.

References

Aguilera Flores, M.M., Ávila Vázquez, V., Medellín Castillo, N.A. et al. (2021). Ibuprofen degradation and energy generation in a microbial fuel cell using a bioanode fabricated from devil fish bone char. *Journal of Environmental Science and Health* 56 (A): 874–885.

Ahmad, A., Das, S., and Ghangrekar, M. (2020). Removal of xenobiotics from wastewater by electrocoagulation: a mini-review. *Journal of the Indian Chemical Society* 97: 493–500.

Ahmad, A., Priyadarshani, M., Das, S., and Ghangrekar, M.M. (2021). Role of bioelectrochemical systems for the remediation of emerging contaminants from wastewater: a review. *Journal of Basic Microbiology* 62: 201–222.

Akagunduz, D., Cebecioglu, R., Ozdemir, M., and Catal, T. (2021). Removal of psychoactive pharmaceuticals from wastewaters using microbial electrolysis cells producing hydrogen. *Water Science and Technology* 84: 931–940.

Amin, M.M., Arvin, A., Hosseini, M. et al. (2021). The degradation and simultaneous influence of bisphenol A on methane production in a bio-anode single-chamber microbial electrolysis cell. *Biochemical Engineering Journal* 176: 108219.

Arulmani, S.R.B., Dai, J., Li, H. et al. (2021). Efficient reduction of antimony by sulfate-reducer enriched bio-cathode with hydrogen production in a microbial electrolysis cell. *Science of the Total Environment* 774: 145733.

Ayyaru, S. and Ahn, Y.-H. (2021). Enhanced performance of sulfonated GO in SPEEK proton-exchange membrane for microbial fuel-cell application. *Journal of Environmental Engineering* 147: 04020153.

Bejjanki, D., Muthukumar, K., Radhakrishnan, T. et al. (2021). Simultaneous bioelectricity generation and water desalination using *Oscillatoria* sp. as biocatalyst in photosynthetic microbial desalination cell. *Science of the Total Environment* 754: 142215.

Berk, R.S. and Canfield, J.H. (1964). Bioelectrochemical energy conversion. *Applied Microbiology* 12: 10–12.

Bhowmick, G., Das, S., Verma, H. et al. (2019). Improved performance of microbial fuel cell by using conductive ink printed cathode containing Co_3O_4 or Fe_3O_4. *Electrochimica Acta* 310: 173–183.

Boretti, A. and Rosa, L. (2019). Reassessing the projections of the world water development report. *NPJ Clean Water* 2: 1–6.

Call, D. and Logan, B.E. (2008). Hydrogen production in a single chamber microbial electrolysis cell lacking a membrane. *Environmental Science and Technology* 42: 3401–3406.

Carlo, S., Catia, A., Benjamin, E., and Ioannis, I. (2017). Microbial fuel cells: from fundamentals to applications: a review. *Journal of Power Sources* 356: 225–244.

Catal, T., Yavaser, S., Enisoglu-Atalay, V. et al. (2018). Monitoring of neomycin sulfate antibiotic in microbial fuel cells. *Bioresource Technology* 268: 116–120.

Chakraborty, I., Das, S., Dubey, B., and Ghangrekar, M. (2020a). Novel low cost proton exchange membrane made from sulphonated biochar for application in microbial fuel cells. *Materials Chemistry and Physics* 239: 122025.

Chakraborty, I., Sathe, S., Khuman, C., and Ghangrekar, M. (2020b). Bioelectrochemically powered remediation of xenobiotic compounds and heavy metal toxicity using microbial fuel cell and microbial electrolysis cell. *Materials Science for Energy Technologies* 3: 104–115.

Chang, T.-J., Chang, Y.-H., Chao, W.-L. et al. (2018). Effect of hydraulic retention time on electricity generation using a solid plain-graphite plate microbial fuel cell anoxic/oxic process for treating pharmaceutical sewage. *Journal of Environmental Science and Health* 53 (A): 1185–1197.

Chaturvedi, V. and Verma, P. (2016). Microbial fuel cell: a green approach for the utilization of waste for the generation of bioelectricity. *Bioresources and Bioprocessing* 3: 1–14.

Chaurasia, A.K., Goyal, H., and Mondal, P. (2020). Hydrogen gas production with Ni, Ni–Co and Ni–Co–P electrodeposits as potential cathode catalyst by microbial electrolysis cells. *International Journal of Hydrogen Energy* 45: 18250–18265.

Chen, G., Zhang, F., Logan, B.E., and Hickner, M.A. (2013). Poly(vinyl alcohol) separators improve the coulombic efficiency of activated carbon cathodes in microbial fuel cells. *Electrochemistry Communications* 34: 150–152.

Chen, J., Hu, Y., Huang, W. et al. (2018). Biodegradation of oxytetracycline and electricity generation in microbial fuel cell with *in situ* dual graphene modified bioelectrode. *Bioresource Technology* 270: 482–488.

Chen, W., Liu, Z., Li, Y. et al. (2021). Improved electricity generation, coulombic efficiency and microbial community structure of microbial fuel cells using sodium citrate as an effective additive. *Journal of Power Sources* 482: 228947.

Cheng, D., Ngo, H.H., Guo, W. et al. (2020). Performance of microbial fuel cell for treating swine wastewater containing sulfonamide antibiotics. *Bioresource Technology* 311: 123588.

Cheng, D., Ngo, H.H., Guo, W. et al. (2021). Evaluation of a continuous flow microbial fuel cell for treating synthetic swine wastewater containing antibiotics. *Science of the Total Environment* 756: 144133.

Chia, M.Y., San Thiam, H., Leong, L.K. et al. (2020). Study on improvement of the selectivity of proton exchange membrane via incorporation of silicotungstic acid-doped silica into SPEEK. *International Journal of Hydrogen Energy* 45: 22315–22323.

Chiranjeevi, P., Kumar, A.N., and Mohan, S.V. (2014). Critical assessment of biofilm and suspended growth reactor configurations for acidogenic biohydrogen production using wastewater as a function of redox microenvironment. *International Journal of Hydrogen Energy* 39: 7561–7571.

Choudhury, P., Ray, R.N., Bandyopadhyay, T.K. et al. (2021). Process engineering for stable power recovery from dairy wastewater using microbial fuel cell. *International Journal of Hydrogen Energy* 46: 3171–3182.

Cotterill, S.E., Dolfing, J., Jones, C. et al. (2017). Low temperature domestic wastewater treatment in a microbial electrolysis cell with 1 m2Anodes: towards system scale-up. *Fuel Cells* 17: 584–592.

Cusick, R.D., Bryan, B., Parker, D.S. et al. (2011). Performance of a pilot-scale continuous flow microbial electrolysis cell fed winery wastewater. *Applied Microbiology and Biotechnology* 89: 2053–2063.

Cusick, R.D., Kiely, P.D., and Logan, B.E. (2010). A monetary comparison of energy recovered from microbial fuel cells and microbial electrolysis cells fed winery or domestic wastewaters. *International Journal of Hydrogen Energy* 35: 8855–8861.

Das, I., Das, S., Dixit, R., and Ghangrekar, M. (2020a). Goethite supplemented natural clay ceramic as an alternative proton exchange membrane and its application in microbial fuel cell. *Ionics* 26: 3061–3072.

Das, I., Das, S., and Ghangrekar, M. (2020b). Application of bimetallic low-cost CuZn as oxygen reduction cathode catalyst in lab-scale and field-scale microbial fuel cell. *Chemical Physics Letters* 751: 137536.

Das, I., Das, S., Sharma, S., and Ghangrekar, M. (2020c). Ameliorated performance of a microbial fuel cell operated with an alkali pre-treated clayware ceramic membrane. *International Journal of Hydrogen Energy* 45: 16787–16798.

Das, I., Ghangrekar, M.M., Satyakam, R. et al. (2020d). On-site sanitary wastewater treatment system using 720-L stacked microbial fuel cell: case study. *Journal of Hazardous, Toxic, and Radioactive Waste* 24: 04020025.

Das, S., Das, S., Das, I., and Ghangrekar, M. (2019a). Application of bioelectrochemical systems for carbon dioxide sequestration and concomitant valuable recovery: a review. *Materials Science for Energy Technologies* 2: 687–696.

Das, S., Das, S., and Ghangrekar, M. (2019b). Quorum-sensing mediated signals: a promising multi-functional modulators for separately enhancing algal yield and power generation in microbial fuel cell. *Bioresource Technology* 294: 122138.

Das, S., Diels, L., Pant, D. et al. (2020e). Review: microbial electrosynthesis: a way towards the production of electro-commodities through carbon sequestration with microbes as biocatalysts. *Journal of the Electrochemical Society* 167: 155510.

Das, S., Mishra, A., and Ghangrekar, M. (2020f). Concomitant production of bioelectricity and hydrogen peroxide leading to the holistic treatment of wastewater in microbial fuel cell. *Chemical Physics Letters* 759: 137986.

Dessì, P., Rovira-Alsina, L., Sánchez, C. et al. (2020). Microbial electrosynthesis: towards sustainable biorefineries for production of green chemicals from CO_2 emissions. *Biotechnology Advances* 46: 107675.

Duarte, K.D. and Kwon, Y. (2020). *In situ* carbon felt anode modification via codeveloping *Saccharomyces cerevisiae* living-template titanium dioxide nanoclusters in a yeast-based microbial fuel cell. *Journal of Power Sources* 474: 228651.

Elmaadawy, K., Liu, B., Hu, J. et al. (2020). Performance evaluation of microbial fuel cell for landfill leachate treatment: research updates and synergistic effects of hybrid systems. *Journal of Environmental Sciences* 96: 1–20.

Gajda, I., Greenman, J., and Ieropoulos, I.A. (2018a). Recent advancements in real-world microbial fuel cell applications. *Current Opinion in Electrochemistry* 11: 78–83.

Gajda, I., Stinchcombe, A., Merino-Jimenez, I. et al. (2018b). Miniaturized ceramic-based microbial fuel cell for efficient power generation from urine and stack development. *Frontiers in Energy Research* 6: 84.

Ge, Z. and He, Z. (2016). Long-term performance of a 200 liter modularized microbial fuel cell system treating municipal wastewater: treatment, energy, and cost. *Environmental Science: Water Research & Technology* 2: 274–281.

Ge, Z., Li, J., Xiao, L. et al. (2014). Recovery of electrical energy in microbial fuel cells: brief review. *Environmental Science & Technology Letters* 1: 137–141.

Ghadge, A.N. and Ghangrekar, M.M. (2015). Performance of low cost scalable air–cathode microbial fuel cell made from clayware separator using multiple electrodes. *Bioresource Technology* 182: 373–377.

Ghasemi, M., Daud, W.R.W., Hassan, S.H. et al. (2016). Carbon nanotube/polypyrrole nanocomposite as a novel cathode catalyst and proper alternative for Pt in microbial fuel cell. *International Journal of Hydrogen Energy* 41: 4872–4878.

Ghorai, A., Roy, S., Das, S. et al. (2021). Preparation of sulfonated polytriazoles with a phosphaphenanthrene unit via click polymerization: fabrication of membranes and properties thereof. *ACS Applied Polymer Materials* 3: 4127–4138.

Goren, A.Y. and Okten, H.E. (2021). Energy production from treatment of industrial wastewater and boron removal in aqueous solutions using microbial desalination cell. *Chemosphere* 285: 131370.

Gul, H., Raza, W., Lee, J. et al. (2021). Progress in microbial fuel cell technology for wastewater treatment and energy harvesting. *Chemosphere* 281: 130828.

Guo, K., Prévoteau, A., Patil, S.A., and Rabaey, K. (2015). Engineering electrodes for microbial electrocatalysis. *Current Opinion in Biotechnology* 33: 149–156.

Gupta, A., Das, S., and Ghangrekar, M. (2020a). Optimal cathodic imposed potential and appropriate catalyst for the synthesis of hydrogen peroxide in microbial electrolysis cell. *Chemical Physics Letters* 754: 137690.

Gupta, S., Srivastava, P., Patil, S.A., and Yadav, A.K. (2020b). A comprehensive review on emerging constructed wetland coupled microbial fuel cell technology: potential applications and challenges. *Bioresource Technology* 320: 124376.

Gurav, R., Bhatia, S.K., Choi, T.-R. et al. (2021). Seafood processing chitin waste for electricity generation in a microbial fuel cell using halotolerant catalyst *Oceanisphaera arctica* YHY1. *Sustainability* 13: 8508.

Han, X., Qu, Y., Li, D. et al. (2021). Combined microbial electrolysis cell–iron-air battery system for hydrogen production and swine wastewater treatment. *Process Biochemistry* 101: 104–110.

Heidrich, E., Curtis, T., and Dolfing, J. (2011). Determination of the internal chemical energy of wastewater. *Environmental Science and Technology* 45: 827–832.

Heidrich, E., Dolfing, J., Scott, K. et al. (2013). Production of hydrogen from domestic wastewater in a pilot-scale microbial electrolysis cell. *Applied Microbiology and Biotechnology* 97: 6979–6989.

Hindatu, Y., Annuar, M.S.M., and Gumel, A.M. (2017). Mini-review: anode modification for improved performance of microbial fuel cell. *Renewable and Sustainable Energy Reviews* 73: 236–248.

Hou, Y., Tu, L., Qin, S. et al. (2021). Dye wastewater treatment and hydrogen production in microbial electrolysis cells using MoS_2-graphene oxide cathode: effects of dye concentration, co-substrate and buffer solution. *Process Biochemistry* 102: 51–58.

Hou, Y., Zhang, R., Luo, H. et al. (2015). Microbial electrolysis cell with spiral wound electrode for wastewater treatment and methane production. *Process Biochemistry* 50: 1103–1109.

Hua, T., Li, S., Li, F. et al. (2019). Microbial electrolysis cell as an emerging versatile technology: a review on its potential application, advance and challenge. *Journal of Chemical Technology and Biotechnology* 94: 1697–1711.

Huang, W., Chen, J., Hu, Y., and Zhang, L. (2018). Enhancement of Congo red decolorization by membrane-free structure and bio-cathode in a microbial electrolysis cell. *Electrochimica Acta* 260: 196–203.

Ieropoulos, I.A., Stinchcombe, A., Gajda, I. et al. (2016). Pee power urinal: microbial fuel cell technology field trials in the context of sanitation. *Environmental Science: Water Research & Technology* 2: 336–343.

Ismail, Z.Z. and Habeeb, A.A. (2017). Experimental and modeling study of simultaneous power generation and pharmaceutical wastewater treatment in microbial fuel cell based on mobilized biofilm bearers. *Renewable Energy* 101: 1256–1265.

Jadhav, D., Ghadge, A., Mondal, D., and Ghangrekar, M. (2014). Comparison of oxygen and hypochlorite as cathodic electron acceptor in microbial fuel cells. *Bioresource Technology* 154: 330–335.

Jadhav, G. and Ghangrekar, M. (2009). Performance of microbial fuel cell subjected to variation in pH, temperature, external load and substrate concentration. *Bioresource Technology* 100: 717–723.

Jafary, T., Daud, W.R.W., Ghasemi, M. et al. (2015). Biocathode in microbial electrolysis cell; present status and future prospects. *Renewable and Sustainable Energy Reviews* 47: 23–33.

Jain, A. and He, Z. (2018). Cathode-enhanced wastewater treatment in bioelectrochemical systems. *NPJ Clean Water* 1: 1–5.

Jayabalan, T., Matheswaran, M., Radhakrishnan, T., and Mohamed, S.N. (2021). Influence of nickel molybdate nanocatalyst for enhancing biohydrogen production in microbial electrolysis cell utilizing sugar industrial effluent. *Bioresource Technology* 320: 124284.

Jung, S., Lee, J., Park, Y.K., and Kwon, E.E. (2020). Bioelectrochemical systems for a circular bioeconomy. *Bioresource Technology* 300: 122748.

Jwa, E., Yun, Y.-M., Kim, H. et al. (2019). Domestic wastewater treatment in a tubular microbial electrolysis cell with a membrane electrode assembly. *International Journal of Hydrogen Energy* 44: 652–660.

Kadier, A., Simayi, Y., Kalil, M.S. et al. (2014). A review of the substrates used in microbial electrolysis cells (MECs) for producing sustainable and clean hydrogen gas. *Renewable Energy* 71: 466–472.

Karuppiah, T., Uthirakrishnan, U., Sivakumar, S.V. et al. (2021). Processing of electroplating industry wastewater through dual chambered microbial fuel cells (MFC) for simultaneous treatment of wastewater and green fuel production. *International Journal of Hydrogen Energy* 47: 37569–37576.

Kodali, M., Herrera, S., Kabir, S. et al. (2018). Enhancement of microbial fuel cell performance by introducing a nano-composite cathode catalyst. *Electrochimica Acta* 265: 56–64.

Kubannek, F., Block, J., Munirathinam, B., and Krull, R. (2022). Reaction kinetics of anodic biofilms under changing substrate concentrations: uncovering shifts in Nernst-Monod curves via substrate pulses. *Engineering in Life Sciences* 22: 152–164.

Leicester, D.D., Amezaga, J.M., Moore, A., and Heidrich, E.S. (2020). Optimising the hydraulic retention time in a pilot-scale microbial electrolysis cell to achieve high volumetric treatment rates using concentrated domestic wastewater. *Molecules* 25: 2945.

Li, H., Zhang, S., Yang, X.-L. et al. (2019). Enhanced degradation of bisphenol A and ibuprofen by an up-flow microbial fuel cell-coupled constructed wetland and analysis of bacterial community structure. *Chemosphere* 217: 599–608.

Li, M., Zhou, M., Tian, X. et al. (2018). Microbial fuel cell (MFC) power performance improvement through enhanced microbial electrogenicity. *Biotechnology Advances* 36: 1316–1327.

Li, N., An, J., Zhou, L. et al. (2016). A novel carbon black graphite hybrid air-cathode for efficient hydrogen peroxide production in bioelectrochemical systems. *Journal of Power Sources* 306: 495–502.

Li, X., Jin, X., Zhao, N. et al. (2017). Efficient treatment of aniline containing wastewater in bipolar membrane microbial electrolysis cell-Fenton system. *Water Research* 119: 67–72.

Liang, B., Guo, S., Zhao, Y. et al. (2020). Single iron atoms anchored on activated carbon as active centres for highly efficient oxygen reduction reaction in air–cathode microbial fuel cell. *Journal of Power Sources* 450: 227683.

Liang, B., Li, K., Liu, Y., and Kang, X. (2019). Nitrogen and phosphorus dual-doped carbon derived from chitosan: an excellent cathode catalyst in microbial fuel cell. *Chemical Engineering Journal* 358: 1002–1011.

Liang, B., Zhang, X., Zhong, M. et al. (2021). Transition metal (Fe, Co, Ni) and sulfur codoped nitrogen-enriched hydrothermal carbon as high-performance cathode catalyst for microbial fuel cell. *Journal of Power Sources* 506: 230178.

Liang, P., Duan, R., Jiang, Y. et al. (2018). One-year operation of 1000-L modularized microbial fuel cell for municipal wastewater treatment. *Water Research* 141: 1–8.

Lim, S.S., Daud, W.R.W., Jahim, J.M. et al. (2012). Sulfonated poly (ether ether ketone)/poly (ether sulfone) composite membranes as an alternative proton exchange membrane in microbial fuel cells. *International Journal of Hydrogen Energy* 37: 11409–11424.

Lim, S.S., Fontmorin, J.-M., Pham, H.T. et al. (2021). Zinc removal and recovery from industrial wastewater with a microbial fuel cell: experimental investigation and theoretical prediction. *Science of the Total Environment* 776: 145934.

Lim, S.S., Yu, E.H., Daud, W.R.W. et al. (2017). Bioanode as a limiting factor to biocathode performance in microbial electrolysis cells. *Bioresource Technology* 238: 313–324.

Lin, X.-Q., Li, Z.-L., Nan, J. et al. (2021). Biodegradation and metabolism of tetrabromobisphenol A in microbial fuel cell: behaviors, dynamic pathway and the molecular ecological mechanism. *Journal of Hazardous Materials* 417: 126104.

Liu, D., Chang, Q., Gao, Y. et al. (2020). High performance of microbial fuel cell afforded by metallic tungsten carbide decorated carbon cloth anode. *Electrochimica Acta* 330: 135243.

Liu, S.-H., Huang, W.-J., Lin, C.-W., and Zhu, T.-J. (2021). Copper removal and microbial community analysis in a single medium sediment microbial fuel cell. *Journal of Water Process Engineering* 44: 102348.

Logan, B.E. (2009). Exoelectrogenic bacteria that power microbial fuel cells. *Nature Reviews Microbiology* 7: 375–381.

Logan, B.E. (2010). Scaling up microbial fuel cells and other bioelectrochemical systems. *Applied Microbiology and Biotechnology* 85: 1665–1671.

Logan, B.E., Rossi, R., Ragab, A., and Saikaly, P.E. (2019). Electroactive microorganisms in bioelectrochemical systems. *Nature Reviews Microbiology* 17: 307–319.

Logan, B.E., Wallack, M.J., Kim, K.-Y. et al. (2015). Assessment of microbial fuel cell configurations and power densities. *Environmental Science & Technology Letters* 2: 206–214.

Lu, M., Chen, S., Babanova, S. et al. (2017). Long-term performance of a 20-L continuous flow microbial fuel cell for treatment of brewery wastewater. *Journal of Power Sources* 356: 274–287.

Lusk, B.G., Peraza, I., Albal, G. et al. (2018). pH dependency in anode biofilms of Thermincola ferriacetica suggests a proton-dependent electrochemical response. *Journal of the American Chemical Society* 140: 5527–5534.

Mani, P.,Bowman, K., Keshavarz, T.et al. (2020). Development of an electroactive aerobic biocathode for microbial fuel cell applications. *Environmental Microbiology Reports* 12: 607–612.

Maxwell, S. (2016). The water-energy nexus in water and wastewater treatment. *Journal American Water Works Association* 108: 20–24.

Min, B. and Logan, B.E. (2004). Continuous electricity generation from domestic wastewater and organic substrates in a flat plate microbial fuel cell. *Environmental Science and Technology* 38: 5809–5814.

Modin, O., Fuad, N., and Rauch, S. (2017). Microbial electrochemical recovery of zinc. *Electrochimica Acta* 248: 58–63.

Mohamed, S.N., Thomas, N., Tamilmani, J. et al. (2020). Bioelectricity generation using iron (II) molybdate nanocatalyst coated anode during treatment of sugar wastewater in microbial fuel cell. *Fuel* 277: 118119.

Monteil, H., Pechaud, Y., Oturan, N., and Oturan, M.A. (2019). A review on efficiency and cost effectiveness of electro-and bio-electro-Fenton processes: application to the treatment of pharmaceutical pollutants in water. *Chemical Engineering Journal* 376: 119577.

Morvarid, K.Z., Hamid-Reza, K., and Manouchehr, V. (2017). Electricity generation, desalination and microalgae cultivation in a biocathode-microbial desalination cell. *Journal of Environmental Chemical Engineering* 5: 843–848.

Mu, Y., Rabaey, K., Rozendal, R.A. et al. (2009). Decolorization of azo dyes in bioelectrochemical systems. *Environmental Sciency and Technology* 43: 5137–5143.

Nagulapally, S.R., Ahmad, A., Henry, A. et al. (2009). Occurrence of ciprofloxacin-, trimethoprim-sulfamethoxazole-, and vancomycin-resistant bacteria in a municipal wastewater treatment plant. *Water Environment Research* 81: 82–90.

Nancharaiah, Y.V., Mohan, S.V., and Lens, P.N. (2015). Metals removal and recovery in bioelectrochemical systems: a review. *Bioresource Technology* 195: 102–114.

Nancharaiah, Y.V., Mohan, S.V., and Lens, P.N. (2016). Recent advances in nutrient removal and recovery in biological and bioelectrochemical systems. *Bioresource Technology* 215: 173–185.

Nath, D., Das, S., and Ghangrekar, M. (2021). High throughput techniques for the rapid identification of electroactive microorganisms. *Chemosphere* 285: 131489.

Neethu, B., Bhowmick, G., and Ghangrekar, M. (2019). A novel proton exchange membrane developed from clay and activated carbon derived from coconut shell for application in microbial fuel cell. *Biochemical Engineering Journal* 148: 170–177.

Nimje, V.R., Chen, C.-Y., Chen, H.-R. et al. (2012). Comparative bioelectricity production from various wastewaters in microbial fuel cells using mixed cultures and a pure strain of *Shewanella oneidensis*. *Bioresource Technology* 104: 315–323.

Noori, M.T., Bhowmick, G., Tiwari, B. et al. (2018). Utilisation of waste medicine wrappers as an efficient low-cost electrode material for microbial fuel cell. *Environmental Technology* 41 (10): 1209–1218.

Ondon, B.S., Li, S., Zhou, Q., and Li, F. (2020). Simultaneous removal and high tolerance of norfloxacin with electricity generation in microbial fuel cell and its antibiotic resistance genes quantification. *Bioresource Technology* 304: 122984.

Pant, D., Singh, A., Van Bogaert, G. et al. (2012). Bioelectrochemical systems (BES) for sustainable energy production and product recovery from organic wastes and industrial wastewaters. *RSC Advances* 2: 1248–1263.

Pareek, A., Sravan, J.S., and Mohan, S.V. (2019). Exploring chemically reduced graphene oxide electrode for power generation in microbial fuel cell. *Materials Science for Energy Technologies* 2: 600–606.

Penteado, E.D., Fernandez-Marchante, C.M., Zaiat, M. et al. (2016). Energy recovery from winery wastewater using a dual chamber microbial fuel cell. *Journal of Chemical Technology and Biotechnology* 91: 1802–1808.

Pinto, R., Srinivasan, B., Guiot, S., and Tartakovsky, B. (2011). The effect of real-time external resistance optimization on microbial fuel cell performance. *Water Research* 45: 1571–1578.

Pophali, A., Singh, S., and Verma, N. (2020). Simultaneous hydrogen generation and COD reduction in a photoanode-based microbial electrolysis cell. *International Journal of Hydrogen Energy* 45: 25985–25995.

Potter, M.C. (1911). Electrical effects accompanying the decomposition of organic compounds. *Proceedings of the Royal Society of London. Series B* 84: 260–276.

Priyadarshini, M., Ahmad, A., Das, S., and Ghangrekar, M.M. (2021). Application of microbial electrochemical technologies for the treatment of petrochemical wastewater with concomitant valuable recovery: a review. *Environmental Science and Pollution Research* 29: 61783–61802.

Pugazhendi, A., Al-Mutairi, A.E., Jamal, M.T. et al. (2020). Treatment of seafood industrial wastewater coupled with electricity production using air–cathode microbial fuel cell under saline conditions. *International Journal of Energy Research* 44: 12535–12545.

Qin, B., Luo, H., Liu, G. et al. (2012). Nickel ion removal from wastewater using the microbial electrolysis cell. *Bioresource Technology* 121: 458–461.

Rahimnejad, M., Adhami, A., Darvari, S. et al. (2015). Microbial fuel cell as new technology for bioelectricity generation: a review. *Alexandria Engineering Journal* 54: 745–756.

Rahimnejad, M., Bakeri, G., Najafpour, G. et al. (2014). A review on the effect of proton exchange membranes in microbial fuel cells. *Biofuel Research Journal* 1: 7–15.

Rahimnejad, M., Ghoreyshi, A.A., Najafpour, G., and Jafary, T. (2011). Power generation from organic substrate in batch and continuous flow microbial fuel cell operations. *Applied Energy* 88: 3999–4004.

Raj, R., Tripathi, A., Das, S., and Ghangrekar, M. (2021). Removal of caffeine from wastewater using electrochemical advanced oxidation process: a mini review. *Case Studies in Chemical and Environmental Engineering* 4: 100129.

Rashid, T., Sher, F., Hazafa, A. et al. (2021). Design and feasibility study of novel paraboloid graphite based microbial fuel cell for bioelectrogenesis and pharmaceutical wastewater treatment. *Journal of Environmental Chemical Engineering* 9: 104502.

Rusli, S.F.N., Bakar, M.H.A., Loh, K.S., and Mastar, M.S. (2019). Review of high-performance biocathode using stainless steel and carbon-based materials in microbial fuel cell for electricity and water treatment. *International Journal of Hydrogen Energy* 44: 30772–30787.

Sadhasivam, T., Dhanabalan, K., Roh, S.-H. et al. (2017). A comprehensive review on unitized regenerative fuel cells: crucial challenges and developments. *International Journal of Hydrogen Energy* 42: 4415–4433.

Saeed, H.M., Husseini, G.A., Yousef, S. et al. (2015). Microbial desalination cell technology: a review and a case study. *Desalination* 359: 1–13.

Samsudeen, N., Radhakrishnan, T., and Matheswaran, M. (2015). Bioelectricity production from microbial fuel cell using mixed bacterial culture isolated from distillery wastewater. *Bioresource Technology* 195: 242–247.

Santoro, C., Kodali, M., Shamoon, N. et al. (2019). Increased power generation in supercapacitive microbial fuel cell stack using FeNC cathode catalyst. *Journal of Power Sources* 412: 416–424.

Sarmin, S., Ethiraj, B., Islam, M.A. et al. (2019). Bio-electrochemical power generation in petrochemical wastewater fed microbial fuel cell. *Science of the Total Environment* 695: 133820.

Sathe, S., Bhowmick, G., Dubey, B., and Ghangrekar, M. (2020). Surfactant removal from wastewater using photo-cathode microbial fuel cell and laterite-based hybrid treatment system. *Bioprocess and Biosystems Engineering* 43: 2075–2084.

Sathe, S., Chakraborty, I., Dubey, B., and Ghangrekar, M. (2022). Microbial fuel cell coupled Fenton oxidation for the cathodic degradation of emerging contaminants from wastewater: applications and challenges. *Environmental Research* 204: 112135.

Sevda, S. and Abu-Reesh, I.M. (2019). Improved petroleum refinery wastewater treatment and seawater desalination performance by combining osmotic microbial fuel cell and up-flow microbial desalination cell. *Environmental Technology* 40: 888–895.

Shabani, M., Younesi, H., Pontié, M. et al. (2020). A critical review on recent proton exchange membranes applied in microbial fuel cells for renewable energy recovery. *Journal of Cleaner Production* 264: 121446.

Shabani, M., Pontié, M., Younesi, H. et al. (2021). Biodegradation of acetaminophen and its main by-product 4-aminophenol by *Trichoderma harzianum* versus mixed biofilm of *Trichoderma harzianum/Pseudomonas fluorescens* in a fungal microbial fuel cell. *Journal of Applied Electrochemistry* 51: 581–596.

Shahgaldi, S., Ghasemi, M., Daud, W.R.W. et al. (2014). Performance enhancement of microbial fuel cell by PVDF/Nafion nanofibre composite proton exchange membrane. *Fuel Processing Technology* 124: 290–295.

Shreeram, D.D., Panmanee, W., McDaniel, C.T. et al. (2018). Effect of impaired twitching motility and biofilm dispersion on performance of *Pseudomonas aeruginosa*-powered microbial fuel cells. *Journal of Industrial Microbiology and Biotechnology* 45: 103–109.

Singh, A. and Kaushik, A. (2021). Removal of Cd and Ni with enhanced energy generation using biocathode microbial fuel cell: insights from molecular characterization of biofilm communities. *Journal of Cleaner Production* 315: 127940.

Singha, S., Jana, T., Modestra, J.A. et al. (2016). Highly efficient sulfonated polybenzimidazole as a proton exchange membrane for microbial fuel cells. *Journal of Power Sources* 317: 143–152.

Sirajudeen, A.A.O., Annuar, M.S.M., Ishak, K.A. et al. (2021). Innovative application of biopolymer composite as proton exchange membrane in microbial fuel cell utilizing real wastewater for electricity generation. *Journal of Cleaner Production* 278: 123449.

Sogani, M., Pankan, A.O., Dongre, A. et al. (2021). Augmenting the biodegradation of recalcitrant ethinylestradiol using *Rhodopseudomonas palustris* in a hybrid photo-assisted microbial fuel cell with enhanced bio-hydrogen production. *Journal of Hazardous Materials* 408: 124421.

Soltani, F., Navidjouy, N., Khorsandi, H. et al. (2021). A novel bio-electro-Fenton system with dual application for the catalytic degradation of tetracycline antibiotic in wastewater and bioelectricity generation. *RSC Advances* 11: 27160–27173.

Son, S., Koo, B., Chai, H. et al. (2021). Comparison of hydrogen production and system performance in a microbial electrolysis cell containing cathodes made of non-platinum catalysts and binders. *Journal of Water Process Engineering* 40: 101844.

Stratford, J.P., Beecroft, N.J., Slade, R.C. et al. (2014). Anodic microbial community diversity as a predictor of the power output of microbial fuel cells. *Bioresource Technology* 156: 84–91.

Su, S.-G., Cheng, H.-Y., Zhu, -T.-T. et al. (2018). Kinetic competition between microbial anode respiration and nitrate respiration in a bioelectrochemical system. *Bioelectrochemistry (Amsterdam, Netherlands)* 123: 241–247.

Sund, C.J., McMasters, S., Crittenden, S.R. et al. (2007). Effect of electron mediators on current generation and fermentation in a microbial fuel cell. *Applied Microbiology and Biotechnology* 76: 561–568.

Suzuki, K., Kato, Y., Yui, A. et al. (2018). Bacterial communities adapted to higher external resistance can reduce the onset potential of anode in microbial fuel cells. *Journal of Bioscience and Bioengineering* 125: 565–571.

Tawalbeh, M., Al-Othman, A., Singh, K. et al. (2020). Microbial desalination cells for water purification and power generation: a critical review. *Energy* 209: 118493.

Tenca, A., Cusick, R.D., Schievano, A. et al. (2013). Evaluation of low cost cathode materials for treatment of industrial and food processing wastewater using microbial electrolysis cells. *International Journal of Hydrogen Energy* 38: 1859–1865.

Tian, X., Song, Y., Shen, Z. et al. (2020). A comprehensive review on toxic petrochemical wastewater pretreatment and advanced treatment. *Journal of Cleaner Production* 245: 118692.

Torres, C.I., Marcus, A.K., Parameswaran, P., and Rittmann, B.E. (2008). Kinetic experiments for evaluating the Nernst–Monod model for anode-respiring bacteria (ARB) in a biofilm anode. *Environmental Science & Technology* 42: 6593–6597.

Tursun, H., Liu, R., Li, J. et al. (2016). Carbon material optimized biocathode for improving microbial fuel cell performance. *Frontiers in Microbiology* 7: 6.

van Hees, W. (1965). A bacterial methane fuel cell. *Journal of the Electrochemical Society* 112: 258.

Wang, Q., Lv, R., Rene, E.R. et al. (2020a). Characterization of microbial community and resistance gene (CzcA) shifts in up-flow constructed wetlands–microbial fuel cell treating Zn (II) contaminated wastewater. *Bioresource Technology* 302: 122867.

Wang, Y., Xu, A., Cui, T. et al. (2020b). Construction and application of a 1-liter upflow-stacked microbial desalination cell. *Chemosphere* 248: 126028.

Wen, H., Zhu, H., Xu, Y. et al. (2021). Removal of sulfamethoxazole and tetracycline in constructed wetlands integrated with microbial fuel cells influenced by influent and operational conditions. *Environmental Pollution* 272: 115988.

Wilberforce, T., Sayed, E.T., Abdelkareem, M.A. et al. (2021). Value added products from wastewater using bioelectrochemical systems: current trends and perspectives. *Journal of Water Process Engineering* 39: 101737.

Wu, D., Sun, F., Chua, F.J.D., and Zhou, Y. (2020). Enhanced power generation in microbial fuel cell by an agonist of electroactive biofilm: Sulfamethoxazole. *Chemical Engineering Journal* 384: 123238.

Wu, Y., Zhao, X., Jin, M. et al. (2018). Copper removal and microbial community analysis in single-chamber microbial fuel cell. *Bioresource Technology* 253: 372–377.

Xiao, L., Ge, Z., Kelly, P. et al. (2014). Evaluation of normalized energy recovery (NER) in microbial fuel cells affected by reactor dimensions and substrates. *Bioresource Technology* 157: 77–83.

Xue, W., Li, F., and Zhou, Q. (2019a). Degradation mechanisms of sulfamethoxazole and its induction of bacterial community changes and antibiotic resistance genes in a microbial fuel cell. *Bioresource Technology* 289: 121632.

Xue, W., Zhou, Q., and Li, F. (2019b). Bacterial community changes and antibiotic resistance gene quantification in microbial electrolysis cells during long-term sulfamethoxazole treatment. *Bioresource Technology* 294: 122170.

Yadav, G., Sharma, I., Ghangrekar, M., and Sen, R. (2020). A live bio-cathode to enhance power output steered by bacteria–microalgae synergistic metabolism in microbial fuel cell. *Journal of Power Sources* 449: 227560.

Yang, W., Wang, X., Son, M., and Logan, B.E. (2020). Simultaneously enhancing power density and coulombic efficiency with a hydrophobic Fe–N_4/activated carbon air cathode for microbial fuel cells. *Journal of Power Sources* 465: 228264.

Yang, X., Zou, R., Tang, K. et al. (2021). Degradation of metoprolol from wastewater in a bio-electro-Fenton system. *Science of the Total Environment* 771: 145385.

Yaqoob, A.A., Ibrahim, M.N.M., and Rodríguez-Couto, S. (2020). Development and modification of materials to build cost-effective anodes for microbial fuel cells (MFCs): an overview. *Biochemical Engineering Journal* 164: 107779.

Yu, J., Seon, J., Park, Y. et al. (2012). Electricity generation and microbial community in a submerged-exchangeable microbial fuel cell system for low-strength domestic wastewater treatment. *Bioresource Technology* 117: 172–179.

Yu, Z., Leng, X., Zhao, S. et al. (2018). A review on the applications of microbial electrolysis cells in anaerobic digestion. *Bioresource Technology* 255: 340–348.

Zhang, L., Wang, J., Fu, G., and Zhang, Z. (2020). Simultaneous electricity generation and nitrogen and carbon removal in single-chamber microbial fuel cell for high-salinity wastewater treatment. *Journal of Cleaner Production* 276: 123203.

Zhang, P., Liu, J., Qu, Y. et al. (2017). Enhanced performance of microbial fuel cell with a bacteria/multi-walled carbon nanotube hybrid biofilm. *Journal of Power Sources* 361: 318–325.

Zhang, Q., Zhang, L., Wang, H. et al. (2018a). Simultaneous efficient removal of oxyfluorfen with electricity generation in a microbial fuel cell and its microbial community analysis. *Bioresource Technology* 250: 658–665.

Zhang, S., You, J., Kennes, C. et al. (2018b). Current advances of VOCs degradation by bioelectrochemical systems: a review. *Chemical Engineering Journal* 334: 2625–2637.

Zhao, H. and Kong, C.-H. (2018). Elimination of pyraclostrobin by simultaneous microbial degradation coupled with the Fenton process in microbial fuel cells and the microbial community. *Bioresource Technology* 258: 227–233.

Zhao, H. and Zhang, Q. (2021). Performance of electro-Fenton process coupling with microbial fuel cell for simultaneous removal of herbicide mesotrione. *Bioresource Technology* 319: 124244.

Zheng, T., Li, J., Ji, Y. et al. (2020). Progress and prospects of bioelectrochemical systems: electron transfer and its applications in the microbial metabolism. *Frontiers in Bioengineering and Biotechnology* 8: 10.

Zhou, H., Zhang, D., Zhang, Y. et al. (2019). Magnetic cathode stimulates extracellular electron transfer in bioelectrochemical systems. *ACS Sustainable Chemistry & Engineering* 7: 15012–15018.

Zikmund, E., Kim, K.-Y., and Logan, B.E. (2018). Hydrogen production rates with closely-spaced felt anodes and cathodes compared to brush anodes in two-chamber microbial electrolysis cells. *International Journal of Hydrogen Energy* 43: 9599–9606.

Zinadini, S., Zinatizadeh, A., Rahimi, M. et al. (2017). High power generation and COD removal in a microbial fuel cell operated by a novel sulfonated PES/PES blend proton exchange membrane. *Energy* 125: 427–438.

Zou, R., Angelidaki, I., Yang, X. et al. (2020). Degradation of pharmaceuticals from wastewater in a 20-L continuous flow bio-electro-Fenton (BEF) system. *Science of the Total Environment* 727: 138684.

Zou, S. and He, Z. (2018). Efficiently "pumping out" value-added resources from wastewater by bioelectrochemical systems: a review from energy perspectives. *Water Research* 131: 62–73.

19

Hybrid Treatment Solutions for Removal of Micropollutant From Wastewaters

Monali Priyadarshini[1], *S. M. Sathe*[2], *and Makarand M. Ghangrekar*[1,2]

[1]*School of Environmental Science and Engineering, Indian Institute of Technology Kharagpur, West Bengal, India*
[2]*Department of Civil Engineering, Indian Institute of Technology Kharagpur, West Bengal, India*

19.1 Background of Hybrid Treatment Processes

Improvement in living standards, industrialization, and urbanization have resulted in an increase in the discharge of complex microconstituents (MCs) in domestic wastewater. Earlier, the majority of MCs were confined to industrial wastewaters; however, recently, MCs are now commonly detected in domestic wastewater, with concentrations ranging from ng L^{-1} to μg L^{-1}. The MCs commonly detected in domestic wastewater include antibiotics, endocrine disrupting chemicals, heavy metals, hydrocarbons, pharmaceuticals, personal care products, preservatives, surfactants, etc. (Petrović et al. 2003; Rout et al; 2021). The primary and secondary biological wastewater treatments are not capable of completely removing such non-biodegradable contaminants from wastewater. Tertiary treatments, such as adsorption, advanced oxidation processes (AOPs), membrane filtration, electrochemical oxidation (EO), photocatalysis, etc., have been widely used for elimination of MCs from wastewater (Dhangar and Kumar 2020). Though successful, these tertiary treatment processes are energy and chemical intensive. This makes the overall wastewater treatment process costly. Even then, for the removal of MCs, tertiary treatments are an ideal choice of environmental engineers. The advantages and disadvantages of individual treatment technologies that used for the removal of MCs, which are also called as emerging contaminants, (ECs) are listed in Table 19.1.

More recently, in order to overcome the shortcomings associated with individual treatments, researchers have tried combining more than one treatment technique in a single setup for wastewater treatment. Such hybrid processes combine more than one treatment technique to enhance the pollutant removal efficiency. The hybrid processes offer advantages of both individual systems, which not only yield better pollutant removal but also complement both processes thereby reducing operational cost. This chapter discusses the application of hybrid systems for wastewater treatment with more specific emphasis on MCs removal. Current research gaps and future direction of research in this field are also identified.

Microconstituents in the Environment: Occurrence, Fate, Removal, and Management, First Edition.
Edited by Rao Y. Surampalli, Tian C. Zhang, Chih-Ming Kao, Makarand M. Ghangrekar, Puspendu Bhunia, Manaswini Behera, and Prangya R. Rout.

Table 19.1 Advantages and disadvantages of individual treatment systems treating MCs.

Treatment	Advantages	Disadvantages
Adsorption	• High removal of MCs; • Possible use of waste-based materials as adsorbent.	• Need of frequent regeneration; • Proper disposal of exhausted adsorbent is challenging; • Requirement of prior primary and secondary treatments for long operational life.
Constructed wetland.	• Low operation and maintenance cost; • Low energy demand; • Simultaneous removal of MCs and organic matter, nutrients, suspended solids.	• Large area requirement; • Clogging due to high suspended solids; • Longer contact time for MCs degradation.
Membrane filtration.	• Shorter retention time for removal of MCs; • High removal of MCs; • Simultaneous removal of MCs.	• High operation and maintenance cost; • Requirement of prior primary and secondary treatments for long operational life.
Bio-electrochemical systems.	• Low operation and maintenance cost; • Simultaneous bioenergy and valuables recovery; • Less sludge production.	• Longer contact time for microconstituents degradation; • Limitation in upscaling a single system of larger volume.
Advanced oxidation processes.	• High removal of MCs; • Shorter retention time for MCs removal; • Capability to handle shock loading; • Less area requirement.	• High operation and maintenance cost; • Requirement of chemicals for pH changes; • Requirement of prior primary and secondary treatments for long operational life.
Electrocoagulation.	• No requirement of external chemical addition; • Removal of small colloidal particles; • Less area requirement.	• Continuous requirement of electricity; • Frequent electrode replacement requirement; • Need of electrolyte with high conductivity.

19.2 Types of Hybrid Processes for Microconstituents Removal

Hybrid processes are a combination of various unit operations and unit processes intended to improve the overall efficiency of wastewater treatment. Owing to the integration of two or more treatment technologies in a single set-up, the disadvantages of individual systems can be overcome. As a result, the removal efficiency of MCs and other pollutants in a hybrid system is significantly improved. However, based on the type of target pollutant, the selection of hybrid treatment can be different. In wastewater treatment, physical treatments are used for the removal of suspended solids, biological treatments for the removal of biodegradable organic matter and nutrients, and tertiary treatments for the removal of complex organic pollutants or MCs. Since wastewater comprises of more than one class of pollutants (organic matter, nutrients, suspended solids, pathogens, MCs, etc.), the hybrid systems are beneficial as they can eliminate all classes of pollutants. Broadly, the hybrid processes can be classified into the following four classes: i) physical and biological processes; ii) physico-chemical processes; iii) chemical and biological processes; and iv) combined physical, biological, and chemical processes (Tee et al. 2016). Some of the combinations of such hybrid treatment systems for MCs removal are discussed below.

19.2.1 Constructed Wetlands

Constructed wetlands (CWs) are the artificially designed wastewater remediation technology that mimics the water purification process occurring in a natural wetland system. These systems consist of vegetation, microorganisms, wastewater, and substrates, and utilize chemical, physical, and biological mechanisms to remove the organic pollutants and MCs from aqueous solution or wastewater (Figure 19.1). These CWs are typically classified into the four following types based on their flow patterns: i) horizontal flow CW (HF-CW); ii) vertical flow CW (VF-CW); and based on wetland hydrology (iii) subsurface flow CW (SSF-CW); and iv) free water surface flow CW (FWS-CW). The FWS-CW is most widely used for the removal of MCs and suspended solids in tropical areas, whereas SSF-CW is commonly used in colder areas, where the atmospheric temperature frequently falls below freezing. In recent years, to enhance the pollutant removal efficiency in CWs, modified CWs, such as step feeding CW, artificial aerated CW, baffled flow CW, hybrid tower CW, circular flow corridor CW, have been proposed for the abatement of pollutants (Wu et al. 2014). The performance of CWs for treating MCs containing wastewater strongly relies on the operating conditions, such as plant selection, root depth of plant, wastewater type, liquid depth, hydraulic retention time (HRT), substrate type,

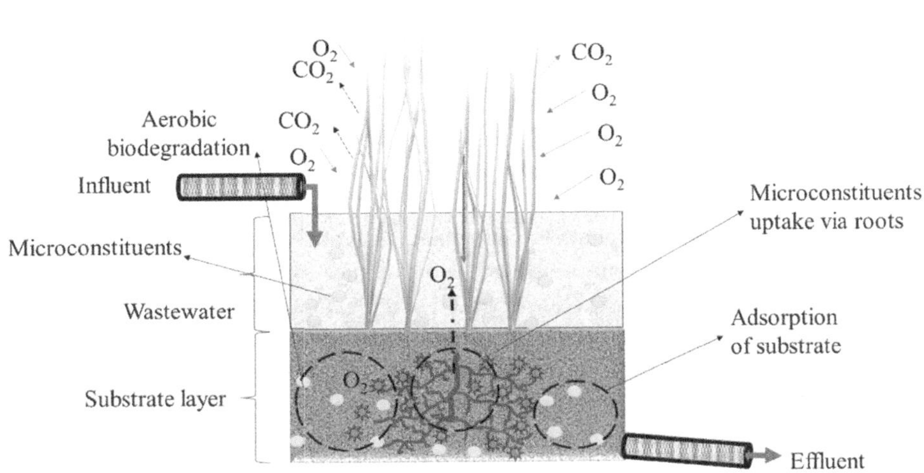

Figure 19.1 A schematic of a constructed wetland for MCs removal from wastewater.

mode of operation, pollutant loading rate. These parameters need to be optimized to successfully scale up CWs for efficient removal of MCs from wastewater to produced a reusable quality of water (Wu et al. 2015).

19.2.1.1 Applications

Over the last few years, CWs have been implemented successfully for the removal of recalcitrant MCs, such as pharmaceuticals, personal care products, surfactants, dyes, pesticides, and other complex organic compounds, from wastewater. Additionally, in the recent past, nutrients and heavy metals such as cadmium, chromium, zinc, lead, copper, and nickel have also been successfully removed using CWs. For example, in an investigation, 80% removal of propranolol and naproxen with initial concentrations of 3.6 and 7 ng L^{-1}, respectively, was observed in a field-scale VF-CW planted with the common reed macrophyte species and fed with hospital wastewater (Vystavna et al. 2017). Similarly, in another investigation, successful removal of various pharmaceuticals and personal care products, such as ibuprofen, diclofenac, acetaminophen, tonalide, oxybenzone, and triclosan, was demonstrated by a field-scale three-stage hybrid CW (VF-SSF, HF-SSF, FWS). The hybrid CW was planted with *Phragmites australis* species and almost 89% of chemical oxygen demand (COD), 99% of biochemical oxygen demand (BOD), 98% of NH_4–N, and >80% removal of all MCs were achieved after 7 days of HRT (Ávila et al. 2015), thereby affirming the efficacy of CWs in removal of MCs.

Several investigations have reported that different sorbents like light expanded clay aggregates, biochar, quartz sand, anthracite, biological ceramsite, and granulated cork, etc., have also been utilized in CWs as media for the removal of MCs (Dordio et al. 2009; Jiang et al. 2014). In an investigation, 96, 97, and 75% removal efficiencies for ibuprofen, carbamazepine, and clofibric acid, respectively, were achieved under summer conditions at an HRT of 7 days in SSF-CW planted with *Typha* species and using light expanded clay aggregates as sorbent media. However, the pollutant removal efficiency decreased in the winter season under similar operating conditions (81.9, 88.2, and 47.3% for ibuprofen, carbamazepine, and clofibric acid, respectively) (Dordio et al. 2010). This showed the dependency of external factors, such as temperature, on the MCs removal efficiency in CW. Despite the good results obtained in lab- and field-scale experiments, their practical application is still not feasible because of larger land requirements and due to the requirement of higher detention period. Hence, to overcome this bottleneck, integration of other technologies such as microbial fuel cell (MFC) with CW has been commenced to enhance the hydraulic loading rate as well as the pollutant removal efficiency.

19.2.1.2 Constructed Wetland Coupled With Microbial Fuel Cell

The basic set-up of a MFC comprises of two distinct zones: anaerobic (anodic compartment) and aerobic (cathodic compartment). In the anodic compartment, electrochemically active microorganisms oxidize the organic matters and release electrons and protons, whereas in the cathodic compartment, the electrons and protons generated in the anodic compartment during the oxidation process react with oxygen to produce water via oxygen reduction reaction. A typical CW also consists of an aerobic and anaerobic zone, where oxidation and reduction processes take place. The area near the roots of a plant is likely to be aerobic, while away from the roots is the anaerobic zone. Plants possess the unique ability to release oxygen through their roots. This functional

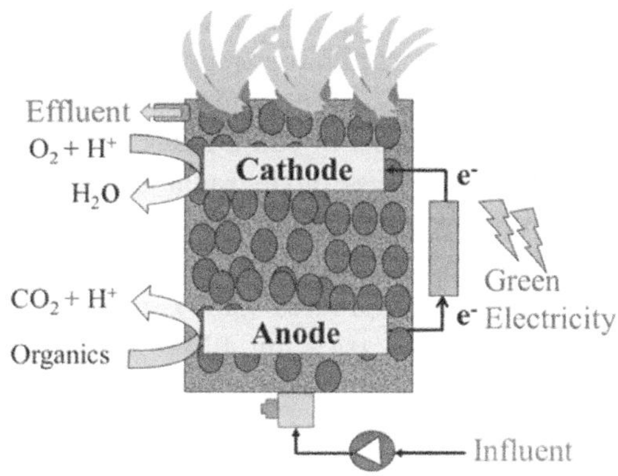

Figure 19.2 Schematic diagram of hybrid constructed wetland coupled microbial fuel cell.

similarity of CW and MFC leads to hybridization of two different systems in a single set-up known as CW-MFC (Yadav et al. 2012). This approach is gaining a lot of attention because it can act as a wastewater treatment unit and bioremediation unit for contaminated sites while simultaneously producing electricity (Figure 19.2).

Removal of sulfadiazine with average effluent concentrations of 7.58 to 15.62 μg L^{-1} was observed in coupled up-flow CW-MFCs planted with *Ipomoea aquatica*, which achieved 94% of COD and 99.8% removal of sulfadiazine using granular activated carbon as anode and cathode. Decrease in power density from 15.41 to 12.84 mW m^{-2} was observed when HRT of CW-MFC was reduced from 3 to 1.5 days (Song et al. 2018). Similarly, 97% removal of sulfamethoxazole with an initial concentration of 30 mg L^{-1} was achieved in a hybrid CW-MFC, which generated a maximum power density of 57.8 mW m^{-2} when the CW-MFC was planted with *Oenanthe javanica* and operated with HRT of 2.5 days (Ebrahimi et al. 2021). In another investigation, 93.9% removal of nitrobenzene was achieved at an HRT of 3 days using CW-MFC planted with *Scirpus Validus* and operated with graphite felts as anode and cathode. Moreover, the degradation of nitrobenzene was carried out with three different plants (*Scirpus validus*, *Typha orientalis*, and *Iris pseudacorus*). It was observed that, as the concentration of nitrobenzene increased from 20 to 200 mg L^{-1}, parts of *Typha orientalis* and *Iris pseudacorus* showed intolerance and died, while *Scirpus validus* grew well, showing a better capability of handling nitrobenzene loading (Di et al. 2020).

The investigations point toward the fact that, depending upon the target pollutants, the vegetation should be selected so as to have better performance of CWs. These investigations demonstrated that coupled CW-MFC can be an effective technology for the removal of MCs and bioelectricity generation. Owing to the hybridization and use of two technologies, CW-MFCs have an edge over standalone CWs. However, low power output compared to other renewable energy sources is the major shortcoming of the CW-MFC system. Therefore, more scaling-up trials using more efficient electrode materials are necessary for a better understanding of operational parameters in order to achieve better electrical energy output.

19.2.2 Combined Biological and Advanced Oxidation Processes

Advanced Oxidation Processes (AOPs) are a set of chemical treatment processes where degradation of MCs occurs by reactive oxygen species (ROS) such as hydroxyl radical ($OH^•$), superoxide ($O_2^{•-}$), hydroperoxyl ($HO_2^•$), and sulphate anions ($SO_4^{•-}$). These ROS are highly reactive because of their electrophilic nature and have the potential to break down the larger chain MCs into smaller fragments, which can be subsequently degraded into simple by-products or completely mineralized into end products like CO_2 and H_2O. Generally, these ROS can be produced by using ozone (O_3), ultrasonication, H_2O_2, and ultraviolet (UV) radiation. However, high-energy consumption of ultrasonication treatment, UV lights, and continuous consumption of H_2O_2 and O_3 make AOPs costlier and limit the field-scale applications of this technology. Additionally, AOPs in some cases could result in more toxic intermediates that are more toxic and less biodegradable than the original target MCs (Oller et al. 2011). Moreover, use of a high quantity of oxidants can also create toxicity toward microorganisms. Therefore, many researchers have proposed biological treatments like activated sludge process (ASP), membrane bioreactor (MBR), moving-bed biofilm reactor (MBBR), trickling filter, rotating biological contactor, aerobic bioreactor, anaerobic bioreactor, microalgae bioreactor, fungal bioreactor, and enzyme treatment prior to AOPs to enhance biodegradability. The biologically treated effluent can be then treated in AOPs with considerably low cost and low HRT.

19.2.2.1 Activated Sludge Process Coupled With Advanced Oxidation Process

Over the past years, integrated ASP-AOP technology have demonstrated great potential toward the effective removal of MCs from wastewater. For example, in an investigation, removal of nine pharmaceuticals, namely acetaminophen, antipyrine, caffeine, carbamazepine, diclofenac, hydrochlorothiazide, ketorolac, metoprolol, and sulfamethoxazole, were achieved by sequential ASP and catalytic ozonation systems. During the ASP treatment only acetaminophen and caffeine were completely eliminated; whereas partial removal (~50%) of metoprolol, sulfamethoxazole, and hydrochlorothiazide was achieved along with a low elimination of antipyrine, carbamazepine, diclofenac, and ketorolac under an HRT of 24 h. However, in the O_3/Fe_3O_4 process, ~100% mineralization of all the 9 pharmaceuticals was achieved in 10 min by employing 13 mg L^{-1} of O_3 dosage and 150 mg L^{-1} doseage of Fe_3O_4 (Espejo et al. 2014). Similarly, the degradation of 200 mg L^{-1} of sulfolane or 2,3,4,5- tetrahydrothiophene-1,1–dioxide was performed in an ASP integrated UV/H_2O_2 system and almost 81% mineralization of sulfolane was achieved in <24 h using the ASP-UV/H_2O_2 system. However, it was observed that mineralization of ~16.7% was achieved in only the ASP system, which was 4.8-fold lower than the mineralization obtained by the combined ASP-UV/H_2O_2 system (Khan et al. 2019).

19.2.2.2 Moving Bed Biofilm Reactor Coupled With Advanced Oxidation Process

The integrated MBBR-AOP have been successfully reported in the literature for decontamination of MCs such as pesticides, herbicides, pharmaceuticals, dyes, and chlorinated phenols from wastewater. In an experiment, a comparative performance evaluation of combined MBBR-MBR-UV/H_2O_2 and MBR-UV/H_2O_2 systems was undertaken for removal of 100 μg L^{-1} of carbamazepine, 10 μg L^{-1} of ciprofloxacin, and 100 μg L^{-1} of ibuprofen from wastewater. The hybrid MBBR-MBR-UV/H_2O_2 system showed complete removal of pharmaceuticals after 5 min of UV/H_2O_2 treatment, when the MBBR-MBR system was operated at 10 h of HRT and 6500 mg L^{-1} of mixed liquor suspended solids (MLSS) concentration, whereas without the MBBR system, the removal efficiency was compromised (Monteoliva-García et al. 2020).

In another work, removal of five persistent pharmaceuticals (clofibric acid, carbamazepine, naproxen, ketoprofen, diclofenac) with an initial concentration of 1 μg L^{-1}, was assessed in a hybrid MBBR coupled with a hydrodynamic cavitation/H_2O_2/UV system. The finding of the investigation revealed that in the MBBR system, no removal of carbamazepine was achieved along with 28, 63, 80, and 85% removal of clofibric acid, ketoprofen, naproxen, and diclofenac, respectively in 48 h of HRT, whereas complete removal of naproxen and ketoprofen along with close to 98% removal of carbamazepine and diclofenac was achieved after 10 h in the combined MBBR-HC/H_2O_2/UV treatment (Zupanc et al. 2013). Apart from these, several other combined biological wastewater treatment systems and AOPs for degradation of MCs have been reported in the literature, which are summarized in Table 19.2.

These findings revealed that integrating a biological process and AOPs can effectively remove MCs from aqueous solution and simultaneously reduce the treatment time by more than half compared to standalone biological treatments. Additionally, AOPs as pretreatment to biological treatment processes can lead to the degradation of MCs, thereby enhancing the biodegradability of the wastewater. However, in hybrid biological and AOP treatment systems, the treatment efficiencies rely heavily upon the selection of appropriate types of AOPs (Fenton reagent, O_3/H_2O_2, O_3/UV, H_2O_2/UV, TiO_2/UV), biological processes, chemical oxidants, physical and chemical properties of target MCs, use of monospecific cultures, and toxic intermediates. Thus, evaluation of the individual biological and chemical oxidation processes is necessary for finding the optimal operating conditions for the hybrid process to enhance the destruction of persistent MCs (Oller et al. 2011).

19.2.2.3 Bio-Electrochemical Systems and Advanced Oxidation Processes

Bio-electrochemical systems (BES) are promising technologies that utilize microorganisms as biocatalysts to convert chemical energy of organic matter into electrical energy and/or other valuables products such as hydrogen and methane. The organic matter used by the microorganisms includes domestic wastewater, industrial wastewater, lignocellulosic biomass, etc. Due to the capability of simultaneous wastewater treatment and valuable recovery, the BES have globally gained a lot of interest from researchers (Pant et al. 2010). The two classes of BES, namely MFC and microbial electrolysis cell (MEC), are discussed in this section pertaining to their coupling with AOPs for the removal of MCs.

The MFC is a type of BES that generates electricity via oxidation of organic matter by anaerobic microorganisms in the anodic chamber, whereas in the case of MEC, an external potential is applied to promote the production of valuables in the cathodic chamber of BES. Although in MEC external power is required, the required energy is less than with a conventional electrolysis process. In MFC, the anaerobic microorganisms in the anodic chamber act as a biocatalyst for the degradation of organic matter to release H^+, e^-, and CO_2. The H^+ ions (protons) migrate toward the cathode through a proton exchange membrane, while electrons transfer toward the cathode through an external circuit (Logan et al. 2006, 2008). In MFC, the anode acts as a continuous electron acceptor, thus making the anodic degradation process independent of the dissolved terminal electron acceptors in wastewater. The degradation of MCs can be accomplished in the anodic chamber as well as the cathodic chamber of BES. The degradation of MCs in the anodic chamber is discussed in detail in Chapter 18. This section deals with the cathodic degradation of MCs via integration of BES with AOPs in a single set-up.

Table 19.2 Treatment of MCs by using different hybrid biological-AOP technologies.

Type of System	Wastewater Type	Target Pollutant	Initial Concentration	HRT	Efficiency (%)	References
Activated sludge process-ozonation.	Pharmaceutical wastewater.	Bisphenol-A 17α- ethinylestradiol, Erythromycin and Roxithromycin.	400 ng L^{-1} 2 ng L^{-1} 170 ng L^{-1} 27 ng L^{-1}	10 days	90	(Schaar et al. 2010)
Membrane bioreactor-ozonation.	Pharmaceutical wastewater.	Acyclovir, Diacetylacyclovir.	154 mg L^{-1} 0.4 mg L^{-1}	10 days	99	(Mascolo et al. 2010)
Aspergillus sp. – electrochemical oxidation.	Food industry wastewater.	Phenols.	–	3 days	96	(Kyriacou et al. 2005)
Immobilized biomass reactor- Solar photo-Fenton.	Pharmaceutical wastewater.	Nalidixic acid.	38 mg L^{-1}	4 days	96	(Sirtori et al. 2009)
Activated sludge process-TiO_2-photocatalysis.	High salinity media.	Phenols.	200 mg L^{-1}	25 h	98	(L'Amour et al. 2008)
Activated sludge process-ozonation.	Mixture of domestic and industrial wastewater.	Acetaminophen, Caffeine, Galaxolide, Ibuprofen, Nicotine.	–	–	100	(Rosal et al. 2010)
Membrane bioreactor-UV/H_2O_2.	Pharmaceutical wastewater.	Carbamazepine, Ciprofloxacin, Ibuprofen.	5000 μg L^{-1} 500 μg L^{-1} 5000 μg L^{-1}	16 h	98.16 100 99.14	(Monteoliva-García et al. 2019)

In the cathodic chamber of the MFC, the incoming electrons are reduced via terminal electron acceptors, which is oxygen in the majority of the cases. Based on the cathode catalyst and cathode material, the cathodic oxygen reduction reaction can be progressed via a two- or four-electron pathway. The two- and four-electron oxygen reduction leads to *in-situ* synthesis of H_2O_2 and H_2O, respectively, as the end product. Although from the view point of power generation, the four-electron pathway is preferred, the H_2O_2 synthesized via the two-electron pathway is a strong oxidizer that can be used as a disinfectant and/or oxidizing agent (Liew et al. 2014). The *in-situ* synthesized H_2O_2 in the cathodic chamber can be used for the tertiary treatment of wastewater. Various metal-based cathode catalysts have been successfully implemented in BES for *in-situ* synthesis of H_2O_2 (Das et al. 2020). Compared to the MFC mode of operation, higher H_2O_2 yield can be obtained in the MEC mode, thus it can be considered as a potential treatment option for the tertiary treatment of wastewater (Chung et al. 2020).

19.2.2.4 Bio-Electro Fenton-Based Advanced Oxidation Processes

The Fenton oxidation is a well-established AOP, which has been widely used for the degradation of MCs, such as dyes, pharmaceuticals, antibiotics, surfactants, etc. In the Fenton process, using iron as the catalyst, $OH^{\bullet}$ is generated that has an oxidizing potential of 2.8 eV. This makes the process highly effective for the degradation and mineralization of a wide range of MCs (Neyens and Baeyens 2003; Asghar et al. 2015). The bio-electro-Fenton (BEF) process is a merger of Fenton oxidation with BES, in which the *in-situ* synthesized H_2O_2 is dissociated via an iron source (heterogeneous or homogenous) to form $OH^{\bullet}$ in the cathodic chamber of BES.

The first successful application of this hybridization technique was undertaken by Zhu and Ni (2009) using scrap iron as an Fe source in the cathodic chamber of MFC for the degradation of *p*-nitrophenol. The results of the investigation demonstrated complete removal of *p*-nitrophenol in 12 h, simultaneously harvesting a maximum power density of 143 mW m^{-2}. In another investigation, paracetamol was degraded with a removal efficiency of 70% in 9 h of contact time using a BEF-MFC set-up. In the same investigation, the researchers highlighted the importance of acidic pH for the BEF process. The reported concentration of $OH^{\bullet}$ in the acidic pH was 85 μM (pH of 2.0), which was ~4 times higher compared to that at the pH of 3.5 (Zhang et al. 2015b). This shows that the *in-situ* generation $OH^{\bullet}$ is highly dependent on pH and the yield of $OH^{\bullet}$ is higher under acidic pH.

A similar observation of accelerated pollutant removal using BEF under acidic pH is concluded by other researchers, as well as during the cathodic degradation of other MCs (Tao et al. 2013; Xu et al. 2020c; Sathe et al. 2021). This might be due to better stability of the precursors (H_2O_2 and Fe in the dissolved state) in the acidic regime. In another investigation, estrogens, namely 17β-estradiol and 17α-ethynyl-estradiol, were removed using BEF-MFC with removal efficiency of 81 and 56%, respectively, in 10 h of contact time (Xu et al. 2013). Some of the researchers have highlighted the importance of external resistance on the pollutant removal efficiency using BEF. For instance, Tao et al. (2013) reported that the *p*-nitrophenol removal decreased from 96 to 64% with increase in external resistance from 20 to 10,000 Ω. At lower external resistance, the input flux of incoming electrons from anode toward cathode is higher, and as a result, the H_2O_2 yield at the cathode is higher. The H_2O_2 being one of the precursors of the Fenton oxidation can be a rate limiting factor for the synthesis of $OH^{\bullet}$.

Although H_2O_2 can be generated at the cathode in MFC, its yield is much less for its replication on a pilot scale. Instead, enhanced and comparatively stable H_2O_2 yield is possible in the MEC mode. This has prompted researchers to initiate cathodic Fenton oxidation in an MEC set-up. In comparison with conventional electro-Fenton, the main difference in BEF-MEC is the presence of electro-active microorganisms, which also supplement some part of the electricity requirement. This significantly reduces the amount of external electrical energy required in BEF compared to conventional electro-Fenton oxidation. The same is validated by Li et al. (2017), in which a BEF-MEC system required energy consumption of 0.728 kWh for the degradation of 1 kg of aniline, which was close to 100-fold less than that required in a conventional electro-Fenton process (74 kWh kg^{-1}). In the same investigation, using BEF-MEC, aniline degradation of 97.1% was accomplished in 144 h of retention time at an applied potential of 0.5 V (against Ag/AgCl). The degradation of non-steroidal anti-inflammatory drugs that are resilient to biological treatments was explored in the BEF-MEC process. At optimal contact time of 5 h, removal of ketoprofen, diclofenac, ibuprofen, and naproxen was 59–61, 87–97, 80–86, and 75–81%, respectively. The optimal operating conditions were observed at external applied potential of 0.3 V (against Ag/AgCl) and pH of 2.0 (Nadais et al. 2018).

As a replacement for an external power source, MFC can be used to supply electrical power to the electro-Fenton process. Such integration of electro-Fenton powered by MFC will reduce the operating cost of the wastewater treatment. In an investigation, a single chamber MFC was used to power the electro-Fenton oxidation system that was operated for the degradation of phenol. It was reported that after 22 h of treatment in the electro-Fenton system, the phenol was completely degraded into simple organic acids, such as oxalic and formic acid, while demonstrating a total organic carbon (TOC) removal of 75% (Zhu and Logan 2013). A similar set-up was investigated by Zhang et al. (2015a) in which a single chamber MFC was used to power an aerobic electrochemical oxidation set-up. The methyl orange was used as a target pollutant in the aerobic electrochemical oxidation system. The methyl orange decolorization in the electrochemical system reached 90.4% in a contact time of 6 h under a fixed voltage of 700 mV. In comparison with conventional electro-Fenton, the hybrid system of MFC coupled electro-Fenton can be a cost-effective alternative. However, one shortcoming of this process is the need of acidic pH to initiate Fenton oxidation. Hence, it can be considered as suitable for the treatment of specific pollutants released from a confined source. As for such cases, the pH adjustments could be practical. For implementation in domestic wastewater treatment, a more reliable solution should be explored, which can be practiced at near neutral pH.

19.2.2.5 Photo-Electrocatalyst-Based Advanced Oxidation Process

Similar to BEF, another application of integration of AOPs in BES is via implementation of a photocatalyst as cathode catalyst. When irradiated using a light source, the photocatalysts adsorb photons to generate photoelectrons that react with oxygen molecules to generate $O_2^{\bullet-}$. Moreover, the photo-induced holes created in the valence band of the photocatalyst have the ability to generate H_2O_2 and $OH^{\bullet}$, successively following reaction with H_2O. When such photocatalysts are used as cathode catalysts in MFC, the cathodic oxygen reduction reaction can be improved owing to the generated photoelectrons on the photocatalyst surface. In this manner, the number of electrons at the cathode can be increased in addition to the generation of ROS following implementation of photocatalysts as a cathode catalyst (Bhowmick et al. 2019; Xu et al. 2020b).

In an investigation, Jiang et al. (2016) used a composite of zero valent iron and TiO_2 as cathode catalyst in MFC for the cathodic degradation of tetracycline in the presence of a light source. The COD removal efficiency (tetracycline measured in terms of COD) in the cathodic chamber was 70 and 55% in 2h of contact time with and without light, respectively. This improvement in pollutant removal in the presence of light might be due to the formation of ROS. The TiO_2 is a well proven photocatalyst that is widely used in water and wastewater treatment. A hybrid system comprising of up-flow MFC, laterite filter, and TiO_2-based photo-electrocatalysis was developed for the removal of organic matter and surfactant from the wastewater. The secondary treatment of wastewater was accomplished in the anodic chamber of MFC followed by laterite-based filter, together achieving close to 60 and 90% of COD and SDS removal, respectively. The cathodic treatment using TiO_2 as photo-electrocatalyst caused 50 to 55% of sodium dodecyl sulphate removal with an initial concentration of <1 mg L^{-1} in the cathodic chamber of MFC (Sathe et al. 2020).

Ibuprofen is a most widely used pharmaceutical as an anti-inflammatory drug; however, its presence in water bodies is potentially toxic to aquatic flora and fauna. In a separate investigation, a photo-cathode MFC coated with $CuInS_2$ as acathode catalyst was used for the cathodic degradation of ibuprofen. The results demonstrated that ibuprofen removal reached 75.94% with a pseudo-first-order reaction kinetic rate of 0.0419 h^{-1} at an optimal pH of 5.0 (Xu et al. 2020b). The same group of scientists, in another investigation, explored cathodic degradation of ofloxacin using $LiNbO_3$ as a photo-electrocatalyst catalyst. The MFC achieved an ofloxacin removal efficiency of 86.5% in 7h of contact time, simultaneously harvesting a maximum power density of 0.546 W m^{-2} (Xu et al. 2020a). The results of these investigations concluded that use of photocatalysts on a cathode not only demonstrated improved power generation but also cathodic pollutant degradation owing to the generation of ROS at the cathode. Moreover, the photocatalysts are immobilized on the cathode surface itself, which prevents the requirement of an additional photocatalyst recovery system.

19.2.3 Membrane Bioreactor

The MBR is the hybrid treatment process comprising of suspended growth biological process and membrane filtration, which is categorized as combined physical and biological processes. In MBR, the membrane filtration unit consists of microfiltration and ultrafiltration membranes. Over the past decade, MBR is a well researched topic and it is considered as a capable wastewater treatment technology owing to its reliable performance, low sludge generation, ability to operate at near neutral pH, low carbon footprint, and decent effluent quality (Iorhemen et al. 2016). The MBR is a proven treatment technology for the removal of persistent organic pollutants at lab-scale levels and has been effectively replicated in field-scale applications for the tertiary treatment of wastewater. In comparison with other biological treatments, MBR has high concentrations of biomass in suspension that have high solid retention time. Moreover, the sieving effect during membrane filtration prevents the MCs with high molecular weight from entering into the permeate. These two are the prime reasons responsible for achieving high MCs removal in MBR (Goswami et al. 2018).

Since inception, MBR has been widely implemented for the removal of various classes of MCs such as pharmaceuticals, personal care products, endocrine disturbing chemicals, pesticides, etc. (Goswami et al. 2018). In an investigation, the performance of lab-scale MBR was evaluated for the removal of eight different endocrine disturbing chemicals from wastewater. The removal efficiency of greater than 95% was obtained using MBR, which consisted of a submerged hollow fiber polyethylene membrane of 0.4 μm nominal pore size. It was observed that a higher sludge retention time (SRT) greater than 10 days was reflected in higher pollutant removal compared with short SRTs of 10 or 5 days (Zhou et al. 2011).

A similar observation on the importance of high SRT was also confirmed by another group of scientists during the degradation of antibiotics using MBR (Xia et al. 2012). The high SRT in the MBR causes increased contact of pollutants with sludge biomass, which results in better adsorption and successive biodegradation. Moreover, high SRT gave enough scope for slow growing microorganisms to grow, thus leading to better MCs removal via enhanced diversity of the microorganisms (Wijekoon et al. 2013). Apart from sole MBR treatment, merging of some other treatments is also undertaken by researchers, which have the capability not only to enhance pollutant removal efficiency but also reduce membrane fouling, thus reducing the operating cost of the MBR process.

19.2.3.1 Granular Sludge Membrane Bioreactor

Stabilized aerobic granular sludge is converted into dominant biomass that can have a mean diameter of up to 800 μm. Such compact granular biomass has a flocculent nature, which can be easily settled (Tay et al. 2007). Simultaneous removal of pharmaceuticals along with organic matter and nutrients was investigated in an aerobic granular sludge MBR set-up by Zhao et al. (2014). In the MBR, the removal of prednisolone was highest (98.46%), followed by naproxen (84.02%) and ibuprofen (63.32%). Concurrent COD and nutrient (NH_3-N and total phosphorus) removal of 87.98 and 90%, respectively, was also accomplished in MBR. Similarly, in another investigation, the removal of five pharmaceuticals and personal care products was evaluated using granular sludge MBR operated with a polyvinylidene fluoride hollow fiber membrane (0.22 μm pore size). Due to higher quantum of biomass present in the MBR (mixed liquor volatile suspended solids concentration of 5.1 g L^{-1}), the pharmaceuticals might have been easily adsorbed on the sludge particles, which were then finally removed via separation and biodegradation in MBR (Wang et al. 2016). However, some of the MCs, such as sulfamethoxazole, have lower adsorption capacity in the sludge; hence, for such pollutants, biodegradation might be the dominant removal mechanism. On the other hand, ibuprofen possesses positive surface charge; thus, it can be easily adsorbed onto the negatively charged biomass (Wang et al. 2016). This confirms that although adsorption, biodegradation, membrane separation, and membrane interception are the principal phenomena involved in the removal of pollutants using granular sludge MBR, the extent of each process might vary based on the nature of the pollutant.

19.2.3.2 Advanced Oxidation Process Coupled Membrane Bioreactor

The AOPs can offer effective treatment for elimination of the majority of MCs from wastewater; however, the suspended organic matter or biomass flocs can be scavengers of the

ROS generated during the AOPs. In this regard, a pre-treatment in MBR can offer a vital treatment in terms of suspended solids and flocs removal. However, the majority of the MCs can be converted into biodegradable intermediates using AOPs. In such cases, the MBR can offer a post treatment to the effluent of AOPs, thus achieving the complete mineralization of the partially degraded MCs. In this manner, the integrated systems of AOP followed by MBR can be an effective hybridized treatment technique for the mineralization of the MCs.

An MBR coupled UV/H_2O_2 system was operated for the removal of ciprofloxacin, carbamazepine, and ibuprofen. An MBR operated with 16 h of HRT and 4.25 g L^{-1} concentration of MLSS demonstrated partial target pollutants removal; however, further treatment in the UV/H_2O_2 system caused 95.04, 100, and 98.29% removal for carbamazepine, ciprofloxacin, and ibuprofen, respectively (Monteoliva-García et al. 2019). In another investigation, MBR followed by the EO process achieved ~97% removal of 4 pharmaceuticals in 18 h and 40 min retention time in MBR and EO, respectively (Ouarda et al. 2018). Additionally, MBR coupled with electrocoagulation (Bani-Melhem and Smith 2012), MBR coupled with solar Fenton oxidation (Karaolia et al. 2017), MBR coupled ozonation and sonolysis (Prado et al. 2017), and MBR coupled with photocatalytic oxidation (Deveci et al. 2016), etc. have also been effective for the removal of MCs from wastewater.

In an interesting finding, nalidixic acid, which is an antibacterial drug, was completely removed in a treatment train comprising of MBR followed by either ozonation or UV/H_2O_2. In the integrated system, the permeate from MBR was treated by ozonation or UV/H_2O_2 and then recirculated back into the MBR. The effect of recirculation was evaluated on the nalidixic acid removal. It was observed that, although nalidixic acid was completely removed in both the systems (with and without recirculation), the removal of 34 out of 55 degradation by-products was higher in recirculated hybrid systems (Laera et al. 2012). Similarly, a positive effect of recirculation on the effective elimination of intermediates was observed in another case employing MBR with ozonation for the removal of an anti-viral drug from wastewater (Mascolo et al. 2010). In the absence of such a recirculated system, a separate downstream treatment process is required for the removal of intermediates. Hence, an integrated system of AOP-MBR might be a cost-effective approach for the effective mineralization of the MCs. Overall, the results of these investigations validated that MBR coupled AOP can offer reliable and consistent results for the degradation and mineralization of MCs, thus offering effective tertiary treatment of wastewater.

19.2.3.3 Membrane Bioreactor Coupled With Microbial Fuel Cell

The integration of ASP and MFC was implemented for wastewater treatment in which a single-chamber MFC was immersed in the ASP. The anodic effluent was diverted in the cathodic chamber for the polishing treatment via ASP (Cha et al. 2010). However, the hybridization of MFC with ASP requires a downstream clarifier for the separation of aerobic sludge. Instead, use of aerobic MBR in the cathodic chamber can demonstrate retention of biomass within the aeration tank, thus averting the need for secondary clarifiers. The combination of MFC with MBR can be used for enhancing pollutant removal efficiency by offering an effective polishing treatment to the wastewater. The involvement of MFC in the MBR process can also help in reducing membrane fouling, which is the main disadvantage of the MBR process.

Using polypyrrole-anthraquinone-2-sulfonate modified stainless steel mesh as the cathode with a membrane filter, the MFC achieved a maximum power density of 15.87 mW m^{-2}, while demonstrating reduced membrane fouling. The MFC-MBR system achieved >90% methylene blue removal in 1 h of cathodic retention time (Li et al. 2014). Tan et al. (2017) reported that incorporation of MFC into anaerobic MBR reduced membrane fouling compared with sole anaerobic MBR. The hybrid system demonstrated the COD removal of 93.39% during the treatment of palm oil mill effluent with an HRT of 12.5 h and temperature of 45°C. Although limited investigations are available on this topic, the initial results confirmed the capability of MFC-MBR systems in achieving higher removal of pollutants. Owing to reduced membrane fouling and reduced energy consumption, the MFC-MBR hybrid systems can be considered as a promising option for the tertiary treatment of wastewater. The MFCs can provide effective secondary treatment of wastewater in the anodic MCs. The effluent from the anodic chamber can then be circulated to the cathodic chamber, wherein the aerobic MBR system can be implemented for polishing treatment to the wastewater.

19.2.4 Electrocoagulation

Electrocoagulation is a physiochemical process, where *in-situ* formation of coagulants occurs to neutralize the charge on the colloidal particles for their destabilization, thereby facilitating the removal of pollutants through co-precipitation and adsorption on the metal hydroxide flocs. The basic idea of electrocoagulation was first proposed by A.E. Dietrich in the early nineteenth century to treat bilge wastewater from ships. Also, in 1909, J.T. Harries patented the use of aluminum and iron as sacrificial anodes for treating wastewater through electrocoagulation. Thus, through the past few decades, electrocoagulation technology has gained much attention from researchers for the efficient treatment of wastewater. The basic mechanism of electrocoagulation is based on the theory of electrolysis, wherein the current passes between a pair of electrodes in the presence of a conductive solution. As a result, the oppositely charge ions are attracted toward the electrodes. Thus, oxidation at anode and reduction at cathode takes place, which helps in the formation of metal hydroxide complexes within an electrocoagulation set-up, as shown in Eqs (19.1) to (19.5).

At anode:

$$M(s) \rightarrow M^{n+}(aq) + ne^{-} \quad (19.1)$$

$$2H_2O(l) \rightarrow 4H^{+}(aq) + O_2(g) + 4e^{-} \quad (19.2)$$

At cathode:

$$M^{n+}(aq) + ne^{-} \rightarrow M(s) \quad (19.3)$$

$$2H_2O(l) + 2e^{-} \rightarrow H_2(g) + 2OH^{-} \quad (19.4)$$

where M can be any metal like iron, aluminum, magnesium, and copper. Additionally, the mass of metal coagulants produced in the solution from the anode is given by Faraday's law, following Eq. (19.5) (Moussa et al. 2017).

$$m = (I \times t \times M_w) / (z \times \mathrm{F}) \tag{19.5}$$

where m is the mass of metal ions produced (g); I is the external supplied current (A); t is the electrolysis time (s); M_w is the molecular weight of the anode material (g mol^{-1}); F is Faraday's constant (96,485 C/mol); and z is the number of electrons involved in the electrolysis reaction.

The conventional electrocoagulation set-up consists of a pair of electrodes, which are externally connected to a direct current (DC) power supply (Figure 19.3). The anode is connected to the positive terminal and the cathode is connected to the negative terminal of the DC power supply. Through applying current between the electrodes, positive ions are released by the anode and hydroxide ions along with hydrogen gas are liberated at the cathode. The positive ions in the solution neutralize the surface charge of pollutants, and also react with hydroxide ions to form metal hydroxide complexes. Thus, the neutralized pollutants stick to the surface of metal complexes and aggregates are formed, which also attach the pollutants to their surface and settle under gravity. The hydrogen gas helps to float the lighter pollutant particles. As a result, scum can be formed on the top of the treating solution. Furthermore, either sedimentation or filtration is required to separate the settled particles from the solution. The performance efficiency of electrocoagulation in terms of pollutant removal and operating costs is influenced by various operating parameters such as current density, electrolyte concentration, electrolysis time, charge loading, inter-electrode distance, initial concentration of pollutants, electrode material, etc.

The electrocoagulation process has been employed to treat wastewater arising from industries, such as textile, dairy, pharmaceuticals, slaughterhouse, restaurant, and many

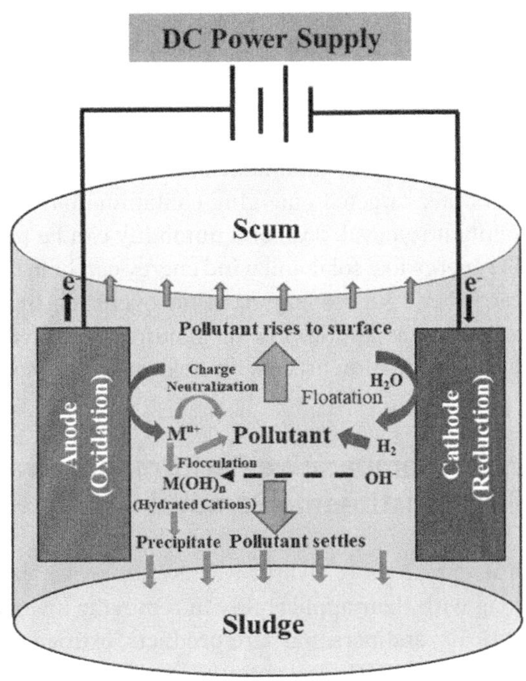

Figure 19.3 A typical electrocoagulation set-up and mechanism involved therein.

others. Moreover, heavy metals, emerging contaminants, and nutrients have also been effectively removed from contaminated water through electrocoagulation. In this regard, Ahmadzadeh et al. (2017) evaluated the performance of electrocoagulation with aluminum electrodes to remove ciprofloxacin from hospital wastewater. The investigation was carried out at optimum current density of 12.5 mA cm^{-2}, electrolyte concentration of 0.07 M, electrolysis time of 20 min, and an inter-electrode gap of 1 cm. The results revealed that about 88.57% of ciprofloxacin removal was achieved at an initial concentration of 32.5 mg L^{-1}. Moreover, 66.80 g m^{-3} of electrode and 0.61 kWh m^{-3} of energy consumption were recorded at optimum operating conditions for ciprofloxacin removal.

Similarly, Shadmehr et al. (2019) applied electrocoagulation for the treatment of propiconazole fungicide from wastewater and ~79.83% of propiconazole removal with initial pollutant concentration of 25.93 mg L^{-1}, current density of 10.95 mA cm^{-2}, and electrolysis time of 36.44 min was reported. Most researchers have used aluminum and iron as the anode material owing to low cost and easy availability. However, in some cases, either low removal efficiencies of pollutants or high sludge generation was observed with these electrodes; therefore, a few other anodes like zinc and magnesium were used in electrocoagulation processes. For instance, the zinc anode was used in treating phenol from wastewater, and 84.2% phenol removal was achieved at current density of 25 mA cm^{-2}, pH of 3.2, and electrolysis time of 120 min (Fajardo et al. 2015).

Similarly, removal of indigo carmine dye using the magnesium anode was also investigated in electrocoagulation processes and the result revealed that 96% of indigo carmine dye removal was achieved with initial dye concentration of 100 mg L^{-1} at current density of 5 mA cm^{-2} in 4 h of treatment. The results also revealed that increasing the current density from 5 to 8 mA cm^{-2} did not show any further improvement in removal efficiency, which can be due to the formation of passive film on the anode surface, which eventually can increase the resistance. As a result of this, a slow amount of metal hydroxide is formed for the precipitation of pollutants, even at high current density (Donneys-Victoria et al. 2019). Thus, not much improvement in performance was observed at high current density. Nonetheless, electrodes such as zinc and magnesium also obtained efficacious removal of phenol and dye contaminants. New electrode material must be explored in future studies with other targeted emerging contaminants so that comparative assessment in terms of pollutant removal, cost, and durability can be performed. Additionally, the use of renewable energy like solar and wind energy can help to overcome the disadvantages of electrical scarcity and high energy cost for operating the electrocoagulation process. Thus, more investigations and efforts are required to develop a renewable energy-based efficacious electrocoagulation process for field-scale applications.

19.3 Comparative Performance Evaluation of Hybrid Systems for Microconstituents Removal

Different kinds of hybrid wastewater treatment systems are discussed in this chapter along with their applicability in removing a variety of MCs, namely antibiotics, pharmaceuticals, and personal care products, estrogens, surfactants, etc. Various AOPs, such as ozonation, UV/H_2O_2, Fenton oxidation and its modifications, EO, photocatalysis, etc. are highly effective for the speedy degradation of MCs owing to the generation of ROS. Compared with these chemical processes, biological processes, such as CWs, MBR, MBBR, etc., offer slower treatment of MCs. The required HRT in chemical processes is generally a few minutes, whereas in the case of biological treatments, it can extend for some hours to days. Moreover, the biological processes being dependent on the action of microorganisms are readily affected by the initial concentration of the target pollutants. For instance, the toxic MCs beyond threshold concentrations can adversely affect the biological treatments and once affected, it might require a long time to achieve normalcy. Thus, in cases where the concentration of MCs is high or there is a frequent possibility of shock loading of such pollutants, first-stage chemical treatments can be preferred followed by biological processes. However, in the case of domestic wastewater, chemical treatments are ideal to be placed as tertiary treatments owing to the presence of a high concentration of biodegradable organic matter.

The chemical processes and other hybrid systems involving AOPs might not be feasible for the complete mineralization of a large quantity of wastewater at a low cost. Instead, it might be an ideal proposition to use AOPs to reduce the concentration of MCs below the toxic limit or convert MCs into biodegradable intermediates and then use biological treatments for complete mineralization. Despite the generation of ROS having high redox potentials, some of the degradation by-products such as carboxylic acids are resistant to AOPs. This avoids the complete mineralization of the MCs in a shorter retention time. In such cases, a downstream biological treatment might be effective as most of these short-chain degradation by-products are biodegradable. Hence, based on the potential MCs and the possible reuse of the treated effluent, the type and sequence of a hybrid system can be adopted.

19.4 Conclusions and Future Directions

Wastewater from a variety of sources is released into the environment through sources like industries, municipalities, and agriculture practices. Since the wastewater contains many harmful and bio-toxic MCs, the treatment of such pollutant laden wastewater is essential before its discharge into the receiving water bodies. Hence, a cost-effective method is required to treat a wide range of pollutants, and as compared to conventional treatment methods, hybrid systems provide a relatively robust treatment option that can reduce the issues related to energy, cost, and treatment time. However, there are still a few bottlenecks for adopting hybrid systems, which need to be overcome for successful commercialization of the technology. The evaluation for providing efficacious treatment technology for removal of these pollutants must rely on several important factors, such as flexibility, reliability, environmental compatibility, and operating and maintenance costs of the treatment technique. Thus, more field-scale demonstrations should be conducted in the near future, which would help to choose the best technology for the treatment of MC-laden wastewaters. Additionally, prior to the implementation of field-scale hybrid technologies, optimization of process parameters, reactions kinetics within a treatment system, a degradation mechanism, by-products formation, and reactor design need to be explored in future investigations.

Additionally, transformation by-products of these contaminants is not generally explored due to difficulty in detection and lack of knowledge of their bio-toxicity and persistency. Therefore, adverse effects of long-term exposure of these pollutants on the environment must be investigated in future studies. Also, as these pollutants are detected in complex wastewater matrices, such as sludge, new protocols and detection methods should be developed for the extraction and identification of their toxic intermediate by-products. However, owing to the frequent occurrence of MCs at trace concentrations and their treatment difficulties, it is vital to reduce the concentration of these pollutants at source, which can be controlled through public awareness, legislation, and by imposing restricted environmental laws related to the release of these contaminants.

The hybrid systems are effective in the removal of MCs from the wastewater as they offer advantages of more than one system in a single set-up. Moreover, such treatments are more stable and reliable compared to a single treatment. Hybrid systems comprising of biological treatment as a core treatment along with other treatments are considered as cost-effective hybrid technologies. Thus, the applications of hybrid systems have immense potential in wastewater treatment that can be explored in future field-scale applications.

References

Ahmadzadeh, S., Asadipour, A., Pournamdari, M. et al. (2017). Removal of ciprofloxacin from hospital wastewater using electrocoagulation technique by aluminum electrode: optimization and modelling through response surface methodology. *Process Safety and Environmental Protection* 109: 538–547.

Asghar, A., Raman, A.A.A., and Daud, W.M.A.W. (2015). Advanced oxidation processes for *in-situ* production of hydrogen peroxide/hydroxyl radical for textile wastewater treatment: a review. *Journal of Cleaner Production* 87 (1): 826–838.

Ávila, C., Bayona, J.M., Martín, I. et al. (2015). Emerging organic contaminant removal in a full-scale hybrid constructed wetland system for wastewater treatment and reuse. *Ecological Engineering* 80: 108–116.

Bani-Melhem, K. and Smith, E. (2012). Grey water treatment by a continuous process of an electrocoagulation unit and a submerged membrane bioreactor system. *Chemical Engineering Journal* 198–199: 201–210.

Bhowmick, G.D., Chakraborty, I., Ghangrekar, M.M., and Mitra, A. (2019). TiO_2/activated carbon photo cathode catalyst exposed to ultraviolet radiation to enhance the efficacy of integrated microbial fuel cell-membrane bioreactor. *Bioresource Technology Reports* 7: 100303.

Cha, J., Choi, S., Yu, H. et al. (2010). Directly applicable microbial fuel cells in aeration tank for wastewater treatment. *Bioelectrochemistry* 78 (1): 72–79.

Chung, T.H., Meshref, M.N.A., Hai, F.I. et al. (2020). Microbial electrochemical systems for hydrogen peroxide synthesis: critical review of process optimization, prospective environmental applications, and challenges. *Bioresource Technology* 313: 123727.

Das, S., Mishra, A., and Ghangrekar, M.M. (2020). Production of hydrogen peroxide using various metal-based catalysts in electrochemical and bioelectrochemical systems: mini review. *Journal of Hazardous, Toxic, and Radioactive Waste* 24 (3): 06020001.

Deveci, E.Ü., Dizge, N., Yatmaz, H.C., and Tansel, B. (2016). Degradation of recalcitrant textile dyes by coupling fungal and photocatalytic membrane reactors. *Clean: Soil, Air, Water* 44 (10): 1345–1351.

Dhangar, K. and Kumar, M. (2020). Tricks and tracks in removal of emerging contaminants from the wastewater through hybrid treatment systems: a review. *Science of the Total Environment* 738: 140320.

Di, L., Li, Y., Nie, L. et al. (2020). Influence of plant radial oxygen loss in constructed wetland combined with microbial fuel cell on nitrobenzene removal from aqueous solution. *Journal of Hazardous Materials* 394: 122542.

Donneys-Victoria, D., Bermúdez-Rubio, D., Torralba-Ramírez, B. et al. (2019). Removal of indigo carmine dye by electrocoagulation using magnesium anodes with polarity change. *Environmental Science and Pollution Research* 26 (7): 7164–7176.

Dordio, A., Carvalho, A.J.P., Teixeira, D.M. et al. (2010). Removal of pharmaceuticals in microcosm constructed wetlands using *Typha* spp. and LECA. *Bioresource Technology* 101 (3): 886–892.

Dordio, A., Pinto, J., Dias, C.B. et al. (2009). Atenolol removal in microcosm constructed wetlands. *International Journal of Environmental Analytical Chemistry* 89 (8–12): 835–848.

Ebrahimi, A., Sivakumar, M., McLauchlan, C. et al. (2021). A critical review of the symbiotic relationship between constructed wetland and microbial fuel cell for enhancing pollutant removal and energy generation. *Journal of Environmental Chemical Engineering* 9 (1): 105011.

Espejo, A., Aguinaco, A., Amat, A.M., and Beltrán, F.J. (2014). Some ozone advanced oxidation processes to improve the biological removal of selected pharmaceutical contaminants from urban wastewater. *Journal of Environmental Science and Health: Part A Toxic/Hazardous Substances and Environmental Engineering* 49 (4): 410–421.

Fajardo, A.S., Rodrigues, R.F., Martins, R.C. et al. (2015). Phenolic wastewaters treatment by electrocoagulation process using Zn anode. *Chemical Engineering Journal* 275: 331–341.

Goswami, L., Vinoth Kumar, R., Borah, S.N. et al. (2018). Membrane bioreactor and integrated membrane bioreactor systems for micropollutant removal from wastewater: a review. *Journal of Water Process Engineering* 26: 314–328.

Iorhemen, O.T., Hamza, R.A., and Tay, J.H. (2016). Membrane bioreactor (MBR) technology for wastewater treatment and reclamation: membrane fouling. *Membranes* 6 (2): 33.

Jiang, C., Jia, L., Zhang, B. et al. (2014). Comparison of quartz sand, anthracite, shale and biological ceramsite for adsorptive removal of phosphorus from aqueous solution. *Journal of Environmental Sciences (China)* 26 (2): 466–477.

Jiang, C., Liu, L., and Crittenden, J.C. (2016). An electrochemical process that uses an Fe^0/TiO_2 cathode to degrade typical dyes and antibiotics and a bio-anode that produces electricity. *Frontiers of Environmental Science and Engineering* 10 (4): 15.

Karaolia, P., Michael-Kordatou, I., Hapeshi, E. et al. (2017). Investigation of the potential of a Membrane BioReactor followed by solar Fenton oxidation to remove antibiotic-related microcontaminants. *Chemical Engineering Journal* 310: 491–502.

Khan, M.F., Yu, L., Achari, G., and Tay, J.H. (2019). Degradation of sulfolane in aqueous media by integrating activated sludge and advanced oxidation process. *Chemosphere* 222: 1–8.

Kyriacou, A., Lasaridi, K.E., Kotsou, M. et al. (2005). Combined bioremediation and advanced oxidation of green table olive processing wastewater. *Process Biochemistry* 40 (3–4): 1401–1408.

L'Amour, R.J.A., Azevedo, E.B., Leite, S.G.F., and Dezotti, M. (2008). Removal of phenol in high salinity media by a hybrid process (activated sludge + photocatalysis). *Separation and Purification Technology* 60 (2): 142–146.

Laera, G., Cassano, D., Lopez, A. et al. (2012). Removal of organics and degradation products from industrial wastewater by a membrane bioreactor integrated with ozone or UV/H_2O_2 treatment. *Environmental Science and Technology* 46 (2): 1010–1018.

Li, X., Jin, X., Zhao, N. et al. (2017). Efficient treatment of aniline containing wastewater in bipolar membrane microbial electrolysis cell-Fenton system. *Water Research* 119: 67–72.

Li, Y., Liu, L., Liu, J. et al. (2014). PPy/AQS (9, 10-anthraquinone-2-sulfonic acid) and PPy/ARS (Alizarin Red's) modified stainless steel mesh as cathode membrane in an integrated MBR/MFC system. *Desalination* 349: 94–101.

Liew, K.B., Daud, W.R.W., Ghasemi, M. et al. (2014). Non-Pt catalyst as oxygen reduction reaction in microbial fuel cells: a review. *International Journal of Hydrogen Energy* 39 (10): 4870–4883.

Logan, B.E., Call, D., Cheng, S. et al. (2008). Microbial electrolysis cells for high yield hydrogen gas production from organic matter. *Environmental Science and Technology* 42 (23): 8630–8640.

Logan, B.E., Hamelers, B., Rozendal, R. et al. (2006). Microbial fuel cells: methodology and technology. *Environmental Science and Technology* 40 (17): 5181–5192.

Mascolo, G., Laera, G., Pollice, A. et al. (2010). Effective organics degradation from pharmaceutical wastewater by an integrated process including membrane bioreactor and ozonation. *Chemosphere* 78 (9): 1100–1109.

Monteoliva-García, A., Martín-Pascual, J., Muñío, M.M., and Poyatos, J.M. (2019). Removal of a pharmaceutical mix from urban wastewater coupling membrane bioreactor with advanced oxidation processes. *Journal of Environmental Engineering* 145 (9): 04019055.

Monteoliva-García, A., Martín-Pascual, J., Muñío, M.M., and Poyatos, J.M. (2020). Effects of carrier addition on water quality and pharmaceutical removal capacity of a membrane bioreactor: advanced oxidation process combined treatment. *Science of the Total Environment* 708: 135104.

Moussa, D.T., El-Naas, M.H., Nasser, M., and Al-Marri, M.J. (2017). A comprehensive review of electrocoagulation for water treatment: potentials and challenges. *Journal of Environmental Management* 186: 24–41.

Nadais, H., Li, X., Alves, N. et al. (2018). Bio-electro-Fenton process for the degradation of non-steroidal anti-inflammatory drugs in wastewater. *Chemical Engineering Journal* 338: 401–410.

Neyens, E. and Baeyens, J. (2003). A review of classic Fenton's peroxidation as an advanced oxidation technique. *Journal of Hazardous Materials* 98 (1–3): 33–50.

Oller, I., Malato, S., and Sánchez-Pérez, J.A. (2011). Combination of advanced oxidation processes and biological treatments for wastewater decontamination: a review. *Science of the Total Environment* 409 (20): 4141–4166.

Ouarda, Y., Tiwari, B., Azaïs, A. et al. (2018). Synthetic hospital wastewater treatment by coupling submerged membrane bioreactor and electrochemical advanced oxidation process: kinetic study and toxicity assessment. *Chemosphere* 193: 160–169.

Pant, D., Van Bogaert, G., Diels, L., and Vanbroekhoven, K. (2010). A review of the substrates used in microbial fuel cells (MFCs) for sustainable energy production. *Bioresource Technology* 101 (6): 1533–1543.

Petrović, M., Gonzalez, S., and Barceló, D. (2003). Analysis and removal of emerging contaminants in wastewater and drinking water. *TrAC: Trends in Analytical Chemistry* 22 (10): 685–696.

Prado, M., Borea, L., Cesaro, A. et al. (2017). Removal of emerging contaminant and fouling control in membrane bioreactors by combined ozonation and sonolysis. *International Biodeterioration and Biodegradation* 119: 577–586.

Rosal, R., Rodríguez, A., Perdigón-Melón, J.A. et al. (2010). Occurrence of emerging pollutants in urban wastewater and their removal through biological treatment followed by ozonation. *Water Research* 44 (2): 578–588.

Rout, P.R., Zhang, T.C., Bhunia, P., and Surampalli, R.Y. (2021). Treatment technologies for emerging contaminants in wastewater treatment plants: a review. *Science of the Total Environment* 753: 141990.

Sathe, S.M., Bhowmick, G.D., Dubey, B.K., and Ghangrekar, M.M. (2020). Surfactant removal from wastewater using photo-cathode microbial fuel cell and laterite-based hybrid treatment system. *Bioprocess and Biosystems Engineering* 43 (11): 2075–2084.

Sathe, S.M., Chakraborty, I., Sankar Cheela, V.R. et al. (2021). A novel bio-electro-Fenton process for eliminating sodium dodecyl sulphate from wastewater using dual chamber microbial fuel cell. *Bioresource Technology* 341: 125850.

Schaar, H., Clara, M., Gans, O., and Kreuzinger, N. (2010). Micropollutant removal during biological wastewater treatment and a subsequent ozonation step. *Environmental Pollution* 158 (5): 1399–1404.

Shadmehr, J., Mirsoleimani-azizi, S.M., Zeinali, S., and Setoodeh, P. (2019). Electrocoagulation process for propiconazole elimination from wastewater: experimental design for correlative modeling and optimization. *International Journal of Environmental Science and Technology* 16 (10): 5409–5420.

Sirtori, C., Zapata, A., Oller, I. et al. (2009). Solar photo-fenton as finishing step for biological treatment of a pharmaceutical wastewater. *Environmental Science and Technology* 43 (4): 1185–1191.

Song, H.L., Li, H., Zhang, S. et al. (2018). Fate of sulfadiazine and its corresponding resistance genes in up-flow microbial fuel cell coupled constructed wetlands: effects of circuit operation mode and hydraulic retention time. *Chemical Engineering Journal* 350: 920–929.

Tan, S.P., Kong, H.F., Bashir, M.J.K. et al. (2017). Treatment of palm oil mill effluent using combination system of microbial fuel cell and anaerobic membrane bioreactor. *Bioresource Technology* 245: 916–924.

Tao, H.C., Wei, X.Y., Zhang, L.J. et al. (2013). Degradation of p-nitrophenol in a BES-Fenton system based on limonite. *Journal of Hazardous Materials* 254–255 (1): 236–241.

Tay, J.H., Yang, P., Zhuang, W.Q. et al. (2007). Reactor performance and membrane filtration in aerobic granular sludge membrane bioreactor. *Journal of Membrane Science* 304 (1–2): 24–32.

Tee, P.F., Abdullah, M.O., Tan, I.A.W. et al. (2016). Review on hybrid energy systems for wastewater treatment and bio-energy production. *Renewable and Sustainable Energy Reviews* 54: 235–246.

Vystavna, Y., Frkova, Z., Marchand, L. et al. (2017). Removal efficiency of pharmaceuticals in a full-scale constructed wetland in East Ukraine. *Ecological Engineering* 108: 50–58.

Wang, X.-C., Shen, J.-M., Chen, Z.-L. et al. (2016). Removal of pharmaceuticals from synthetic wastewater in an aerobic granular sludge membrane bioreactor and determination of the bioreactor microbial diversity. *Applied Microbiology and Biotechnology* 100 (18): 8213–8223.

Wijekoon, K.C., Hai, F.I., Kang, J. et al. (2013). The fate of pharmaceuticals, steroid hormones, phytoestrogens, UV-filters and pesticides during MBR treatment. *Bioresource Technology* 144: 247–254.

Wu, H., Zhang, J., Ngo, H.H. et al. (2015). A review on the sustainability of constructed wetlands for wastewater treatment: design and operation. *Bioresource Technology* 175: 594–601.

Wu, S., Kuschk, P., Brix, H. et al. (2014). Development of constructed wetlands inperformance intensifications for wastewater treatment: a nitrogen and organic matter targeted review. *Water Research* 57: 40–55.

Xia, S., Jia, R., Feng, F. et al. (2012). Effect of solids retention time on antibiotics removal performance and microbial communities in an A/O-MBR process. *Bioresource Technology* 106: 36–43.

Xu, N., Zhang, Y., Tao, H. et al. (2013). Bio-electro-Fenton system for enhanced estrogens degradation. *Bioresource Technology* 138: 136–140.

Xu, P., Zheng, D., He, Q., and Yu, J. (2020a). The feasibility of ofloxacin degradation and electricity generation in photo-assisted microbial fuel cells with $LiNbO_3$/CF photocatalytic cathode. *Separation and Purification Technology* 250: 117106.

Xu, P., Zheng, D., Xie, Z. et al. (2020b). The degradation of ibuprofen in a novel microbial fuel cell with PANi@CNTs/SS bio-anode and $CuInS_2$ photocatalytic cathode: property, efficiency and mechanism. *Journal of Cleaner Production* 265: 121872.

Xu, P., Zheng, D., Xie, Z. et al. (2020c). The mechanism and oxidation efficiency of bio-electro-Fenton system with Fe@Fe_2O_3/ACF composite cathode. *Separation and Purification Technology* 234: 116103.

Yadav, A., Dash, P., Mohanty, A. et al. (2012). Performance assessment of innovative constructed wetland–microbial fuel cell for electricity production and dye removal. *Ecological Engineering* 47: 126–131.

Zhang, B., Wang, Z., Zhou, X. et al. (2015a). Electrochemical decolorization of methyl orange powered by bioelectricity from single-chamber microbial fuel cells. *Bioresource Technology* 181: 360–362.

Zhang, L., Yin, X., and Li, S.F.Y. (2015b). Bio-electrochemical degradation of paracetamol in a microbial fuel cell-Fenton system. *Chemical Engineering Journal* 276: 185–192.

Zhao, X., Chen, Z.-L., Wang, X.-C. et al. (2014). PPCPs removal by aerobic granular sludge membrane bioreactor. *Applied Microbiology and Biotechnology* 98 (23): 9843–9848.

Zhou, Y., Huang, X., Zhou, H. et al. (2011). Removal of typical endocrine disrupting chemicals by membrane bioreactor: in comparison with sequencing batch reactor. *Water Science and Technology* 64 (10): 2096–2102.

Zhu, X. and Logan, B.E. (2013). Using single-chamber microbial fuel cells as renewable power sources of electro-Fenton reactors for organic pollutant treatment. *Journal of Hazardous Materials* 252–253: 198–203.

Zhu, X. and Ni, J. (2009). Simultaneous processes of electricity generation and p-nitrophenol degradation in a microbial fuel cell. *Electrochemistry Communications* 11 (2): 274–277.

Zupanc, M., Kosjek, T., Petkovšek, M. et al. (2013). Removal of pharmaceuticals from wastewater by biological processes, hydrodynamic cavitation and UV treatment. *Ultrasonics Sonochemistry* 20 (4): 1104–1112.

Part V

Aspects of Sustainability and Environmental Management

20

Regulatory Framework of Microconstituents

Wei-Han Lin[1], Jiun-Hau Ou[2], Ying-Liang Yu[2,4], Pu-Fong Liu[2], Rao Y. Surampalli[3], and Chih-Ming Kao[2,4]

[1]*School of Environment, Tsinghua University, Beijing, PR China*
[2]*Institute of Environmental Engineering, National Sun Yat-sen University, Kaohsiung, Taiwan*
[3]*Global Institute for Energy, Environment and Sustainability, Lenexa, KS, USA*
[4]*Civil and Environmental Engineering Department, College of Engineering, University of Nebraska, Lincoln, Omaha, NE, USA*

20.1 Introduction

Due to rapid economic growth, industrialization, and urbanization, environmental pollution has become a major issue of public interest, especially in many developing countries, which leads to overexploitation of natural resources and serious threats to ecosystems (Ajibade et al. 2021). Microconstituents (MCs) are used to describe artificial or natural compounds (inorganic or organic chemicals and elements) which are detected in different environmental media, and cause potential impacts on human health and other species. MCs are usually produced by the degradation of organic compounds, resulting in the formation and accumulation of persistent metabolites in the natural environment. Due to their special natures and properties, traditional pollution treatment methods usually cannot effectively control MCs, although the activities, routes, and trends for MCs production are similar to traditional pollutants (Köhl et al. 2019; Aghalari et al. 2020). Pharmaceuticals and personal care products (PPCPs) are the massive sources of MCs in the environment. PPCPs are pervasive in our lives, and masses of PPCPs become disposed of as wastes each year when they pass their expiration date, become detrimental, or are unwanted. Ongoing studies disclose that pharmaceuticals that escape into the environment act as endocrine disruptors, which have been linked to major abnormalities and impaired reproductive performance in some species. Prevention can be one of the best effective mechanisms for avoiding MCs from entering the environment. The management framework and regulations of MCs pollution are discussed in this chapter.

20.2 Management and Regulatory Framework of Microconstituents

The establishment of an MCs pollution management and regulatory framework needs to consider the following three major aspects:

Microconstituents in the Environment: Occurrence, Fate, Removal, and Management, First Edition.
Edited by Rao Y. Surampalli, Tian C. Zhang, Chih-Ming Kao, Makarand M. Ghangrekar, Puspendu Bhunia, Manaswini Behera, and Prangya R. Rout.

1) The damages of MCs to the environment is not just limited to the release area, as their influences and impacts are global (Gong and Xie 2020). Thus, the management and control of MC pollution needs the cooperation activities of all countries. The participation of international conferences and government agencies can strengthen international communication and learn from the cutting-edge research, treatment methods, regulatory framework, and policy guidelines of MC management and control. MC pollution is a cross-cutting problem involving many disciplines and fields, and thus it needs to gain full attention from the advantages of various departments to formulate scientific and effective preventive measures that jointly control MCs pollution (Lützhøft et al. 2012; Gong and Xie 2020).
2) The source control is an essential means for MC pollution reduction. Legislation needs to be strengthened for MCs source control of trace elements and production prohibition. The usage of MC contained-products may result in further MC production and pollution. A good example is that the United States issued the Microbead-Free Waters Act of 2015. This Act prohibits the production and sale of cleaning cosmetics with plastic microbeads, which effectively reduces the release of plastic microbeads to the ecosystems (Auta et al. 2017). Moreover, the public's awareness of environmental hazards in MCs is still superficial, so it is a necessity to strengthen publicity and raise society's awareness of risks and hazards caused by MCs. The establishment of the related regulations can reduce the MCs-contained wastes, and also improve the public's environmental awareness. Additionally, recycling of MC-contained products and treatment of MC-contained wastes in a centralized and circular-economic way will be an important measure for the reduction of MC-contained wastes in the environment. This will also slow the accumulation of MCs in the environment (Sun et al. 2020).
3) The application of biological techniques for MCs removal has attracted wide attention. Some microbial species can biodegrade or biotransform MCs into non-toxic end products (Ball 2017). Moreover, development of biodegradable MC-contained products (e.g., biodegradable plastics) is also an effective way to solve this environmental problem. After being discarded, MCs can be biodegraded by intrinsic microorganisms, and this can reduce the risks to the environment (Carvalho et al. 2015; Ball 2017).

20.3 Regulations on Microconstituents

20.3.1 Pharmaceuticals and Personal Care Products (PPCPs)

Most standards of MCs do not exist, although some regions or countries have implemented laws and regulations on some MCs (Luo et al. 2014). For example, Directive 2008/105/EC has already agreed on a set of environmental quality standards for MCs (e.g., nonylphenol, bisphenol A, diuron) (Directive 2008). Moreover, nonylphenol and its ethoxylates are toxic compounds based on the definition of Section 64 of the Canadian Environmental Protection Act of 1999. Other MCs, such as steroid hormones and PPCPs, have also been added to the list of controlled substances. To set monitoring limits of MCs, additional studies on the acute and chronic effects (biological reactions) of these compounds need to be conducted (Luo et al. 2014).

Conduction of investigation studies is crucial to obtain key data related to the pollution status of MCs in the environment (Al-Odaini et al. 2010). For example, Al-Qaim (2014) conducted a study to investigate the presence of human drugs and synthetic hormones in the Tanks River in Malaysia. He found several compounds in the water samples (e.g., carbamazepine, caffeine, levonorgestrel). These have been approved as the most common compounds in Malaysia (Al-Qaim et al. 2014). A survey of more than 100 compounds in a drinking water treatment plant in the United States found that there were 22 compounds in treated water (Sharmin and Chowdhury 2016).

In 1998, the regulations of the pharmaceutical industry were reviewed by the US Environmental Protection Agency to control air emissions and sewage discharges. The first European Union's (EU's) Guidelines for Environmental Risk Assessment of Human Drugs was issued by The European Medicines Agency (Ågerstrand 2015). The health risk assessments of new drugs are required under the Pharmaceutical Administration Law (Health and Services 1997). The environmental risk assessments of PPCPs have not been standardized. PPCPs environmental risk assessment regulations and guidelines should be formulated by all countries in the future (Liu et al. 2013). It is a direct and effective strategy to reduce the disposal of PPCPs by raising the awareness of PPCPs through source control (Ewadh et al. 2017). In Australia, the registration of new drugs should be assessed for pharmaceutical environmental risks according to the regulations by the Australian Therapeutic Goods Administration (TGA 2008). In 2010, the laws and regulations on environmental management of hazardous chemicals included the Environmental Management Method for New Chemicals promulgated in China (MEP 2010), and the Management and Registration Measures for Environmental Management of Hazardous Chemicals became effective in March 2013 (MEP 2012).

Except for the new laws for PPCPs management, expired drugs should be collected by making drug management and recovery plans. Moreover, PPCPs may enter the subsurface environment (soil and groundwater environment) and so further investigation and evaluation of the sewage, landfill waste, and reclaimed irrigation water are required. The traditional wastewater treatment plants (WWTPs) facilities need to be upgraded or improved to effectively prevent the escape of MCs from the WWTPs, and illegal wastewater discharges are strictly forbidden (Liu and Wong 2013).

20.3.2 Microplastics

The United Nations Environmental Programme (UNEP) called for immediate action to eliminate microplastics in the oceans, because microplastics were consumed by a significant amount of marine organisms, resulting in their physical and chemical harm (Caruso 2015). In 2018, a global review of national laws and regulations on single-use plastics was released by the UNEP and the World Resources Institute (WRI). The report calls for a more comprehensive regulatory approach, integrating the life cycle of plastic products, including their production, distribution, use, and disposal. The purpose of this is to provide a reference for countries and other stakeholders who are interested in learning about the methods currently used to solve the plastic pollution problem.

Approximately 127 countries have passed legislation for the regulation of plastic bags, while only 8 countries (United States, Canada, Italy, France, New Zealand, Sweden, South Korea, and United Kingdom) have formulated national regulations or laws for the prohibition of microbeads (enacted microbeads bans). Among these countries, New Zealand's regulations prohibit the use of microbeads in personal products as well as automotive, household, and industrial cleaning products. Moreover, India, Belgium, Ireland, and Brazil have initiated new laws or regulations to prohibit microbeads at the national level, and the EU has proposed a procedure to restrict microplastics addition in consumer and professional products. Additionally, companies, non-governmental organizations (NGPs), and governments, promote voluntary phase-outs to reduce the number of products containing microbeads.

The UNEP proposed a program involving more than 40 million people from 120 countries, and formulated educational measures to raise awareness and promote the reduction of plastic use, encourage recycling, and evaluate disposal (UNEP 2014; Caruso 2015). The UPEP/Mediterranean Action Plan, Oslo/Paris convention for the protection of the marine environment of the North-East Atlantic, and the Baltic Marine Environment Protection Commission-Helsinki Commission have formulated guidelines for marine debris (including microplastics) assessment (Lassen et al. 2012). The plan included the organization of some seminars to encourage capacity building and the dissemination of good practices among individuals. In 2011, the plastics industry established the "Joint Declaration of the Global Plastics Association on Solving the Problem of Marine Garbage," which included a plan to reduce garbage and a commitment to support multiple garbage assessments. NGOs have come up with programs to raise awareness and help the investigation and quantification of the extent of microplastics pollution and its effects at regional, national, and international scale (Auta et al. 2017; Nielsen et al. 2020).

Except for the above stated guidelines proposed by different international originations, regulations for microplastics pollution control have been established from governmental levels. In 2015, the microbeads ban in California (Assembly Bill No. 888, 2015) was passed. The ban was set up to provide the strongest protection for the state to prevent pollution by plastic microbeads. The bill encourages enterprises to use natural substitutes such as walnut shells, sea salt, and almonds. This has been followed by the US House of Representatives Legislation (Microbead-free Water Act of 2015, H.R. 1321) prohibiting the manufacture, sale, or distribution of personal care products containing plastic microbeads beginning on 1 July 2017 (Strifling 2016; Xanthos and Walker 2017). Based on the amendment of Canada's Environmental Protection Act, new regulations prohibited the production and import of microbeads from the beginning of 2018, and the sales ban was implemented after six months. Moreover, natural health products and non-prescription drugs containing microbeads were banned from July 2019 (Canada's Environmental Protection Act 2017).

In 2014, the EU stipulated that washing products containing microplastics were no longer allowed to be labeled with eco-labels (EU Commission Decision 2014/893/EU) (Kijek 2015). In 2017, in the United Kingdom, the Department for Environment, Food and Rural Affairs (DEFRA) issued a regulation to ban the sale and manufacture of microplastics in personal care products and cosmetics (DEFRA 2016). In France, a law was published in 2017 (French Decree, 2017-291) to prohibit the use of solid plastic particles in washing, exfoliating, and cleaning cosmetics from January 2018. The law affects plastic cotton swabs, which were banned in January 2020. Moreover, the Swedish Chemicals Agency (KEMI) proposed to ban products containing microbeads in 2016 (supported by the Danish Ministry of Environment and Food) (Kentin and Kaarto 2018; Hurley et al. 2019).

Many countries/regions have implemented regulations or laws to prohibit the use, production, and sale of microplastics (microbeads) in personal care products and cosmetics to reduce microplastics in environment. To effectively manage and control the impacts of microplastics on the ecosystem, various methods and measures need to be adopted to raise our awareness of the daily use of plastic materials by encouraging companies and all users to use plastics more effectively and wisely, and thus, a safer environment for marine/aquatic life and human can be created. The Joint Group of Experts on the Scientific Aspects of Marine Environmental Protection encourages all countries to adopt a circular economy strategy, and make efforts to reduce the amount of plastics entering the ocean, because this is a more economical way to reduce the number of microplastics entering the ocean (Kershaw and Rochman 2015; Kershaw et al. 2019).

20.3.3 Persistent Organic Pollutants (POPs) and Persistent Bioaccumulated Toxics (PBTs)

The descriptions of regulatory frameworks and international agreements related to POPs and PBTs are summarized in chronological order (van Wijk et al. 2009; Moermond et al. 2012; Matthies et al. 2016). Although the Rotterdam Convention (on prior informed consent) and the Basel Convention (on waste) can solve the distribution and final disposal of POPs and PBTs, the main global legally binding instrument is the Stockholm Convention (Matthies et al. 2016). The jurisdictions of international agreements and regulatory frameworks are different (Sands and Galizzi 2004). The Protocol on Persistent Organic Pollutants (Aarhus) of the United Nations Economic Commission for Europe includes Canada, Russia, Europe, and the US. The Regulation on Registration, Evaluation, Authorization and Restriction of Chemicals and EC 1107/2009 includes the EU (Solomon et al. 2013). Mexico, Canada, and the US have participated in the initiative of sound management of chemicals of the North American Agreement on Environmental Cooperation. Canada's Environmental Protection Act or the Chemical Management Plan under the Toxic Substances Control Act in the US are national laws and regulations. China, Australia, New Zealand, Korea, and Japan apply very similar approaches to those used in the EU and North America.

It needs to be pointed out that Canada's pesticide regulations based on the Pest Control Products Act and the US Federal Insecticide, Bactericide and Rodenticide Act, do not include a specific evaluation of PBTs or POPs (Moermond et al. 2012; Matthies et al. 2016; Westwood et al. 2019). Based on the Stockholm International Convention on POPs, each participating country must take administrative and legislative actions to prevent the environmental impact of POPs. The requirements of the Stockholm Convention were incorporated into EU law by Regulation (EU) 2019/1021 and Irish law through Directive No.146 of 2020. According to Article 7 of the Convention, as new POPs are restricted or prohibited, the Republic of Ireland must formulate and regularly update the National Implementation Plan for POPs. As the competent department, the Environmental Protection Agency reviewed and updated the National Implementation Plan of POPs in 2018. The Ireland Environmental Protection Agency included detailed action plans to support the measures for PDPs control (Westwood et al. 2019).

20.3.4 Endocrine-Disrupting Chemicals (EDCs)

To reduce the environmental impact associated with the discharges of EDCs from wastewater, South Africa has formulated regulations and guidelines to improve wastewater treatment processes. This is the focus of organizations including the National Water Law No. 36 of 1998 and South Africa's national implementation plan formulated according to Article 7 of the Stockholm Convention (Burger and Nel 2008).

Within the EU's standardized regulatory framework system, environmental pollution discharges are handled within the frameworks of biocidal products, industrial chemicals, water pollutants, and plant protection products. These frameworks are bound by the following provisions: i) *Plant protection products*: Regulation (EC) 1107/2009 on the placing of plant protection products on the market; ii) *Industrial chemicals*: Regulation (EC) 1907/2006, concerning the Registration, Evaluation, Authorization and Restriction of Chemicals; iii) *Water pollutants*: Directive 2000/60/EC and Decision 2015/495 (Rosales et al. 2019), establishing a framework for Community action in the field of water policy; iv) *US's pesticide regulatory framework includes two regulations*: the Federal Insecticide, Fungicide and Rodenticide Act (FIFRA) and the Federal Food, Drug and Cosmetic Act (FD&C Act) (USEPA 2010); and v) *Biocidal products*: Regulation (EU) 528/2012, concerning making available on the market and use of biocidal products (Biocidal Products Regulation).

Several regulatory frameworks in Canada explicitly deal with environmental data processing, including those proposed in the Canadian Environmental Protection Act of 1999 (CEPA), the Pest Control Products Act (PCPA), and the Food and Drugs Act (F&DA). To reduce the harmful effects of EDCs in discharged wastewater on the environment and human health, South Africa has formulated regulations and guidelines that provide the legal requirements for WWTPs. Although many wastewater treatment plants do not meet the sewage discharge requirements, there is currently no standard legal restrictions applicable to EDCs in South Africa. In Brazil, the environmental pollution problems have been dealt with by the regulatory framework on EDCs and the policy initiatives to formulate national industrial chemicals legislation. The National Environmental Management Act (NEMA) No. 107 of 1998 reaffirmed the provisions of Section 24 of the South African Constitution, which promotes and guarantees the internationally recognized principles of sustainable development (Bray 1999). In Australia, under the framework of industrial chemicals assessment, namely the National Industrial Chemicals Notification and Assessment Scheme (NICNAS), the environmental pollution problem is explicitly dealt with. In China, environmental design has been carried out under the background of environmental protection policy initiatives and industry standards.

20.4 Concluding Remarks

Science must be the basis for formulating MCs management and control regulatory framework and policies. Many fragmented laws, regulations, and rules need to be replaced by comprehensive and holistic systems, which also need to be explicitly authorized by designated agencies. It is a necessity to incorporate MCs into the current environmental monitoring program, and this is expected by decision-makers, NGOs, and scientists. In the near future, environmental quality standards need be determined for more MCs. It should be a priority to better understand MCs occurrence in time and space. This chapter puts forward a feasible management and regulatory framework, and summarizes various international regulations and laws on pollution of MCs. In the future, researchers and governmental agencies need to deepen the investigation and evaluation of single and mixed toxicity of MCs, and authorities also need to update relevant legislation and establish the limits for specific MCs.

References

Ågerstrand, M. (2015). Improving environmental risk assessment of human pharmaceuticals. *Environmental Science & Technology* 49 (9): 5336–5345.

Aghalari, Z., Dahms, H.-U., Sillanpää, M. et al. (2020). Effectiveness of wastewater treatment systems in removing microbial agents: a systematic review. *Globalization and Health* 16 (1): 1–11.

Ajibade, F.O., Adelodun, B., Lasisi, K.H. et al. (2021). Environmental pollution and their socioeconomic impacts. In: *Microbe Mediated Remediation of Environmental Contaminants* (eds. Kumar, A., Singh, V.K., Singh, P., and Mishra, V.K.), Chapter 25, 321–354. Woodhead Publishing.

Al-Odaini, N.A., Zakaria, M.P., Yaziz, M.I., and Surif, S. (2010). Multi-residue analytical method for human pharmaceuticals and synthetic hormones in river water and sewage effluents by solid-phase extraction and liquid chromatography–tandem mass spectrometry. *Journal of Chromatography A* 1217 (44): 6791–6806.

Al-Qaim, F.F., Abdullah, M.P., Othman, M.R. et al. (2014). A validation method development for simultaneous LC-ESI-TOF/MS analysis of some pharmaceuticals in Tangkas River, Malaysia. *Journal of the Brazilian Chemical Society* 25: 271–281.

Auta, H.S., Emenike, C., and Fauziah, S. (2017). Distribution and importance of microplastics in the marine environment: a review of the sources, fate, effects, and potential solutions. *Environment International* 102: 165–176.

Ball, P. (2017). Plastics on the menu. *Nature Materials* 16 (6): 606–606.

Bray, E. (1999). Co-operative governance in the context of the National Environmental Management Act 107 of 1998. *South African Journal of Environmental Law and Policy* 6 (1): 1–12.

Burger, A. and Nel, A. (2008). *Scoping Study to Determine the Potential Impact of Agricultural Chemical Substances (Pesticides) with Endocrine Disruptor Properties on the Water Resources of South Africa*. Water Research Commission Pretoria.

Canada's Environmental Protection Act (CEP) (2017). Microbeads in toiletries regulations. *Canada Gazette* 151: 12–14.

Caruso, G. (2015). Microplastics in marine environments: possible interactions with the microbial assemblage. *Journal of Pollution Effects Cont.* 3: e111.

Carvalho, R.N., Ceriani, L., Ippolito, A., and Lettieri, T. (2015). *Development of the First Watch List Under the Environmental Quality Standards Directive*. JRC Science Report. Report EUR 27142, Ispra, Italy.

DEFRA (2016). UK plant health risk register entry for anoplophora glabripennis. Available at: https://secure.fera.defra.gov.uk/phiw/riskRegister/viewPestRisks.cfm?cslref=1786 (accessed 18 April 2016).

Directive SF (2008). Directive 2008/56/EC of the European Parliament and of the Council. *Journal Council Decision.*

Ewadh, H.M., Abdullah, S.R., Anwar, N., and Hasan, H.A. (2017). Pharmaceuticals and personal care products: sources, toxicity in the environment, regulations and removal technologies. *Journal of Chemical and Pharmaceutical Sciences* 10: 1180–1187.

Gong, J. and Xie, P. (2020). Research progress in sources, analytical methods, eco-environmental effects, and control measures of microplastics. *Chemosphere* 254: 126790. doi: 10.1016/j.chemosphere.2020.126790.

Health U.D. Services of (1997). *Centre for drug evaluation and research (CDER).* Washington, DC: US Government Printing Office.

Hurley, C., Janssen, J., Schreuder, M. et al. (2019) Memo on national legislation and initiatives banning microplastics. Leiden: Universiteit Leiden, 1–10.

Kentin, E. and Kaarto, H. (2018). An EU ban on microplastics in cosmetic products and the right to regulate. *Review of European, Comparative & International Environmental Law* 27 (3): 254–266.

Kershaw, P. and Rochman, C. (2015). Sources, fate and effects of microplastics in the marine environment. Part 2 of a global assessment. Reports and Studies-IMO/FAO/Unesco-IOC/WMO/IAEA/UN/UNEP Joint Group of Experts on the Scientific Aspects of Marine Environmental Protection (GESAMP) Eng. No. 93.

Kershaw, P., Turra, A., and Galgani, F. eds. (2019). *Guidelines for the Monitoring and Assessment of Plastic Litter and Microplastics in the Ocean.* London: GESAMP.

Kijek, T. (2015). Modelling of eco-innovation diffusion: the EU eco-label. *Comparative Economic Research* 18 (1): 65–79.

Köhl, J., Booij, K., Kolnaar, R., and Ravensberg, W.J. (2019). Ecological arguments to reconsider data requirements regarding the environmental fate of microbial biocontrol agents in the registration procedure in the European Union. *BioControl* 64 (5): 469–487.

Lassen, C., Hansen, S.F., Magnusson, K. et al. (2012). Microplastics: occurrence, effects and sources of releases to the environment in Denmark. *Significance* 2 (2).

Liu, J.-L. and Wong, M.-H. (2013). Pharmaceuticals and personal care products (PPCPs): a review on environmental contamination in China. *Environment International* 59: 208–224.

Luo, Y., Guo, W., Ngo, H.H. et al. (2014). A review on the occurrence of micropollutants in the aquatic environment and their fate and removal during wastewater treatment. *Science of the Total Environment* 473: 619–641.

Lützhøft, H.-C.H., Donner, E., Wickman, T. et al. (2012). A source classification framework supporting pollutant source mapping, pollutant release prediction, transport and load forecasting, and source control planning for urban environments. *Environmental Science and Pollution Research* 19 (4): 1119–1130.

Matthies, M., Solomon, K., Vighi, M. et al. (2016). The origin and evolution of assessment criteria for persistent, bioaccumulative and toxic (PBT) chemicals and persistent organic pollutants (POPs). *Environmental Science: Processes & Impacts* 18 (9): 1114–1128.

MEP (2010). Decree No. 7 of the Ministry of Environmental Protection of the People's Republic of China, Measures on Environmental Management of New Chemical Substances, 19 January 2010 (in Chinese).

MEP (2012). MEP (Ministry of Environmental Production of China) Environmental management and registration method for hazardous chemicals. http://www.zhb.gov.cn/gkml/hbb/bl/201210/t20121016_238481.htm (2012).

Moermond, C.T., Janssen, M.P., de Knecht, J.A. et al. (2012). PBT assessment using the revised Annex XIII of REACH: a comparison with other regulatory frameworks. *Integrated Environmental Assessment and Management* 8 (2): 359–371.

Nielsen, T.D., Hasselbalch, J., Holmberg, K., and Stripple, J. (2020). Politics and the plastic crisis: a review throughout the plastic life cycle. *Wiley Interdisciplinary Reviews: Energy and Environment* 9 (1): e360.

Rosales, E., Diaz, S., Pazos, M., and Sanromán, M.A. (2019). Comprehensive strategy for the degradation of anti-inflammatory drug diclofenac by different advanced oxidation processes. *Separation and Purification Technology* 208: 130–141.

Sands, P. and Galizzi, P. (2004). *Documents in International Environmental Law*. Cambridge UK: Cambridge University Press.

Sharmin, R. and Chowdhury, S. (2016). A review on the removal of pharmaceuticals and personal care products from wastewater. *International Journal of Scientific Research* 33–35.

Solomon, K., Matthies, M., and Vighi, M. (2013). Assessment of PBTs in the European Union: a critical assessment of the proposed evaluation scheme with reference to plant protection products. *Environmental Sciences Europe* 25 (1): 1–17.

Strifling, D.A. (2016). The Microbead-Free Waters Act of 2015. *Journal of Land Use & Environmental Law* 32 (1): 151–166.

Sun, Q., Ren, S.-Y., and Ni, H.-G. (2020). Incidence of microplastics in personal care products: an appreciable part of plastic pollution. *Science of the Total Environment* 742: 140218.

TGA (2008). Ageing. Australian Therapeutic Goods Administration. Regulation of complementary medicines.

UNEP, I. and TNC, W. (2014). Green infrastructure: guide for water management.

USEPA (2010). Integrated Risk Information System (IRIS). Arsenic, inorganic (CASRN 7440-38-2), Cadmium (CASRN 7440-43-9). http://cfpub.epa.gov/00A0;

van Wijk, D., Chénier, R., Henry, T. et al. (2009). Integrated approach to PBT and POP prioritization and risk assessment. *Integrated Environmental Assessment and Management: An International Journal* 5 (4): 697–711.

Westwood, A., Olszynski, M., Fox, C. et al. (2019). The role of science in contemporary Canadian environmental decision-making: the example of environmental assessment. *UBCL Review* 52: 243.

Xanthos, D. and Walker, T.R. (2017). International policies to reduce plastic marine pollution from single-use plastics (plastic bags and microbeads): a review. *Marine Pollution Bulletin* 118 (1–2): 17–26.

21

Laboratory to Field Application of Technologies for Effective Removal of Microconstituents From Wastewaters

Indrajit Chakraborty[1], *Manikanta M. Doki*[1], *and Makarand M. Ghangrekar*[1,2]

[1]*Department of Civil Engineering, Indian Institute of Technology Kharagpur, West Bengal, India*
[2]*School of Environmental Science and Engineering, Indian Institute of Technology Kharagpur, West Bengal, India*

21.1 Introduction

Recalcitrance of microconstituents (MCs) reduces treatability by conventional treatment processes. The removal of these xenobiotic compounds requires special attention owing to the high bond strength that requires higher energy for dissociation and also the biotoxic nature of these compounds that impedes the biochemical processes in bioreactors used in effluent treatment plants (Shcherbakova et al. 1999). Hence, in the cases wherein biological treatment is implemented, special attention is required toward both selection of appropriate detection methods, as well as identification of optimum operating parameters for the biodegradation of these slow degrading organic compounds. The physico-chemical processes, although providing a surer path to MC degradation, require identification of the degradation by-products owing to incomplete mineralization of the target pollutant to be ensured to ascertain the sustainability of the process (Fatta-Kassinos et al. 2011; Ike et al. 2019).

Although different investigations on full-scale installed bioreactors in the field report partial or considerable removal of MCs, the removal from the aqueous phase is often due to the partitioning of the MCs to the settled sludge (Radjenović et al. 2009; Fijalkowski et al. 2017). Apart from the adsorption/absorption onto sludge, the biochemical degradation of MCs also has been reported using different types of technologies. In field-scale technologies, the activated sludge process (ASP), membrane bioreactors (MBR), phyto-remediation, and electrically conductive biofilters are popular, which are claimed to achieve a certain level of MC degradation. To understand the biochemical reaction kinetics and the degradation pathway, a pragmatic approach is to investigate lab-scale/bench-scale bioreactors, which can provide valuable inputs for deciding the optimum operating parameters in future field-scale reactors.

In the case of physico-chemical processes, the advanced oxidation processes (AOPs) followed by membrane processes or adsorption media can be an effective technique for the removal of MCs from wastewater. The field-scale AOPs that are implemented for the

Microconstituents in the Environment: Occurrence, Fate, Removal, and Management, First Edition.
Edited by Rao Y. Surampalli, Tian C. Zhang, Chih-Ming Kao, Makarand M. Ghangrekar, Puspendu Bhunia, Manaswini Behera, and Prangya R. Rout.

removal of the MCs have been time tested for disinfection of both water and wastewater. Application of ozone, UV, chlorination, electrochemical oxidation, and in certain cases hydrogen peroxide, has been well explored for disinfection purposes (Miklos et al. 2018). Implementation of these AOPs for removal of MCs is still in its experimental stage owing to the uncertainty in terms of deciding the process parameters, fate, and identification of different degradation by-products and optimum reactor engineering (Giwa et al. 2021). In this preview, different lab-scale set-ups have been investigated in the past using standalone or combined/hybrid AOPs for the removal and mineralization of the MCs.

Apart from the AOPs, membrane separation techniques are also shown as effective techniques for the removal of the MCs from the environment (Khanzada et al. 2020). Issues such as biofouling and pore clogging pose operating problems in the case of membrane filtration, hence demand further research in terms of material synthesis for membrane development and for innovative reactor configurations. Sufficient literature is available on the membrane operation for the direct membrane filtration of wastewater; however, a fewer number of the reported works focus on the biofouling strategies that can be adopted for extending the longevity of the membrane utilized. The implementation of different strategies, either in terms of membrane development or reactor configuration, have to be evaluated at lab-scale, the outcome of which can be further translated to field-scale with suitable assumptions. In addition to single standalone technologies that are being implemented, the hybrid technologies, such as membrane bioreactors, and photo-catalytic membrane reactors coupling photo-catalysis on the membrane surface to facilitate concomitant photolysis of the target pollutant (majorly MCs) and membrane filtration of the effluent, etc., are innovative lab-scale implementations that are currently being explored.

21.1.1 Microconstituent Origin and Type

Microconstituents (MCs) in wastewater can originate from different domestic, institutional, or industrial sources. The MCs include pharmaceutically active compounds, personal care products, pesticides, biocidal compounds, steroid hormones, surfactants, and industrial chemicals under a broad classification. Although the broad classification allows a certain degree of distinction among the different classes of xenobiotic compounds, detection of MCs is a costly and tedious process owing to which tracing the fate and transport of individual MCs becomes a difficult task. MCs can enter the water matrix through different pathways. Pharmaceutical compounds, such as antibiotics, anti-depressants, analgesics, and steroidal compounds, originate from human urine and from the expired medicines that are often disposed of into wastewater due to lack of awareness. Bisphenol A, a widely regulated endocrine disrupting chemical (EDC), and its homologs originate from the leaching of paints and polycarbonate plastics and are prevalently found in wastewater as well as sludge. Pesticides originating from agricultural runoff often enter rivers and other fresh water bodies and eventually find their way into the water matrix. Hence, the occurrence of pesticides in wastewater treatment plants (WWTPs) can be traced back to the insurgence of these compounds via the aforementioned pathway.

In a unique review work, the authors compiled the literature data available for the drugs of abuse that were detected by different researchers in wastewater samples in Europe (Postigo et al. 2008). The most prevalently studied abusive drug was the cocainic group of compounds, with cocaine being the most abundant drug found in the range of 40 to 860 ng L^{-1} in sewage treatment plant (STP) effluent and 6 to 40 ng L^{-1} in surface water (Postigo et al. 2008). Another important class of pharmaceutical compounds, namely psychiatric drugs, such as diazepam (a type of benzodiazepine), fluoxetine, sertraline (antidepressants), meprobamate (used to treat anxiety disorders), etc., were found in the ng L^{-1} range in different surface water, potable water, treated and untreated wastewater, landfill leachate, etc. in different study locations of the USA and Europe (Calisto and Esteves 2009). Antibiotics are another class of pharmaceutical compounds that can be sourced from natural biota, discharged from an antibiotic manufacturing facility, and from domestic households and hospitals. Antibiotics, such as streptomycin and β-lactams, are naturally occurring antibiotics that are produced by soil bacteria (Kümmerer 2009). It was observed that, for Western developed nations, the load of antibiotic discharged through sewage was only 30%, the remaining 70% being from the diffuse/non-point source of domestic households (Schuster et al. 2008).

21.1.2 Refractory Nature and Corresponding Degradation Barriers of Microconstituents

The presence of MCs in the surface water and wastewaters is a function of continuous introduction of the same from diffused and point sources as well as the recalcitrant nature of these compounds. Although different stages of a wastewater treatment scheme facilitate a certain degree of removal of MCs (Figure 21.1), the treatment technologies need to be evaluated in terms of holistic removal of these compounds. The recalcitrance can be traced to the strong chemical bonds of these compounds, as well as their associated biotoxic nature.

For instance, Batt et al. (2007) examined the fate of antibiotics in four full-scale WWTPs that used different treatment technologies. The first wastewater treatment plant comprised of a two-stage conventional ASP. Stage one had a shorter solid retention time (SRT), and stage two had a longer SRT, allowing for effective nitrification. The second WWTP employed extended aeration ASP with auxiliary chemical addition, such as $FeCl_3$. For the third and fourth systems, WWTPs were equipped with a rotating biological contactor reactor and

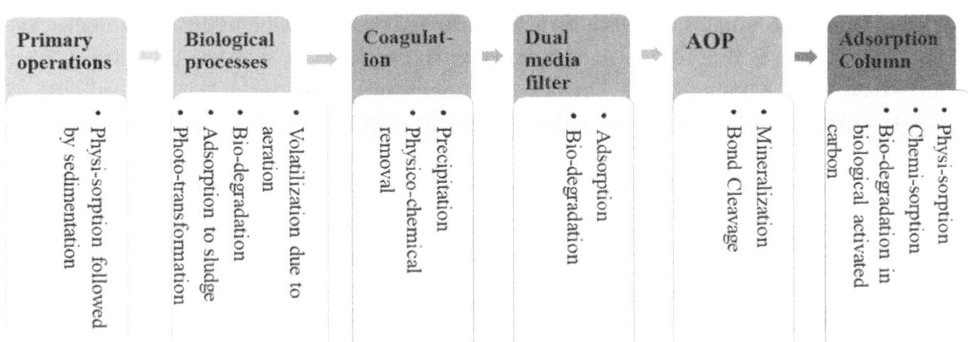

Figure 21.1 Removal mechanisms of MCs in different stages of wastewater treatment.

pure oxygenated activated sludge system in the covered aeration tank, respectively. Except for the antibiotic trimethoprim, the concentrations of tetracycline, ciprofloxacin, and sulfamethoxazole were significantly reduced in the first plant. The removal of trimethoprim was observed to be primarily due to microbial degradation with limited or no sorption on sludge. Thus, the investigation indicated requirement of longer SRT, suggesting the proliferation of nitrifying bacteria and their participation in trimethoprim breakdown. The concentrations of sulfamethoxazole and trimethoprim in the second plant were significantly reduced, which was exclusively attributed to the microbial degradation. Addition of $FeCl_3$ to the aeration tank also aided in the precipitation of tetracycline, which forms complexes with Fe^{+3} ions.

Antibiotic levels were significantly reduced in the third and fourth plants, using a rotating biological contactor and pure oxygen activated sludge systems. For the third selected technology using a rotating biological contactor, a downstream ultraviolet (UV-C) disinfection unit was augmented to realize further degradation of the antibiotics. However, the effluent from UV-C indicated no effect of UV irradiation on the antibiotic compound. This investigation exhibited that degradation of MCs requires longer SRT, pure oxygen generators, and chemicals to break down antibiotics, all of which increases the cost of treatment. It also demonstrated that antibiotics are present at all stages of the WWTP, resulting in MCs in the effluents, which further establishes the need for additional removal strategies of these MCs to avoid the negative consequences. Combined physico-chemical and biological processes can also efficiently degrade MCs.

A different investigation studied the influence of pH, H_2O_2, and initial water physico-chemical characteristics on the kinetics of MC transformation utilizing UV/H_2O_2 and low-pressure UV photolysis (Baeza and Knappe 2011). The investigation compared removal efficiency of MCs in three different sources of water, namely the effluent from a WWTP, lake water from North Carolina Lake in the USA, and ultrapure water. Antibiotics such as sulfamethoxazole, sulfamethazine, sulfadiazine, and trimethoprim and EDCs such as diclofenac and bisphenol A, spiked at an initial concentration of $4 \pm 1\,\mu M$. It was found that for sulpha-medications, such as sulfamethoxazole, sulfadiazine, and sulfamethazine solution, the pH has a direct influence on photolysis rates. However, pH demonstrated no effect on hydroxyl radical oxidation rates. The neutral form of sulfamethoxazole photolyses more easily than the anionic form, and the opposite is true for sulfamethazine and sulfadiazine.

Trimethoprim's cationic form has a higher rate of hydroxyl radical oxidation under acidic condition ($pH = 3.6$) than the near neutral medium ($pH = 7.85$). The hydroxyl radical scavenging rate increases due to dissolved organic carbon, chloride, sulphate, alkalinity, and nitrate (Baeza and Knappe 2011). The percentage removal of these MCs varied from 31 to 97% in the experiments carried out with the effluent of WWTP spiked with the mixture of EDCs and antibiotics ($4 \pm 1\,\mu M$). Since the local lake water had lower scavenging ions in the water matrix, the percentage removal of spiked EDCs and antibiotics varied from 43 to 98% (Baeza and Knappe 2011). The transformation efficiencies were higher in the order of diclofenac > sulfamethoxazole > sulfamethazine > sulfadiazine > trimethoprim > bisphenol A (Baeza and Knappe 2011).

As mentioned earlier, the disadvantages of the AOPs include tendency of by-product formation that can be more toxic than the parental compounds and inadequacy of standalone AOPs. These drawbacks can be avoided by combining processes to provide the lowest possible concentrations in effluents (Sanches et al. 2013). However, biological processes, such as ASP and anaerobic SBR, are already on a field-scale and in such cases, selection of optimum operating procedure for removal of xenobiotic compounds is required (Figure 21.2). For instance, removal efficiencies of MCs, namely acyclovir, bezafibrate, atenolol, carbendazim, sulfamethoxazole, trimethoprim, DHH-carbamazepine, metoprolol, benzotriazole, 3-OH-carbamazepine, tramadol, 2-OH-carbamazepine, diclofenac, venlafaxine, oxazepam, acesulfame, primidone, iopromide, terbutryn, and carbamazepine were investigated by combining aerobic and anaerobic digestion in attached and suspended growth processes under different operating conditions under different substrate availability (Falås et al. 2016). Although the mentioned investigation was on a lab-scale ASPs of 500 mL volume, the outcome from the experiment could be translated to pilot-scale installations as evident from the outcome of further experimentation with real wastewater using the acclimatized sludge containing 31 types of xenobiotic compounds. The xenobiotic compounds such as bezafibrate, atenolol, and acyclovir were significantly removed in the ASP fed with municipal wastewater. The oxic biofilm reactor removed diuron and diclofenac, while the anaerobic reactor removed venlafaxine, diatrizoate, and tramadol (Falås et al. 2016).

An important conclusion proposed in this investigation was that the degradation of the MCs did not have any strong correlation with the sludge age (Falås et al. 2016). It was further reported that when the sludge from the synthetic wastewater fed lab-scale ASPs was used as an inoculum source for the real wastewater treatment, the microorganisms were able to degrade the MC, namely acyclovir, bezafibrate, atenolol, carbendazim, sulfamethoxazole, trimethoprim, DHH-carbamazepine, metoprolol, benzotriazole, 3-OH-carbamazepine, tramadol, 2-OH-carbamazepine, diclofenac, venlafaxine, oxazepam, acesulfame, primidone, iopromide, terbutryn, and carbamazepine, without any lag/acclimatization period. This is contrary to the popular hypothesis of acclimatization that often claims that the microbes require a certain acclimatization

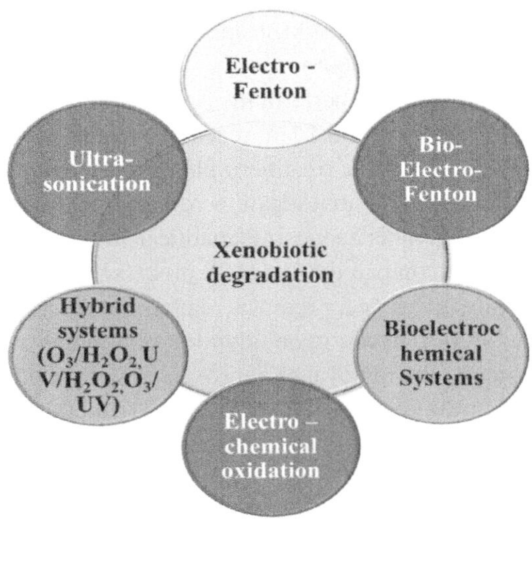

Figure 21.2 Different technologies having potential application for MC degradation.

period and higher sludge age for removal of xenobiotic compounds. These observations indicate that the lab-scale bioreactors can provide vital information in deciding the operational strategies for the field-scale reactors and emphasises the necessity of having a closer scrutiny of these technologies.

21.2 Case Studies for Lab to Field Applications

To understand the scenario of different treatment options, the lab-scale investigations are revisited for both emerging as well as established processes. The following sections evaluate the past investigations on the basis of scalability and reliability of the novel technologies and the operating methodologies for the established technologies.

21.2.1 Conventional Treatment Methods

Among the established technologies, ASP, membrane filtration, trickling filters, anaerobic filters (in some cases with conductive media), and constructed wetlands are popular choices. The ASP is the most explored technology owing to the prevalence of this technology in STPs. In a particular investigation carried out in Germany, ibuprofen, a non-steroidal anti-inflammatory drug with an initial concentration range of 4.9 to 12.3 $\mu g\ L^{-1}$, was effectively removed up to 97.3 ± 1.3% in the ASP in a full-scale WWTP. The removal of diclofenac, another similar class of compound, was also reported in this investigation; wherein, the influent and the effluent concentration of the diclofenac ranged between 1.6 and 4.4 $\mu g\ L^{-1}$, and 1.0 and 2.15 $\mu g\ L^{-1}$, respectively. In this case, the removal efficiency ranged between 0 and 60%. Similarly, removal of Clofibric acid was between 0 and 50% with an influent concentration of 0.06 to 0.15 $\mu g\ L^{-1}$. Such lower removal efficiencies reverberate the initial discussion on the compromised efficiency of the most conventional systems to remediate the MCs. Alternatives to conventional ASP, such as membrane bioreactors, can well deliver higher removal efficiency. In the same investigation, the removal of diclofenac (28–78%) and Clofibric (below limit of detection of 0.05 $\mu g\ L^{-1}$) was higher for a lab-scale MBR inoculated with the activated sludge from the full-scale WWTP (Bernhard et al. 2006).

Comparing the performance of MBR and ASP further, 31 MCs were investigated in pilot-scale MBRs (3.6 and 4.7 m^3 reactor volume, respectively) and the results were compared with ASP in a full-scale treatment plant receiving an average flow of 42,000 $m^3\ d^{-1}$ (Radjenović et al. 2009). The investigation revealed that contaminants, such as indomethacin, pravastatin, propyphenazone, mefenamic acid, gemfibrozil, and diclofenac, were poorly removed in ASP as compared to the MBR process that effectively eliminated these residuals. The MBR was also effective for removal of pharmaceuticals, such as β-blockers, ranitidine, famotidine, and erythromycin, upon adoption of a higher retention time (Radjenović et al. 2009). For example, the removal of naproxen was reported to be 70% in ASP, which was enhanced to 90% in the case of pilot-scale MBR. Trimethoprim removal in ASP was estimated as 40%, while that in the MBRs, a removal efficiency of 70% was reported.

A common mechanism of removal was assumed to be the phenomenum of sorption on the sludge in both the processes. Pharmaceutical compounds, such as ketoprofen, ibuprofen, ofloxacin, diclofenac, and azithromycin, were identified in the sewage sludge at concentrations up to 336.3, 741.1, 454.7, 380.7, and 299.6 ng/g of dry weight, respectively. While the sorption theory could well define the performance of the ASP in terms of effluent concentrations of the pharmaceuticals and quantity of pharma compounds adsorbed per gram of dried sludge, the higher removal efficiency for the MBR process was owing to different phenomena. To understand the higher removal mechanism in MBR, the hypothesis of sludge adsorption was further investigated. The sorption rate of pharmaceuticals onto sludge was observed to be high for the MBR sludge. However, the pharmaceutical compound content per gram of dried sludge was lower in MBR sludge than the ASP. A plausible explanation would be enhanced biodegradation in the MBR, hence less adsorption on the sludge of MBR operated under higher SRT compared to the ASP (Radjenović et al. 2009). Although MBR has been popularly explored for such enhanced biodegradation efficiency, the MBR technology has different limitations, which are being explored further in lab- and pilot-scale investigations. The MBR technology is a hybrid variant of biological processes that combines the microbial degradation and membrane technology together. The details of the technology are elaborated on in the next section.

Among other conventional processes, aerobic and anaerobic SBR, constructed wetlands, and trickling filters also have been explored for removal of MCs. In one investigation, ibuprofen and ketoprofen were removed in an aerobic SBR (Abu Hasan et al. 2016). In this investigation, a maximum concentration of ibuprofen and ketoprofen was kept as 39.3 and 2.01 $\mu g\ L^{-1}$, respectively. The investigation also reported three specific strains of microbes, namely *Bacillus pseudomycoides, Rhodococcus ruber*, and *Vibrio mediterranei*, resistant to the pharmaceutical compounds, which effectively participated in the biodegradation of the chosen antibiotics (Abu Hasan et al. 2016).

Interestingly, the degradation kinetics of a mixture of phenolic compounds was investigated in SBR with the phenolic derivatives as the carbon source (Tomei and Annesini 2008). The investigation indicated that the degradation of the 4-nitrophenol powered the further breakdown of the highly refractory phenol derivative 3,4-dimethyl phenol (Tomei and Annesini 2008). This investigation was carried out in a lab-scale set-up of 5 liter volume and operated at an HRT of 16 h. The investigation provided a proof of concept regarding the acclimatization and ability of the aerobic sludge developed in the biological reactor to degrade high concentrations of phenolic compounds. The experimental results were in corroboration with the Haldane kinetics model, which further validated the substrate inhibition theory.

The above discussion emphasises the importance of removal kinetics and so it is imperative to further understand the removal mechanisms of different MCs. Removal of chiral pharmaceuticals (alprenolol, bisoprolol, metoprolol, propranolol, venlafaxine, salbutamol, fluoxetine, norfluoxetine), each with an initial concentration of 1.3 $\mu g\ L^{-1}$, was investigated in an aerobic granular sludge reactor (Amorim et al. 2016). The investigation indicated that among the chiral pharma compounds selected, the removal of norfluoxetine was considerably higher as compared to the others. However, the investigation did not describe the exclusive degradation mechanism and only reported sorption of the selected pharmaceuticals in the granular sludge (Amorim et al. 2016).

In a different investigation, 2-fluorophenol was degraded in an aerobic sequencing batch type reactor using acetate as a co-substrate (Duque et al. 2011). The complete mineralization of the MC was indicated by fluoride release in stoichiometric proportions as predicted. The highest concentration of 2-fluoorophenol degraded in the reactor was 0.44 mM using a higher abundance of acetate (5.9 mM) and two specific aerobic strains of microbes were identified, which were associated with the degradation of the 2-fluorophenol. The investigation indicated that higher concentrations of such toxic compounds could be degraded in a co-substrate augmented condition, with a sludge age showing prolonged exposure to such toxic contaminant laden wastewater. The investigations described above are lab/small pilot-scale prototypes that revealed important information regarding the co-substrate availability/necessity, sludge age conditions, and degradation/removal mechanisms of chosen MCs. Such information is vital toward the field-scale implementation of the SBR for concomitant removal of MCs as well as conventional pollutants (Duque et al. 2011).

In addition to suspended growth processes, attached growth conventional processes, such as tricking filter and MBBR, are also explored with lab-scale prototypes. Among the attached growth processes, electrically conductive biofilters are hybrid variants that make use of the electroactive microorganism's exoelectrogenic behavior and the symbiotic communities of other bacteria. Moreover, these mechanisms are responsible for decontaminating MCs via direct interspecies electron transfer and extracellular electron transfer pathways to provide the necessary energy for cleaving the complicated structures. This is dealt with in the next section with a special emphasis on the type of configurations and media used. In addition to the biological processes, the AOP variants, such as ozonation, photocatlytic investigations, and ultra-violet irradiation, also have been explored for removal of MCs.

As discussed previously, ozonation was primarily employed as a disinfection technology for water treatment schemes and for sewage treatment plants targeting reuse (Xu et al. 2002). Additionally, industrial wastewater containing a high fraction of refractory organics also has been implementing ozonation as a standard AOP prior to effluent discharge/reuse (Rice 1996; Arslan-Alaton and Alaton 2007). As the understanding of the MCs in the water matrix has been enhanced, probing technologies relevant to the degradation of such contaminants has become a popular topic. Ozonation is one of such technologies that is being implemented for this use (Sumikura et al. 2007). Sulfamethaoxazole drug, with an initial concentration of 200 mg L^{-1} was removed by 60 min ozonation in a lab-scale set-up (1.2 L volume). The ozone treatment did not completely mineralize the compound, but enhanced the biodegradability index (BOD_5/COD) ratio from 0 to 0.28 (Dantas et al. 2008). The toxicity test in the same investigation indicated low acute toxicity levels.

Application of ozonation to treat 17β estradiol has also been demonstrated in a few investigations. For example, the higher effectiveness of ozonation is highlighted in an investigation, wherein similar estrogenicity reduction was achieved by 10 min ozonation as compared to chlorination that took 120 min (Alum et al. 2004). The same investigation also reported complete mineralization of bisphenol A, 17α-ethynyl estradiol during the ozonation. Ozonation has also been effective in the removal of recalcitrant surfactants (Ikehata and El-Din 2010). The majority of investigations have targeted removal of linear alky benzene sulfonate and alkyl phenol ethoxyates. In a particular investigation, LAS with an initial concentration of 4.63 to 5.30 mg L^{-1} was degraded to the extent of 67 to 90% with a total applied dose of 15.9 to 16.7 mg-ozone L^{-1} and a contact time of 30 min. However, surfactants are typical refractory compounds that occur at the ppm level, which implies that the reaction kinetics defined in the case of these compounds might not be applicable for mapping the degradation of MCs having lower trace concentrations.

An investigation conducted by Mathon et al. (2017), where 12 xenobiotic compounds were spiked in treated secondary effluent, demonstrated the oxidation rate kinetics determination, classification of the xenobiotics in terms of oxidation feasibility, and the optimum ozone dose in terms of per gram of DOC removed. An ozone dose of 0.2 to 0.4 g-O_3/g DOC was estimated to be effective for the removal of the targeted MCs. Furthermore, the effect of pH and different components of the water matrix on ozonation of triclosan as a model pollutant (1–5 mg L^{-1}) was examined in a lab-scale ozonation system (borosilicate glass cylinder of 1.2 L volume) for surface water collected from a lake in Turkey (Orhon et al. 2017). The investigation aimed to determine the effect of different scavenging moieties during ozonation of pollutants. It could be observed in this investigation that with a contact time of 10 min and transferred ozone concentration of 5 mg L^{-1}, triclosan and its by-products were completely mineralized. It was also emphasised in the investigation that the natural organic matter present in the surface water produced a scavenging effect on the generated hydroxyl radical (Orhon et al. 2017).

In a different investigation, the effect of the water matrix on triclosan and three other MCs were further elaborated on (Hernández-Leal et al. 2011). The research focused on the degradation of bisphenol-A, hexylcinnamic aldehyde, 4-methylbenzylidene-camphor (4MBC), benzophenone-3 (BP3), triclosan, galaxolide, and ethylhexyl methoxycinnamate within a concentration range of 100 to 1600 μg L^{-1}. Similar to Orhan et al. (2017), this investigation also demonstrated a negative effect of the water matrix moieties on the ozone and subsequent hydroxyl radical generation. The removal of the compounds was observed to be >99% in deionized water, while the removal was attenuated for experiments conducted with aerobically treated gray water spiked with MCs (Hernández-Leal et al. 2011). The different ozonation experiments indicate that the technique of ozonation is an effective technique toward the degradation of high- as well as low-strength of refractory compounds. The application of physical processes, such as UV radiation or ultrasonic vibrations, for removal of xenobiotic compounds from secondary treated sewage, are not as effective as compared to standalone chemical processes (Miklos et al. 2018). However, the application of these physical techniques in conjunction with the chemical species activation is an effective technique, which is described in the next section.

21.2.2 Hybrid Treatment Methods

21.2.2.1 Hybrid Biochemical Processes

The hybrid technologies can be classified into biological-based processes and physico-chemical processes, and in some cases hydraulically connected multistage bio-physico-chemical or physico-chemical-biological processes. Among the bio-based hybrid processes, the membrane bioreactor is one of the well tested technologies that have been explored in different configurations in past investigations. Riso-Miguel et al. (2021) investigated the influence of concentrations and hydraulic retention time (HRT) on eliminating pharmaceutical chemicals, such as acetaminophen, fluoxetine, metoprolol, diclofenac, metformin, and carbamazepine, in a membrane bioreactor inoculated with activated sludge from municipal WWTPs. The concentrations in the influent were raised from 4–800 nM at regular intervals of every 2 weeks after an acclimatization period of 180 days at an HRT of 3.5 days and SRT of 15 days in the first portion. It was discovered that the removal percentages did not align with influent concentrations and have remained relatively stable. The MC acetaminophen was completely removed in the MBR (100% removal), while the removal efficiency of the other MCs, such as metoprolol and diclofenac, varied between 20 and 40%.

For the pharmaceutical compound carbamazepine, the removal efficiency unusually varied between 0 and 20% owing to the reconversion of the undetectable conjugate form of carbamazepine after biological treatment. The removal efficiency of metformin varied between 10 and 30%, and that of fluoxetine varied between 25 and 70%. Although the higher loading rates of the antibiotics exhibited a proportional increase in the removal rates, percentage removal was not proportionately enhanced. Examination of the compounds on an individual basis revealed larger solid–water distribution coefficient ($k_d > 0.5\,L\,g^{-1}$) and lesser Henry's law constants resulted in fluoxetine removal as a sorption mechanism, with some reduction also occurring via nitrifiers and heterotrophic bacteria. Metoprolol and diclofenac removal was facilitated by heterotrophic bacteria. Negative removals of carbamazepine and metformin compounds were reported owing to the reconversion of the undetectable conjugated forms of the same to the original detectable form post-biochemical reactions (Rios-Miguel et al. 2021). In response to increased pharmaceutical concentrations, the microbial community structure also changed. The predominance of *Nitrospira* and *Planctomycetes* declined, while *Bacteroidetes* and *Acidobacteria* soared.

In the second part of the investigation, the HRT was varied to 1, 3, and 5 days to monitor the effect of the varying contact period of the MCs with the microbial community. During the experiments pertaining to this HRT variation, the influent concentration was kept at 800 nM, and other functional parameters stayed unaltered (Rios-Miguel et al. 2021). It was observed that the HRT had no impact on the removal percentages of acetaminophen, carbamazepine, fluoxetine, diclofenac, and metoprolol. Metformin was the primary constituent eliminated by expanding the HRT to 5 days, recommending that the reaction time was liable for the deficient expulsion of metformin. Since the biomass concentration increased by three times at HRT of 1 day, it brought about higher removal percentages of metformin of ~80% in the reactor, which indicates that the metformin removal is a function of biomass concentration and the HRT.

Tadkaew et al. (2010) examined the impact of pH of mixed liquor on the expulsion of MCs, ionizable compounds such as sulfamethoxazole, ketoprofen, ibuprofen, and diclofenac, and non-ionizable components such as bisphenol A and carbamazepine in the reactor and evaluated the performance of the reactor under different pH for removal of MCs. When the pH of the reactor is acidic or basic, the biological performance of the MBR is affected, resulting in reduced removal efficiencies of TOC or TN. The TOC and TN removal drastically reduced to 60 and 90% and marginally reduced to 80 and 90% at pH above 8.0 and below 5.0, respectively. The optimum performance was observed under a pH of 7.0, wherein the TOC and TN removal was ascertained to be 90 and 100% (Tadkaew et al. 2010).

It was revealed that ionizable compounds exhibited a substantial removal efficiency at an acidic pH of 5.0, which is linked to their speciation behavior. Most of these chemicals are hydrophobic at this pH, resulting in adsorption on the sludge and increased removal efficiency compared to the mixed liquors with acidic pH. This might be attributable to the mixed liquor's physico-chemical features rather than the biological conditions in the reactor. Non-ionizable chemical removal efficiencies, on the other hand, are independent of the pH of the mixed liquor and attributed to the biodegradability of chemicals (Tadkaew et al. 2010).

The performance of the pilot-scale MBR loaded with varied amoxicillin (AMX) concentrations of 5, 10, 20, 40, 70, and 100 mg L^{-1} was evaluated (Rezaei et al. 2020). The reactor attained a steady state after 70 days of operation under an HRT of 12 h and a flow rate of 1.6 L h^{-1}. Amoxicillin removal efficiency was reported to be up to 95% for influent concentrations in the range of 5 to 20 mg L^{-1}, which dropped to only 72% for influent concentration of 40 mg L^{-1}. However, removal efficiencies of 63 and 61% were reached at doses of 70 and 100 mg L^{-1}, respectively. This indicated the inhibitory effect of the chosen antibiotic on the microbes present and hence affected the performance of MBR. Although the performance of the MBR was reduced to a certain extent at higher concentrations, the biomass exhibited resilience toward the high concentration of antibiotic indicating the applicability of the MBR as a pre-treatment to AOPs during treatment of hospital and pharmaceutical wastewater.

The performance of the MBR in terms of COD removal efficiency for different concentrations of AMX was >95%, except for 70 and 100 mg L^{-1} for which the COD removal was roughly 94 and 92%, respectively. The nitrification efficiency was initially 53.6% and declined with time and reached its lowest level of 29.6% at 24 h after initial shock loading of the reactor with an AMX concentration of 100 mg L^{-1} from an initial total nitrogen concentration of 35 mg L^{-1} (Rezaei et al. 2020). After AMX entered the bioreactor, the average MLSS growth rate significantly dropped. Furthermore, the sludge volume index increased with AMX shock, resulting in a reduction in sedimentation and the drop in the specific oxygen uptake rate indicating antibiotic toxicity for microorganisms and a decrease in oxygen consumption.

Another hybrid biological process that takes advantage of the bio-electrochemical reaction pathways is the electrically conductive biofilters. The primary use of microbial electrochemical framework assisted biofilter was shown by Aguirre-Sierra et al. (2016) utilizing coke as a conductive media. At 4-fold organic loading rates, the coke-based horizontal flow anaerobic biofilter demonstrated equivalent pollutant removal efficiency to the gravel-based biofilter. Bacterial dynamics indicated distinct speciation of the microbial consortia in the two biofilters, suggesting that the conductivity of media material has an active role in the proliferation of the microbial species (Aguirre-Sierra et al. 2016). The experiment exhibited that replacing the inert medium with electrically conductive media in a biofilter improved pollutant removal efficiency. The presence of conductive media promotes faster electron transfer via the exoelectrogenic pathway as well as facilitates direct inter-species electron transfer. Both these electron transfer mechanisms are reported to boost metabolic activities and degradation of contaminants under *ex-situ* as well as *in-situ* conditions.

The enhancement of performance by introduction of conductive media in biofilters was further demonstrated in an experiment involving degradation of pharmaceutical compounds (Pun et al. 2019). Implementation of an electrically conductive biofilter resulted in 90% removal of 13 selected pharmaceuticals within an HRT of 24 h. The robustness of the biofilter was evident by considering the fact that the TOC removal efficiency remained unaltered in both phases of operation, namely with and without the presence of pharmaceuticals in the influent wastewater. The robust performance of the biofilter was achieved in the entire considered range of organic loading rate (3–14 g-TOC m^{-2} d^{-1}) and there was no discernible change in the elimination of organic materials in the presence or absence of pharmaceuticals (Pun et al. 2019).

The cited investigation demonstrated the capacity of electrically conductive biofilters to accommodate higher loading of emerging organic pollutants without impacting the biofilter's overall treatment effectiveness. Tertiary treatment systems are expensive, hence employing such biofilters for removal of complex organic MCs during secondary treatment can help to lower the cost of tertiary treatment. Furthermore, biological treatments are vulnerable to complex organic MCs. The better performance of electrically conductive biofilters in the presence of pharmaceuticals suggests that they may withstand shock loading better than conventional biological treatments.

Conductive coke granules as media were investigated in downflow biofilters, which have prominent aerobic conditions due to passive aeration (Aguirre-Sierra et al. 2020). The biofilter using coke as media delivered more than 90% of removal efficiency at a hydraulic loading rate of 0.28 to 0.56 m^3 m^{-2} d^{-1}, and 97 and 81% of ammonia removal efficiency at a hydraulic loading rate of 0.28 and 1.12 m^3 m^{-2} d^{-1}, respectively. The aerobic conditions in the downflow filter created a favorable environment for the growth of nitrifiers, which could be proved by the removal of ammonia and microbial dynamics (Aguirre-Sierra et al. 2020). Overall, the downflow type aerobic biofilter outperformed conventional biofilters using gravel as the inert material for COD and nitrogen removal.

21.2.2.2 Hybrid Advanced Oxidation Processes

Photo-catalysis is a novel technology that employs the electro-magnetic radiation in the ultraviolet or visible light spectra to excite photo-catalysts that are capable of generating free radicals in excited state. In a particular investigation, application of a TiO_2-based photo-catalyst for removal of MCs from secondary treated sewage was evaluated (Choi et al. 2014). The investigation demonstrated that TiO_2/UV photo-catalyst combination enhanced the rate kinetics to 0.1154 ± 0.0135 min^{-1} as compared to standalone UV radiation (0.0032 ± 0.0004 min^{-1}) for acetaminophen. The investigation further established the negative correlation between the dissolved organic matter content and degradation rate kinetics. The ·OH scavenging action of the dissolved organic matter reduced the performance of the TiO_2/UV-A, as well as TiO_2/UV-C systems.

The effects of the scavenging species on the degradation rate kinetics was further demonstrated in a different investigation employing carbon nanotube-TiO_2 (CNT-TiO_2) composites (Awfa et al. 2020). The effect of natural organic matter (NOM) on the removal of the chosen compound carbamazepine was modeled by choosing five different types of NOM comprising of secondary treated wastewater, river water, and three surrogates of NOM, namely reverse osmosis isolate, humic acid, and fluvic acid. The results indicated that the NOM with higher molecular weight and low carbon/oxygen content ratio had the severe scavenging effect on the reactive oxygen species. In the same investigation, comparing photolysis and photo-catalysis, carbamazepine was rapidly degraded with degradation rate kinetic values of 7.0 ± 0.08 × 10^{-2} min^{-1} for UV-C/CNT-TiO_2, 4.75 ± 0.07 × 10^{-2} min^{-1} for UV-C/TiO_2 combination, 3.95 ± 0.07 × 10^{-2} min^{-1} for solar/TiO_2, and 5.5 ± 0.06 × 10^{-2} min^{-1} for solar/CNT-TiO_2 (Awfa et al. 2020). It can be observed from the above values of degradation rate constants that the dissociation of carbamazepine was higher in the case of photo-catalyst combinations as compared to photolysis experiments with standalone UV lamps. It can be further observed that the photo-catalytic activity of the CNT-TiO_2, that incorporates innovative carbon structure with high porosity and active functional sites, enhanced the photocatalytic activity as compared to base photo-catalyst TiO_2.

The efficacy of carbon-based catalysts has been well explored in the form of carbon nitride photo-catalysts that have a unique layered graphitic structure with uniform spatially distributed graphitic and pyridinic nitrogen groups. The band gap of graphitic carbon nitride is 2.70 eV corresponding to a wavelength of 460 nm (Wang et al. 2012). Graphitic carbon nitride-based photo-catalysts demonstrated promising performance owing to their highly tuneable structure and excellent electronic properties (Wang et al. 2017). Application of exfoliated graphitic carbon nitride was evaluated for complete mineralization of nine MCs in simulated wastewater samples in a lab-scale reactor at 417 nm using a light emitting diode as light source. The selected MCs, namely, isoproturon, diclofenac, fluoxetine, carbamazepine, atenolol, bezafibrate, tramadol, venlafaxine, and clopidogrel, were degraded with graphitic carbon nitride immobilized over glass rings within a residence time of 10 min (Amorim et al. 2016). The degradation rate kinetics were determined for each of the compounds, such as atenolol (initial concentration, 12.5 ± 1.2 ng L^{-1}), bezafibrate (38.7 ± 6.0 ng L^{-1}), carbamazepine (763 ± 18 ng L^{-1}), clopidogrel (93.2 ± 10.7 ng L^{-1}), diclofenac (1102 ± 31 ng L^{-1}), fluoxetine (21.7 ± 4.5 ng L^{-1}), isoproturon (84.6 ± 1.2 ng L^{-1}), tramadol (3930 ± 244 ng L^{-1}), and venlafaxine (249 ± 1.2 ng L^{-1}) and the values were 74.3 ± 2.3 × 10^{-2} min^{-1}, 51.9 ± 1.2 × 10^{-2} min^{-1}, 238 ± 1 × 10^{-2} min^{-1}, 108 ± 2 × 10^{-2} min^{-1}, 90.6 ± 2.6 × 10^{-2} min^{-1}, 27.4 ± 4.5 × 10^{-2} min^{-1}, 110 ± 8 × 10^{-2} min^{-1}, 44.6 ± 3.6 × 10^{-2} min^{-1}, and 41.3 ± 1.2 × 10^{-2} min^{-1}, respectively. The investigation demonstrated successful degradation of chosen MCs and also catered to the problem of photo-catalyst wash-off during batch experiments (Amorim et al. 2016).

The application of graphitic carbon nitride as a Fenton-like reagent has become a focus that has been further experimented on in the past decade (Yang et al. 2020). The graphitic carbon nitride facilitates H_2O_2 production upon photo-excitation or as an electrocatalyst, while the Fe source catalyses the dissociation of the produced H_2O_2 to produce reactive oxygen species. The application of such photocatalytic degradation of MCs using graphitic carbon nitride-based photo-catalysts has been elaborated on in a comprehensive review described in the past, which may be referred to for understanding the hybrid photo-catalytic process (Yang et al. 2020).

A different variant of Fenton, the electro-Fenton, is a promising technology that couples the anodic electrochemical oxidation with cathodic H_2O_2 generation and subsequent dissociation by an Fe source. The Fe can be sourced from the catholyte or can be coated onto the cathode surface. The advantage of electro-Fenton is that the Fe^{+2} consumed in the catalytic dissociation of H_2O_2 is regenerated by an accepting electron from the cathode surface (Sathe et al. 2022). Moxifloxacin with an initial concentration of 0.15 mM was degraded in an electro-Fenton lab-scale set-up within a retention period of 60, 30, 25, and 11 min for 60, 100, 300, and 400 mA imposed current values, respectively (Yahya et al. 2017). The degradation kinetics for moxifloxacin degradation followed pseudo-first order kinetics and it was observed to increase from 0.05 to 0.64 min^{-1} when the current was increased from 60 to 400 mA. However, the degradation rate kinetics was estimated to decrease when the current was further increased to 500 mA. The reason behind the decrease of the rate kinetics at higher current values was owing to the four-electron reduction of oxygen, hydrogen evolution reaction at the cathode, and oxidation of H_2O_2 (Yahya et al. 2017).

A step ahead in the direction of electro-Fenton is photo-electro-Fenton, wherein a photo assisted anode is coupled with an electro-Fenton reaction at the cathode. Degradation of terbutryn (63%), clorfenvinphos (57%), and diclofenac (90%), with initial concentration of 500 μg L^{-1} in a photo-electrochemical oxidation cell using TiO_2 nanotubes (TiO_2-NT) as anode and carbon felt as cathode, concluded that the hybrid photo-electrochemical oxidation system was better than photocatalysis alone for degradation of xenobiotics (Salmerón et al. 2021b). Similar to other AOPs, the effect of dissolved organic matter was also profound in this work (Salmerón et al. 2021b).

However, the promising performance of the lab-scale set-ups becomes highly reduced in pilot-scale investigations of electro-Fenton processes. In a pilot-scale investigation, the performance of the electrodes was drastically reduced within a short interval of operational cycle in terms of H_2O_2 generation, which is a vital step in electro-Fenton treatment (Salmerón et al. 2021a). For a fresh carbon-polytetrafluoroethylene gas diffusion electrode (carbon-PTFE-GDE), the H_2O_2 production was estimated as 43 mg L^{-1} in 30 min and Coulombic efficiency (CE) was ascertained as 46%. The deposition of salts on the cathode surface led to a decrease in H_2O_2 electro-generation to 16 mg L^{-1} after 30 min, with a corresponding lowering of CE to 21%. Furthermore, in the same investigation, the electrode regeneration effect was also demonstrated (Salmerón et al. 2021a). It was found that the performance of the regenerated carbon-PTFE-GDE electrode was lowered further, contrary to the initial hypothesis. In the case of a regenerated electrode, the accumulated H_2O_2 concentration for an interval of 30 min was found to be 12 mg L^{-1} with a CE of 15% (Salmerón et al. 2021a). The bio-electro-Fenton (BEF) process is another innovative derivative of the electro-Fenton, which has been extensively explored in past investigations (Sathe et al. 2022). The BEF process has been elaborated on in Chapter 19 and hence not included here.

The different technologies described in this current section are selected from different domains that deal with different forms of MCs. While the described investigations are not exhaustive in nature, they represent major kinds of techniques used for the degradation of recalcitrant compounds. In addition to these technologies, certain other forms of removal strategies, such as adsorption using innovative and highly activated adsorbents, and biochemical and/or physico-chemical processes in the form of bio-electrochemical systems, are also well explored in the domain of MC removal (Table 21.1). The bio-electrochemical systems are detailed in Chapter 18 and a few representative investigations are referred to in Table 21.1.

Table 21.1 Performance of different technologies for MC removal.

Sl. No	Technology Used	Xenobiotic	Initial Concentration	Removal Percentage	References
1	Adsorption on to Carbon cloth	Ametryn	6.5×10^{-5} M	98.52	Ayranci and Hoda (2005)
		Aldicarb		33.85	
		Dinoseb		86.92	
		Diuron		49.23	
	Adsorption on to Maize straw decorated with sulphide	Tylosin	20 mg L^{-1}	~80	Guo et al. (2018)
	Adsorption of Triazine pesticides on magnetically recoverable Fe_3O_4/graphene nanocomposite	Ametryn	10 mg L^{-1}	93.61	Boruah et al. (2017)
		Prometryn		91.34	
		Simazine		88.55	
		Simeton		81.22	
		Atrazine		75.24	
	Air cathode single chambered MFC	Penicillin	50 mg L^{-1}	98	Wen et al. (2011)
2	Double chambered MFC	Chloramphenicol	50 mg L^{-1}	84	Zhang et al. (2017)
	Single chambered air cathode MFC	Neomycin sulphate	20 mg L^{-1}	54	Catal et al. (2018)
3	Bio-electro-Fenton (BEF) system	Diclofenac	500 µg L^{-1}	100	Zou et al. (2020)
		Ibuprofen			
		Ketoprofen			
		Naproxen			
		Carbamazepine			
		Clorfibric acid			
		Vanillic acid	20 mg L^{-1}	100	B. Li et al. (2021)
		Syringic acid		94.32	
		Hydroxybenzoic acid		100	
		Sulfamethoxazole	11 mg L^{-1}	94.66	S. Li et al. (2020)
4	UV	2,4-dichlorophenoxyacetic acid	100 mg L^{-1}	~66	Adak et al. (2019)
	UV-H_2O_2			97	
5	Electrochemical oxidation	Thiamethoxam	2 mg L^{-1}	76	Lebik-Elhadi et al. (2018)
		Diclofenac	175 mg L^{-1}	93	Brillas et al. (2010)
		Bisphenol A	20 mg L^{-1}	100	Murugananthan et al. (2008)

21.3 Future Outlook

This chapter describes the implementation strategies for the different technologies for removal of MCs from wastewaters. The strategies described encompass different mechanisms for removal of the compounds including biodegradation, physi-sorption, and/or chemi-sorption onto sludge and coagulants, oxidation by reactive oxygen and chlorine species, and photolysis and ultrasonic cleavage of refractory MCs, to name a few. Typically, the biological processes can be tailored to suit MC removal by different strategies, such as provision of co-substrate in case of high-strength wastewaters, increased sludge age, recirculation strategies, and introducing conductive media for augmenting the available energy yield during metabolism. Among the different strategies, the implementation of anaerobic biological processes coupled with a membrane bioreactor or by introducing conductive media are most promising and can be further explored for establishing standard operating protocols. However, owing to demographic variations at a global scale and also the differential nature of scavenging ions due to different geographical locations, the decision on the type of strategy to be implemented should be at least region specific, if not case specific.

Similar to biological processes, the AOPs and the physico-chemical transformations also beget similar strategies of test run prior to real implementation. The lab-scale investigations described in this chapter would thus provide a template for understanding the different factors that needs to be considered in future pilot and consequent field-scale implementations. Furthermore, the application of AOPs should also be integrated with downstream biological and/or adsorption processes. One innovative approach would be to implement multistage activated carbon filter units including a bacteriologically active carbon (BAC) followed by a granular activated carbon (GAC). Such a two-stage filter would ensure adsorption followed by biodegradation of biodegradable by-products of the MCs and adsorption of the MCs in BAC and GAC, respectively. However, this is just one of the strategies that can be implemented to reduce the toxicity of the MCs further.

A prominent area of research for the future would be the identification of the MCs and detection of the treatment by-products of such MCs. Understanding fate and transport of the MCs through analytical methods is an important step owing to the structural and chemical composition variation of the MCs. Such variation leads to uncertainty in analysis and often makes it difficult to identify the class of the compound, thereby reducing the effectiveness of the monitoring methods. Hence, further research should also focus on the detection of such compounds in real wastewater.

21.4 Conclusions

This chapter gives an overview of the different novel lab-scale applications as well as explores the operating mechanisms of the conventional processes that are to be adopted for removal of microconstituents (MCs). The removal of MCs from wastewater is facilitated by different strategies and technologies. As evident from the investigations, a combination of multistage treatment is necessary for effective removal of MCs. The conclusive remarks pertaining to the different described methods will provide vital clues to the scientific community as to the selectivity of these technologies and aid in paving the way for future research. The hybrid processes exhibit great potential and can be further adopted in pilot investigations after successful testing in lab set-ups, as described in this chapter.

References

Abu Hasan, H., Sheikh Abdullah, S.R., Al-Attabi, A.W.N. et al. (2016). Removal of ibuprofen, ketoprofen, COD and nitrogen compounds from pharmaceutical wastewater using aerobic suspension-sequencing batch reactor (ASSBR). *Separation and Purification Technology* 157: 215–221.

Adak, A., Das, I., Mondal, B. et al. (2019). Degradation of 2,4-dichlorophenoxyacetic acid by UV 253.7 and UV-H_2O_2: reaction kinetics and effects of interfering substances. *Emerging Contaminants* 5: 53–60.

Aguirre-Sierra, A., Bacchetti-De Gregoris, T., Berná, A. et al. (2016). Microbial electrochemical systems outperform fixed-bed biofilters in cleaning up urban wastewater. *Environmental Science: Water Research and Technology* 2 (6): 984–993.

Aguirre-Sierra, A., Bacchetti-De Gregoris, T., Salas, J.J. et al. (2020). A new concept in constructed wetlands: assessment of aerobic electroconductive biofilters. *Environmental Science: Water Research and Technology* 6 (5): 1312–1323.

Alum, A., Yoon, Y., Westerhoff, P., and Abbaszadegan, M. (2004). Oxidation of bisphenol A, 17β-estradiol, and 17α-ethynyl estradiol and byproduct estrogenicity. *Environmental Toxicology* 19 (3): 257–264.

Amorim, C.L., Moreira, I.S., Ribeiro, A.R. et al. (2016). Treatment of a simulated wastewater amended with a chiral pharmaceuticals mixture by an aerobic granular sludge sequencing batch reactor. *International Biodeterioration & Biodegradation* 115: 277–285.

Arslan-Alaton, I. and Alaton, I. (2007). Degradation of xenobiotics originating from the textile preparation, dyeing, and finishing industry using ozonation and advanced oxidation. *Ecotoxicology and Environmental Safety* 68 (1): 98–107.

Awfa, D., Ateia, M., Fujii, M., and Yoshimura, C. (2020). Photocatalytic degradation of organic micropollutants: inhibition mechanisms by different fractions of natural organic matter. *Water Research* 174: 115643.

Ayranci, E. and Hoda, N. (2005). Adsorption kinetics and isotherms of pesticides onto activated carbon-cloth. *Chemosphere* 60 (11): 1600–1607.

Baeza, C. and Knappe, D.R.U. (2011). Transformation kinetics of biochemically active compounds in low-pressure UV photolysis and UV/H_2O_2 advanced oxidation processes. *Water Research* 45 (15): 4531–4543.

Batt, A.L., Kim, S., and Aga, D.S. (2007). Comparison of the occurrence of antibiotics in four full-scale wastewater treatment plants with varying designs and operations. *Chemosphere* 68 (3): 428–435.

Bernhard, M., Müller, J., and Knepper, T.P. (2006). Biodegradation of persistent polar pollutants in wastewater: comparison of an optimised lab-scale membrane bioreactor and activated sludge treatment. *Water Research* 40 (18): 3419–3428.

Boruah, P.K., Sharma, B., Hussain, N., and Das, M.R. (2017). Magnetically recoverable Fe_3O_4/graphene nanocomposite towards efficient removal of triazine pesticides from aqueous solution: investigation of the adsorption phenomenon and specific ion effect. *Chemosphere* 168: 1058–1067.

Brillas, E., Garcia-Segura, S., Skoumal, M., and Arias, C. (2010). Electrochemical incineration of diclofenac in neutral aqueous medium by anodic oxidation using Pt and boron-doped diamond anodes. *Chemosphere* 79 (6): 605–612.

Calisto, V. and Esteves, V.I. (2009). Psychiatric pharmaceuticals in the environment. *Chemosphere* 77 (10): 1257–1274.

Catal, T., Yavaser, S., Enisoglu-Atalay, V. et al. (2018). Monitoring of neomycin sulfate antibiotic in microbial fuel cells. *Bioresource Technology* 268: 116–120.

Choi, J., Lee, H., Choi, Y. et al. (2014). Heterogeneous photocatalytic treatment of pharmaceutical micropollutants: effects of wastewater effluent matrix and catalyst modifications. *Applied Catalysis B: Environmental* 147: 8–16.

Dantas, R.F., Contreras, S., Sans, C., and Esplugas, S. (2008). Sulfamethoxazole abatement by means of ozonation. *Journal of Hazardous Materials* 150 (3): 790–794.

Duque, A.F., Bessa, V.S., Carvalho, M.F. et al. (2011). 2-Fluorophenol degradation by aerobic granular sludge in a sequencing batch reactor. *Water Research* 45 (20): 6745–6752.

Falås, P., Wick, A., Castronovo, S. et al. (2016). Tracing the limits of organic micropollutant removal in biological wastewater treatment. *Water Research* 95: 240–249.

Fatta-Kassinos, D., Vasquez, M.I., and Kümmerer, K. (2011). Transformation products of pharmaceuticals in surface waters and wastewater formed during photolysis and advanced oxidation processes: degradation, elucidation of byproducts and assessment of their biological potency. *Chemosphere* 85 (5): 693–709.

Fijalkowski, K., Rorat, A., Grobelak, A., and Kacprzak, M.J. (2017). The presence of contaminations in sewage sludge: the current situation. *Journal of Environmental Management* 203: 1126–1136.

Giwa, A., Yusuf, A., Balogun, H.A. et al. (2021). Recent advances in advanced oxidation processes for removal of contaminants from water: a comprehensive review. *Process Safety and Environmental Protection* 146: 220–256.

Guo, X., Yin, Y., Yang, C., and Dang, Z. (2018). Maize straw decorated with sulfide for tylosin removal from the water. *Ecotoxicology and Environmental Safety* 152: 16–23.

Hernández-Leal, L., Temmink, H., Zeeman, G., and Buisman, C.J.N. (2011). Removal of micropollutants from aerobically treated grey water via ozone and activated carbon. *Water Research* 45 (9): 2887–2896.

Ike, I.A., Karanfil, T., Cho, J., and Hur, J. (2019). Oxidation byproducts from the degradation of dissolved organic matter by advanced oxidation processes: a critical review. *Water Research* 164: 114929.

Ikehata, K. and El-Din, M.G. (2010). Degradation of recalcitrant surfactants in wastewater by ozonation and advanced oxidation processes: a review. *Ozone Science & Engineering* 26 (4): 327–343.

Khanzada, N.K., Farid, M.U., Kharraz, J.A. et al. (2020). Removal of organic micropollutants using advanced membrane-based water and wastewater treatment: a review. *Journal of Membrane Science* 598: 117672.

Kümmerer, K. (2009). Antibiotics in the aquatic environment: a review. Part I: *Chemosphere* 75 (4): 417–434.

Lebik-Elhadi, H., Frontistis, Z., Ait-Amar, H. et al. (2018). Electrochemical oxidation of pesticide thiamethoxam on boron doped diamond anode: role of operating parameters and matrix effect. *Process Safety and Environmental Protection* 116: 535–541.

Li, B., Sun, J.D., Tang, C. et al. (2021). A novel core-shell Fe@Co nanoparticles uniformly modified graphite felt cathode (Fe@Co/GF) for efficient bio-electro-Fenton degradation of phenolic compounds. *Science of the Total Environment* 760: 143415.

Li, S., Hua, T., Yuan, C.S. et al. (2020). Degradation pathways, microbial community and electricity properties analysis of antibiotic sulfamethoxazole by bio-electro-Fenton system. *Bioresource Technology* 298: 122501.

Mathon, B., Coquery, M., Miege, C. et al. (2017). Removal efficiencies and kinetic rate constants of xenobiotics by ozonation in tertiary treatment. *Water Science and Technology* 75 (12): 2737–2746.

Miklos, D.B., Remy, C., Jekel, M. et al. (2018). Evaluation of advanced oxidation processes for water and wastewater treatment: a critical review. *Water Research* 139: 118–131.

Murugananthan, M., Yoshihara, S., Rakuma, T., and Shirakashi, T. (2008). Mineralization of bisphenol A (BPA) by anodic oxidation with boron-doped diamond (BDD) electrode. *Journal of Hazardous Materials* 154 (1–3): 213–220.

Orhon, K.B., Orhon, A.K., Dilek, F.B., and Yetis, U. (2017). Triclosan removal from surface water by ozonation: kinetics and by-products formation. *Journal of Environmental Management* 204: 327–336.

Postigo, C., Lopez de Alda, M.J., and Barceló, D. (2008). Analysis of drugs of abuse and their human metabolites in water by LC-MS2: a non-intrusive tool for drug abuse estimation at the community level. *Trends in Analytical Chemistry* 27 (11): 1053–1069.

Pun, Á., Boltes, K., Letón, P., and Esteve-Nuñez, A. (2019). Detoxification of wastewater containing pharmaceuticals using horizontal flow bioelectrochemical filter. *Bioresource Technology Reports* 7: 100296.

Radjenović, J., Petrović, M., and Barceló, D. (2009). Fate and distribution of pharmaceuticals in wastewater and sewage sludge of the conventional activated sludge (CAS) and advanced membrane bioreactor (MBR) treatment. *Water Research* 43 (3): 831–841.

Rezaei, J., Bouteh, E., Torabian, A., and Ghulami, H. (2020). Evaluation of membrane bioreactor-hollow fiber (MBR-HF) pilot performance in the treatment of wastewater facing with different concentrations of amoxicillin (AMX) as shock loads. *Journal of Environmental Chemical Engineering* 8 (4): 103944.

Rice, R.G. (1996). Applications of ozone for industrial wastewater treatment: a review. *Ozone: Science and Engineering* 18 (6): 477–515.

Rios-Miguel, A.B., Jetten, M.S.M., and Welte, C.U. (2021). Effect of concentration and hydraulic reaction time on the removal of pharmaceutical compounds in a membrane bioreactor inoculated with activated sludge. *Microbial Biotechnology* 14 (4): 1707–1721.

Salmerón, I., Oller, I., Plakas, K.V., and Malato, S. (2021a). Carbon-based cathodes degradation during electro-Fenton treatment at pilot scale: changes in H_2O_2 electrogeneration. *Chemosphere* 275: 129962.

Salmerón, I., Sharma, P.K., Polo-López, M.I. et al. (2021b). Electrochemically assisted photocatalysis for the simultaneous degradation of organic micro-contaminants and inactivation of microorganisms in water. *Process Safety and Environmental Protection* 147: 488–496.

Sanches, S., Penetra, A., Rodrigues, A. et al. (2013). Removal of pesticides from water combining low pressure UV photolysis with nanofiltration. *Separation and Purification Technology* 115: 73–82.

Sathe, S.M., Chakraborty, I., Dubey, B.K., and Ghangrekar, M.M. (2022). Microbial fuel cell coupled Fenton oxidation for the cathodic degradation of emerging contaminants from wastewater: applications and challenges. *Environmental Research* 204: 112135.

Schuster, A., Hädrich, C., and Kümmerer, K. (2008). Flows of active pharmaceutical ingredients originating from health care practices on a local, regional, and nationwide level in Germany-is hospital effluent treatment an effective approach for risk reduction? *Water, Air, and Soil Pollution: Focus* 8 (5–6): 457–471.

Shcherbakova, V.A., Laurinavichius, K.S., and Akimenko, V.K. (1999). Toxic effect of surfactants and probable products of their biodegradation on methanogenesis in an anaerobic microbial community. *Chemosphere* 39 (11): 1861–1870.

Sumikura, M., Hidaka, M., Murakami, H. et al. (2007). Ozone micro-bubble disinfection method for wastewater reuse system. *Water Science and Technology* 56: 53–61.

Tadkaew, N., Sivakumar, M., Khan, S.J. et al. (2010). Effect of mixed liquor pH on the removal of trace organic contaminants in a membrane bioreactor. *Bioresource Technology* 101 (5): 1494–1500.

Tomei, M.C. and Annesini, M.C. (2008). Biodegradation of phenolic mixtures in a sequencing batch reactor: a kinetic study. *Environmental Science and Pollution Research* 15: 3.

Wang, A., Wang, C., Fu, L. et al. (2017). Recent advances of graphitic carbon nitride-based structures and applications in catalyst, sensing, imaging, and leds. *Nano-Micro Letters* 9 (4): 1–21.

Wang, X., Blechert, S., and Antonietti, M. (2012). Polymeric graphitic carbon nitride for heterogeneous photocatalysis. *ACS Catalysis* 2 (8): 1596–1606.

Wen, Q., Kong, F., Zheng, H. et al. (2011). Electricity generation from synthetic penicillin wastewater in an air-cathode single chamber microbial fuel cell. *Chemical Engineering Journal* 168 (2): 572–576.

Xu, P., Janex, M.L., Savoye, P. et al. (2002). Wastewater disinfection by ozone: main parameters for process design. *Water Research* 36 (4): 1043–1055.

Yahya, M.S., Beqqal, N., Guessous, A. et al. (2017). Degradation and mineralization of moxifloxacin antibiotic in aqueous medium by electro-Fenton process: kinetic assessment and oxidation products. *Cogent Chemistry* 3 (1): 1290021. http://www.editorialmanager.com/cogentchem.

Yang, Y., Li, X., Zhou, C. et al. (2020). Recent advances in application of graphitic carbon nitride-based catalysts for degrading organic contaminants in water through advanced oxidation processes beyond photocatalysis: a critical review. *Water Research* 184: 116200.

Zhang, Q., Zhang, Y., and Li, D. (2017). Cometabolic degradation of chloramphenicol via a meta-cleavage pathway in a microbial fuel cell and its microbial community. *Bioresource Technology* 229: 104–110.

Zou, R., Angelidaki, I., Yang, X. et al. (2020). Degradation of pharmaceuticals from wastewater in a 20-L continuous flow bio-electro-Fenton (BEF) system. *Science of the Total Environment* 727: 138684.

22

Sustainability Outlook

Green Design, Consumption, and Innovative Business Model

Tsai Chi Kuo

National Taiwan University of Science and Technology, Taipei, Taiwan

22.1 Introduction

The pursuit of sustainability is increasingly recognized as an effective strategy to deal with some of the contemporary challenges facing global supply chains. The definition of sustainable strategies (Giannakis and Papadopoulos 2016) could be considered as the level of future uncertainty and therefore the risks that decisions may impose on the natural and social environments, in addition to the investment costs that are required to make supply chains more sustainable. Enterprises have been trying to become sustainable through different business strategies such as cost effectiveness, lean, high-quality, on-time delivery, and agile, resilient, and world-class manufacturing. Recently, as environmental awareness has been growing worldwide, many enterprises begin to think seriously about integrating their products and service sector with sustainability and cutting supply chain expenses to attain a competitive edge over others (Luthra et al. 2016). The World Commission on Environment and Development (WCED 1987) has defined sustainability as an integration of social, environmental, and economic issues. It is usually operated through the triple bottom line (TBL): economic, social, and environment (Gold et al. 2013). As a result, research on a sustainable/green supply chain has been attracting interest across academic practices in the last few years.

Generally, the scope of the green/sustainable supply chain is broad thinking. It covers the members of upstream and downstream and integrates with the issues of life-cycle management: design, manufacturing, transportation, use, and disposal. Although the concept of green design (GD) was proposed initially (Kuo 2000), it had not achieved a favorable position until green supply chain management was introduced and bloomed. With this systematic concept, more theories and practices are developed and integrated under the concept of sustainability of supply chain management.

Microconstituents in the Environment: Occurrence, Fate, Removal, and Management, First Edition.
Edited by Rao Y. Surampalli, Tian C. Zhang, Chih-Ming Kao, Makarand M. Ghangrekar, Puspendu Bhunia, Manaswini Behera, and Prangya R. Rout.

A number of literature reviews on green or sustainable supply chain management have been published in the past few years. Table 22.1 lists the published review papers. These researchers have identified various topical issues covered from general models to specific aspects, and analytical methods.

1) General models (Srivastava 2007; Seuring and Müller 2008),
2) Specific aspects (Igarashi et al. 2013; Taticchi et al. 2013; Heckmann et al. 2015; Kamalahmadi and Parast 2016), or
3) Analytical methods (Hassini et al. 2012; Seuring 2013; Brandenburg et al. 2014; Fahimnia et al. 2015; Govindan et al. 2015; Ntabe et al. 2015).

Although the above research covers the emergent research literature in the green/sustainable supply chain, the scope of a sustainable supply chain is increasing, expanded from design, business model, to sustainable consumption and smart living. Also, the strategies and supported technologies are adapting fast. Therefore, it is worth having a review that covers the product design, business strategies and operations, and consumptions based on the triple bottom line (TBL) of supply chain sustainability.

The aim of this chapter is to summarize existing research on green products, business models, and consumption by using TBL for the whole supply chains. This chapter provides insights into future research directions and needs. More than 150 research papers were selected and analyzed from SCI or SSCI databases. Papers were identified by means of a

Table 22.1 The previous literature reviews of sustainable or green supply chain.

	This Research	(1)	(2)	(3)	(4)	(5)	(6)	(7)	(8)	(9)	(10)	(11)
Design/Product	v	v		v								
Process/ Operations	v	v				v						
Methodologies	v	v					v	v			v	
Modeling	v	v	v			v	v	v			v	
Tools	v	v				v	v					
Trigger	v			v								
Supplier management	v			v				v	v			
Resilience	v				v							
By authors, areas, industry sectors										v	v	v
Business strategy	v											
Consumptions	v											
Technologies	v											

(1) Srivastava (2007), (2) Heckmann et al. (2015), (3) Seuring and Müller (2008), (4) Kamalahmadi and Parast (2016), (5) Igarashi et al. (2013), (6) Seuring (2013), (7) Brandenburg et al. (2014), (8) Govindan et al. (2015), (9) Fahimnia et al. (2015), (10) Ntabe et al. (2015), (11) Hassini et al. (2012).
Source: Adapted from Srivastava, S.K., 2007,Heckmann et al. 2015, Seuring and Müller 2008.

structured keyword search on major databases and publisher websites (Ebsco, Springerlink, Wiley Interscience, Elsevier ScienceDirect, Emerald Insight). Keywords such as "design," "recycle," "resource," "manufacturing," "regulation," "supply," "supply chain," "logistics/ logistical," "innovative/ innovation," and "product service," were combined with sustainability related ones, such as "sustainable/sustainability," "sustainable development," "environment(al)," "green," "social," and "ethics/ethical" (Table 22.1).

This chapter is structured as follows: as the study deals with a literature review, a classical section labeled as such is not provided. Instead, the chapter starts by outlining the content analysis method as applied to the research process. Next, some descriptive background on the papers (e.g., years of publication, major journals) is presented. Furthermore, the findings from the content analysis are discussed, with a particular focus on the sustainability dimensions and modeling approaches. This will lead to the discussion of the findings and brief conclusions.

22.2 Sustainable/Green Supply Chain

Traditionally, the supply chain can be separated as forward supply chain and reverse supply chain. With reverse logistics, the waste can be reduced. However, the enterprises not only develop corporate environmental proactivity, but also move toward green/sustainable supply chain management (Wu et al. 2014). The implementation of environmental management involves numerous managerial challenges that are related to the organizational complexities, collaboration, and performance.

22.2.1 Collaboration

Kuo et al. (2012) applied the IDEF (integration definition) tool set that represents the architecture of the collaborative model and managed the quantity and quality of the supply network of the motorcycle industry. With collaboration, the suppliers and customers could cooperate together to increase their competitivities. Chen et al. (2015a) also established a web-based green collaboration management system, which incorporates environment protection requests into product development and green supply chain activities. With this system, the green products, brand leadership, and environment protection endeavor can be assured.

22.2.2 System Improvements

Since the green product system is different from the non-green product system, the system should be improved. Lin et al. (2015a) integrated a new product development (NPD) framework that includes the following two stages: i) quality function deployment (QFD), fuzzy interpretive structural modeling (FISM), and fuzzy analytic network process (FANP); and ii) fuzzy failure mode and effects analysis (FFMEA) to determine the importance of Electronic Commerces (ECs) with respect to risk control. Tseng et al. (2012) developed an intelligent system for sustainable product design at the concept development stage. In the initial design stage, designers are allowed to establish the concept without time and distance limits because of its web-based structure. Chiang and Roy (2012) used a back-propagation neural network (BPNN) to create the analysis of approximate life-cycle inventory (LCI) and to identify hazardous chemical substances. Chen et al. (2012b) used a two-stage network Data Envelopment Analysis (DEA) to improve the design efficiency of sustainable product design performances. Lin et al. (2010) used conjoint analysis to evaluate the design scheme based on customer preferences and to seek the optimal niche green technology. Lin et al. (2010) analyzed the certified enterprises and firms that qualify for environmental protection standards, ISO 14001, in Taiwan, showing that they have a significant effect on environmental design implementation.

22.2.3 Supplier Evaluations

Tseng and Chiu (2013) identified the firm's criteria and supplier selection to improve the firm's performance. In an earlier study, Lu et al. (2006) presented an analytical hierarchy process (AHP) decision-making method and fuzzy logic process for green supply chain management (GSCM) to measure and evaluate suppliers' performance. Wu et al. (2015) combined fuzzy set theory and the Decision-Making Trial and Evaluation Laboratory (DEMATEL) method for a hybrid approach to investigate the effects of each criterion within GSCP. To select the business strategy based on life-cycle management, specifically design, purchasing, manufacturing, and marketing and service, Chen et al. (2012a) conducted an analytic network process (ANP). Hsu et al. (2013) also applied the DEMATEL to deal with the importance and evaluation of supplier selection. Furthermore, Lin (2013) used fuzzy DEMATEL to evaluate the influential factors among the green supply chain management practices. The artificial neural network and MADA methods have also been used for green supplier selection (Kuo et al. 2010). Another study by Hsu et al. (2012) used DEMATEL based on ANP (called DANP) with VIKOR to find the vendor of recycled materials among the aluminum composite panel (ACP) industry. Wong et al. (2012) examined the boundary spanning role of graphic oxide (GO) and investigated the influence of environmental management capability (EMC) of suppliers on firm performance and pollution reduction.

22.2.4 Performance and Uncertainty

To increase the green supply chain performance, some research presented "top-down" green efforts by policy-makers and "bottom-up" efforts by companies in the supply chain (Chiu and Teng 2013; Tseng et al. 2013). Tsai (2012) found that process stewardship has a positive influence on performance outcomes and that the EMC of suppliers moderates the relationship between process stewardship and financial performance. Furthermore, Yang et al. (2013), based on the data from a survey of Taiwan firms, confirmed that internal green practices and external green collaboration have positive effects on green performance, which in turn helps to enhance firm competitiveness. Chiou et al. (2011) developed a questionnaire-based survey data collected from 124 companies from 8 industrial sectors in Taiwan and analyzed using Structural Equation Modeling to verify the significance of the proposed relationships. Gong and Chen (2012) proposed several critical control processes by using Unified Modeling Language (UML) sequence diagrams to enhance the effectiveness of product management based on International Quality Management System, IECQ

QC080000 HSPM. Sheu and Chen (2012) analyzed the effects of governmental financial intervention on green supply chain competition using a three-stage game-theoretical model. In a previous study, Chen and Sheu (2009) found that a proper design of environmental-regulation pricing strategies is able to promote Extended Product Responsibility for green supply chain firms in a competitive market.

22.3 Environmental Sustainability: Innovative Design and Manufacturing

Several studies have identified the characteristics of companies with the most advanced environmental design utilizing raw materials, energy, recyclability, product life-cycle assessment, and packaging optimization (Tien et al. 2002; Chung and Tsai 2006; Shang et al. 2010; Tsai et al. 2013). The green design method could be categorized as described below.

22.3.1 Design Improvements

Several design improvemnet technologies have been raised: i) design for disassembly and recyclability; ii) module design; and iii) life-cycle design.

22.3.1.1 Disassembly and Recyclability

The problems of disassembly and recycling have received much attention based on green design. For disassembly and recycling, the critical problem is to find a disassembly sequence and to feedback to the designer for design change. Kuo et al. (2000) presented a graph-based heuristic approach (disassembly tree) to perform disassembly and recycling analysis. Smith et al. (2012) proposed a new "disassembly sequence structure graph" (DSSG) model for multiple-target selective disassembly sequence planning, an approach for creating DSSGs, and methods for searching DSSGs. Tseng et al. (2010) proposed a disassembly-oriented assessment method for product modular design that considers the economy. Furthermore, Smith and Chen (2011) used a rule-based recursive method to find a near-optimal heuristic selective disassembly sequence for green design. Fan et al. (2013) evaluated the recycling rates and costs, as well as the disassembly time of a notebook at its end-of-life stage. Kuo (2010) also used case-based reasoning to provide a recyclability index of a product. Therefore, the product design could incorporate recycling planning during the design stage. Shih et al. (2006) proposed an intelligent evaluation approach that incorporates case-based reasoning models, economic analysis models, and domain expertise. Huang et al. (2012) analyzed and identified modules and disassembly patterns to enhance 3R-abilities (reuse, recycle, and recovery).

22.3.1.2 Modularity Design

Modularity design integrates different components and/or subassemblies that share the same physical relationship and similar functions. With modularized product design, the waste is minimized, since it can be easily disassembly or replaced. Luh et al. (2010) developed a generic modularized product architecture that facilitates data management of green product development. To increase the reuse rate, Ji et al. (2013) leveraged technical system modularity (TSM) and material reuse modularity (MRM) within a coherent framework. Tseng et al. (2008) used modular design to support green life-cycle engineering. Furthermore, Chang et al. (2013b) developed a QFD and design structure matrix (DSM) system for green design in modular product development. Kuo (2013b) constructed a collaborative design framework for the enterprises to collect and calculate product carbon footprints in a ready and timely manner throughout the entire supply chain. Chang et al. (2014) systematically integrated life-cycle analysis in product development, from concept design, part design, and process design to decision-making. Smith and Yen (2010) developed an innovative method that uses the concepts of atomic theory to solve design modularization problems of green product design. With this method, products can be modularized based upon given green constraint (e.g., material compatibility, part recyclability, part disassemblability).

22.3.1.3 Life-Cycle Design

With the increased emphasis on green design, an increasing number of enterprises conduct the life-cycle assessment (LCA) to evaluate the environmental effects of products. The life-cycle inventory (LCI) data is needed to perform LCA. The LCI is the data on raw material, manufacturing, transport, use, and disposal that is collected and compiled into a database. The LCA software system named "Do-It-Pro" had been developed by the Industrial Technology Research Institute (ITRI 2015). The LCA could also be applied to the analysis of the environmental effects of an industry. Liu et al. (2010) investigated the major environmental effects of the DRAM products in Taiwan's semiconductor industry and determined which LCA method is more applicable. Su and Lee (2009) conducted the life-cycle inventory analyses of biofuels and found positive energy benefits of producing biofuels. Lu et al. (2006) used cost–benefit analysis and formal LCA to revise the current recycling policies. Trappey et al. (2012) developed the system dynamics modeling of product carbon footprint life cycles for collaborative green supply chains. Trappey's model uses an economic input–output life-cycle assessment approach to evaluate the carbon emissions of new products.

22.3.2 Green Manufacturing

Green technology aspects of precision engineering and manufacturing are becoming even more important in current and future technologies. Green manufacturing could be seen as a manufacturing method that minimizes waste and pollution. Tsai et al. (2011) used the activity-based costing (ABC) system to justify the capital investments of green manufacturing systems (GMSs). Therefore, some research proposed a lean-green concept (Chiarini 2014; Dhingra et al. 2014; Galeazzo et al. 2014; Johansson and Sundin 2014; Pampanelli et al. 2014; Verrier et al. 2014; Garza-Reyes 2015).

22.3.2.1 Green Manufacturing Process and System Development

Chuang (2014) used the six sigma approach and particle swarm optimization to improve the green performance of global footwear manufacturing processes. Chiang et al. (2011a) used the analytic hierarchy process (AHP) method to analyze green manufacturing indicators relatively influencing environmental performance. The results showed that the choice of lead-free substitute materials, soil heavy metal pollution, and compliance with environmental laws and regulations are the three important indicators of the environmental performance of lead-free manufacturing. Lin and Tseng (2012) replaced the easy-to-make-mistake methods of predicting kinetics and the consumption energy of the transitional self-accelerating decomposition temperature (SADT) tests using simply thermal analysis combined with kinetic and thermal hazard simulation. Yang et al. (2012) applied QFD to build a green manufacturing system. Chuang and Yang (2014) implemented the analytical network process (ANP) to find the key success factors of implementing a green-manufacturing system.

22.3.2.2 Recycling Technology

The remanufacturing and recycling processes of end-of-life products are very important to reduce waste. Hong et al. (2012) examined the effect of exogenous subsidies on recycled material flows in a decentralized recycling system, where each entity acts according to its own interests. Lee et al. (2015) analyzed the related knowledge of the management of end-of-life fluorescent lamp tubes, the recycling or reprocessing technologies, and policies in Taiwan and other nations. Using pineapple leaf fiber and recycled disposable chopsticks, Shih et al. (2014) developed hybrid fibers and biodegradable polymers.

22.3.2.3 Hazard Material Control

Hsu and Hu (2009) presented an analytic network process (ANP) approach to incorporate the issue of hazardous substance management (HSM) into supplier selection. Later, Kuo and Chu (2013) conducted a failure mode and effects analysis (FMEA) and X-bar control chart for component risk based on component types. Furthermore, Koh et al. (2012) developed a conceptual model that outlines the antecedents of a successful embeddedness of environmental directives, Waste Electrical and Electronic Equipment (WEEE), and Restriction of the use of certain Hazardous Substances (RoHS), on greening a supply chain.

22.3.2.4 Remanufacturing and Inventory Model

Chung and Wee (2008) utilized green-component life-cycle value design and reverse manufacturing in a semi-closed supply chain. Additionally, Wee and Chung (2009) analyzed the optimized replenishment policy for an integrated production inventory deteriorating model considering green component-value design and remanufacturing. Later, Chung and Wee (2010) discussed the green product design value and information technology investment on replenishment model with remanufacturing. They analyzed remanufacturing of the short life-cycle deteriorating product in a green supply chain inventory control system (Chung and Wee 2011). Wu (2012) formulated a two-period model to investigate the OEM's product-design strategy and the remanufacturer's pricing strategy in an extensive-form game in which the equilibrium decisions of the resulting scenarios are derived. Wu (2013) further formulated a two-period supply chain model consisting of two chain members, an OEM and a remanufacturer, to investigate the product design decision of the OEM and both chain members' competitive pricing strategies.

22.3.3 Summary of Environmental Sustainability

Green design could trigger innovation. The core ideas of green design are less energy and resource usage, lower pollution, and higher recyclability. Wang et al. (2010) used an improved TRIZ method and eco-innovative design method to develop a new green fire proof system for kitchen equipment. Yang and Chen (2011) enhanced the design level of new products to achieve eco-innovation under the case-based reasoning (CBR) framework. Trappey et al. (2011) adopted LCA, quality function deployment for environment (QFDE), theory of inventive problem solving (TRIZ), and back-propagation network (BPN) to achieve eco- and inno-design objectives. Chen and Liu (2001) used TRIZ inventive principles to support the designer to build eco-innovative design. Yeh et al. (2011) integrated the QFD and TRIZ to construct a contradiction matrix to find green design solutions. Chou (2014) proposed an Algorithm for Inventive-Problem Solving (ARIZ) based on a life-cycle engineering (LCE) model to develop eco-designs of products. Through this ARIZ model, the modular analysis of alternative attributes associated with the morphological approach was developed. Chen et al. (2006) found that the performances of the green product innovation and green process innovation positively correlated with the corporate competitive advantage.

22.4 Economical Sustainability: Innovation Business Model

The environmental problem needs to be emphasised for green design to bloom. However, the green design products have not achieved a favorable position in the marketplace as would be expected, even though they appear to be more environmentally-friendly. Green products are designed not only to meet environmental objectives, such as resource and energy conservation and environmental burden reduction, but also to consider cost effectiveness, market demand, and multi-functionality requirements (Lee et al. 2001; Kuo et al. 2009).

With the innovative green design, the business model has also been renewed. Some researchers have proposed the product service system to eliminate the environmental effects (Kuo et al. 2010; Kuo 2013a). Chiu et al. (2015) used a multiple criteria decision-making (MCDM) tool to integrate both Analytic Hierarchy Process (AHP) and technique for Order Preference by Similarity to Ideal Solution (TOPSIS) to present new business market opportunities. Kuo (2011) developed a simulation model to compare the purchase or rental decision-making based on a product service system. Moreover, Chou et al. (2015a) used the concept of sustainable product-service efficiency to explore the relationship between product-service value and sustainability effect.

22.4.1 Business Model and Performance

Hai and Sung (2010) used SWOT to analyze the green competitive strategy of enterprises. Chen and Wu (2015) conducted a survey and found that the perceived benefits of green business, as reported by service businesses, have a significant positive influence on implementation intentions; however, the perceived risk has a significant negative influence on the intention to implement green business. Other research has analyzed the effects of low carbon economy strategies and policies for the enterprises (Hu et al. 2013). Hung et al. (2013) used the modified Delphi with end-user participation to provide strategic foresight for firms in Taiwan's PC industry ecosystem. Chen et al. (2015b) also found that eco-organizational innovation has the strongest effect on business performance. Huang et al. (2011) developed a driving force-pressure-state-impact-response (DPSIR) framework and found that some enterprises added the GHG-related indictors to increase the enterprises competitivies. Huang and Shih (2009) used the case study of China Steel Corporation and found that the enterprises could make a profit and reduce costs through energy sold, by-products, and recycling.

22.4.2 Summary of Economic Sustainability

Green design sometime conflicts with the other design criteria, such as cost, quality, and price. It can be seen as a trade-off problem for an enterprise–multi-criteria evaluation and optimization. Chiang and Che (2015) developed a decision-making methodology for low-carbon electronic product design based on the carbon footprint. Additionally, Su et al. (2012) considered the price, quality of product, and environmental quality, and they developed mathematical models to maximize profit. Chiang et al. (2011b) used a back-propagation neural network (BPNN) model and a technique for order preference by similarity to the ideal solution (TOPSIS) method for estimating quantities of hazardous chemical substances and energy consumption. Tsai (2012) calculated the corresponding weights for each factor and then proceeded with fuzzy multiple attribute decision-making (FMADM) and developed a checklist evaluation model for green design. Tsai et al. (2012) applied a mathematical programming approach for a green product mix decision that incorporates capacity expansion features.

22.5 Social Sustainability

The effects of domestic and international environmental regulations generated tremendous ripples and effects for the enterprises to reform their business strategies and green design (Koh et al. 2012). As the enterprises are under the pressures of environmental regulations, they change their reaction to an aggressive attitude to reform their business strategy. An enterprise used the corporate social responsibility (CSR) report to communicate with its stakeholder about their environmental efforts. Additionally, some enterprises integrated the environmental performance into their business performances.

22.5.1 Corporate Social Responsibility

Currently, enterprises begin to adopt the CSR as their business strategy and to communicate with their stakeholders. Tang et al. (2005) suggested that public participation in environmental management would enhance the effectiveness of Environmental Impact Assessment (EIA). Chen et al. (2015b) applied structural equation modeling and showed that green shared vision positively influences green mindfulness, green self-efficacy, and green creativity. Lin et al. (2015b) developed an analytical hierarchy process (ANP) and found the highest priority index among the three components of TBL, namely social development, environmental protection, and economic development. Guo et al. (2015) focused on green product (GP) and employed a novel decision-making trial and evaluation laboratory method to evaluate and investigate green corporate social responsibility (GCSR) indicators. Chiouy et al. (2011) conducted sustainable supplier selection and assessment for the information and electronics industry based on literature reviews and industry expert input.

22.5.2 Sustainable Consumption

With the development of green design and green supply chain management, further research found that the larger gaps existing between consumers' expectations and their perceptions of environmental attributes of green information products (Tseng et al. 2013; Tseng and Hung 2013). Wu and Lin (2014) used structural equation modeling and explored the influence of green marketing strategies on business performance. Their model showed that organic farms' corporate image could be enhanced through green marketing strategies, thereby improving their business performance. Customers are still concerned about the prices of green products. Lee (2014) examined the effects of pricing strategies (i.e., market-skimming pricing vs. market-penetration pricing) and advertising strategies (functional advertisement vs. emotional advertisement) on consumer-perceived quality, perceived risk, perceived value, and adoption in the context of green product innovation. Yang et al. (2014) used Kano's model to classify attributes as an attractive quality for green marketing. Tsaur (2015) explored the price strategy of reused personal computers that are discarded and sold on the secondary market. In an earlier studies, Hwang et al. (2010) conducted a structural equation model (SEM) and performed a study on the relationship between the PDCA (plan-do-check-action) cycle of green purchasing and the Supply-Chain Operations Reference (SCOR) model.

22.5.3 Brief Summary of Social Sustainability

Social sustainability is emphasised by academics and practices. It not only covers more enterprises toward the many researchers but begins to explore the issue of Bottom of Pyramid (BOP) consumer groups. BOP population groups are those in which individuals have daily incomes of less than 2 USD. The total number of global BOP consumers is 4 billion. These groups are also referred to as "rising stars" or "rising powers." In "The Fortune at the Bottom of the Pyramid," Prahalad and Hart (2002) explained that enterprises should develop the BOP market while reducing or eradicating poverty through technology and business model innovation. They further developed and analyzed this idea through their progressive models BOP 1.0 (selling to the poor), BOP 2.0 (business co-venturing), and BOP 3.0 (inclusive business) (Simanis and Hart 2008; Hart and Cañeque 2015).

22.6 Conclusions and Future Research Development

This chapter aims to provide a critical review of the literature on Eco design, business model innovation, and sustainable consumption. The current review indicates that the economic issues significantly affect environmental protection and social equal issues. When considering adoption of sustainability, an enterprise will face pressures from its stakeholders, such as government, customers, competitors, the media, non-government organizations (NGOs), and the like, to achieve its sustainable goal. Future research development should address the following issues.

22.6.1 Future Research Development

Based on the statistics, carbon dioxide (CO_2) emissions from the consumption of energy in Taiwan were 293 million tons in 2011, accounting for 0.9% of global emissions (EIA 2012). In response to the international increase in energy prices and under environmental pressure to reduce global emissions of greenhouse gasses, the promotion of solar water heaters (SWHs) has become a crucial aspect of the Taiwanese government's energy saving policies (Chang et al. 2013a). Additionally, several studies have been conducted on Taiwan's renewable energy polices (Tsai 2014), including solar photovoltaic policy (Chou et al. 2015b), wind power (Liou 2011), and carbon taxes (Hua and Wu 2000).

22.6.2 Industry 4.0 in Sustainable Life

As industry 4.0 is proposed, more research used it in the sustainable supply chain. The contents of industry include internet of things (IoT), cloud computing, robotics, smart factories, 3D printing, and big data. For example, the IoT application has been used in the mobile taxi for quick response to customer needs (Lanza et al. 2015). Singh et al. (2015) used cloud computing technologies to calculate the beef carbon footprint. With cloud computing and big data technologies, it is easy to collect data and to analyze it. Gebler et al. (2014) have shown that 3D printing could reduce carbon emissions.

22.6.3 Conclusions

With the enterprise toward supply chain sustainability, the enterprise needs to formulate its business strategies to fulfill its stakeholders demands. The business strategies should include two perspectives: internal and external management. However, different strategies will lead to uncertainties and risks:

1) *Pressure*: With the environment deteriorating, people have put pressure on businesses to improve the environmental and resource consequences of their products and processes in recent decades.
2) *Strategy*: Based on the literature review, the content of sustainable strategy should include top managers commitment, accountability of the whole supply chain, R&D activities, key performance indicators (KPI), incentives, and motivation of the enterprise.
3) *Internal management*: This includes managing the process of continuous improvement, minimizing environmental impacts, and enhancing efficiency.
4) *External management*: It is an essential influential factor in adopting sustainable practices.
5) *Uncertainty and risk management*: The uncertainty considerations involve having sufficient knowledge of the whole supply chain, minimizing risks, and preventing risks, among others. Once a suitable framework is found to fit a particular enterprise, knowledge about risks can be reinforced by cooperating with others externally and by dealing with risks rooted in social and environmental effects and, at the same time, by realizing that risks are constantly evolving.

References

Brandenburg, M., Govindan, K., Sarkis, J., and Seuring, S. (2014). Quantitative models for sustainable supply chain management: developments and directions. *European Journal of Operational Research* 233: 299–312.

Chang, D., Lee, C.K.M., and Chen, C.-H. (2014). Review of life cycle assessment towards sustainable product development. *Journal of Cleaner Production* 83: 48–60.

Chang, P.-L., Ho, S.-P., and Hsu, C.-W. (2013a). Dynamic simulation of government subsidy policy effects on solar water heaters installation in Taiwan. *Renewable and Sustainable Energy Reviews* 20: 385–396.

Chang, T.-R., Wang, C.-S., and Wang, C.-C. (2013b). A systematic approach for green design in modular product development. *The International Journal of Advanced Manufacturing Technology* 68: 2729–2741.

Chen, C.-C., Shih, H.-S., Shyur, H.-J., and Wu, K.-S. (2012a). A business strategy selection of green supply chain management via an analytic network process. *Computers & Mathematics with Applications* 64: 2544–2557.

Chen, C., Zhu, J., Yu, J.-Y., and Noori, H. (2012b). A new methodology for evaluating sustainable product design performance with two-stage network data envelopment analysis. *European Journal of Operational Research* 221: 348–359.

Chen, J.-H. and Wu, S.-I. (2015). A comparison of green business relationship models between industry types. *Total Quality Management & Business Excellence* 26: 778–792.

Chen, J.L. and Liu, C.-C. (2001). An eco-innovative design approach incorporating the TRIZ method without contradiction analysis. *The Journal of Sustainable Product Design* 1: 263–272.

Chen, Y.-C., Chu, C., Sun, H.-M. et al. (2015a). Application of green collaboration operation on network industry. *International Journal of Precision Engineering and Manufacturing-Green Technology* 2: 73–83.

Chen, Y.J. and Sheu, J.-B. (2009). Environmental-regulation pricing strategies for green supply chain management. *Transportation Research Part E: Logistics and Transportation Review* 45: 667–677.

Chen, Y.-S., Chang, C.-H., Yeh, S.-L., and Cheng, H.-I. (2015b). Green shared vision and green creativity: the mediation roles of green mindfulness and green self-efficacy. *Quality & Quantity* 49: 1169–1184.

Chen, Y.-S., Lai, S.-B., and Wen, C.-T. (2006). The influence of green innovation performance on corporate advantage in Taiwan. *Journal of Business Ethics* 67: 331–339.

Chiang, S.-Y., Wei, C.-C., Chiang, T.-H., and Chen, W.-L. (2011a). How can electronics industries become green manufacturers in Taiwan and Japan. *Clean Technologies and Environmental* 13: 37–47.

Chiang, T.-A. and Che, Z.H. (2015). A decision-making methodology for low-carbon electronic product design. *Decision Support Systems* 71: 1–13.

Chiang, T.-A., Che, Z.H., and Wang, T.-T. (2011b). A design for environment methodology for evaluation and improvement of derivative consumer electronic product development. *Journal of Systems Science and Systems Engineering* 20: 260–274.

Chiang, T.-A. and Roy, R. (2012). An intelligent benchmark-based design for environment system for derivative electronic product development. *Computers in Industry* 63: 913–929.

Chiarini, A. (2014). Sustainable manufacturing-greening processes using specific Lean Production tools: an empirical observation from European motorcycle component manufacturers. *Journal of Cleaner Production* 85: 226–233.

Chiou, T.-Y., Chan, H.K., Lettice, F., and Chung, S.H. (2011). The influence of greening the suppliers and green innovation on environmental performance and competitive advantage in Taiwan. *Transportation Research Part E: Logistics and Transportation Review* 47: 822–836.

Chiouy, C.-Y., Chou, S.-H., and Yeh, C.-Y. (2011). Using fuzzy AHP in selecting and prioritizing sustainable supplier on CSR for Taiwan's electronics industry. *Journal of Information and Optimization Sciences* 32: 1135–1153.

Chiu, M.-C., Kuo, M.-Y., and Kuo, T.-C. (2015). A new methodology to develop business model of product service system. *International Journal of Industrial Engineering: Theory, Applications and Practice* 22: 369–381.

Chiu, M.-C. and Teng, L.-W. (2013). Sustainable product and supply chain design decisions under uncertainties. *International Journal of Precision Engineering and Manufacturing* 14: 1953–1960.

Chou, C.-J., Chen, C.-W., and Conley, C. (2015a). An approach to assessing sustainable product-service systems. *Journal of Cleaner Production* 86: 277–284.

Chou, J.-R. (2014). An ARIZ-based life cycle engineering model for eco-design. *Journal of Cleaner Production* 66: 210–223.

Chou, S.-Y., Nguyen, T.A.T., Yu, T.H.-K., and Phan, N.K.P. (2015b). Financial assessment of government subsidy policy on photovoltaic systems for industrial users: a case study in Taiwan. *Energy Policy* 87: 505–516.

Chuang, S.-P. (2014). Assessing and improving the green performance using a compound approach. *Flexible Services and Manufacturing Journal* 26: 69–91.

Chuang, S.-P. and Yang, C.-L. (2014). Key success factors when implementing a green-manufacturing system. *Production Planning & Control* 25: 923–937.

Chung, C.-J. and Wee, H.-M. (2008). Green-component life-cycle value on design and reverse manufacturing in semi-closed supply chain. *International Journal of Production Economics* 113: 528–545.

Chung, C.-J. and Wee, H.-M. (2010). Green-product-design value and information-technology investment on replenishment model with remanufacturing. *International Journal of Computer Integrated Manufacturing* 23: 466–485.

Chung, C.-J. and Wee, H.-M. (2008). Green-component life-cycle value on design and reverse manufacturing in semi-closed supply chain, International Journal of Production Economics 113 (2): 528–545, ISSN 0925–5273, https://doi.org/10.1016/j.ijpe.2007.10.020.

Chung, Y.C. and Tsai, C.H. (2006). Research on environmental design implementation in Taiwan's Manufacturers Certified by ISO 14001. *Asian Journal on Quality* 7: 177–194.

Dhingra, R., Kress, R., and Upreti, G. (2014). Does lean mean green? *Journal of Cleaner Production* 85: 1–7.

EIA (2012). *International Energy Statistics*. US Energy Information Administration (EIA).

Fahimnia, B., Sarkis, J., and Davarzani, H. (2015). Green supply chain management: a review and bibliometric analysis. *International Journal of Production Economics* 162: 101–114.

Fan, S.-K.S., Fan, C., Yang, J.-H., and Liu, K.F.-R. (2013). Disassembly and recycling cost analysis of waste notebook and the efficiency improvement by re-design process. *Journal of Cleaner Production* 39: 209–219.

Galeazzo, A., Furlan, A., and Vinelli, A. (2014). Lean and green in action: interdependencies and performance of pollution prevention projects. *Journal of Cleaner Production* 85: 191–200.

Garza-Reyes, J.A. (2015). Lean and green: a systematic review of the state-of-the-art literature. *Journal of Cleaner Production* 102: 18–29.

Gebler, M., Schoot Uiterkamp, A.J.M., and Visser, C. (2014). A global sustainability perspective on 3D printing technologies. *Energy Policy* 74: 158–167.

Giannakis, M. and Papadopoulos, T. (2016). Supply chain sustainability: a risk management approach. *International Journal of Production Economics* 171 (4): 455–470.

Gold, S., Hahn, R., and Seuring, S. (2013). Sustainable supply chain management in "Base of the Pyramid" food projects: a path to triple bottom line approaches for multinationals? *International Business Review* 22: 784–799.

Gong, D.-C. and Chen, J.-L. (2012). Critical control processes to fulfil environmental requirements at the product development stage. *International Journal of Computer Integrated Manufacturing* 25: 457–472.

Govindan, K., Rajendran, S., Sarkis, J., and Murugesan, P. (2015). Multi criteria decision-making approaches for green supplier evaluation and selection: a literature review. *Journal of Cleaner Production* 98: 66–83.

Guo, W.-F., Zhou, J., Yu, C.-L. et al. (2015). Evaluating the green corporate social responsibility of manufacturing corporations from a green industry law perspective. *International Journal of Production Research* 53: 665–674.

Hai, H.-L. and Sung, W.-L. (2010). Assessing the green competitive strategy of enterprises in Taiwan under the Gore effect. *Journal of Information and Optimization Sciences* 31: 837–856.

Hart, S.L. and Cañeque, F.C. (2015). *Base of the Pyramid 3.0: Sustainable Development through Innovation and Entrepreneurship*. (ed. Hart, S.L.). Greenleaf Publishing.

Hassini, E., Surti, C., and Searcy, C. (2012). A literature review and a case study of sustainable supply chains with a focus on metrics. *International Journal of Production Economics* 140: 69–82.

Heckmann, I., Comes, T., and Nickel, S. (2015). A critical review on supply chain risk – definition, measure and modeling. *Omega* 52: 119–132.

Hong, I.H., Ammons, J.C., and Realff, M.J. (2012). Examining subsidy impacts on recycled WEEE material flows. In: *Waste Electrical and Electronic Equipment (WEEE) Handbook* (eds. Goodship, V. and Stevels,A.), 466–490. Woodhead Publishing.

Hsu, C.H., Wang, F.-K., and Tzeng, G.-H. (2012). The best vendor selection for conducting the recycled material based on a hybrid MCDM model combining DANP with VIKOR. *Resources, Conservation and Recycling* 66: 95–111.

Hsu, C.-W. and Hu, A.H. (2009). Applying hazardous substance management to supplier selection using analytic network process. *Journal of Cleaner Production* 17: 255–264.

Hsu, C.-W., Kuo, T.-C., Chen, S.-H., and Hu, A.H. (2013). Using DEMATEL to develop a carbon management model of supplier selection in green supply chain management. *Journal of Cleaner Production* 56: 164–172.

Hu, W.-C., Chien, H.-C., Shih, R., and Lien, C.-A. (2013). Mobilizing business for low-carbon economy: Taiwan's vision, pathway, and challenge. *Greenhouse Gas Measurement and Management* 3: 166–178.

Hua, J. and Wu, Y. (2000). The effect of carbon taxes on energy use by the manufacturing sector of Taiwan. *Water, Air, & Soil Pollution* 124: 245–257.

Huang, C.-C., Liang, W.-Y., Chuang, H.-F., and Chang, Z.-Y. (2012). A novel approach to product modularity and product disassembly with the consideration of 3R-abilities. *Computers & Industrial Engineering* 62: 96–107.

Huang, H.-F., Kuo, J., and Lo, S.-L. (2011). Review of PSR framework and development of a DPSIR model to assess greenhouse effect in Taiwan. *Environmental Monitoring and Assessment* 177: 623–635.

Huang, P.-S. and Shih, L.-H. (2009). Effective environmental management through environmental knowledge management. *International Journal of Environmental Science and Technology* 6: 35–50.

Hung, C.-Y., Lee, W.-Y., and Wang, D.-S. (2013). Strategic foresight using a modified Delphi with end-user participation: a case study of the iPad's impact on Taiwan's PC ecosystem. *Technological Forecasting and Social Change* 80: 485–497.

Hwang, Y.-D., Wen, Y.-F., and Chen, M.-C. (2010). A study on the relationship between the PDSA cycle of green purchasing and the performance of the SCOR model. *Total Quality Management & Business Excellence* 21: 1261–1278.

Igarashi, M., de Boer, L., and Fet, A.M. (2013). What is required for greener supplier selection? A literature review and conceptual model development. *Journal of Purchasing and Supply Management* 19: 247–263.

Industrial Technology Research Institute (ITRI) (2015). *Do it Pro*. ITRI, http://scmp.itri.org.tw/Frontend/WebPage/index.aspx#.

Ji, Y., Jiao, R.J., Chen, L., and Wu, C. (2013). Green modular design for material efficiency: a leader–follower joint optimization model. *Journal of Cleaner Production* 41: 187–201.

Johansson, G. and Sundin, E. (2014). Lean and green product development: two sides of the same coin? *Journal of Cleaner Production* 85: 104–121.

Kamalahmadi, M. and Parast, M.M. (2016). A review of the literature on the principles of enterprise and supply chain resilience: major findings and directions for future research. *International Journal of Production Economics* 171(1): 116–133.

Koh, S.C.L., Gunasekaran, A., and Tseng, C.S. (2012). Cross-tier ripple and indirect effects of directives WEEE and RoHS on greening a supply chain. *International Journal of Production Economics* 140: 305–317.

Kuo, T. (2013a). Mass customization and personalization software development: a case study eco-design product service system. *Journal of Intelligent Manufacturing* 24: 1019–1031.

Kuo, T. and Chu, C.-H. (2013). Risk management of hazardous substances in selection of green suppliers. *International Journal of Precision Engineering and Manufacturing* 14: 1057–1063.

Kuo, T., Ma, H.-Y., Huang, S. et al. (2010). Barrier analysis for product service system using interpretive structural model. *The International Journal of Advanced Manufacturing Technology* 49: 407–417.

Kuo, T.C. (2000). Disassembly sequence and cost analysis for electromechanical products. *Robotics and Computer-Integrated Manufacturing* 16: 43–54.

Kuo, T.C. (2010). Combination of case-based reasoning and analytical hierarchy process for providing intelligent decision support for product recycling strategies. *Expert Systems With Applications* 37: 5558–5563.

Kuo, T.C. (2011). Simulation of purchase or rental decision-making based on product service system. *The International Journal of Advanced Manufacturing Technology* 52: 1239–1249.

Kuo, T.C. (2013b). The construction of a collaborative framework in support of low carbon product design. *Robotics and Computer-Integrated Manufacturing* 29: 174–183.

Kuo, T.C., Hsu, C.W., Ku, K.C. et al. (2012). A collaborative model for controlling the green supply network in the motorcycle industry. *Advanced Engineering Informatics* 26: 941–950.

Kuo, T.-C., Wu, H.-H., and Shieh, J.-I. (2009). Integration of environmental considerations in quality function deployment by using fuzzy logic. *Expert Systems with Applications* 36: 7148–7156.

Kuo, T.C., Zhang, H.C., and Huang, S.H. (2000). Disassembly analysis for electromechanical products: a graph-based heuristic approach. *International Journal of Production Research* 38: 993–1007.

Lanza, J., Sánchez, L., Muñoz, L. et al. (2015). Large-scale mobile sensing enabled internet-of-things testbed for smart city services. *International Journal of Distributed Sensor Networks* 2015: 15.

Lee, B.C.Y. (2014). Critical decisions in new product launch: pricing and advertising strategies on consumer adoption of green product innovation. *Asian Journal of Technology Innovation* 22: 16–32.

Lee, C.-H., Popuri, S., Peng, Y.-H. et al. (2015). Overview on industrial recycling technologies and management strategies of end-of-life fluorescent lamps in Taiwan and other developed countries. *Journal of Material Cycles and Waste Management* 17: 312–323.

Lee, S.G., Lye, S.W., and Khoo, M.K. (2001). A multi-objective methodology for evaluating product end-of-life options and disassembly. *The International Journal of Advanced Manufacturing Technology* 18: 148–156.

Lin, C.-P. and Tseng, J.-M. (2012). Green technology for improving process manufacturing design and storage management of organic peroxide. *Chemical Engineering Journal* 180: 284–292.

Lin, C.-Y., Lee, A.H.I., and Kang, H.-Y. (2015a). An integrated new product development framework: an application on green and low-carbon products. *International Journal of Systems Science* 46: 733–753.

Lin, C., Madu, C.N., Kuei, C.-H. et al. (2015b). Developing an assessment framework for managing sustainability programs: analytic network process approach. *Expert Systems with Applications* 42: 2488–2501.

Lin, K.H., Shih, L.H., and Lee, S.C. (2010). Optimization of product line design for environmentally conscious technologies in notebook industry. *International Journal of Environmental Science and Technology* 7: 473–484.

Lin, R.-J. (2013). Using fuzzy DEMATEL to evaluate the green supply chain management practices. *Journal of Cleaner Production* 40: 32–39.

Liou, H.M. (2011). Wind power in Taiwan: policy and development challenges. *Energy Policy* 39: 3238–3251.

Liu, C.H., Lin, S.J., and Lewis, C. (2010). Life cycle assessment of DRAM in Taiwan's semiconductor industry. *Journal of Cleaner Production* 18: 419–425.

Lu, L.-T., Wernick, I.K., Hsiao, T.-Y. et al. (2006). Balancing the life cycle impacts of notebook computers: Taiwan's experience. *Resources, Conservation and Recycling* 48: 13–25.

Luh, Y.-P., Chu, C.-H., and Pan, -C.-C. (2010). Data management of green product development with generic modularized product architecture. *Computers in Industry* 61: 223–234.

Luthra, S., Garg, D., and Haleem, A. (2016). The impacts of critical success factors for implementing green supply chain management towards sustainability: an empirical

investigation of Indian automobile industry. *Journal of Cleaner Production* 121: 142–158. ISSN 0959-6526, https://doi.org/10.1016/j.jclepro.2016.01.095.
Ntabe, E.N., LeBel, L., Munson, A.D., and Santa-Eulalia, L.A. (2015). A systematic literature review of the supply chain operations reference (SCOR) model application with special attention to environmental issues. *International Journal of Production Economics* 169: 310–332.
Pampanelli, A.B., Found, P., and Bernardes, A.M. (2014). A Lean & Green Model for a production cell. *Journal of Cleaner Production* 85: 19–30.
Prahalad, C.K. and Hart, S.L. (2002). The fortune at the bottom of the pyramid. *Strategy and Business 26: 1–15.*.
Seuring, S. (2013). A review of modeling approaches for sustainable supply chain management. *Decision Support Systems* 54: 1513–1520.
Seuring, S. and Müller, M. (2008). From a literature review to a conceptual framework for sustainable supply chain management. *Journal of Cleaner Production* 16: 1699–1710.
Shang, K.-C., Lu, C.-S., and Li, S. (2010). A taxonomy of green supply chain management capability among electronics-related manufacturing firms in Taiwan. *Journal of Environmental Management* 91: 1218–1226.
Sheu, J.-B. and Chen, Y.J. (2012). Impact of government financial intervention on competition among green supply chains. *International Journal of Production Economics* 138: 201–213.
Shih, L.-H., Chang, Y.-S., and Lin, Y.-T. (2006). Intelligent evaluation approach for electronic product recycling via case-based reasoning. *Advanced Engineering Informatics* 20: 137–145.
Shih, Y.-F., Chang, W.-C., Liu, W.-C. et al. (2014). Pineapple leaf/recycled disposable chopstick hybrid fiber-reinforced biodegradable composites. *Journal of the Taiwan Institute of Chemical Engineers* 45: 2039–2046.
Simanis, E.N. and Hart, S.L. (2008). *The base of the pyramid protocol: toward next-generation BoP strategy center for sustainable global enterprise*. Johnson School of Management, Cornell University.
Singh, A., Mishra, N., Ali, S.I. et al. (2015). Cloud computing technology: reducing carbon footprint in beef supply chain. *International Journal of Production Economics* 164: 462–471.
Smith, S., Smith, G., and Chen, W.-H. (2012). Disassembly sequence structure graphs: an optimal approach for multiple-target selective disassembly sequence planning. *Advanced Engineering Informatics* 26: 306–316.
Smith, S. and Yen, C.-C. (2010). Green product design through product modularization using atomic theory. *Robotics and Computer-Integrated Manufacturing* 26: 790–798.
Smith, S.S. and Chen, W.-H. (2011). Rule-based recursive selective disassembly sequence planning for green design. *Advanced Engineering Informatics* 25: 77–87.
Srivastava, S.K. (2007). Green supply-chain management: a state-of-the-art literature review. *International Journal of Management Reviews* 9: 53–80.
Su, C.-L. and Lee, Y.-M. (2009). Development status and life cycle inventory analysis of biofuels in Taiwan. *Energy Policy* 37: 754–758.
Su, J.C.P., Wang, L., and Ho, J.C. (2012). The impacts of technology evolution on market structure for green products. *Mathematical and Computer Modelling* 55: 1381–1400.
Tang, S.-Y., Tang, C.-P., and Lo, C.W.-H. (2005). Public participation and environmental impact assessment in Mainland China and Taiwan: political foundations of environmental management. *The Journal of Development Studies* 41: 1–32.

Taticchi, P., Tonelli, F., and Pasqualino, R. (2013). Performance measurement of sustainable supply chains: a literature review and a research agenda. *International Journal of Productivity and Performance Management* 62: 782–804.

Tien, S.-W., Chung, Y.-C., and Tsai, C.-H. (2002). Environmental design implementation in Taiwan's industries. *Environmental Impact Assessment Review* 22: 685–702.

Trappey, A.C., Ou, J.R., Lin, G.P., and Chen, M.-Y. (2011). An eco- and inno-product design system applying integrated and intelligent QFDE and TRIZ methodology. *Journal of Systems Science and Systems Engineering* 20: 443–459.

Trappey, A.J.C., Trappey, C.V., Hsiao, C.-T. et al. (2012). System dynamics modelling of product carbon footprint life cycles for collaborative green supply chains. *International Journal of Computer Integrated Manufacturing* 25: 934–945.

Tsai, C.-C. (2012). A research on selecting criteria for new green product development project: taking Taiwan consumer electronics products as an example. *Journal of Cleaner Production* 25: 106–115.

Tsai, W.-H., Chen, H.-C., Leu, J.-D. et al. (2013). A product-mix decision model using green manufacturing technologies under activity-based costing. *Journal of Cleaner Production* 57: 178–187.

Tsai, W.-H., Chen, H.-C., Liu, J.-Y. et al. (2011). Using activity-based costing to evaluate capital investments for green manufacturing systems. *International Journal of Production Research* 49: 7275–7292.

Tsai, W.-H., Lin, W.-R., Fan, Y.-W. et al. (2012). Applying a mathematical programming approach for a green product mix decision. *International Journal of Production Research* 50: 1171–1184.

Tsai, W.T. (2014). An analysis of shifting to a low-carbon society through energy policies and promotion measures in Taiwan. *Energy Sources, Part B: Economics, Planning, and Policy* 9: 391–397.

Tsaur, R.-C. (2015). Green product pricing decision analysis with application to personal computers. *International Journal of Production Research* 53: 307–320.

Tseng, H.-E., Chang, C.-C., and Cheng, C.-J. (2010). Disassembly-oriented assessment methodology for product modularity. *International Journal of Production Research* 48: 4297–4320.

Tseng, H.-E., Chang, C.-C., and Li, J.-D. (2008). Modular design to support green life-cycle engineering. *Expert Systems with Applications* 34: 2524–2537.

Tseng, K.C., Lin, B., and Han, C.-M. (2012). An intelligent system for sustainable product design at the concept development stage. *Computer-Aided Design and Applications* 9: 397–408.

Tseng, M.-L. and Chiu, A.S.F. (2013). Evaluating firm's green supply chain management in linguistic preferences. *Journal of Cleaner Production* 40: 22–31.

Tseng, M.-L., Chiu, S.F., Tan, R.R., and Siriban-Manalang, A.B. (2013). Sustainable consumption and production for Asia: sustainability through green design and practice. *Journal of Cleaner Production* 40: 1–5.

Tseng, S.-C. and Hung, S.-W. (2013). A framework identifying the gaps between customers' expectations and their perceptions in green products. *Journal of Cleaner Production* 59: 174–184.

Verrier, B., Rose, B., Caillaud, E., and Remita, H. (2014). Combining organizational performance with sustainable development issues: the Lean and Green project benchmarking repository. *Journal of Cleaner Production* 85: 83–93.

Wang, M.-C., Lee, Y.-D., and Tseng, H.-C. (2010). Integrating TRIZ method and eco-innovative design to develop a new green fire-proof system. *Journal of Information and Optimization Sciences* 31: 1061–1070.

Wee, H.M. and Chung, C.J. (2009). Optimising replenishment policy for an integrated production inventory deteriorating model considering green component-value design and remanufacturing. *International Journal of Production Research* 47: 1343–1368.

Wong, C.W.Y., Lai, K.-H., Shang, K.-C. et al. (2012). Green operations and the moderating role of environmental management capability of suppliers on manufacturing firm performance. *International Journal of Production Economics* 140: 283–294.

World Commission on Environment and Development (WCED) (1987). *Our Common Future*. Oxford, UK: Oxford University Press.

Wu, C.-H. (2012). Product-design and pricing strategies with remanufacturing. *European Journal of Operationl Research* 222: 204–215.

Wu, C.-H. (2013). OEM product design in a price competition with remanufactured product. *Omega* 41: 287–298.

Wu, K.-J., Liao, C.-J., Tseng, M.-L., and Chiu, A.S.F. (2015). Exploring decisive factors in green supply chain practices under uncertainty. *International Journal of Production Economics* 159: 147–157.

Wu, S.-I. and Lin, S.-R. (2016). The effect of green marketing strategy on business performance: a study of organic farms in Taiwan. *Total Quality Management & Business Excellence* 27 (1–2): 141–156. doi: 10.1080/14783363.2014.959256

Wu, T., Jim Wu, Y.-C., Chen, Y.J., and Goh, M. (2014). Aligning supply chain strategy with corporate environmental strategy: a contingency approach. *International Journal of Production Economics* 147 (B): 220–229.

Yang, C.J. and Chen, J.L. (2011). Accelerating preliminary eco-innovation design for products that integrates case-based reasoning and TRIZ method. *Journal of Cleaner Production* 19: 998–1006.

Yang, C.-L., Huang, R.-H., and Ke, W.-C. (2012). Applying QFD to build green manufacturing system. *Production Planning & Control* 23: 145–159.

Yang, C.-S., Lu, C.-S., Haider, J.J., and Marlow, P.B. (2013). The effect of green supply chain management on green performance and firm competitiveness in the context of container shipping in Taiwan. *Transportation Research Part E: Logistics and Transportation Review* 55: 55–73.

Yang, W.-G., Huang, Y.-F., Kuo, C.-H., and Lin, J.-H. (2014). Using Kano's model and green marketing viewpoint to investigate the customer demand of organic specialty store in Taiwan. *Journal of Information and Optimization Sciences* 35: 25–45.

Yeh, C.H., Huang, J.Y., and Yu, C.K. (2011). Integration of four-phase QFD and TRIZ in product R&D: a notebook case study. *Research in Engineering Design* 22: 125–141.

List of Abbreviations

Acronym	Definition
ABC	activity-based costing
AC	activated carbon
ACCS	activated carbon derived from coconut shell
ACP	aluminum composite panel
ACR	anaerobic contact reactor
AD	anaerobic digestion
AEM	anion exchange membrane
AF	Amyloid-Fibril
AHP	analytical hierarchy process
AHTN	-acetyl-1,1,3,4,4,6-hexamethyl-1,2,3,4-tetrahydronapthalene, Toxalide
ALK	alkyd resin
AMX	amoxicillin
AnMBR	anaerobic membrane reactors
AMP	analytic network process
ANP	analytical network process
AO	anodic oxidation
AO-cocoamido	Cocamidopropylamine oxi
AOP	advanced oxidation process
AO-R_{12}	lauramine oxide
AO-R_{14}	myristamine oxide
APAM	anion polyacrylamide
API	Application Programming Interface
API	atmospheric pressure ionization
ARG	antibiotic resistance gene
ARIZ	Algorithm for Inventive-Problem Solving
AS	alcohol sulphates
ASP	activated sludge process
Au	gold
A/V	area to volume ratio
BAC	bacteriologically active carbon

Microconstituents in the Environment: Occurrence, Fate, Removal, and Management, First Edition.
Edited by Rao Y. Surampalli, Tian C. Zhang, Chih-Ming Kao, Makarand M. Ghangrekar, Puspendu Bhunia, Manaswini Behera, and Prangya R. Rout.

BASINS	Better Assessment Science Integrating Point and Nonpoint Sources
BC	biochar
BDL	below detection limit
BDR-15	4,4'-dibromodiphenyl ether
BDST	bed depth service time
BEF	bio-electro-Fenton
BES	bio-electrochemical system
BFR	brominated fire retardant
BP3	benzophenone-3
BPA	bisphenol A
BPN	back-propagation network
BPS	biophenol S
BEF	bio-electro-Fenton
BET	Brunauer-Emmett-Teller
BMP	best management practice
BOP	bottom of pyramid
BPNN	back-propagation neural network
Br-DBP	brominated DBP
C_{10}-LAS	C_{10}-linear alkylbenzene sulfonate
C_{13}-LAS	C_{13}-linear alkylbenzene sulfonate
CA	cellular acetate
CAFO	concentrated animal feeding operations
CAGR	compound annual growth rate
carbon-PTFE-GDE	carbon-polytetrafluoroethylene gas diffusion electrode
CB	Cucurbituril
CCL	contaminant candidate list
CE	coulombic efficiency
CEB	chemically enhanced backwashing
CEC	contaminants of emerging concern
CEC	cation exchange capacity
CEM	cation exchange membrane
CeO_2	cerium oxide
CeO_2-rGO	cerium oxide (CeO_2) and reduced graphene oxide
CEPA	Canadian Environmental Protection Act
Cl-DBP	chlorinated disinfection by-products
CMPO	octylphenyl-N,N-di-isobutyl carbamoylphosphine oxide
CN	cellulose nitrate
CNT	carbon nanotubes
CNT/PPy	carbon nanotube polypyrrole nanocomposite
CNT-TiO_2	carbon nanotube-TiO_2
CuZn	copper and zinc bimetallic catalyst
CO	carbon monoxide
CO_2	carbon dioxide
COD	chemical oxygen demand
CPAM	cation polyacrylamide
CSR	corporate social responsibility

CTA	cellulose triacetate
CW	constructed wetland
CW-MFC	constructed wetland integrated MFC
CWWTP	conventional wastewater treatment plant
D	dialysis
DBP	disinfection by-products
DBP	dibutyl phthalate
DC	direct current
DCF	diclofenac
D2EHPA	di(2-ethylhexyl) phosphoric acid
DDD	1,1-dichloro-2,2-bis(p-chloro-phenyl) ethane
DDE	1,1-dichloro-2,2-bis (p-chlorophenyl) ethylene
DDT	dichlorodiphenyltrichloroethane
DEBP	di(2-ethylhexyl)phthalate
DEET	*N,N'*-diethyltoluamide
DEFRA	Department for Environment, Food and Rural Affairs
DEMATEL	Decision-Making Trial and Evaluation Laboratory
DEHMPA	di(2-ethylhexyl) methanediphosphonic acid
DEHP	Di(2-ethylhexyl)phthalate
DEM	Digital Elevation Model
DG	digestate
DGT	diffusion gradients in thin films
DHS	downflow aerobic hanging sponge system
DIET	direct interspecies electron transfer
DMAC	dodecyl trimethyl ammonium chloride
DMAD	dimethyl adipate
DMG	dimethylglyoxim
DNAP	DEMATEL-based on analytical network process
DNAPL	dense non-aqueous phase liquids
DO	dissolved oxygen
DOC	dissolved organic carbon
DOP	dioctyl phthalate
DSM	design structure matrix
DSSG	disassembly sequence structure graph
DTA	differential thermal analysis
DTMPPA	di(2,4,4trimethylpentyl) phosphinic acid
DVB	divinylbenzene
DWTP	drinking water treatment plant
E1	estrone
E2	estradiol
E3	estriol
EBCT	empty bed contact time
EC	electrochemical
ECH/DMA	epichlorohydrin/dimethylamine
ED	electrodialysis
EDC	endocrine disrupting compound

EDI	electrodeionization
EDTA	ethylenediamine tetraacetic acid
EE2	17α-ethynylestradiol
EEC	electrical energy consumption
EHDP	2-ethylhexyl diphenyl phosphate
EHP	electron-hole pair
EIA	Environmental Impact Assessment
ELISA	enzyme-linked immunosorbent assay
EMC	electrode material consumption
EMC	environmental management capability
EMR	electromagnetic radiation
EO	electrochemical oxidation
EOC	emerging organic contaminants
EP	epoxy resin
EPC	ethylene propylene copolymer
EPDM	ethylene propylene diene rubber
EVAC	ethylene vinyl acetate copolymer
μFTIR	micro-Fourier transform infrared spectroscopy
FD&C Act	Federal Food, Drug and Cosmetic Act
FIFRA	Federal Insecticide, Fungicide and Rodenticide Act
FEMA	failure mode and effects analysis
Fe–AAPyr–GNS	iron aminoantipyrine and graphene nanosheet derived catalyst
Fe–NSC	iron and sulphur co-doped nitrogen-enriched hydrothermal carbon catalyst
FISM	fuzzy interpretive structural modeling
F/M	food to microorganisms
FMFC	fungal microbial fuel cell
FNAP	fuzzy analytic network process
FO	forward osmosis
EP	emerging pollutant
FR	flame retardant
FTIR	Fourier Transform infrared spectroscopy
FUNG	fungicide
FWSCW	free water surface constructed wetland
G-5	5% Goethite supplemented natural clay ceramic
GAC	granular activated carbon
GC×GC-TOFMS	two-dimensional gas chromatography coupled with time-of-flight mass spectrometry
GC/MS	gas chromatography with mass spectrometry
GCSR	green corporate social response
GIS	geographical information system
GMS	green manufacturing systems
GO	graphene oxide
GP	green product
GSCM	green supply chain management
H_2O_2	hydrogen peroxide
HA	humic acid

HC	hydrochar
HDPE	high density polyethylene
HF-CW	horizontal flow constructed wetland
HHA	haloacetic acid
HHCB	3,4,6,7,8-hexahydro-4,6,6,7,8,8-hexamethyl-cyclopenta-[γ]-2-benzopyran, Galaxolide
HI	hazard index
HMS	homosalate
HPLC-MS/MS	high-performance liquid chromatography/tandem mass spectrometry
HRAP	high-rate algal pond
HRT	hydraulic retention time
HRU	hydrologic response unit
HSM	hazardous substance management
HSPF	Hydrological Simulation Program Fortran
HSSFCE	horizontal subsurface flow constructed wetland
HWW	hospital wastewater
HYSPLIT	Hybrid Single Particle Lagrangian Integrated Trajectory Model
IDEF	integration definition
IDW	inverse distance weighting
IER	ion exchange resin
INSE	insecticide
IOT	internet of things
IrO_2	iridium dioxide
K_{bio}	biomass normalized rate concentration
K_d	distribution coefficient
K_P	soil adsorption coefficient
KEMI	Swedish Chemicals Agency
K_{OW}	octanol-water partitioning coefficient
LAE	leaf area index
LAS	linear alkylbenzene sulfonate
LCI	life cycle inventory
LC/MS	liquid chromatography with mass spectrometry
LC-MS/MS	liquid chromatography with
LDPE	low-density polyethylene
LED	light-emitting diode
LLE	liquid-liquid extraction
LNAPL	light non-aqueous phase liquids
LUB	length of the unused bed
LULU	Land use Land cover
MAP	magnesium ammonium phosphate
MBBR	moving bed biofilm reactor
4MBC	4-methyl-benzilidine-camphor
MBR	membrane bioreactor
MBST	membrane-based separation technology
MC	microconstituent
MCC	microbial carbon-capture cell

MCDM	multiple criteria decision-making
MD	membrane distillation
MDC	microbial desalination cell
MDG	membrane degasification
MDR	modified dose-response
ME	membrane extraction
Me-NP	metal nanoparticle
MEC	microbial electrolysis cell
MEC-F	Fenton-assisted microbial electrolysis cell
MES	microbial electrosynthesis
MF	microfiltration
MFC	microbial fuel cell
Mg	magnesium
MHW	moderately hard water
MLE	modified Ludzack-Ettinger
MLR	multiple linear regression
MLSS	mixed liquor suspended solids
MFC	microbial fuel cell
MFC-CW	constructed wetland integrated microbial fuel cell
MFC-EF	electro Fenton-assisted microbial fuel cell
mg/L	milligrams per liter
MM	molecular mass
MoS_2-GO NF	molybdenum disulphide, graphene oxide and nickel foam composite electrode
MP	microplastics
MP	micropollutant
MRM	material reuse modularity
MS	mass spectrometry
MSFD	Marine Strategy Framework Directive
MSS	multi spectral scanner
MTBE	methyl tert-butyl ether
MTZ	mass transfer zone
MUSCL	monotone upwind schemes for conservation laws
MWCNT	multi-wall carbon nanotube
MWCO	molecular weight cut-off
MWNT	multi-walled nanotube
MX	multispectral
N-DBP	nitrogenous DBP
NDPES	national pollutant discharge elimination system
NEMA	National Environmental Management Act
NER	normalized energy recovery
NERc	unit of COD removed
NERv	volume of wastewater being treated
NF	nanofiltration
ng/L	nanograms per liter
NGP	non-governmental organization
NH	ammonium

NICNAS	National Industrial Chemicals Notification and Assessment Scheme
Ni/Co	nickel and cobalt catalyst
Ni–Co–P	Ni, cobalt and phosphorous catalyst
$NiMoO_4$	Nickel molybdate nanocatalyst
Ni–Pd	nickel and palladium catalyst
NM	nanomaterial
NMR	nuclear magnetic resonance
NOM	natural organic matter
NP	nanoparticle
N–P/chitosan	nitrogen and phosphorus dual-doped chitosan derived carbon catalyst
NPD	new product development
NPEC	nonylphenoxy carboxylic acids
NPEO	nonylphenol ethoxylates
NPS	non-point source
NSAIDs	nonsteroidal anti-inflammatory drugs
NSE	Nash-Sutcliffe Efficiency
O	oxygen
O_3	ozone
OC	organochlorine
OCP	organochlorine pesticide
ODPABA	octyl dimethyl-p-aminobenzoic acid
OECD	Organisation for Economic Cooperation and Development
OH·	hydroxyl radical
OLR	organic loading rate
OP	olive pomace
OP	organophosphate
OPF	oil palm fiber
OPFR	organophosphorus flame retardant
95PPU	uncertainty factor P-factor
P	phosphorus
PA	polyamide
PAA	polyacrylic acid
PA-DVB	polyacrylic divinylbenzene
PAE	phthalate ester
PAH	polycyclic aromatic hydrocarbons
PAK	polyacrylate
PAM	polyacrylamide
PAN	polyacrylonitrile
PAN-AA	polyacrylonitrile-acrylic acid
PAV	polyvinyl acetate
PBC	polychlorinated biphenyl
PBDE	polybrominated diphenyl ether
PBIAS	percent bias
PbO_2	lead oxide
PBT	persistent bioaccumulate toxin
PBT	Polybutylene Terephthalate

PC	polycarbonate
PC	principal components
PCA	principal component analysis
PCM	polycyclic musk
PCE	tetrachloroethene
PCP	personal care products
PD	palladium
PDADMAC	poly (diallyldimethylammonium chloride)
PDB	passive diffusion bag
PDCA	plan-do-check-action
PE	polyethylene
PEC	predicted environmental concentrations
PECA	Pest Control Products Act
PEM	proton exchange membrane
PEP	phosphoenolpyruvic acid
PES	polyethersulphone
PET	polyethylene terephthalate
PEVA	poly (ethylene-co-vinyl acetate)
PFC	polyfluorinated chemical
PFAS	perfluorinated alkylated substances
PFOA	perfluorooctanoic acid
PFOS	perfluorooctane sulfonic acid
PHA	olyhydroxyalkanoate
PhAC	pharmaceutically active compound
PhATE	Pharmaceutical Assessment and Transport Evaluation
PHB	olyhydroxyalkanoate
PhE	phthalate ester
Phe	phenylalanine
PIMS	passive integrative mercury sampler
pKa	acid dissociation constant
PLI	pollutant load index
PLM	permeation liquid membrane
PLOAD	pollutant load
PMA	poly (*N*-methyl acrylamide)
PMFC	photo-assisted M
PMS	permonosulfate
PNEC	predicted no effect concentrations
PO	phosphate
POCIS	polar organic chemical integrative sampler
POM	polyoxymethylene
POP	persistent organic pollutant
POTW	publicly owned treatment works
PP	polypropylene
ppb	parts per billion
PPCP	pharmaceutical and personal care products
PPS	polyphenylenesulfide

PR	phenoxy resin
PS	polysulphone/polysulphate
PS	polystyrene
PTFE	polytetrafluoroethylene
PULSE	PUSLE support practice factor
PUR	polyurethane
PVC	polyvinyl chloride
Pt	platinum
PTFE	polytetrafluoroethylene
PV	pervaporation
PVA	poly(vinyl acetate)
PVC	polyvinyl chloride
PVDF	polyvinylidene fluoride
PVS	poly(vinyl acetate)
Pyr-GCMS	pyrolysis gas chromatography mass spectroscopy
QA-QC	quality-assurance and quality-control
QC	quality control
QD	quantum dot
QFD	quality function deployment
QFDE	quality function deployment for environment
QMSA	quantitative molecular similarity assessment
QqTOF	quadrupole-time-of-flight
QSAR	quantitative structures-activity relationship
μRaman	microRaman spectroscopy
R&D	Research & Development
RAIDAR	Risk Identification and Ranking
RI	remedial investigation
rGO	reduced graphene oxide
RM	red mud
RO	reverse osmosis
RoHS	restriction of the use of certain hazardous substances
ROS	reactive oxidizing species
RQ	risk quotient
RuO_2	ruthenium dioxide
RS	remote sensing
S	sulphur
SA	surface area
SABF	submerged aerated biological filters
SADT	self-accelerating decomposition temperature
SAR	Synthetic Aperture Radar
SAS	secondary alkyl sulfonate
SBB	sugar beet bagasse
SBC	sulfonated biochar
SBR	sequencing batch reactor
SBR	styrene butadiene rubber
SCOR	Supply-Chain Operations Reference

SDS	sodium dodecyl sulfate
SEC	supporting electrolyte consumption
SEM	structural equation model
SEM-EDS/X	scanning electron microscopy with energy dispersive X-ray spectroscopy
SGO	sulfonated graphene oxide
SHBG	sex hormone-binding globulin
SLMD	stabilized liquid membrane device
SMFC	sediment microbial fuel cell
SMX	sulfamethoxazole
S/N	signal-to-noise ratio
SnO_2	tin dioxide
S-OBPI	sulfonated oxy-polybenzimidazole
SPD	solid-phase denitrification
SPE	solid-phase extraction
SPEEK	sulfonated polyether ether ketone
SPES	sulfonated polyethersulfone
SPMD	semi-permeable membrane devices
SRT	solid retention time
SS	sludge survey
SSE	sum of the squares of error
SSF-CW	subsurface flow constructed wetland
STP	sewage treatment plant
SWAT	soil and water assessment tool
SWAT-CUP	soil and water assessment tool calibration and uncertainty procedures
SWCNT	single-walled carbon nanotube
SWH	solar water heater
SWMM	Storm Water Management Model
SWRRB	Simulator for Water Resources in Rural Basics
TBBPA	3,3′, 5,5′ tetrabromobisphenol A
TBOEP	tris(2-butoxyethy) phosphate
TBP	tri-n-butyl phosphate
TCC	triclocarban
TDS-GCMS	thermal desorption system gas chromatography-mass spectroscopy
TECP	tris(2-chloroethyl) phosphate
TCPP	tris(1-chloro-2-propyl) phosphate
TCS	triclosan
TDCPP	tris(1,3-dichloropropyl) phosphate
TF	trickling filter
T_g	glass transition temperature
TGA	thermo-gravitometric analysis
THM	trihalomethane
TiO_2	titanium dioxide
TiO_2-NT	TiO_2 nanotubes
TM	Thermatic Mapper
TMAC	tetradecyl trimethyl ammonium chloride
TMDL	total maximum daily load

TMP	transmembrane pressure
TMPP	tris(methylphenyl) phosphate
TNBP	tri(butyl) phosphate
TNSSS	Targeted National Sewage Sludge Survey
TOF	time-of-flight
TOPSIS	technique for Order Preference by Similarity to Ideal Solution
TP	transformation product
TPHP	tris(phenyl) phosphate
TrOCs	trace organic compounds
TSM	technical system modularity
UASB	up-flow anaerobic sludge blanket
UF	ultrafiltrtaion
UML	Unified Modeling Language
UNEP	United Nations Environmental Programme
UHPLC-TOFMS)	Ultra-High Performance Liquid Chromatography coupled with Time Of Flight Mass Spectrometry
UPC	Un-Plastic Collective
U_r	adsorbent usage rate
USDA-ARC	United States Department of Agriculture - Agricultural Research Center
USEPA	United States Environmental Protection Agency
USGS	United States Geological Survey
USLE	Universal Soil Loss Equation
UV	ultra-violet
UV-C	downstream ultraviolet
VF	vermifiltration
VFA	volatile fatty acid
VFCW	vertical flow constructed wetland
VOC	volatile organic compound
WAP	waste stabilization pond
WASP	Water Quality Analysis Simulation Program
WEEE	waste electrical and electronic equipment
WEF	Water Environment Federation
WO_3	tungsten oxide
WRF	white-rot fungi
WRI	World Resources Institute
WS_2	tungsten sulphide
WWTP	wastewater treatment plant
XRD	X-Ray diffraction
ZnO	zinc oxide
ZrO_2	zirconium dioxide

Index

a

Microconstituents in the Environment: Occurrence, Fate, Removal, and Management, First Edition.
Edited by Rao Y. Surampalli, Tian C. Zhang, Chih-Ming Kao, Makarand M. Ghangrekar, Puspendu Bhunia, Manaswini Behera, and Prangya R. Rout.

t

u

v